Michael Ruzhansky • Durvudkhan Suragan

Hardy Inequalities on Homogeneous Groups

100 Years of Hardy Inequalities

Michael Ruzhansky
Department of Mathematics
Imperial College London
London, UK

Department of Mathematics: Analysis, Logic
and Discrete Mathematics
Ghent University
Ghent, Belgium

School of Mathematical Sciences
Queen Mary University of London
London, UK

Durvudkhan Suragan
Department of Mathematics
Nazarbayev University
Astana, Kazakhstan

https://doi.org/10.1007/978-3-030-02895-4

Mathematics Subject Classification (2010): 22E30, 22E60, 31B10, 31B25, 35A08, 35A23, 31C99, 35H10, 35J70, 35R03, 43A80, 43A99, 46E35, 53C17, 53C35

This book is an open access publication.

Ferran Sunyer i Balaguer (1912–1967) was a self-taught Catalan mathematician who, in spite of a serious physical disability, was very active in research in classical mathematical analysis, an area in which he acquired international recognition. His heirs created the Fundació Ferran Sunyer i Balaguer inside the Institut d'Estudis Catalans to honor the memory of Ferran Sunyer i Balaguer and to promote mathematical research.

Each year, the Fundació Ferran Sunyer i Balaguer and the Institut d'Estudis Catalans award an international research prize for a mathematical monograph of expository nature. The prize-winning monographs are published in this series. Details about the prize and the Fundació Ferran Sunyer i Balaguer can be found at

http://ffsb.espais.iec.cat/en

This book has been awarded the Ferran Sunyer i Balaguer 2018 prize.

The members of the scientific commitee of the 2018 prize were:

Antoine Chambert-Loir
Université Paris-Diderot (Paris 7)

Rafael de la Llave
Georgia Institute of Technology

Jiang-Hua Lu
The University of Hong Kong

Joan Porti
Universitat Autònoma de Barcelona

Kristian Seip
Norwegian University of Science and Technology

Ferran Sunyer i Balaguer Prize winners since 2007:

2007 Rosa Miró-Roig
Determinantal Ideals, PM 264

2008 Luis Barreira
Dimension and Recurrence in Hyperbolic Dynamics, PM 272

2009 Timothy D. Browning
Quantitative Arithmetic of Projective Varieties, PM 277

2010 Carlo Mantegazza
Lecture Notes on Mean Curvature Flow, PM 290

2011 Jayce Getz and Mark Goresky
Hilbert Modular Forms with Coefficients in Intersection Homology and Quadratic Base Change, PM 298

2012 Angel Cano, Juan Pablo Navarrete and José Seade
Complex Kleinian Groups, PM 303

2013 Xavier Tolsa
Analytic capacity, the Cauchy transform, and non-homogeneous Calderón–Zygmund theory, PM 307

2014 Veronique Fischer and Michael Ruzhansky
Quantization on Nilpotent Lie Groups, Open Access, PM 314

2015 The scientific committee decided not to award the prize

2016 Vladimir Turaev and Alexis Virelizier
Monoidal Categories and Topological Field Theory, PM 322

2017 Antoine Chambert-Loir, Johannses Nicaise and Julien Sebag
Motivic Integration, PM 325

Contents

Preface

The subject of Hardy inequalities has now been a fascinating subject of continuous research by numerous mathematicians for exactly[1] one century, 1918–2018.

It appears to have been inspired by D. Hilbert's investigations in the theory of integral equations where he came across a beautiful fact that the series

$$\sum_{m,n=1}^{\infty} \frac{a_m a_n}{m+n}$$

with positive entries $a_n \geq 0$ is convergent whenever $\sum_{m=1}^{\infty} a_m^2$ is convergent.

In a few years period at least four different proofs of this fact have been published: the original proof of Hilbert given by his doctoral student H. Weyl in 1908 in his Inaugural-Dissertation [Wey08a, Page 83] also appearing in [Wey08b], a proof by F. Wiener [Wie10] in 1910, and two proofs by I. Schur [Sch11] in 1911. All these proofs including Wiener's proof in the paper bearing the title "Elementarer Beweis eines Reihensatzes von Herrn Hilbert" were still not considered elementary enough by G.H. Hardy, so he came up in 1918 with yet another proof in [Har19] which seemed to him "to lack nothing in simplicity". In fact, there, he derived Hilbert's theorem as a simple 3-line corollary to the following statement: if the series $\sum_{m=1}^{\infty} a_m^2$ is convergent and we set $A_n := a_1 + \cdots + a_n$, then also the series

$$\sum_{n=1}^{\infty} \left(\frac{A_n}{n}\right)^2$$

is convergent. Thus, this moment could be considered as the birth of what is now known as Hardy's inequalities, although Hardy himself reservedly commented on his theorem with "it seems to be of some interest in itself".

After G.H. Hardy communicated his proof to Marcel Riesz, at once Riesz came up with another argument leading to the following generalization of Hardy's

[1]The original inequality was published by G.H. Hardy in "Notes on some points in the integral calculus (51)", Messenger of Mathematics, 48 (1918), pp. 107–112, see the note in [Har20, Footnote 4] for a historic remark.

result: if $\varkappa > 1$ and $\sum_{m=1}^{\infty} a_m^{\varkappa}$ is convergent, then also the series

$$\sum_{n=1}^{\infty} \left(\frac{A_n}{n}\right)^{\varkappa}$$

is convergent. Thus, this can be also regarded as the birth of what is now known as L^p-Hardy's inequalities (but should be probably then called Hardy–Riesz inequalities). The proof of Riesz and the historical account of this matter was then published as a short note[2] by Hardy in [Har20]. Hardy also gave the exact value of the best constant in the inequality, together with its extension to the integral formulation in the form of

$$\int_a^{\infty} \left(\frac{\int_a^x f(t)dt}{x}\right)^{\varkappa} dx \leq \left(\frac{\varkappa}{\varkappa-1}\right)^{\varkappa} \int_a^{\infty} f^{\varkappa}(x)dx,$$

where a and f are positive. Interestingly, Hardy called his own proof for the best constant "unnecessarily complicated", so in [Har20] he gave another simpler proof that was "sent to him by Prof. Schur by letter".

Over the last 100 years the subject of Hardy inequalities and related analysis has been a topic of intensive research: currently MathSciNet lists more than 800 papers containing words 'Hardy inequality' in the title, and almost 3500 papers containing words 'Hardy inequality' in the abstract or in the review. In view of this wealth of information *we apologize for the inevitability of missing to mention many important contributions to the subject.*

Nevertheless, the Hardy inequalities with many references have been already presented in several monographs and reviews; here we can mention excellent presentations by Opic and Kufner [OK90] in 1990, Davies [Dav99] in 1999, Kufner and Persson [KP03] (and with Samko [KPS17]), Edmunds and Evans [EE04] in 2004, part of Mazya's books [Maz85, Maz11], Ghoussoub and Moradifam [GM13] in 2013, and Balinsky, Evans and Lewis [BEL15] in 2015, as well as books on different areas related to Hardy inequalities: Hardy inequalities on time scales [AOS16], Hardy inequalities with general kernels [KHPP13], weighted Hardy inequalities [KP03], Hardy inequalities and sequence spaces [GE98]. The history and prehistory of Hardy inequalities were discussed in [KMP07] and in [KMP06], respectively, also with 'what should have happened if Hardy had discovered this' considerations [PS12].

However, all of these presentations are largely confined to the Euclidean part of the available wealth of information on this subject.

At the same time there is another layer of intensive research over the recent years related to Hardy and related inequalities in subelliptic settings motivated

[2]It seems Hardy liked publishing such notes as, according to MathSciNet, 51 of his papers start with the words "A note on...", together with papers titled "Additional note on..." or "A further note on..."

by their applications to problems involving sub-Laplacians. This is complemented by the more general anisotropic versions of the theory.

In this direction, the subelliptic ideas of the analysis on the Heisenberg group, significantly advanced by Folland and Stein in [FS74], were subsequently consistently developed by Folland [Fol75] leading to the foundations for analysis on stratified groups (or homogeneous Carnot groups). Furthermore, in their fundamental book [FS82] in 1982 titled "**Hardy spaces on homogeneous groups**", Folland and Stein laid down foundations for the 'anisotropic' analysis on general homogeneous groups, i.e., Lie groups equipped with a compatible family of dilations. Such groups are necessarily nilpotent, and the realm of homogeneous groups almost exhausts the whole class of nilpotent Lie groups including the classes of stratified, and more generally, graded groups. Happily, *the title of our monograph pays tribute to G.H. Hardy as well as to Folland and Stein's book.*

Among many, one of the motivations behind doing analysis on homogeneous groups is the "distillation of ideas and results of harmonic analysis depending only on the group and dilation structures".

The place where Hardy inequalities and homogeneous groups meet is a beautiful area of mathematics which was not consistently treated in the book form. We took it as an incentive to write this monograph to collect and deepen the understanding of Hardy inequalities and closely related topics from the point of view of Folland and Stein's homogeneous groups. While we describe the general theory of Hardy, Rellich, Caffarelli–Kohn–Nirenberg, Sobolev, and other inequalities in the setting of general homogeneous groups, a particular attention is paid to the special class of stratified groups. In this setting the theory of Hardy inequalities becomes intricately intertwined with the properties of sub-Laplacians and subelliptic partial differential equations.

These topics constitute the core of this book with the material complemented with additional closely related topics such as uncertainty principles, function spaces on homogeneous groups, the potential theory for stratified groups, and elements of the potential theory and related Hardy–Rellich inequalities for general Hörmander's sums of squares and their fundamental solutions.

We tried to make the exposition self-contained as much as possible, giving relevant references for further material. In general, for an extensive discussion of the background material related to the general theory of homogeneous and stratified groups we can refer the reader to the monograph [FR16] of which the current book is also a natural outgrowth.

The authors would like to thank our collaborators on different topics also reflected in this book: Nicola Garofalo, Ari Laptev, Tohru Ozawa, Aidyn Kassymov, Bolys Sabitbek and Nurgissa Yessirkegenov. In addition, we would like to thank Aidyn Kassymov, Bolys Sabitbek and Nurgissa Yessirkegenov for helping to proofread the preliminary version of the manuscript. We are also grateful to Gerald Folland for a positive reaction and remarks.

Finally, it is our pleasure to acknowledge the financial support by EPSRC (grants EP/K039407/1 and EP/R003025/1), by the Leverhulme Trust (grant RPG-2017-151), and by the FWO Odysseus project at different stages of preparing this monograph.

Michael Ruzhansky
Durvudkhan Suragan

London, Astana, 2018

Introduction

The present book is devoted to the exposition of the research developments at the intersection of two active fields of mathematics: Hardy inequalities and related analysis, and the noncommutative analysis in the setting of nilpotent Lie groups of different types. Both subjects are very broad and deserve separate monograph presentations on their own, and many good books are already available.

However, the recent active research in the area does allow one to make a consistent treatment of 'anisotropic' Hardy inequalities, their numerous features, and a number of related topics. This brings many new insights to the subject, also allowing to underline the interesting character of its subelliptic features. The progress in this field is facilitated by the rapid developments in both areas of Hardy inequalities and related topics, and in the noncommutative analysis on Folland and Stein's homogeneous groups.

We will now give some short insights into both fields and into the scope of this book. Here we only give a general overview, with more detailed references and explanations of different features presented throughout the monograph.

Hardy inequalities and related topics

The classical L^2-Hardy inequality in the modern literature in the Euclidean space $\mathbb{R}^n$ with $n \geq 3$ can be written in the form

$$\left\| \frac{f}{|x|_E} \right\|_{L^2(\mathbb{R}^n)} \leq \frac{2}{n-2} \|\nabla f\|_{L^2(\mathbb{R}^n)}, \tag{1}$$

where ∇ is the standard gradient in $\mathbb{R}^n$,

$$|x|_E = \sqrt{x_1^2 + \cdots + x_n^2}$$

is the Euclidean norm, $f \in C_0^\infty(\mathbb{R}^n)$, and where the constant $\frac{2}{n-2}$ is known to be sharp. In addition to references in the preface, the multidimensional version of the Hardy inequality was proved by J. Leray [Ler33].

It has numerous applications in different fields, for example in the spectral theory, leading to the lower bounds for the quadratic form associated to the Laplacian operator. It is also related to many other areas and fields, notably to

M. Ruzhansky, D. Suragan, *Hardy Inequalities on Homogeneous Groups*,
Progress in Mathematics 327, https://doi.org/10.1007/978-3-030-02895-4_1

the uncertainty principles. The uncertainty principle in physics is a fundamental concept going back to Heisenberg's work on quantum mechanics [Hei27, Hei85], as well as to its mathematical justification by Hermann Weyl [Wey50]. In the simplest Euclidean setting it can be stated as the inequality

$$\left(\int_{\mathbb{R}^n} |\nabla\phi|^2 dx\right)\left(\int_{\mathbb{R}^n} |x|_E^2 \phi^2 dx\right) \geq \frac{n^2}{4}\left(\int_{\mathbb{R}^n} \phi^2 dx\right)^2, \tag{2}$$

for all real-valued functions $\phi \in C_0^\infty(\mathbb{R}^n)$, where the constant $\frac{n^2}{4}$ is sharp. It can be shown to be a consequence of (1). There are good surveys on the mathematical aspects of uncertainty principles by Fefferman [Fef83] and by Folland and Sitaram [FS97]. We also note that the uncertainty principle can be also obtained without Hardy inequalities, see, e.g., Ciatti, Ricci and Sundari [CRS07].

The inequality (1) can be extended to L^p-spaces, taking the form

$$\left\| \frac{f}{|x|_E} \right\|_{L^p(\mathbb{R}^n)} \leq \frac{p}{n-p} \|\nabla f\|_{L^p(\mathbb{R}^n)}, \quad n \geq 2, \quad 1 \leq p < n, \tag{3}$$

where $f \in C_0^\infty(\mathbb{R}^n)$, and where the constant $\frac{p}{n-p}$ is known to be sharp.

As mentioned in the preface, such inequalities go back to Hardy [Har19], and have been evolving and growing over the years. In fact, the subject is so deep and broad at the same time that it would be impossible to give justice to all the authors who have made their contributions. To this end, we can refer to several extensive presentations of the subject in the books and surveys and the references therein: Opic and Kufner [OK90] in 1990, Davies [Dav99] in 1999, Edmunds and Evans [EE04] in 2004, part of Mazya's books [Maz85, Maz11], Ghoussoub and Moradifam [GM13] in 2013, and the recent book by Balinsky, Evans and Lewis [BEL15]. Hardy type inequalities have been very intensively studied, see, e.g., also Davies and Hinz [DH98], Davies [Dav99] as well as Ghoussoub and Moradifam [GM11] for reviews and applications.

One further extension of the Hardy inequality is the now classical result by Rellich appearing at the 1954 ICM in Amsterdam [Rel56] with the inequality

$$\left\| \frac{f}{|x|_E^2} \right\|_{L^2(\mathbb{R}^n)} \leq \frac{4}{n(n-4)} \|\Delta f\|_{L^2(\mathbb{R}^n)}, \quad n \geq 5, \tag{4}$$

with the sharp constant. We can refer, for example, to Davies and Hinz [DH98] (see also Brézis and Vázquez [BV97]) for further history and later extensions, including the derivation of sharp constants.

Higher-order Hardy inequalities have been also intensively investigated. Some of such results go back to 1961 to Birman [Bir61, p. 48] who has shown, for functions $f \in C_0^k(0,\infty)$, the family of inequalities

$$\left\| \frac{f}{x^k} \right\|_{L^2(0,\infty)} \leq \frac{2^k}{(2k-1)!!} \left\| f^{(k)} \right\|_{L^2(0,\infty)}, \quad k \in \mathbb{N}, \tag{5}$$

where $(2k-1)!! = (2k-1)\cdot(2k-3)\cdots 3\cdot 1$. For $k=1$ and $k=2$ this reduces to one-dimensional Hardy and Rellich inequalities, respectively. Such one-dimensional inequalities have recently found new life and one can find their historical discussion by Gesztesy, Littlejohn, Michael and Wellman in [GLMW17].

There is now a whole scope of related inequalities playing fundamental roles in different branches of mathematics, in particular, in the theory of linear and nonlinear partial differential equations. For example, the analysis of more general weighted Hardy–Sobolev type inequalities has also a long history, initiated by Caffarelli, Kohn and Nirenberg [CKN84] as well as by Brézis and Nirenberg in [BN83], and then Brézis and Lieb [BL85] with a mixture with Sobolev inequalities, Brézis and Vázquez in [BV97, Section 4], also [BM97], with many subsequent works in this direction. We also refer to more recent paper of Hoffmann-Ostenhof and Laptev [HOL15] on this subject and to further references therein. Many of these inequalities will be also appearing in the present book.

Of course, there are many more aspects to Hardy inequalities. In particular, working in domains, one can establish inequalities under certain boundary conditions. For example, for Hardy inequalities for Robin Laplacians and p-Laplacians see [KL12] and [EKL15], respectively, or [LW99, BLS04] for magnetic versions, or [BM97, HOHOL02] for versions involving the distance to the boundary. For Hardy inequalities for discrete Laplacians see, e.g., [KL16], or [HOHOLT08] for many-particle versions.

Homogeneous groups of different types

The harmonic analysis on homogeneous groups goes back to 1982 to Folland and Stein who in their book [FS82] laid down the foundations of 'anisotropic' harmonic analysis, that is, the harmonic analysis that depends only on the group and dilation structures of the group.

Such homogeneous groups are necessarily nilpotent, and provide a unified framework including many well-known classes of (nilpotent) Lie groups: *the Euclidean space, the Heisenberg group, H-type groups, polarizable Carnot groups, stratified groups (homogeneous Carnot groups), graded groups*. All of these groups are homogeneous and have the rational weights for their dilations.

The class of homogeneous groups is closer to the classical analysis than one might first think: in fact, any homogeneous group can be identified with some space $\mathbb{R}^n$ with a polynomial group law. The simplest examples are $\mathbb{R}^n$ itself, where the group law is linear, or the Heisenberg group, where the group law is quadratic in the last variable.

An important feature of homogeneous groups is that they do not have to allow for homogeneous hypoelliptic left invariant partial differential operators. In fact, if such an operator exists, the group has to be *graded* and its weights of dilations are rational. The class of stratified groups is a particularly important class of graded groups allowing for a homogeneous second-order sub-Laplacian. In general, nilpotent Lie groups provide local models for many questions in subelliptic analysis and sub-Riemannian geometry, their importance widely recognized since the essential role they played in deriving sharp subelliptic estimates for differential operators on manifolds, starting from the seminal paper by Rothschild and Stein [RS76] (see also [Fol77, Rot83]).

In order to facilitate the exposition in the sequel, in Chapter 1 we will recall all the necessary facts needed for the analysis in this book.

We note, however, that the general scope of techniques available on such groups is much more extensive than presented in Chapter 1. The fundamental paper by Folland [Fol75] developed the rich functional analysis on stratified groups. Further functional spaces (e.g., of Besov type) on stratified groups have been analysed by Saka [Sak79]. There are many sources with rather comprehensive and deep treatments of general nilpotent Lie groups, for example, the books by Goodman [Goo76] or Corwin and Greanleaf [CG90]. Good sources of information are the notes by Fulvio Ricci [Ric] and Folland's books [Fol89, Fol95, Fol16].

As a side remark we can note that there is also a number of recent works developing function spaces on graded groups extending Folland and Saka's constructions in the stratified case, see [FR17] and [FR16] for Sobolev, and [CR16] and [CR17] for Besov spaces, respectively.

In our presentation and approach to the basic analysis on homogeneous groups of different types we mostly rely on the recent *open access* book [FR16]. Moreover, the exposition of the topics in this book is done more in the spirit of the classical potential theory, without much reference to the Fourier analysis. However, here we should mention that the noncommutative Fourier analysis on nilpotent Lie groups is extremely rich, with many powerful approaches available, such as Kirillov's orbit method [Kir04], Mackey general description of the unitary dual, or the von Neumann algebra approaches of Dixmier [Dix77, Dix81]. We can refer to [FR16, Appendix B] for a workable summary of these methods.

The recently developed noncommutative quantization theories on nilpotent Lie groups, in particular, the global theory of pseudo-differential operators on graded groups, indeed heavily rely on such Fourier analysis. We refer the interested reader to [FR16] for the thorough exposition and application of such methods. A good exposition of the analysis of questions not requiring the Fourier analysis, in the setting of stratified groups, can be found in the book [BLU07] by Bonfiglioli, Lanconelli and Uguzzoni.

Hardy inequalities and potential theory on stratified groups

The study of the subelliptic Hardy inequalities has also begun more than 40 years ago due to their importance for many questions involving subelliptic partial differential equations, unique continuation, sub-Riemannian geometry, subelliptic spectral theory, etc. Not surprisingly, here the work started with the most important example of the Heisenberg group, where we can mention a fundamental contribution by Garofalo and Lanconelli [GL90].

There is a deep link with the properties of the fundamental solutions for the sub-Laplacian on stratified groups. In general, the understanding of the fundamental solutions of differential operators is one of the keys for solving boundary value problems for differential equations in a domain, and this idea has a long history dating back to the works of mathematicians such as Gauss [Gau77, Gau29] and Green [Gre28].

In general, the sub-Laplacians on stratified (and on more general graded) groups play important roles not only in theoretical settings (see, e.g., Gromov [Gro96] or Danielli, Garofalo and Nhieu [DGN07] for general expositions from different points of view), but also in applications of mathematics, for example in mathematical models of crystal material and human vision (see, for example, [Chr98] and [CMS04]).

The fundamental solution for the sub-Laplacian on stratified groups behaves well and was already understood by Folland [Fol75]. In particular, one always has its existence and uniqueness, an advantageous feature when compared to higher-order operators on stratified groups, or more general hypoelliptic operators on graded groups, see Geller [Gel83], and an exposition in [FR16, Section 3.2.7].

Roughly speaking, there are three versions of Hardy type inequalities on stratified groups available in the literature:

(A) Using the homogeneous quasi-norm, sometimes called the $\mathcal{L}$-gauge, given by the appropriate power of the fundamental solution of the sub-Laplacian $\mathcal{L}$. Thus, if $d(x)$ is the $\mathcal{L}$-gauge, then $d(x)^{2-Q}$ is a constant multiple of Folland's [Fol75] fundamental solution of the sub-Laplacian $\mathcal{L}$, with Q being the homogeneous dimension of the stratified group $\mathbb{G}$; these will be discussed in Chapter 7.

(B) Using the Carnot–Carathéodory distance, i.e., the control distance associated to the sub-Laplacian.

(C) Using the Euclidean distance on the first stratum of the group.

One can note that if one is not interested in best constants in such inequalities one can work with any of these equivalent quasi-norms. In fact, in such a case one can also work with fractional-order derivatives expressed as arbitrary powers of the sub-Laplacian, see, e.g., Ciatti, Cowling and Ricci [CCR15] for such an analysis on stratified groups, as well as Yafaev [Yaf99] for some Euclidean considerations also with best constants, or Hoffmann-Ostenhof and Laptev [HOL15] and references therein.

However, the best constants in the corresponding inequalities in cases (A)–(C) above may depend on the quasi-norm that one is using.

Thus, in the case (A) there is an extensive literature on Hardy inequalities and related topics on stratified groups relating them to the fundamental solution to the sub-Laplacian. Here we can briefly mention some papers [GL90, GK08, Gri03, NZW01, HN03, D'A04b, WN08, DGP11, Kom10, JS11, Lia13, Yan13, CCR15,

Yen16, GKY17], with more details and acknowledgements given throughout the book. Here, the Hardy inequality typically takes the form

$$\left\| \frac{f}{d(x)} \right\|_{L^p(\mathbb{G})} \leq \frac{p}{Q-p} \|\nabla_H f\|_{L^p(\mathbb{G})}, \quad Q \geq 3, \ 1 < p < Q, \tag{6}$$

where Q is the homogeneous dimension of the stratified group $\mathbb{G}$, ∇_H is the horizontal gradient, and $d(x)$ is the so-called $\mathcal{L}$-gauge related to the fundamental solution of the sub-Laplacian, and the constant is sharp. The analysis in the case (A) in terms of the fundamental solution of the sub-Laplacian will be the subject of Chapter 11 of this book.

The results on Hardy and other inequalities for the case (B) are less extensive, mostly devoted to the case of the Heisenberg group. However, the case (C) has recently attracted a lot of attention due to its geometrically clear nature and importance for questions in partial differential equations, see, e.g., [BT02a] and [D'A04b]. A typical horizontal Hardy inequality would take the form

$$\left\| \frac{f}{|x'|} \right\|_{L^p(\mathbb{G})} \leq \frac{p}{N-p} \|\nabla_H f\|_{L^p(\mathbb{G})}, \quad 1 < p < N, \tag{7}$$

where N is the dimension of the first stratum, x' denotes the variables in the first stratum of $\mathbb{G}$, and

$$|x'| = \sqrt{x_1'^2 + \cdots + x_N'^2}$$

is the Euclidean norm on the first stratum of $\mathbb{G}$, which can be identified with $\mathbb{R}^N$. The constant $\frac{p}{N-p}$ in (7) is also sharp.

In Chapter 6 we aim at giving a comprehensive treatment of such horizontal estimates based on the divergence relations and on the potential theory on the stratified groups.

Another ingredient that we find to be missing in the literature in the setting of stratified groups is the *classical style potential theory* working with layer potential operators. Indeed, nowadays the appearing boundary layer operators and elements of the potential theory serve as the main apparatus for the analysis and construction of solutions to boundary value problems. That have led to a vast literature concerning modern theory of boundary layer operators and potential theory in $\mathbb{R}^n$ as well as their important applications. In the subelliptic setting we can mention the works of Jerison [Jer81] and Romero [Rom91] on the Heisenberg group. We refer to [GL03], [GN88], [GW92], [LU97], [RS17d] and [WN16] as well as to references therein for more general Green function analysis of second order subelliptic (and weighted degenerate) operators. In this book, we follow a more geometric approach of our recent paper [RS17c] to present such a subject in the setting of general stratified groups, and to give its applications to several questions, such as boundary value problems for the sub-Laplacian, traces of Newton potential operators, and Hardy inequalities with boundary terms. All these topics will be the subject of discussions in Chapter 11.

The boundary value problems in the subelliptic settings are substantially more complicated than in the elliptic case due to the appearance of so-called *characteristic points* at the boundary – some problems of such a type are well explained, e.g., in [DGN06]. However, there is still a particular type of boundary conditions (Kac' boundary value problem) which can be viewed as a subelliptic version of M. Kac's question: *is there any boundary value problem for the Laplacian which is explicitly solvable in the classical sense for any smooth domain?*

An answer to M. Kac's question was given in [RS16c] for the Heisenberg group and in [RS17c] for general stratified groups. The appearing boundary conditions are, however, nonlocal and the corresponding boundary value problem can be called Kac's boundary value problem. One interesting fact is that the explicit solutions that one constructs for Kac's boundary value problem for the sub-Laplacian work also well in the presence of characteristic points on the boundary.

In Section 11.5 we also discuss another version of such a question: *is there a class of domains in which the Dirichlet boundary value problem for the sub-Laplacian is explicitly solvable in the classical sense?* This is discussed in the setting of H-type groups following our recent paper [GRS17] with Nicola Garofalo.

Furthermore, in Chapter 12 we will give an exposition of the potential theory and related Hardy–Rellich inequalities for more general Hörmander's sums of squares, based on the properties of the fundamental solutions rather than those of the $\mathcal{L}$-gauge. This has a definitive advantage of eliminating the need to use Folland's formula relating the $\mathcal{L}$-gauge with the fundamental solution. As a result, we can extend the analysis to more general settings, also those without any group structure, dealing with Hörmander's sums of squares beyond the setting of the stratified groups.

In general, there are several ways to obtain improvements of Hardy inequalities by including boundary terms. For the Laplacians such problems have been considered in [ACR02] by using variational method and in [WZ03] by using conformal transformation method. The methods described in this book are based on the potential theory and, compared to other approaches, do not rely so much on the particular structure of the Euclidean space. Certain Hardy and Rellich inequalities for sums of squares have been considered by Grillo [Gri03], compared to which our approach provides refinements from several points of view, based on the generalized representation formulae (Green's formulae) of non-subharmonic functions for improved Hardy and Rellich type inequalities with boundary terms.

Hardy inequalities and related topics on homogeneous groups

The lack of homogeneous hypoelliptic left invariant differential operators on general homogeneous groups is compensated by several other advantageous properties, such as a good polar decomposition which, combined with dilations, still allows one to explore the radial structure of the group. For this purpose, we extensively work with the radial derivative operator $\mathcal{R}$ and the Euler operator $\mathbb{E}$ which we describe in Section 1.3.

This fits well with the structure of Hardy and other inequalities as their maximisers are often achieved on radial functions. An advantage of such an approach is that one can also work with arbitrary quasi-norms and anisotropic structures still yielding similar properties and best constants in the inequalities. In addition, in Section 1.3.3 we demonstrate how the Hardy type inequalities for radial functions often imply similar inequalities for functions of general (non-radial) type.

Thus, Chapter 2 is devoted to Hardy inequalities on homogeneous groups, their weighted and critical versions, stability and remainder estimates. Furthermore, Chapter 3 is devoted to Rellich, Caffarelli–Kohn–Nirenberg and Sobolev type inequalities on homogeneous groups. In Chapter 9 we present different versions of uncertainty principles on homogeneous groups. We follow an abstract approach by defining abstract position and momentum operators satisfying minimal structural properties, already allowing one to establish a number of uncertainty type relations. Consequently, different choices of such abstract position and momentum operators are possible based on the additional structural properties available on the group.

In Chapter 10 we discuss different function spaces on homogeneous groups, with or without differentiability properties. The spaces involving radial derivatives are the Euler–Hilbert–Sobolev and Sobolev–Lorentz–Zygmund spaces. We investigate their basic properties and embeddings. The spaces not involving the differentiable structure are the Morrey and Campanato spaces. There, we discuss the boundedness of integral operators, namely, the Hardy–Littlewood maximal operator, Bessel–Riesz operators, generalized Bessel–Riesz operators, generalized fractional integral operators and Olsen type inequalities in generalized Morrey spaces on homogeneous groups.

Incidentally, all these and other results on homogeneous groups in this book give new statements already in the Euclidean setting of $\mathbb{R}^n$ when we are working with anisotropic differential structure in view of the arbitrariness of the choice of any homogeneous quasi-norm.

Let us consider, for instance, the following Bessel–Riesz operators

$$I_{\alpha,\gamma}f(x) = \int_{\mathbb{R}^n} \frac{|x-y|_E^{\alpha-n}}{(1+|x-y|_E)^\gamma} f(y)dy, \tag{8}$$

where $f \in L^p_{\mathrm{loc}}(\mathbb{R}^n)$, $p \geq 1$, $\gamma \geq 0$ and $0 < \alpha < n$. Classical results on the Bessel–Riesz operators are due to Hardy, Littlewood and Sobolev, precisely, the boundedness of the Bessel–Riesz operators on Lebesgue spaces was shown by Hardy and Littlewood in [HL27], [HL32] and by Sobolev in [Sob38]. In the case of $\mathbb{R}^n$, the Hardy–Littlewood maximal operator, the Riesz potential $I_{\alpha,0} = I_\alpha$, the generalized fractional integral operators, which are a generalized form of the Riesz potential $I_{\alpha,0} = I_\alpha$, Bessel–Riesz operators and Olsen type inequalities are widely analysed on Lebesgue spaces, Morrey spaces and generalized Morrey spaces. For further discussions in this direction we refer to [Ada75, CF87, BNC14, Nak94, EN04, Eri02, KNS99, Nak01, Nak02, GE09, SST12, IGLE15, IGE16], as well as to

[Bur13] for a recent survey. Morrey spaces for non-Euclidean distances find their applications in many problems, see, e.g., [GS15a, GS15b] and [GS16]. A natural analogue of the Bessel–Riesz operator (8) on homogeneous groups is the operator

$$I_{\alpha,\gamma}f(x) := \int_{\mathbb{G}} \frac{|xy^{-1}|^{\alpha-Q}}{(1+|xy^{-1}|)^{\gamma}} f(y)dy.$$

In the setting of graded Lie groups the connections between these operators and the Sobolev spaces have been investigated in [FR16, Chapter 4] using the heat kernel methods. Here, in Chapter 10 we concentrate on the harmonic analysis aspects in the framework of Morrey and Campanato spaces on homogeneous groups.

In addition, we can mention an overview of constructions for Morrey–Campanato spaces in [RSS13] by Rafeiro, N. Samko and S. Samko, or in [RT15] by Rosenthal and Triebel. It is worth noting that Morrey–Campanato spaces can be interpolated [VS14]. One also considered variable exponent versions of Morrey spaces and maximal and singular operators there, see [GS13, GS16] and references therein.

In Chapter 4 we look at the Hardy inequalities from the point of view of operators of fractional orders. Certainly, fractional powers of Laplacians and sub-Laplacians can be defined in different ways, e.g., using the Fourier or spectral analysis. However, here, we first adopt the integral representation that turns out to make perfect sense on general homogeneous groups. More specifically, for $p > 1$ and $s \in (0,1)$, we consider the *fractional p-sub-Laplacian* $(-\Delta_p)^s$ on a general homogeneous group $\mathbb{G}$ defined by the formula

$$(-\Delta_p)^s u(x) := 2 \lim_{\delta \searrow 0} \int_{\mathbb{G}\setminus B(x,\delta)} \frac{|u(x)-u(y)|^{p-2}(u(x)-u(y))}{|y^{-1}x|^{Q+sp}} dy, \quad x \in \mathbb{G},$$

where $B(x,\delta) = B_{|\cdot|}(x,\delta)$ is a quasi-ball with respect to the quasi-norm $|\cdot|$, with radius δ centred at $x \in \mathbb{G}$. It turns out that this operator has many advantageous properties similar to those exhibited by the usual p-Laplacians on the Euclidean spaces, and in Chapter 4 we present their analysis and some applications to 'partial differential' functional equations and related spectral questions. Consequently, we look at operators of fractional orders from a different point of view, and in Section 4.7 we discuss the boundedness of the operator

$$T_\alpha f(x) := |x|^{-\alpha}\mathcal{L}^{-\alpha/2} f(x),$$

on L^p-spaces on stratified Lie groups, where $\mathcal{L}$ is a sub-Laplacian. The analysis is based on the Riesz kernel representation of such operators, and we also supplement it with several versions of the Landau–Kolmogorov inequalities.

In Chapter 5 we discuss integral versions of Hardy inequalities. In fact, such a point of view goes back to one of the original versions of such inequalities by Hardy [Har20], where he has shown the inequality

$$\int_b^\infty \left(\frac{\int_b^x f(t)dt}{x}\right)^p dx \leq \left(\frac{p}{p-1}\right)^p \int_b^\infty f(x)^p dx,$$

where $p>1$, $b>0$, and $f\geq 0$ is a non-negative function. It turns out that such an inequality can be also put in the framework of general homogeneous groups, especially since it does not involve derivatives, so that one does not need to specify one's analysis to a particular choice of the gradient. Thus, in Chapter 5 we present inequalities of such a type in weighted and unweighted settings, actually providing characterizations of weights for which integral Hardy inequalities hold true. We also present inequalities in the convolution form which, in turn, can be used for the derivation of Hardy–Littlewood–Sobolev and Stein–Weiss inequalities. The latter can be then established both on general homogeneous groups as well as on stratified/graded groups using Riesz kernels of hypoelliptic differential operators.

In Chapter 8 we discuss the so-called geometric versions of Hardy inequalities. By this one usually means Hardy inequalities on domains when the distance to the boundary enters the inequality as a weight. For example, if Ω is a convex open set of the Euclidean space $\mathbb{R}^n$, then a geometric version of the Hardy inequality can take a form

$$\int_\Omega |\nabla u|^2 dx \geq \frac{1}{4}\int_\Omega \frac{|u|^2}{\operatorname{dist}(x,\partial\Omega)^2}dx,$$

for all $u\in C_0^\infty(\Omega)$, with the sharp constant $1/4$. In the case of half-spaces and, more generally, convex domains, in Chapter 8 we present such inequalities in the setting of stratified groups.

Chapter 1

Analysis on Homogeneous Groups

In this chapter we provide preliminaries for the analysis on homogeneous groups to make the use of the monograph more self-sufficient. We make a selection of topics which will be playing a role in the subsequent analysis. Thus, we first discuss relevant properties of general Lie groups and algebras and then concentrate on properties of homogeneous groups required for our further analysis. Lastly, we introduce the notion of the Euler operator on homogeneous groups and establish its main properties.

This chapter is not intended to be a comprehensive treatise of homogeneous groups but rather a description of a collection of tools used throughout the book. The theory of homogeneous groups for their use in analysis was developed by Folland and Stein [FS82]. A recent rather comprehensive description of homogeneous groups and their place among nilpotent Lie groups have appeared in [FR16]. We refer to both books for the expositions devoted specifically to homogeneous groups. For some related information we may also refer to Ricci's notes [Ric].

There are many sources with rather comprehensive and deep treatments of nilpotent Lie groups, for example the books by Goodman [Goo76] or Corwin and Greenleaf [CG90]. There are also many books on groups or Lie groups, we can refer for example to [RT10, Part III] for a basic introduction. Therefore, we assume the reader to have some familiarity with the concepts of the Lie groups and Lie algebras.

1.1 Homogeneous groups

In this section we discuss nilpotent Lie algebras and groups in the spirit of Folland and Stein's book [FS82] as well as introduce homogeneous (Lie) groups. For more analysis and details in this direction we refer to the recent open access book [FR16].

Let $\mathfrak{g}$ be a Lie algebra (always assumed real and finite-dimensional), and let $\mathbb{G}$ be the corresponding connected and simply-connected Lie group. The lower

M. Ruzhansky, D. Suragan, *Hardy Inequalities on Homogeneous Groups*, Progress in Mathematics 327, https://doi.org/10.1007/978-3-030-02895-4_2

central series of $\mathfrak{g}$ is defined inductively by

$$\mathfrak{g}_{(1)} := \mathfrak{g}, \quad \mathfrak{g}_{(j)} := [\mathfrak{g}, \mathfrak{g}_{(j-1)}].$$

If the lower central series of a Lie algebra $\mathfrak{g}$ terminates at 0 in a finite number of steps then this Lie algebra is called *nilpotent.* Moreover, if $\mathfrak{g}_{(s+1)} = \{0\}$ and $\mathfrak{g}_{(s)} \neq \{0\}$, then $\mathfrak{g}$ is said to be *nilpotent of step s*. A Lie group $\mathbb{G}$ is *nilpotent* (of step s) whenever its Lie algebra is nilpotent (of step s). If $\exp : \mathfrak{g} \to \mathbb{G}$ is the exponential map, by the Campbell–Hausdorff formula for $X, Y \in \mathfrak{g}$ sufficiently close to 0 we have

$$\exp X \exp Y = \exp H(X, Y),$$

where $H(X, Y)$, the Campbell–Hausdorff series, is an infinite linear combination of X and Y and their iterated commutators and H is universal, i.e., independent of $\mathfrak{g}$, and that

$$H(X, Y) = X + Y + \frac{1}{2}[X, Y] + \cdots,$$

where the dots indicate terms of order ≥ 3.

If $\mathfrak{g}$ is nilpotent, the Campbell–Hausdorff series terminates after finitely many terms and defines a polynomial map from $V \times V$ to V, where V is the underlying vector space of $\mathfrak{g}$.

Altogether, we have the following useful properties:

Proposition 1.1.1 (Exponential mapping and Haar measure). *Let $\mathbb{G}$ be a connected and simply-connected nilpotent Lie group with Lie algebra $\mathfrak{g}$. Then:*

(i) *The exponential map* exp *is a diffeomorphism from $\mathfrak{g}$ to $\mathbb{G}$. Moreover, if $\mathbb{G}$ is identified with $\mathfrak{g}$ via* exp, *then the group law $(x, y) \mapsto xy$ is a polynomial map.*

(ii) *If λ denotes a Lebesgue measure on $\mathfrak{g}$, then $\lambda \circ \exp^{-1}$ is a bi-invariant Haar measure on $\mathbb{G}$.*

Proof of Proposition 1.1.1. Part (i) is a direct consequence of the fact that $\mathbb{G}$ is uniquely (up to isomorphism) determined by $\mathfrak{g}$, see, e.g., [FS82, Proposition 1.2], [CG90, Section 1.2] or [FR16, Proposition 1.6.6].

Let us give an argument for Part (ii). Let us denote the lower central series for $\mathfrak{g}$ by

$$\mathfrak{g}_{(1)}, \ldots, \mathfrak{g}_{(m)}, \mathfrak{g}_{(m+1)} = \{0\}$$

and denote

$$n := \dim \mathfrak{g} \quad \text{and} \quad n_j := \dim \mathfrak{g}_{(j)}.$$

Let $X_{n-n_m+1}, \ldots, X_n$ be a basis for $\mathfrak{g}_{(m)}$, and we extend it to a basis

$$X_{n-n_{m-1}+1}, \ldots, X_n$$

for $\mathfrak{g}_{(m-1)}$, and so forth obtaining eventually a basis $X_1, \ldots, X_n$ for $\mathfrak{g}$. Let $\xi_1, \ldots, \xi_n$ be the dual basis for $\mathfrak{g}^*$, and let $\eta_k := \xi_k \circ \exp^{-1}$. These $\eta_1, \ldots, \eta_n$ are a system of global coordinates on $\mathbb{G}$. By using the Campbell–Hausdorff formula and the construction of the η_k's we obtain

$$\eta_k(xy) = \eta_k(x) + \eta_k(y) + P_k(x, y),$$

where $P_k(x, y)$ depends only on the coordinates $\eta_i(x), \eta_i(y)$ with $i < k$. Thus, with respect to the coordinates η_k, the differentials of the maps $x \mapsto xy$ with fixed y and $y \mapsto xy$ with fixed x are given by lower triangular matrices with only 1 elements on the diagonal, and therefore, each of the determinants is equal to one. This implies that the volume form $d\eta_1 \cdots d\eta_n$ on $\mathbb{G}$, which corresponds to the Lebesgue measure on $\mathfrak{g}$, is left and right invariant. □

Definition 1.1.2 (Dilations on a Lie algebra)**.** A family of *dilations* of a Lie algebra $\mathfrak{g}$ is a family of linear mappings

$$\{\delta_r : r > 0\}$$

from $\mathfrak{g}$ to itself which satisfies:

- the mappings are of the form

$$\delta_r = \exp(A \log r),$$

  where A is a diagonalisable linear operator on $\mathfrak{g}$ with positive eigenvalues.
- In particular, $\delta_{rs} = \delta_r \delta_s$ for all $r, s > 0$. If $\alpha > 0$ and $\{\delta_r\}$ is a family of dilations on $\mathfrak{g}$, then so is $\{\widetilde{\delta_r}\}$, where

$$\widetilde{\delta_r} := \delta_{r^\alpha} = \exp(\alpha A \log r).$$

  By adjusting α we can always assume that the minimum eigenvalue of A is equal to 1.

Let $\mathcal{A}$ be the set of eigenvalues of A and denote by $W_a \subset \mathfrak{g}$ the corresponding eigenfunction space of A, where $a \in \mathcal{A}$. Then we have

$$\delta_r X = r^a X \quad \text{for} \quad X \in W_a.$$

If $X \in W_a$ and $Y \in W_b$, then

$$\delta_r[X, Y] = [\delta_r X, \delta_r Y] = r^{a+b}[X, Y]$$

and thus $[W_a, W_b] \subset W_{a+b}$. In particular, since $a \geq 1$ for $a \in \mathcal{A}$, we see that $\mathfrak{g}_{(j)} \subset \bigoplus_{a \geq j} W_a$. Since the set $\mathcal{A}$ is finite, it follows that $\mathfrak{g}_{(j)} = \{0\}$ for j sufficiently large. Thus, we obtain:

Proposition 1.1.3 (Lie algebras with dilations are nilpotent)**.** *If a Lie algebra $\mathfrak{g}$ admits a family of dilations then it is nilpotent.*

However, not all nilpotent Lie algebras admit a dilation structure: an example of a (nine-dimensional) nilpotent Lie algebra that does not allow any compatible family of dilations was constructed by Dyer [Dye70].

Definition 1.1.4 (Graded Lie algebras and groups). A Lie algebra $\mathfrak{g}$ is called *graded* if it is endowed with a vector space decomposition (where all but finitely many of the V_k's are 0)

$$\mathfrak{g} = \oplus_{j=1}^{\infty} V_j \quad \text{such that} \quad [V_i, V_j] \subset V_{i+j}.$$

Consequently, a Lie group is called *graded* if it is a connected simply-connected Lie group whose Lie algebra is graded.

Definition 1.1.5 (Stratified Lie algebras and groups). A graded Lie algebra $\mathfrak{g}$ is called *stratified* if V_1 generates $\mathfrak{g}$ as an algebra. In this case, if $\mathfrak{g}$ is nilpotent of step m we have

$$\mathfrak{g} = \oplus_{j=1}^{m} V_j, \quad [V_j, V_1] = V_{j+1},$$

and the natural dilations of $\mathfrak{g}$ are given by

$$\delta_r\left(\sum_{k=1}^{m} X_k\right) = \sum_{k=1}^{m} r^k X_k, \quad (X_k \in V_k).$$

Consequently, a Lie group is called *stratified* if it is a connected simply-connected Lie group whose Lie algebra is stratified.

Definition 1.1.6 (Homogeneous groups). Let δ_r be dilations on $\mathbb{G}$. We say that a Lie group $\mathbb{G}$ is a *homogeneous group* if:

a. It is a connected and simply-connected nilpotent Lie group $\mathbb{G}$ whose Lie algebra $\mathfrak{g}$ is endowed with a family of dilations $\{\delta_r\}$.
b. The maps $\exp \circ \delta_r \circ \exp^{-1}$ are group automorphism of $\mathbb{G}$.

Since the exponential mapping exp is a global diffeomorphism from $\mathfrak{g}$ to $\mathbb{G}$ by Proposition 1.1.1, (i), it induces the corresponding family on $\mathbb{G}$ which we may still call the dilations on $\mathbb{G}$ and denote by δ_r. Thus, for $x \in \mathbb{G}$ we will write $\delta_r(x)$ or abbreviate it writing simply rx.

The origin of $\mathbb{G}$ will be usually denoted by 0.

Now let us give some well-known examples of homogeneous groups.

Example 1.1.7 (Abelian groups). The Euclidean space $\mathbb{R}^n$ is a homogeneous group with dilation given by the scalar multiplication.

Example 1.1.8 (Heisenberg groups). If n is a positive integer, the Heisenberg group $\mathbb{H}^n$ is the group whose underlying manifold is $\mathbb{C}^n \times \mathbb{R}$ and whose multiplication is given by

$$(z_1, \ldots, z_n, t)(z_1', \ldots, z_n', t') = \left(z_1 + z_1', \ldots, z_n + z_n', t + t' + 2\mathrm{Im}\sum_{k=1}^{n} z_k \overline{z}_k'\right).$$

The Heisenberg group $\mathbb{H}^n$ is a homogeneous group with dilations

$$\delta_r(z_1, \ldots, z_n, t) = (rz_1, \ldots, rz_n, r^2 t).$$

Example 1.1.9 (Upper triangular groups). Let $\mathbb{G}$ be the group of all $n \times n$ real matrices (a_{ij}) such that $a_{ii} = 1$ for $1 \leq i \leq n$ and $a_{ij} = 0$ when $i > j$. Then $\mathbb{G}$ is a homogeneous group with dilations

$$\delta_r(a_{ij}) = r^{j-i} a_{ij}.$$

These Examples 1.1.7, 1.1.8 and 1.1.9 are all examples of the stratified groups. It is also possible to define other families of dilations on these groups. For instance, on $\mathbb{R}^n$ we can define

$$\delta_r(x_1, \ldots, x_n) = (r^{d_1} x_1, \ldots, r^{d_n} x_n),$$

where $1 = d_1 \leq d_2 \leq \cdots \leq d_n$, and on $\mathbb{H}^n$ we can define

$$\delta_r(x_1 + iy_1, \ldots, x_n + iy_n, t) = (r^{a_1} x_1 + ir^{b_1} y_1, \ldots, r^{a_n} x_n + ir^{b_n} y_n, r^c t),$$

where $\min\{a_1, \ldots, a_n, b_1, \ldots, b_n\} = 1$ and $a_j + b_j = c$ for all j. In general, these dilations do not have to be stratified. However, when we refer to $\mathbb{R}^n$ or $\mathbb{H}^n$ we shall assume that they are equipped with the natural dilations defined in Examples 1.1.7, 1.1.8 unless we state otherwise.

Let $d_1, \ldots, d_n$ be the eigenvalues of A, enumerated in nondecreasing order according to their multiplicity, and let $\overline{d} = \max d_k$. The mappings $\{\delta_r = \exp(A \log r)\}$ give the dilation structure to an n-dimensional homogeneous group $\mathbb{G}$, with

$$1 = d_1 \leq d_2 \leq \cdots \leq d_n = \overline{d}. \tag{1.1}$$

Let us fix a basis $\{X_k\}_{k=1}^n$ of the Lie algebra $\mathfrak{g}$ of the Lie group $\mathbb{G}$ such that

$$AX_k = d_k X_k$$

for each k. Then one can define a standard Euclidean norm $\|\cdot\|$ on $\mathfrak{g}$ by declaring the X_k's to be orthonormal. This norm can be also considered as a function on $\mathbb{G}$ by the formula

$$\|x\| = \|\exp^{-1} x\|. \tag{1.2}$$

The number

$$Q := \sum_{k=1}^{n} d_k = \operatorname{Tr}(A) \tag{1.3}$$

is called the *homogeneous dimension* of $\mathbb{G}$. From now on Q will always denote the homogeneous dimension of $\mathbb{G}$.

1.2 Properties of homogeneous groups

In this section we discuss properties of homogeneous groups that are important for their understanding and that will be also useful for our further analysis. For different further properties of homogeneous and graded groups we can refer the reader to the open access book [FR16, Chapter 3].

1.2.1 Homogeneous quasi-norms

We start by the definition of a quasi-norm.

Definition 1.2.1 (Quasi-norms). Let us define a *homogeneous quasi-norm* on a homogeneous group $\mathbb{G}$ to be a continuous function $x \mapsto |x|$ from $\mathbb{G}$ to $[0, \infty)$ that satisfies

(a) for all $x \in \mathbb{G}$ and $r > 0$:

$$|x^{-1}| = |x| \quad \text{and} \quad |rx| = r|x|.$$

(b) The non-degeneracy:

$$|x| = 0 \quad \text{if and only if} \quad x = 0.$$

Here and elsewhere we denote by $rx = \delta_r x$ the dilation of x induced by the dilations on the Lie algebra through the exponential mapping.

There always exist homogeneous quasi-norms on homogeneous groups. Moreover, there always exist quasi-norms that are C^∞-smooth on $\mathbb{G}\backslash\{0\}$. Let us give such an example. Observe that

$$X = \sum_{k=1}^{n} c_k X_k \in \mathfrak{g} \quad \text{implies} \quad \|\delta_r X\| = \left(\sum_{k=1}^{n} c_k^2 r^{2d_k} \right)^{1/2},$$

where $\|\cdot\|$ is the Euclidean norm from (1.2). We can notice that for $X \neq 0$ the function $\|\delta_r X\|$ is a strictly increasing function of r, and it tends to 0 and ∞ as $r \to 0$ and $r \to \infty$, respectively. Now, for $x = \exp X$, we can define a homogeneous quasi-norm on $\mathbb{G}$ by setting

$$|0| := 0 \quad \text{and} \quad |x| := 1/r \quad \text{for} \quad x \neq 0,$$

where $r = r(X) > 0$ is the unique number such that

$$\|\delta_{r(X)} X\| = 1.$$

By the implicit function theorem and the fact that the Euclidean unit sphere is a C^∞ manifold we see that this function is C^∞ on $\mathbb{G}\backslash\{0\}$.

If $x \in \mathbb{G}$ and $r > 0$ we define the ball of radius r about x by

$$B(x,r) := \{y \in \mathbb{G} : |x^{-1}y| < r\}.$$

It can be noticed that $B(x,r)$ is the left translate by x of $B(0,r)$, which in turn is the image under δ_r of $B(0,1)$.

Lemma 1.2.2 (Closed quasi-balls are compact). *$\overline{B(x,r)}$ is compact for any $x \in \mathbb{G}$ and $r > 0$.*

Proof. Let us define

$$\rho(x) := \sum_{k=1}^{n} \frac{|c_k|}{d_k} \quad \text{for} \quad x = \exp\left(\sum_{k=1}^{n} c_k X_k\right),$$

where d_k are as in (1.1). Then ρ satisfies all the properties of a homogeneous quasi-norm. Obviously $\{x : \rho(x) = 1\}$ is compact and does not contain 0, so the function $x \mapsto |x|$ attains a positive minimum η on it. Since $|rx| = r|x|$ and $\rho(rx) = r\rho(x)$, it follows that $|x| \geq \eta\rho(x)$ for all x and for some $\eta > 0$, and hence that

$$\overline{B(0,\eta)} \subset \{x : \rho(x) \leq 1\}.$$

Thus, $\overline{B(0,\eta)}$ is compact, and it follows by dilation and translation that $\overline{B(x,r)}$ is compact for all $r > 0$, $x \in \mathbb{G}$. □

We can compare the quasi-norms with each other and with the Euclidean norm (1.2).

Proposition 1.2.3 (Quasi-norms and the Euclidean norm). *We have the following properties:*

(1) *Any two homogeneous quasi-norms on a homogeneous group are equivalent.*
(2) *There are the constants $C_1, C_2 > 0$ such that*

$$C_1\|x\| \leq |x| \leq C_2\|x\|^{1/\overline{d}} \quad \textit{for all} \quad |x| \leq 1.$$

Proof. Proof of Part (2). When $y = \exp(\sum c_k X_k)$ we have $\|ry\| = (\sum c_k^2 r^{2d_k})^{1/2}$ and hence

$$r^{\overline{d}}\|y\| \leq \|ry\| \leq r\|y\|$$

for $r \leq 1$. A positive maximum C_1^{-1} and a positive minimum $C_2^{-\overline{d}}$ on $\{y : |y| = 1\}$ are attained by the Euclidean norm $\|y\|$ in view of the compactness in Lemma 1.2.2. Any $x \neq 0$ can be written as $x = |x|y$ where $|y| = 1$, so that for $|x| \leq 1$,

$$\|x\| \leq |x|\|y\| \leq C_1^{-1}|x|, \quad \|x\| \geq |x|^{\overline{d}}\|y\| \geq C_2^{-\overline{d}}|x|^{\overline{d}},$$

completing the proof of Part (2).

Proof of Part (1). Let $|\cdot|_1$ and $|\cdot|_2$ be two homogeneous quasi-norms. By a similar argument to Part (2), we observe that since the ball $\overline{B(0,1)}$ with respect to $|\cdot|_1$ is compact by Lemma 1.2.2, and $|\cdot|_2$ is continuous, we have

$$\left|\frac{x}{|x|_1}\right|_2 \leq C < \infty$$

for all $x \neq 0$. By homogeneity it follows that $|x|_2 \leq C|x|_1$ for all $x \in \mathbb{G}$. Switching the roles of $|\cdot|_1$ and $|\cdot|_2$ we obtain the statement. □

The reason why the homogeneous quasi-norms have the prefix 'quasi' becomes clear from the following proposition that shows that in general the triangle inequality is satisfied only with some constant:

Proposition 1.2.4 (Triangle inequality with constant). *Let $\mathbb{G}$ be a homogeneous group. Then we have the following properties:*

(1) *If $|\cdot|$ is a homogeneous quasi-norm on $\mathbb{G}$, there exists $C > 0$ such that for every $x, y \in \mathbb{G}$, we have*

$$|xy| \leq C(|x| + |y|).$$

(2) *There always exists a homogeneous quasi-norm $|\cdot|$ on $\mathbb{G}$ which satisfies the triangle inequality (with constant $C = 1$):*

$$|xy| \leq |x| + |y| \tag{1.4}$$

for all $x, y \in \mathbb{G}$.

Proof. Let us prove Part (1). The function $(x, y) \mapsto |xy|$ attains a finite maximum $C > 0$ on the set $\{(x, y) \in \mathbb{G} \times \mathbb{G} : |x| + |y| = 1\}$ which is compact by Lemma 1.2.2. Then, given any $x, y \in \mathbb{G}$, set $r = |x| + |y|$. It follows that

$$|xy| = r|r^{-1}(xy)| = r|(r^{-1}x)(r^{-1}y)| \leq Cr = C(|x| + |y|),$$

completing the proof.

We leave Part (2) without proof, referring to [FR16, Proposition 3.1.38 and Theorem 3.1.39] for the complete argument. □

Proposition 1.2.5. *There exists a constant $C > 0$ such that for every $x \in \mathbb{G}$ and $s \in [0, 1]$, we have*

$$|\exp(s \log(x))| \leq C|x|. \tag{1.5}$$

Proof. Let $x \neq 0$, otherwise (1.5) is trivial. Using the fact that $|\cdot|$ is homogeneous of degree 1, we have

$$\frac{|\exp(s\log(x))|}{|x|} = |\delta_{1/|x|}(\exp(s\log(x))| = |\exp(s\log(\delta_{1/|x|}(x))|.$$

With $|\delta_{1/|x|}(x)| = 1$, it follows that

$$\frac{|\exp(s\log(x))|}{|x|} \leq \max_{\xi\in\mathbb{G}:|\xi|=1, s\in[0,1]} \rho(\exp(s\log(\xi))) =: C.$$

Note that C is finite, since the set $\{\xi : |\xi| = 1\}$ is compact (see Lemma 1.2.2) as well as $|\cdot|$, exp and log are all continuous functions. □

The bi-invariant Haar measure on $\mathbb{G}$ comes from the Lebesgue measure on $\mathfrak{g}$ by Proposition 1.1.1. Fixing the normalisation of the Haar measure on $\mathbb{G}$ we require that the Haar measure of $B(0,1)$ is 1. (Thus, if $\mathbb{G} = \mathbb{R}^n$ with the usual Lebesgue measure, our Haar measure is $\Gamma((n+2)/2)/\pi^{n/2}$ times the Lebesgue measure.) The measure of any measurable set $E \subset \mathbb{G}$ will be denoted by $|E|$, and we shall denote the integral of a function f with respect to this measure by $\int_{\mathbb{G}} f dx$ or by $\int_{\mathbb{G}} f(x)dx$, or simply by $\int f$ or by $\int f(x)dx$.

Recalling (1.3), the homogeneous dimension of $\mathbb{G}$ is

$$Q = \sum_{k=1}^{n} d_k = \mathrm{Tr}(A),$$

and we have

$$|\delta_r(E)| = r^Q|E|, \quad d(rx) = r^Q dx. \tag{1.6}$$

In particular, we have $|B(x,r)| = r^Q$ for all $r > 0$ and $x \in \mathbb{G}$.

Definition 1.2.6 (Homogeneous functions and operators). A function f on $\mathbb{G}\backslash\{0\}$ is said to be *homogeneous of degree* λ if it satisfies

$$f \circ \delta_r = r^\lambda f \text{ for all } r > 0.$$

We note that for f and g, we have the formula

$$\int_{\mathbb{G}} f(x)\,(g\circ\delta_r)(x)dx = r^{-Q}\int_{\mathbb{G}} (f\circ\delta_{1/r})(x)g(x)dx,$$

given that the integrals exist. Hence we can extend the mapping $f \mapsto f\circ\delta_r$ to distributions by defining, for any distribution f and any test function ϕ, the distribution $f\circ\delta_r$ by

$$\langle f\circ\delta_r, \phi\rangle = r^{-Q}\langle f, \phi\circ\delta_{1/r}\rangle,$$

where $\langle\cdot,\cdot\rangle$ denotes the usual duality between functions and distributions. The distribution f is called *homogeneous of degree* λ if it satisfies

$$f \circ \delta_r = r^\lambda f \text{ for all } r > 0.$$

Also, a linear operator D on $\mathbb{G}$ is called *homogeneous of degree* λ if it satisfies

$$D(f\circ\delta_r) = r^\lambda (Df)\circ\delta_r \text{ for all } r > 0,$$

for any f. If D is a linear operator homogeneous of degree λ and f is a homogeneous function of degree μ, then Df is homogeneous of degree $\mu - \lambda$.

The following extension of the reverse triangle inequality is often useful:

Proposition 1.2.7 (Reverse triangle inequality). *Let f be a homogeneous function of degree λ and of class C^1 on $\mathbb{G}\backslash\{0\}$. Then there is a constant $C > 0$ such that we have*

$$|f(xy) - f(x)| \leq C|y||x|^{\lambda-1} \quad \textit{for all} \quad |y| \leq |x|/2.$$

Proof. Suppose that $|x| = 1$ and $|y| \leq 1/2$, and we use the fact that both sides of the desired inequality are homogeneous of degree λ. In this case x and xy are bounded, and also bounded away from zero, and the map $y \mapsto xy$ is C^1, so by the usual mean value theorem and Proposition 1.2.3, we obtain

$$|f(xy) - f(x)| \leq C\|y\| \leq C'|y| = C'|y||x|^{\lambda-1},$$

using that both sides of the desired inequality are homogeneous functions of the same degree λ. □

In particular, this proposition can be applied to C^1 homogeneous quasi-norms. Specifically, the combination of Proposition 1.2.4 and Proposition 1.2.7 leads to a constant $\gamma > 0$ such that we have

$$|xy| \leq \gamma(|x| + |y|) \quad \text{for all} \quad x, y \in \mathbb{G}, \tag{1.7}$$

$$||xy| - |x|| \leq \gamma|y| \quad \text{for all} \quad x, y \in \mathbb{G} \quad \text{with} \quad |y| \leq |x|/2. \tag{1.8}$$

Henceforth, γ will always be called the minimal constant satisfying (1.7) and (1.8). Obviously, $\gamma \geq 1$. We will be using (1.7) and (1.8) without comment in the sequel. The following simple fact will also be useful later:

Lemma 1.2.8 (Peetre type inequality). *For every $x, y \in \mathbb{G}$ and $s > 0$, we have*

$$(1 + |x|)^s(1 + |y|)^{-s} \leq \gamma^s(1 + |xy^{-1}|)^s.$$

Proof. Because of $|x| \leq \gamma(|xy^{-1}| + |y|)$ we have

$$1 + |x| \leq \gamma(1 + |xy^{-1}|)(1 + |y|),$$

and we obtain the needed inequality by raising both sides to the sth power. □

Let us now fix the notation for some common function spaces on $\mathbb{G}$. Let $\Omega \subset \mathbb{G}$, and let $C(\Omega)$ ($C_0(\Omega)$) be the space of continuous functions on $\mathbb{G}$ (continuous functions with compact support, respectively). If Ω is open, then $C^{(k)}(\Omega)$ is called the class of k times continuously differentiable functions on Ω,

$$C^\infty(\Omega) = \bigcap_{k=1}^{\infty} C^{(k)}(\Omega) \quad \text{and} \quad C_0^\infty(\Omega) = C^\infty(\Omega) \cap C_0(\Omega).$$

When $\Omega = \mathbb{G}$ we shall usually omit mentioning it. If $0 < p \leq \infty$, then L^p will denote the usual Lebesgue space on $\mathbb{G}$. For $0 < p < \infty$ we write

$$\|f\|_p := \left(\int_{\mathbb{G}} |f(x)|^p dx \right)^{1/p},$$

despite the fact that this is not a norm for $p < 1$. However, the map $(f, g) \mapsto \|f - g\|_p^p$ is a metric on L^p for $p < 1$. We recall that if f is a measurable function on $\mathbb{G}$, its *distribution function* $\lambda_f : [0, \infty] \to [0, \infty]$ is defined by

$$\lambda_f(\alpha) := |\{x : |f(x)| > \alpha\}|, \tag{1.9}$$

and its *nonincreasing rearrangement* $f^* : [0, \infty) \to [0, \infty)$ is defined by

$$f^*(t) = \inf\{\alpha : \lambda_f(\alpha) \leq t\}. \tag{1.10}$$

Moreover,

$$\int_{\mathbb{G}} |f(x)|^p dx = -\int_0^\infty \alpha^p d\lambda_f(\alpha) = p \int_0^\infty \alpha^{p-1} \lambda_f(\alpha) d\alpha = \int_0^\infty f^*(t)^p dt.$$

For $0 < p < \infty$, the weak-L^p is the space of functions f such that

$$[f]_p := \sup_{\alpha > 0} \alpha^p \lambda_f(\alpha) = \sup_{t > 0} t^{1/p} f^*(t) < \infty.$$

This $[\cdot]_p$ is not a norm but it defines a topology on the weak-L^p space. A subadditive operator which is bounded from L^p to weak L^q is said to be *weak type* (p, q).

1.2.2 Polar coordinates

There is an analogue of polar coordinates on homogeneous groups. We start with the following observation:

Proposition 1.2.9 (Polar decomposition: a special case). *Let f be a locally integrable function on $\mathbb{G}\backslash\{0\}$ and assume that it is homogeneous of degree $-Q$. Then there is a constant μ_f (the 'average value' of f) such that for every $g \in L^1((0, \infty), r^{-1}dr)$, we have*

$$\int_{\mathbb{G}} f(x) g(|x|) dx = \mu_f \int_0^\infty g(r) r^{-1} dr. \tag{1.11}$$

Proof. Define $L_f : (0, \infty) \to \mathbb{C}$ by

$$L_f(r) := \begin{cases} \int_{1 \leq |x| \leq r} f(x) dx & \text{if} \quad r \geq 1, \\ -\int_{r \leq |x| \leq 1} f(x) dx & \text{if} \quad r < 1. \end{cases}$$

By changing the variables $x \mapsto sx$ and using the homogeneity of f, it can be verified that

$$L_f(rs) = L_f(r) + L_f(s)$$

for all $r, s > 0$. From the continuity of L_f, it then follows that

$$L_f = L_f(e) \log r,$$

and we set $\mu_f := L_f(e)$. Then equality (1.11) is obvious when g is the characteristic function of an interval, and it follows in general by taking linear combinations and limits of such functions. $\square$

Proposition 1.2.10 (Polar decomposition). *Let*

$$\wp := \{x \in \mathbb{G} : |x| = 1\} \tag{1.12}$$

be the unit sphere with respect to the homogeneous quasi-norm $|\cdot|$. Then there is a unique Radon measure σ on $\wp$ such that for all $f \in L^1(\mathbb{G})$,

$$\int_{\mathbb{G}} f(x)dx = \int_0^\infty \int_{\wp} f(ry) r^{Q-1} d\sigma(y) dr. \tag{1.13}$$

Proof. Let $\widetilde{f} \in C(\mathbb{G}\backslash\{0\})$ be the homogeneous extension of $f \in C(\wp)$ defined by

$$\widetilde{f}(x) := |x|^{-Q} f(|x|^{-1}x).$$

Then $\widetilde{f}$ satisfies the hypotheses of Proposition 1.2.9. The map $f \mapsto \mu_{\widetilde{f}}$ is clearly a positive linear functional on $C(\wp)$, so it is given by the integration against some Radon measure σ on $\wp$. If $g \in C_0(0,\infty)$ then we have

$$\begin{aligned} \int_{\mathbb{G}} f(|x|^{-1}x) g(|x|) dx &= \int_{\mathbb{G}} \widetilde{f}(x) |x|^Q g(|x|) dx = \mu_{\widetilde{f}} \int_0^\infty r^{Q-1} g(r) dr \\ &= \int_0^\infty \int_{\wp} f(y) g(r) r^{Q-1} d\sigma(y) dr. \end{aligned}$$

Since linear combination of functions of the form $f(|x|^{-1}x)g(|x|)$ are dense in $L^1(\mathbb{G})$, this completes the existence proof, and from the decomposition it follows that such a measure is necessarily unique. $\square$

Corollary 1.2.11. *Let $C := \sigma(\wp)$. Then if $0 < a < b < \infty$ and $\alpha \in \mathbb{C}$, we have*

$$\int_{a<|x|<b} |x|^{\alpha - Q} dx = \begin{cases} C\alpha^{-1}(b^\alpha - a^\alpha) & \text{if } \alpha \neq 0, \\ C \log(b/a) & \text{if } \alpha = 0. \end{cases}$$

Corollary 1.2.12. *Let f be a measurable function on $\mathbb{G}$ such that*

$$f(x) = O(|x|^{\alpha - Q})$$

for some $\alpha \in \mathbb{R}$. If $\alpha > 0$ then f is integrable near 0, and if $\alpha < 0$ then f is integrable near ∞.

These two corollaries will be frequently used without comment in the sequel.

1.2.3 Convolutions

Let f and g be two integrable function on $\mathbb{G}$. Then their *convolution* $f * g$ is well defined by

$$(f * g)(x) := \int_{\mathbb{G}} f(y)g(y^{-1}x)dy = \int_{\mathbb{G}} f(xy^{-1})g(y)dy.$$

The basic facts about convolution of L^p and weak-L^p functions can be formulated in two propositions. For other properties of convolutions on groups we can refer to [FR16, Sections 1.5 and 3.1.10].

Proposition 1.2.13 (Young's inequality). *Suppose*

$$1 \leq p, q, r \leq \infty \quad \textit{and} \quad \frac{1}{p} + \frac{1}{q} = \frac{1}{r} + 1.$$

If $f \in L^p$ and $g \in L^q$, then $f * g \in L^r$ and

$$\|f * g\|_{L^r(\mathbb{G})} \leq \|f\|_{L^p(\mathbb{G})} \|g\|_{L^q(\mathbb{G})}.$$

Proof. First assuming $r = \infty$, in this case p and q are conjugate exponents and the result follows from Hölder's inequality.

Second assuming $r = q$, $p = 1$, let q' be the conjugate exponent to q. By Hölder's inequality,

$$\begin{aligned} |f * g(x)| &\leq \int_{\mathbb{G}} |f(xy^{-1})|^{(1/q)+(1/q')} |g(y)| dy \\ &\leq \left(\int_{\mathbb{G}} |f(xy^{-1})| dy \right)^{1/q'} \left(\int_{\mathbb{G}} |f(xy^{-1})| |g(y)|^q dy \right)^{1/q} \\ &= \|f\|_1^{1/q'} \left(\int_{\mathbb{G}} |f(xy^{-1})| |g(y)|^q dy \right)^{1/q}. \end{aligned}$$

Thus, by Fubini's theorem, we obtain

$$\begin{aligned} \int_{\mathbb{G}} |f * g(x)|^q dx &\leq \|f\|_1^{q/q'} \int_{\mathbb{G}} \int_{\mathbb{G}} |f(xy^{-1})| |g(y)|^q dy dx \\ &= \|f\|_1^{(q/q')+1} \|g\|_q^q, \end{aligned}$$

so that $\|f * g\|_q \leq \|f\|_1 \|g\|_q$. The rest follows by interpolation. □

Proposition 1.2.14 (Young's inequality for weak-L^p spaces). *Suppose*

$$q \leq p < \infty, \quad 1 < q, r < \infty, \quad \textit{and} \quad \frac{1}{p} + \frac{1}{q} = \frac{1}{r} + 1.$$

If $f \in L^p$ and $g \in L^q$ then $f * g \in$ weak-L^r and there exists $C_1 = C_1(p, q)$ such that

$$[f * g]_r \leq C_1 \|f\|_p [g]_q.$$

Moreover, if $p > 1$ *then* $f * g \in L^r$ *and there exists* $C_2 = C_2(p,q)$ *such that*

$$\|f * g\|_r \le C_2 \|f\|_p [g]_q.$$

Proof. By the Marcinkiewicz interpolation theorem one can notice that the strong result for $p > 1$ follows from the weak result. Suppose then that $f \in L^p$ and $g \in L^q$, and (without loss of generality) that $\|f\|_p = [q]_q = 1$. Given $\alpha > 0$ set

$$M := (\alpha/2)^{r/q} (q/r)^{r/qp'},$$

where p' is the conjugate exponent to p. Define $g_1(x) := g(x)$ if $|g(x)| \le M$ and $g_1(x) := 0$ otherwise, and set $g_2 := g - g_1$. Since

$$\lambda_{f*g}(\alpha) \le \lambda_{f*g_1}(\alpha/2) + \lambda_{f*g_2}(\alpha/2),$$

it is enough to show that each term on the right side is bounded by $C\alpha^{-r}$, where C depends only on p and q. On the one hand, since $q^{-1} - (p')^{-1} = r^{-1} > 0$ we have $p'q > 0$ and therefore

$$\begin{aligned} \int_{\mathbb{G}} |g_1(x)|^{p'} dx &= p' \int_0^\infty \alpha^{p'-1} \lambda_{g_1}(\alpha) d\alpha \le p' \int_0^M \alpha^{p'-1} \lambda_g(\alpha) d\alpha \\ &\le p' \int_0^M \alpha^{p'-1-q} d\alpha = \frac{p'}{p'-q} M^{p'-q} = \frac{r}{q} M^{qp'/r} = (\alpha/2)^{p'}. \end{aligned}$$

Thus, for every $x \in \mathbb{G}$, by Hölder's inequality (or by Proposition 1.2.13) we have

$$|f * g_1(x)| \le \|f\|_p \|g_1\|_{p'} \le \alpha/2,$$

which implies that $\lambda_{f*g_1}(\alpha/2) = 0$. On the other hand, since $q > 1$, we have

$$\begin{aligned} \int_{\mathbb{G}} |g_2(x)| dx &= \int_0^\infty \lambda_{g_2}(\alpha) d\alpha = \int_0^M \lambda_g(M) d\alpha + \int_M^\infty \lambda_g(\alpha) d\alpha \\ &\le M \cdot M^{-q} + \int_M^\infty \alpha^{-q} d\alpha = \frac{q}{q-1} M^{1-q}, \end{aligned}$$

and therefore by Proposition 1.2.13,

$$\|f * g_2\|_p \le \|f\|_p \|g_2\|_1 \le q(q-1)^{-1} M^{1-q}.$$

But then

$$\begin{aligned} \lambda_{f*g_2}(\alpha/2) &\le [2\|f * g_2\|_p/\alpha]^p \\ &\le \left(\frac{2}{\alpha}\right)^p \left(\frac{q}{q-1}\right)^p M^{(1-q)p} \\ &= C(p,q)\alpha^{-r}. \end{aligned}$$

This completes the proof. $\square$

Now let us summarize some properties of approximations to the identity in terms of the convolution. The following notation will be used throughout this monograph: if ϕ is a function on $\mathbb{G}$ and $t > 0$, we define ϕ_t by

$$\phi_t := t^{-Q}\phi \circ \delta_{1/t}, \quad \text{that is,} \quad \phi_t(x) := t^{-Q}\phi(x/t). \tag{1.14}$$

We notice that if $\phi \in L^1(\mathbb{G})$ then $\int_{\mathbb{G}} \phi_t(x)dx$ is independent of t.

Proposition 1.2.15 (Approximation of identity). *Let $\phi \in L^1(\mathbb{G})$ and let $a := \int_{\mathbb{G}} \phi(x)dx$. Then we have the following properties:*

(i) *If $f \in L^p(\mathbb{G})$ for $1 \leq p < \infty$, then $\|f * \phi_t - af\|_p \to 0$ as $t \to 0$.*

(ii) *If f is bounded and right uniformly continuous, then $\|f * \phi_t - af\|_\infty \to 0$ as $t \to 0$.*

(iii) *If f is bounded on $\mathbb{G}$ and continuous on an open set $\Omega \subset \mathbb{G}$, then $f * \phi_t - af \to 0$ uniformly on compact subsets of Ω as $t \to 0$.*

Proof. For a function f on $\mathbb{G}$ and $y \in \mathbb{G}$, let us define

$$f^y(x) := f(xy^{-1}).$$

If $f \in L^p$ for $1 \leq p < \infty$, then it can be shown that

$$\|f^y - f\|_p \to 0 \quad \text{as } y \to 0, \tag{1.15}$$

for example, using the fact that C_0 is dense in L^p. If $p = \infty$, property (1.15) holds if and only if f is (almost everywhere equal to) a right uniformly continuous function. We now observe that

$$\begin{aligned} f * \phi_t(x) - af(x) &= \int_{\mathbb{G}} f(xy^{-1})t^{-Q}\phi(y/t)dy - af(x) \\ &= \int_{\mathbb{G}} f(x(tz)^{-1})\phi(z)dz - af(x) \\ &= \int_{\mathbb{G}} [f(x(tz)^{-1}) - f(x)]\phi(z)dz. \end{aligned}$$

Hence by Minkowski's inequality,

$$\|f * \phi_t - af\|_p \leq \int_{\mathbb{G}} \|f^{tz} - f\|_p |\phi(z)|dz.$$

Since $\|f^{tz} - f\|_p \leq 2\|f\|_p$, under the hypothesis of (i) or (ii) it follows from (1.15) and the dominated convergence theorem that $\|f * \phi_t \ - af\| \to 0$. The routine modification of this argument (with $p = \infty$) needed to establish (iii) is left to the reader. □

1.2.4 Polynomials

The Lie algebra $\mathfrak{g}$ of $\mathbb{G}$ can be understood from different prospectives:

- as tangent vectors at the origin,
- as left invariant (and right invariant) vector fields.

While in this book we will not make much use of the first interpretation, we will need the second. Consequently, let us denote by $\mathfrak{g}_L$ and $\mathfrak{g}_R$ the spaces of left invariant and right invariant vector fields on $\mathbb{G}$.

Let us fix a basis $X_1, \ldots, X_n$ for $\mathfrak{g}$ consisting of eigenvectors for the dilations δ_r with eigenvalues $r^{d_1}, \ldots, r^{d_n}$, i.e., such that

$$\delta_r X_k = r^{d_k} X_k.$$

In other words, first, we consider X_k as left invariant differential operators on $\mathbb{G}$ and we denote by $Y_1, \ldots, Y_n$ the corresponding basis for $\mathfrak{g}_R$: that is, Y_k is the element of $\mathfrak{g}_R$ such that $Y_k|_0 = X_k|_0$ and for $f \in C^1$ we have

$$X_k f(y) = \frac{d}{dt} f(y \cdot \exp(tX_k))|_{t=0},$$
$$Y_k f(y) = \frac{d}{dt} f(\exp(tX_k) \cdot y)|_{t=0}.$$

Then X_k and Y_k are the differential operators homogeneous of degree d_k since

$$\begin{aligned} X_k(f \circ \delta_r)(y) &= \frac{d}{dt} f((ry)\exp(r^{d_k} tX_k))|_{t=0} \\ &= r^{d_k} \frac{d}{dt} f((ry)\exp(tX_k))|_{t=0} \\ &= r^{d_k} (X_k f \circ \delta_r)(y), \end{aligned}$$

and similarly for Y_k. For $I = (i_1, \ldots, i_n) \in \mathbb{N}^n$, we use the notation

$$X^I = X_1^{i_1} X_2^{i_2} \cdots X_n^{i_n}, \quad Y^I = Y_1^{i_1} Y_2^{i_2} \cdots Y_n^{i_n}.$$

According to the Poincaré–Birkhoff–Witt theorem, the operators X^I give a basis for the algebra of left invariant differential operators on the Lie group $\mathbb{G}$. In addition, we also use the notations

$$|I| := i_1 + i_2 + \cdots + i_n, \quad d(I) := d_1 i_1 + d_2 i_2 + \cdots + d_n i_n. \tag{1.16}$$

Here $|I|$ is called an order of the differential operators X^I and Y^I, and $d(I)$ is their degree of homogeneity or the homogeneous degree. If we denote by Δ the set of all numbers $d(I)$ as I ranges over $\mathbb{N}^n$ then we have $\mathbb{N} \subset \Delta$ as $d_1 = 1$.

There are two useful facts. On the one hand, left translations are isometries on $L^2(\mathbb{G})$, and the operators X_k and Y_k are formally skew-adjoint. Therefore,

$$\int_{\mathbb{G}} (X^I f)g = (-1)^{|I|} \int_{\mathbb{G}} f(X^I g), \quad \int_{\mathbb{G}} f(Y^I g) = (-1)^{|I|} \int_{\mathbb{G}} (Y^I f)g,$$

for all smooth functions f and g for which the integrands decay suitably at infinity. On the other hand, the operators X^I and Y^I interact with convolutions by the formulae

$$X^I(f*g) = f*(X^I g), \quad Y^I(f*g) = (Y^I f)*g, \quad (X^I f)*g = f*(Y^I g).$$

Except for the last one these equalities are direct consequences of differentiating, and the third can be obtained by integration by parts:

$$\begin{aligned}(X^I f)*g(x) &= \int_{\mathbb{G}} X^I f(xy) g(y^{-1}) dy = (-1)^{|I|} \int_{\mathbb{G}} f(xy) X^I [g(y^{-1})] dy \\ &= \int_{\mathbb{G}} f(xy)(Y^I g)(y^{-1}) dy = f*(Y^I g)(x).\end{aligned}$$

Definition 1.2.16 (Polynomials on the homogeneous group $\mathbb{G}$). A function P on $\mathbb{G}$ will be called a *polynomial* if $P \circ \exp$ is a polynomial on $\mathfrak{g}$.

We can form a global coordinate system on $\mathbb{G}$ and generate the algebra of polynomials on $\mathbb{G}$ by setting

$$\eta_k = \xi_k \circ \exp^{-1},$$

where $\eta_1, \ldots, \eta_n$ are polynomials on $\mathbb{G}$, and $\xi_1, \ldots, \xi_n$ are the basis for the linear forms on $\mathfrak{g}$ dual to the basis $X_1, \ldots, X_n$ for $\mathfrak{g}$. Therefore, each polynomial on $\mathbb{G}$ can be defined uniquely in the form

$$P = \sum_I a_I \eta^I,$$

where

$$\eta^I = \eta_1^{i_1} \cdots \eta_n^{i_n},$$

$a_I \in \mathbb{C}$, and all but finitely many of the coefficients a_I vanish. Since η^I is homogeneous of degree $d(I)$, the set of possible homogeneous degrees for polynomials coincides with the set Δ. The *isotropic degree* of a polynomial P is

$$\max\{|I| : a_I \neq 0\}.$$

And the *homogeneous degree* of a polynomial P is

$$\max\{d(I) : a_I \neq 0\}.$$

By $\mathcal{P}_N^{\text{iso}}$, for $N \in \mathbb{N}$, we denote the space of polynomials of isotropic degree $\leq N$, and by $\mathcal{P}_a$, for $a \in \Delta$, we denote the space of polynomials of homogeneous degree $\leq a$. Since $1 \leq d_k \leq \bar{d}$ for $k = 1, \ldots, n$, we observe that

$$\mathcal{P}_N \subset \mathcal{P}_N^{\text{iso}} \subset \mathcal{P}_{\bar{d}N}$$

for $N \in \mathbb{N}$.

There is a more explicit description of the group law in terms of the coordinates η_k since the map

$$(x, y) \mapsto \eta_k(xy)$$

is a polynomial on $\mathbb{G} \times \mathbb{G}$. Thus, we have

$$\eta_k((rx)(ry)) = r^{d_k}\eta_k(xy),$$

that is, it is jointly homogeneous of degree d_k and, according to the Baker–Campbell–Hausdorff formula, we have $\eta_k(xy) = \eta_k(x) + \eta_k(y)$ modulo terms of isotropic degree ≥ 2. It follows that

$$\eta_k(xy) = \eta_k(x) + \eta_k(y) + \sum_{I\neq 0, J\neq 0, d(I)+d(J)=d_k} C_k^{IJ}\eta^I(x)\eta^J(y), \tag{1.17}$$

where C_k^{IJ} are constants. It is easy to see that the monomials η^I, η^J can only involve coordinates with homogeneous degree less than d_k, since the multi-indices I and J in (1.17) must satisfy $d(I) < d_k$ and $d(J) < d_k$. In particular, only the coordinates $\eta_1, \ldots, \eta_{j-1}$ can be involved, for instance:

$$\begin{aligned} d_k = 1: &\quad \eta_k(xy) = \eta_k(x) + \eta_k(y), \\ d_k = 2: &\quad \eta_k(xy) = \eta_k(x) + \eta_k(y) + \sum_{d_j=d_l=1} C_k^{jl}\eta_j(x)\eta_l(y). \end{aligned}$$

Proposition 1.2.17 (Polynomials are translation invariant). *For any $a \in \Delta$, $\mathcal{P}_a$ is left translation invariant.*

Proof. According the formula (1.17), it is easy to see that $\eta_k(xy)$ is in $\mathcal{P}_{d_k}$ (as a function of x for each y, and also as a function of y for each x). On the other hand, the η_k's generate all polynomials, therefore $\mathcal{P}_a$ is left translation invariant for all $a \in \Delta$. □

Definition 1.2.18 (Coordinate functions on the group). For $x \in \mathbb{G}$ and $k = 1, \ldots, n$, we can think of

$$x_k := \eta_k(x)$$

as the *coordinates* of the variable x. Thus, each x_k becomes a polynomial of homogeneous degree k.

We now establish a link between left and right invariant differential operators and derivatives with respect to coordinate functions on the group.

Proposition 1.2.19 (Formulae for invariant derivatives). *We have*

$$X_k = \sum P_{kj}(\partial/\partial x_j), \quad Y_k = \sum Q_{kj}(\partial/\partial x_j), \tag{1.18}$$

where $P_{kk} = Q_{kk} = 1$, $P_{kj} = Q_{kj} = 0$ if $d_j < d_k$ or if $d_j = d_k$ and $j \neq k$, and P_{kj}, Q_{kj} are homogeneous polynomials of degree $d_k - d_j$ if $d_j > d_k$.

Proof. Let us define the operator $L_x : \mathbb{G} \to \mathbb{G}$ by $L_x(y) := xy$ for $x \in \mathbb{G}$. Then, using the fact that X_k agrees with $\partial/\partial x_k$ at 0, for each differentiable function f on $\mathbb{G}$ and $x \in \mathbb{G}$, we have

$$X_k f(x) = (X_k f) \circ L_x(0) = X_k(f \circ L_x)(0) = (\partial/\partial x_k)(f \circ L_x)(0).$$

Therefore, by the chain rule, we obtain

$$X_k f(x) = \sum_{k=1}^{n} \frac{\partial f}{\partial x_k}(x) \frac{\partial [x_k \circ L_x]}{\partial x_k}(0).$$

But by formula (1.17) it follows that

$$\frac{\partial [x_j \circ L_x]}{\partial x_k}(0) = \delta_{kj} + \sum_{d(I)=d_j - d_k} C_j^{I[k]} \eta^I(x),$$

where $[k]$ is the multi-index with 1 in the kth place and zeros elsewhere. The desired result for X_k follows from this, and for Y_k it can be proved in a similar way. □

There are also similar expressions for $\partial/\partial x_k$ in terms of X_j or Y_j:

$$\partial/\partial x_k = \sum P'_{kj} X_j = \sum Q'_{kj} Y_j,$$

where P'_{kj}, Q'_{kj} are of the same form as P_{kj}, Q_{kj} in (1.18). Above formulae can be directly obtained from (1.18) with $j = n$, that is, we have

$$\begin{aligned}
X_n &= \partial/\partial x_n, \\
X_{n-1} &= \partial/\partial x_{n-1} + P_{(n-1)n} \partial/\partial x_n, \\
X_{n-2} &= \partial/\partial x_{n-2} + P_{(n-2)(n-1)} \partial/\partial x_{n-1} + P_{(n-2)n} \partial/\partial x_n,
\end{aligned}$$

therefore, we obtain

$$\begin{aligned}
\partial/\partial x_n &= X_n, \\
\partial/\partial x_{n-1} &= X_{n-1} - P_{(n-1)n} \partial/\partial x_n, \\
\partial/\partial x_{n-2} &= X_{n-2} - P_{(n-2)(n-1)} \partial/\partial x_{n-1} - P_{(n-2)n} \partial/\partial x_n,
\end{aligned}$$

and so on. Similarly, one can obtain expressions for higher-order derivatives. For instance,

$$X^I = \sum_{|J| \le |I|, d(J) \ge d(I)} P_{IJ} (\partial/\partial x)^J, \tag{1.19}$$

where P_{IJ} is a homogeneous polynomial of degree $d(J) - d(I)$. Analogously, we obtain formulae for Y^I in terms of $(\partial/\partial x)^J$ and for $(\partial/\partial x)^I$ in terms of X^J or

Y^J, that is,

$$X^I = \sum_{|J|\le|I|, d(J)\le d(I)} P_{IJ} Y^J,$$
$$Y^I = \sum_{|J|\le|I|, d(J)\le d(I)} Q_{IJ} X^J,$$

where P_{IJ} and Q_{IJ} are homogeneous polynomials of degree $d(J) - d(I)$.

Proposition 1.2.20 (Determination of invariant differential operators). *Let $a \in \Delta$ and let $\mu := \dim P_a$. Then the following maps are linear isomorphisms from P_a to $\mathbb{C}^\mu$:*

(i) $P \to ((\partial/\partial x)^I P(0))_{d(I)\le a}$,

(ii) $P \to (X^I P(0))_{d(I)\le a}$,

(iii) $P \to (Y^I P(0))_{d(I)\le a}$.

Proof. Note that Case (i) is a simple consequence of Taylor's theorem. Also, in view of (1.19), since P_{IJ} is a constant function when $d(I) = d(J)$ and $P_{IJ}(0) = 0$ when $d(J) > d(I)$, we have

$$X^I|_0 = \sum_{|J|\le|I|, d(J)=d(I)} P_{IJ} (\partial/\partial x)^J|_0,$$

and similarly for the other formulae relating X^I, Y^I and $(\partial/\partial\eta)^I$. Cases (ii) and (iii) follow easily from this observation together with Case (i). □

The properties above motivate the following:

Definition 1.2.21 (Taylor polynomials). Let $x \in \mathbb{G}$, $a \in \Delta$, and let f be a function whose (distributional) derivatives $X^I f$ (resp. $Y^I f$) are continuous functions in a neighborhood of x for $d(I) \le a$. The left (resp. right) *Taylor polynomial* of f at x of homogeneous degree a is the unique $P \in P_a$ such that $X^I P(0) = X^I f(x)$ (resp. $Y^I P(0) = Y^I f(x)$) for all I such that $d(I) \le a$.

Now we provide simple proofs of an explicit expression of the Taylor formula and the Taylor inequality in the spirit of [Bon09].

Let $X \in \mathfrak{g}$ be given, and suppose $\gamma(t)$ is any integral curve of X, i.e.

$$\dot\gamma = X(\gamma(t))$$

for all $t \in \mathbb{R}$. If $m \in \mathbb{N} \cup \{0\}$ and $u \in C^{m+1}(\mathbb{G})$ is real-valued, then since

$$\frac{d^k}{dt^k}(u(\gamma(t))) = (X^k u)(\gamma(t)),$$

for each $k \in \mathbb{N}\cup\{0\}$, by applying the usual Taylor formula (with integral reminder) to $t \mapsto u(\gamma(t))$, we obtain

$$u(\gamma(t)) = \sum_{k=0}^{m} \frac{t^k}{k!}(X^k u)(\gamma(0)) + \frac{1}{m!}\int_0^t (t-s)^m (X^{m+1}u)(\gamma(s))ds. \tag{1.20}$$

Moreover, it is easy to see that the integral curve γ of $\log h$ starting at x is $s \mapsto \gamma(s) = x\exp(s\log h)$. With $\gamma(0) = x, \gamma(1) = xh$, we get

$$u(xh) = \sum_{k=0}^{m} \frac{1}{k!}((\log h)^k u)(x) + \frac{1}{m!}\int_0^1 (1-s)^m ((\log h)^{m+1}u)(x\exp(s\log h))ds. \tag{1.21}$$

On the other hand, there always exist (polynomial) functions $\mathbb{G} \ni h \mapsto \zeta_i(h) \in \mathbb{R}$ such that

$$\log h = \zeta_1(h)X_1 + \cdots + \zeta_n(h)X_n,$$

for all $h \in \mathbb{G}$, where $\{X_1, \ldots, X_n\}$ is a basis of the Lie algebra of $\mathbb{G}$. Thus, we have

$$(\log h)^k = \left(\sum_{i=1}^{n} \zeta_i(h)X_i\right)^k = \sum_{i_1,\ldots,i_k=1}^{n} \zeta_{i_1}(h)\cdots\zeta_{i_k}(h)X_{i_1}\cdots X_{i_k},$$

for every $k \in \mathbb{N}$.

Therefore, (1.21) implies that

$$\begin{aligned} u(xh) = u(x) &+ \sum_{k=1}^{m} \sum_{I=(i_1,\ldots,i_k),i_1,\ldots,i_k\leq n} \frac{X_I u(x)}{k!}\zeta_{i_1}(h)\cdots\zeta_{i_k}(h) \\ &+ \sum_{I=(i_1,\ldots,i_{m+1}),i_1,\ldots,i_{n+1}\leq n} \zeta_{i_1}(h)\cdots\zeta_{i_{m+1}}(h) \\ &\times \int_0^1 (X_I u)\left(x\exp\left(\sum_{i\leq n} s\zeta_i(h)X_i\right)\right)\frac{(1-s)^m}{m!}ds, \end{aligned} \tag{1.22}$$

where $X_I = X_{i_1}\cdots X_{i_k}$ and $I = (i_1, \ldots, i_k)$ with $i_1, \ldots, i_k \in \{1, \ldots, n\}$.

For a multi-index α, we will be using the notations (1.16), i.e.,

$$|\alpha| = \alpha_1 + \cdots + \alpha_n, \quad d(\alpha) = d_1\alpha_1 + \cdots + d_n\alpha_n,$$

where d_j is the homogeneous degree of X_k; these are called the Euclidean length and the homogeneous length of α, respectively. One also sets

$$\mathfrak{G} := \{d(\alpha) : \alpha \in (\mathbb{N}\cup\{0\})^n\}.$$

As usual, $[\beta]$ below is the integer part of the real number β. Now we are in a position to state the Taylor formula on homogeneous groups.

Theorem 1.2.22 (Taylor formula). *Let $\mathbb{G}$ be a homogeneous group (identified with $\mathbb{R}^n$ as a topological space). Suppose $\{X_1, \dots, X_n\}$ is the Jacobian basis for its Lie algebra, $m \in \mathfrak{G}$ and $u \in C^{[m]+1}(\mathbb{G})$. Let also $x_0 \in \mathbb{G}$ be fixed. Then, for every $x \in \mathbb{G}$ we have*

$$\begin{aligned} u(x) &= P_m(u, x_0)(x) + R_m(x, x_0) \\ &= u(x_0) + \sum_{k=1}^{[m]} \sum_{\substack{I=(i_1,\dots,i_k),\\ i_1,\dots,i_k \le n,\\ d(I)\le m}} \frac{X_I u(x_0)}{k!} \zeta_{i_1}(x_0^{-1}x) \cdots \zeta_{i_k}(x_0^{-1}x) + R_m(x, x_0), \end{aligned} \tag{1.23}$$

where the reminder term $R_m(x, x_0)$ is given by

$$\begin{aligned} R_m(x, x_0) = &\sum_{k=1}^{[m]} \sum_{\substack{I=(i_1,\dots,i_k),\\ i_1,\dots,i_k \le n,\\ d(I)>m}} \frac{X_I u(x_0)}{k!} \zeta_{i_1}(x_0^{-1}x) \cdots \zeta_{i_k}(x_0^{-1}x) \\ &+ \sum_{\substack{I=(i_1,\dots,i_{[n]+1}),\\ i_1,\dots,i_{[m]+1} \le n}} \zeta_{i_1}(x_0^{-1}x) \cdots \zeta_{i_{[m]+1}}(x_0^{-1}x) \\ &\times \int_0^1 (X_I u)\left(x_0 \exp\left(\sum_{i\le n} s\zeta_i(x_0^{-1}x) X_i\right)\right) \frac{(1-s)^{[m]}}{[m]!} ds. \end{aligned}$$

Proof of Theorem 1.2.22. If $x_0 \in \mathbb{G}$ is any fixed element, by replacing x and h in the formula (1.22) by respectively x_0 and $x_0^{-1}x$, we obtain the following:

$$\begin{aligned} u(x) = u(x_0) + &\sum_{k=1}^{[m]} \sum_{\substack{I=(i_1,\dots,i_k),\\ i_1,\dots,i_k \le n,\\ d(I)>m}} \frac{X_I u(x_0)}{k!} \zeta_{i_1}(x_0^{-1}x) \cdots \zeta_{i_k}(x_0^{-1}x) \\ &+ \sum_{\substack{I=(i_1,\dots,i_{[m]+1}),\\ i_1,\dots,i_{[m]+1} \le n}} \zeta_{i_1}(x_0^{-1}x) \cdots \zeta_{i_{[m]+1}}(x_0^{-1}x) \\ &\times \int_0^1 (X_I u)\left(x_0 \exp\left(\sum_{i\le n} s\zeta_i(x_0^{-1}x) X_i\right)\right) \frac{(1-s)^{[m]}}{[m]!} ds. \end{aligned} \tag{1.24}$$

Since a polynomial $\zeta_i(x)$ is homogeneous of degree σ_i, there exists $C_1 > 0$ such that

$$C_1^{-1}|x|^{\sigma_i} \le |\zeta_i(x)| \le C_1 |x|^{\sigma_i}, \quad \forall x \in \mathbb{G}, \quad i = 1, \dots, n.$$

As a consequence, for every $k \in \mathbb{N}$, $\zeta_{i_1} \cdots \zeta_{i_k}$ is a homogeneous polynomial of degree $d_{i_1} + \dots + d_{i_k}$. Similarly, $\zeta_{i_1} \cdots \zeta_{i_{[m]+1}}$ is homogeneous of degree $\ge [m] + 1$ (since the d_i's are ≥ 1) and, there appear only derivatives $X_I u$ with $d(I) \geq [m]+1 > m$ in the integral summands.

We restate (1.24) emphasizing out the polynomial of degree $\leq m$ in the right-hand side:

$$\begin{aligned} u(x) = u(x_0) &+ \sum_{k=1}^{[m]} \sum_{\substack{I=(i_1,\dots,i_k),\\ i_1,\dots,i_k\leq n,\\ d(I)\leq m}} \frac{X_I u(x_0)}{k!} \zeta_{i_1}(x_0^{-1}x)\cdots\zeta_{i_k}(x_0^{-1}x) \\ &+ \sum_{k=1}^{[m]} \sum_{\substack{I=(i_1,\dots,i_k),\\ i_1,\dots,i_k\leq n,\\ d(I)> m}} \frac{X_I u(x_0)}{k!} \zeta_{i_1}(x_0^{-1}x)\cdots\zeta_{i_k}(x_0^{-1}x) \\ &+ \sum_{\substack{I=(i_1,\dots,i_{[m]+1}),\\ i_1,\dots,i_{[m]+1}\leq n}} \zeta_{i_1}(x_0^{-1}x)\cdots\zeta_{i_{[m]+1}}(x_0^{-1}x) \\ &\times \int_0^1 (X_I u)\left(x_0 \exp\left(\sum_{i\leq n} s\zeta_i(x_0^{-1}x)X_i\right)\right)\frac{(1-s)^{[m]}}{[m]!}ds \\ &=: P_m(u,x_0)(x) + R_m(x,x_0). \end{aligned}$$

By construction, $P_m(u,x_0)(x)$ is a polynomial of homogeneous degree $\leq m$. $\square$

Theorem 1.2.23 (Taylor inequality). *Assume the hypotheses of Theorem 1.2.22. Then for every fixed homogeneous norm $|\cdot|$ on $\mathbb{G}$ and every $m \in \mathfrak{G}$, there exists $C > 0$ (depending on $\mathbb{G}$ and $|\cdot|$) such that*

$$|R_m(x,x_0)| \leq \sum_{k=1}^{[m]+1} \frac{C^k}{k!} \sum_{\substack{I=(i_1,\dots,i_k),\\ i_1,\dots,i_k\leq n,\\ d(I)> m}} |x_0^{-1}x|^{d(I)} \sup_{|y|\leq C|x_0^{-1}x|} |X_I u(x_0^{-1}y)|. \quad (1.25)$$

Moreover, an explicit formula for the Taylor polynomial $P_m(u,x_0)$ of degree $m \in \mathfrak{G}$ related to u about x_0 is

$$\begin{aligned} P_m(u,x_0)(x) &= u(x_0) \\ &+ \sum_{k=1}^{[m]} \sum_{I=(i_1,\dots,i_k),i_1,\dots,i_k\leq n,d(I)\leq m} \frac{X_I u(x_0)}{k!}\zeta_{i_1}(x_0^{-1}x)\cdots\zeta_{i_k}(x_0^{-1}x). \quad (1.26) \end{aligned}$$

Proof of Theorem 1.2.23. Since

$$\sum_{i=1}^{n} s\zeta_i(x)X_i = s\log(x)$$

by Proposition 1.2.5 we obtain

$$\left|(X_I u)\left(\exp\left(\sum_{i\le n} s\zeta_i(x)X_i\right)\right)\right| \le \sup_{|y|\le C_0|x|} |X_I u(y)|.$$

This implies that

$$\begin{aligned}|R_m(x_0^{-1}x)| &\le \sum_{k=1}^{[m]+1} \sum_{\substack{I=(i_1,\dots,i_k),\\ i_1,\dots,i_k\le n,\\ d(I)>m}} \sup_{|y|\le C_0|x_0^{-1}x|} |X_I u(x_0^{-1}y)| \frac{\zeta_{i_1}(x_0^{-1}x)\cdots\zeta_{i_k}(x_0^{-1}x)}{k!} \\ &\le \sum_{k=1}^{[m]+1} \frac{C_1^k}{k!} \sum_{\substack{I=(i_1,\dots,i_k),\\ i_1,\dots,i_k\le n,\\ d(I)>m}} |x_0^{-1}x|^{\sigma(I)} \sup_{|y|\le C_0|x_0^{-1}x|} |X_I u(x_0^{-1}y)|.\end{aligned}$$

Choosing $C := \max\{C_0, C_1\}$ we complete the proof of the Taylor inequality (1.25).

First, note that the above estimate of R_m gives $R_m(x) = \mathcal{O}(\rho^{m+\varepsilon}(x))$ as $x \to 0$, where

$$\varepsilon := \min_{k=1,\dots,[m]+1} \{d(I) - m : I = (i_1,\dots,i_k), i_1,\dots,i_k \le n, d(I) > m\}. \tag{1.27}$$

Thus, the Taylor formula (1.23) can be rewritten as $u(x)=P_m(x)+\mathcal{O}(|x|^{m+\varepsilon})$ as $x \to 0$, with $\varepsilon > 0$ as in (1.27).

Now let us see that there exists at most one polynomial function P on $\mathbb{G}$, with degree $\le m$, such that, for some $\varepsilon > 0$ (depending on P and m) it holds

$$u(x) = P(x) + \mathcal{O}_{x\to x_0}(|x_0^{-1}x|^{m+\varepsilon}). \tag{1.28}$$

Indeed, suppose there are two such polynomials, A and B (with related $\varepsilon_1, \varepsilon_2 > 0$). Then setting $\varepsilon := \min\{\varepsilon_1, \varepsilon_2\}$ we have

$$Q(x) := B - A = \mathcal{O}_{x\to x_0}(|x_0^{-1}x|^{m+\varepsilon}).$$

Setting $\widetilde{Q}(z) := Q(x_0 z)$, this is equivalent to

$$\widetilde{Q}(z) = \mathcal{O}_{z\to 0}(|z|^{m+\varepsilon}). \tag{1.29}$$

The fact that Q is a polynomial of degree at most m and the ith component function of $x_0 z$ is a polynomial in z of degree at most d_i, it follows that $Q(x_0 z)$ is a polynomial in z of degree at most m.

Therefore, (1.29) is valid if and only if $\widetilde{Q} \equiv 0$, that is, $Q(x_0 z) = 0$ for all $z \in \mathbb{G}$. This is in turn equivalent to $Q \equiv 0$, i.e., $A \equiv B$. Note that the equivalence of all homogeneous norms (cf. Proposition 1.2.3) implies that a polynomial P as in (1.28) is independent of $|\cdot|$. Thus, P_m is the Taylor polynomial of degree m related to u, which has the explicit formula (1.26). $\square$

1.3 Radial and Euler operators

An important tool for working on homogeneous groups will be an extensive use of radial and Euler operators. We now discuss them in some detail and establish a number of properties used throughout the book.

1.3.1 Radial derivative

First we introduce a radial derivative (acting on a differentiable function f) on a homogeneous group $\mathbb{G}$ by

$$\mathcal{R}f(x) := \frac{df(x)}{d|x|}, \tag{1.30}$$

where $|x|$ is a homogeneous quasi-norm of $\mathbb{G}$. Note that the homogeneous quasi-norm $|x|$ in the formula (1.30) can be arbitrary, that is, in general the radial operator $\mathcal{R}$ depends on a chosen homogeneous quasi-norm.

Let $\{X_1, \ldots, X_n\}$ be a basis of the Lie algebra $\mathfrak{g}$ of $\mathbb{G}$ such that we have

$$AX_k = \nu_k X_k \quad \text{for every} \ \ k = 1, \ldots, n.$$

Then the matrix A can be taken to be $A = \operatorname{diag}(\nu_1, \ldots, \nu_n)$ and each X_k is homogeneous of degree ν_k. By decomposing the vector $\exp_{\mathbb{G}}^{-1}(x)$ in $\mathfrak{g}$ with respect to the basis $\{X_1, \ldots, X_n\}$, we get the vector

$$e(x) = (e_1(x), \ldots, e_n(x))$$

given by the formula

$$\exp_{\mathbb{G}}^{-1}(x) = e(x) \cdot \nabla \equiv \sum_{j=1}^{n} e_j(x) X_j,$$

where

$$\nabla = (X_1, \ldots, X_n)$$

is the full gradient. It gives the equality

$$x = \exp_{\mathbb{G}} \left(e_1(x) X_1 + \cdots + e_n(x) X_n \right). \tag{1.31}$$

By homogeneity and denoting $x = ry$, with $y \in \wp$ being on the quasi-sphere (1.12), we get

$$e(x) = e(ry) = (r^{\nu_1} e_1(y), \ldots, r^{\nu_n} e_n(y)).$$

Indeed, since each X_k is homogeneous of degree ν_k, from (1.31) we get that

$$rx = \exp_{\mathbb{G}} \left(r^{\nu_1} e_1(x) X_1 + \cdots + r^{\nu_n} e_n(x) X_n \right),$$

and hence

$$e(rx) = (r^{\nu_1} e_1(x), \ldots, r^{\nu_n} e_n(x)).$$

Thus, since $r > 0$ is arbitrary, without loss of generality taking $|x| = 1$, we can write

$$\frac{d}{d|rx|} f(rx) = \frac{d}{dr} f(\exp_{\mathbb{G}}(r^{\nu_1} e_1(x) X_1 + \cdots + r^{\nu_n} e_n(x) X_n)). \tag{1.32}$$

So, summarizing, one obtains

$$\frac{d}{d|x|}(f(x)) = \frac{d}{dr}(f(ry)) = \frac{d}{dr}(f(\exp_{\mathbb{G}}(r^{\nu_1} e_1(y) X_1 + \cdots + r^{\nu_n} e_n(y) X_n))). \tag{1.33}$$

Throughout this book we will be often abbreviate the notation by writing

$$\mathcal{R} := \frac{d}{dr}, \tag{1.34}$$

meaning that the derivative is taken with respect to the radial direction with respect to the quasi-norm $|\cdot|$.

We can also observe that for any differentiable function f we have

$$\frac{d}{d|x|} f(x) = \frac{d}{d|x|} f\left(\frac{x}{|x|}|x|\right) = \frac{x}{|x|}\frac{d}{dx} f(x) = \frac{x \cdot \nabla_E}{|x|} f(x), \tag{1.35}$$

since for $x \in \mathbb{G}$, we have that $\frac{x}{|x|}$ does not depend on $|x|$, and where

$$\nabla_E = \left(\frac{\partial}{\partial x_1}, \ldots, \frac{\partial}{\partial x_n}\right)$$

is an anisotropic (Euclidean) gradient on $\mathbb{G}$ consisting of partial derivatives with respect to coordinate functions.

Although x_j and $\frac{\partial}{\partial x_j}$ may have degrees of homogeneity depending on j, the operator

$$\mathcal{R} = \frac{x \cdot \nabla_E}{|x|} = \frac{d}{d|x|} \tag{1.36}$$

is homogeneous of degree -1.

1.3.2 Euler operator

Given the radial derivative operator $\mathcal{R}$, we define the *Euler operator* on $\mathbb{G}$ by

$$\mathbb{E} := |x|\mathcal{R}. \tag{1.37}$$

Since $\mathcal{R}$ is homogeneous of degree -1, the operator $\mathbb{E}$ is homogeneous of degree 0.

We can note the following useful property shedding some more light on the link between the radial derivative and the Euler operators, also clarifying how to take derivatives with respect to points that are not on the quasi-sphere $\wp$. Thus, for $x \in \mathbb{G}$, we can write $x = ry$ with $y \in \wp$. Then, denoting $\rho := e^t r$ for $t \in \mathbb{R}$, we have

$$\frac{d}{dt}(f(e^t x)) = \frac{d}{dt}(f(e^t ry)) = \frac{d}{dt}(f(\rho y)) = \rho\frac{d}{d\rho}(f(\rho y)) = \mathbb{E}f(\rho y) = \mathbb{E}f(e^t x),$$

that is,

$$\frac{d}{dt}f(e^t x) = \mathbb{E}f(e^t x). \tag{1.38}$$

The Euler operator has the following useful properties, also justifying the name of Euler associated to this operator.

Proposition 1.3.1 (Properties of the Euler operator). *We have the following properties:*

(i) *Let $\nu \in \mathbb{R}$. If $f : \mathbb{G}\backslash\{0\} \to \mathbb{R}$ is differentiable, then*

$$\mathbb{E}(f) = \nu f \quad \textit{if and only if} \quad f(rx) = r^\nu f(x) \quad (\forall r > 0, x \neq 0).$$

(ii) *The formal adjoint operator of $\mathbb{E}$ has the form*

$$\mathbb{E}^* = -Q\mathbb{I} - \mathbb{E}, \tag{1.39}$$

where $\mathbb{I}$ is the identity operator.

(iii) *For all complex-valued functions $f \in C_0^\infty(\mathbb{G}\backslash\{0\})$ we have*

$$\|\mathbb{E}f\|_{L^2(\mathbb{G})} = \|\mathbb{E}^* f\|_{L^2(\mathbb{G})} \,. \tag{1.40}$$

Proof. Part (i). If a function f is positively homogeneous of order ν, that is, if

$$f(rx) = r^\nu f(x)$$

holds for all $r > 0$ and $x := \rho y \neq 0$, $y \in \wp$, then using (1.30) for such an f, it follows that

$$\mathbb{E}f = \nu f(x).$$

Conversely, let us fix $x \neq 0$ and define

$$g(r) := f(rx).$$

Using (1.30), the equality $\mathbb{E}f(rx) = \nu f(rx)$ means that

$$g'(r) = \frac{d}{dr}f(rx) = \frac{1}{r}\mathbb{E}f(rx) = \frac{\nu}{r}f(rx) = \frac{\nu}{r}g(r).$$

Consequently, $g(r) = g(1)r^\nu$, i.e., $f(rx) = r^\nu f(x)$ and thus f is positively homogeneous of order ν.

Part (ii). We can calculate the formal adjoint operator of $\mathbb{E}$ on $C_0^\infty(\mathbb{G}\backslash\{0\})$ as follows:

$$\begin{aligned}\int_{\mathbb{G}} \mathbb{E}f(x)\overline{g(x)}dx &= \int_0^\infty \int_{\wp} \frac{d}{dr} f(ry)\overline{g(ry)} r^Q d\sigma(y)dr \\ &= -\int_0^\infty \int_{\wp} f(ry)\left(Qr^{Q-1}\overline{g(ry)} + r^Q \frac{d}{dr}\overline{g(ry)}\right) d\sigma(y)dr \\ &= -\int_{\mathbb{G}} f(x)(Q+\mathbb{E})\overline{g(x)}dx,\end{aligned}$$

by the polar decomposition in Proposition 1.2.10 and the integration by parts using formula (1.30).

Part (iii). By using the representation of $\mathbb{E}^*$ in (1.39), we get

$$\begin{aligned}\|\mathbb{E}^* f\|^2_{L^2(\mathbb{G})} &= \|(-Q\mathbb{I} - \mathbb{E})f\|^2_{L^2(\mathbb{G})} \\ &= Q^2 \|f\|^2_{L^2(\mathbb{G})} + 2Q\mathrm{Re}\int_{\mathbb{G}} f(x)\overline{\mathbb{E}f(x)}dx + \|\mathbb{E}f\|^2_{L^2(\mathbb{G})}.\end{aligned} \tag{1.41}$$

Then we have

$$\begin{aligned}2Q\mathrm{Re}\int_{\mathbb{G}} f(x)\overline{\mathbb{E}f(x)}dx &= 2Q\mathrm{Re}\int_0^\infty \int_{\wp} f(ry)\frac{d}{dr}\overline{f(ry)} r^Q d\sigma(y)dr \\ &= Q\int_0^\infty r^Q \int_{\wp} \frac{d}{dr}(|f(ry)|^2)d\sigma(y)dr \\ &= -Q^2 \int_0^\infty \int_{\wp} |f(ry)|^2 r^{Q-1} d\sigma(y)dr \\ &= -Q^2 \|f\|^2_{L^2(\mathbb{G})}.\end{aligned} \tag{1.42}$$

Combining this with (1.41) we obtain (1.40). ☐

Let us introduce the following operator that will be of importance in the sequel,

$$\mathbb{A} := \mathbb{E}\mathbb{E}^*.$$

It is easy to see that this operator is formally self-adjoint, that is,

$$\mathbb{A} = \mathbb{E}\mathbb{E}^* = \mathbb{E}^*\mathbb{E} = \mathbb{A}^*,$$

where we can use Proposition 1.3.1, Part (ii), to also write

$$\mathbb{E}\mathbb{E}^* = \mathbb{E}^*\mathbb{E} = -Q\mathbb{E} - \mathbb{E}^2.$$

Then by replacing f by $\mathbb{E}f$ in (1.40), we obtain the equality

$$\|\mathbb{A}f\|_{L^2(\mathbb{G})} = \left\|\mathbb{E}^2 f\right\|_{L^2(\mathbb{G})} \tag{1.43}$$

for all complex-valued functions $f \in C_0^\infty(\mathbb{G}\backslash\{0\})$. Moreover, the operator $\mathbb{A}$ is *Komatsu-non-negative* in $L^2(\mathbb{G})$, which means that $(-\infty, 0)$ is included in the resolvent set $\rho(\mathbb{A})$ of $\mathbb{A}$ and we have the property

$$\exists M > 0,\ \forall \lambda > 0, \quad \|(\lambda + \mathbb{A})^{-1}\|_{L^2(\mathbb{G})\to L^2(\mathbb{G})} \leq M\lambda^{-1}.$$

Indeed, and more precisely, we have the following:

Lemma 1.3.2 ($\mathbb{A} = \mathbb{E}\mathbb{E}^*$ is Komatsu-non-negative). *The operator $\mathbb{A} = \mathbb{E}\mathbb{E}^*$ is Komatsu-non-negative in $L^2(\mathbb{G})$:*

$$\|(\lambda + \mathbb{A})^{-1}\|_{L^2(\mathbb{G})\to L^2(\mathbb{G})} \leq \lambda^{-1} \text{ for all } \lambda > 0. \tag{1.44}$$

Proof. We start with $f \in C_0^\infty(\mathbb{G}\backslash\{0\})$. Using Proposition 1.3.1, Part (ii), a direct calculation shows that we have the equality

$$\begin{aligned}
&\|(\lambda\mathbb{I} + \mathbb{A})f\|^2_{L^2(\mathbb{G})} = \|(\lambda\mathbb{I} - \mathbb{E}(Q\mathbb{I} + \mathbb{E}))f\|^2_{L^2(\mathbb{G})} \\
&\quad = \lambda^2 \|f\|^2_{L^2(\mathbb{G})} + \|\mathbb{E}(Q\mathbb{I} + \mathbb{E})f\|^2_{L^2(\mathbb{G})} - 2\lambda \mathrm{Re}\int_{\mathbb{G}} f(x)\overline{Q\mathbb{E}f + \mathbb{E}^2 f}dx.
\end{aligned} \tag{1.45}$$

Since

$$\begin{aligned}
\mathrm{Re}\int_{\mathbb{G}} f(x)\overline{\mathbb{E}^2 f}dx &= \mathrm{Re}\int_0^\infty \int_{\wp} f(ry)\overline{\frac{d}{dr}(\mathbb{E}f(ry))r^Q}d\sigma(y)dr \\
&= -\mathrm{Re}\int_0^\infty \int_{\wp} \overline{(\mathbb{E}f(ry))}\left(r^Q\frac{d}{dr}f(ry) + Qr^{Q-1}f(ry)\right)d\sigma(y)dr \\
&= -\|\mathbb{E}f\|^2_{L^2(\mathbb{G})} - Q\mathrm{Re}\int_{\mathbb{G}} \overline{\mathbb{E}f(x)}f(x)dx,
\end{aligned}$$

we have

$$\begin{aligned}
&-2\lambda\mathrm{Re}\int_{\mathbb{G}} f(x)\overline{Q\mathbb{E}f(x) + \mathbb{E}^2 f(x)}dx \\
&= -2\lambda Q\mathrm{Re}\int_{\mathbb{G}} f(x)\overline{\mathbb{E}f(x)}dx - 2\lambda\mathrm{Re}\int_{\mathbb{G}} f(x)\overline{\mathbb{E}^2 f(x)}dx = 2\lambda \|\mathbb{E}f\|^2_{L^2(\mathbb{G})}.
\end{aligned} \tag{1.46}$$

Combining (1.45) with (1.46), we obtain the equality

$$\|(\lambda\mathbb{I} - \mathbb{E}(Q\mathbb{I} + \mathbb{E}))f\|^2_{L^2(\mathbb{G})} = \lambda^2 \|f\|^2_{L^2(\mathbb{G})} + 2\lambda \|\mathbb{E}f\|^2_{L^2(\mathbb{G})} + \|\mathbb{E}(Q\mathbb{I} + \mathbb{E})f\|^2_{L^2(\mathbb{G})}.$$

By dropping positive terms, it follows that

$$\|(\lambda\mathbb{I} - \mathbb{E}(Q\mathbb{I} + \mathbb{E}))f\|^2_{L^2(\mathbb{G})} \geq \lambda^2 \|f\|^2_{L^2(\mathbb{G})},$$

which implies (1.44). □

We can refer to [FR16, Section A.3] for more details on general further properties of Komatsu-non-negative operators and their use in the theory of fractional powers of operators.

1.3.3 From radial to non-radial inequalities

Now we show that the Euler operator and, consequently, also the radial derivative operator, have a very useful property that in order to prove certain inequalities on $\mathbb{G}$ is may be enough to prove them only for radially symmetric functions. We summarize it in the following proposition. Such ideas will be of use, for example, in the analysis of remainder estimates in Theorem 2.3.1 or in Theorem 3.2.6.

As usual, we will say that a function $f = f(x)$ on $\mathbb{G}$ is *radial*, or *radially symmetric*, if it depends only on $|x|$; clearly, this notion depends on the seminorm that we are using.

Proposition 1.3.3 (Radialisation of functions). *Let $\phi_1, \phi_2, \phi_3 \in L^1_{\rm loc}(\mathbb{G})$ be arbitrary radially symmetric functions. For $f \in L^p_{\rm loc}(\mathbb{G})$, define its radial average by*

$$\widetilde{f}(|x|) := \left(\frac{1}{|\wp|}\int_{\wp} |f(|x|y)|^p d\sigma(y)\right)^{1/p}. \tag{1.47}$$

Then for any $f \in L^p_{\rm loc}(\mathbb{G})$ and $1 < p < \infty$ we have the equality

$$\int_{\mathbb{G}} \phi_1(x) \left|\widetilde{f}(|x|)\right|^p dx = \int_{\mathbb{G}} \phi_1(x)\, |f(x)|^p\, dx. \tag{1.48}$$

Moreover, if $\phi_2, \phi_3 \geq 0$, we have the inequalities

$$\int_{\mathbb{G}} \phi_2(x) \left|\mathbb{E}^k \widetilde{f}(|x|)\right|^p dx \leq \int_{\mathbb{G}} \phi_2(x) \left|\mathbb{E}^k f(x)\right|^p dx \tag{1.49}$$

and

$$\int_{\mathbb{G}} \phi_3(x) \left|\mathcal{R}^k \widetilde{f}(|x|)\right|^p dx \leq \int_{\mathbb{G}} \phi_3(x) \left|\mathcal{R}^k f(x)\right|^p dx, \tag{1.50}$$

for all $1 < p < \infty$, any $k \in \mathbb{N}$ and all $f \in L^p_{\rm loc}(\mathbb{G})$ such that $\mathbb{E}^k f \in L^p_{\rm loc}(\mathbb{G})$ or $\mathcal{R}^k f \in L^p_{\rm loc}(\mathbb{G})$, respectively. The constants in these inequalities are sharp, and are attained when $f = \widetilde{f}$.

Proof. Using definition (1.47) and the polar decomposition formula in Proposition 1.2.10, we have

$$\begin{aligned}\int_{\mathbb{G}} |\widetilde{f}(|x|)|^p \phi_1(x) dx &= |\wp| \int_0^\infty |\widetilde{f}(r)|^p \phi_1(r) r^{Q-1} dr \\ &= |\wp| \int_0^\infty \frac{1}{|\wp|}\int_{\wp} |f(ry)|^p d\sigma(y) \phi_1(r) r^{Q-1} dr = \int_{\mathbb{G}} |f(x)|^p \phi_1(x) dx,\end{aligned} \tag{1.51}$$

which proves the identity (1.48).

To prove (1.49) let us show first that

$$|\mathbb{E}^k \widetilde{f}| \leq \left(\frac{1}{|\wp|}\int_{\wp} \left|\mathbb{E}^k f(ry)\right|^p d\sigma(y)\right)^{\frac{1}{p}}, \quad r = |x|, \tag{1.52}$$

holds for any $k \in \mathbb{N}$. We use the induction. For $k = 1$, by the Hölder inequality we obtain

$$
\begin{aligned}
|\mathbb{E}\widetilde{f}| &= r\left(\frac{1}{|\wp|}\int_{\wp}|f(ry)|^p d\sigma(y)\right)^{\frac{1}{p}-1}\frac{1}{|\wp|}\left|\int_{\wp}|f(ry)|^{p-2}f(ry)\overline{\frac{d}{dr}f(ry)d\sigma(y)}\right| \\
&\leq \left(\frac{1}{|\wp|}\int_{\wp}|f(ry)|^p d\sigma(y)\right)^{\frac{1-p}{p}}\frac{1}{|\wp|}\int_{\wp}|f(ry)|^{p-1}\left|r\frac{d}{dr}f(ry)\right|d\sigma(y) \\
&\leq \left(\frac{1}{|\wp|}\int_{\wp}|f(ry)|^p d\sigma(y)\right)^{\frac{1-p}{p}}\frac{1}{|\wp|}\left(\int_{\wp}\left|r\frac{d}{dr}f(ry)\right|^p d\sigma(y)\right)^{\frac{1}{p}}\left(\int_{\wp}|f(ry)|^p d\sigma(y)\right)^{\frac{p-1}{p}} \\
&= \left(\frac{1}{|\wp|}\int_{\wp}|\mathbb{E}f(ry)|^p\, d\sigma(y)\right)^{\frac{1}{p}}.
\end{aligned}
$$

For the induction step, we assume that for some $\ell \in \mathbb{N}$ we have

$$
|\mathbb{E}^{\ell}\widetilde{f}| \leq \left(\frac{1}{|\wp|}\int_{\wp}|\mathbb{E}^{\ell}f(ry)|^p\, d\sigma(y)\right)^{1/p}, \tag{1.53}
$$

and we want to prove that it then follows that

$$
|\mathbb{E}^{\ell+1}\widetilde{f}| \leq \left(\frac{1}{|\wp|}\int_{\wp}|\mathbb{E}^{\ell+1}f(ry)|^p\, d\sigma(y)\right)^{1/p}.
$$

So, using (1.53), similarly to the case $\ell = 1$ above, we calculate

$$
\begin{aligned}
|\mathbb{E}^{\ell+1}\widetilde{f}(r)| &\leq \left|\mathbb{E}\left(\left(\frac{1}{|\wp|}\int_{\wp}|\mathbb{E}^{\ell}f(ry)|^p\, d\sigma(y)\right)^{\frac{1}{p}}\right)\right| \\
&= r\left(\frac{1}{|\wp|}\int_{\wp}|\mathbb{E}^{\ell}f(ry)|^p\, d\sigma(y)\right)^{\frac{1}{p}-1}\left|\frac{1}{|\wp|}\int_{\wp}|\mathbb{E}^{\ell}f(ry)|^{p-2}\,\mathbb{E}f(ry)\overline{\frac{d}{dr}(\mathbb{E}^{\ell}f(ry))d\sigma(y)}\right| \\
&\leq \left(\frac{1}{|\wp|}\int_{\wp}|\mathbb{E}^{\ell}f(ry)|^p\, d\sigma(y)\right)^{\frac{1}{p}-1}\left(\frac{1}{|\wp|}\int_{\wp}|\mathbb{E}^{\ell}f(ry)|^{p-1}\,|\mathbb{E}^{\ell+1}f(ry)|\, d\sigma(y)\right) \\
&\leq \left(\frac{1}{|\wp|}\int_{\wp}|\mathbb{E}^{\ell}f(ry)|^p d\sigma(y)\right)^{\frac{1}{p}-1}\frac{1}{|\wp|}\left(\int_{\wp}|\mathbb{E}^{\ell+1}f(ry)|^p\, d\sigma(y)\right)^{\frac{1}{p}} \\
&\quad\times\left(\int_{\wp}|\mathbb{E}^{\ell}f(ry)|^p d\sigma(y)\right)^{\frac{p-1}{p}} \\
&= \left(\frac{1}{|\wp|}\int_{\wp}|\mathbb{E}^{\ell+1}f(ry)|^p\, d\sigma(y)\right)^{\frac{1}{p}}.
\end{aligned}
$$

Here in the last line we have used Hölder's inequality. It proves (1.52).

Now (1.52) yields

$$\begin{aligned}\int_{\mathbb{G}} \left|\mathbb{E}^k \widetilde{f}(x)\right|^p \phi_2(x)dx &= |\wp| \int_0^\infty \left|\mathbb{E}^k \widetilde{f}(r)\right|^p \phi_2(r) r^{Q-1} dr \\ &\leq |\wp| \int_0^\infty \frac{1}{|\wp|} \int_{\wp} \left|\mathbb{E}^k f(ry)\right|^p \phi_2(r) r^{Q-1} d\sigma(y) dr \\ &= \int_{\mathbb{G}} \left|\mathbb{E}^k f(x)\right|^p \phi_2(x)dx.\end{aligned}$$

This completes the proof of (1.49). The proof of (1.50) is similar. □

1.3.4 Euler semigroup $e^{-t\mathbb{E}^*\mathbb{E}}$

Here we will describe the operator semigroup $\{e^{-t\mathbb{E}^*\mathbb{E}}\}_{t>0}$ associated with the Euler operator on homogeneous groups.

Theorem 1.3.4 (Euler semigroup). *Let $\mathbb{G}$ be a homogeneous group of homogeneous dimension Q. Let $x \in \mathbb{G}$, $x \neq 0$, and let $y := \frac{x}{|x|}$, and $t > 0$. Then the semigroup $e^{-t\mathbb{E}^*\mathbb{E}}$ is given by*

$$(e^{-t\mathbb{E}^*\mathbb{E}} f)(x) = \frac{e^{-tQ^2/4}}{\sqrt{4\pi t}} |x|^{-Q/2} \int_0^\infty e^{-\frac{(\ln|x| - \ln s)^2}{4t}} s^{-Q/2} f(sy) s^{Q-1} ds. \tag{1.54}$$

Before we prove formula (1.54), let us introduce some notation that will be useful in the sequel. Thus, let us define the map $F : L^2(\mathbb{G}) \to L^2(\mathbb{R} \times \wp)$ by the formula

$$(Ff)(s,y) := e^{sQ/2} f(e^s y), \tag{1.55}$$

for $y \in \wp$ and $s \in \mathbb{R}$. Its inverse map $F^{-1} : L^2(\mathbb{R} \times \wp) \to L^2(\mathbb{G})$ can be given by the formula

$$(F^{-1} g)(x) := r^{-Q/2} g(\ln r, y), \tag{1.56}$$

and one can readily check that F preserves the L^2 norm. The map F can be also described as

$$(Ff)(s,y) = (U(s)f)(y)$$

for all $y \in \wp$ and $s \in \mathbb{R}$, with the dilation mapping $U(t)$ defined by

$$U(t)f(x) := e^{tQ/2} f(e^t x). \tag{1.57}$$

We then immediately have

$$(F(U(t)f))(s,y) = (U(s)(U(t)f))(y) = (U(s+t)f)(y) = (Ff)(s+t,y). \tag{1.58}$$

The dilations $U(t)$ can be linked to the Euler operator through the relation

$$\frac{d}{dt} f(e^t x) = \mathbb{E} f(e^t x), \quad x \in \mathbb{G},$$

see (1.38). It then follows that these dilations can be also seen as a group of unitary operators $U(t) = e^{iAt}$ with the generator

$$Af = \frac{1}{i}\frac{d}{dt}U(t)f|_{t=0} = \frac{1}{i}\left(\mathbb{E} + \frac{Q}{2}\right)f = -i\mathbb{E}f - i\frac{Q}{2}f. \tag{1.59}$$

Since $\mathbb{E}^* = -Q\mathbb{I} - \mathbb{E}$ by Proposition 1.3.1, Part (ii), the formula (1.59) implies that

$$A = A^* = -i\mathbb{E} - i\frac{Q}{2}, \tag{1.60}$$

which yields the relation

$$\mathbb{A} = \mathbb{E}^*\mathbb{E} = \left(-iA - \frac{Q}{2}\right)\left(iA - \frac{Q}{2}\right) = A^2 + \frac{Q^2}{4}. \tag{1.61}$$

The family $U(t)$ is mapped to the multiplication by exponents $e^{it\tau}$ through the Mellin transformation $M : L^2(\mathbb{G}) \to L^2(\mathbb{R} \times \wp)$ defined by the formula $M = \mathcal{F} \circ F$, where $\mathcal{F}$ is the Fourier transform on $\mathbb{R}$, that is,

$$(Mf)(\tau, y) := \frac{1}{\sqrt{2\pi}}\int_{\mathbb{R}} e^{-is\tau}(Ff)(s, y)ds. \tag{1.62}$$

Indeed, using (1.58) and changing variables, we have

$$\begin{aligned}(MU(t)f)(\tau, y) &= \frac{1}{\sqrt{2\pi}}\int_{\mathbb{R}} e^{-is\tau}(Ff)(s + t, y)ds \\ &= \frac{e^{it\tau}}{\sqrt{2\pi}}\int_{\mathbb{R}} e^{-is\tau}(Ff)(s, y)ds = e^{it\tau}(Mf)(\tau, y).\end{aligned} \tag{1.63}$$

Before finally proving Theorem 1.3.4, let us point out that it implies the following representation of the semigroup e^{-tA^2}.

Corollary 1.3.5 (Semigroup e^{-tA^2}). *Let F and F^{-1} be mappings as in (1.55) and (1.56), respectively. Then we have*

$$Fe^{-tA^2}F^{-1}f(r, y) = \frac{1}{\sqrt{4\pi t}}\int_{\mathbb{R}} \exp\left(-\frac{(r-s)^2}{4t}\right)f(sy)ds. \tag{1.64}$$

Proof of Corollary 1.3.5. Setting $e^{-tA^2} = e^{tQ^2/4}e^{-t\mathbb{E}^*\mathbb{E}}$ as well as combining (1.55) and (1.56), and using (1.54) we get

$$\begin{aligned}&Fe^{-tA^2}F^{-1}f(r, y) \\ &= F\left(e^{tQ^2/4}\frac{e^{-tQ^2/4}r^{-Q/2}}{\sqrt{4\pi t}}\int_0^\infty e^{-\frac{(\ln r - \ln s)^2}{4t}}s^{Q/2-1}(s^{-Q/2}f(\ln s, y))ds\right) \\ &= e^{rQ/2}\frac{e^{-rQ/2}}{\sqrt{4\pi t}}\int_0^\infty e^{-\frac{(r-\ln s)^2}{4t}}\frac{f(\ln s, y)}{s}ds = \frac{1}{\sqrt{4\pi t}}\int_{-\infty}^\infty e^{-\frac{(r-s_1)^2}{4t}}f(s_1 y)ds_1,\end{aligned}$$

which is (1.64), where we have used the new variable $s = e^{s_1}$ in the last line. □

Let us finally prove Theorem 1.3.4.

Proof of Theorem 1.3.4. Noting that from the definition we have $iAe^{itA} = \partial_t U(t)$, we can calculate

$$(MiAe^{iAt}f)(\tau, y) = (M\partial_t U(t)f)(\tau, y) = \partial_t(MU(t)f)(\tau, y).$$

Now using (1.63) we get from above that

$$(MiAe^{iAt}f)(\tau, y) = \partial_t e^{it\tau}(Mf)(\tau, y) = i\tau e^{it\tau}(Mf)(\tau, y),$$

which (after setting $t = 0$) implies

$$(MAf)(\tau, y) = \tau(Mf)(\tau, y) \tag{1.65}$$

for f in the domain $\mathcal{D}(A)$. It follows that the condition $f \in \mathcal{D}(A)$ can be described by the property that the function $(\tau, y) \mapsto \tau(Mf)(\tau, y) \in L^2(\mathbb{R} \times \wp)$.

So, first we prove that

$$(Me^{-tA^2}f)(\tau, y) = e^{-t\tau^2}(Mf)(\tau, y). \tag{1.66}$$

We have

$$(Me^{-tA^2}f)(\tau, y) = \sum_{k=0}^{\infty} \frac{(-t)^k}{k!}(MA^{2k}f)(\tau, y). \tag{1.67}$$

Moreover, by iterating (1.65), it follows that

$$(MA^{2k}f)(\tau, y) = \tau^{2k}(Mf)(\tau, y), \quad k = 0, 1, 2, \ldots.$$

Combining this with (1.67), we get

$$(Me^{-tA^2}f)(\tau, y) = \sum_{k=0}^{\infty} \frac{(-t)^k}{k!}\tau^{2k}(Mf)(\tau, y) = e^{-t\tau^2}(Mf)(\tau, y).$$

That is, we have showed that (1.66) holds. Thus, it follows that

$$e^{-tA^2} = M^{-1}e^{-t\tau^2}M.$$

Here by using that $M = \mathcal{F} \circ F$, we have

$$e^{-tA^2} = F^{-1} \circ \mathcal{F}^{-1}(e^{-t\tau^2}\mathcal{F} \circ F). \tag{1.68}$$

Furthermore, we calculate

$$\begin{aligned}
\mathcal{F}^{-1}(e^{-t\tau^2}Mf)(\lambda, y) &= \mathcal{F}^{-1}(e^{-t\tau^2}\mathcal{F} \circ F)(\lambda, y) \\
&= \frac{1}{2\pi}\int_{\mathbb{R}}\int_{\mathbb{R}} e^{i\lambda\tau}e^{-t\tau^2}e^{-is\tau}(Ff)(s, y)dsd\tau \\
&= \frac{1}{2\pi}\int_{\mathbb{R}}\left(\int_{\mathbb{R}} e^{-t\tau^2+i(\lambda-s)\tau}d\tau\right)(Ff)(s, y)ds \\
&= \frac{1}{\sqrt{4\pi t}}\int_{\mathbb{R}} e^{-\frac{(\lambda-s)^2}{4t}}(Ff)(s, y)ds =: \varphi_t(\lambda, y),
\end{aligned}$$

since

$$\int_{\mathbb{R}} e^{-t\tau^2+i(\lambda-s)\tau}d\tau = \sqrt{\frac{\pi}{t}}e^{-\frac{(\lambda-s)^2}{4t}}.$$

From this and (1.68), using (1.55), (1.56) and $M = \mathcal{F} \circ F$ with $x = |x|y$, we compute

$$\begin{aligned}(e^{-tA^2}f)(|x|y) &= (F^{-1}\varphi_t)(|x|y) = r^{-Q/2}\varphi_t(\ln|x|, y)\\ &= \frac{1}{\sqrt{4\pi t}}|x|^{-Q/2}\int_{\mathbb{R}} e^{-\frac{(\ln|x|-s)^2}{4t}}(Ff)(s,y)ds\\ &= \frac{1}{\sqrt{4\pi t}}|x|^{-Q/2}\int_0^\infty e^{-\frac{(\ln|x|-\ln z)^2}{4t}}z^{\frac{Q}{2}-1}f(zy)dz,\end{aligned}$$

where we have used the change of variables $z = e^s$ in the last line.

Since we have $e^{-t\mathbb{E}^*\mathbb{E}} = e^{-tQ^2/4}e^{-tA^2}$ by (1.61), we arrive at

$$\begin{aligned}(e^{-t\mathbb{E}^*\mathbb{E}}f)(|x|y) &= e^{-tQ^2/4}(e^{-tA^2}f)(|x|y)\\ &= \frac{1}{\sqrt{4\pi t}}|x|^{-Q/2}e^{-tQ^2/4}\int_0^\infty e^{-\frac{(\ln|x|-\ln z)^2}{4t}}z^{\frac{Q}{2}-1}f(zy)dz\\ &= \frac{1}{\sqrt{4\pi t}}|x|^{-Q/2}e^{-tQ^2/4}\int_0^\infty e^{-\frac{(\ln|x|-\ln z)^2}{4t}}z^{-\frac{Q}{2}}f(zy)z^{Q-1}dz,\end{aligned}$$

completing the proof of (1.54). □

Remark 1.3.6.

1. The representation of the Euler semigroup in Theorem 1.3.4 becomes instrumental in deriving several forms of the Hardy–Sobolev and Gagliardo–Nirenberg type inequalities for the Euler operator, as we will show in the sequel, see, e.g., Section 10.4.
2. In the Euclidean case $\mathbb{R}^n$, the results of this section have been obtained in [BEHL08]. For general homogeneous groups, our presentation followed the results obtained in [RSY18a].

1.4 Stratified groups

An important special case of homogeneous groups is that of stratified groups introduced in Definition 1.1.5. Because this is an important class that will be analysed in Chapter 6 from the point of view of Hardy and other inequalities, here we will provide more details on it and fix the corresponding notation.

1.4.1 Stratified Lie groups

We recall the definition of stratified groups.

Definition 1.4.1 (Stratified groups). A Lie group $\mathbb{G} = (\mathbb{R}^n, \circ)$ is called a *stratified group* (or a *homogeneous Carnot group*) if it satisfies the following conditions:

(a) For some natural numbers $N + N_2 + \cdots + N_r = n$, that is $N = N_1$, the decomposition $\mathbb{R}^n = \mathbb{R}^N \times \cdots \times \mathbb{R}^{N_r}$ is valid, and for every $\lambda > 0$ the dilation $\delta_\lambda : \mathbb{R}^n \to \mathbb{R}^n$ given by

$$\delta_\lambda(x) \equiv \delta_\lambda(x', x^{(2)}, \ldots, x^{(r)}) := (\lambda x', \lambda^2 x^{(2)}, \ldots, \lambda^r x^{(r)})$$

is an automorphism of the group $\mathbb{G}$. Here $x' \equiv x^{(1)} \in \mathbb{R}^N$ and $x^{(k)} \in \mathbb{R}^{N_k}$ for $k = 2, \ldots, r$.

(b) Let N be as in (a) and let $X_1, \ldots, X_N$ be the left invariant vector fields on $\mathbb{G}$ such that $X_k(0) = \frac{\partial}{\partial x_k}|_0$ for $k = 1, \ldots, N$. Then

$$\operatorname{rank}(\operatorname{Lie}\{X_1, \ldots, X_N\}) = n,$$

for every $x \in \mathbb{R}^n$, i.e., the iterated commutators of $X_1, \ldots, X_N$ span the Lie algebra of $\mathbb{G}$.

The number r is called the *step of* $\mathbb{G}$ and the left invariant vector fields $X_1, \ldots, X_N$ are called the (Jacobian) generators of $\mathbb{G}$. The homogeneous dimension of a stratified Lie group $\mathbb{G}$ is given by

$$Q = \sum_{k=1}^{r} k N_k, \quad N_1 = N.$$

The second-order differential operator

$$\mathcal{L} = \sum_{k=1}^{N} X_k^2 \tag{1.69}$$

is called the (*canonical*) *sub-Laplacian* on $\mathbb{G}$. The sub-Laplacian $\mathcal{L}$ is a left invariant homogeneous hypoelliptic differential operator and it is elliptic if and only if the step of $\mathbb{G}$ is equal to 1.

The *hypoellipticity* of $\mathcal{L}$ means that for a distribution $f \in \mathcal{D}'(\Omega)$ in any open set Ω, if $\mathcal{L}f \in C^\infty(\Omega)$ then $f \in C^\infty(\Omega)$. It is a special case of Hörmander's sum of squares theorem [Hör67].

The left invariant vector field X_k has an explicit form given in Proposition 1.2.19, namely,

$$X_k = \frac{\partial}{\partial x'_k} + \sum_{l=2}^{r} \sum_{m=1}^{N_l} a_{k,m}^{(l)}(x', \ldots, x^{(l-1)}) \frac{\partial}{\partial x_m^{(l)}}, \tag{1.70}$$

where $a_{k,m}^{(l)}$ is a homogeneous (with respect to δ_λ) polynomial function of degree $l-1$. We will also use the following notation for the *horizontal gradient*

$$\nabla_H := (X_1, \ldots, X_N),$$

for the *horizontal divergence*

$$\operatorname{div}_H v := \nabla_H \cdot v,$$

and for the *horizontal p-Laplacian (or p-sub-Laplacian)*

$$\mathcal{L}_p f := \operatorname{div}_H(|\nabla_H f|^{p-2} \nabla_H f), \quad 1 < p < \infty. \tag{1.71}$$

Denoting the Euclidean distance by

$$|x'| = \sqrt{x_1'^2 + \cdots + x_N'^2}$$

for the Euclidean norm on $\mathbb{R}^N$, the representation (1.70) for derivatives leads to the identities

$$|\nabla_H |x'|^\gamma| = \gamma |x'|^{\gamma-1}, \tag{1.72}$$

and

$$\operatorname{div}_H \left(\frac{x'}{|x'|^\gamma} \right) = \frac{\sum_{j=1}^N |x'|^\gamma X_j x'_j - \sum_{j=1}^N x'_j \gamma |x'|^{\gamma-1} X_j |x'|}{|x'|^{2\gamma}} = \frac{N-\gamma}{|x'|^\gamma} \tag{1.73}$$

for all $\gamma \in \mathbb{R}$, $|x'| \neq 0$.

It was shown by Folland [Fol75] that the sub-Laplacian $\mathcal{L}$ in (1.69) on a general stratified group $\mathbb{G}$ has a unique *fundamental solution* ε, that is,

$$\mathcal{L}\varepsilon = \delta, \tag{1.74}$$

where δ is the delta-distribution at the unit element of $\mathbb{G}$. Moreover, the function ε is homogeneous of degree $2-Q$.

The function

$$d(x) := \begin{cases} \varepsilon(x)^{\frac{1}{2-Q}}, & \text{for } x \neq 0, \\ 0, & \text{for } x = 0, \end{cases} \tag{1.75}$$

is called the *$\mathcal{L}$-gauge* on $\mathbb{G}$. It is a *homogeneous quasi-norm* on $\mathbb{G}$, that is, it is a continuous function $d : \mathbb{G} \to [0, \infty)$, smooth away from the origin, which satisfies the conditions

$$d(\lambda x) = \lambda d(x),\ d(x^{-1}) = d(x) \text{ and } d(x) = 0 \text{ if only if } x = 0. \tag{1.76}$$

We refer to the original paper [Fol75] by Folland as well as to a recent presentation in [FR16, Section 3.2.7] for further details and properties of these fundamental solutions.

For future use, we record the action of $\mathcal{L}$ on d and its powers. Since $\mathcal{L}d^{2-Q} = 0$ in $\mathbb{G}\backslash\{0\}$, a straightforward calculation shows that for $Q \geq 3$ we have

$$\mathcal{L}d = (Q-1)\frac{|\nabla_H d|^2}{d} \text{ in } \mathbb{G}\backslash\{0\}, \tag{1.77}$$

as well as, consequently, for all $\alpha \in \mathbb{R}$,

$$\mathcal{L}d^\alpha = \alpha(\alpha + Q - 2)d^{\alpha-2}|\nabla_H d|^2 \text{ in } \mathbb{G}\backslash\{0\}. \tag{1.78}$$

1.4.2 Extended sub-Laplacians

In general, most of the results described in this book in the setting of stratified groups can be extended to any second-order hypoelliptic differential operators which are "equivalent" to the sub-Laplacian $\mathcal{L}$. Let us very briefly discuss this matter in the sprit of [BLU07].

Let $A = (a_{k,j})_{1\leq k,j \leq N_1}$ be a positive-definite symmetric matrix. Consider the following second-order hypoelliptic differential operator based on the matrix A and the vector fields $\{X_1, \ldots, X_{N_1}\}$ from the first stratum, given by

$$\mathcal{L}_A = \sum_{k,j=1}^{N_1} a_{k,j} X_k X_j. \tag{1.79}$$

For instance, in the Euclidean case, that is, for $\mathbb{G} = (\mathbb{R}^N, +)$ and $N_1 = N$, the constant coefficients second-order elliptic operator

$$\Delta_A = \sum_{k,j=1}^{N} a_{k,j} \frac{\partial^2}{\partial x_k \partial x_j}$$

is transformed into the Laplacian

$$\Delta = \sum_{k=1}^{N} \frac{\partial^2}{\partial x_k^2}$$

under a linear change of coordinates in $\mathbb{R}^N$. Thus, the operator Δ_A is "equivalent" to the operator Δ by a linear change of the coordinate system.

In general, to apply the above argument to transform $\mathcal{L}_A$ to the sub-Laplacian $\mathcal{L}$ it is not enough to change the basis by a linear transformation. However, it is enough in the setting of free stratified groups. We say that a stratified group $\mathbb{G}$ is a *free stratified group* if its Lie algebra is (isomorphic to) a free Lie algebra. For instance, the Heisenberg group $\mathbb{H}^1$ is a free stratified group. In this case we have the following result.

Theorem 1.4.2 ([BLU07]). *Let $\mathbb{G}$ be a free stratified group and let A be a given positive-definite symmetric matrix. Let $X = \{X_1, \ldots, X_{N_1}\}$ be left invariant vector fields in the first stratum of the Lie algebra of $\mathbb{G}$. Let*

$$Y_k := \sum_{j=1}^{N_1} \left(A^{\frac{1}{2}}\right)_{k,j} X_j, \quad k = 1, \ldots, N_1.$$

Consider the related second-order differential operator

$$\mathcal{L}_A = \sum_{k=1}^{N_1} Y_k^2 = \sum_{k,j=1}^{N_1} a_{k,j} X_k X_j.$$

Then there exists a Lie group automorphism T_A of $\mathbb{G}$ such that

$$Y_k(u \circ T_A) = (X_k u) \circ T_A, \quad k = 1, \ldots, N_1,$$
$$\mathcal{L}_A(u \circ T_A) = (\mathcal{L}u) \circ T_A,$$

for every smooth function $u : \mathbb{G} \to \mathbb{R}$. Moreover, T_A has polynomial component functions and commutes with the dilations of $\mathbb{G}$.

Remark 1.4.3. The automorphism T_A may not exist when $\mathbb{G}$ is not a free stratified group. However, for any stratified group $\mathbb{G}$ one can find a different stratified group $\mathbb{G}_* = (\mathbb{R}^N, *, \delta_\lambda)$, that is, the stratified group with the same underlying manifold $\mathbb{R}^N$ and the same group of dilations δ_λ as $\mathbb{G}$, and a Lie-group isomorphism from $\mathbb{G}$ to $\mathbb{G}_*$ turning the extended sub-Laplacian $\mathcal{L}_A$ on $\mathbb{G}$ into the sub-Laplacian $\mathcal{L}$ on $\mathbb{G}_*$, see [BLU07, Chapter 16.3].

1.4.3 Divergence theorem

Here we discuss the divergence theorem on stratified Lie groups that will be useful for our analysis at different places of the book.

Let $d\nu$ denote the volume element on $\mathbb{G}$ corresponding to the first stratum on $\mathbb{G}$:

$$d\nu := d\nu(x) = \bigwedge_{j=1}^{N} dx_j. \tag{1.80}$$

However, for simplicity of the exposition, we will mainly use the notation $dx := d\nu(x)$. Regarding it as a differential form, let $\langle X_k, d\nu \rangle$ denote the natural pairing between vector fields and differential forms. As it will follow from the proof of Theorem 1.4.5, using formula (1.70) for the left invariant operators X_k expressing them in terms of the Euclidean derivatives, the pairing $\langle X_k, d\nu\rangle$ can be also expressed in terms of the differential forms corresponding to the Euclidean coordinates in the form

$$\langle X_k, d\nu(x)\rangle = \bigwedge_{j=1, j\neq k}^{N_1} dx_j^{(1)} \bigwedge_{l=2}^{r} \bigwedge_{m=1}^{N_l} \theta_{l,m}, \tag{1.81}$$

with

$$\theta_{l,m} = -\sum_{k=1}^{N_1} a_{k,m}^{(l)}(x^{(1)}, \ldots, x^{(l-1)}) dx_k^{(1)} + dx_m^{(l)}, \tag{1.82}$$

for $l = 2, \ldots, r$ and $m = 1, \ldots, N_l$, where $a_{k,m}^{(l)}$ is a homogeneous polynomial of degree $l-1$ from (1.70).

Definition 1.4.4 (Admissible domains). A bounded open set $\Omega \subset \mathbb{G}$ will be called an *admissible domain* if its boundary $\partial\Omega$ is piecewise smooth and simple, that is, it has no self-intersections. The condition for the boundary to be simple amounts to $\partial\Omega$ being orientable.

The following divergence theorem can be regarded as a consequence of the abstract Stokes formula. However, we give a detailed local proof which will also lead to the explicit representation formula (1.82) that will be of use in the sequel.

Theorem 1.4.5 (Divergence formula). *Let $\Omega \subset \mathbb{G}$ be an admissible domain. Let $f_k \in C^1(\Omega) \bigcap C(\overline{\Omega})$, $k = 1, \ldots, N_1$. Then for each $k = 1, \ldots, N_1$, we have*

$$\int_{\Omega} X_k f_k d\nu = \int_{\partial\Omega} f_k \langle X_k, d\nu\rangle. \tag{1.83}$$

Consequently, we also have

$$\int_{\Omega} \sum_{k=1}^{N_1} X_k f_k d\nu = \int_{\partial\Omega} \sum_{k=1}^{N_1} f_k \langle X_k, d\nu\rangle. \tag{1.84}$$

Proof of Theorem 1.4.5. Using (1.70), for any function f we obtain the following differentiation formula

$$\begin{aligned}
df &= \sum_{k=1}^{N_1} \frac{\partial f}{\partial x_k^{(1)}} dx_k^{(1)} + \sum_{l=2}^{r} \sum_{m=1}^{N_l} \frac{\partial f}{\partial x_m^{(l)}} dx_m^{(l)} \\
&= \sum_{k=1}^{N_1} X_k f dx_k^{(1)} - \sum_{k=1}^{N_1} \sum_{l=2}^{r} \sum_{m=1}^{N_l} a_{k,m}^{(l)}(x^{(1)}, \ldots, x^{(l-1)}) \frac{\partial f}{\partial x_m^{(l)}} dx_k^{(l)} \sum_{l=2}^{r} \sum_{m=1}^{N_l} \frac{\partial f}{\partial x_m^{(l)}} dx_m^{(l)} \\
&= \sum_{k=1}^{N_1} X_k f dx_k^{(1)} + \sum_{l=2}^{r} \sum_{m=1}^{N_l} \frac{\partial f}{\partial x_m^{(l)}} \left(-\sum_{k=1}^{N_1} a_{k,m}^{(l)}(x^{(1)}, \ldots, x^{(l-1)}) dx_k^{(1)} + dx_m^{(l)} \right) \\
&= \sum_{k=1}^{N_1} X_k f dx_k^{(1)} + \sum_{l=2}^{r} \sum_{m=1}^{N_l} \frac{\partial f}{\partial x_m^{(l)}} \theta_{l,m},
\end{aligned}$$

where for each $l = 2, \dots, r$ and $m = 1, \dots, N_l$,

$$\theta_{l,m} = -\sum_{k=1}^{N_1} a_{k,m}^{(l)}(x^{(1)}, \dots, x^{(l-1)})dx_k^{(1)} + dx_m^{(l)}.$$

That is, we can write

$$df = \sum_{k=1}^{N_1} X_k f dx_k^{(1)} + \sum_{l=2}^{r}\sum_{m=1}^{N_l} \frac{\partial f}{\partial x_m^{(l)}}\theta_{l,m}. \tag{1.85}$$

It is simple to see that

$$\langle X_s, dx_j^{(1)}\rangle = \frac{\partial}{\partial x_s^{(1)}} dx_j^{(1)} = \delta_{sj},$$

where δ_{sj} is the Kronecker delta. Moreover, we have

$$\begin{aligned}
&\langle X_s, \theta_{l,m}\rangle \\
&\quad= \left(\frac{\partial}{\partial x_s^{(1)}} + \sum_{h=2}^{r}\sum_{g=1}^{N_h} a_{s,g}^{(h)}(x^{(1)}, \dots, x^{(h-1)})\frac{\partial}{\partial x_g^{(h)}}\right) \\
&\qquad\times\left(-\sum_{k=1}^{N_1} a_{k,m}^{(l)}(x^{(1)}, \dots, x^{(l-1)})dx_k^{(1)} + dx_m^{(l)}\right) \\
&\quad= -\sum_{k=1}^{N_1}\left(\frac{\partial}{\partial x_s^{(1)}} a_{k,m}^{(l)}(x^{(1)}, \dots, x^{(l-1)})\right) dx_k^{(1)} \\
&\qquad- \sum_{k=1}^{N_1} a_{k,m}^{(l)}(x^{(1)}, \dots, x^{(l-1)})\frac{\partial}{\partial x_s^{(1)}} dx_k^{(1)} + \frac{\partial}{\partial x_s^{(1)}} dx_m^{(l)} \\
&\qquad- \sum_{k=1}^{N_1}\sum_{h=2}^{r}\sum_{g=1}^{N_h} a_{s,g}^{(h)}(x^{(1)}, \dots, x^{(h-1)})\left(\frac{\partial}{\partial x_g^{(h)}} a_{k,m}^{(l)}(x^{(1)}, \dots, x^{(l-1)})\right) dx_k^{(1)} \\
&\qquad- \sum_{k=1}^{N_1}\sum_{h=2}^{r}\sum_{g=1}^{N_h} a_{s,g}^{(h)}(x^{(1)}, \dots, x^{(h-1)}) a_{k,m}^{(l)}(x^{(1)}, \dots, x^{(l-1)})\frac{\partial}{\partial x_g^{(h)}} dx_k^{(1)} \\
&\qquad+ \sum_{h=2}^{r}\sum_{g=1}^{N_h} a_{s,g}^{(h)}(x^{(1)}, \dots, x^{(h-1)})\frac{\partial}{\partial x_g^{(h)}} dx_m^{(l)} \\
&\quad= -\sum_{k=1}^{N_1}\left(\frac{\partial}{\partial x_s^{(1)}} a_{k,m}^{(l)}(x^{(1)}, \dots, x^{(l-1)})\right) dx_k^{(1)} - \sum_{k=1}^{N_1} a_{k,m}^{(l)}(x^{(1)}, \dots, x^{(l-1)})\delta_{sk} \\
&\qquad- \sum_{k=1}^{N_1}\sum_{h=2}^{r}\sum_{g=1}^{N_h} a_{s,g}^{(h)}(x^{(1)}, \dots, x^{(h-1)})\left(\frac{\partial}{\partial x_g^{(h)}} a_{k,m}^{(l)}(x^{(1)}, \dots, x^{(l-1)})\right) dx_k^{(1)}
\end{aligned}$$

$$
\begin{aligned}
&+\sum_{h=2}^{r}\sum_{g=1}^{N_h} a_{s,g}^{(h)}(x^{(1)},\dots,x^{(h-1)})\delta_{gm}\delta_{hl}\\
&=-\sum_{k=1}^{N_1}\sum_{h=2}^{r}\sum_{g=1}^{N_h} a_{s,g}^{(h)}(x^{(1)},\dots,x^{(h-1)})\left(\frac{\partial}{\partial x_g^{(h)}}a_{k,m}^{(l)}(x^{(1)},\dots,x^{(l-1)})\right)dx_k^{(1)}\\
&-\sum_{k=1}^{N_1}\left(\frac{\partial}{\partial x_s^{(1)}}a_{k,m}^{(l)}(x^{(1)},\dots,x^{(l-1)})\right)dx_k^{(1)}\\
&=-\sum_{k=1}^{N_1}\left[\sum_{h=2}^{r}\sum_{g=1}^{N_l} a_{s,g}^{(h)}(x^{(1)},\dots,x^{(h-1)})\left(\frac{\partial}{\partial x_g^{(h)}}a_{k,m}^{(l)}(x^{(1)},\dots,x^{(l-1)})\right)\right.\\
&\left.+\frac{\partial}{\partial x_s^{(1)}}a_{k,m}^{(l)}(x^{(1)},\dots,x^{(l-1)})\right]dx_k^{(1)}.
\end{aligned}
$$

That is, we have

$$\langle X_s, dx_j^{(1)}\rangle = \delta_{sj},$$

for $s, j = 1, \dots, N_1$, and

$$\langle X_s, \theta_{l,m}\rangle = \sum_{k=1}^{N_1} \mathcal{C}_k dx_k^{(1)},$$

for $s = 1, \dots, N_1,\ l = 2, \dots, r,\ m = 1, \dots, N_l$. Here we used the notation

$$
\begin{aligned}
\mathcal{C}_k = &-\sum_{h=2}^{r}\sum_{g=1}^{N_l} a_{s,g}^{(h)}(x^{(1)},\dots,x^{(h-1)})\frac{\partial}{\partial x_g^{(h)}}a_{k,m}^{(l)}(x^{(1)},\dots,x^{(l-1)})\\
&-\frac{\partial}{\partial x_s^{(1)}}a_{k,m}^{(l)}(x^{(1)},\dots,x^{(l-1)}).
\end{aligned}
$$

By (1.80) we have

$$d\nu = d\nu(x) = \bigwedge_{j=1}^{N} dx_j = \bigwedge_{j=1}^{N_1} dx_j^{(1)} \bigwedge_{l=2}^{r}\bigwedge_{m=1}^{N_l} dx_m^{(l)} = \bigwedge_{j=1}^{N_1} dx_j^{(1)} \bigwedge_{l=2}^{r}\bigwedge_{m=1}^{N_l} \theta_{l,m},$$

so that we obtain

$$\langle X_k, d\nu(x)\rangle = \bigwedge_{j=1, j\neq k}^{N_1} dx_j^{(1)} \bigwedge_{l=2}^{r}\bigwedge_{m=1}^{N_l} \theta_{l,m}.$$

Therefore, using (1.85) we get

$$
\begin{aligned}
d(f_s\langle X_s, d\nu(x)\rangle) &= df_s \wedge \langle X_s, d\nu(x)\rangle\\
&= \sum_{k=1}^{N_1} X_k f_s dx_k^{(1)} \wedge \langle X_s, d\nu(x)\rangle + \sum_{l=2}^{r}\sum_{m=1}^{N_l}\frac{\partial f_s}{\partial x_m^{(l)}}\theta_{l,m}\wedge\langle X_s, d\nu(x)\rangle\\
&= X_s f_s d\nu(x),
\end{aligned}
$$

that is, we have

$$d(\langle f_k X_k, d\nu(x)\rangle) = X_k f_k d\nu(x), \quad k = 1, \ldots, N_1.$$

Now using the classical Stokes theorem (see, e.g., [DFN84, Theorem 26.3.1]) we obtain (1.83). Taking a sum over k we also obtain (1.84). □

1.4.4 Green's identities for sub-Laplacians

In this section we prove Green's first and second formulae for the sub-Laplacian on stratified groups. These formulae will be useful throughout the book when we will be dealing with inequalities and with the potential theory on stratified groups. We will formulate them in admissible domains in the sense of Definition 1.4.4.

Theorem 1.4.6 (Green's first and second identities). *Let $\mathbb{G}$ be a stratified group and let $\Omega \subset \mathbb{G}$ be an admissible domain.*

(1) *Green's first identity: Let $v \in C^1(\Omega) \bigcap C(\overline{\Omega})$ and $u \in C^2(\Omega) \bigcap C^1(\overline{\Omega})$. Then*

$$\int_\Omega \left((\widetilde{\nabla} v)u + v\mathcal{L}u\right) d\nu = \int_{\partial\Omega} v\langle \widetilde{\nabla} u, d\nu\rangle, \tag{1.86}$$

where $\mathcal{L}$ is the sub-Laplacian on $\mathbb{G}$ and where the vector field $\widetilde{\nabla} u$ is defined by

$$\widetilde{\nabla} u := \sum_{k=1}^{N_1} (X_k u)\, X_k. \tag{1.87}$$

(2) *Green's second identity: Let $u, v \in C^2(\Omega) \bigcap C^1(\overline{\Omega})$. Then*

$$\int_\Omega (u\mathcal{L}v - v\mathcal{L}u) d\nu = \int_{\partial\Omega} (u\langle \widetilde{\nabla} v, d\nu\rangle - v\langle \widetilde{\nabla} u, d\nu\rangle). \tag{1.88}$$

Remark 1.4.7.

1. The definition (1.87) means that $\widetilde{\nabla} u$ is a vector field. Consequently, the expression $(\widetilde{\nabla} v)u$ is a scalar, given by

$$\left(\widetilde{\nabla} v\right) u = \widetilde{\nabla} v u = \sum_{k=1}^{N_1} (X_k v)\,(X_k u) = \sum_{k=1}^{N_1} X_k v X_k u.$$

At the same time the expression $\widetilde{\nabla}(vu)$ is a vector field, also understood as an operator.

2. Although we formulate Green's identities in bounded domains, they are still applicable in unbounded domains for functions with necessary decay rates at infinity. It can be readily shown by the standard argument using quasi-balls with radii $R \to \infty$.

3. The version (1.86) of Green's first identity was proved for the ball in [Gav77] and for any smooth domain of the complex Heisenberg group in [Rom91]. Other analogues have been also obtained in [BLU07] and [CGN08] but using different terminologies. Also, the group structure is not needed for it, see Proposition 12.2.1, with other versions also known, see, e.g., [CGL93]. The version given in Theorem 1.4.6 was obtained in [RS17c].

Proof of Theorem 1.4.6. Part (1). Let $f_k := vX_k u$, so that

$$\sum_{k=1}^{N_1} X_k f_k = (\widetilde{\nabla} v)u + v\mathcal{L}u.$$

By using the divergence formula in Theorem 1.4.5 we obtain

$$\begin{aligned}\int_\Omega \left(\widetilde{\nabla} vu + v\mathcal{L}u\right) d\nu &= \int_\Omega \sum_{k=1}^{N_1} X_k f_k d\nu = \int_{\partial\Omega} \sum_{k=1}^{N_1} \langle f_k X_k, d\nu\rangle \\ &= \int_{\partial\Omega} \sum_{k=1}^{N_1} \langle vX_k u X_k, d\nu\rangle = \int_{\partial\Omega} v\langle \widetilde{\nabla} u, d\nu\rangle,\end{aligned}$$

yielding (1.86).

Part (2). Rewriting (1.86) we have

$$\int_\Omega \left((\widetilde{\nabla} u)v + u\mathcal{L}v\right) d\nu = \int_{\partial\Omega} u\langle \widetilde{\nabla} v, d\nu\rangle,$$

$$\int_\Omega \left((\widetilde{\nabla} v)u + v\mathcal{L}u\right) d\nu = \int_{\partial\Omega} v\langle \widetilde{\nabla} u, d\nu\rangle.$$

By subtracting the second identity from the first one and using

$$(\widetilde{\nabla} u)v = (\widetilde{\nabla} v)u,$$

we obtain (1.88). □

Taking $v = 1$ in Theorem 1.4.6 we obtain the following analogue of Gauss' mean value formula for harmonic functions:

Corollary 1.4.8 (Gauss' mean value formula). *If $\mathcal{L}u = 0$ in an admissible domain $\Omega \subset \mathbb{G}$, then*

$$\int_{\partial\Omega} \langle \widetilde{\nabla} u, d\nu\rangle = 0.$$

As in the classical theory, we can approximate functions with (weak) singularities such as smooth functions because the Green formulae are still valid for them. In this sense, without further justification and using these Green formulae, in particular, we apply them to the fundamental solution ε of the sub-Laplacian $\mathcal{L}$ as in (1.74). We define the function

$$\varepsilon(x, y) := \varepsilon(x^{-1}y). \tag{1.89}$$

The properties of the $\mathcal{L}$-gauge imply that $\varepsilon(x, y) = \varepsilon(y, x)$.

Thus, for $x \in \Omega$, taking $v = 1$ and $u(y) = \varepsilon(x, y)$ we can record the following consequence of Theorem 1.4.6, (1):

Corollary 1.4.9. *If $\Omega \subset \mathbb{G}$ is an admissible domain, and $x \in \Omega$, then*

$$\int_{\partial\Omega} \langle \widetilde{\nabla}\varepsilon(x, y), d\nu(y)\rangle = 1,$$

where ε is the fundamental solution of the sub-Laplacian $\mathcal{L}$.

Putting the fundamental solution ε instead of v in (1.88) we obtain the following representation formulae.

Corollary 1.4.10 (Representation formulae for functions on stratified groups)**.** *Let $\mathbb{G}$ be a stratified group and let $\Omega \subset \mathbb{G}$ be an admissible domain.*

(1) *Let $u \in C^2(\Omega)\bigcap C^1(\overline{\Omega})$. Then for $x \in \Omega$ we have*

$$\begin{aligned} u(x) = &\int_{\Omega} \varepsilon(x, y)\mathcal{L}u(y)d\nu(y) \\ &+ \int_{\partial\Omega} u(y)\langle \widetilde{\nabla}\varepsilon(x, y), d\nu(y)\rangle - \int_{\partial\Omega} \varepsilon(x, y)\langle \widetilde{\nabla}u(y), d\nu(y)\rangle. \end{aligned}$$

(2) *Let $u \in C^2(\Omega)\bigcap C^1(\overline{\Omega})$ and $\mathcal{L}u = 0$ on Ω, then for $x \in \Omega$ we have*

$$u(x) = \int_{\partial\Omega} u(y)\langle \widetilde{\nabla}\varepsilon(x, y), d\nu(y)\rangle - \int_{\partial\Omega} \varepsilon(x, y)\langle \widetilde{\nabla}u(y), d\nu(y)\rangle.$$

(3) *Let $u \in C^2(\Omega)\bigcap C^1(\overline{\Omega})$ and $u(x) = 0,\ x \in \partial\Omega$, then*

$$u(x) = \int_{\Omega} \varepsilon(x, y)\mathcal{L}u(y)d\nu(y) - \int_{\partial\Omega} \varepsilon(x, y)\langle \widetilde{\nabla}u(y), d\nu(y)\rangle.$$

(4) *Let $u \in C^2(\Omega)\bigcap C^1(\overline{\Omega})$ and $\sum_{j=1}^{N_1} X_j u\langle X_j, d\nu\rangle = 0$ on $\partial\Omega$, then*

$$u(x) = \int_{\Omega} \varepsilon(x, y)\mathcal{L}u(y)d\nu(y) + \int_{\partial\Omega} u(y)\langle \widetilde{\nabla}\varepsilon(x, y), d\nu(y)\rangle.$$

1.4.5 Green's identities for p-sub-Laplacians

In this section we show how Green's first and second formulae for the sub-Laplacian from Section 1.4.4 on stratified groups can be extended to the p-sub-Laplacian for all $1 < p < \infty$. As before, we will formulate them in admissible domains in the sense of Definition 1.4.4. We recall the definition of the p-sub-Laplacian from (1.71) as

$$\mathcal{L}_p f := \operatorname{div}_H(|\nabla_H f|^{p-2}\nabla_H f), \quad 1 < p < \infty.$$

Theorem 1.4.11 (Green's first and second identities for p-sub-Laplacian). *Let $\mathbb{G}$ be a stratified group and let $\Omega \subset \mathbb{G}$ be an admissible domain. Let $1 < p < \infty$.*

(1) *Green's first identity: Let $v \in C^1(\Omega)\bigcap C(\overline{\Omega})$ and $u \in C^2(\Omega)\bigcap C^1(\overline{\Omega})$. Then*

$$\int_\Omega \left((|\nabla_{\mathbb{G}} u|^{p-2}\widetilde{\nabla} v)u + v\mathcal{L}_p u\right) d\nu = \int_{\partial\Omega} |\nabla_{\mathbb{G}} u|^{p-2} v\langle \widetilde{\nabla} u, d\nu\rangle, \tag{1.90}$$

where

$$\widetilde{\nabla} u = \sum_{k=1}^{N_1} (X_k u)\, X_k.$$

(2) *Green's second identity: Let $u, v \in C^2(\Omega)\bigcap C^1(\overline{\Omega})$. Then*

$$\begin{aligned}\int_\Omega \left(u\mathcal{L}_p v - v\mathcal{L}_p u + (|\nabla_{\mathbb{G}} v|^{p-2} - |\nabla_{\mathbb{G}} u|^{p-2})(\widetilde{\nabla} v)u\right) d\nu \\ = \int_{\partial\Omega} (|\nabla_{\mathbb{G}} v|^{p-2} u\langle \widetilde{\nabla} v, d\nu\rangle - |\nabla_{\mathbb{G}} u|^{p-2} v\langle \widetilde{\nabla} u, d\nu\rangle).\end{aligned} \tag{1.91}$$

Proof of Theorem 1.4.11. Part (1). Let $f_k := v|\nabla_{\mathbb{G}} u|^{p-2} X_k u$, then

$$\sum_{k=1}^{N_1} X_k f_k = (|\nabla_{\mathbb{G}} u|^{p-2}\widetilde{\nabla} v)u + v\mathcal{L}_p u.$$

By integrating both sides of this equality over Ω and using Proposition 1.4.5 we obtain

$$\begin{aligned}\int_\Omega \left((|\nabla_{\mathbb{G}} u|^{p-2}\widetilde{\nabla} v)u + v\mathcal{L}_p u\right) d\nu = \int_\Omega \sum_{k=1}^{N_1} X_k f_k d\nu = \int_{\partial\Omega} \sum_{k=1}^{N_1} \langle f_k X_k, d\nu\rangle \\ = \int_{\partial\Omega} \sum_{k=1}^{N_1} \langle v|\nabla_{\mathbb{G}} u|^{p-2} X_k u X_k, d\nu\rangle = \int_{\partial\Omega} |\nabla_{\mathbb{G}} u|^{p-2} v\langle \widetilde{\nabla} u, d\nu\rangle,\end{aligned}$$

showing (1.90).

Part (2). Using (1.90) we have

$$\int_{\Omega}\left((|\nabla_{\mathbb{G}}v|^{p-2}\widetilde{\nabla}u)v+u\mathcal{L}_p v\right)d\nu=\int_{\partial\Omega}|\nabla_{\mathbb{G}}v|^{p-2}u\langle\widetilde{\nabla}v,d\nu\rangle,$$

$$\int_{\Omega}\left((|\nabla_{\mathbb{G}}u|^{p-2}\widetilde{\nabla}v)u+v\mathcal{L}_p u\right)d\nu=\int_{\partial\Omega}|\nabla_{\mathbb{G}}u|^{p-2}v\langle\widetilde{\nabla}u,d\nu\rangle.$$

By subtracting the second identity from the first one, the equality

$$(\widetilde{\nabla}u)v=(\widetilde{\nabla}v)u$$

implies (1.91). □

Taking $v=1$ in Theorem 1.4.11 we get the following analogue of Gauss' mean value formula for p-harmonic functions:

Corollary 1.4.12 (Gauss' mean value formula for p-harmonic functions). *If $1<p<\infty$ and $\mathcal{L}_p u=0$ in an admissible domain $\Omega\subset\mathbb{G}$, then*

$$\int_{\partial\Omega}|\nabla_{\mathbb{G}}u|^{p-2}\langle\widetilde{\nabla}u,d\nu\rangle=0.$$

1.4.6 Sub-Laplacians with drift

In this section we briefly describe the so-called sub-Laplacians with drift. While such operators can be analysed on more general groups, we restrict our presentation to stratified groups $\mathbb{G}$ only since this will be the setting where we will be using these operators.

Definition 1.4.13 (Sub-Laplacian with drift). The (extended) positive sub-Laplacian with drift is defined on $C_0^\infty(\mathbb{G})$ as the operator

$$\mathcal{L}_X:=-\sum_{i,j=1}^{N}a_{i,j}X_iX_j-\gamma X, \tag{1.92}$$

where $\gamma\in\mathbb{R}$, the matrix $(a_{i,j})_{i,j=1}^N$ is real, symmetric, positive definite, and $X\in\mathfrak{g}$ is a left invariant vector field on $\mathbb{G}$.

Similar to Section 1.4.2 the operator (1.92) can be transformed to the *(positive) sub-Laplacian with drift* of the form

$$\mathcal{L}_X=-\sum_{j=1}^{N}X_j^2-\gamma X:=\mathcal{L}_0-\gamma X, \tag{1.93}$$

where $\mathcal{L}_0$ is the positive sub-Laplacian on $\mathbb{G}$ defined by

$$\mathcal{L}_0=-\sum_{j=1}^{N}X_j^2. \tag{1.94}$$

The details of such a transformation can be found in [HMM05] or [MOV17].

If $X = \sum_{j=1}^N a_j X_j$, then we denote $\|X\| := \left(\sum_{j=1}^N a_j^2\right)^{1/2}$ and

$$b_X := \frac{\|X\|}{2}, \tag{1.95}$$

where $a_j \in \mathbb{R}$ for $j = 1, \ldots, N$.

Let us collect the following spectral properties of (positive) sub-Laplacians with drift (1.93) which are true on more general groups than the stratified ones. In the case $\gamma = 1$ this was shown in [HMM05, Proposition 3.1] while here we follow [RY18b] for general $\gamma \in \mathbb{R}$:

Proposition 1.4.14 (Spectral properties of sub-Laplacians with drift). *Let $\mathbb{G}$ be a connected Lie group with unit e, $X_1, \ldots, X_N$ an algebraic basis of $\mathfrak{g}$ and let $X \in \mathfrak{g}\backslash\{0\}$. Let $\gamma \in \mathbb{R}$. Then we have for the operator $\mathcal{L}_X$, with domain $C_0^\infty(\mathbb{G})$, the following properties:*

(i) *the operator $\mathcal{L}_X$ is symmetric on $L^2(\mathbb{G}, \mu)$ for some positive measure μ on $\mathbb{G}$ if and only if there exists a positive character χ of $\mathbb{G}$ and a constant C such that $\mu = C\mu_X$ and $\nabla_H \chi|_e = \gamma X|_e$, where μ_X is the measure absolutely continuous with respect to the Haar measure μ with density χ;*

(ii) *assume that $\nabla_H \chi|_e = \gamma X|_e$ for some positive character χ of $\mathbb{G}$. Then the operator $\mathcal{L}_X$ is essentially self-adjoint on $L^2(\mathbb{G}, \mu_X)$ and its spectrum is contained in the interval $[\gamma^2 b_X^2, \infty)$.*

Proof of Proposition 1.4.14. Let μ be a positive measure on $\mathbb{G}$. Then for all test functions $\phi, \psi \in C_0^\infty(\mathbb{G})$ we can calculate

$$\begin{aligned}
\int_{\mathbb{G}} (\mathcal{L}_X \phi)\psi d\mu &= -\sum_{j=1}^N \left(\int_{\mathbb{G}} (X_j^2 \phi)\psi d\mu\right) - \gamma \int_{\mathbb{G}} \psi X \phi d\mu \\
&= \sum_{j=1}^N \left(\int_{\mathbb{G}} X_j \phi X_j \psi d\mu + \int_{\mathbb{G}} \psi X_j \phi X_j \mu\right) + \gamma \int_{\mathbb{G}} \phi X \psi d\mu + \gamma \int_{\mathbb{G}} \phi \psi X \mu \\
&= \sum_{j=1}^N \left(-\int_{\mathbb{G}} \phi X_j^2 \psi d\mu - 2\int_{\mathbb{G}} \phi X_j \psi X_j \mu - \int_{\mathbb{G}} \phi \psi X_j^2 \mu\right) \\
&\quad + \gamma \int_{\mathbb{G}} \phi X \psi d\mu + \gamma \int_{\mathbb{G}} \phi \psi X \mu \\
&= \int_{\mathbb{G}} \phi(\mathcal{L}_X \psi) d\mu + 2\gamma \int_{\mathbb{G}} \phi X \psi d\mu - 2\int_{\mathbb{G}} \phi \nabla_H \psi \nabla_H \mu + \int_{\mathbb{G}} \phi \psi (\mathcal{L}_0 + \gamma X)\mu \\
&= \int_{\mathbb{G}} \phi(\mathcal{L}_X \psi) d\mu + \langle \phi, 2\gamma (X\psi)\mu - 2\nabla_H \psi \nabla_H \mu + \psi(\mathcal{L}_0 + \gamma X)\mu \rangle \\
&=: \int_{\mathbb{G}} \phi(\mathcal{L}_X \psi) d\mu + \mathfrak{I}(\phi, \psi, \mu),
\end{aligned} \tag{1.96}$$

where $\mathcal{L}_0$ is defined in (1.94), $\nabla_H = (X_1, \ldots, X_N)$ and $\langle \cdot, \cdot \rangle$ is the pairing between distributions and test functions on $\mathbb{G}$. From this we see that $\mathcal{L}_X$ is symmetric on $L^2(\mathbb{G}, \mu)$ if and only if $\mathfrak{I}(\phi, \psi, \mu) = 0$ for all functions ϕ and ψ, that is,

$$(\mathcal{L}_0 + \gamma X)\mu = 0, \quad \gamma(X\psi)\mu - \nabla_H \psi \nabla_H \mu = 0, \quad \forall \psi \in C_0^\infty(\mathbb{G}). \tag{1.97}$$

The vector fields $X_1, \ldots, X_N$ satisfy Hörmander's condition, so that $\mathcal{L}_0 + \gamma X$ is hypoelliptic, which implies with the condition $(\mathcal{L}_0 + \gamma X)\mu = 0$ that μ has a smooth density ω with respect to the Haar measure. Then, as in [HMM05, Proof of Proposition 3.1], we show that

$$X = \sum_{j=1}^{N} a_j X_j, \tag{1.98}$$

for some coefficients $a_1, \ldots, a_N$. Using the fact that $X_1, \ldots, X_N$ are linearly independent and the second equation of (1.97), we obtain that

$$X_k \omega = \gamma a_k \omega, \tag{1.99}$$

where $k = 1, \ldots, N$. The solution of (1.99) is given by

$$\omega(x) = \omega(e) \exp\left(\gamma \int_0^1 \sum_{k=1}^{N} a_k \vartheta_k(t) dt\right),$$

which is a positive and uniquely determined by its value at the identity, where $\vartheta_k(t)$ is the piecewise C^1 path. By normalizing ω, we get that $\omega(e) = 1$, and that it is a character of $\mathbb{G}$. Then, we see that the function $x \mapsto \omega(xy)/\omega(y)$ is a solution of (1.99) for any y in $\mathbb{G}$. Since the value of this function at the identity is 1, we have $\omega(xy) = \omega(x)\omega(y)$ for any $x, y \in \mathbb{G}$, and ω is a character of $\mathbb{G}$. From (1.98) and (1.99), we get $\nabla_H \chi|_e = \gamma X|_e$ with $\chi = \omega$. This proves Part (i) of Proposition 1.4.14.

As in the case $\gamma = 1$ (see [HMM05, Proposition 3.1]), by considering the isometry $\mathcal{U}_2 f = \chi^{-1/2} f$ of $L^2(\mathbb{G}, \mu)$ onto $L^2(\mathbb{G}, \mu_X)$, we have

$$\chi^{\frac{1}{2}} \mathcal{L}_X (\chi^{-\frac{1}{2}} f) = (\mathcal{L}_0 + \gamma^2 b_X^2) f, \tag{1.100}$$

which is an essentially self-adjoint operator on $L^2(\mathbb{G}, \mu)$, where b_X is defined in (1.95). Since the spectrum of this operator is contained in $[\gamma^2 b_X^2, \infty)$, we obtain that $\mathcal{L}_X$ is essentially self-adjoint on $L^2(\mathbb{G}, \mu_X)$ and its spectrum is contained in $[\gamma^2 b_X^2, \infty)$.

This completes the proof of Proposition 1.4.14. □

As a corollary of Proposition 1.4.14 let us collect the properties that will be important for us in the sequel.

Corollary 1.4.15 (Transformation of sub-Laplacian with drift). *Let $\mathbb{G}$ be a stratified group and assume conditions of Proposition 1.4.14. Assume that there exists a positive character χ of $\mathbb{G}$ such that*

$$\nabla_H \chi|_e = \gamma X|_e.$$

Then the operator $\mathcal{L}_X$ is formally self-adjoint with respect to the positive measure $\mu_X = \chi\mu$, where μ is the Haar measure of $\mathbb{G}$. The operator $\mathcal{L}_X$ is self-adjoint on $L^2(\mathbb{G}, \mu_X)$ and the mapping

$$L^2(\mathbb{G}, \mu) \ni f \mapsto \chi^{-1/2} f \in L^2(\mathbb{G}, \mu_X) \tag{1.101}$$

is an isometric isomorphism.

For a detailed discussion about more properties of the sub-Laplacians with drift we refer to [HMM04], [HMM05] and [MOV17].

1.4.7 Polarizable Carnot groups

In (1.74) we recalled the result of Folland that the sub-Laplacian $\mathcal{L}$ on general stratified groups always has a unique fundamental solution ε. The explicit formula (1.75) relating the fundamental solution to the $\mathcal{L}$-gauge turns out to be useful in many explicit calculations.

In applications to nonlinear partial differential equations, a natural question arises to express the fundamental solution of the p-sub-Laplacian (1.71) in terms of the fundamental solution of the sub-Laplacian or, equivalently, in terms of the $\mathcal{L}$-gauge. One of the largest classes of stratified Lie groups, for which the fundamental solution of the p-sub-Laplacian is known to be expressed explicitly in terms of the $\mathcal{L}$-gauge are the so-called polarizable Carnot groups which we now briefly discuss.

A Lie group $\mathbb{G}$ is called a *polarizable Carnot group* if the $\mathcal{L}$-gauge d satisfies the following ∞-sub-Laplacian equality

$$\mathcal{L}_\infty d := \frac{1}{2}\langle \nabla_H(|\nabla_H d|^2), \nabla_H d\rangle = 0 \quad \text{in } \mathbb{G}\backslash\{0\}. \tag{1.102}$$

It is known that the Euclidean space, the Heisenberg group $\mathbb{H}^n$ and Kaplan's H-type groups are polarizable Carnot groups.

It was shown by Balogh and Tyson in [BT02b] that if $\mathbb{G}$ is a polarizable Carnot group, then the fundamental solutions of the p-sub-Laplacian (1.71) are given by the explicit formulae

$$\varepsilon_p := \begin{cases} c_p d^{\frac{p-Q}{p-1}}, & \text{if } p \neq Q, \\ -c_Q \log d, & \text{if } p = Q. \end{cases} \tag{1.103}$$

This class of groups also admits an advantageous version of the polar coordinates decomposition. In particular, it can be shown (see [BT02b, Proposition

2.20]) that (1.103) implies a useful identity

$$\mathcal{L}_\infty u = \frac{Q-1}{Q-2}\frac{|\nabla_H u|^4}{u},$$

which can be also written as

$$\sum_{j=1}^{N} \frac{uX_j u X_j |\nabla_H u|}{|\nabla_H u|^3} = \frac{Q-1}{Q-2}. \tag{1.104}$$

It can be shown that the $\mathcal{L}$-gauge d on polarizable Carnot groups satisfies a number of further useful relations. For example, the following formula established in [BT02b] will be useful for some calculations in the sequel:

$$\nabla_H \left(\frac{d}{|\nabla_H d|^2} \nabla_H d \right) = Q \text{ in } \mathbb{G}\backslash \mathcal{Z}, \tag{1.105}$$

where the set

$$\mathcal{Z} := \{0\} \bigcup \{x \in \mathbb{G}\backslash\{0\} : \nabla_H d = 0\}$$

has Haar measure zero, and we have $\nabla_H d \neq 0$ for a.e. $x \in \mathbb{G}$.

As usual, the Green identities are still valid for functions with (weak) singularities provided we can approximate them by smooth functions. Thus, for example, for $x \in \Omega$ in a polarizable Carnot group, taking $v = 1$ and $u(y) = \varepsilon_p(x, y)$ we have the following corollary of Theorem 1.4.11 as an extension of Corollary 1.4.9:

Corollary 1.4.16. *Let Ω be an admissible domain in a polarizable Carnot group $\mathbb{G}$ and let $x \in \Omega$. Then we have*

$$\int_{\partial\Omega} |\nabla_{\mathbb{G}} \varepsilon_p|^{p-2} \langle \widetilde{\nabla} \varepsilon_p(x, y), d\nu(y)\rangle = 1.$$

Note that there are stratified Lie groups other than polarizable Carnot groups where the fundamental solution of the sub-Laplacian can be expressed explicitly (see, e.g., [BT02b, Section 6]).

In particular, since on the polarizable Carnot groups we have the fundamental solution ε_p, putting it instead of v in (1.91) we get the following representation type formulae extending those for $p = 2$ from Corollary 1.4.10:

Corollary 1.4.17 (Representation formulae for functions on polarizable Carnot groups)**.** *Let Ω be an admissible domain in a polarizable Carnot group $\mathbb{G}$.*

1. *Let $u \in C^2(\Omega) \bigcap C^1(\overline{\Omega})$. Then for $x \in \Omega$ we have*

$$u(x) = \int_\Omega \varepsilon_p \mathcal{L}_p u - (|\nabla_{\mathbb{G}} \varepsilon_p|^{p-2} - |\nabla_{\mathbb{G}} u|^{p-2})(\widetilde{\nabla} \varepsilon_p) u d\nu$$
$$+ \int_{\partial\Omega} (|\nabla_{\mathbb{G}} \varepsilon_p|^{p-2} u \langle \widetilde{\nabla} \varepsilon_p, d\nu\rangle - |\nabla_{\mathbb{G}} u|^{p-2} \varepsilon_p \langle \widetilde{\nabla} u, d\nu\rangle).$$

2. *Let* $u \in C^2(\Omega)\bigcap C^1(\overline{\Omega})$ *and* $\mathcal{L}_p u = 0$ *on* Ω*, then for* $x \in \Omega$ *we have*

$$u(x) = \int_\Omega (|\nabla_{\mathbb{G}} u|^{p-2} - |\nabla_{\mathbb{G}} \varepsilon_p|^{p-2})(\widetilde{\nabla}\varepsilon_p) u d\nu + \int_{\partial\Omega} (|\nabla_{\mathbb{G}} \varepsilon_p|^{p-2} u \langle \widetilde{\nabla}\varepsilon_p, d\nu\rangle - |\nabla_{\mathbb{G}} u|^{p-2}\varepsilon_p \langle \widetilde{\nabla} u, d\nu\rangle).$$

3. *Let* $u \in C^2(\Omega)\bigcap C^1(\overline{\Omega})$ *and*

$$u(x) = 0, \;\; x \in \partial\Omega,$$

then

$$u(x) = \int_\Omega \varepsilon_p \mathcal{L}_p u - (|\nabla_{\mathbb{G}} \varepsilon_p|^{p-2} - |\nabla_{\mathbb{G}} u|^{p-2})(\widetilde{\nabla}\varepsilon_p) u d\nu - \int_{\partial\Omega} |\nabla_{\mathbb{G}} u|^{p-2}\varepsilon_p \langle \widetilde{\nabla} u, d\nu\rangle.$$

4. *Let* $u \in C^2(\Omega)\bigcap C^1(\overline{\Omega})$ *and* $\sum_{j=1}^{N_1} X_j u \langle X_j, d\nu\rangle = 0$ *on* $\partial\Omega$*, then*

$$u(x) = \int_\Omega \varepsilon_p \mathcal{L}_p u - (|\nabla_{\mathbb{G}} \varepsilon_p|^{p-2} - |\nabla_{\mathbb{G}} u|^{p-2})(\widetilde{\nabla}\varepsilon_p) u d\nu + \int_{\partial\Omega} |\nabla_{\mathbb{G}} \varepsilon_p|^{p-2} u \langle \widetilde{\nabla}\varepsilon_p, d\nu\rangle.$$

1.4.8 Heisenberg group

One of the important examples of the stratified groups is the Heisenberg group that was introduced in Example 1.1.8. Here we collect several of its basic properties that will be of use later in the book. We will give both real and complex descriptions of the Heisenberg group as both will be of use to us in the sequel.

Real description of the Heisenberg group. The Heisenberg group $\mathbb{H}^n$ is the manifold $\mathbb{R}^{2n+1}$ but with the group law given by

$$(x^{(1)}, y^{(1)}, t^{(1)})(x^{(2)}, y^{(2)}, t^{(2)}) := \left(x^{(1)} + x^{(2)}, y^{(1)} + y^{(2)}, t^{(1)} + t^{(2)} + \frac{1}{2}(x^{(1)} \cdot y^{(2)} - x^{(2)} \cdot y^{(1)})\right), \tag{1.106}$$

for $(x^{(1)}, y^{(1)}, t^{(1)}), (x^{(2)}, y^{(2)}, t^{(2)}) \in \mathbb{R}^n \times \mathbb{R}^n \times \mathbb{R} \sim \mathbb{H}^n$, where $x^{(1)} \cdot y^{(2)}$ and $x^{(2)} \cdot y^{(1)}$ are the usual scalar products on $\mathbb{R}^n$. The canonical basis of the Lie algebra $\mathfrak{h}^n$ of the Heisenberg group $\mathbb{H}^n$ is given by the left invariant vector fields

$$X_j = \partial_{x_j} - \frac{y_j}{2}\partial_t, \;\; Y_j = \partial_{y_j} + \frac{x_j}{2}\partial_t, \;\; j = 1, \dots, n, \;\; T = \partial_t. \tag{1.107}$$

It follows that the basis elements X_j, Y_j, T, $j = 1, \ldots, n$, have the following commutator relations,

$$[X_j, Y_j] = T, \quad j = 1, \ldots, n,$$

with all the other commutators being zero. The Heisenberg (Lie) algebra $\mathfrak{h}^n$ is stratified via the decomposition

$$\mathfrak{h}^n = V_1 \oplus V_2,$$

where V_1 is linearly spanned by the X_j's and Y_j's, and $V_2 = \mathbb{R}T$. Therefore, the natural dilations on $\mathfrak{h}^n$ are given by

$$\delta_r(X_j) = rX_j, \;\; \delta_r(Y_j) = rY_j, \;\; \delta_r(T) = r^2 T.$$

On the level of the Heisenberg group $\mathbb{H}^n$ this can be expressed as

$$\delta_r(x, y, t) = r(x, y, t) = (rx, ry, r^2 t), \;\; (x, y, t) \in \mathbb{H}^n, \; r > 0.$$

Consequently, $Q = 2n + 2$ is the homogeneous dimension of the Heisenberg group $\mathbb{H}^n$. The (negative) sub-Laplacian on $\mathbb{H}^n$ is given by

$$\mathcal{L} := \sum_{j=1}^{n} (X_j^2 + Y_j^2) = \sum_{j=1}^{n} \left(\partial_{x_j} - \frac{y_j}{2}\partial_t\right)^2 + \left(\partial_{y_j} + \frac{x_j}{2}\partial_t\right)^2,$$

corresponding to the horizontal gradient

$$\nabla_H := (X_1, \ldots, X_n, Y_1, \ldots, Y_n).$$

We can also write

$$\mathcal{L} = \Delta_{x,y} + \frac{|x|^2 + |y|^2}{4}\partial_t^2 + Z\partial_t, \qquad \text{with } Z = \sum_{j=1}^{n} (x_j \partial_{y_j} - y_j \partial_{x_j}),$$

where $\Delta_{x,y}$ is the Euclidean Laplacian with respect to x, y, and Z is the tangential derivative in the (x, y)-variables.

Complex description of the Heisenberg group. There is an alternative description of the Heisenberg group using complex rather than real variables. It is easy to see that both descriptions are equivalent.

The Heisenberg group $\mathbb{H}^n$ is the space $\mathbb{C}^n \times \mathbb{R}$ with the group operation given by

$$(\zeta, t) \circ (\eta, \tau) = (\zeta + \eta, t + \tau + 2\,\mathrm{Im}\,\zeta\eta), \tag{1.108}$$

for $(\zeta, t), (\eta, \tau) \in \mathbb{C}^n \times \mathbb{R}$.

Comparing (1.106) with (1.108) we can note the change of the constant from $\frac{1}{2}$ to 2. As a result, we are getting different constants in the group law and in the formulae for the left invariant vector fields. We chose to give two descriptions with different constants since the adaptation of such constants in real and complex descriptions of the Heisenberg group seems to be happening in most of the literature. As a result, it will make it more convenient to refer to the relevant literature when needed. This should lead to no confusion since we will never be using these two descriptions at the same time. We note that in general, one can put any constant instead of $\frac{1}{2}$ or 2, the appearing objects are all isomorphic. We can refer the reader to [FR16, Section 6.1.1] for a detailed discussion on the choice of constants in the descriptions of the Heisenberg group.

Writing $\zeta = x + iy$ with x_j, y_j, $j = 1, \ldots, n$, the real coordinates on $\mathbb{H}^n$, the left invariant vector fields

$$\begin{aligned}
\tilde{X}_j &= \frac{\partial}{\partial x_j} + 2y_j \frac{\partial}{\partial t}, \quad j = 1, \ldots, n, \\
\tilde{Y}_j &= \frac{\partial}{\partial y_j} - 2x_j \frac{\partial}{\partial t}, \quad j = 1, \ldots, n, \\
T &= \frac{\partial}{\partial t},
\end{aligned}$$

form a basis for the Lie algebra $\mathfrak{h}^n$ of $\mathbb{H}^n$; again, this can be compared to (1.107).

At the same time, $\mathbb{H}^n$ can be seen as the boundary of the Siegel upper half-space in $\mathbb{C}^{n+1}$,

$$\mathbb{H}^n = \{(\zeta, z_{n+1}) \in \mathbb{C}^{n+1} : \operatorname{Im} z_{n+1} = |\zeta|^2, \zeta = (z_1, \ldots, z_n)\}.$$

Again, we can refer the reader, e.g., to [FR16, Section 6.1.1] for more details on different descriptions of the Heisenberg group.

Parametrising $\mathbb{H}_n$ by $z = (\zeta, t)$ where $t = \operatorname{Re} z_{n+1}$, a basis for the complex tangent space of $\mathbb{H}_n$ at the point z is given by the left invariant vector fields

$$X_j = \frac{\partial}{\partial z_j} + i\overline{z}\frac{\partial}{\partial t}, \quad j = 1, \ldots, n.$$

We denote their conjugates by

$$X_{\overline{j}} \equiv \overline{X}_j = \frac{\partial}{\partial \overline{z}_j} - iz\frac{\partial}{\partial t}, \quad j = 1, \ldots, n.$$

The operator

$$\mathcal{L}_{a,b} = \sum_{j=1}^{n} (aX_j X_{\overline{j}} + bX_{\overline{j}} X_j), \quad a + b = n, \tag{1.109}$$

is a left invariant, rotation invariant differential operator that is homogeneous of degree two. We can refer to the book of Folland and Kohn [FK72] for further properties of such operators. However, we can note that this operator is a slight generalization of the standard sub-Laplacian or Kohn-Laplacian $\mathcal{L}_b$ on the

Heisenberg group $\mathbb{H}^n$ which, when acting on the coefficients of a $(0,q)$-form can be written as

$$\mathcal{L}_b = -\frac{1}{n}\sum_{j=1}^{n}\left((n-q)X_jX_{\bar{j}} + qX_{\bar{j}}X_j\right).$$

Folland and Stein [FS74] obtained the fundamental solution of the operator $\mathcal{L}_{a,b}$ as a constant multiple of the function

$$\varepsilon_{a,b}(z) = \varepsilon(z) = \varepsilon(\zeta,t) = \frac{1}{(t+i|\zeta|^2)^a(t-i|\zeta|^2)^b}. \tag{1.110}$$

More precisely, the distribution $\frac{1}{c_{a,b}}\varepsilon$ is the fundamental solution of $\mathcal{L}_{a,b}$ since ε from (1.110) satisfies the equation

$$\mathcal{L}_{a,b}\varepsilon = c_{a,b}\delta. \tag{1.111}$$

The constant $c_{a,b}$ is zero if a and $b = -1,-2,\ldots,n,n+1,\ldots$, and $c_{a,b} \neq 0$ if a or $b \neq -1,-2,\ldots,n,n+1,\ldots$. In fact, then we can take

$$c_{a,b} = \frac{2(a^2+b^2)\mathrm{Vol}(B_1)}{(2i)^{n+1}}\frac{n!}{a(a-1)\cdots(a-n)}(1-\exp(-2ia\pi)) \tag{1.112}$$

for $a \notin \mathbb{Z}$, see Romero [Rom91, Proof of Theorem 1.6]. We will use the above description of the Heisenberg group and of the (rescaled) fundamental solution (1.110) to $\mathcal{L}_{a,b}$ in Section 11.3.3.

1.4.9 Quaternionic Heisenberg group

In this section we describe the basics of the quaternionic version of the Heisenberg group. We start by recalling the notion of quaternions and summarizing their main properties. As the space of quaternions is usually denoted by $\mathbb{H}$, we keep this notation here as well. There should be no notational confusion with the Heisenberg group since the quaternionic notation will be mostly localized to this section only.

Let $\mathbb{H}$ be the set of quaternions

$$x := x_0 + x_1 i_1 + x_2 i_2 + x_3 i_3,$$

where $(x_0,x_1,x_2,x_3) \in \mathbb{R}^4$, and $1, i_1, i_2, i_3$ are the basis elements of $\mathbb{H}$ with the following rules of multiplication:

$$i_1^2 = i_2^2 = i_3^2 = i_1 i_2 i_3 = -1,$$
$$i_1 i_2 = -i_2 i_1 = i_3,\ i_2 i_3 = -i_3 i_2 = i_1,\ i_3 i_1 = -i_1 i_3 = i_2.$$

The usual convention is that the real part of $x \in \mathbb{H}$ is the real number x_0 and its imaginary part is the point $(x_1,x_2,x_3) \in \mathbb{R}^3$. And so, the real and imaginary parts of x can be denoted by $\Re x$ and $\Im x$, respectively. In addition, we use more precise notations for the imaginary parts as

$$\Im_1 x := x_1, \quad \Im_2 x := x_2, \quad \Im_3 x := x_3.$$

The conjugate of x is denoted by

$$\overline{x} := x_0 - x_1 i_1 - x_2 i_2 - x_3 i_3,$$

and the modulus $|x|$ is defined by

$$|x|^2 := x\overline{x} = \sum_{j=0}^{3} x_j^2.$$

The Grassmanian product (or the quaternion product) of x and y is defined by

$$xy := (x_0 y_0 - \Im x \cdot \Im y) + (x_0 \Im y + y_0 \Im x + \Im x \times \Im y),$$

where

$$\Im x \times \Im y := \det \begin{pmatrix} i_1 & i_2 & i_3 \\ x_1 & x_2 & x_3 \\ y_1 & y_2 & y_3 \end{pmatrix}.$$

Let us denote $\mathbb{H}_q := \mathbb{H} \times \mathbb{R}^3$, it is called the *quaternion Heisenberg group.* Then $\mathbb{H}_q$ becomes a non-commutative group with the group law

$$\begin{aligned}&(x, t_1, t_2, t_3) \circ (y, \tau_1, \tau_2, \tau_3)\\ &\quad := (x + y, t_1 + \tau_1 - 2\Im_1(\overline{y}x), t_2 + y_2 - 2\Im_2(\overline{y}x), t_3 + \tau_3 - 2\Im_3(\overline{y}x)),\end{aligned}$$

for all $(x, t), (y, \tau) \in \mathbb{H}_q$. We note that $e = (0, 0, 0, 0)$ is the identity element of $\mathbb{H}_q$ and the inverse of every element $(x, t_1, t_2, t_3) \in \mathbb{H}_q$ is $(-x, -t_1, -t_2, -t_3)$.

The Haar measure on $\mathbb{H}_q$ coincides with the Lebesgue measure on $\mathbb{H} \times \mathbb{R}^3$ which is denoted by $d\nu = dxdt$. Let $\mathfrak{h}_q$ be the Lie algebra of left invariant vector fields on $\mathbb{H}_q$. A basis of $\mathfrak{h}_q$ is given by $\{X_0, X_1, X_2, X_3\}$ and $\{T_1, T_2, T_3\}$, where

$$\begin{aligned}
X_0 &= \frac{\partial}{\partial x_0} - 2x_1 \frac{\partial}{\partial t_1} - 2x_2 \frac{\partial}{\partial t_2} - 2x_3 \frac{\partial}{\partial t_3},\\
X_1 &= \frac{\partial}{\partial x_1} + 2x_0 \frac{\partial}{\partial t_1} - 2x_3 \frac{\partial}{\partial t_2} + 2x_2 \frac{\partial}{\partial t_3},\\
X_2 &= \frac{\partial}{\partial x_2} + 2x_3 \frac{\partial}{\partial t_1} + 2x_0 \frac{\partial}{\partial t_2} - 2x_1 \frac{\partial}{\partial t_3},\\
X_3 &= \frac{\partial}{\partial x_3} - 2x_2 \frac{\partial}{\partial t_1} + 2x_1 \frac{\partial}{\partial t_2} + 2x_0 \frac{\partial}{\partial t_3},
\end{aligned}$$

and

$$T_k = \frac{\partial}{\partial t_k}, \quad k = 1, 2, 3.$$

The Lie brackets of these vector fields are given by

$$[X_0, X_1] = [X_3, X_2] = 4T_1,$$
$$[X_0, X_2] = [X_1, X_3] = 4T_2,$$
$$[X_0, X_3] = [X_2, X_1] = 4T_3.$$

Thus, the sub-Laplacian on $\mathbb{H}_q$ is given by

$$\mathcal{L} = \sum_{j=0}^{3} X_j^2 = \Delta_x - 4|x|^2\Delta_t - 4\sum_{k=1}^{3}(i_k x \cdot \nabla_x)\frac{\partial}{\partial t_k}, \tag{1.113}$$

where

$$\Delta_x = \sum_{k=0}^{3} \frac{\partial^2}{\partial x_k^2}, \quad \text{and} \quad \Delta_t = \sum_{k=1}^{3} \frac{\partial^2}{\partial t_k^2}.$$

Note that the fundamental solution of the sub-Laplacian $\mathcal{L}$ on $\mathbb{H}_q$ was found by Tie and Wong in [TW09]. We restate their results in the following theorem.

Theorem 1.4.18 (Fundamental solutions for sub-Laplacian on quaternion Heisenberg groups)**.** *The fundamental solution $\Gamma(\xi)$ of the sub-Laplacian $\mathcal{L}$ on the quaternion Heisenberg group $\mathbb{H}_q$ is given by*

$$\Gamma(\xi) := \Gamma(|x|, t) = \frac{2}{(2\pi)^{7/2}|x|^2}\int_{\mathbb{S}^2} \frac{1}{(|x|^2 - i(t\cdot n))^3} d\sigma, \tag{1.114}$$

where $\xi = (x, t) \in \mathbb{H}_q$, $n = (n_1, n_2, n_3)$ is a point on the unit sphere $\mathbb{S}^2$ in $\mathbb{R}^3$ with centre at the origin, and $d\sigma$ is the surface measure on $\mathbb{S}^2$. That is,

$$\mathcal{L}\Gamma_\zeta = -\delta_\zeta, \tag{1.115}$$

where $\Gamma_\zeta(\xi) = \Gamma(\zeta^{-1} \circ \xi)$ and δ_ζ is the Dirac distribution at $\zeta \equiv (y, \tau) \in \mathbb{H}_q$.

The quaternion Heisenberg group is a special case of the model step two nilpotent Lie group. It is a homogeneous group with respect to the dilation

$$\delta_\lambda : \mathbb{R}^7 \to \mathbb{R}^7, \quad \delta_\lambda = (\lambda x, \lambda^2 t).$$

Thus,

$$d(\xi) = \frac{1}{\Gamma^{1/8}(\xi)}, \quad \xi = (x, t) \in \mathbb{H}_q, \tag{1.116}$$

is a homogeneous quasi-norm on $\mathbb{H}_q$ with respect to the dilation δ_λ (see, e.g., [Cyg81]).

1.4.10 *H*-type groups

The H-type groups are a special family of stratified groups with a similar structure to that of the Heisenberg group; one of their important features is that the fundamental solutions to the sub-Laplacian are known explicitly.

We briefly recall the main notions related to this family of groups adopting the notation from [BLU07]; we refer to it for further details.

Definition 1.4.19 (Prototype H-type groups)**.** The space $\mathbb{R}^{m+n}$ equipped with the group law

$$(x,t)\circ(y,\tau)=\begin{pmatrix} x_k+y_k, \quad k=1,\ldots,m \\ t_k+\tau_k+\frac{1}{2}\langle A^{(k)}x,y\rangle, \quad k=1,\ldots,n \end{pmatrix} \tag{1.117}$$

and with the dilations

$$\delta_\lambda(x,t)=(\lambda x,\lambda^2 t)$$

is called a *prototype H-type group*. Here $A^{(k)}$ is an $m\times m$ skew-symmetric orthogonal matrix, such that,

$$A^{(k)}A^{(l)}+A^{(l)}A^{(k)}=0$$

for all $k,l\in\{1,\ldots,n\}$ with $k\neq l$.

Clearly, the Euclidean (Abelian) group and the Heisenberg group are examples of prototype H-type groups.

We leave aside the general H-type groups since it can be shown that any (abstract) H-type group is naturally isomorphic to a prototype H-group (see [BLU07, Theorem 18.2.1]).

It can be directly checked that prototype H-groups are two step nilpotent Lie groups in which the identity of the group is the origin $(0,0)$ and the inverse of (x,t) is

$$(x,t)^{-1}=(-x,-t).$$

It can be also verified that the vector field in the Lie algebra $\mathfrak{g}$ of $\mathbb{G}$ that agrees at the origin with $\frac{\partial}{\partial x_j}$, $j=1,\ldots,m$, is given by

$$X_j=\frac{\partial}{\partial x_j}+\frac{1}{2}\sum_{k=1}^{n}\left(\sum_{i=1}^{m}a_{j,i}^k x_i\right)\frac{\partial}{\partial t_k}, \tag{1.118}$$

where $a_{j,i}^k$ is the (j,i)th element of the matrix $A^{(k)}$.

The prototype H-type groups are stratified with a basis of the first stratum given by these vector fields $X_1,\ldots,X_m$. Thus, the (negative) sub-Laplacian on a prototype H-type group $\mathbb{G}$ is given by

$$\mathcal{L}=\sum_{j=1}^{m}X_j^2=\Delta_x+\frac{1}{4}|x|^2\Delta_t+\sum_{k=1}^{n}\langle A^{(k)}x,\nabla_x\rangle\frac{\partial}{\partial t_k}, \tag{1.119}$$

where Δ and ∇ are the Euclidean Laplacian and the Euclidean gradient, respectively. There is no restriction to suppose that, if ϱ is the centre of the Lie algebra $\mathfrak{g}$ of $\mathbb{G}$, $\varrho^\perp$ is the orthogonal complement of ϱ and

$$m = \dim(\varrho^\perp), \quad n = \dim(\varrho).$$

So, the $\mathbb{R}^n$-component of the prototype H-type group $\mathbb{G} \simeq \mathbb{R}^{m+n}$ can be thought of as of its centre.

We have that the homogeneous dimension of the group is

$$Q = m + 2n.$$

We note that since for H-type groups we have $m \geq 2$ and $n \geq 1$, we actually always have $Q \geq 4$.

Now using a generic coordinate $\xi \equiv (x,t)$, $x \in \mathbb{R}^m$, $t \in \mathbb{R}^n$, let us introduce the following functions on $\mathbb{G}$:

$$v : \mathbb{G} \to \varrho^\perp, \quad v(\xi) := \sum_{j=1}^{m} \langle \exp_{\mathbb{G}}^{-1}(\xi), X_j \rangle X_j,$$

where $\{X_1, \ldots, X_m\}$ is an orthogonal basis of $\varrho^\perp$,

$$z : \mathbb{G} \to \varrho, \quad z(\xi) := \sum_{j=1}^{n} \langle \exp_{\mathbb{G}}^{-1}(\xi), Z_j \rangle Z_j,$$

where $\{Z_1, \ldots, Z_n\}$ is an orthogonal basis of ϱ. Thus, by the definition of v and z, for any $\xi \in \mathbb{G}$, one has

$$\xi = \exp(v(\xi) + z(\xi)), \quad v(\xi) \in \varrho^\perp, \quad z(\xi) \in \varrho,$$

and by a direct calculation we have (see, e.g., [BLU07, Proof of Remark 18.3.3]) that

$$|v(\xi)| = |x|, \quad |z(\xi)| = |t|.$$

The fundamental solutions for the sub-Laplacian on abstract H-type groups were found by A. Kaplan in [Kap80]. Such results boil down to the following statement.

Theorem 1.4.20 (Fundamental solutions for sub-Laplacian on H-type groups). *There exists a positive constant c such that*

$$\Gamma(\xi) := c \left(|x|^4 + 16|t|^2\right)^{(2-Q)/4}$$

is the fundamental solution of the sub-Laplacian $\mathcal{L}$, that is,

$$\mathcal{L}\Gamma_\zeta = -\delta_\zeta, \tag{1.120}$$

where $\Gamma_\zeta(\xi) = \Gamma(\zeta^{-1} \circ \xi)$ and δ_ζ is the Dirac distribution at $\zeta \equiv (y, \tau) \in \mathbb{G}$.

For future use in Section 11.5, we will prefer to have the appearing function Γ positive, which leads to the appearance of the minus sign in (1.120).

For further details and analysis on H-type and related groups we may refer the reader to Kohn–Nirenberg [KN65], Folland [Fol75], Kaplan [Kap80], as well as to a more detailed exposition in [BLU07].

Chapter 2

Hardy Inequalities on Homogeneous Groups

This chapter is devoted to Hardy inequalities and the analysis of their remainders in different forms. Moreover, we discuss several related inequalities such as Rellich inequalities and uncertainty principles.

In this chapter we will use all the notations given in Chapter 1 concerning homogeneous groups and the operators defined on it. In particular, $\mathbb{G}$ *is always a homogeneous group of homogeneous dimension* $Q \geq 1$. Some statements will hold for $Q \geq 2$ or for $Q \geq 3$ but we will be specifying this explicitly in formulations when needed.

2.1 Hardy inequalities and sharp remainders

In this section we analyse the anisotropic version of the classical L^p-Hardy inequality

$$\left\| \frac{f}{|x|_E} \right\|_{L^p(\mathbb{R}^n)} \leq \frac{p}{n-p} \|\nabla f\|_{L^p(\mathbb{R}^n)}, \quad n \geq 2, \quad 1 \leq p < n, \tag{2.1}$$

where ∇ is the standard gradient in $\mathbb{R}^n$, $|x|_E = \sqrt{x_1^2 + \cdots + x_n^2}$ is the Euclidean norm, $f \in C_0^\infty(\mathbb{R}^n)$, and the constant $\frac{p}{n-p}$ is known to be sharp. We also discuss in detail its critical cases and remainder estimates. As consequences, we derive Rellich type inequalities and the corresponding uncertainty principles.

2.1.1 Hardy inequality and uncertainty principle

First we establish the L^p-Hardy inequality and derive a formula for the remainder on a homogeneous group $\mathbb{G}$ of homogeneous dimension $Q \geq 2$. The radial operator $\mathcal{R}$ from (1.30) is entering the appearing expressions.

Theorem 2.1.1 (Hardy inequalities on homogeneous groups). *Let* $|\cdot|$ *be any homogeneous quasi-norm on* $\mathbb{G}$.

M. Ruzhansky, D. Suragan, *Hardy Inequalities on Homogeneous Groups*,
Progress in Mathematics 327, https://doi.org/10.1007/978-3-030-02895-4_3

(i) *Let $f \in C_0^\infty(\mathbb{G}\backslash\{0\})$ be a complex-valued function. Then we have*

$$\left\| \frac{f}{|x|} \right\|_{L^p(\mathbb{G})} \leq \frac{p}{Q-p} \|\mathcal{R}f\|_{L^p(\mathbb{G})}, \quad 1 < p < Q, \tag{2.2}$$

where the constant $\frac{p}{Q-p}$ is sharp. Moreover, the equality in (2.2) is attained if and only if $f = 0$.

(ii) *For a real-valued function $f \in C_0^\infty(\mathbb{G}\backslash\{0\})$ and with the notations*

$$u := u(x) = -\frac{p}{Q-p}\mathcal{R}f(x),$$
$$v := v(x) = \frac{f(x)}{|x|},$$

we have

$$\|u\|^p_{L^p(\mathbb{G})} - \|v\|^p_{L^p(\mathbb{G})} = p \int_{\mathbb{G}} I_p(v,u)|v-u|^2 dx, \tag{2.3}$$

where

$$I_p(h,g) = (p-1)\int_0^1 |\xi h + (1-\xi)g|^{p-2}\xi d\xi. \tag{2.4}$$

(iii) *For $Q \geq 3$, for a complex-valued function $f \in C_0^\infty(\mathbb{G}\backslash\{0\})$ we have*

$$\|\mathcal{R}f\|^2_{L^2(\mathbb{G})} = \left(\frac{Q-2}{2}\right)^2 \left\|\frac{f}{|x|}\right\|^2_{L^2(\mathbb{G})} + \left\|\mathcal{R}f + \frac{Q-2}{2}\frac{f}{|x|}\right\|^2_{L^2(\mathbb{G})}, \tag{2.5}$$

that is, when $p = 2$, (2.3) holds for complex-valued functions as well.

Remark 2.1.2.

1. In the case of $\mathbb{G} = \mathbb{R}^n$ and $|x| = |x|_E = \sqrt{x_1^2 + \cdots + x_n^2}$ the Euclidean norm, we have $Q = n$ and $\mathcal{R} = \partial_r$ is the usual radial derivative, and (2.2) implies the classical Hardy inequality (2.1). Indeed, in this case for $1 < p < n$ inequality (2.2) yields

$$\begin{aligned} \left\|\frac{f}{|x|_E}\right\|_{L^p(\mathbb{R}^n)} &\leq \frac{p}{n-p}\|\mathcal{R}f\|_{L^p(\mathbb{R}^n)} = \frac{p}{n-p}\|\partial_r f\|_{L^p(\mathbb{R}^n)} \\ &= \frac{p}{n-p}\left\|\frac{x}{|x|_E}\cdot\nabla f\right\|_{L^p(\mathbb{R}^n)} \leq \frac{p}{n-p}\|\nabla f\|_{L^p(\mathbb{R}^n)}, \end{aligned} \tag{2.6}$$

in view of the Cauchy–Schwarz inequality for the Euclidean norm.

An interesting feature of the Hardy inequality in Part (i) is that the constant in (2.2) is sharp for any homogeneous quasi-norm $|\cdot|$.

2. In the setting of Part 1 above the remainder formula (2.3) for the Euclidean norm $|\cdot|_E$ in $\mathbb{R}^n$ was analysed by Ioku, Ishiwata and Ozawa [IIO17].

3. In Theorem 2.1.1, Part (ii) implies Part (i). To show it one can notice that the right-hand side of (2.3) is non-negative, which implies that

$$\left\|\frac{f}{|x|}\right\|_{L^p(\mathbb{G})} \leq \frac{p}{Q-p}\|\mathcal{R}f\|_{L^p(\mathbb{G})}, \quad 1<p<Q, \tag{2.7}$$

for any real-valued $f \in C_0^\infty(\mathbb{G}\backslash\{0\})$. Moreover, by using the following identity we obtain the same inequality for all complex-valued functions: for all $z \in \mathbb{C}$ we have

$$|z|^p = \left(\int_{-\pi}^{\pi} |\cos\theta|^p d\theta\right)^{-1} \int_{-\pi}^{\pi} |\text{Re}(z)\cos\theta + \text{Im}(z)\sin\theta|^p \, d\theta, \tag{2.8}$$

which is a consequence of the decomposition of a complex number $z = r(\cos\phi + i\sin\phi)$.

That is, we obtain inequality (2.2), and also that the constant $\frac{p}{Q-p}$ is sharp, in view of the remainder formula. Now let us show that this constant is attained only for $f = 0$. Identity (2.8) says that it is sufficient to look only for real-valued functions f. If the right-hand side of (2.3) vanishes, then we must have $u = v$, that is,

$$-\frac{p}{Q-p}\mathcal{R}f(x) = \frac{f(x)}{|x|}.$$

This also means that $\mathbb{E}f = -\frac{Q-p}{p}f$. Lemma 1.3.1 implies that f is positively homogeneous of order $-\frac{Q-p}{p}$, i.e., there exists a function $h : \wp \to \mathbb{C}$ such that

$$f(x) = |x|^{-\frac{Q-p}{p}} h\left(\frac{x}{|x|}\right), \tag{2.9}$$

where $\wp$ is the unit sphere for the quasi-norm $|\cdot|$. It confirms that f cannot be compactly supported unless it is identically zero.

4. The identity (2.8) has been often used in similar estimates for passing from real-valued to complex-valued functions, see, e.g., Davies [Dav80, p. 176].
5. Let us denote by $H^1_{\mathcal{R}}(\mathbb{G})$ the functional space of the functions $f \in L^2(\mathbb{G})$ with $\mathcal{R}f \in L^2(\mathbb{G})$. Then Theorem 2.1.1 can be extended for functions in $H^1_{\mathcal{R}}(\mathbb{G})$, that is, the proof of (2.2) given above works in this case. As for the sharpness and the equality in (2.2), having (2.9) also implies that $\frac{f(x)}{|x|} = |x|^{-\frac{Q}{p}} h\left(\frac{x}{|x|}\right)$ is not in $L^p(\mathbb{G})$ unless $h = 0$ and $f = 0$.

Remark 2.1.2, Part 2, shows that (2.3) implies Part (i) of Theorem 2.1.1, that is, we only need to prove Parts (ii) and (iii). However, we now give an independent proof of (2.2) for complex-valued functions without relying on the formula (2.8). We see that this argument will be also useful in the proof of Part (ii).

Proof of Theorem 2.1.1. Proof of Part (i). Using the polar decomposition from Proposition 1.2.10, a direct calculation shows that

$$\begin{aligned}\int_{\mathbb{G}} \frac{|f(x)|^p}{|x|^p} dx &= \int_0^\infty \int_{\wp} \frac{|f(ry)|^p}{r^p} r^{Q-1} d\sigma(y) dr \\ &= -\frac{p}{Q-p} \int_0^\infty r^{Q-p} \operatorname{Re} \int_{\wp} |f(ry)|^{p-2} f(ry) \overline{\frac{df(ry)}{dr}} d\sigma(y) dr \\ &= -\frac{p}{Q-p} \operatorname{Re} \int_{\mathbb{G}} \frac{|f(x)|^{p-2} f(x)}{|x|^{p-1}} \overline{\frac{df(x)}{d|x|}} dx. \end{aligned} \tag{2.10}$$

Now by the Hölder inequality with $\frac{1}{p} + \frac{1}{q} = 1$ we obtain

$$\begin{aligned}\int_{\mathbb{G}} \frac{|f(x)|^p}{|x|^p} dx &= -\frac{p}{Q-p} \operatorname{Re} \int_{\mathbb{G}} \frac{|f(x)|^{p-2} f(x)}{|x|^{p-1}} \overline{\frac{df(x)}{d|x|}} dx \\ &\leq \frac{p}{Q-p} \left(\int_{\mathbb{G}} \left| \frac{|f(x)|^{p-2} f(x)}{|x|^{p-1}} \right|^q dx \right)^{\frac{1}{q}} \left(\int_{\mathbb{G}} \left| \frac{df(x)}{d|x|} \right|^p dx \right)^{\frac{1}{p}} \\ &= \frac{p}{Q-p} \left(\int_{\mathbb{G}} \frac{|f(x)|^p}{|x|^p} dx \right)^{1-\frac{1}{p}} \left\| \frac{df(x)}{d|x|} \right\|_{L^p(\mathbb{G})}. \end{aligned}$$

This proves inequality (2.2) in Part (i).

Proof of Part (ii). Since

$$u := u(x) = -\frac{p}{Q-p} \mathcal{R} f, \quad \text{and} \quad v := v(x) = \frac{f(x)}{|x|},$$

the formula (2.10) can be restated as

$$\|v\|^p_{L^p(\mathbb{G})} = \operatorname{Re} \int_{\mathbb{G}} |v|^{p-2} v \overline{u} dx. \tag{2.11}$$

For a real-valued f the formula (2.10) becomes

$$\int_{\mathbb{G}} \frac{|f(x)|^p}{|x|^p} dx = -\frac{p}{Q-p} \int_{\mathbb{G}} \frac{|f(x)|^{p-2} f(x)}{|x|^{p-1}} \frac{df(x)}{d|x|} dx$$

and (2.11) becomes

$$\|v\|^p_{L^p(\mathbb{G})} = \int_{\mathbb{G}} |v|^{p-2} v u dx. \tag{2.12}$$

Moreover, for any L^p-integrable real-valued functions u and v, we have

$$\begin{aligned} &\|u\|^p_{L^p(\mathbb{G})} - \|v\|^p_{L^p(\mathbb{G})} + p \int_{\mathbb{G}} (|v|^p - |v|^{p-2} v u) dx \\ &= \int_{\mathbb{G}} (|u|^p + (p-1)|v|^p - p|v|^{p-2} v u) dx = p \int_{\mathbb{G}} I_p(v,u) |v-u|^2 dx, \end{aligned} \tag{2.13}$$

where

$$I_p(v,u) = (p-1)\int_0^1 |\xi v + (1-\xi)u|^{p-2}\xi d\xi.$$

To show the last equality in (2.13), we observe the identity, for real numbers $u \neq v$,

$$\begin{aligned}
&\frac{1}{p}|u|^p + \left(1-\frac{1}{p}\right)|v|^p - |v|^{p-2}vu \\
&= \left(1-\frac{1}{p}\right)(|v|^p - |u|^p) - u(|v|^{p-2}v - |u|^{p-2}u) \\
&= (p-1)\int_0^1 |\xi v + (1-\xi)u|^{p-2}(\xi v + (1-\xi)u)d\xi\,(v-u) \\
&\quad - (p-1)\int_0^1 |\xi v + (1-\xi)u|^{p-2}d\xi\, u(v-u) \\
&= (p-1)\int_0^1 |\xi v + (1-\xi)u|^{p-2}\xi d\xi\,(v-u)^2,
\end{aligned}$$

using the integral expression for the remainder in the Taylor expansion formula.

Combining (2.13) with (2.12) we arrive at

$$\|u\|^p_{L^p(\mathbb{G})} - \|v\|^p_{L^p(\mathbb{G})} = p\int_{\mathbb{G}} I_p(v,u)|v-u|^2 dx.$$

It completes the proof of Part (ii).

Proof of Part (iii). When $p = 2$, the equality (2.11) for complex-valued functions reduces to

$$\|v\|^2_{L^2(\mathbb{G})} = \operatorname{Re}\int_{\mathbb{G}} v\overline{u}dx.$$

Then we have

$$\begin{aligned}
\|u\|^2_{L^2(\mathbb{G})} - \|v\|^2_{L^2(\mathbb{G})} &= \|u\|^2_{L^2(\mathbb{G})} - \|v\|^2_{L^2(\mathbb{G})} + 2\int_{\mathbb{G}}(|v|^2 - \operatorname{Re} v\overline{u})dx \\
&= \int_{\mathbb{G}}(|u|^2 + |v|^2 - 2\operatorname{Re} v\overline{u})dx = \int_{\mathbb{G}} |u-v|^2 dx,
\end{aligned}$$

that is, (2.5) is proved. □

As a direct consequence of the inequality (2.2) we obtain the corresponding uncertainty principle:

Corollary 2.1.3 (Uncertainty principle on homogeneous groups). *For every complex-valued function $f \in C_0^\infty(\mathbb{G}\backslash\{0\})$ we have*

$$\left(\int_{\mathbb{G}}\left|\frac{df(x)}{d|x|}\right|^p dx\right)^{1/p}\left(\int_{\mathbb{G}}|x|^q|f|^q dx\right)^{1/q} \geq \frac{Q-p}{p}\int_{\mathbb{G}}|f|^2 dx. \tag{2.14}$$

Here $Q \geq 2$, $|\cdot|$ is an arbitrary homogeneous quasi-norm on $\mathbb{G}$, $1 < p < Q$ and $\frac{1}{p} + \frac{1}{q} = 1$.

Proof. The inequality (2.7) and the Hölder inequality imply that

$$\left(\int_{\mathbb{G}}\left|\frac{df(x)}{d|x|}\right|^p dx\right)^{\frac{1}{p}}\left(\int_{\mathbb{G}}|x|^q|f|^q dx\right)^{\frac{1}{q}}$$
$$\geq \frac{Q-p}{p}\left(\int_{\mathbb{G}}\frac{|f|^p}{|x|^p}dx\right)^{\frac{1}{p}}\left(\int_{\mathbb{G}}|x|^q|f|^q dx\right)^{\frac{1}{q}} \geq \frac{Q-p}{p}\int_{\mathbb{G}}|f|^2 dx,$$

This shows (2.14). □

Remark 2.1.4. In the Abelian case $\mathbb{G} = (\mathbb{R}^n, +)$ with the standard Euclidean distance $|x|_E$, we have $Q = n$, so that (2.14) with $p = q = 2$ and $n \geq 3$ implies the uncertainty principle

$$\int_{\mathbb{R}^n}\left|\frac{x}{|x|_E}\cdot\nabla u(x)\right|^2 dx\int_{\mathbb{R}^n}|x|_E^2|u(x)|^2 dx \geq \left(\frac{n-2}{2}\right)^2\left(\int_{\mathbb{R}^n}|u(x)|^2 dx\right)^2, \quad (2.15)$$

which in turn implies the classical uncertainty principle for $\mathbb{G} \equiv \mathbb{R}^n$:

$$\int_{\mathbb{R}^n}|\nabla u(x)|^2 dx\int_{\mathbb{R}^n}|x|_E^2|u(x)|^2 dx \geq \left(\frac{n-2}{2}\right)^2\left(\int_{\mathbb{R}^n}|u(x)|^2 dx\right)^2, \quad n \geq 3.$$

2.1.2 Weighted Hardy inequalities

In this section $\mathbb{G}$ is a homogeneous group of homogeneous dimension $Q \geq 3$. Let $|\cdot|$ be an arbitrary homogeneous quasi-norm on $\mathbb{G}$. Here, we are going to discuss weighted Hardy inequalities on $\mathbb{G}$ which are the consequences of exact equalities.

Theorem 2.1.5 (Weighted Hardy identity in $L^2(\mathbb{G})$)**.** *Let $\mathbb{G}$ be a homogeneous group of homogeneous dimension $Q \geq 3$ and let $|\cdot|$ be a homogeneous quasi-norm on $\mathbb{G}$. Then for every complex-valued function $f \in C_0^\infty(\mathbb{G}\backslash\{0\})$ and for any $\alpha \in \mathbb{R}$ we have the equality*

$$\left\|\frac{1}{|x|^\alpha}\mathcal{R}f\right\|^2_{L^2(\mathbb{G})} = \left(\frac{Q-2}{2}-\alpha\right)^2\left\|\frac{f}{|x|^{\alpha+1}}\right\|^2_{L^2(\mathbb{G})} + \left\|\frac{1}{|x|^\alpha}\mathcal{R}f + \frac{Q-2-2\alpha}{2|x|^{\alpha+1}}f\right\|^2_{L^2(\mathbb{G})}. \quad (2.16)$$

The equality (2.16) implies many different inequalities. For instance, by taking $\alpha = 1$ and simplifying its coefficient, for any $Q \geq 3$ we obtain the identity

$$\left\|\frac{1}{|x|}\mathcal{R}f\right\|^2_{L^2(\mathbb{G})} = \left(\frac{Q-4}{2}\right)^2\left\|\frac{f}{|x|^2}\right\|^2_{L^2(\mathbb{G})} + \left\|\frac{1}{|x|}\mathcal{R}f + \frac{Q-4}{2|x|^2}f\right\|^2_{L^2(\mathbb{G})}. \quad (2.17)$$

By dropping the last term in (2.16) which is non-negative we obtain:

Corollary 2.1.6 (Weighted Hardy inequality in $L^2(\mathbb{G})$). *Let $Q \geq 3$ and let $\alpha \in \mathbb{R}$ be such that $Q - 2 - 2\alpha \neq 0$. Then for all complex-valued functions $f \in C_0^\infty(\mathbb{G}\backslash\{0\})$ we have*

$$\left\| \frac{f}{|x|^{\alpha+1}} \right\|_{L^2(\mathbb{G})} \leq \frac{2}{|Q-2-2\alpha|} \left\| \frac{1}{|x|^\alpha} \mathcal{R} f \right\|_{L^2(\mathbb{G})}. \tag{2.18}$$

Here the constant $\frac{2}{|Q-2-2\alpha|}$ is sharp and it is attained if and only if $f = 0$.

Remark 2.1.7.

1. It is interesting to note that the constant $\frac{2}{|Q-2-2\alpha|}$ in (2.18) is sharp for any homogeneous quasi-norm $|\cdot|$ on $\mathbb{G}$.
2. When $\alpha = 1$, (2.17) or (2.18) also imply that

$$\left\| \frac{f}{|x|^2} \right\|_{L^2(\mathbb{G})} \leq \frac{2}{Q-4} \left\| \frac{1}{|x|} \mathcal{R} f \right\|_{L^2(\mathbb{G})}, \quad Q \geq 5, \tag{2.19}$$

again with $\frac{2}{Q-4}$ being the sharp constant.

3. If $\alpha = 0$, the identity (2.16) recovers Part (iii) of Theorem 2.1.1. However, we will use Part (iii) of Theorem 2.1.1 in the proof of Theorem 2.1.5.
4. In the Abelian case $\mathbb{G} = (\mathbb{R}^n, +)$, $n \geq 3$, we have $Q = n$, so for any homogeneous quasi-norm $|\cdot|$ on $\mathbb{R}^n$ identity (2.16) implies the following inequality with the optimal constant:

$$\frac{|n-2-2\alpha|}{2} \left\| \frac{f}{|x|^{\alpha+1}} \right\|_{L^2(\mathbb{R}^n)} \leq \left\| \frac{1}{|x|^\alpha} \frac{x}{|x|} \cdot \nabla f \right\|_{L^2(\mathbb{R}^n)} \quad \text{for all } \alpha \in \mathbb{R}.$$

In the case of the Euclidean distance $|x|_E = \sqrt{x_1^2 + \cdots + x_n^2}$, by the Cauchy–Schwarz inequality we obtain the following estimate:

$$\frac{|n-2-2\alpha|}{2} \left\| \frac{f}{|x|_E^{\alpha+1}} \right\|_{L^2(\mathbb{R}^n)} \leq \left\| \frac{1}{|x|_E^\alpha} \nabla f \right\|_{L^2(\mathbb{R}^n)} \tag{2.20}$$

for all $\alpha \in \mathbb{R}$ and for any $f \in C_0^\infty(\mathbb{R}^n\backslash\{0\})$. The sharpness of the constant $\frac{|n-2-2\alpha|}{2}$ in the Euclidean case of $\mathbb{R}^n$ with the Euclidean norm, (2.20) was shown in [CW01, Theorem 1.1. (ii)].

5. Hardy inequalities with homogeneous weights have been also considered by Hoffmann-Ostenhof and Laptev [HOL15]. There are also further many-particle versions of such inequalities, see [HOHOLT08] and many further references therein. We will discuss some of such inequalities in Section 6.11 and Section 6.12.
6. Theorem 2.1.5 was established in [RS17b]. Its extension from L^2 to L^p spaces presented in Theorem 2.1.8 was made in [Ngu17].

Proof of Theorem 2.1.5. We first observe the equality

$$\frac{1}{|x|^{\alpha}}\mathcal{R}f = \mathcal{R}\frac{f}{|x|^{\alpha}} + \alpha\frac{f}{|x|^{\alpha+1}} \tag{2.21}$$

for any $\alpha \in \mathbb{R}$, which follows from

$$\mathcal{R}\frac{f}{|x|^{\alpha}} = \frac{1}{|x|^{\alpha}}\mathcal{R}f + f\mathcal{R}\frac{1}{|x|^{\alpha}}$$

and hence, by using (1.30), we have

$$\mathcal{R}\frac{1}{|x|^{\alpha}} = \frac{d}{dr}\frac{1}{r^{\alpha}} = -\alpha\frac{1}{r^{\alpha+1}} = -\alpha\frac{1}{|x|^{\alpha+1}}, \quad r = |x|.$$

Then using (2.21) we can write

$$\begin{aligned}\left\|\frac{1}{|x|^{\alpha}}\mathcal{R}f\right\|^2_{L^2(\mathbb{G})} &= \left\|\mathcal{R}\frac{f}{|x|^{\alpha}} + \frac{\alpha f}{|x|^{\alpha+1}}\right\|^2_{L^2(\mathbb{G})} \\ &= \left\|\mathcal{R}\frac{f}{|x|^{\alpha}}\right\|^2_{L^2(\mathbb{G})} + 2\alpha\mathrm{Re}\int_{\mathbb{G}}\mathcal{R}\left(\frac{f}{|x|^{\alpha}}\right)\frac{\overline{f}}{|x|^{\alpha+1}}dx + \left\|\frac{\alpha f}{|x|^{\alpha+1}}\right\|^2_{L^2(\mathbb{G})}.\end{aligned}$$

By applying (2.5) to the function $\frac{f}{|x|^{\alpha}}$ and using (2.21) we have that

$$\left\|\mathcal{R}\frac{f}{|x|^{\alpha}}\right\|^2_{L^2(\mathbb{G})} = \left(\frac{Q-2}{2}\right)^2\left\|\frac{f}{|x|^{1+\alpha}}\right\|^2_{L^2(\mathbb{G})} + \left\|\frac{1}{|x|^{\alpha}}\mathcal{R}f + \frac{Q-2-2\alpha}{2|x|^{\alpha+1}}f\right\|^2_{L^2(\mathbb{G})}.$$

In addition, a direct calculation using the polar decomposition in Proposition 1.2.10 shows that

$$\begin{aligned}2\alpha\mathrm{Re}\int_{\mathbb{G}}\mathcal{R}\left(\frac{f}{|x|^{\alpha}}\right)\frac{\overline{f}}{|x|^{\alpha+1}}dx &= 2\alpha\mathrm{Re}\int_0^{\infty}r^{Q-2}\int_{\wp}\frac{d}{dr}\left(\frac{f(ry)}{r^{\alpha}}\right)\frac{\overline{f(ry)}}{r^{\alpha}}d\sigma(y)dr \\ &= \alpha\int_0^{\infty}r^{Q-2}\int_{\wp}\frac{d}{dr}\left(\frac{|f(ry)|^2}{r^{2\alpha}}\right)d\sigma(y)dr = -\alpha(Q-2)\left\|\frac{f}{|x|^{\alpha+1}}\right\|^2_{L^2(\mathbb{G})}.\end{aligned}$$

In conclusion, combining these identities we arrive at

$$\left\|\frac{1}{|x|^{\alpha}}\mathcal{R}f\right\|^2_{L^2(\mathbb{G})} = \left(\frac{Q-2}{2}-\alpha\right)^2\left\|\frac{f}{|x|^{\alpha+1}}\right\|^2_{L^2(\mathbb{G})} + \left\|\frac{1}{|x|^{\alpha}}\mathcal{R}f + \frac{Q-2-2\alpha}{2|x|^{\alpha+1}}f\right\|^2_{L^2(\mathbb{G})},$$

yielding (2.16). □

To present a weighted L^p-Hardy inequality on $\mathbb{G}$ we will use the following function R_p in analogy to I_p in (2.4). For $\xi, \eta \in \mathbb{C}$ we denote

$$R_p(\xi,\eta) := \frac{1}{p}|\eta|^p + \frac{p-1}{p}|\xi|^p - \mathrm{Re}(|\xi|^{p-2}\xi\overline{\eta}). \tag{2.22}$$

By the convexity of the function $z \mapsto |z|^p$ we see that $R_p(\xi,\eta) \geq 0$ is non-negative and $R_p(\xi,\eta)=0$ and if and only if $\xi=\eta$. If $\xi,\eta\in\mathbb{R}$, we then have

$$R_p(\xi,\eta)=(p-1)\int_0^1 |t\xi+(1-t)\eta|^{p-2}tdt\,|\xi-\eta|^2,$$

analogous to I_p in (2.4).

Theorem 2.1.8 (Weighted Hardy identity in $L^p(\mathbb{G})$). *Let $\mathbb{G}$ be a homogeneous group of homogeneous dimension Q. Let $1<p<Q$ and $\alpha\in\mathbb{R}$. Then for any homogeneous quasi-norm $|\cdot|$ on $\mathbb{G}$ and for all complex-valued functions $f\in C_0^\infty(\mathbb{G}\backslash\{0\})$ we have*

$$\begin{aligned}\left\|\frac{\mathcal{R}f}{|x|^\alpha}\right\|^p_{L^p(\mathbb{G})} &= \left|\frac{Q-p(1+\alpha)}{p}\right|^p\int_{\mathbb{G}}\frac{|f|^p}{|x|^{p(1+\alpha)}}dx\\ &+p\int_{\mathbb{G}}\frac{1}{|x|^{p\alpha}}R_p\left(-\frac{Q-p(1+\alpha)}{p}\frac{f}{|x|},\mathcal{R}f\right)dx.\end{aligned} \tag{2.23}$$

By dropping the last term in (2.23) which is non-negative we obtain:

Corollary 2.1.9 (Weighted Hardy inequality in $L^p(\mathbb{G})$). *Let $1<p<Q$ and let $\alpha\in\mathbb{R}$. Then for all complex-valued functions $f\in C_0^\infty(\mathbb{G}\backslash\{0\})$ we have*

$$\frac{|Q-p(1+\alpha)|}{p}\left\|\frac{f}{|x|^{1+\alpha}}\right\|_{L^p(\mathbb{G})}\leq\left\|\frac{\mathcal{R}f}{|x|^\alpha}\right\|_{L^p(\mathbb{G})}. \tag{2.24}$$

If $Q-p(1+\alpha)\neq 0$, then the constant $\frac{|Q-p(1+\alpha)|}{p}$ in (2.24) is sharp and it is attained if and only if $f=0$.

Proof of Theorem 2.1.8. We can assume that $Q-p(1+\alpha)\neq 0$, otherwise there is nothing to prove. A direct calculation gives

$$\begin{aligned}\int_{\mathbb{G}}\frac{|f(x)|^p}{|x|^{p(1+\alpha)}}dx &= \int_0^\infty r^{Q-p(1+\alpha)-1}\int_{\wp}|f(ry)|^p d\sigma(y)dr\\ &=\frac{1}{Q-p(1+\alpha)}\int_0^\infty (r^{Q-p(1+\alpha)})'\int_{\wp}|f(ry)|^p d\sigma(y)dr\\ &=-\frac{1}{Q-p(1+\alpha)}\mathrm{Re}\int_0^\infty r^{Q-p(1+\alpha)}\int_{\wp}|f(ry)|^{p-2}f(ry)\overline{\mathcal{R}f(ry)}d\sigma(y)dr\\ &=-\frac{1}{Q-p(1+\alpha)}\mathrm{Re}\int_{\mathbb{G}}\frac{|f(x)|^{p-2}f(x)}{|x|^{(p-1)(1+\alpha)}}\frac{\overline{\mathcal{R}f(x)}}{|x|^\alpha}dx\\ &=\frac{p-1}{p}\int_{\mathbb{G}}\frac{|f|^p}{|x|^{p(1+\alpha)}}dx+\frac{1}{p}\left(\frac{p}{|Q-p(1+\alpha)|}\right)^p\int_{\mathbb{G}}\frac{|\mathcal{R}f|^p}{|x|^{p\alpha}}dx\\ &-\int_{\mathbb{G}}R_p\left(\frac{f}{|x|^{1+\alpha}},-\frac{p}{Q-p(1+\alpha)}\frac{\mathcal{R}f}{|x|^\alpha}\right)dx,\end{aligned}$$

which implies the identity (2.23). □

Proof of Corollary 2.1.9. The equality (2.23) implies inequality (2.24) since the last term in (2.23) is non-negative. Let us now show the sharpness of the constant. For this we approximate the function $r^{-(Q-p(1+\alpha))/p}$ by smooth compactly supported functions, for details of such an argument see also the proof of Theorem 3.1.4. Using (2.23) it follows that the equality in (2.24) holds if and only if

$$R_p\left(-\frac{Q-p(1+\alpha)}{p}\frac{f}{|x|}, \mathcal{R}f\right) = 0,$$

or, equivalently,

$$\mathcal{R}f = -\frac{Q-p(1+\alpha)}{p}\frac{f}{|x|}.$$

In turn, this is equivalent to

$$\mathbb{E}f = -\frac{Q-p(1+\alpha)}{p}f.$$

By Proposition 1.3.1 it follows that f is positively homogeneous of order $-(Q-p(1+\alpha))/p$. Since $|f|/|x|^{1+\alpha}$ is in $L^p(\mathbb{G})$, it follows that $f=0$. □

2.1.3 Hardy inequalities with super weights

In this section we discuss sharp L^p-Hardy type inequalities with *super weights*, i.e., with weights of the form

$$\frac{(a+b|x|^{\alpha})^{\frac{\beta}{p}}}{|x|^m}. \tag{2.25}$$

Such weights are sometimes called the super weights because of the arbitrariness of the choice of any homogeneous quasi-norm as well as a wide range of parameters. However, all the inequalities can be obtained with best constants.

Theorem 2.1.10 (Hardy inequalities with super weights). *Let $\mathbb{G}$ be a homogeneous group of homogeneous dimension $Q \geq 1$. Let $a, b > 0$ and $1 < p < \infty$. Then we have the following inequalities:*

(i) *If $\alpha\beta > 0$ and $pm \leq Q-p$, then for all $f \in C_0^{\infty}(\mathbb{G}\backslash\{0\})$ we have*

$$\frac{Q-pm-p}{p}\left\|\frac{(a+b|x|^{\alpha})^{\frac{\beta}{p}}}{|x|^{m+1}}f\right\|_{L^p(\mathbb{G})} \leq \left\|\frac{(a+b|x|^{\alpha})^{\frac{\beta}{p}}}{|x|^{m}}\mathcal{R}f\right\|_{L^p(\mathbb{G})}. \tag{2.26}$$

If $Q \neq pm+p$ then the constant $\frac{Q-pm-p}{p}$ is sharp.

(ii) *If $\alpha\beta < 0$ and $pm-\alpha\beta \leq Q-p$, then for all $f \in C_0^{\infty}(\mathbb{G}\backslash\{0\})$ we have*

$$\frac{Q-pm+\alpha\beta-p}{p}\left\|\frac{(a+b|x|^{\alpha})^{\frac{\beta}{p}}}{|x|^{m+1}}f\right\|_{L^p(\mathbb{G})} \leq \left\|\frac{(a+b|x|^{\alpha})^{\frac{\beta}{p}}}{|x|^{m}}\mathcal{R}f\right\|_{L^p(\mathbb{G})}. \tag{2.27}$$

If $Q \neq pm+p-\alpha\beta$ then the constant $\frac{Q-pm+\alpha\beta-p}{p}$ is sharp.

Proof of Theorem 2.1.10. Proof of Part (i). We can assume $Q \neq pm + p$ since in the case $Q = pm + p$ there is nothing to prove. As usual with $(r, y) = (|x|, \frac{x}{|x|}) \in (0, \infty) \times \wp$ on $\mathbb{G}$, where

$$\wp := \{x \in \mathbb{G} : |x| = 1\},$$

using the polar decomposition in Proposition 1.2.10 and integrating by parts, we obtain

$$\int_{\mathbb{G}} \frac{(a + b|x|^{\alpha})^{\beta}}{|x|^{pm+p}} |f(x)|^p dx = \int_0^{\infty} \int_{\wp} \frac{(a + br^{\alpha})^{\beta}}{r^{pm+p}} |f(ry)|^p r^{Q-1} d\sigma(y) dr. \tag{2.28}$$

Since $a, b > 0$, $\alpha\beta > 0$ and $m < \frac{Q-p}{p}$ we obtain

$$\begin{aligned}
&\int_{\mathbb{G}} \frac{(a + b|x|^{\alpha})^{\beta}}{|x|^{pm+p}} |f(x)|^p dx \\
&\leq \int_0^{\infty} \int_{\wp} (a + br^{\alpha})^{\beta} r^{Q-1-pm-p} \left(\frac{\alpha\beta b r^{\alpha}}{(a + br^{\alpha})(Q - pm - p)} + 1 \right) |f(ry)|^p d\sigma(y) dr \\
&= \int_0^{\infty} \int_{\wp} \frac{d}{dr} \left(\frac{(a + br^{\alpha})^{\beta} r^{Q-pm-p}}{Q - pm - p} \right) |f(ry)|^p d\sigma(y) dr \\
&= -\frac{p}{Q - pm - p} \int_0^{\infty} (a + br^{\alpha})^{\beta} r^{Q-pm-p} \operatorname{Re} \int_{\wp} |f(ry)|^{p-2} f(ry) \overline{\frac{df(ry)}{dr}} d\sigma(y) dr \\
&\leq \left| \frac{p}{Q - pm - p} \right| \int_{\mathbb{G}} \frac{(a + b|x|^{\alpha})^{\beta} |\mathcal{R}f(x)| |f(x)|^{p-1}}{|x|^{pm+p-1}} dx \\
&= \frac{p}{Q - pm - p} \int_{\mathbb{G}} \frac{(a + b|x|^{\alpha})^{\frac{\beta(p-1)}{p}} |f(x)|^{p-1}}{|x|^{(m+1)(p-1)}} \frac{(a + b|x|^{\alpha})^{\frac{\beta}{p}}}{|x|^m} |\mathcal{R}f(x)| dx.
\end{aligned}$$

Now by using Hölder's inequality we arrive at the inequality

$$\begin{aligned}
&\int_{\mathbb{G}} \frac{(a + b|x|^{\alpha})^{\beta}}{|x|^{pm+p}} |f(x)|^p dx \\
&\leq \frac{p}{Q - pm - p} \left(\int_{\mathbb{G}} \frac{(a + b|x|^{\alpha})^{\beta}}{|x|^{pm+p}} |f(x)|^p dx \right)^{\frac{p-1}{p}} \left(\int_{\mathbb{G}} \frac{(a + b|x|^{\alpha})^{\beta}}{|x|^{pm}} |\mathcal{R}f(x)|^p dx \right)^{\frac{1}{p}},
\end{aligned}$$

which gives (2.26).

We need to check the equality condition in the above Hölder inequality in order to show the sharpness of the constant. Setting

$$g(x) = |x|^C,$$

where $C \in \mathbb{R}, C \neq 0$ and $Q \neq pm + p$ a direct calculation shows

$$\left| \frac{1}{C} \right|^p \left(\frac{(a + b|x|^{\alpha})^{\frac{\beta}{p}} |\mathcal{R}g(x)|}{|x|^m} \right)^p = \left(\frac{(a + b|x|^{\alpha})^{\frac{\beta(p-1)}{p}} |g(x)|^{p-1}}{|x|^{(m+1)(p-1)}} \right)^{\frac{p}{p-1}},$$

which satisfies the equality condition of the Hölder inequality. This gives the sharpness of the constant $\frac{Q-pm-p}{p}$ in the inequality (2.26).

Proof of Part (ii). Here we can assume that $Q \neq pm+p-\alpha\beta$ since for $Q = pm+p-\alpha\beta$ there is nothing to prove. Using the polar decomposition in Proposition 1.2.10, as before we have the equality (2.28). Since $\alpha\beta < 0$ and $pm-\alpha\beta < Q-p$ we obtain

$$\begin{aligned}
&\int_{\mathbb{G}} \frac{(a+b|x|^{\alpha})^{\beta}}{|x|^{pm+p}}|f(x)|^p dx \\
&\leq \int_0^{\infty}\int_{\wp}(a+br^{\alpha})^{\beta} r^{Q-1-pm-p} \\
&\quad \times \left(\frac{br^{\alpha}}{a+br^{\alpha}}+\frac{a}{a+br^{\alpha}}\cdot\frac{Q-pm-p}{Q-pm-p+\alpha\beta}\right)|f(ry)|^p d\sigma(y)dr \\
&= \int_0^{\infty}\int_{\wp}\frac{(a+br^{\alpha})^{\beta}r^{Q-1-pm-p}}{Q-pm-p+\alpha\beta}\left(\frac{\alpha\beta br^{\alpha}}{a+br^{\alpha}}+Q-pm-p\right)|f(ry)|^p d\sigma(y)dr \\
&= \int_0^{\infty}\int_{\wp}\frac{d}{dr}\left(\frac{(a+br^{\alpha})^{\beta}r^{Q-pm-p}}{Q-pm-p+\alpha\beta}\right)|f(ry)|^p d\sigma(y)dr \\
&= -\frac{p}{Q-pm-p+\alpha\beta} \\
&\quad \times \int_0^{\infty}(a+br^{\alpha})^{\beta}r^{Q-pm-p}\,\mathrm{Re}\int_{\wp}|f(ry)|^{p-2}f(ry)\overline{\frac{df(ry)}{dr}}d\sigma(y)dr \\
&\leq \left|\frac{p}{Q-pm-p+\alpha\beta}\right|\int_{\mathbb{G}}\frac{(a+b|x|^{\alpha})^{\beta}|\mathcal{R}f(x)||f(x)|^{p-1}}{|x|^{pm+p-1}}dx \\
&= \frac{p}{Q-pm-p+\alpha\beta}\int_{\mathbb{G}}\frac{(a+b|x|^{\alpha})^{\frac{\beta(p-1)}{p}}|f(x)|^{p-1}}{|x|^{(m+1)(p-1)}}\frac{(a+b|x|^{\alpha})^{\frac{\beta}{p}}}{|x|^m}|\mathcal{R}f(x)|dx.
\end{aligned}$$

By Hölder's inequality, it follows that

$$\begin{aligned}
\int_{\mathbb{G}}\frac{(a+b|x|^{\alpha})^{\beta}}{|x|^{pm+p}}|f(x)|^p dx \leq & \frac{p}{Q-pm-p+\alpha\beta}\left(\int_{\mathbb{G}}\frac{(a+b|x|^{\alpha})^{\beta}}{|x|^{pm+p}}|f(x)|^p dx\right)^{(p-1)/p} \\
&\times\left(\int_{\mathbb{G}}\frac{(a+b|x|^{\alpha})^{\beta}}{|x|^{pm}}|\mathcal{R}f(x)|^p dx\right)^{1/p},
\end{aligned}$$

which gives (2.27).

To show the sharpness of the constant we will check the equality condition in the above Hölder inequality. Thus, by taking

$$h(x) = |x|^C,$$

where $C \in \mathbb{R}, C \neq 0$ and $Q \neq pm+p-\alpha\beta$, we get

$$\left|\frac{1}{C}\right|^p\left(\frac{(a+b|x|^{\alpha})^{\frac{\beta}{p}}|\mathcal{R}h(x)|}{|x|^m}\right)^p = \left(\frac{(a+b|x|^{\alpha})^{\frac{\beta(p-1)}{p}}|h(x)|^{p-1}}{|x|^{(m+1)(p-1)}}\right)^{p/(p-1)},$$

which satisfies the equality condition in Hölder's inequality. This gives the sharpness of the constant $\frac{Q-pm-p+\alpha\beta}{p}$ in (2.27). □

2.1.4 Hardy inequalities of higher order with super weights

The iteration process gives the following higher-order L^p-Hardy type inequalities with super weights. Here as before, $\mathbb{G}$ is a homogeneous group of homogeneous dimension $Q \geq 1$ and $|\cdot|$ is a homogeneous quasi-norm on $\mathbb{G}$.

Theorem 2.1.11 (Higher-order Hardy inequalities with super weights). *Let $a, b > 0$ and $1 < p < \infty$, $Q \geq 1$, $k \in \mathbb{N}$. Then we have the following inequalities.*

(i) *If $\alpha\beta > 0$ and $pm \leq Q - p$, then for all $f \in C_0^\infty(\mathbb{G}\backslash\{0\})$ we have*

$$\left[\prod_{j=0}^{k-1}\left(\frac{Q-p}{p}-(m+j)\right)\right]\left\|\frac{(a+b|x|^\alpha)^{\frac{\beta}{p}}}{|x|^{m+k}}f\right\|_{L^p(\mathbb{G})} \leq \left\|\frac{(a+b|x|^\alpha)^{\frac{\beta}{p}}}{|x|^m}\mathcal{R}^k f\right\|_{L^p(\mathbb{G})}. \tag{2.29}$$

(ii) *If $\alpha\beta < 0$ and $pm - \alpha\beta \leq Q - p$, then for all $f \in C_0^\infty(\mathbb{G}\backslash\{0\})$ we have*

$$\left[\prod_{j=0}^{k-1}\left(\frac{Q-p+\alpha\beta}{p}-(m+j)\right)\right]\left\|\frac{(a+b|x|^\alpha)^{\frac{\beta}{p}}}{|x|^{m+k}}f\right\|_{L^p(\mathbb{G})} \leq \left\|\frac{(a+b|x|^\alpha)^{\frac{\beta}{p}}}{|x|^m}\mathcal{R}^k f\right\|_{L^p(\mathbb{G})}. \tag{2.30}$$

Remark 2.1.12. 1. In the case of $k = 1$, (2.29) gives inequality (2.26), and (2.30) gives inequality (2.27).

2. In the Euclidean case $\mathbb{G} = \mathbb{R}^n$ and $|\cdot| = |\cdot|_E$ the Euclidean norm, the super weights in the form (2.25) have appeared in [GM08], together with some applications to problems for differential equations. The case of homogeneous groups, as well as the iterative higher order estimates as in Theorem 2.1.11 were analysed in [RSY17b, RSY18b].

Proof of Theorem 2.1.11. We can iterate (2.26). That is, we start with

$$\frac{Q-pm-p}{p}\left\|\frac{(a+b|x|^\alpha)^{\frac{\beta}{p}}}{|x|^{m+1}}f\right\|_{L^p(\mathbb{G})} \leq \left\|\frac{(a+b|x|^\alpha)^{\frac{\beta}{p}}}{|x|^m}\mathcal{R}f\right\|_{L^p(\mathbb{G})}. \tag{2.31}$$

In (2.31) replacing f by $\mathcal{R}f$ we obtain

$$\frac{Q-pm-p}{p}\left\|\frac{(a+b|x|^\alpha)^{\frac{\beta}{p}}}{|x|^{m+1}}\mathcal{R}f\right\|_{L^p(\mathbb{G})} \leq \left\|\frac{(a+b|x|^\alpha)^{\frac{\beta}{p}}}{|x|^m}\mathcal{R}^2 f\right\|_{L^p(\mathbb{G})}. \tag{2.32}$$

On the other hand, replacing m by $m+1$, (2.31) gives

$$\frac{Q-p(m+1)-p}{p}\left\|\frac{(a+b|x|^{\alpha})^{\frac{\beta}{p}}}{|x|^{m+2}}f\right\|_{L^p(\mathbb{G})} \le \left\|\frac{(a+b|x|^{\alpha})^{\frac{\beta}{p}}}{|x|^{m+1}}\mathcal{R}f\right\|_{L^p(\mathbb{G})}.$$

Combining this with (2.32) we obtain

$$\begin{aligned}\left(\frac{Q-pm-p}{p}\right)\left(\frac{Q-p(m+1)-p}{p}\right)&\left\|\frac{(a+b|x|^{\alpha})^{\frac{\beta}{p}}}{|x|^{m+2}}f\right\|_{L^p(\mathbb{G})}\\ &\le \left\|\frac{(a+b|x|^{\alpha})^{\frac{\beta}{p}}}{|x|^{m}}\mathcal{R}^2f\right\|_{L^p(\mathbb{G})}.\end{aligned}$$

This iteration process gives

$$\prod_{j=0}^{k-1}\left(\frac{Q-p}{p}-(m+j)\right)\left\|\frac{(a+b|x|^{\alpha})^{\frac{\beta}{p}}}{|x|^{m+k}}f\right\|_{L^p(\mathbb{G})} \le \left\|\frac{(a+b|x|^{\alpha})^{\frac{\beta}{p}}}{|x|^{m}}\mathcal{R}^kf\right\|_{L^p(\mathbb{G})}.$$

Similarly, we have for $\alpha\beta<0$, $pm-\alpha\beta\le Q-2$ and $f\in C_0^{\infty}(\mathbb{G}\backslash\{0\})$ that

$$\begin{aligned}\prod_{j=0}^{k-1}\left(\frac{Q-p+\alpha\beta}{p}-(m+j)\right)&\left\|\frac{(a+b|x|^{\alpha})^{\frac{\beta}{p}}}{|x|^{m+k}}f\right\|_{L^p(\mathbb{G})}\\ &\le \left\|\frac{(a+b|x|^{\alpha})^{\frac{\beta}{p}}}{|x|^{m}}\mathcal{R}^kf\right\|_{L^p(\mathbb{G})},\end{aligned}$$

completing the proof. □

2.1.5 Two-weight Hardy inequalities

In this section, using the method of factorization of differential expressions, we obtain Hardy type inequalities with two general weights $\phi(x)$ and $\psi(x)$. The idea of the factorization method can be best illustrated by the following example of an estimate due to Gesztesy and Littlejohn [GL17].

Example 2.1.13 (Gesztesy and Littlejohn two-parameter inequality). Let $\alpha,\beta\in\mathbb{R}$, $x\in\mathbb{R}^n\backslash\{0\}$ and $n\ge 2$. Let us define the operator

$$T_{\alpha,\beta}:=-\Delta+\alpha|x|_E^{-2}x\cdot\nabla+\beta|x|_E^{-2}.$$

One readily checks that its formal adjoint is given by

$$T^{+}_{\alpha,\beta}:=-\Delta-\alpha|x|_E^{-2}x\cdot\nabla+(\beta-\alpha(n-2))|x|_E^{-2}.$$

Using the non-negativity of the operator $T^+_{\alpha,\beta}T_{\alpha,\beta}$ on $C_0^\infty(\mathbb{R}^n\backslash\{0\})$, for all $f \in C_0^\infty(\mathbb{R}^n\backslash\{0\})$ we can deduce that

$$\begin{aligned}\int_{\mathbb{R}^n} |(\Delta f)(x)|^2 dx \geq &((n-4)\alpha - 2\beta) \int_{\mathbb{R}^n} |x|_E^{-2}|(\nabla f)(x)|^2 dx \\ &- \alpha(\alpha-4) \int_{\mathbb{R}^n} |x|_E^{-4}|x\cdot(\nabla f)(x)|^2 dx \\ &+ \beta((n-4)(\alpha-2)-\beta) \int_{\mathbb{R}^n} |x|_E^{-4}|f(x)|^2 dx.\end{aligned} \tag{2.33}$$

By choosing particular values of α and β it can be checked that this inequality yields classical Rellich and Hardy–Rellich type inequalities as special cases, see [GL17] for details.

On the other hand, using the non-negativity of the operator $T_{\alpha,\beta}T^+_{\alpha,\beta}$, it was shown in [RY17] that for $\alpha, \beta \in \mathbb{R}$ and $n \geq 2$, and for all $f \in C_0^\infty(\mathbb{R}^n\backslash\{0\})$ we can deduce another two-parameter inequality

$$\begin{aligned}&\int_{\mathbb{R}^n} |(\Delta f)(x)|^2 dx \\ &\geq (n\alpha - 2\beta)\int_{\mathbb{R}^n} |x|_E^{-2}|(\nabla f)(x)|^2 dx - \alpha(\alpha+4)\int_{\mathbb{R}^n} |x|_E^{-4}|x\cdot(\nabla f)(x)|^2 dx \\ &\quad + (2(n-4)(\alpha(n-2)-\beta) - 2\alpha^2(n-2) + \alpha\beta n - \beta^2)\int_{\mathbb{R}^n} |x|_E^{-4}|f(x)|^2 dx.\end{aligned} \tag{2.34}$$

The following result is a two-weight inequality on general homogeneous groups with general weights.

Theorem 2.1.14 (Two-weight Hardy inequality). *Let $\mathbb{G}$ be a homogeneous group of homogeneous dimension $Q \geq 3$ and let $|\cdot|$ be a homogeneous quasi-norm on $\mathbb{G}$. Let $\phi, \psi \in L^2_{\mathrm{loc}}(\mathbb{G}\backslash\{0\})$ be any real-valued functions such that $\mathcal{R}\phi, \mathcal{R}\psi \in L^2_{\mathrm{loc}}(\mathbb{G}\backslash\{0\})$. Let $\alpha \in \mathbb{R}$. Then for all complex-valued functions $f \in C_0^\infty(\mathbb{G}\backslash\{0\})$ we have inequalities*

$$\begin{aligned}&\int_{\mathbb{G}} (\phi(x))^2 |\mathcal{R}f(x)|^2 dx \\ &\quad \geq \alpha \int_{\mathbb{G}} (\phi(x)\mathcal{R}\psi(x) + \psi(x)\mathcal{R}\phi(x))\,|f(x)|^2 dx \\ &\quad\quad + \alpha(Q-1)\int_{\mathbb{G}} \frac{\phi(x)\psi(x)}{|x|}|f(x)|^2 dx - \alpha^2 \int_{\mathbb{G}} (\psi(x))^2 |f(x)|^2 dx\end{aligned} \tag{2.35}$$

and

$$\begin{aligned}&\int_{\mathbb{G}} (\phi(x))^2 |\mathcal{R}f(x)|^2 dx \\ &\quad \geq \alpha \int_{\mathbb{G}} (\psi(x)\mathcal{R}\phi(x) - \phi(x)\mathcal{R}\psi(x))\,|f(x)|^2 dx\end{aligned}$$

$$+\alpha(Q-1)\int_{\mathbb{G}}\frac{\phi(x)\psi(x)}{|x|}|f(x)|^2dx-\alpha^2\int_{\mathbb{G}}(\psi(x))^2|f(x)|^2dx$$
$$-(Q-1)\int_{\mathbb{G}}\frac{(\phi(x))^2}{|x|^2}|f(x)|^2dx+(Q-1)\int_{\mathbb{G}}\frac{\phi(x)\mathcal{R}\phi(x)}{|x|}|f(x)|^2dx$$
$$+\int_{\mathbb{G}}\phi(x)\mathcal{R}^2\phi(x)|f(x)|^2dx. \tag{2.36}$$

From Theorem 2.1.14 one can get different weighted Hardy inequalities. Let us point out several examples.

Remark 2.1.15.

1. If we take $\phi(x)\equiv 1$ in Theorem 2.1.14, we obtain for all $\alpha\in\mathbb{R}$ and for all $f\in C_0^\infty(\mathbb{G}\backslash\{0\})$ families of inequalities

$$\int_{\mathbb{G}}|\mathcal{R}f(x)|^2dx\geq\int_{\mathbb{G}}\left(\alpha\mathcal{R}\phi(x)+\alpha(Q-1)\frac{\psi(x)}{|x|}-\alpha^2(\psi(x))^2\right)|f(x)|^2dx$$

   and

$$\int_{\mathbb{G}}|\mathcal{R}f(x)|^2dx$$
$$\geq\int_{\mathbb{G}}\left(-\alpha\mathcal{R}\phi(x)+\alpha(Q-1)\frac{\psi(x)}{|x|}-\alpha^2(\psi(x))^2-\frac{Q-1}{|x|^2}\right)|f(x)|^2dx.$$

2. If we take $\phi(x)=|x|^{-a}$ and $\psi(x)=|x|^{-b}$ for $a,b\in\mathbb{R}$, then (2.35) implies that

$$\int_{\mathbb{G}}\frac{|\mathcal{R}f(x)|^2}{|x|^{2a}}dx\geq\alpha(Q-a-b-1)\int_{\mathbb{G}}\frac{|f(x)|^2}{|x|^{a+b+1}}dx-\alpha^2\int_{\mathbb{G}}\frac{|f(x)|^2}{|x|^{2b}}dx.$$

   In the case when we take $b=a+1$, we get

$$\int_{\mathbb{G}}\frac{|\mathcal{R}f(x)|^2}{|x|^{2a}}dx\geq(\alpha(Q-2a-2)-\alpha^2)\int_{\mathbb{G}}\frac{|f(x)|^2}{|x|^{2a+2}}dx.$$

   Then, by maximizing the constant $(\alpha(Q-2a-2)-\alpha^2)$ with respect to α we obtain the weighted Hardy inequality from Corollary 2.1.6, namely,

$$\int_{\mathbb{G}}\frac{|\mathcal{R}f(x)|^2}{|x|^{2a}}dx\geq\frac{(Q-2a-2)^2}{4}\int_{\mathbb{G}}\frac{|f(x)|^2}{|x|^{2a+2}}dx, \tag{2.37}$$

   for which it is known that the constant in (2.37) is sharp.

3. If we take $\phi(x) = |x|^{-a}(\log|x|)^c$ and $\psi(x) = |x|^{-b}(\log|x|)^d$ for $a, b, c, d \in \mathbb{R}$, then we obtain from (2.35) the inequality

$$\begin{aligned}\int_{\mathbb{G}} \frac{(\log|x|)^{2c}}{|x|^{2a}}|\mathcal{R}f(x)|^2dx \\ \geq \alpha \int_{\mathbb{G}} \left(\frac{(c+d)(\log|x|)^{c+d-1} + (Q-1-a-b)(\log|x|)^{c+d}}{|x|^{a+b+1}}\right)|f(x)|^2dx \\ - \alpha^2 \int_{\mathbb{G}} \frac{(\log|x|)^{2d}}{|x|^{2b}}|f(x)|^2dx.\end{aligned}$$

If we take $a = \frac{Q-2}{2}$, $b = \frac{Q}{2}$, $c = 1$ and $d = 0$, it follows that

$$\int_{\mathbb{G}} \frac{(\log|x|)^2}{|x|^{Q-2}}|\mathcal{R}f(x)|^2dx \geq (\alpha - \alpha^2)\int_{\mathbb{G}} \frac{|f(x)|^2}{|x|^Q}dx.$$

After maximizing the above constant with respect to α we obtain the critical Hardy inequality

$$\int_{\mathbb{G}} \frac{(\log|x|)^2}{|x|^{Q-2}}|\mathcal{R}f(x)|^2dx \geq \frac{1}{4}\int_{\mathbb{G}} \frac{|f(x)|^2}{|x|^Q}dx, \tag{2.38}$$

recovering the critical inequality in Theorem 2.2.4. There, it is shown that the constant $\frac{1}{4}$ in (2.38) is sharp.

4. We can refer to [GL17] for a thorough discussion of the factorization method, its history and different features. We also refer to [GP80] for obtaining the Hardy inequality and to [Ges84] for logarithmic refinements by this factorization method.

The inequalities in Theorem 2.1.14 were obtained in [RY17] which we follow for the proof.

Proof of Theorem 2.1.14. Let us introduce the one-parameter differential expression

$$T_\alpha := \phi(x)\mathcal{R} + \alpha\psi(x).$$

One can readily calculate for the formal adjoint operator of T_α on $C_0^\infty(\mathbb{G}\backslash\{0\})$:

$$\begin{aligned}&\int_{\mathbb{G}} \phi(x)\mathcal{R}f(x)\overline{g(x)}dx + \alpha\int_{\mathbb{G}} \psi(x)f(x)\overline{g(x)}dx \\ &= \int_0^\infty \int_{\wp} \phi(ry)\frac{d}{dr}(f(ry))\overline{g(ry)}r^{Q-1}d\sigma(y)dr + \alpha\int_{\mathbb{G}} \psi(x)f(x)\overline{g(x)}dx \\ &= -\int_0^\infty \int_{\wp} \phi(ry)f(ry)\overline{\frac{d}{dr}(g(ry))}r^{Q-1}d\sigma(y)dr \\ &\quad - (Q-1)\int_0^\infty \int_{\wp} \phi(ry)f(ry)\overline{g(ry)}r^{Q-2}d\sigma(y)dr + \alpha\int_{\mathbb{G}} \psi(x)f(x)\overline{g(x)}dx\end{aligned}$$

$$
\begin{aligned}
&= \int_{\mathbb{G}} f(x)\left(-\phi(x)\overline{\mathcal{R}g(x)}\right)dx - (Q-1)\int_{\mathbb{G}} f(x)\left(\frac{\phi(x)}{|x|}\overline{g(x)}\right)dx \\
&\quad - \int_{\mathbb{G}} f(x)\left(\mathcal{R}\phi(x)\overline{g(x)}\right)dx + \alpha\int_{\mathbb{G}} f(x)\left(\psi(x)\overline{g(x)}\right)dx.
\end{aligned}
$$

Thus, the formal adjoint operator of T_α has the form

$$
T_\alpha^+ := -\phi(x)\mathcal{R} - \frac{Q-1}{|x|}\phi(x) - \mathcal{R}\phi(x) + \alpha\psi(x),
$$

where $x \neq 0$. Then we have

$$
\begin{aligned}
(T_\alpha^+ T_\alpha f)(x) = &- \phi(x)\mathcal{R}(\phi(x)\mathcal{R}f(x)) - \alpha\phi(x)\mathcal{R}(f(x)\psi(x)) - \frac{Q-1}{|x|}(\phi(x))^2\mathcal{R}f(x) \\
&- \frac{\alpha(Q-1)}{|x|}\phi(x)\psi(x)f(x) - \phi(x)\mathcal{R}\phi(x)\mathcal{R}f(x) - \alpha\psi(x)f(x)\mathcal{R}\phi(x) \\
&+ \alpha\phi(x)\psi(x)\mathcal{R}f(x) + \alpha^2(\psi(x))^2 f(x).
\end{aligned}
$$

By the non-negativity of $T_\alpha^+ T_\alpha$, introducing polar coordinates $(r,y) = (|x|, \frac{x}{|x|}) \in (0,\infty)\times\wp$ on $\mathbb{G}$, where $\wp$ is the quasi-sphere as in (1.12), and using Proposition 1.2.10 one calculates

$$
\begin{aligned}
0 &\le \int_{\mathbb{G}} |(T_\alpha f)(x)|^2 dx = \int_{\mathbb{G}} f(x)\overline{(T_\alpha^+ T_\alpha f)(x)}dx \\
&= \operatorname{Re}\int_{\mathbb{G}} f(x)\overline{(T_\alpha^+ T_\alpha f)(x)}dx =: I_1 + I_2 + I_3 + I_4 + I_5 + I_6,
\end{aligned}
\tag{2.39}
$$

where we set

$$
I_1 = -\operatorname{Re}\int_0^\infty\int_\wp f(ry)\phi(ry)\overline{\frac{d}{dr}\left(\phi(ry)\frac{d}{dr}(f(ry))\right)}r^{Q-1}d\sigma(y)dr, \tag{2.40}
$$

$$
\begin{aligned}
I_2 &= -\alpha\operatorname{Re}\int_0^\infty\int_\wp f(ry)\phi(ry)\overline{\frac{d}{dr}(f(ry)\psi(ry))}r^{Q-1}d\sigma(y)dr, \\
I_3 &= -(Q-1)\operatorname{Re}\int_0^\infty\int_\wp f(ry)\frac{(\phi(ry))^2\overline{\frac{d}{dr}(f(ry))}}{r}r^{Q-1}d\sigma(y)dr, \\
I_4 &= -\alpha(Q-1)\int_{\mathbb{G}}\frac{\phi(x)\psi(x)}{|x|}|f(x)|^2dx - \alpha\int_{\mathbb{G}}\mathcal{R}\phi(x)\psi(x)|f(x)|^2dx \\
&\quad + \alpha^2\int_{\mathbb{G}}(\psi(x))^2|f(x)|^2dx, \\
I_5 &= -\operatorname{Re}\int_0^\infty\int_\wp f(ry)\frac{d}{dr}(\phi(ry))\phi(ry)\overline{\frac{d}{dr}(f(ry))}r^{Q-1}d\sigma(y)dr
\end{aligned}
$$

and

$$
I_6 = \alpha\operatorname{Re}\int_0^\infty\int_\wp f(ry)\phi(ry)\psi(ry)\overline{\frac{d}{dr}(f(ry))}r^{Q-1}d\sigma(y)dr.
$$

Now we simplify the sum of the terms I_1, I_2, I_3, I_5 and I_6. By a direct calculation we obtain

$$
\begin{aligned}
I_1 = &(Q-1)\mathrm{Re}\int_0^\infty\int_\wp (\phi(ry))^2 f(ry)\overline{\frac{d}{dr}(f(ry))}r^{Q-2}d\sigma(y)dr\\
&+\mathrm{Re}\int_0^\infty\int_\wp (\phi(ry))^2\frac{d}{dr}(f(ry))\overline{\frac{d}{dr}(f(ry))}r^{Q-1}d\sigma(y)dr\\
&+\mathrm{Re}\int_0^\infty\int_\wp \phi(ry)\frac{d}{dr}(\phi(ry))\overline{\frac{d}{dr}f(ry)}f(ry)r^{Q-1}d\sigma(y)dr\\
=&\frac{Q-1}{2}\int_0^\infty\int_\wp (\phi(ry))^2\frac{d}{dr}|f(ry)|^2 r^{Q-2}d\sigma(y)dr\\
&+\int_0^\infty\int_\wp (\phi(ry))^2\left|\frac{d}{dr}(f(ry))\right|^2 r^{Q-1}d\sigma(y)dr\\
&+\frac{1}{2}\int_0^\infty\int_\wp \phi(ry)\frac{d}{dr}(\phi(ry))\frac{d}{dr}|f(ry)|^2 r^{Q-1}d\sigma(y)dr\\
=&\int_0^\infty\int_\wp (\phi(ry))^2\left|\frac{d}{dr}(f(ry))\right|^2 r^{Q-1}d\sigma(y)dr\\
&-\frac{(Q-1)(Q-2)}{2}\int_0^\infty\int_\wp (\phi(ry))^2|f(ry)|^2 r^{Q-3}d\sigma(y)dr\\
&-(Q-1)\int_0^\infty\int_\wp \phi(ry)\frac{d}{dr}(\phi(ry))|f(ry)|^2 r^{Q-2}d\sigma(y)dr\\
&-\frac{1}{2}\int_0^\infty\int_\wp \left(\frac{d}{dr}(\phi(ry))\right)^2|f(ry)|^2 r^{Q-1}d\sigma(y)dr\\
&-\frac{1}{2}\int_0^\infty\int_\wp \phi(ry)\frac{d^2}{dr^2}(\phi(ry))|f(ry)|^2 r^{Q-1}d\sigma(y)dr\\
&-\frac{Q-1}{2}\int_0^\infty\int_\wp \phi(ry)\frac{d}{dr}(\phi(ry))|f(ry)|^2 r^{Q-2}d\sigma(y)dr\\
=&\int_{\mathbb{G}}(\phi(x))^2|\mathcal{R}f(x)|^2dx-\frac{(Q-1)(Q-2)}{2}\int_{\mathbb{G}}\frac{(\phi(x))^2}{|x|^2}|f(x)|^2dx\\
&-(Q-1)\int_{\mathbb{G}}\frac{\phi(x)\mathcal{R}\phi(x)}{|x|}|f(x)|^2dx-\frac{1}{2}\int_{\mathbb{G}}(\mathcal{R}\phi(x))^2|f(x)|^2dx\\
&-\frac{1}{2}\int_{\mathbb{G}}\mathcal{R}^2\phi(x)\phi(x)|f(x)|^2dx-\frac{Q-1}{2}\int_{\mathbb{G}}\frac{\phi(x)\mathcal{R}\phi(x)}{|x|}|f(x)|^2dx.
\end{aligned}
$$

Now we calculate I_2 as follows,

$$
I_2 = -\alpha\mathrm{Re}\int_0^\infty\int_\wp \phi(ry)\psi(ry)f(ry)\overline{\frac{d}{dr}(f(ry))}r^{Q-1}d\sigma(y)dr
$$

$$
\begin{aligned}
&-\alpha\int_0^\infty\int_\wp \phi(ry)\frac{d}{dr}(\psi(ry))|f(ry)|^2r^{Q-1}d\sigma(y)dr\\
=&-\frac{\alpha}{2}\int_0^\infty\int_\wp \phi(ry)\psi(ry)\frac{d}{dr}|f(ry)|^2r^{Q-1}d\sigma(y)dr\\
&-\alpha\int_0^\infty\int_\wp \phi(ry)\frac{d}{dr}(\psi(ry))|f(ry)|^2r^{Q-1}d\sigma(y)dr\\
=&-\alpha\int_0^\infty\int_\wp \phi(ry)\frac{d}{dr}(\psi(ry))|f(ry)|^2r^{Q-1}d\sigma(y)dr\\
&+\frac{\alpha}{2}\int_0^\infty\int_\wp \psi(ry)\frac{d}{dr}(\phi(ry))|f(ry)|^2r^{Q-1}d\sigma(y)dr\\
&+\frac{\alpha}{2}\int_0^\infty\int_\wp \phi(ry)\frac{d}{dr}(\psi(ry))|f(ry)|^2r^{Q-1}d\sigma(y)dr\\
&+\frac{\alpha(Q-1)}{2}\int_0^\infty\int_\wp \phi(ry)\psi(ry)|f(ry)|^2r^{Q-2}d\sigma(y)dr\\
=&\frac{\alpha}{2}\int_{\mathbb{G}}\psi(x)\mathcal{R}\phi(x)|f(x)|^2dx+\frac{\alpha}{2}\int_{\mathbb{G}}\phi(x)\mathcal{R}\psi(x)|f(x)|^2dx\\
&+\frac{\alpha(Q-1)}{2}\int_{\mathbb{G}}\frac{\phi(x)\psi(x)}{|x|}|f(x)|^2dx-\alpha\int_{\mathbb{G}}\phi(x)\mathcal{R}\psi(x)|f(x)|^2dx.
\end{aligned}
$$

For I_3, one has

$$
\begin{aligned}
I_3=&-(Q-1)\mathrm{Re}\int_0^\infty\int_\wp(\phi(ry))^2f(ry)\overline{\frac{d}{dr}(f(ry))}r^{Q-2}d\sigma(y)dr\\
=&-\frac{Q-1}{2}\int_0^\infty\int_\wp(\phi(ry))^2\frac{d}{dr}|f(ry)|^2r^{Q-2}d\sigma(y)dr\\
=&(Q-1)\int_0^\infty\int_\wp\phi(ry)\frac{d}{dr}(\phi(ry))|f(ry)|^2r^{Q-2}d\sigma(y)dr\\
&+\frac{(Q-1)(Q-2)}{2}\int_0^\infty\int_\wp(\phi(ry))^2|f(ry)|^2r^{Q-3}d\sigma(y)dr\\
=&(Q-1)\int_{\mathbb{G}}\frac{\phi(x)\mathcal{R}\phi(x)}{|x|}|f(x)|^2dx+\frac{(Q-1)(Q-2)}{2}\int_{\mathbb{G}}\frac{(\phi(x))^2}{|x|^2}|f(x)|^2dx.
\end{aligned}
$$

For I_5, we have

$$
\begin{aligned}
I_5=&-\mathrm{Re}\int_0^\infty\int_\wp f(ry)\frac{d}{dr}(\phi(ry))\phi(ry)\overline{\frac{d}{dr}(f(ry))}r^{Q-1}d\sigma(y)dr\\
=&-\frac{1}{2}\int_0^\infty\int_\wp\frac{d}{dr}(\phi(ry))\phi(ry)\frac{d}{dr}|f(ry)|^2r^{Q-1}d\sigma(y)dr\\
=&\frac{1}{2}\int_0^\infty\int_\wp\phi(ry)\frac{d^2}{dr^2}(\phi(ry))|f(ry)|^2r^{Q-1}d\sigma(y)dr
\end{aligned}
$$

$$
\begin{aligned}
&+\frac{1}{2}\int_0^\infty\int_\wp\left(\frac{d}{dr}(\phi(ry))\right)^2|f(ry)|^2r^{Q-1}d\sigma(y)dr\\
&+\frac{Q-1}{2}\int_0^\infty\int_\wp\frac{\phi(ry)\frac{d}{dr}(\phi(ry))}{r}|f(ry)|^2r^{Q-1}d\sigma(y)dr\\
&=\frac{1}{2}\int_{\mathbb{G}}\mathcal{R}^2\phi(x)\phi(x)|f(x)|^2dx+\frac{1}{2}\int_{\mathbb{G}}(\mathcal{R}\phi(x))^2|f(x)|^2dx\\
&+\frac{Q-1}{2}\int_{\mathbb{G}}\frac{\mathcal{R}\phi(x)\phi(x)}{|x|}|f(x)|^2dx.
\end{aligned}
$$

Finally, for I_6 we obtain

$$
\begin{aligned}
I_6&=\alpha\mathrm{Re}\int_0^\infty\int_\wp f(ry)\phi(ry)\psi(ry)\overline{\frac{d}{dr}(f(ry))}r^{Q-1}d\sigma(y)dr \qquad (2.41)\\
&=\frac{\alpha}{2}\int_0^\infty\int_\wp\phi(ry)\psi(ry)\frac{d}{dr}|f(ry)|^2r^{Q-1}d\sigma(y)dr\\
&=-\frac{\alpha}{2}\int_0^\infty\int_\wp\frac{d}{dr}(\phi(ry))\psi(ry)|f(ry)|^2r^{Q-1}d\sigma(y)dr\\
&-\frac{\alpha}{2}\int_0^\infty\int_\wp\frac{d}{dr}(\psi(ry))\phi(ry)|f(ry)|^2r^{Q-1}d\sigma(y)dr\\
&-\frac{\alpha(Q-1)}{2}\int_0^\infty\int_\wp\phi(ry)\psi(ry)\frac{|f(ry)|^2}{r}r^{Q-1}d\sigma(y)dr\\
&=-\frac{\alpha}{2}\int_{\mathbb{G}}\mathcal{R}\phi(x)\psi(x)|f(x)|^2dx-\frac{\alpha}{2}\int_{\mathbb{G}}\phi(x)\mathcal{R}\psi(x)|f(x)|^2dx\\
&-\frac{\alpha(Q-1)}{2}\int_{\mathbb{G}}\frac{\phi(x)\psi(x)}{|x|}|f(x)|^2dx.
\end{aligned}
$$

Putting (2.40)–(2.41) in (2.39), we obtain that

$$
\begin{aligned}
&\int_{\mathbb{G}}(\phi(x))^2|\mathcal{R}f(x)|^2dx\\
&-\alpha\int_{\mathbb{G}}\left(\phi(x)\mathcal{R}\psi(x)+\psi(x)\mathcal{R}\phi(x)+(Q-1)\frac{\phi(x)\psi(x)}{|x|}\right)|f(x)|^2dx\\
&+\alpha^2\int_{\mathbb{G}}(\psi(x))^2|f(x)|^2dx\geq 0,
\end{aligned}
$$

which implies (2.35).

Thus, we have obtained (2.35) using the non-negativity of $T_\alpha^+T_\alpha$. Now we can obtain (2.36) using the non-negativity of $T_\alpha T_\alpha^+$. Similar to the above we calculate

$$
(T_\alpha T_\alpha^+f)(x)=-\phi(x)\mathcal{R}(\phi(x)\mathcal{R}f(x))+\alpha\phi(x)\mathcal{R}(f(x)\psi(x))-\frac{Q-1}{|x|}(\phi(x))^2\mathcal{R}f(x)
$$

$$
\begin{aligned}
&-\frac{\alpha(Q-1)}{|x|}\phi(x)\psi(x)f(x)-\phi(x)\mathcal{R}\phi(x)\mathcal{R}f(x)-\alpha\psi(x)f(x)\mathcal{R}\phi(x)\\
&-\alpha\phi(x)\psi(x)\mathcal{R}f(x)+\alpha^2(\psi(x))^2f(x)\\
&-(Q-1)\phi(x)f(x)\mathcal{R}\left(\frac{\phi(x)}{|x|}\right)-\phi(x)f(x)\mathcal{R}^2\phi(x).
\end{aligned}
$$

Using the non-negativity of $T_\alpha T_\alpha^+$ we get

$$
\begin{aligned}
0 &\le \int_{\mathbb{G}} |(T_\alpha f)(x)|^2 dx = \int_{\mathbb{G}} f(x)\overline{(T_\alpha T_\alpha^+ f)(x)}dx\\
&= \operatorname{Re}\int_{\mathbb{G}} f(x)\overline{(T_\alpha T_\alpha^+ f)(x)}dx = \widetilde{I}_1+\widetilde{I}_2+\widetilde{I}_3+\widetilde{I}_4+\widetilde{I}_5+\widetilde{I}_6,
\end{aligned}
$$

where

$$
\widetilde{I}_1 = I_1,\ \widetilde{I}_2=-I_2,\ \widetilde{I}_3=I_3,\ \widetilde{I}_4=I_4+A,\ \widetilde{I}_5=I_5,\ \widetilde{I}_6=-I_6
$$

with

$$
A=-(Q-1)\int_{\mathbb{G}}\phi(x)|f(x)|^2\mathcal{R}\left(\frac{\phi(x)}{|x|}\right)dx-\int_{\mathbb{G}}\phi(x)\mathcal{R}^2\phi(x)|f(x)|^2dx.
$$

Taking into account these and (2.40)–(2.41), we obtain (2.36). □

To finish this section, we observe another version of weighted Hardy inequality with the radial derivative. Anticipating the material presented in later chapters, the proof will be based on the integral Hardy inequality from Theorem 5.1.1.

Theorem 2.1.16 (Weighted Hardy inequality for radial functions). *Let $\mathbb{G}$ be a homogeneous group of homogeneous dimension Q. Let $\phi>0$, $\psi>0$ be positive weight functions on $\mathbb{G}$ and let $1<p\le q<\infty$. Then there exists a positive constant $C>0$ such that*

$$
\left(\int_{\mathbb{G}}\phi(x)|f(x)|^q dx\right)^{1/q}\le C\left(\int_{\mathbb{G}}\psi(x)|\mathcal{R}f(x)|^p dx\right)^{1/p} \tag{2.42}
$$

holds for all radial functions f with $f(0)=0$ if and only if

$$
\sup_{R>0}\left(\int_{|x|\ge R}\phi(x)dx\right)^{1/q}\left(\int_0^R\left(\int_{\wp} r^{Q-1}\psi(ry)d\sigma(y)\right)^{1-p'}dr\right)^{1/p'}<\infty. \tag{2.43}
$$

Remark 2.1.17. In the Abelian case $\mathbb{G}=(\mathbb{R}^n,+)$ and $Q=n$, (2.42) was obtained in [DHK97] and in [Saw84]. On homogeneous groups it was observed in [RY18a] and we follow the proof there.

Proof of Theorem 2.1.16. For $r=|x|$, let us denote $\tilde{f}(r)=f(x)$. We also denote

$$
\Phi(r):=\int_{\wp} r^{Q-1}\phi(ry)d\sigma(y),\quad \Psi(r):=\int_{\wp} r^{Q-1}\psi(ry)d\sigma(y).
$$

Consequently, using $\tilde{f}(0)=0$, we have

$$\begin{aligned}\left(\int_{\mathbb{G}}\phi(x)|f(x)|^q dx\right)^{1/q} &= \left(\int_{\wp}\int_0^\infty r^{Q-1}\phi(ry)|\tilde{f}(r)|^q dr d\sigma(y)\right)^{1/q}\\ &= \left(\int_0^\infty \Phi(r)|\tilde{f}(r)|^q dr\right)^{1/q}\\ &= \left(\int_0^\infty \Phi(r)\left|\int_0^r \mathcal{R}\tilde{f}(r)dr\right|^q dr\right)^{1/q}\\ &\leq C\left(\int_0^\infty \Psi(r)\left|\mathcal{R}\tilde{f}(r)\right|^p dr\right)^{1/p}\\ &= C\left(\int_{\mathbb{G}}\psi(x)|\mathcal{R}f(x)|^p dx\right)^{1/p}\end{aligned}$$

if and only if the condition (2.43) holds by Theorem 5.1.1, namely by (5.2) and (5.3). □

2.2 Critical Hardy inequalities

In this section we discuss critical Hardy inequalities. The critical behaviour may be manifested with respect to different parameters. For example, one major critical case arises when we have $p=Q$ in (2.2). In this case, in the Euclidean case of $\mathbb{R}^n$ with $p=Q=n$ it is known that the Hardy inequality (2.1) fails for any constant, see, e.g., [ET99] and [IIO16a], and references therein. In such critical cases it is natural to expect the appearance of the logarithmic terms.

One version of such a critical case is the inequality

$$\sup_{R>0}\left\|\frac{f-f_R}{|x|\log\frac{R}{|x|}}\right\|_{L^Q(\mathbb{G})}\leq\frac{Q}{Q-1}\|\mathcal{R}f\|_{L^Q(\mathbb{G})},\quad Q\geq 2, \tag{2.44}$$

for all $f\in C_0^\infty(\mathbb{G}\backslash\{0\})$, where we denote

$$f_R(x):=f\left(R\frac{x}{|x|}\right)\quad\text{for}\quad x\in\mathbb{G}\quad\text{and}\quad R>0.$$

In fact, this inequality is a special case (with $p=Q$) of the following more general family of critical inequalities derived in Theorem 2.2.1, namely,

$$\sup_{R>0}\left\|\frac{f-f_R}{|x|^{\frac{Q}{p}}\log\frac{R}{|x|}}\right\|_{L^p(\mathbb{G})}\leq\frac{p}{p-1}\left\|\frac{1}{|x|^{\frac{Q}{p}-1}}\mathcal{R}f\right\|_{L^p(\mathbb{G})}, \tag{2.45}$$

for all $1<p<\infty$. Here the constant $\frac{p}{p-1}$ in (2.45) is sharp but is in general unattainable.

The inequalities (2.45) are all critical with respect to their weight $|x|^{-\frac{Q}{p}}$ in L^p-space because its pth power gives $|x|^{-Q}$ which is the critical order for the integrability at zero and at infinity in $L^p(\mathbb{G})$.

Moreover, we show another type of a critical Hardy inequality on general homogeneous groups with the logarithm being on the right-hand side:

$$\left\| \frac{f}{|x|} \right\|_{L^Q(\mathbb{G})} \leq Q \|(\log|x|)\mathcal{R}f\|_{L^Q(\mathbb{G})}. \tag{2.46}$$

Furthermore, we also give improved versions of the Hardy inequality (2.1) on quasi-balls of homogeneous (Lie) groups, the so-called Hardy–Sobolev type inequalities.

2.2.1 Critical Hardy inequalities

First we discuss a family of generalized critical Hardy inequalities on the homogeneous group $\mathbb{G}$ mentioned in (2.45). In this section, $\mathbb{G}$ is a homogeneous group of homogeneous dimension $Q \geq 2$ and $|\cdot|$ is an arbitrary homogeneous quasi-norm on $\mathbb{G}$.

Theorem 2.2.1 (A family of critical Hardy inequalities). *Let $f \in C_0^\infty(\mathbb{G}\backslash\{0\})$ and denote $f_R(x) := f(R\frac{x}{|x|})$ for $x \in \mathbb{G}$ and $R > 0$. Then we have the inequalities*

$$\sup_{R>0} \left\| \frac{f - f_R}{|x|^{\frac{Q}{p}} \log\frac{R}{|x|}} \right\|_{L^p(\mathbb{G})} \leq \frac{p}{p-1} \left\| \frac{1}{|x|^{\frac{Q}{p}-1}} \mathcal{R}f \right\|_{L^p(\mathbb{G})}, \quad 1 < p < \infty, \tag{2.47}$$

where the constant $\frac{p}{p-1}$ is sharp. Moreover, denoting

$$u_R(x) := \frac{f(x) - f_R(x)}{|x|^{\frac{Q}{p}} \log\frac{R}{|x|}} \quad \text{and} \quad v(x) := \frac{1}{|x|^{\frac{Q}{p}-1}} \mathcal{R}f(x),$$

for each $R > 0$ we have the following expression for the remainder:

$$\begin{aligned} \left\| \frac{f - f_R}{|x|^{\frac{Q}{p}} \log\frac{R}{|x|}} \right\|^p_{L^p(\mathbb{G})} &= \left(\frac{p}{p-1}\right)^p \left\| \frac{1}{|x|^{\frac{Q}{p}-1}} \mathcal{R}f \right\|^p_{L^p(\mathbb{G})} \\ &\quad - p \int_{\mathbb{G}} I(u_R, -\frac{p}{p-1}v) \left| \frac{p}{p-1}v + u_R \right|^2 dx, \end{aligned} \tag{2.48}$$

where I is defined by

$$I(u,g) := \left(\frac{1}{p}|g|^p + \left(1 - \frac{1}{p}\right)|u|^p - |u|^{p-2}\operatorname{Re}(u\overline{g})\right)|u-g|^{-2} \geq 0, \quad u \neq g,$$

$$I(g,g) := \frac{p-1}{2}|g|^{p-2}.$$

Remark 2.2.2.

1. For $p = Q$ the inequality (2.47) becomes

$$\sup_{R>0}\left\|\frac{f-f_R}{|x|\log\frac{R}{|x|}}\right\|_{L^Q(\mathbb{G})}\le\frac{Q}{Q-1}\|\mathcal{R}f\|_{L^Q(\mathbb{G})},\quad Q\ge 2,\tag{2.49}$$

which gives the mentioned critical estimate (2.44).

2. In the Euclidean isotropic case $\mathbb{G}=\mathbb{R}^n$ with the Euclidean norm $|\cdot|_E$, similar to Remark 2.1.2, Part 1, inequalities (2.47) yields inequalities for the usual gradient ∇ in $\mathbb{R}^n$ for all $1<p<\infty$:

$$\sup_{R>0}\left\|\frac{f-f_R}{|x|_E^{\frac{n}{p}}\log\frac{R}{|x|_E}}\right\|_{L^p(\mathbb{R}^n)}\le\frac{p}{p-1}\left\|\frac{1}{|x|_E^{\frac{n}{p}-1}}\nabla f\right\|_{L^p(\mathbb{R}^n)},\tag{2.50}$$

where $f_R(x):=f(R\frac{x}{|x|_E})$ for $x\in\mathbb{R}^n$ and $R>0$. The critical inequality (2.49) becomes

$$\sup_{R>0}\left\|\frac{f-f_R}{|x|_E\log\frac{R}{|x|_E}}\right\|_{L^n(\mathbb{R}^n)}\le\frac{n}{n-1}\|\nabla f\|_{L^n(\mathbb{R}^n)},\quad n\ge 2.\tag{2.51}$$

3. For the sharpness of the constant $\frac{p}{p-1}$ in (2.47) we refer to the Euclidean case with the Euclidean norm where this constant is known to be sharp but is in general unattainable. This was shown in [IIO16a]. In the case of general homogeneous groups this follows from the expression (2.48) for the remainder.

4. For $p=2$, in the Euclidean case $\mathbb{G}=\mathbb{R}^n$ with the Euclidean norm, the estimate (2.50) was shown in [MOW17a]. In principle, this case can be also obtained from [MOW15a, Theorem 1.1]. Inequality (2.51) in bounded domains of $\mathbb{R}^n$ was analysed in [II15].

Proof of Theorem 2.2.1. Using the polar decomposition in Proposition 1.2.10, a straightforward calculation shows

$$\begin{aligned}
&\int_{B(0,R)}\frac{|f(x)-f_R(x)|^p}{|x|^Q|\log\frac{R}{|x|}|^p}dx\\
&=\int_0^R\int_{\wp}\frac{|f(ry)-f(Ry)|^p}{r^Q\left(\log\frac{R}{r}\right)^p}r^{Q-1}d\sigma(y)dr\\
&=\int_0^R\frac{d}{dr}\left(\frac{1}{p-1}\frac{1}{\left(\log\frac{R}{r}\right)^{p-1}}\int_{\wp}|f(ry)-f(Ry)|^p d\sigma(y)\right)dr
\end{aligned}$$

$$-\frac{p}{p-1}\mathrm{Re}\int_0^R\left(\frac{1}{(\log\frac{R}{r})^{p-1}}\int_{\wp}|f(ry)-f(Ry)|^{p-2}(f(ry)-f(Ry))\overline{\frac{df(ry)}{dr}}d\sigma(y)\right)dr$$

$$=-\frac{p}{p-1}\mathrm{Re}\int_0^R\left(\frac{1}{(\log\frac{R}{r})^{p-1}}\int_{\wp}|f(ry)-f(Ry)|^{p-2}(f(ry)-f(Ry))\overline{\frac{df(ry)}{dr}}d\sigma(y)\right)dr,$$

where σ is the Borel measure on $\wp$ and the contribution on the boundary at $r=R$ vanishes due to the inequalities

$$|f(ry)-f(Ry)|\le C(R-r),\qquad \frac{R-r}{R}\le\log\frac{R}{r}.$$

By using the formula (1.30) we obtain

$$\int_{B(0,R)}\frac{|f(x)-f_R(x)|^p}{|x|^Q|\log\frac{R}{|x|}|^p}dx$$

$$=-\frac{p}{p-1}\mathrm{Re}\int_0^R\frac{1}{(\log\frac{R}{r})^{p-1}}\int_{\wp}|f(ry)-f(Ry)|^{p-2}(f(ry)-f(Ry))\overline{\frac{df(ry)}{dr}}d\sigma(y)dr$$

$$=-\frac{p}{p-1}\mathrm{Re}\int_{B(0,R)}\left|\frac{f(x)-f_R(x)}{|x|^{\frac{Q}{p}}\log\frac{R}{|x|}}\right|^{p-2}\frac{(f(x)-f_R(x))}{|x|^{\frac{Q}{p}}\log\frac{R}{|x|}}\frac{1}{|x|^{\frac{Q}{p}-1}}\overline{\frac{df(x)}{d|x|}}dx.$$

Similarly, one has

$$\int_{B^c(0,R)}\frac{|f(x)-f_R(x)|^p}{|x|^Q|\log\frac{R}{|x|}|^p}dx=\int_R^\infty\int_{\wp}\frac{|f(ry)-f(Ry)|^p}{r^Q\left(\log\frac{r}{R}\right)^p}r^{Q-1}d\sigma(y)dr$$

$$=-\int_R^\infty\frac{d}{dr}\left(\frac{1}{p-1}\frac{1}{(\log\frac{r}{R})^{p-1}}\int_{\wp}|f(ry)-f(Ry)|^pd\sigma(y)\right)dr$$

$$+\frac{p}{p-1}\mathrm{Re}\int_R^\infty\left(\frac{1}{(\log\frac{r}{R})^{p-1}}\int_{\wp}|f(ry)-f(Ry)|^{p-2}(f(ry)-f(Ry))\overline{\frac{df(ry)}{dr}}d\sigma(y)\right)dr$$

$$=-\frac{p}{p-1}\mathrm{Re}\int_{B^c(0,R)}\left|\frac{f(x)-f_R(x)}{|x|^{\frac{Q}{p}}\log\frac{R}{|x|}}\right|^{p-2}\frac{(f(x)-f_R(x))}{|x|^{\frac{Q}{p}}\log\frac{R}{|x|}}\frac{1}{|x|^{\frac{Q}{p}-1}}\overline{\frac{df(x)}{d|x|}}dx.$$

This implies that

$$\int_{\mathbb{G}}\frac{|f(x)-f_R(x)|^p}{|x|^Q|\log\frac{R}{|x|}|^p}dx$$

$$=-\frac{p}{p-1}\mathrm{Re}\int_{\mathbb{G}}\left|\frac{f(x)-f_R(x)}{|x|^{\frac{Q}{p}}\log\frac{R}{|x|}}\right|^{p-2}\frac{(f(x)-f_R(x))}{|x|^{\frac{Q}{p}}\log\frac{R}{|x|}}\frac{1}{|x|^{\frac{Q}{p}-1}}\overline{\frac{df(x)}{d|x|}}dx$$

$$=\left(\frac{p}{p-1}\right)^p\|v\|^p_{L^p(\mathbb{G})}-p\int_{\mathbb{G}}I(u,-\frac{p}{p-1}v)\left|\frac{p}{p-1}v+u\right|^2dx,$$

with

$$u = \frac{f(x) - f_R(x)}{|x|^{\frac{Q}{p}} \log \frac{R}{|x|}}, \qquad v = \frac{1}{|x|^{\frac{Q}{p}-1}} \frac{df(x)}{d|x|},$$

and I is defined by

$$I(f,g) := \left(\frac{1}{p}|g|^p + \frac{1}{p'}|f|^p - |f|^{p-2}\operatorname{Re}(f\overline{g})\right)|f-g|^{-2} \geq 0,\ f \neq g,\ \frac{1}{p} + \frac{1}{p'} = 1,$$

$$I(g,g) := \frac{p-1}{2}|g|^{p-2}.$$

Thus, we establish

$$\|u\|^p_{L^p(\mathbb{G})} = \left(\frac{p}{p-1}\right)^p \|v\|^p_{L^p(\mathbb{G})} - p\int_{\mathbb{G}} I\left(u, -\frac{p}{p-1}v\right)\left|\frac{p}{p-1}v + u\right|^2 dx.$$

This proves the equality (2.48) and the inequality (2.47) since the last term is non-positive. □

We have the following consequence of Theorem 2.2.1:

Corollary 2.2.3 (Critical uncertainty type principles). *Let $1 < p < \infty$ and $f \in C_0^\infty(\mathbb{G}\backslash\{0\})$. Then for any $R > 0$ and $\frac{1}{p} + \frac{1}{q} = \frac{1}{2}$ with $q > 1$, we have*

$$\left\|\frac{1}{|x|^{\frac{Q}{p}-1}}\mathcal{R}f\right\|_{L^p(\mathbb{G})} \|f\|_{L^q(\mathbb{G})} \geq \frac{p-1}{p}\left\|\frac{f(f-f_R)}{|x|^{\frac{Q}{p}}\log\frac{R}{|x|}}\right\|_{L^2(\mathbb{G})} \tag{2.52}$$

and also

$$\left\|\frac{1}{|x|^{\frac{Q}{p}-1}}\mathcal{R}f\right\|_{L^p(\mathbb{G})} \left\|\frac{f-f_R}{|x|^{\frac{Q}{p'}}\log\frac{R}{|x|}}\right\|_{L^{p'}(\mathbb{G})} \geq \frac{p-1}{p}\left\|\frac{f-f_R}{|x|^{\frac{Q}{2}}\log\frac{R}{|x|}}\right\|^2_{L^2(\mathbb{G})} \tag{2.53}$$

for $\frac{1}{p} + \frac{1}{p'} = 1$.

Proof of Corollary 2.2.3. By Theorem 2.2.1 and Hölder's inequality we have

$$\begin{aligned}
\left\|\frac{1}{|x|^{\frac{Q}{p}-1}}\mathcal{R}f\right\|_{L^p(\mathbb{G})} \|f\|_{L^q(\mathbb{G})} &\geq \frac{p-1}{p}\left\|\frac{f-f_R}{|x|^{\frac{Q}{p}}\log\frac{R}{|x|}}\right\|_{L^p(\mathbb{G})} \|f\|_{L^q(\mathbb{G})} \\
&= \frac{p-1}{p}\left(\int_{\mathbb{G}}\left|\frac{f-f_R}{|x|^{\frac{Q}{p}}\log\frac{R}{|x|}}\right|^{2\frac{p}{2}} dx\right)^{\frac{1}{2}\frac{2}{p}} \left(\int_{\mathbb{G}}|f|^{2\frac{q}{2}}dx\right)^{\frac{1}{2}\frac{2}{q}} \\
&\geq \frac{p-1}{p}\left(\int_{\mathbb{G}}\left|\frac{f(f-f_R)}{|x|^{\frac{Q}{p}}\log\frac{R}{|x|}}\right|^2 dx\right)^{1/2} = \frac{p-1}{p}\left\|\frac{f(f-f_R)}{|x|^{\frac{Q}{p}}\log\frac{R}{|x|}}\right\|_{L^2(\mathbb{G})}.
\end{aligned}$$

The formula (2.52) is proved. The proof of (2.53) is similar so we can omit it. □

2.2.2 Another type of critical Hardy inequality

In this section we analyse another type of a critical Hardy inequality when the logarithmic term appears on the other side of the inequality. This extends inequality (2.38) that was obtained for $p = 2$ by a factorization method.

Theorem 2.2.4 (Critical Hardy inequality). *Let $\mathbb{G}$ be a homogeneous group of homogeneous dimension $Q \geq 1$ and let $|\cdot|$ be a homogeneous quasi-norm on $\mathbb{G}$. Let $1 < p < \infty$. Then for any complex-valued function $f \in C_0^\infty(\mathbb{G}\backslash\{0\})$ we have*

$$\int_{\mathbb{G}} \frac{|f(x)|^p}{|x|^Q} dx \leq p^p \int_{\mathbb{G}} \frac{|\log|x||^p}{|x|^{Q-p}} |\mathcal{R}f(x)|^p dx, \tag{2.54}$$

and the constant p^p in this inequality is sharp.

Remark 2.2.5.

1. Inequality (2.54) was mentioned in (2.46) as another type of a critical Hardy inequality in the critical case of $p = Q$, in which case we have

$$\int_{\mathbb{G}} \frac{|f(x)|^Q}{|x|^Q} dx \leq Q^Q \int_{\mathbb{G}} |(\log|x|)\mathcal{R}f(x)|^Q dx. \tag{2.55}$$

2. In the Euclidean case $\mathbb{G} = (\mathbb{R}^n, +)$ we have $Q = n$, so for any quasi-norm $|\cdot|$ on $\mathbb{R}^n$ the inequality (2.54) implies a new inequality with the optimal constant: For each $f \in C_0^\infty(\mathbb{R}^n\backslash\{0\})$, we have

$$\left\| \frac{f}{|x|} \right\|_{L^n(\mathbb{R}^n)} \leq n \left\| (\log|x|) \frac{x}{|x|} \cdot \nabla f \right\|_{L^n(\mathbb{R}^n)}. \tag{2.56}$$

   If we take now the standard Euclidean distance $|x|_E = \sqrt{x_1^2 + \cdots + x_n^2}$, it follows that we have

$$\left\| \frac{f}{|x|_E} \right\|_{L^n(\mathbb{R}^n)} \leq n \left\| (\log|x|_E) \nabla f \right\|_{L^n(\mathbb{R}^n)}, \tag{2.57}$$

   for all $f \in C_0^\infty(\mathbb{R}^n\backslash\{0\})$, where ∇ is the standard gradient in $\mathbb{R}^n$. The constants in the above inequalities are sharp.

3. The inequality (2.57) and its consequence are analogous to the critical Hardy inequality of Edmunds and Triebel [ET99] that they showed in $\mathbb{R}^n$ for the Euclidean norm $|\cdot|_E$ in bounded domains $B \subset \mathbb{R}^n$:

$$\left\| \frac{f(x)}{|x|_E(1 + \log\frac{1}{|x|_E})} \right\|_{L^n(B)} \leq \frac{n}{n-1} \|\nabla f\|_{L^n(B)}, \quad n \geq 2, \tag{2.58}$$

   with sharp constant $\frac{n}{n-1}$, which was also discussed in [AS06]. This inequality was also shown to be equivalent to the critical case of the Sobolev–Lorentz inequality. However, a different feature of (2.57) compared to (2.58) is that the logarithmic term enters the other side of the inequality. Inequalities of this type have been investigated in [RS16a].

Proof of Theorem 2.2.4. Let $R > 0$ be such that supp$f \subset B(0, R)$. A direct calculation using the polar decomposition in Proposition 1.2.10 with integration by parts yields

$$\begin{aligned}
\int_{B(0,R)} \frac{|f(x)|^p}{|x|^Q} dx &= \int_0^R \int_{\wp} |f(\delta_r(y))|^p r^{Q-1-Q} d\sigma(y) dr \\
&= -p \int_0^R \log r \operatorname{Re} \int_{\wp} |f(\delta_r(y))|^{p-2} f(\delta_r(y)) \overline{\frac{df(\delta_r(y))}{dr}} d\sigma(y) dr \\
&\le p \int_{B(0,R)} \frac{|\mathcal{R}f(x)||f(x)|^{p-1}}{|x|^{Q-1}} |\log|x|| dx \\
&= p \int_{B(0,R)} \frac{|\mathcal{R}f(x)||\log|x||}{|x|^{\frac{Q}{p}-1}} \frac{|f(x)|^{p-1}}{|x|^{\frac{Q(p-1)}{p}}} dx,
\end{aligned}$$

and by using the Hölder inequality, we obtain that

$$\int_{B(0,R)} \frac{|f(x)|^p}{|x|^Q} dx \le p \left(\int_{B(0,R)} \frac{|\mathcal{R}f(x)|^p |\log|x||^p}{|x|^{Q-p}} dx \right)^{\frac{1}{p}} \left(\int_{B(0,R)} \frac{|f(x)|^p}{|x|^Q} dx \right)^{\frac{p-1}{p}},$$

which gives (2.54).

Now it remains to show the optimality of the constant, so we need to check the equality condition in the above Hölder inequality. Let us consider the test function

$$h(x) = \log|x|.$$

Thus, we have

$$\left(\frac{|\mathcal{R}h(x)||\log|x||}{|x|^{\frac{Q}{p}-1}} \right)^p = \left(\frac{|h(x)|^{p-1}}{|x|^{\frac{Q(p-1)}{p}}} \right)^{\frac{p}{p-1}},$$

which satisfies the equality condition in Hölder's inequality. This gives the optimality of the constant p^p in (2.54). $\square$

As usual, the Hardy inequality implies the corresponding uncertainty principle:

Corollary 2.2.6 (Another type of critical uncertainty principle). *Let $\mathbb{G}$ be a homogeneous group of homogeneous dimension $Q \ge 2$. Let $|\cdot|$ be a homogeneous quasi-norm on $\mathbb{G}$. Then for each $f \in C_0^\infty(\mathbb{G}\backslash\{0\})$ we have*

$$\left(\int_{\mathbb{G}} |(\log|x|)\mathcal{R}f|^Q dx \right)^{\frac{1}{Q}} \left(\int_{\mathbb{G}} |x|^{\frac{Q}{Q-1}} |f|^{\frac{Q}{Q-1}} dx \right)^{\frac{Q-1}{Q}} \ge \frac{1}{Q} \int_{\mathbb{G}} |f|^2 dx. \tag{2.59}$$

Proof. From the inequality (2.54) we get

$$\left(\int_{B(0,R)} |(\log|x|)\mathcal{R}f|^Q\,dx\right)^{\frac{1}{Q}} \left(\int_{B(0,R)} |x|^{\frac{Q}{Q-1}}|f|^{\frac{Q}{Q-1}}\,dx\right)^{\frac{Q-1}{Q}}$$

$$\geq \frac{1}{Q}\left(\int_{B(0,R)} \frac{|f|^Q}{|x|^Q}\,dx\right)^{\frac{1}{Q}} \left(\int_{B(0,R)} |x|^{\frac{Q}{Q-1}}|f|^{\frac{Q}{Q-1}}\,dx\right)^{\frac{Q-1}{Q}} \geq \frac{1}{Q}\int_{B(0,R)} |f|^2 dx,$$

where we have used the Hölder inequality in the last line. This shows (2.59). □

2.2.3 Critical Hardy inequalities of logarithmic type

In this section we present yet another logarithmic type of critical Hardy inequalities on the homogeneous group $\mathbb{G}$ of homogeneous dimension $Q \geq 1$. As usual, let $|\cdot|$ be a homogeneous quasi-norm on $\mathbb{G}$.

Theorem 2.2.7 (Another family of logarithmic Hardy inequalities). *Let $1 < \gamma < \infty$ and $\max\{1, \gamma - 1\} < p < \infty$. Then for all complex-valued functions $f \in C_0^\infty(\mathbb{G}\backslash\{0\})$ and all $R > 0$ we have the inequality*

$$\left\| \frac{f - f_R}{|x|^{\frac{Q}{p}}\left(\log\frac{R}{|x|}\right)^{\frac{\gamma}{p}}} \right\|_{L^p(\mathbb{G})} \leq \frac{p}{\gamma - 1}\left\| |x|^{\frac{p-Q}{p}}\left(\log\frac{R}{|x|}\right)^{\frac{p-\gamma}{p}} \mathcal{R}f \right\|_{L^p(\mathbb{G})}, \tag{2.60}$$

where $f_R(x) := f\left(R\frac{x}{|x|}\right)$, and the constant $\frac{p}{\gamma-1}$ is sharp.

Proof of Theorem 2.2.7. For a quasi-ball $B(0,R)$ we have using the polar decomposition in Proposition 1.2.10 that

$$\int_{B(0,R)} \frac{|f(x) - f_R(x)|^p}{|x|^Q\left|\log\frac{R}{|x|}\right|^\gamma}dx = \int_0^R \int_\wp \frac{|f(ry) - f(Ry)|^p}{r^Q\left(\log\frac{R}{r}\right)^\gamma} r^{Q-1} d\sigma(y)dr$$

$$= \int_0^R \frac{d}{dr}\left(\frac{1}{(\gamma-1)\left(\log\frac{R}{r}\right)^{\gamma-1}}\int_\wp |f(ry) - f(Ry)|^p d\sigma(y)\right)dr$$

$$- \frac{p}{\gamma-1}\mathrm{Re}\int_0^R \left(\log\frac{R}{r}\right)^{-\gamma+1}$$

$$\times \int_\wp |f(ry) - f(Ry)|^{p-2}(f(ry) - f(Ry))\overline{\frac{df(ry)}{dr}}d\sigma(y)dr$$

$$= -\frac{p}{\gamma-1}\mathrm{Re}\int_0^R \left(\log\frac{R}{r}\right)^{-\gamma+1}$$

$$\times \int_\wp |f(ry) - f(Ry)|^{p-2}(f(ry) - f(Ry))\overline{\frac{df(ry)}{dr}}d\sigma(y)dr,$$

where $p-\gamma+1>0$, so that the boundary term at $r=R$ vanishes due to inequalities

$$|f(ry)-f(Ry)|\leq C(R-r), \qquad \log\frac{R}{r}\geq\frac{R-r}{R}.$$

Then by the Hölder inequality we get

$$\begin{aligned}
&\int_0^R\int_{\wp}\frac{|f(ry)-f(Ry)|^p}{r\left(\log\frac{R}{r}\right)^{\gamma}}d\sigma(y)dr\\
&=-\frac{p}{\gamma-1}\mathrm{Re}\int_0^R\left(\log\frac{R}{r}\right)^{-\gamma+1}\int_{\wp}|f(ry)-f(Ry)|^{p-2}(f(ry)-f(Ry))\overline{\frac{df(ry)}{dr}}d\sigma(y)dr\\
&\leq\frac{p}{\gamma-1}\int_0^R\left(\log\frac{R}{r}\right)^{-\gamma+1}\int_{\wp}|f(ry)-f(Ry)|^{p-1}\left|\frac{df(ry)}{dr}\right|d\sigma(y)dr\\
&\leq\frac{p}{\gamma-1}\left(\int_0^R\int_{\wp}\frac{|f(ry)-f(Ry)|^p}{r\left(\log\frac{R}{r}\right)^{\gamma}}d\sigma(y)dr\right)^{(p-1)/p}\\
&\quad\times\left(\int_0^R\int_{\wp}r^{p-1}\left(\log\frac{R}{r}\right)^{p-\gamma}\left|\frac{df(ry)}{dr}\right|^p d\sigma(y)dr\right)^{1/p}.
\end{aligned}$$

Thus, we obtain

$$\begin{aligned}
&\left(\int_{B(0,R)}\frac{|f(x)-f_R(x)|^p}{|x|^Q\left|\log\frac{R}{|x|}\right|^{\gamma}}dx\right)^{1/p}\\
&\quad\leq\frac{p}{\gamma-1}\left(\int_{B(0,R)}|x|^{p-Q}\left|\log\frac{R}{|x|}\right|^{p-\gamma}|\mathcal{R}f(x)|^p\,dx\right)^{1/p}.
\end{aligned}\tag{2.61}$$

Similarly, we have

$$\begin{aligned}
&\left(\int_{B^c(0,R)}\frac{|f(x)-f_R(x)|^p}{|x|^Q\left|\log\frac{R}{|x|}\right|^{\gamma}}dx\right)^{1/p}\\
&\quad\leq\frac{p}{\gamma-1}\left(\int_{B^c(0,R)}|x|^{p-Q}\left|\log\frac{R}{|x|}\right|^{p-\gamma}|\mathcal{R}f(x)|^p\,dx\right)^{1/p}.
\end{aligned}\tag{2.62}$$

The inequalities (2.61) and (2.62) imply (2.60). Furthermore, the optimality of the constant in (2.60) is proved exactly in the same way as in the Euclidean case (see [MOW15b, Section 3]). ☐

Corollary 2.2.8 (Another family of logarithmic uncertainty type principles). *Let $1<p<\infty$ and $q>1$ be such that $\frac{1}{p}+\frac{1}{q}=\frac{1}{2}$. Let $1<\gamma<\infty$ and $\max\{1,\gamma-1\}<p<\infty$. Then for any $R>0$ and $f\in C_0^{\infty}(\mathbb{G}\backslash\{0\})$ we have*

$$\left\||x|^{\frac{p-Q}{p}}\left(\log\frac{R}{|x|}\right)^{(p-\gamma)/p}\mathcal{R}f\right\|_{L^p(\mathbb{G})}\|f\|_{L^q(\mathbb{G})}\geq\frac{\gamma-1}{p}\left\|\frac{f(f-f_R)}{|x|^{\frac{Q}{p}}\left(\log(R/|x|)\right)^{\gamma/p}}\right\|_{L^2(\mathbb{G})}.$$

Moreover,

$$\left\| |x|^{\frac{p-Q}{p}} \left(\log \frac{R}{|x|}\right)^{\frac{p-\gamma}{p}} \mathcal{R}f \right\|_{L^p(\mathbb{G})} \left\| \frac{f - f_R}{|x|^{\frac{Q}{p'}} \left(\log \frac{R}{|x|}\right)^{2-\frac{\gamma}{p}}} \right\|_{L^{p'}(\mathbb{G})} \geq \frac{\gamma - 1}{p} \left\| \frac{f - f_R}{|x|^{\frac{Q}{2}} \log \frac{R}{|x|}} \right\|^2_{L^2(\mathbb{G})} \tag{2.63}$$

holds for $\frac{1}{p} + \frac{1}{p'} = 1$.

Proof of Corollary 2.2.8. By (2.60), we have

$$\left\| |x|^{\frac{p-Q}{p}} \left(\log \frac{R}{|x|}\right)^{\frac{p-\gamma}{p}} \mathcal{R}f \right\|_{L^p(\mathbb{G})} \|f\|_{L^q(\mathbb{G})} \geq \frac{\gamma - 1}{p} \left\| \frac{f - f_R}{|x|^{\frac{Q}{p}} \left(\log \frac{R}{|x|}\right)^{\frac{\gamma}{p}}} \right\|_{L^p(\mathbb{G})} \|f\|_{L^q(\mathbb{G})}$$

$$= \frac{\gamma - 1}{p} \left(\int_{\mathbb{G}} \left| \frac{f(x) - f_R(x)}{|x|^{\frac{Q}{p}} \left(\log \frac{R}{|x|}\right)^{\frac{\gamma}{p}}} \right|^{2\frac{p}{2}} dx \right)^{\frac{1}{2}\frac{2}{p}} \left(\int_{\mathbb{G}} |f(x)|^{2\frac{q}{2}} dx \right)^{\frac{1}{2}\frac{2}{q}},$$

and using Hölder's inequality, we obtain

$$\left\| |x|^{\frac{p-Q}{p}} \left(\log \frac{R}{|x|}\right)^{\frac{p-\gamma}{p}} \mathcal{R}f \right\|_{L^p(\mathbb{G})} \|f\|_{L^q(\mathbb{G})}$$

$$\geq \frac{\gamma - 1}{p} \left(\int_{\mathbb{G}} \left| \frac{f(x)(f(x) - f_R(x))}{|x|^{\frac{Q}{p}} \left(\log \frac{R}{|x|}\right)^{\gamma/p}} \right|^2 dx \right)^{1/2} = \frac{\gamma - 1}{p} \left\| \frac{f(f - f_R)}{|x|^{\frac{Q}{p}} \left(\log \frac{R}{|x|}\right)^{\gamma/p}} \right\|_{L^2(\mathbb{G})}.$$

Similarly, one can prove (2.63). ☐

Remark 2.2.9. When $\gamma = p$, the statement in Theorem 2.2.7 appeared in [RS16a, Theorem 3.1] in the form

$$\left\| \frac{f - f_R}{|x|^{\frac{Q}{p}} \log \frac{R}{|x|}} \right\|_{L^p(\mathbb{G})} \leq \frac{p}{p-1} \left\| |x|^{\frac{p-Q}{p}} \mathcal{R}f \right\|_{L^p(\mathbb{G})}, \quad 1 < p < \infty, \tag{2.64}$$

for all $R > 0$. For general p and γ it was analysed in [RS16a]. In the Euclidean case with the Euclidean distance such inequalities have been analysed in [MOW15b, Section 3].

2.3 Remainder estimates

In this section we analyse the remainder estimates for L^p-weighted Hardy inequalities with sharp constants on homogeneous groups. In addition, other refined versions involving a distance and the critical case $p = Q = 2$ on the quasi-ball are discussed.

The analysis of remainder terms in Hardy inequalities has a long history initiated by Brézis and Nirenberg in [BN83], with subsequent works by Brézis and Lieb [BL85] for Hardy–Sobolev inequalities, Brézis and Vázquez in [BV97, Section 4]. Nowadays there is a lot of literature on this subject and this section will contain some further references on this subject.

2.3.1 Remainder estimates for L^p-weighted Hardy inequalities

Let $\mathbb{G}$ be a homogeneous group of homogeneous dimension $Q \geq 3$ and let $|\cdot|$ be a homogeneous quasi-norm on $\mathbb{G}$. We now present a family of remainder estimates for the weighted L^p-Hardy inequalities, with a freedom of choosing the parameter $b \in \mathbb{R}$.

Theorem 2.3.1 (Remainder estimates for L^p-weighted Hardy inequalities). *Let*

$$2 \leq p < Q, \quad -\infty < \alpha < \frac{Q-p}{p},$$

and let

$$\delta_1 = Q - p - \alpha p - \frac{Q+pb}{p},$$
$$\delta_2 = Q - p - \alpha p - \frac{bp}{p-1},$$

for $b \in \mathbb{R}$. Then for all complex-valued functions $f \in C_0^\infty(\mathbb{G}\backslash\{0\})$ we have

$$\begin{aligned}\int_{\mathbb{G}} \frac{|\mathcal{R}f(x)|^p}{|x|^{\alpha p}} dx - \left(\frac{Q-p-\alpha p}{p}\right)^p \int_{\mathbb{G}} \frac{|f(x)|^p}{|x|^{p(\alpha+1)}} dx \\ \geq C_p \frac{\left(\int_{\mathbb{G}} |f(x)|^p |x|^{\delta_1} dx\right)^p}{\left(\int_{\mathbb{G}} |f(x)|^p |x|^{\delta_2} dx\right)^{p-1}},\end{aligned} \tag{2.65}$$

where the constant $C_p = c_p \left|\frac{Q(p-1)-pb}{p^2}\right|^p$ is sharp, with

$$c_p = \min_{0<t\leq 1/2} ((1-t)^p - t^p + pt^{p-1}). \tag{2.66}$$

Due to the positivity of the last remainder term, Theorem 2.3.1 implies the L^p-weighted Hardy inequalities with the radial derivative. In the case of the Euclidean norm, they reduce to the usual L^p-weighted Hardy estimates:

Remark 2.3.2 (L^p-weighted Hardy inequalities).

1. If we take $b = \frac{Q(p-1)}{p}$ we have $C_p = 0$, and then inequality (2.65) gives the L^p-weighted Hardy inequalities with sharp constant on $\mathbb{G}$:
$$\int_{\mathbb{G}} \frac{|\mathcal{R}f(x)|^p}{|x|^{\alpha p}} dx \geq \left(\frac{Q-p-\alpha p}{p}\right)^p \int_{\mathbb{G}} \frac{|f(x)|^p}{|x|^{p(\alpha+1)}} dx, \quad -\infty < \alpha < \frac{Q-p}{p}, \quad 2 \leq p < Q, \tag{2.67}$$
for all complex-valued functions $f \in C_0^\infty(\mathbb{G}\backslash\{0\})$. Such inequalities on homogeneous groups have been investigated in [RS17b], and their remainders have been analysed in [RSY18b].
2. If $\mathbb{G} = (\mathbb{R}^n, +)$ with $Q = n$, the inequality (2.67) gives the L^p-weighted Hardy inequalities with sharp constant for any quasi-norm on $\mathbb{R}^n$: For any complex-valued function $f \in C_0^\infty(\mathbb{R}^n\backslash\{0\})$ we obtain
$$\int_{\mathbb{R}^n} \left|\frac{x}{|x|} \cdot \nabla f(x)\right|^p |x|^{-\alpha p} dx \geq \left(\frac{n-p-\alpha p}{p}\right)^p \int_{\mathbb{R}^n} \frac{|f(x)|^p}{|x|^{p(\alpha+1)}} dx,$$
where $-\infty < \alpha < \frac{n-p}{p}$ and $2 \leq p < n$, and where ∇ is the standard gradient in $\mathbb{R}^n$. Now, if we take the Euclidean norm $|x|_E = \sqrt{x_1^2 + x_2^2 + \cdots + x_n^2}$, by using the Schwarz inequality we obtain the Euclidean form of the L^p-weighted Hardy inequalities with sharp constants:
$$\int_{\mathbb{R}^n} \frac{|\nabla f(x)|^p}{|x|_E^{\alpha p}} dx \geq \left(\frac{n-p-\alpha p}{p}\right)^p \int_{\mathbb{R}^n} \frac{|f(x)|^p}{|x|_E^{p(\alpha+1)}} dx, \quad -\infty < \alpha < \frac{n-p}{p}, \quad 2 \leq p < n, \tag{2.68}$$
for any complex-valued function $f \in C_0^\infty(\mathbb{R}^n\backslash\{0\})$.
3. Moreover, for any function $f \in C_0^\infty(\mathbb{R}^n\backslash\{0\})$ and for any $b \in \mathbb{R}$, we have
$$\begin{aligned} &\int_{\mathbb{R}^n} \left|\frac{x}{|x|} \cdot \nabla f(x)\right|^p |x|^{-\alpha p} dx - \left(\frac{n-p-\alpha p}{p}\right)^p \int_{\mathbb{R}^n} \frac{|f(x)|^p}{|x|^{p(\alpha+1)}} dx \\ &\geq C_p \frac{\left(\int_{\mathbb{R}^n} |f(x)|^p |x|^{\delta_1} dx\right)^p}{\left(\int_{\mathbb{R}^n} |f(x)|^p |x|^{\delta_2} dx\right)^{p-1}}, \quad 2 \leq p < n, \ -\infty < \alpha < \frac{n-p}{p}, \end{aligned} \tag{2.69}$$
where ∇ is the standard gradient in $\mathbb{R}^n$. As in Part 2 above, by the Schwarz inequality with the usual Euclidean distance $|\cdot|_E$, we obtain
$$\begin{aligned} &\int_{\mathbb{R}^n} \frac{|\nabla f(x)|^p}{|x|_E^{\alpha p}} dx - \left(\frac{n-p-\alpha p}{p}\right)^p \int_{\mathbb{R}^n} \frac{|f(x)|^p}{|x|_E^{p(\alpha+1)}} dx \\ &\geq C_p \frac{\left(\int_{\mathbb{R}^n} |f(x)|^p |x|_E^{\delta_1} dx\right)^p}{\left(\int_{\mathbb{R}^n} |f(x)|^p |x|_E^{\delta_2} dx\right)^{p-1}}, \quad 2 \leq p < n, \ -\infty < \alpha < \frac{n-p}{p}, \end{aligned} \tag{2.70}$$

for all complex-valued functions $f \in C_0^\infty(\mathbb{R}^n\backslash\{0\})$ and for any $b \in \mathbb{R}$, where the constant C_p is sharp.

4. In $\mathbb{R}^n$ a variant of inequality (2.65) is known for radially symmetric functions. Let $n \geq 3$, $2 \leq p < n$ and $-\infty < \alpha < \frac{n-p}{p}$. Let $N \in \mathbb{N}$, $t \in (0,1)$, $\gamma < \min\{1-t, \frac{p-N}{p}\}$ and $\delta = N - n + \frac{N}{1-t-\gamma}\left(\gamma + \frac{n-p-\alpha p}{p}\right)$. Then there exists a constant $C > 0$ such that the inequality
$$\int_{\mathbb{R}^n} \frac{|\nabla f|^p}{|x|_E^{\alpha p}} dx - \left(\frac{n-p-\alpha p}{p}\right)^p \int_{\mathbb{R}^n} \frac{|f|^p}{|x|_E^{p(\alpha+1)}} dx$$
$$\geq C \frac{\left(\int_{\mathbb{R}^n} |f|^{\frac{N}{1-t-\gamma}} |x|_E^{\delta} dx\right)^{\frac{p(1-t-\gamma)}{Nt}}}{\left(\int_{\mathbb{R}^n} |f|^p |x|_E^{-\alpha p} dx\right)^{\frac{1-t}{t}}}$$
holds for all radially symmetric functions $f \in W_{0,\alpha}^{1,p}(\mathbb{R}^n)$, $f \neq 0$, where $W_{0,\alpha}^{1,p}(\mathbb{R}^n)$ is an appropriate Sobolev type space. For $\alpha = 0$ this was shown by Sano and Takahashi [ST17] and then extended in [ST18a] for any $-\infty < \alpha < \frac{n-p}{p}$.

5. The constant c_p in (2.66) appears in view of the following result that will be also of use in the determination of this constant:

   Lemma 2.3.3 ([FS08]). *Let $p \geq 2$ and let a, b be real numbers. Then there exists $c_p > 0$ such that*
$$|a-b|^p \geq |a|^p - p|a|^{p-2}ab + c_p|b|^p$$
   holds, where $c_p = \min\limits_{0<t\leq 1/2}((1-t)^p - t^p + pt^{p-1})$ is sharp in this inequality.

6. Remainder estimates of different forms are possible. In general, it is known from Ghoussoub and Moradifam [GM08] that there are no strictly positive functions $V \in C^1(0,\infty)$ such that the inequality
$$\int_{\mathbb{R}^n} |\nabla f|^2 dx \geq \left(\frac{n-2}{2}\right)^2 \int_{\mathbb{R}^n} \frac{|f|^2}{|x|_E^2} dx + \int_{\mathbb{R}^n} V(|x|_E)|f|^2 dx$$
holds for all Sobolev space functions $f \in W^{1,2}(\mathbb{R}^n)$. At the same time, Cianchi and Ferone showed in [CF08] that for all $1 < p < n$ there exists a constant $C = C(p,n)$ such that
$$\int_{\mathbb{R}^n} |\nabla f|^p dx \geq \left(\frac{n-p}{p}\right)^p \int_{\mathbb{R}^n} \frac{|f|^p}{|x|_E^p} dx\, (1 + C d_p(f)^{2p^*})$$
holds for all real-valued weakly differentiable functions f in $\mathbb{R}^n$ such that f and $|\nabla f| \in L^p(\mathbb{R}^n)$ go to zero at infinity, where
$$d_p(f) = \inf_{c\in\mathbb{R}} \frac{\|f - c|x|_E^{-\frac{n-p}{p}}\|_{L^{p^*,\infty}(\mathbb{R}^n)}}{\|f\|_{L^{p^*,p}(\mathbb{R}^n)}}$$

with $p^* = \frac{np}{n-p}$, and $L^{\tau,\sigma}(\mathbb{R}^n)$ is the Lorentz space for $0 < \tau \le \infty$ and $1 \le \sigma \le \infty$. In the case of a bounded domain Ω, Wang and Willem [WW03] for $p = 2$ and Abdellaoui, Colorado and Peral [ACP05] for $1 < p < \infty$ investigated other expressions of remainders, see also [ST17] and [ST18a] for more details.

In the following proof we will rely on a useful feature that some estimates involving radial derivatives of the Euler operator can be proved first for radial functions, and then extended to non-radial ones by a more abstract argument, see Section 1.3.3.

Proof of Theorem 2.3.1. Let $f \in C_0^\infty(\mathbb{G}\backslash\{0\})$ be a radial function, then f can be represented as $f(x) = \widetilde{f}(|x|)$. By using Brézis–Vázquez' idea ([BV97]), we define

$$\widetilde{g}(r) := r^{\frac{Q-p-\alpha p}{p}} \widetilde{f}(r). \tag{2.71}$$

Since $\widetilde{f} = \widetilde{f}(r) \in C_0^\infty(0,\infty)$ and $\alpha < \frac{Q-p}{p}$, we have $\widetilde{g}(0) = 0$ and $\widetilde{g}(+\infty) = 0$. We set

$$g(x) := \widetilde{g}(|x|)$$

for $x \in \mathbb{G}$. Introducing polar coordinates $(r,y) = (|x|, \frac{x}{|x|}) \in (0,\infty) \times \wp$ on $\mathbb{G}$, by Proposition 1.2.10 we have

$$\begin{aligned}
J &:= \int_{\mathbb{G}} \frac{|\mathcal{R}f(x)|^p}{|x|^{\alpha p}} dx - \left(\frac{Q-p-\alpha p}{p}\right)^p \int_{\mathbb{G}} \frac{|f(x)|^p}{|x|^{p(\alpha+1)}} dx \\
&= |\wp| \int_0^\infty \left|\frac{d}{dr}\widetilde{f}(r)\right|^p r^{-\alpha p+Q-1} dr - |\wp| \left(\frac{Q-p-\alpha p}{p}\right)^p \int_0^\infty |\widetilde{f}(r)|^p r^{-p(\alpha+1)+Q-1} dr \\
&= |\wp| \int_0^\infty \left|\left(\frac{Q-p-\alpha p}{p}\right) r^{-\frac{Q-\alpha p}{p}} \widetilde{g}(r) - r^{-\frac{Q-p-\alpha p}{p}} \frac{d}{dr}\widetilde{g}(r)\right|^p r^{Q-1-\alpha p} dr \\
&\quad - |\wp| \left(\frac{Q-p-\alpha p}{p}\right)^p \int_0^\infty |\widetilde{g}(r)|^p r^{-1} dr,
\end{aligned}$$

where $|\wp|$ is the $Q-1$-dimensional surface measure of the unit sphere. Here applying Lemma 2.3.3 to the integrand of the first term in the last expression above, we get

$$\begin{aligned}
&\left|\left(\frac{Q-p-\alpha p}{p}\right) r^{-\frac{Q-\alpha p}{p}} \widetilde{g}(r) - r^{-\frac{Q-p-\alpha p}{p}} \frac{d}{dr}\widetilde{g}(r)\right|^p r^{Q-1-\alpha p} \\
&\quad \ge \left(\left(\frac{Q-p-\alpha p}{p}\right)^p r^{-Q+\alpha p} |\widetilde{g}(r)|^p\right) r^{Q-1-\alpha p} \\
&\quad - p\left(\frac{Q-p-\alpha p}{p}\right)^{p-1} |\widetilde{g}(r)|^{p-2}\widetilde{g}(r) \frac{d}{dr}\widetilde{g}(r) r^{-(\frac{Q-\alpha p}{p})(p-1)} r^{-(\frac{Q-p-\alpha p}{p})} r^{Q-1-\alpha p} \\
&\quad + c_p \left|\frac{d}{dr}\widetilde{g}(r)\right|^p r^{-Q+p+\alpha p} r^{Q-1-\alpha p}
\end{aligned}$$

$$= \left(\frac{Q-p-\alpha p}{p}\right)^p r^{-1}|\widetilde{g}(r)|^p - p\left(\frac{Q-p-\alpha p}{p}\right)^{p-1} |\widetilde{g}(r)|^{p-2}\widetilde{g}(r)\frac{d}{dr}\widetilde{g}(r)$$
$$+ c_p\left|\frac{d}{dr}\widetilde{g}(r)\right|^p r^{p-1}.$$

Since $\widetilde{g}(0) = \widetilde{g}(+\infty) = 0$ and $p \geq 2$, we note that

$$p\int_0^\infty |\widetilde{g}(r)|^{p-2}\widetilde{g}(r)\frac{d}{dr}\widetilde{g}(r)dr = \int_0^\infty \frac{d}{dr}(|\widetilde{g}(r)|^p)dr = 0.$$

This gives a so-called ground state representation of the Hardy difference J:

$$J \geq c_p|\wp|\int_0^\infty \left|\frac{d}{dr}\widetilde{g}(r)\right|^p r^{p-1}dr = c_p\int_{\mathbb{G}} |\mathcal{R}g(x)|^p|x|^{p-Q}dx. \tag{2.72}$$

Putting $a = \frac{Q-p}{p}$ in Lemma 2.3.3, we obtain for any $b \in \mathbb{R}$ that

$$\left|\frac{Q(p-1)-pb}{p^2}\right|\int_{\mathbb{G}} |g|^p|x|^{-\frac{Q+pb}{p}}dx$$
$$\leq \left(\int_{\mathbb{G}} |\mathcal{R}g|^p|x|^{p-Q}dx\right)^{\frac{1}{p}}\left(\int_{\mathbb{G}} |g|^p|x|^{-\frac{bp}{p-1}}dx\right)^{\frac{p-1}{p}}.$$

It gives the estimate

$$\begin{aligned} J &\geq c_p\int_{\mathbb{G}} |\mathcal{R}g(x)|^p|x|^{p-Q}dx \\ &\geq c_p\left|\frac{Q(p-1)-pb}{p^2}\right|^p \frac{\left(\int_{\mathbb{G}} |g|^p|x|^{-\frac{Q+pb}{p}}dx\right)^p}{\left(\int_{\mathbb{G}} |g|^p|x|^{-\frac{bp}{p-1}}dx\right)^{p-1}}. \end{aligned} \tag{2.73}$$

Taking into account that $g(x) = \widetilde{g}(|x|)$, $x \in \mathbb{G}$, and (2.71), one calculates

$$\begin{aligned} \int_{\mathbb{G}} |x|^{-\frac{Q+pb}{p}}|g(x)|^p dx &= |\wp|\int_0^\infty r^{Q-p-\alpha p}|\widetilde{f}(r)|^p r^{-\frac{Q+pb}{p}} r^{Q-1}dr \\ &= \int_{\mathbb{G}} |f(x)|^p|x|^{Q-p-\alpha p-\frac{Q+pb}{p}}dx = \int_{\mathbb{G}} |f(x)|^p|x|^{\delta_1}dx. \end{aligned}$$

On the other hand,

$$\begin{aligned} \int_{\mathbb{G}} |x|^{-\frac{bp}{p-1}}|g(x)|^p dx &= |\wp|\int_0^\infty r^{Q-p-\alpha p}|\widetilde{f}(r)|^p r^{-\frac{bp}{p-1}} r^{Q-1}dr \\ &= \int_{\mathbb{G}} |f(x)|^p|x|^{Q-p-\alpha p-\frac{bp}{p-1}}dx = \int_{\mathbb{G}} |f(x)|^p|x|^{\delta_2}dx. \end{aligned}$$

Putting these equalities into (2.73), we obtain

$$J \geq c_p \left| \frac{Q(p-1)-pb}{p^2} \right|^p \frac{\left(\int_{\mathbb{G}} |f(x)|^p |x|^{\delta_1} dx\right)^p}{\left(\int_{\mathbb{G}} |f(x)|^p |x|^{\delta_2} dx\right)^{p-1}}.$$

Now let us prove the statement for non-radial functions. For a non-radial function f we consider the radial one obtained as its spherical average with respect to the homogeneous quasi-norm $|\cdot|$:

$$U(r) := \left(\frac{1}{|\wp|} \int_{\wp} |f(ry)|^p d\sigma(y) \right)^{\frac{1}{p}}. \tag{2.74}$$

Using Hölder's inequality, we calculate

$$\begin{aligned}
\frac{d}{dr} U(r) &= \frac{1}{p} \left(\frac{1}{|\wp|} \int_{\wp} |f(ry)|^p d\sigma(y) \right)^{\frac{1}{p}-1} \frac{1}{|\wp|} \int_{\wp} p |f(ry)|^{p-2} f(ry) \overline{\frac{d}{dr} f(ry)} d\sigma(y) \\
&\leq \left(\frac{1}{|\wp|} \int_{\wp} |f(ry)|^p d\sigma(y) \right)^{\frac{1}{p}-1} \frac{1}{|\wp|} \int_{\wp} |f(ry)|^{p-1} \left| \frac{d}{dr} f(ry) \right| d\sigma(y) \\
&\leq \left(\frac{1}{|\wp|} \int_{\wp} |f(ry)|^p d\sigma(y) \right)^{\frac{1}{p}-1} \frac{1}{|\wp|} \left(\int_{\wp} \left| \frac{d}{dr} f(ry) \right|^p d\sigma(y) \right)^{\frac{1}{p}} \left(\int_{\wp} |f(ry)|^p d\sigma(y) \right)^{\frac{p-1}{p}} \\
&= \left(\frac{1}{|\wp|} \int_{\wp} \left| \frac{d}{dr} f(ry) \right|^p d\sigma(y) \right)^{\frac{1}{p}}.
\end{aligned}$$

Here we note that since there exists the function $h(x) = e^{-|x|}$ which satisfies the equality

$$\left| \frac{d}{dr} h(x) \right|^p = (|h(x)|^{p-1})^{\frac{p}{p-1}},$$

the equality condition in the above Hölder inequality holds. Thus, we have

$$\frac{d}{dr} U(r) \leq \left(\frac{1}{|\wp|} \int_{\wp} \left| \frac{d}{dr} f(ry) \right|^p d\sigma(y) \right)^{\frac{1}{p}}.$$

It follows that

$$\begin{aligned}
|\wp| \int_0^{\infty} \left| \frac{d}{dr} U(r) \right|^p r^{Q-1-\alpha p} dr &\leq |\wp| \int_0^{\infty} \frac{1}{|\wp|} \int_{\wp} \left| \frac{d}{dr} f(ry) \right|^p r^{Q-1-\alpha p} d\sigma(y) dr \\
&= \int_{\mathbb{G}} |\mathcal{R} f|^p \, |x|^{-\alpha p} dx,
\end{aligned}$$

that is,

$$\int_{\mathbb{G}} |\mathcal{R} U|^p \, |x|^{-\alpha p} dx \leq \int_{\mathbb{G}} |\mathcal{R} f|^p \, |x|^{-\alpha p} dx. \tag{2.75}$$

In view of (2.74), we obtain the equality

$$\int_{\mathbb{G}} |U(|x|)|^p |x|^\theta dx = |\wp| \int_0^\infty |U(r)|^p r^{\theta+Q-1} dr \tag{2.76}$$
$$= |\wp| \int_0^\infty \frac{1}{|\wp|} \int_\wp |f(ry)|^p d\sigma(y) r^{\theta+Q-1} dr = \int_{\mathbb{G}} |f(x)|^p |x|^\theta dx,$$

for any $\theta \in \mathbb{R}$. Then, it is easy to see that (2.75) and (2.76) imply that (2.65) holds also for all non-radial functions. ☐

2.3.2 Critical and subcritical Hardy inequalities

Here we discuss the relation between the critical and the subcritical Hardy inequalities on homogeneous groups. We formulate this relation for functions that are radially symmetric with respect to a homogeneous quasi-norm $|\cdot|$ on $\mathbb{G}$.

Proposition 2.3.4 (Critical and subcritical Hardy inequalities). *Let $\mathbb{G}$ be a homogeneous group of homogeneous dimension $Q \geq 3$ and let $\widetilde{\mathbb{G}}$ be a homogeneous group of homogeneous dimension $m \geq 2$, and assume that $Q \geq m+1$. Let $|\cdot|$ denote homogeneous quasi-norms on $\mathbb{G}$ and on $\widetilde{\mathbb{G}}$. Then for any non-negative radially symmetric function $g \in C_0^1(B^m(0,R)\backslash\{0\})$, there exists a non-negative radially symmetric function $f \in C_0^1(B^Q(0,1)\backslash\{0\})$ such that*

$$\int_{B^Q(0,1)} |\mathcal{R}f(x)|^m dx - \left(\frac{Q-m}{m}\right)^m \int_{B^Q(0,1)} \frac{|f(x)|^m}{|x|^m} dx$$
$$= \frac{|\wp|}{|\widetilde{\wp}|} \left(\frac{Q-m}{m-1}\right)^{m-1} \tag{2.77}$$
$$\times \left(\int_{B^m(0,R)} |\mathcal{R}g|^m dz - \left(\frac{m-1}{m}\right)^m \int_{B^m(0,R)} \frac{|g|^m}{|z|^m \left(\log \frac{Re}{|z|}\right)^m} dz \right)$$

holds true, where $|\wp|$ and $|\widetilde{\wp}|$ are $Q-1$- and $m-1$-dimensional surface measures of the unit sphere, respectively.

Proof of Proposition 2.3.4. Let $r = |x|$, $x \in \mathbb{G}$ and $s = |z|$, $z \in \widetilde{\mathbb{G}}$, where $\widetilde{\mathbb{G}}$ is a homogeneous group of homogeneous dimension m. Let us define a radial function $f = f(x) \in C_0^1(B^Q(0,1)\backslash\{0\})$ for a non-negative radial function $g = g(z) \in C_0^1(B^m(0,R)\backslash\{0\})$:

$$f(r) = g(s(r)),$$

where $s(r) = R\exp(1 - r^{-\frac{Q-m}{m-1}})$, that is,

$$r^{-\frac{Q-m}{m-1}} = \log \frac{Re}{s}, \quad s'(r) = \frac{Q-m}{m-1} r^{-\frac{Q-m}{m-1}-1} s(r).$$

Here we see that $s'(r) > 0$ for $r \in [0,1]$ and $s(0) = 0$, $s(1) = R$. Since $g(s) \equiv 0$ near $s = R$, we also note that $f \equiv 0$ near $r = 1$.

Then a direct calculation shows

$$
\begin{aligned}
&\int_{B^Q(0,1)} |\mathcal{R}f|^m dx - \left(\frac{Q-m}{m}\right)^m \int_{B^Q(0,1)} \frac{|f|^m}{|x|^m} dx \\
&= |\wp| \int_0^1 |f'(r)|^m r^{Q-1} dr - \left(\frac{Q-m}{m}\right)^m |\wp| \int_0^1 f^m(r) r^{Q-m-1} dr \\
&= |\wp| \int_0^R |g'(s)s'(r(s))|^m r^{Q-1}(s) \frac{ds}{s'(r(s))} \\
&\quad - \left(\frac{Q-m}{m}\right)^m |\wp| \int_0^R g^m(s) r^{Q-m-1}(s) \frac{ds}{s'(r(s))} \\
&= |\wp| \left(\frac{Q-m}{m-1}\right)^{m-1} \int_0^R |g'(s)|^m s^{m-1} ds \\
&\quad - \left(\frac{Q-m}{m}\right)^m \frac{m-1}{Q-m} |\wp| \int_0^R \frac{g^m(s)}{s\left(\log \frac{Re}{s}\right)^m} ds \\
&= \frac{|\wp|}{|\widetilde{\wp}|} \left(\frac{Q-m}{m-1}\right)^{m-1} \\
&\quad \times \left(\int_{B^m(0,R)} |\mathcal{R}g|^m dz - \left(\frac{m-1}{m}\right)^m \int_{B^m(0,R)} \frac{|g|^m}{|z|^m \left(\log \frac{Re}{|z|}\right)^m} dz\right),
\end{aligned}
$$

yielding (2.77). □

2.3.3 A family of Hardy–Sobolev type inequalities on quasi-balls

Let $\mathbb{G}$ be a homogeneous group of homogeneous dimension $Q \geq 3$. It will be convenient to denote the dilations by $\delta_r(x) = rx$ in the following formulations. Here we discuss another type of Hardy–Sobolev inequalities for functions supported in balls of radius R. As usual, we denote by $B(0,R)$ a quasi-ball of radius R around 0 with respect to the quasi-norm $|\cdot|$.

Theorem 2.3.5 (Another type of Hardy inequalities for $Q \geq 3$). *For each $f \in C_0^\infty(B(0,R)\backslash\{0\})$ and any homogeneous quasi-norm $|\cdot|$ on $\mathbb{G}$ we have*

$$
\left(\int_{B(0,R)} \frac{1}{|x|^2} \left| f(x) - f\left(\frac{\delta_R(x)}{|x|}\right)\right|^2 dx\right)^{1/2} \leq \frac{2}{Q-2} \left(\int_{B(0,R)} |\mathcal{R}f|^2 dx\right)^{1/2}, \tag{2.78}
$$

and

$$
\begin{aligned}
\left(\int_{B(0,R)} \frac{1}{|x|^2} |f(x)|^2 dx\right)^{\frac{1}{2}} &\leq \left(\frac{Q}{Q-2}\right)^{\frac{1}{2}} \frac{1}{R} \left(\int_{B(0,R)} |f(x)|^2 dx\right)^{\frac{1}{2}} \\
&\quad + \frac{2}{Q-2}\left(1 + \left(\frac{Q}{Q-2}\right)^{\frac{1}{2}}\right) \left(\int_{B(0,R)} |\mathcal{R}f|^2 dx\right)^{\frac{1}{2}}.
\end{aligned} \tag{2.79}
$$

Remark 2.3.6.

1. Theorem 2.3.5 could have been formulated for functions $f \in C_0^\infty(\mathbb{G}\backslash\{0\})$ choosing $R > 0$ such that $\operatorname{supp} f \subset B(0,R)$. The introduction of R into the notation is essential here since the dilated function $f\left(\frac{\delta_R(x)}{|x|}\right)$ appears in the inequality (2.78).
2. In the Euclidean setting with the Euclidean norm inequalities in Theorem 2.3.5 have been studied in [MOW13b].

In the case $Q = 2$ we have the following inequalities:

Theorem 2.3.7 (Another type of critical Hardy inequality for $Q = 2$). *Let $\mathbb{G}$ be a homogeneous group of homogeneous dimension $Q = 2$. Then for each $f \in C_0^\infty(B(0,R)\backslash\{0\})$ and any homogeneous quasi-norm $|\cdot|$ on $\mathbb{G}$ we have*

$$\left(\int_{B(0,R)} \frac{1}{|x|^2 \left|\log \frac{R}{|x|}\right|^2} \left|f(x) - f\left(\frac{\delta_R(x)}{|x|}\right)\right|^2 dx\right)^{1/2} \leq 2\left(\int_{B(0,R)} |\mathcal{R} f|^2 dx\right)^{1/2}, \tag{2.80}$$

and

$$\left(\int_{B(0,R)} \frac{|f(x)|^2}{|x|^2\left(1+\left|\log\frac{R}{|x|}\right|^2\right)^2} dx\right)^{1/2} \tag{2.81}$$
$$\leq \frac{\sqrt{2}}{R}\left(\int_{B(0,R)} |f(x)|^2\, dx\right)^{1/2} + 2\left(1+\sqrt{2}\right)\left(\int_{B(0,R)} |\mathcal{R} f|^2 dx\right)^{1/2}.$$

Proof of Theorem 2.3.5. By the polar decomposition from Proposition 1.2.10, we write $(r,y) = (|x|, \frac{x}{|x|}) \in (0,\infty)\times\wp$ on $\mathbb{G}$, where $\wp$ is the unit quasi-sphere, so that

$$\begin{aligned}
&\int_{B(0,R)} \frac{1}{|x|^2}\left|f(x) - f\left(\frac{\delta_R(x)}{|x|}\right)\right|^2 dx \\
&= \int_0^R \int_\wp |f(\delta_r(y)) - f(\delta_R(y))|^2 r^{Q-3} d\sigma(y) dr \\
&= \frac{1}{Q-2} r^{Q-2} \int_\wp |f(\delta_r(y)) - f(\delta_R(y))|^2 d\sigma(y)\Big|_{r=0}^{r=R} \\
&\quad - \frac{1}{Q-2}\int_0^R r^{Q-2}\left(\frac{d}{dr}\int_\wp |f(\delta_r(y)) - f(\delta_R(y))|^2 d\sigma(y)\right) dr \\
&= -\frac{2}{Q-2}\int_0^R r^{Q-2} \operatorname{Re}\int_\wp (f(\delta_r(y)) - f(\delta_R(y)))\overline{\frac{df(\delta_r(y))}{dr}} d\sigma(y) dr.
\end{aligned}$$

Now using Schwarz' inequality, we obtain

$$\int_{B(0,R)} \frac{1}{|x|^2}\left|f(x)-f\left(\frac{\delta_R(x)}{|x|}\right)\right|^2 dx$$
$$\leq \frac{2}{Q-2}\left(\int_0^R\int_{\wp}|f(\delta_r(y))-f(\delta_R(y))|^2 r^{Q-3}d\sigma(y)dr\right)^{1/2}$$
$$\times\left(\int_0^R\int_{\wp}\left|\frac{df(\delta_r(y))}{dr}\right|^2 r^{Q-1}d\sigma(y)dr\right)^{1/2}$$
$$=\frac{2}{Q-2}\left(\int_{B(0,R)}\frac{1}{|x|^2}\left|f(x)-f\left(\frac{\delta_R(x)}{|x|}\right)\right|^2 dx\right)^{1/2}\left(\int_{B(0,R)}|\mathcal{R}f|^2dx\right)^{1/2}.$$

This implies that

$$\left(\int_{B(0,R)}\frac{1}{|x|^2}\left|f(x)-f\left(\frac{\delta_R(x)}{|x|}\right)\right|^2 dx\right)^{1/2}\leq\frac{2}{Q-2}\left(\int_{B(0,R)}|\mathcal{R}f|^2dx\right)^{1/2},$$

that is, the inequality (2.78) is proved. The triangle inequality gives

$$\left(\int_{B(0,R)}\frac{1}{|x|^2}|f|^2dx\right)^{\frac{1}{2}}=\left(\int_{B(0,R)}\frac{1}{|x|^2}\left|f(x)-f\left(\frac{\delta_R(x)}{|x|}\right)+f\left(\frac{\delta_R(x)}{|x|}\right)\right|^2dx\right)^{\frac{1}{2}}$$
$$\leq\left(\int_{B(0,R)}\frac{1}{|x|^2}\left|f(x)-f\left(\frac{\delta_R(x)}{|x|}\right)\right|^2dx\right)^{\frac{1}{2}}+\left(\int_{B(0,R)}\frac{1}{|x|^2}\left|f\left(\frac{\delta_R(x)}{|x|}\right)\right|^2dx\right)^{\frac{1}{2}}. \tag{2.82}$$

Moreover, we have

$$\left(\int_{B(0,R)}\frac{1}{|x|^2}\left|f\left(\frac{\delta_R(x)}{|x|}\right)\right|^2dx\right)^{\frac{1}{2}}=\left(\int_0^R\int_{\wp}|f(\delta_R(y))|^2r^{Q-3}d\sigma(y)dr\right)^{\frac{1}{2}}$$
$$=\left(\frac{R^{Q-2}}{Q-2}\int_{\wp}|f(\delta_R(y))|^2d\sigma(y)\right)^{\frac{1}{2}}$$
$$=\left(\frac{R^{Q-2}}{Q-2}\frac{Q}{R^Q}\int_0^R\int_{\wp}|f(\delta_R(y))|^2r^{Q-1}d\sigma(y)dr\right)^{\frac{1}{2}}$$
$$=\left(\frac{Q}{Q-2}\right)^{\frac{1}{2}}\frac{1}{R}\left(\int_{B(0,R)}\left|f\left(\frac{\delta_R(x)}{|x|}\right)\right|^2dx\right)^{\frac{1}{2}}$$
$$\leq\left(\frac{Q}{Q-2}\right)^{\frac{1}{2}}\frac{1}{R}\left(\left(\int_{B(0,R)}\left|f\left(\frac{\delta_R(x)}{|x|}\right)-f(x)\right|^2dx\right)^{\frac{1}{2}}+\left(\int_{B(0,R)}|f|^2dx\right)^{\frac{1}{2}}\right)$$

$$\leq \left(\frac{Q}{Q-2}\right)^{\frac{1}{2}} \left(\int_{B(0,R)} \frac{1}{|x|^2} \left| f\left(\frac{\delta_R(x)}{|x|}\right) - f(x)\right|^2 dx\right)^{\frac{1}{2}}$$
$$+ \left(\frac{Q}{Q-2}\right)^{\frac{1}{2}} \frac{1}{R} \left(\int_{B(0,R)} |f|^2 dx\right)^{\frac{1}{2}},$$

thus,

$$\left(\int_{B(0,R)} \frac{1}{|x|^2} \left| f\left(\frac{\delta_R(x)}{|x|}\right)\right|^2 dx\right)^{1/2}$$
$$\leq \left(\frac{Q}{Q-2}\right)^{\frac{1}{2}} \left(\int_{B(0,R)} \frac{1}{|x|^2} \left| f\left(\frac{\delta_R(x)}{|x|}\right) - f(x)\right|^2 dx\right)^{1/2} \tag{2.83}$$
$$+ \left(\frac{Q}{Q-2}\right)^{\frac{1}{2}} \frac{1}{R} \left(\int_{B(0,R)} |f|^2 dx\right)^{1/2}.$$

Combining (2.83) with (2.82) we arrive at

$$\left(\int_{B(0,R)} \frac{1}{|x|^2} |f|^2 dx\right)^{1/2}$$
$$\leq \left(1 + \left(\frac{Q}{Q-2}\right)^{1/2}\right) \left(\int_{B(0,R)} \frac{1}{|x|^2} \left| f(x) - f\left(\frac{\delta_R(x)}{|x|}\right)\right|^2 dx\right)^{1/2}$$
$$+ \left(\frac{Q}{Q-2}\right)^{1/2} \frac{1}{R} \left(\int_{B(0,R)} |f|^2 dx\right)^{1/2}.$$

Now by using (2.78) we arrive at (2.79). □

Proof of Theorem 2.3.7. By using polar coordinates $(r, y) = (|x|, \frac{x}{|x|}) \in (0, \infty) \times \wp$ on $\mathbb{G}$, where $\wp$ is the unit quasi-sphere, one calculates

$$\int_{B(0,R)} \frac{1}{|x|^2 |\log(R/|x|)|^2} \left| f(x) - f\left(\frac{\delta_R(x)}{|x|}\right)\right|^2 dx$$
$$= \int_0^R \int_{\wp} |f(\delta_r(y)) - f(\delta_R(y))|^2 \frac{1}{r\,(\log(R/r))^2} d\sigma(y) dr$$
$$= \frac{1}{\log(R/r)} \int_{\wp} |f(\delta_r(y)) - f(\delta_R(y))|^2 d\sigma(y) \Big|_{r=0}^{r=R}$$
$$- \int_0^R \frac{1}{\log(R/r)} \left(\frac{d}{dr} \int_{\wp} |f(\delta_r(y)) - f(\delta_R(y))|^2 d\sigma(y)\right) dr$$

$$= -2\int_0^R \frac{1}{\log(R/r)}\mathrm{Re}\int_{\wp}(f(\delta_r(y)) - f(\delta_R(y)))\overline{\frac{df(\delta_r(y))}{dr}}d\sigma(y)dr.$$

Here we have used the fact that

$$\log(R/r) = \log\left(1+\left(\frac{R}{r}-1\right)\right) \geq \frac{R}{r} - 1 = \frac{R-r}{r},$$

and that

$$|f(\delta_r(y)) - f(\delta_R(y))|^2 \leq C|R-r|^2.$$

Using Schwarz's inequality we obtain

$$\int_{B(0,R)} \frac{1}{|x|^2|\log(R/|x|)|^2}\left|f(x) - f\left(\frac{\delta_R(x)}{|x|}\right)\right|^2 dx$$
$$\leq 2\left(\int_0^R\int_{\wp}\frac{1}{r\left(\log(R/r)\right)^2}|f(\delta_r(y)) - f(\delta_R(y))|^2 d\sigma(y)dr\right)^{\frac{1}{2}}$$
$$\times\left(\int_0^R\int_{\wp}\left|\frac{df(\delta_r(y))}{dr}\right|^2 rd\sigma(y)dr\right)^{\frac{1}{2}}$$
$$= 2\left(\int_{B(0,R)}\frac{1}{|x|^2|\log(R/|x|)|^2}\left|f(x) - f\left(\frac{\delta_R(x)}{|x|}\right)\right|^2 dx\right)^{\frac{1}{2}}\left(\int_{B(0,R)}|\mathcal{R}f|^2dx\right)^{\frac{1}{2}}.$$

It completes the proof of (2.80). To show (2.81) we calculate

$$\left(\int_{B(0,R)}\frac{1}{|x|^2\left(1+|\log(R/|x|)|\right)^2}|f(x)|^2dx\right)^{1/2}$$
$$\leq \left(\int_{B(0,R)}\frac{1}{|x|^2\left(1+|\log(R/|x|)|\right)^2}\left|f(x) - f\left(\frac{\delta_R(x)}{|x|}\right)\right|^2 dx\right)^{1/2} \tag{2.84}$$
$$+\left(\int_{B(0,R)}\frac{1}{|x|^2\left(1+|\log(R/|x|)|\right)^2}\left|f\left(\frac{\delta_R(x)}{|x|}\right)\right|^2 dx\right)^{1/2}.$$

Moreover, we have

$$\left(\int_{B(0,R)}\frac{1}{|x|^2\left(1+|\log(R/|x|)|\right)^2}\left|f\left(\frac{\delta_R(x)}{|x|}\right)\right|^2 dx\right)^{\frac{1}{2}}$$
$$= \left(\int_0^R\int_{\wp}\frac{1}{r\left(1+|\log(R/r)|\right)^2}|f(\delta_R(y))|^2d\sigma(y)dr\right)^{\frac{1}{2}}$$

$$
\begin{aligned}
&= \left(\int_0^R \frac{1}{r\,(1+|\log(R/r)|)^2} dr \int_{\wp} |f(\delta_R(y))|^2 d\sigma(y) \right)^{\frac{1}{2}} \\
&= \left(\frac{1}{1+|\log(R/r)|} \Bigg|_0^R \int_{\wp} |f(\delta_R(y))|^2 d\sigma(y) \right)^{\frac{1}{2}} \\
&= \left(\int_{\wp} |f(\delta_R(y))|^2 d\sigma(y) \right)^{\frac{1}{2}} = \left(\frac{2}{R^2} \int_0^R \int_{\wp} |f(\delta_R(y))|^2 r d\sigma(y) dr \right)^{\frac{1}{2}} \\
&= \left(\frac{2}{R^2} \right)^{\frac{1}{2}} \left(\int_{B(0,R)} \left| f\left(\frac{\delta_R(x)}{|x|} \right) \right|^2 dx \right)^{\frac{1}{2}} \\
&\leq \frac{\sqrt{2}}{R} \left(\left(\int_{B(0,R)} \left| f\left(\frac{\delta_R(x)}{|x|} \right) - f(x) \right|^2 dx \right)^{\frac{1}{2}} + \left(\int_{B(0,R)} |f(x)|^2 dx \right)^{\frac{1}{2}} \right) \\
&\leq \sqrt{2} \left(\int_{B(0,R)} \frac{1}{|x|^2\,(1+|\log(R/|x|)|)^2} \left| f\left(\frac{\delta_R(x)}{|x|} \right) - f(x) \right|^2 dx \right)^{\frac{1}{2}} \\
&+ \frac{\sqrt{2}}{R} \left(\int_{B(0,R)} |f(x)|^2 dx \right)^{\frac{1}{2}},
\end{aligned}
$$

where we use the simple inequality

$$
\frac{1}{R^2} \leq \frac{1}{r^2(1+\log(R/r))^2}, \quad r \in (0,R).
$$

Therefore, we obtain

$$
\begin{aligned}
&\left(\int_{B(0,R)} \frac{1}{|x|^2\,(1+|\log(R/|x|)|)^2} \left| f\left(\frac{\delta_R(x)}{|x|} \right) \right|^2 dx \right)^{\frac{1}{2}} \\
&\quad \leq \sqrt{2} \left(\int_{B(0,R)} \frac{1}{|x|^2\,(1+|\log(R/|x|)|)^2} \left| f\left(\frac{\delta_R(x)}{|x|} \right) - f(x) \right|^2 dx \right)^{\frac{1}{2}} \\
&\quad + \frac{\sqrt{2}}{R} \left(\int_{B(0,R)} |f(x)|^2 dx \right)^{\frac{1}{2}}.
\end{aligned} \tag{2.85}
$$

Combining (2.85) with (2.84) we arrive at

$$
\left(\int_{B(0,R)} \frac{1}{|x|^2\,(1+|\log(R/|x|)|)^2} |f(x)|^2 dx \right)^{\frac{1}{2}}
$$

$$\leq (1+\sqrt{2})\left(\int_{B(0,R)} \frac{1}{|x|^2\left(1+|\log(R/|x|)|\right)^2}\left|f(x)-f\left(\frac{\delta_R(x)}{|x|}\right)\right|^2 dx\right)^{\frac{1}{2}}$$
$$+\frac{\sqrt{2}}{R}\left(\int_{B(0,R)}|f(x)|^2dx\right)^{\frac{1}{2}}.$$

Finally, using (2.80) we obtain (2.81). □

2.3.4 Improved Hardy inequalities on quasi-balls

For $p=Q=2$ and any homogeneous quasi-norm $|\cdot|$ on $\mathbb{G}$ we have the following refinement of Theorem 2.3.7 with an estimate for a remainder.

Theorem 2.3.8 (Remainder estimate in critical Hardy inequality for $Q=2$). *We have*

$$\int_{B^2(0,R)}|\mathcal{R}f|^2dx-\frac{1}{4}\int_{B^2(0,R)}\frac{|f(|x|)|^2}{|x|^2\left(\log\frac{R}{|x|}\right)^2}dx \tag{2.86}$$
$$\geq \frac{4}{R^2|\widehat{\sigma}|}\sup_{\nu>0}\nu^{-4}\left|\int_{B^2(0,R)}f(|x|)\frac{\nu-\log\frac{R}{|x|}}{|x|\left(\log\frac{R}{|x|}\right)^{\frac{1}{2}}}\left(\frac{R}{|x|}\right)^{1-\frac{1}{\nu}}dx\right|^2,\ \forall\nu>0,$$

for all real-valued radial functions $f\in C_0^\infty(B^2(0,R)\backslash\{0\})$, where $B^2(0,R)$ and $|\widehat{\sigma}|$ are 2-dimensional quasi-ball with radius R and 1-dimensional surface measure of the unit sphere, respectively.

Proof of Theorem 2.3.8. Let us define the new function $g=g(x)$ on $B^2(0,R)$ as

$$g(r):=\left(\log\frac{R}{r}\right)^{-\frac{1}{2}}f(r),\ r=|x|,\ x\in B^2(0,R).$$

One calculates

$$I_1:=\int_{B^2(0,R)}|\mathcal{R}f|^2dx-\frac{1}{4}\int_{B^2(0,R)}\frac{|f|^2}{|x|^2\left(\log\frac{R}{|x|}\right)^2}dx$$
$$=|\widehat{\sigma}|\int_0^R(f'(r))^2rdr-\frac{|\widehat{\sigma}|}{4}\int_0^R\frac{|f(r)|^2}{r^2\left(\log\frac{R}{r}\right)^2}rdr$$
$$=|\widehat{\sigma}|\int_0^R\left(-\frac{1}{2}\left(\log\frac{R}{r}\right)^{-\frac{1}{2}}\frac{g(r)}{r}+\left(\log\frac{R}{r}\right)^{\frac{1}{2}}g'(r)\right)^2rdr-\frac{|\widehat{\sigma}|}{4}\int_0^R\frac{|g(r)|^2}{r\log\frac{R}{r}}dr$$
$$=-|\widehat{\sigma}|\int_0^R g(r)g'(r)dr+|\widehat{\sigma}|\int_0^R|g'(r)|^2r\log\frac{R}{r}dr$$

$$= -\frac{|\widehat{\sigma}|}{2}\int_0^R (g^2(r))'dr + \frac{R^2|\widehat{\sigma}|}{4}\left(\int_0^R |g'(r)|^2\frac{4}{R^2}r\log\frac{R}{r}dr\right),$$

and applying $g(0) = g(R) = 0$, we obtain

$$I_1 = \frac{R^2|\widehat{\sigma}|}{4}\left(\int_0^R |g'(r)|^2\frac{4}{R^2}r\log\frac{R}{r}dr\right).$$

Taking into account that $\frac{4}{R^2}\int_0^R r\log\frac{R}{r}dr = 1$ and using Hölder's inequality, we get

$$\left(\int_0^R |g'(r)|^2\frac{4}{R^2}r\log\frac{R}{r}dr\right)^{\frac{1}{2}} = \left(\int_0^R |g'(r)|^2\frac{4}{R^2}r\log\frac{R}{r}dr\right)^{\frac{1}{2}}\left(\frac{4}{R^2}\int_0^R r\log\frac{R}{r}dr\right)^{\frac{1}{2}}$$
$$\geq \int_0^R |g'(r)|\frac{4}{R^2}r\log\frac{R}{r}dr.$$

It follows that

$$I_1 \geq \frac{R^2|\widehat{\sigma}|}{4}\left(\int_0^R |g'(r)|\frac{4}{R^2}r\log\frac{R}{r}dr\right)^2 \geq \frac{4|\widehat{\sigma}|}{R^2}\left|\int_0^R g'(r)r\log\frac{R}{r}dr\right|^2. \tag{2.87}$$

Using $g(0) = g(R) = 0$, we have

$$\int_0^R g'(r)r\log\frac{R}{r}dr = -\int_0^R g(r)\left(\log\frac{R}{r}-1\right)dr = \int_0^R f(r)\frac{1-\log\frac{R}{r}}{\left(\log\frac{R}{r}\right)^{\frac{1}{2}}}dr$$
$$= \frac{1}{|\widehat{\sigma}|}\int_{B^2(0,R)} f(x)\frac{1-\log\frac{R}{|x|}}{|x|\left(\log\frac{R}{|x|}\right)^{\frac{1}{2}}}dx.$$

Putting this in (2.87), we arrive at

$$I_1 \geq \frac{4}{R^2|\widehat{\sigma}|}\left|\int_{B^2(0,R)} f(x)\frac{1-\log\frac{R}{|x|}}{|x|\left(\log\frac{R}{|x|}\right)^{\frac{1}{2}}}dx\right|^2.$$

Now we note that I_1 is invariant under the scaling $f \longmapsto f_\nu(r) = \nu^{-\frac{1}{2}}f(R^{1-\nu}r^\nu)$. Then setting

$$I_2(f) := \int_{B^2(0,R)} f(x)\frac{1-\log\frac{R}{|x|}}{|x|\left(\log\frac{R}{|x|}\right)^{\frac{1}{2}}}dx = |\widehat{\sigma}|\int_0^R f(r)\frac{1-\log\frac{R}{r}}{\left(\log\frac{R}{r}\right)^{\frac{1}{2}}}dr,$$

one obtains

$$I_1 \geq \frac{4}{R^2|\widehat{\sigma}|}|I_2(f_\nu)|^2 \tag{2.88}$$

for any $\nu > 0$. On the other hand, we have

$$I_2(f_\nu) = |\widehat{\sigma}| \int_0^R f_\nu(r) \frac{1-\log\frac{R}{r}}{(\log\frac{R}{r})^{\frac{1}{2}}} dr = |\widehat{\sigma}|\nu^{-\frac{1}{2}} \int_0^R f(r^\nu R^{1-\nu}) \frac{1-\log\frac{R}{r}}{(\log\frac{R}{r})^{\frac{1}{2}}} dr.$$

Using a change of variable $s = r^\nu R^{1-\nu}$, $dr = \frac{1}{\nu}\left(\frac{R}{s}\right)^{\frac{\nu-1}{\nu}} ds$, one calculates

$$\begin{aligned} I_2(f_\nu) &\geq |\widehat{\sigma}|\nu^{-\frac{1}{2}} \int_0^R f(s) \frac{1-\log\left(\frac{R}{s}\right)^{\frac{1}{\nu}}}{\left(\log\left(\frac{R}{s}\right)^{\frac{1}{\nu}}\right)^{\frac{1}{2}}} \frac{1}{\nu}\left(\frac{R}{s}\right)^{1-\frac{1}{\nu}} ds \\ &= |\widehat{\sigma}|\nu^{-2} \int_0^R f(s) \frac{\nu - \log\frac{R}{s}}{(\log\frac{R}{s})^{\frac{1}{2}}} \left(\frac{R}{s}\right)^{1-\frac{1}{\nu}} ds \\ &= \nu^{-2} \int_{B^2(0,R)} f(x) \frac{\nu - \log\frac{R}{|x|}}{|x|\left(\log\frac{R}{|x|}\right)^{\frac{1}{2}}} \left(\frac{R}{|x|}\right)^{1-\frac{1}{\nu}} dx. \end{aligned} \tag{2.89}$$

The estimates (2.88) and (2.89) imply (2.86). ☐

Theorem 2.3.9 (Remainder estimate in critical Hardy inequality). *Let $\mathbb{G}$ be a homogeneous group of homogeneous dimension $Q \geq 2$. Let $|\cdot|$ be a homogeneous quasi-norm on $\mathbb{G}$. Let $q > 0$ be such that*

$$\alpha = \alpha(q, L) := \frac{Q-1}{Q}q + L + 2 \leq Q,$$

for $-1 < L < Q-2$. Then for all real-valued positive non-increasing radial functions $u \in C_0^\infty(B(0,R))$ we have

$$\begin{aligned} &\int_{B(0,R)} |\mathcal{R}u|^Q dx - \left(\frac{Q-1}{Q}\right)^Q \int_{B(0,R)} \frac{|u(x)|^Q}{|x|^Q\left(\log\frac{Re}{|x|}\right)^Q} dx \\ &\geq |\wp|^{1-\frac{Q}{q}} C^{\frac{Q}{q}} \left(\int_{B(0,R)} \frac{|u(x)|^q}{|x|^Q\left(\log\frac{Re}{|x|}\right)^\alpha} dx\right)^{\frac{Q}{q}}, \end{aligned} \tag{2.90}$$

where $|\wp|$ is the measure of the unit quasi-sphere in $\mathbb{G}$ and

$$\begin{aligned} C^{-1} = C(L,Q,q)^{-1} &:= \int_0^1 s^L \left(\log\frac{1}{s}\right)^{\frac{Q-1}{Q}q} ds \\ &= (L+1)^{-\left(\frac{Q-1}{Q}q+1\right)} \Gamma\left(\frac{Q-1}{Q}q+1\right), \end{aligned}$$

where $\Gamma(\cdot)$ is the Gamma function.

Proof of Theorem 2.3.9. As in previous proofs we set

$$v(s)=\left(\log\frac{Re}{r}\right)^{-\frac{Q-1}{Q}}u(r),\quad \text{where}\quad r=|x|, s=s(r)=\left(\log\frac{Re}{r}\right)^{-1},$$
$$s'(r)=\frac{s(r)}{r\log\frac{Re}{r}}\geq 0.$$

We have $v(0)=v(1)=0$ since $u(R)=0$ and, moreover,

$$u'(r)=-\left(\frac{Q-1}{Q}\right)\left(\log\frac{Re}{r}\right)^{-\frac{1}{Q}}\frac{v(s(r))}{r}+\left(\log\frac{Re}{r}\right)^{\frac{Q-1}{Q}}v'(s(r))s'(r)\leq 0.$$

It is straightforward to calculate that

$$\begin{aligned}I&:=\int_{B(0,R)}|\mathcal{R}u|^Q dx-\left(\frac{Q-1}{Q}\right)^Q\int_{B(0,R)}\frac{|u|^Q}{|x|^Q\left(\log\frac{Re}{|x|}\right)^Q}dx\\
&=|\wp|\int_0^R|u'(r)|^Q r^{Q-1}dr-\left(\frac{Q-1}{Q}\right)^Q|\wp|\int_0^R\frac{|u(r)|^Q}{r\left(\log\frac{Re}{r}\right)^Q}dr\\
&=|\wp|\int_0^R\left(\frac{Q-1}{Q}\left(\log\frac{Re}{r}\right)^{-\frac{1}{Q}}\frac{v(s(r))}{r}-\left(\log\frac{Re}{r}\right)^{\frac{Q-1}{Q}}v'(s(r))s'(r)\right)^Q r^{Q-1}dr\\
&\quad-\left(\frac{Q-1}{Q}\right)^Q|\wp|\int_0^R\frac{|u(r)|^Q}{r\left(\log\frac{Re}{r}\right)^Q}dr.\end{aligned}$$

By applying the third relation in Lemma 2.4.2 with

$$a=\frac{Q-1}{Q}\left(\log\frac{Re}{r}\right)^{-\frac{1}{Q}}\frac{v(s(r))}{r}\quad\text{and}\quad b=\left(\log\frac{Re}{r}\right)^{\frac{Q-1}{Q}}v'(s(r))s'(r),$$

and dropping $a^Q\geq 0$ as well as using the boundary conditions $v(0)=v(1)=0$, we get

$$\begin{aligned}I\geq&-|\wp|Q\left(\frac{Q-1}{Q}\right)^{Q-1}\int_0^R v(s(r))^{Q-1}v'(s(r))s'(r)dr \qquad (2.91)\\
&+|\wp|\int_0^R|v'(s(r))|^Q(s'(r))^Q\left(r\log\frac{Re}{r}\right)^{Q-1}dr\\
=&-|\wp|Q\left(\frac{Q-1}{Q}\right)^{Q-1}\int_0^R v(s(r))^{Q-1}v'(s(r))s'(r)dr\\
&+|\wp|\int_0^R|v'(s(r))|^Q\frac{1}{r^Q\left(\log\frac{Re}{r}\right)^{2Q}}\left(r\log\frac{Re}{r}\right)^{Q-1}dr\end{aligned}$$

$$
\begin{aligned}
&= - |\wp| Q \left(\frac{Q-1}{Q}\right)^{Q-1} \int_0^R v(s(r))^{Q-1} v'(s(r)) s'(r) dr \\
&\quad + |\wp| \int_0^R |v'(s(r))|^Q s(r)^{Q-1} s'(r) dr \\
&= - |\wp| Q \left(\frac{Q-1}{Q}\right)^{Q-1} \int_0^1 v(s)^{Q-1} v'(s) ds + |\wp| \int_0^1 |v'(s)|^Q s^{Q-1} ds \\
&= |\wp| \int_0^1 |v'(s)|^Q s^{Q-1} ds.
\end{aligned}
$$

Moreover, by using the inequality

$$
\begin{aligned}
|v(s)| = \left| \int_s^1 v'(t) dt \right| &= \left| \int_s^1 v'(t) t^{\frac{Q-1}{Q} - \frac{Q-1}{Q}} dt \right| \\
&\leq \left(\int_0^1 |v'(t)|^Q t^{Q-1} dt \right)^{\frac{1}{Q}} \left(\log \frac{1}{s} \right)^{\frac{Q-1}{Q}},
\end{aligned}
$$

we obtain

$$
\int_0^1 |v(s)|^q s^L ds \leq \left(\int_0^1 |v'(s)|^Q s^{Q-1} ds \right)^{\frac{q}{Q}} \int_0^1 s^L \left(\log \frac{1}{s} \right)^{\frac{Q-1}{Q} q} ds
$$

for $-1 < L < Q-2$. Thus, we have

$$
\int_0^1 |v'(s)|^Q s^{Q-1} ds \geq C^{\frac{q}{Q}} \left(\int_0^1 |v(s)|^q s^L ds \right)^{\frac{Q}{q}}. \tag{2.92}
$$

Now it follows from (2.91) and (2.92) that

$$
\begin{aligned}
I &\geq |\wp| C^{\frac{Q}{q}} \left(\int_0^1 |v(s)|^q s^L ds \right)^{\frac{Q}{q}} = |\wp| C^{\frac{Q}{q}} \left(\int_0^R \frac{|u(r)|^q}{r \left(\log \frac{Re}{r}\right)^{\alpha}} dr \right)^{\frac{Q}{q}} \\
&= |\wp|^{1-\frac{Q}{q}} C^{\frac{Q}{q}} \left(\int_0^R \frac{|u(x)|^q}{|x|^Q \left(\log \frac{Re}{|x|}\right)^{\alpha}} dx \right)^{\frac{Q}{q}},
\end{aligned}
$$

where $\alpha = \alpha(q, L) = \frac{Q-1}{Q} q + L + 2$. The proof is complete. □

2.4 Stability of Hardy inequalities

Expressions for the remainder in an estimate in terms of the distance function to the set of extremisers are sometimes called the stability estimates in the literature. The purpose of this section is to discuss such estimates for the Hardy inequalities. Several results of such a type have been established in the Euclidean space $\mathbb{R}^n$ in [San18, ST17, ST18a, ST15, ST16]. In the following section our presentation on homogeneous groups follows [RS18].

2.4.1 Stability of Hardy inequalities for radial functions

Let $\mathbb{G}$ be a homogeneous group of homogeneous dimension $Q \geq 3$ and let $|\cdot|$ be a homogeneous quasi-norm on $\mathbb{G}$. We note that although sharp constants in the Hardy inequalities are not achieved the formal extremisers of such inequalities are given by homogeneous functions. To this end let us denote

$$f_\alpha(x) := |x|^{-\frac{Q-p-\alpha p}{p}} \tag{2.93}$$

for $-\infty < \alpha < \frac{Q-p}{p}$, and

$$d_R(f,g) := \left(\int_{\mathbb{G}} \frac{|f(x)-g(x)|^p}{\left|\log \frac{R}{|x|}\right|^p |x|^{p(\alpha+1)}} dx \right)^{\frac{1}{p}} \tag{2.94}$$

for functions f, g for which the integral in (2.94) is finite. We start with the case of radially symmetric functions.

Theorem 2.4.1 (Stability of Hardy inequalities for radially symmetric functions). *Let $\mathbb{G}$ be a homogeneous group of homogeneous dimension Q and let $|\cdot|$ be a homogeneous quasi-norm on $\mathbb{G}$. Let*

$$2 \leq p < Q \quad \text{and} \quad -\infty < \alpha < \frac{Q-p}{p}.$$

Then for all radial complex-valued functions $f \in C_0^\infty(\mathbb{G}\backslash\{0\})$ we have

$$\begin{aligned} \int_{\mathbb{G}} \frac{|\mathcal{R}f(x)|^p}{|x|^{\alpha p}} dx - \left(\frac{Q-p-\alpha p}{p}\right)^p \int_{\mathbb{G}} \frac{|f(x)|^p}{|x|^{p(\alpha+1)}} dx \\ \geq c_p \left(\frac{p-1}{p}\right)^p \sup_{R>0} d_R(f, c_f(R) f_\alpha)^p, \end{aligned} \tag{2.95}$$

where $c_f(R) := R^{\frac{Q-p-\alpha p}{p}} \widetilde{f}(R)$ with $f(x) = \widetilde{f}(r)$, $|x| = r$, $\mathcal{R} := \frac{d}{d|x|}$ is the radial derivative, c_p is defined in Lemma 2.3.3, i.e.,

$$c_p = \min_{0<t\leq 1/2} ((1-t)^p - t^p + pt^{p-1}),$$

and f_α and $d_R(\cdot,\cdot)$ are defined in (2.93) and (2.94), respectively.

Proof of Theorem 2.4.1. Since $p \geq 2$, as in (2.72) in the proof of Theorem 2.3.1, we have

$$J(f) = \int_{\mathbb{G}} \frac{|\mathcal{R}f|^p}{|x|^{\alpha p}} dx - \left(\frac{Q-p-\alpha p}{p}\right)^p \int_{\mathbb{G}} \frac{|f|^p}{|x|^{p(\alpha+1)}} dx$$
$$\geq c_p|\wp| \int_0^\infty \left|\frac{d}{dr}\widetilde{g}\right|^p r^{p-1} dr = c_p \int_{\mathbb{G}} \left|\frac{d}{dr} g\right|^p |x|^{p-Q} dx.$$

By Corollary 2.2.8 with $\gamma = p$, we obtain

$$J(f) \geq c_p \int_{\mathbb{G}} |\mathcal{R}g|^p |x|^{p-Q} dx \geq c_p \left(\frac{p-1}{p}\right)^p \int_{\mathbb{G}} \frac{\left|g(x) - g(\frac{Rx}{|x|})\right|^p}{\left|\log \frac{R}{|x|}\right|^p |x|^Q} dx$$
$$= c_p \left(\frac{p-1}{p}\right)^p \int_{\mathbb{G}} \frac{\left||x|^{\frac{Q-p-\alpha p}{p}} f(x) - R^{\frac{Q-p-\alpha p}{p}} f(\frac{Rx}{|x|})\right|^p}{\left|\log \frac{R}{|x|}\right|^p |x|^Q} dx$$

for any $R > 0$. Here using $f(x) = \widetilde{f}(r)$, $r = |x|$, we can estimate

$$J(f) \geq c_p \left(\frac{p-1}{p}\right)^p \int_{\mathbb{G}} \frac{\left|f(x) - R^{\frac{Q-p-\alpha p}{p}} \widetilde{f}(R)|x|^{-\frac{Q-p-\alpha p}{p}}\right|^p}{\left|\log \frac{R}{|x|}\right|^p |x|^{p(\alpha+1)}} dx$$
$$= c_p \left(\frac{p-1}{p}\right)^p \int_{\mathbb{G}} \frac{\left|f(x) - c_f(R)|x|^{-\frac{Q-p-\alpha p}{p}}\right|^p}{\left|\log \frac{R}{|x|}\right|^p |x|^{p(\alpha+1)}} dx,$$

yielding (2.95). □

2.4.2 Stability of Hardy inequalities for general functions

Here we discuss the stability of L^p-Hardy inequalities for functions which do not have to be radially symmetric. Before discussing the stability estimates for non-radial functions let us recall the following relations.

Lemma 2.4.2. *Let $a, b \in \mathbb{R}$. Then*

(i) *We have*

$$|a-b|^p - |a|^p \geq -p|a|^{p-2}ab, \quad p \geq 1.$$

(ii) *There exists a constant $C = C(p) > 0$ such that*

$$|a-b|^p - |a|^p \geq -p|a|^{p-2}ab + C|b|^p, \quad p \geq 2.$$

(iii) *If $a \geq 0$ and $a - b \geq 0$, then*

$$(a-b)^p + pa^{p-1}b - a^p \geq |b|^p, \quad p \geq 2.$$

Proof of Lemma 2.4.2. One has (see, e.g., [Lin90]) the inequality

$$|d|^p \geq |c|^p + p|c|^{p-2}c(d-c), \quad p \geq 1,$$

which is the condition for the convexity of the function $|x|^p$. Now, taking $a = c$ and $d = a - b$ we get the first relation. Furthermore, consider the function

$$g(s) := |1-s|^p - |s|^p + p|s|^{p-2}s, \quad s \in \mathbb{R}.$$

To show (ii), it is sufficient to prove that $g(s) \geq C > 0$, and then take $s = \frac{a}{b}$. Let $s \geq 1$. Then for $p \geq 2$ we have

$$g'(s) = p((t-1)^{p-1} - t^{p-1}) + p(p-1)s^{p-2} = p(p-1)(s^{p-2} - t^{p-2}) \geq 0,$$

by the mean value theorem for the function x^{p-2}, $x \geq 0$. Thus, we have

$$g(s) \geq g(1) = p - 1$$

for all $s \geq 1$. Similarly, we get $g(s) \geq g(0) = 1$ for all $s \leq 1$. Let now $0 \leq s \leq 1$. Setting

$$C_p := \min_{0 \leq t \leq 1} f(t),$$

we assume that $C_p = f(s_0)$ for some $0 \leq s_0 \leq 1$. The first relation implies that $C_p \geq 0$. If $C_p = 0$, then we have $f(s_0) = 0$ and $\frac{s_0 - 1}{p} f'(s_0) = 0$ which implies $s_0 = 0$, contradicting that $f(0) = 1$. Therefore, we have $g(s) \geq C_p > 0$. This proves (ii). The Taylor formula yields that

$$(a-b)^p + pa^{p-1}b - a^p \geq |b|^p = p(p-1)b^2 \int_0^1 (1-t)(a-\tau b)^{p-2} d\tau.$$

Thus, if $b \leq 0$, then $a - \tau b \geq \tau |b|$, which implies (iii) in this case. On the other hand, if $0 \leq b$, then $a - \tau b \geq (1-\tau)|b|$ which implies (iii) in this case as well. $\square$

To formulate the following result, for $R > 0$, let us set

$$d_H(u; R) := \left(\int_{\mathbb{G}} \frac{\left| u(x) - R^{\frac{Q-p}{p}} u\left(R\frac{x}{|x|}\right) |x|^{-\frac{Q-p}{p}} \right|^p}{|x^p| \log \frac{R}{|x|} |^p} dx \right)^{\frac{1}{p}}.$$

Then we have the following stability property.

Theorem 2.4.3 (Stability of Hardy inequalities for general real-valued functions). *Let $\mathbb{G}$ be a homogeneous group of homogeneous dimension Q and let $|\cdot|$ be a homogeneous quasi-norm on $\mathbb{G}$. Let $2 \leq p < Q$. Then there exists a constant $C > 0$ such that for all real-valued functions $u \in C_0^\infty(\mathbb{G})$ we have*

$$\int_{\mathbb{G}} |\mathcal{R}u|^p \, dx - \left(\frac{Q-p}{p}\right)^p \int_{\mathbb{G}} \frac{|u|^p}{|x|^p} dx \geq C_p \sup_{R>0} d_H^p(u; R). \tag{2.96}$$

Proof of Theorem 2.4.3. Let $x = (r, y) = (|x|, \frac{x}{|x|}) \in (0, \infty) \times \wp$ on $\mathbb{G}$, where $\wp$ is the unit quasi-sphere $\wp := \{x \in \mathbb{G} : |x| = 1\}$, and

$$v(ry) := r^{\frac{Q-p}{p}} u(ry),$$

where $u \in C_0^\infty(\mathbb{G})$. It follows that $v(0) = 0$ and that $\lim_{r\to\infty} v(ry) = 0$ for $y \in \wp$ since u is compactly supported. Using the polar decomposition from Proposition 1.2.10 and integrating by parts, we get

$$\begin{aligned} D &:= \int_{\mathbb{G}} |\mathcal{R}u|^p \, dx - \left(\frac{Q-p}{p}\right)^p \int_{\mathbb{G}} \frac{|u|^p}{|x|^p} dx \\ &= \int_\wp \int_0^\infty \left|-\frac{\partial}{\partial r} u(ry)\right|^p r^{Q-1} - \left(\frac{Q-p}{p}\right)^p |u(ry)|^p r^{Q-p-1} drdy \\ &= \int_\wp \int_0^\infty \left|\frac{Q-p}{p} r^{-\frac{Q}{p}} v(ry) - r^{-\frac{Q-p}{p}} \frac{\partial}{\partial r} v(ry)\right|^p r^{Q-1} \\ &\quad - \left(\frac{Q-p}{p}\right)^p |v(ry)|^p r^{-1} drdy. \end{aligned}$$

Now using relation (ii) in Lemma 2.4.2 with the choice $a = \frac{Q-p}{p} r^{-\frac{Q}{p}} v(ry)$ and $b = r^{-\frac{Q-p}{p}} \frac{\partial}{\partial r} v(ry)$, and using the fact that $\int_0^\infty |v|^{p-2} v \left(\frac{\partial}{\partial r} v\right) dr = 0$, we obtain

$$\begin{aligned} D &\geq \int_\wp \int_0^\infty -p\left(\frac{Q-p}{p}\right)^{p-1} |v(ry)|^{p-2} v(ry) \frac{\partial}{\partial r} v(ry) \\ &\quad + C\left|\frac{\partial}{\partial r} v(ry)\right|^p r^{p-1} drdy \\ &= C \int_{\mathbb{G}} |x|^{p-Q} \, |\mathcal{R}v|^p \, dx. \end{aligned} \tag{2.97}$$

Finally, combining (2.97) and Remark 2.2.9, we arrive at

$$\begin{aligned} D &\geq C_p \int_{\mathbb{G}} \frac{|v(x) - v(R\frac{x}{|x|})|^p}{|x|^Q |\log \frac{R}{|x|}|^p} dx = C \int_\wp \int_0^\infty \frac{|v(ry) - v(Ry)|^p}{r\, |\log \frac{R}{r}|^p} drdy \\ &= C_p \int_\wp \int_0^\infty \frac{|u(ry) - R^{\frac{Q-p}{p}} u(Ry) r^{-\frac{Q-p}{p}}|^p}{r^{1+p-Q} |\log \frac{R}{r}|^p} drdy, \end{aligned}$$

for any $R > 0$. This proves the desired result. □

2.4.3 Stability of critical Hardy inequality

Here we discuss the stability of the critical Hardy inequality, i.e., the L^p-Hardy inequality in the case $p = Q$. For this, let us denote

$$f_{T,R}(x) := T^{\frac{Q-1}{Q}} u\left(Re^{-\frac{1}{T}} \frac{x}{|x|}\right) \left(\log \frac{R}{|x|}\right)^{\frac{Q-1}{Q}}.$$

We also introduce the following 'distance' function:

$$d_{cH}(u;T,R) := \left(\int_{B(0,R)} \frac{|u(x) - f_{T,R}(x)|^Q}{|x|^Q \left|\log \frac{R}{|x|}\right|^Q \left|T \log \frac{R}{|x|}\right|^Q} dx \right)^{\frac{1}{Q}}, \tag{2.98}$$

for some parameter $T > 0$, functions u and $f_{T,R}$ for which the integral in (2.98) is finite.

Theorem 2.4.4 (Stability of critical Hardy inequality). *Let $\mathbb{G}$ be a homogeneous group of homogeneous dimension Q and let $|\cdot|$ be a homogeneous quasi-norm on $\mathbb{G}$. Then there exists a constant $C > 0$ such that for all real-valued functions $u \in C_0^\infty(B(0,R))$ we have*

$$\begin{aligned} \int_{B(0,R)} |\mathcal{R}u(x)|^Q \, dx - \left(\frac{Q-1}{Q}\right)^Q \int_{B(0,R)} \frac{|u(x)|^Q}{|x|^Q (\log \frac{R}{|x|})^Q} dx \\ \geq C_p \sup_{T>0} d_{cH}^Q(u;T,R). \end{aligned} \tag{2.99}$$

Proof of Theorem 2.4.4. With polar coordinates $(r,y) = (|x|, \frac{x}{|x|}) \in (0,\infty) \times \wp$ on $\mathbb{G}$, where $\wp$ is the sphere as in (1.12), we have $u(x) = u(ry) \in C_0^\infty(B(0,R))$. In addition, let us set

$$v(sy) := \left(\log \frac{R}{r}\right)^{-\frac{Q-1}{Q}} u(ry), \quad y \in \wp,$$

where

$$s = s(r) := \left(\log \frac{R}{r}\right)^{-1}.$$

Since $u \in C_0^\infty(B(0,R))$ we have $v(0) = 0$ and v has a compact support. Moreover, it is straightforward that

$$\frac{\partial}{\partial r} u(ry) = -\left(\frac{Q-1}{Q}\right) \left(\log \frac{R}{r}\right)^{-\frac{1}{Q}} \frac{v(sy)}{r} + \left(\log \frac{R}{r}\right)^{\frac{Q-1}{Q}} \frac{\partial}{\partial s} v(sy) s'(r).$$

A direct calculation using the polar decomposition in Proposition 1.2.10 gives

$$\begin{aligned} S &:= \int_{B(0,R)} |\mathcal{R}u|^Q \, dx - \left(\frac{Q-1}{Q}\right)^Q \int_{B(0,R)} \frac{|u|^Q}{|x|^Q \left(\log \frac{R}{|x|}\right)^Q} dx \\ &= \int_\wp \left(\int_0^R \left|\frac{\partial}{\partial r} u(ry)\right|^Q r^{Q-1} - \left(\frac{Q-1}{Q}\right)^Q \frac{|u(ry)|^Q}{r \left(\log \frac{R}{r}\right)^Q} dr \right) dy \end{aligned}$$

$$= \int_{\wp}\int_0^R \left(\left| \left(\frac{Q-1}{Q}\right) \left(r \log \frac{R}{r}\right)^{-\frac{1}{Q}} v(sy) + \left(r \log \frac{R}{r}\right)^{\frac{Q-1}{Q}} \frac{\partial}{\partial s} v(sy) s'(r) \right|^Q \right.$$
$$\left. - \left(\frac{Q-1}{Q}\right)^Q \frac{|v(sy)|^Q}{r \log \frac{R}{r}} \right) drdy.$$

Now by applying relation (ii) from Lemma 2.4.2 with the choice

$$a = \frac{Q-1}{Q} \left(r \log \frac{R}{r}\right)^{-\frac{1}{Q}} v(sy) \quad \text{and} \quad b = \left(r \log \frac{R}{r}\right)^{\frac{Q-1}{Q}} \frac{\partial}{\partial s} v(sy) s'(r),$$

and by using the properties that $v(0) = 0$ and $\lim_{r\to\infty} v(ry) = 0$, we obtain

$$\begin{aligned}
S &\geq \int_{\wp}\int_0^R \Bigg(- Q \left(\frac{Q-1}{Q}\right)^{Q-1} |v(sy)|^{Q-2} v(sy) \frac{\partial}{\partial s} v(sy) s'(r) \\
&\quad + C \left| \frac{\partial}{\partial s} v(sy) \right|^Q (s'(r))^Q \left(r \log \frac{R}{r}\right)^{Q-1} \Bigg) drdy \\
&= \int_{\wp}\int_0^R \Bigg(- Q \left(\frac{Q-1}{Q}\right)^{Q-1} |v(sy)|^{Q-2} v(sy) \frac{\partial}{\partial s} v(sy) s'(r) \\
&\quad + C \left| \frac{\partial}{\partial s} v(sy) \right|^Q \frac{1}{r^Q \left(\log \frac{R}{r}\right)^{2Q}} \left(r \log \frac{R}{r}\right)^{Q-1} \Bigg) drdy \\
&= \int_{\wp}\int_0^R \Bigg(- Q \left(\frac{Q-1}{Q}\right)^{Q-1} |v(sy)|^{Q-2} v(sy) \frac{\partial}{\partial s} v(sy) s'(r) \\
&\quad + C \left| \frac{\partial}{\partial s} v(sy) \right|^Q \frac{1}{\left(\log \frac{R}{r}\right)^{Q-1}} s'(r) \Bigg) drdy \\
&= \int_{\wp}\int_0^R \Bigg(- Q \left(\frac{Q-1}{Q}\right)^{Q-1} |v(sy)|^{Q-2} v(sy) \frac{\partial}{\partial s} v(s) \\
&\quad + C \left| \frac{\partial}{\partial s} v(sy) \right|^Q s^{Q-1} \Bigg) dsdy \\
&= C \int_{\mathbb{G}} |\mathcal{R} v|^Q \, dx,
\end{aligned}$$

that is,

$$S \geq C \int_{\mathbb{G}} |\mathcal{R} v|^Q \, dx. \tag{2.100}$$

According to Remark 2.2.9 for $v \in C_0^\infty(\mathbb{G}\backslash\{0\})$ with $p = Q$ and (2.100), it follows that

$$S \geq C_p \int_{\mathbb{G}} \frac{|v(x) - v(T\frac{x}{|x|})|^Q}{|x|^Q |\log \frac{T}{|x|}|^Q} dx = C_p \int_{\wp}\int_0^\infty \frac{|v(sy) - v(Ty)|^Q}{s |\log \frac{T}{s}|^Q} dsdy$$

$$
= C_p \int_{\wp} \int_0^R \frac{\left| (\log \frac{R}{r})^{-\frac{Q-1}{Q}} u(ry) - T^{\frac{Q-1}{Q}} u(Re^{-\frac{1}{T}} y) \right|^Q}{r(\log \frac{R}{r}) |\log(T \log \frac{R}{r})|^Q} drdy
$$

$$
= C_p \int_{\wp} \int_0^R \frac{\left| u(ry) - T^{\frac{Q-1}{Q}} u(Re^{-\frac{1}{T}} y)(\log \frac{R}{r})^{\frac{Q-1}{Q}} \right|^Q}{r(\log \frac{R}{r})^Q |\log(T \log \frac{R}{r})|^Q} drdy.
$$

Thus, we arrive at

$$
S \geq C_p \int_{B(0,R)} \frac{\left| u(x) - T^{\frac{Q-1}{Q}} u\left(Re^{-\frac{1}{T}} \frac{x}{|x|}\right) \left(\log \frac{R}{|x|}\right)^{\frac{Q-1}{Q}} \right|^Q}{|x|^Q \left|\log \frac{R}{|x|}\right|^Q \left|\log \left(T \log \frac{R}{|x|}\right)\right|^Q} dx
$$

for all $T > 0$. This completes the proof of Theorem 2.4.4. □

Chapter 3

Rellich, Caffarelli–Kohn–Nirenberg, and Sobolev Type Inequalities

This chapter is devoted to other functional inequalities usually associated to the Hardy inequalities. These include Rellich and Caffarell–Kohn–Nirenberg inequalities. We also discuss different aspects of this analysis such as their stability, higher-order inequalities, their weighted and extended versions.

3.1 Rellich inequality

In general, the Rellich type inequalities have the following form

$$\int_{\mathbb{R}^n} \frac{|f(x)|^p}{|x|_E^{\alpha}} dx \leq C \int_{\mathbb{R}^n} \frac{|\Delta f(x)|^p}{|x|_E^{\beta}} dx$$

for certain constants α, β and p. The classical result by Rellich appearing at the 1954 ICM in Amsterdam [Rel56] was the inequality

$$\left\| \frac{f}{|x|_E^2} \right\|_{L^2(\mathbb{R}^n)} \leq \frac{4}{n(n-4)} \|\Delta f\|_{L^2(\mathbb{R}^n)}, \quad n \geq 5. \tag{3.1}$$

To find analogues of (3.1) on the homogeneous groups is an interesting question. The first obstacle to it is that there is neither stratification nor gradation on general homogeneous groups, so there may be no homogeneous left invariant hypoelliptic differential operators on $\mathbb{G}$ at all, to formulate an expression of the form of the right-hand side of (3.1).

However, similar to the results on the Hardy type inequalities before, the inequality (3.1) can be expressed in terms of the radial derivative

$$\partial_r = \frac{x}{|x|_E} \cdot \nabla,$$

M. Ruzhansky, D. Suragan, *Hardy Inequalities on Homogeneous Groups*, Progress in Mathematics 327, https://doi.org/10.1007/978-3-030-02895-4_4

taking the form

$$\left\|\frac{f}{|x|_E^2}\right\|_{L^2(\mathbb{R}^n)} \leq \frac{4}{n(n-4)}\left\|\partial_r^2 f + \frac{n-1}{|x|_E}\partial_r f\right\|_{L^2(\mathbb{R}^n)}, \quad n \geq 5. \tag{3.2}$$

It is well known that in the spherical coordinates on $\mathbb{R}^n$ the Laplacian $\Delta_{\mathbb{R}^n}$ decomposes in the radial and spherical parts as

$$\Delta_{\mathbb{R}^n} f = \frac{\partial^2 f}{\partial r^2} + \frac{n-1}{r}\frac{\partial f}{\partial r} + \frac{1}{r^2}\Delta_{\mathbb{S}^{n-1}} f.$$

So, the operator on the right-hand side in (3.2) is precisely the radial part of the Laplacian on $\mathbb{R}^n$, which can be also expressed as

$$\frac{\partial^2 f}{\partial r^2} + \frac{n-1}{r}\frac{\partial f}{\partial r} = \frac{1}{r^{n-1}}\frac{\partial}{\partial r}\left(r^{n-1}\frac{\partial f}{\partial r}\right). \tag{3.3}$$

Although there is no analogue of the Laplacian on general homogeneous groups, using the radial operator $\mathcal{R}$ from Section 1.3, the expression (3.3) on a homogeneous group of homogeneous dimension Q and homogeneous quasi-norm $|\cdot|$ makes perfect sense in the form of

$$\mathcal{R}^2 f + \frac{Q-1}{|x|}\mathcal{R} f = \frac{1}{|x|^{Q-1}}\mathcal{R}\left(|x|^{Q-1}\mathcal{R} f\right). \tag{3.4}$$

One aim of this section is to show that the Rellich inequality in the form (3.2) extends to general homogeneous groups using the radial operator $\mathcal{R}$, taking the form:

$$\left\|\frac{f}{|x|^2}\right\|_{L^2(\mathbb{G})} \leq \frac{4}{Q(Q-4)}\left\|\mathcal{R}^2 f + \frac{Q-1}{|x|}\mathcal{R} f\right\|_{L^2(\mathbb{G})}, \quad Q \geq 5, \tag{3.5}$$

for all complex-valued functions $f \in C_0^\infty(\mathbb{G}\backslash\{0\})$. The operator on the right-hand side of (3.5) is thus an analogue of the radial part of the Laplacian on $\mathbb{R}^n$.

Thus, in the sequel we will be frequently use the *Rellich type operator* appearing in the right-hand side of (3.8) which we may denote by

$$\tilde{\mathcal{R}} f := \mathcal{R}^2 f + \frac{Q-1}{|x|}\mathcal{R} f. \tag{3.6}$$

Similarly to the Hardy type inequalities from Chapter 2, the expression on the right-hand side of (3.5) appears to be natural since there is no analogue of homogeneous Laplacian or sub-Laplacian on general homogeneous groups to extend (3.1). Moreover, even on a group where such operators exist, this would usually give a refinement of those inequalities since derivatives only in one direction appear.

3.1.1 Rellich type inequalities in L^2

In this section we prove the Rellich type inequality (3.5) and its weighted version. This will be a corollary of the following identity relating different expressions involving the radial derivatives and radially symmetric weights.

Theorem 3.1.1 (Identity leading to Rellich inequality). *Let $\mathbb{G}$ be a homogeneous group of homogeneous dimension $Q \geq 5$. Then for arbitrary homogeneous quasi-norm $|\cdot|$ on $\mathbb{G}$ and every complex-valued function $f \in C_0^\infty(\mathbb{G}\backslash\{0\})$ we have the identity*

$$\left\|\mathcal{R}^2 f + \frac{Q-1}{|x|}\mathcal{R}f + \frac{Q(Q-4)}{4|x|^2}f\right\|^2_{L^2(\mathbb{G})} + \frac{Q(Q-4)}{2}\left\|\frac{1}{|x|}\mathcal{R}f + \frac{Q-4}{2|x|^2}f\right\|^2_{L^2(\mathbb{G})}$$
$$= \left\|\mathcal{R}^2 f + \frac{Q-1}{|x|}\mathcal{R}f\right\|^2_{L^2(\mathbb{G})} - \left(\frac{Q(Q-4)}{4}\right)^2\left\|\frac{f}{|x|^2}\right\|^2_{L^2(\mathbb{G})}. \tag{3.7}$$

Since the left-hand side of (3.7) is non-negative, this implies the following Rellich type inequality on $\mathbb{G}$ for $Q \geq 5$:

Corollary 3.1.2 (Rellich type inequality in $L^2(\mathbb{G})$). *For all complex-valued functions $f \in C_0^\infty(\mathbb{G}\backslash\{0\})$ we have*

$$\left\|\frac{f}{|x|^2}\right\|_{L^2(\mathbb{G})} \leq \frac{4}{Q(Q-4)}\left\|\mathcal{R}^2 f + \frac{Q-1}{|x|}\mathcal{R}f\right\|_{L^2(\mathbb{G})}, \quad Q \geq 5, \tag{3.8}$$

where the constant $\frac{4}{Q(Q-4)}$ is sharp and it is attained if and only if $f = 0$.

Remark 3.1.3. Let us show that the constant $\frac{4}{Q(Q-4)}$ in (3.8) is sharp and is never attained unless $f = 0$. Indeed, if the equality in (3.8) is attained, it follows that both terms on the left-hand side of (3.7) must be zero. In particular, it means that

$$\frac{1}{|x|}\mathcal{R}f + \frac{Q-4}{2|x|^2}f = 0, \tag{3.9}$$

and hence

$$\mathbb{E}f = -\frac{Q-4}{2}f.$$

In view of Proposition 1.3.1, Part (i), the function f must be positively homogeneous of order $-\frac{Q-4}{2}$ which is impossible since $f \in C_0^\infty(\mathbb{G}\backslash\{0\})$ unless $f = 0$, so that the constant is not attained unless $f = 0$.

Furthermore, the first term in (3.7) must be also zero, and by using (3.9) this is equivalent to

$$\mathcal{R}^2 f + \frac{Q-2}{2|x|}\mathcal{R}f = 0$$

which means that $\mathcal{R}f$ is positively homogeneous of order $-\frac{Q-2}{2}$. Thus, by taking an approximation of homogeneous functions f of order $-\frac{Q-4}{2}$, we have that $\mathcal{R}f$ is homogeneous of order $-\frac{Q-2}{2}$, so that the left-hand side of (3.7) converges to zero. Therefore, the constant $\frac{4}{Q(Q-4)}$ in (3.8) is sharp.

Thus, in view of Remark 3.1.3, the statement of Corollary 3.1.2 follows from the identity (3.7), which we will now prove.

Proof of Theorem 3.1.1. Introducing polar coordinates

$$(r,y)=\left(|x|,\frac{x}{|x|}\right)\in(0,\infty)\times\wp$$

on $\mathbb{G}$, using the polar decomposition formula from Proposition 1.2.10, as well as integrating by parts we obtain

$$\begin{aligned}
\int_{\mathbb{G}}\frac{|f(x)|^2}{|x|^4}dx&=\int_0^\infty\int_\wp\frac{|f(ry)|^2}{r^4}r^{Q-1}d\sigma(y)dr\\
&=-\frac{2}{Q-4}\mathrm{Re}\int_0^\infty r^{Q-4}\int_\wp f(ry)\overline{\frac{df(ry)}{dr}}d\sigma(y)dr\\
&=\frac{2}{(Q-3)(Q-4)}\mathrm{Re}\int_0^\infty r^{Q-3}\int_\wp\left(\left|\frac{df(ry)}{dr}\right|^2+f(ry)\overline{\frac{d^2f(ry)}{dr^2}}\right)d\sigma(y)dr\\
&=\frac{2}{(Q-3)(Q-4)}\left(\left\|\frac{1}{|x|}\mathcal{R}f\right\|^2_{L^2(\mathbb{G})}+\mathrm{Re}\int_{\mathbb{G}}\frac{f(x)}{|x|^2}\overline{\mathcal{R}^2f(x)}dx\right).
\end{aligned}\tag{3.10}$$

For the first term, using identity (2.16) with $\alpha=1$, we have identity (2.17), i.e.,

$$\left\|\frac{1}{|x|}\mathcal{R}f\right\|^2_{L^2(\mathbb{G})}=\left(\frac{Q-4}{2}\right)^2\left\|\frac{f}{|x|^2}\right\|^2_{L^2(\mathbb{G})}+\left\|\frac{1}{|x|}\mathcal{R}f+\frac{Q-4}{2|x|^2}f\right\|^2_{L^2(\mathbb{G})}.\tag{3.11}$$

For the second term a direct calculation shows

$$\begin{aligned}
&\mathrm{Re}\int_{\mathbb{G}}\frac{f(x)}{|x|^2}\overline{\mathcal{R}^2f(x)}dx\\
&=\mathrm{Re}\int_{\mathbb{G}}\frac{f(x)}{|x|^2}\overline{\left(\mathcal{R}^2f(x)+\frac{Q-1}{|x|}\mathcal{R}f(x)\right)}dx-(Q-1)\mathrm{Re}\int_{\mathbb{G}}\frac{f(x)}{|x|^3}\overline{\mathcal{R}f(x)}dx\\
&=\mathrm{Re}\int_{\mathbb{G}}\frac{f(x)}{|x|^2}\overline{\left(\mathcal{R}^2f(x)+\frac{Q-1}{|x|}\mathcal{R}f(x)\right)}dx\\
&\quad-\frac{Q-1}{2}\int_0^\infty r^{Q-4}\int_\wp\frac{d|f(ry)|^2}{dr}d\sigma(y)dr\\
&=\mathrm{Re}\int_{\mathbb{G}}\frac{f(x)}{|x|^2}\overline{\left(\mathcal{R}^2f(x)+\frac{Q-1}{|x|}\mathcal{R}f(x)\right)}dx\\
&\quad+\frac{(Q-1)(Q-4)}{2}\int_0^\infty r^{Q-5}\int_\wp|f(ry)|^2d\sigma(y)dr
\end{aligned}$$

$$= \mathrm{Re} \int_{\mathbb{G}} \frac{f(x)}{|x|^2} \overline{\left(\mathcal{R}^2 f(x) + \frac{Q-1}{|x|} \mathcal{R} f(x)\right)} dx$$
$$+ \frac{(Q-1)(Q-4)}{2} \left\| \frac{f}{|x|^2} \right\|^2_{L^2(\mathbb{G})}. \tag{3.12}$$

Combining (3.11) and (3.12) with (3.10) we arrive at

$$\left\| \frac{f}{|x|^2} \right\|^2_{L^2(\mathbb{G})} = \frac{2}{(Q-3)(Q-4)} \left(\left(\frac{Q-4}{2}\right)^2 \left\| \frac{f}{|x|^2} \right\|^2_{L^2(\mathbb{G})} \right.$$
$$+ \left\| \frac{1}{|x|} \mathcal{R} f + \frac{Q-4}{2|x|^2} f \right\|^2_{L^2(\mathbb{G})}$$
$$\left. + \mathrm{Re} \int_{\mathbb{G}} \frac{f(x)}{|x|^2} \overline{\left(\mathcal{R}^2 f(x) + \frac{Q-1}{|x|} \mathcal{R} f(x)\right)} dx + \frac{(Q-1)(Q-4)}{2} \left\| \frac{f}{|x|^2} \right\|^2_{L^2(\mathbb{G})} \right).$$

Collecting same terms, this gives

$$0 = \frac{2Q}{(Q-3)4} \left\| \frac{f}{|x|^2} \right\|^2_{L^2(\mathbb{G})}$$
$$+ \frac{2}{(Q-3)(Q-4)} \left(\mathrm{Re} \int_{\mathbb{G}} \frac{f(x)}{|x|^2} \overline{\left(\mathcal{R}^2 f(x) + \frac{Q-1}{|x|} \mathcal{R} f(x)\right)} dx \right.$$
$$\left. + \left\| \frac{1}{|x|} \mathcal{R} f + \frac{Q-4}{2|x|^2} f \right\|^2_{L^2(\mathbb{G})} \right),$$

that is,

$$\frac{Q(Q-4)}{4} \left\| \frac{f}{|x|^2} \right\|^2_{L^2(\mathbb{G})}$$
$$= -\mathrm{Re} \int_{\mathbb{G}} \frac{f(x)}{|x|^2} \overline{\left(\mathcal{R}^2 f(x) + \frac{Q-1}{|x|} \mathcal{R} f(x)\right)} dx - \left\| \frac{1}{|x|} \mathcal{R} f + \frac{Q-4}{2|x|^2} f \right\|^2_{L^2(\mathbb{G})}.$$

Multiplying both sides by $\frac{4}{Q(Q-4)}$ and simplifying we obtain

$$\mathrm{Re} \int_{\mathbb{G}} \frac{f(x)}{|x|^2} \left(\overline{\frac{f(x)}{|x|^2} + \frac{4}{Q(Q-4)} \left(\mathcal{R}^2 f(x) + \frac{Q-1}{|x|} \mathcal{R} f(x)\right)} \right) dx$$
$$= -\frac{4}{Q(Q-4)} \left\| \frac{1}{|x|} \mathcal{R} f + \frac{Q-4}{2|x|^2} f \right\|^2_{L^2(\mathbb{G})}. \tag{3.13}$$

On the other hand, we also have

$$2\mathrm{Re} \int_{\mathbb{G}} \frac{f(x)}{|x|^2} \left(\overline{\frac{f(x)}{|x|^2} + \frac{4}{Q(Q-4)} \left(\mathcal{R}^2 f(x) + \frac{Q-1}{|x|} \mathcal{R} f(x)\right)} \right) dx$$

$$= \left\| \frac{f}{|x|^2} + \frac{4}{Q(Q-4)} \left(\mathcal{R}^2 f + \frac{Q-1}{|x|} \mathcal{R} f \right) \right\|^2_{L^2(\mathbb{G})} + \left\| \frac{f}{|x|^2} \right\|^2_{L^2(\mathbb{G})} - \left\| \frac{4}{Q(Q-4)} \left(\mathcal{R}^2 f + \frac{Q-1}{|x|} \mathcal{R} f \right) \right\|^2_{L^2(\mathbb{G})}. \tag{3.14}$$

From (3.13) and (3.14) we obtain

$$\begin{aligned} &- \frac{8}{Q(Q-4)} \left\| \frac{1}{|x|} \mathcal{R} f + \frac{Q-4}{2|x|^2} f \right\|^2_{L^2(\mathbb{G})} \\ &= \left(\frac{4}{Q(Q-4)} \right)^2 \left\| \frac{Q(Q-4)}{4} \frac{f}{|x|^2} + \mathcal{R}^2 f + \frac{Q-1}{|x|} \mathcal{R} f \right\|^2_{L^2(\mathbb{G})} + \left\| \frac{f}{|x|^2} \right\|^2_{L^2(\mathbb{G})} \\ &\quad - \left(\frac{4}{Q(Q-4)} \right)^2 \left\| \mathcal{R}^2 f + \frac{Q-1}{|x|} \mathcal{R} f \right\|^2_{L^2(\mathbb{G})}, \end{aligned}$$

thus,

$$\begin{aligned} &\left\| \mathcal{R}^2 f + \frac{Q-1}{|x|} \mathcal{R} f + \frac{Q(Q-4)}{4|x|^2} f \right\|^2_{L^2(\mathbb{G})} + \frac{Q(Q-4)}{2} \left\| \frac{1}{|x|} \mathcal{R} f + \frac{Q-4}{2|x|^2} f \right\|^2_{L^2(\mathbb{G})} \\ &\qquad = \left\| \mathcal{R}^2 f + \frac{Q-1}{|x|} \mathcal{R} f \right\|^2_{L^2(\mathbb{G})} - \left(\frac{Q(Q-4)}{4} \right)^2 \left\| \frac{f}{|x|^2} \right\|^2_{L^2(\mathbb{G})}. \end{aligned}$$

This gives identity (3.7). □

In the Euclidean setting of $\mathbb{R}^n$ the equalities of the type of Theorem 3.1.1 were analysed in [MOW17b]. Theorem 3.1.1 was obtained in [RS17b] and it can be extended to all $\alpha \in \mathbb{R}$. We present these results next, following [Ngu17], also correcting relevant statements. We will be always using the notation

$$\tilde{\mathcal{R}} f := \mathcal{R}^2 f + \frac{Q-1}{|x|} \mathcal{R} f. \tag{3.15}$$

Theorem 3.1.4 (Weighted L^2-Rellich inequalities). *Let $\mathbb{G}$ be a homogeneous group of homogeneous dimension $Q \geq 5$ and let $|\cdot|$ be a homogeneous quasi-norm on $\mathbb{G}$. Then we have the following properties:*

(1) *For any $\alpha \in \mathbb{R}$, for all complex-valued functions $f \in C_0^\infty(\mathbb{G}\backslash\{0\})$ we have the identity*

$$\begin{aligned} \left\| \frac{\tilde{\mathcal{R}} f}{|x|^\alpha} \right\|^2_{L^2(\mathbb{G})} &= C_\alpha^2 \left\| \frac{f}{|x|^{\alpha+2}} \right\|^2_{L^2(\mathbb{G})} + \left\| \frac{\tilde{\mathcal{R}} f}{|x|^\alpha} + C_\alpha \frac{f}{|x|^{\alpha+2}} \right\|^2_{L^2(\mathbb{G})} \\ &\quad + 2C_\alpha \left\| \frac{\mathcal{R} f}{|x|^{1+\alpha}} + \frac{Q-4-2\alpha}{2|x|^{\alpha+2}} f \right\|^2_{L^2(\mathbb{G})}, \end{aligned} \tag{3.16}$$

where

$$C_\alpha = \frac{(Q+2\alpha)(Q-4-2\alpha)}{4}. \tag{3.17}$$

(2) *For any $\alpha \in (-Q/2, (Q-4)/2)$, we have the following weighted Rellich inequality for all complex-valued functions $f \in C_0^\infty(\mathbb{G}\backslash\{0\})$,*

$$C_\alpha \left\| \frac{f}{|x|^{\alpha+2}} \right\|_{L^2(\mathbb{G})} \le \left\| \frac{\tilde{\mathcal{R}} f}{|x|^\alpha} \right\|_{L^2(\mathbb{G})}, \tag{3.18}$$

where the constant $C_\alpha > 0$ as in (3.17) is sharp and it is attained if and only if $f = 0$.

Proof of Theorem 3.1.4. Part (2) is an immediate consequence of Part (1), so we prove Part (1) now.

We can assume that $C_\alpha \neq 0$ since otherwise (3.16) trivially holds. Moreover, we can assume for the proof below that $Q-3-2\alpha \neq 0$. If this is zero, the statement (3.16) would follow by continuity from the same identity for other α's.

As in the proof of Theorem 3.1.1, we calculate

$$\begin{aligned}
\int_{\mathbb{G}} \frac{|f(x)|^2}{|x|^{4+2\alpha}} dx &= \int_0^\infty r^{Q-5-2\alpha} \int_{\wp} |f(ry)|^2 d\sigma(y) dr \\
&= \frac{1}{Q-4-2\alpha} \int_0^\infty (r^{Q-4-2\alpha})' \int_{\wp} |f(ry)|^2 d\sigma(y) dr \\
&= -\frac{2}{Q-4-2\alpha} \mathrm{Re} \int_0^\infty r^{Q-4-2\alpha} \int_{\wp} f(ry) \overline{\mathcal{R} f(ry)} d\sigma(y) dr \\
&= -\frac{2}{(Q-4-2\alpha)(Q-3-2\alpha)} \mathrm{Re} \int_0^\infty (r^{Q-3-2\alpha})' \int_{\wp} f(ry) \overline{\mathcal{R} f(ry)} d\sigma(y) dr \\
&= \frac{2}{(Q-4-2\alpha)(Q-3-2\alpha)} \mathrm{Re} \int_0^\infty r^{Q-3-2\alpha} \\
&\quad \times \int_{\wp} \left(|\mathcal{R} f(ry)|^2 + f(ry) \overline{\mathcal{R}^2 f(ry)} \right) d\sigma(y) dr \\
&= \frac{2}{(Q-4-2\alpha)(Q-3-2\alpha)} \mathrm{Re} \int_{\mathbb{G}} \left(\frac{|\mathcal{R} f(x)|^2}{|x|^{2+2\alpha}} + \frac{f(x) \overline{\mathcal{R}^2 f(x)}}{|x|^{2+2\alpha}} \right) dx.
\end{aligned} \tag{3.19}$$

By Theorem 3.1.1 we have

$$\int_{\mathbb{G}} \frac{|\mathcal{R} f|^2}{|x|^{2+2\alpha}} dx = \frac{(Q-4-2\alpha)^2}{4} \int_{\mathbb{G}} \frac{|f|^2}{|x|^{4+2\alpha}} dx + \int_{\mathbb{G}} \left| \frac{\mathcal{R} f}{|x|^{1+\alpha}} + \frac{Q-4-2\alpha}{2|x|^{2+\alpha}} f \right|^2 dx. \tag{3.20}$$

By integration by parts we obtain

$$\mathrm{Re} \int_{\mathbb{G}} \frac{f \overline{\mathcal{R}^2 f}}{|x|^{2+2\alpha}} dx = \mathrm{Re} \int_{\mathbb{G}} \frac{f \overline{\tilde{\mathcal{R}} f}}{|x|^{2+2\alpha}} dx - (Q-1) \mathrm{Re} \int_{\mathbb{G}} \frac{f \overline{\mathcal{R} f}}{|x|^{3+2\alpha}} dx \tag{3.21}$$

$$= \operatorname{Re} \int_{\mathbb{G}} \frac{f\overline{\tilde{\mathcal{R}} f}}{|x|^{2+2\alpha}} dx + \frac{(Q-1)(Q-4-2\alpha)}{2} \int_{\mathbb{G}} \frac{|f|^2}{|x|^{4+2\alpha}} dx.$$

Plugging (3.20) and (3.21) into (3.19) we get

$$\int_{\mathbb{G}} \frac{|f|^2}{|x|^{4+2\alpha}} dx = -\frac{1}{C_\alpha} \operatorname{Re} \int_{\mathbb{G}} \frac{f\overline{\tilde{\mathcal{R}} f}}{|x|^{2+2\alpha}} dx - \frac{1}{C_\alpha} \int_{\mathbb{G}} \left| \frac{\mathcal{R} f}{|x|^{1+\alpha}} + \frac{Q-4-2\alpha}{2|x|^{2+\alpha}} f \right|^2 dx$$
$$= \frac{1}{2C_\alpha^2} \int_{\mathbb{G}} \frac{|\tilde{\mathcal{R}} f|^2}{|x|^{2\alpha}} dx + \frac{1}{2} \int_{\mathbb{G}} \frac{|f|^2}{|x|^{2\alpha+4}} dx - \frac{1}{2C_\alpha^2} \int_{\mathbb{G}} \left| \frac{\tilde{\mathcal{R}} f}{|x|^{\alpha}} + C_\alpha \frac{f}{|x|^{\alpha+2}} \right|^2 dx$$
$$- \frac{1}{C_\alpha} \int_{\mathbb{G}} \left| \frac{\mathcal{R} f}{|x|^{\alpha+1}} + \frac{Q-4-2\alpha}{2|x|^{\alpha+2}} f \right|^2 dx,$$

which gives equality (3.16).

The inequality (3.18) is a straightforward consequence of (3.16) since $C_\alpha > 0$ for $\alpha \in (-Q/2, (Q-4)/2)$. Let us now show that this constant is sharp. For this, we consider the test function $g(r) := r^{-(Q-4-2\alpha)/2}$ which can be approximated by smooth functions. For example, let $\eta \in C_0^\infty(\mathbb{R})$ be such that $\eta = 1$ on $(-1,1)$ and $\eta = 0$ on $\mathbb{R}\backslash(-2,2)$. For any $\epsilon > 0$, define $f_\epsilon(r) := (1-\eta(r/\epsilon))r^{-\frac{Q-4-2\alpha}{2}}\eta(\epsilon r)$, then

$$\lim_{\epsilon \to 0} f_\epsilon(r) = g(r).$$

We have

$$\frac{\left\| \frac{\tilde{\mathcal{R}} g}{|x|^\alpha} \right\|^2_{L^2(\mathbb{G})}}{\left\| \frac{g}{|x|^{\alpha+2}} \right\|^2_{L^2(\mathbb{G})}} = \frac{\int_{\mathbb{G}} \frac{1}{|x|^{2\alpha}} \left| \mathcal{R}^2 g(|x|) + \frac{Q-1}{|x|} \mathcal{R} g(|x|) \right|^2 dx}{\int_{\mathbb{G}} \frac{|g(|x|)|^2}{|x|^{4+2\alpha}} dx} = C_\alpha^2,$$

which shows the sharpness of C_α. If for some function f, there is equality in (3.18), then from (3.16) we must have

$$\mathcal{R} f + \frac{Q-4-2\alpha}{2|x|} f = 0.$$

In terms of the Euler operator this can be equivalently expressed by

$$\mathbb{E} f = -\frac{Q-4-2\alpha}{2} f.$$

By Proposition 1.3.1 this implies that f is positively homogeneous of order $-(Q-4-2\alpha)/2$, that is, there exists function $h : \wp \to \mathbb{C}$ such that

$$f(x) = |x|^{-(Q-4-2\alpha)/2} h(x/|x|).$$

Since $f(x)/|x|^{2+\alpha}$ is in $L^2(\mathbb{G})$, we must have $h = 0$ on $\wp$ and, consequently, $f = 0$ on $\wp$. $\square$

3.1.2 Rellich type inequalities in L^p

In this section we describe the L^p-versions of the L^2-properties presented in Section 3.1.1, where we have followed the proofs in [RS17b]. In this section we follow [Ngu17]. We start with the identity analogous to that in Theorem 3.1.1.

Theorem 3.1.5 (Identity leading to L^p-Rellich inequality). *Let $\mathbb{G}$ be a homogeneous group of homogeneous dimension Q and let $|\cdot|$ be a homogeneous quasi-norm on $\mathbb{G}$. Let $1 < p < Q$ and $\alpha \in \mathbb{R}$. Then for all complex-valued functions $f \in C_0^\infty(\mathbb{G}\backslash\{0\})$ we have*

$$\int_{\mathbb{G}} \frac{1}{|x|^{p\alpha}} \left| \mathcal{R}f + \frac{Q-1}{|x|} f \right|^p dx = \frac{|Q+p'\alpha|^p}{(p')^p} \int_{\mathbb{G}} \frac{|f|^p}{|x|^{p(1+\alpha)}} dx \tag{3.22}$$
$$+ p \int_{\mathbb{G}} \frac{1}{|x|^{p\alpha}} R_p \left(\frac{Q+p'\alpha}{p'} \frac{f}{|x|}, \mathcal{R}f + \frac{Q-1}{|x|} f \right) dx,$$

where $p' = p/(p-1)$ and R_p is as in (2.22).

Proof of Theorem 3.1.5. First, a direct calculation shows

$$\int_{\mathbb{G}} \frac{|f|^p}{|x|^{p(1+\alpha)}} dx = \int_0^\infty r^{Q-p(1+\alpha)-1} \int_{\wp} |f(ry)|^p d\sigma dr$$
$$= \frac{1}{Q-p(1+\alpha)} \int_0^\infty (r^{Q-p(1+\alpha)})' \int_{\wp} |f(ry)|^p d\sigma dr$$
$$= -\frac{p}{Q-p(1+\alpha)} \mathrm{Re} \int_0^\infty r^{Q-p(1+\alpha)} \int_{\wp} |f(ry)|^{p-2} f(ry) \overline{\mathcal{R}f(ry)} d\sigma(y) dr$$
$$= -\frac{p}{Q-p(1+\alpha)} \mathrm{Re} \int_{\mathbb{G}} \frac{|f|^{p-2} f \overline{\mathcal{R}f}}{|x|^{p(1+\alpha)-1}} dx$$
$$= -\frac{p}{Q-p(1+\alpha)} \left(\mathrm{Re} \int_{\mathbb{G}} \frac{|f|^{p-2} f}{|x|^{(p-1)(1+\alpha)}} \frac{\overline{\mathcal{R}f + \frac{Q-1}{|x|} f}}{|x|^\alpha} - (Q-1) \int_{\mathbb{G}} \frac{|f|^p}{|x|^{p(1+\alpha)}} dx \right).$$

This implies further equalities,

$$\int_{\mathbb{G}} \frac{|f|^p}{|x|^{p(1+\alpha)}} dx = \frac{p'}{Q+p'\alpha} \mathrm{Re} \int_{\mathbb{G}} \frac{|f|^{p-2} f}{|x|^{(p-1)(1+\alpha)}} \frac{\overline{\mathcal{R}f + \frac{Q-1}{|x|} f}}{|x|^\alpha} dx$$
$$= \frac{p-1}{p} \int_{\mathbb{G}} \frac{|f|^p}{|x|^{p(1+\alpha)}} dx + \frac{1}{p} \frac{(p')^p}{|Q+p'\alpha|^p} \left| \mathcal{R}f + \frac{Q-1}{|x|} f \right|^2 dx$$
$$- \int_{\mathbb{G}} R_p \left(\frac{f}{|x|^{1+\alpha}}, \frac{p'}{Q+p'\alpha} \frac{\mathcal{R}f + \frac{Q-1}{|x|} f}{|x|^\alpha} \right) dx,$$

which proves (3.22). ∎

Since the remainder term on the right-hand side of (3.22) is non-negative, this implies the following L^p-version of the Rellich type inequality on $\mathbb{G}$:

Corollary 3.1.6 (Rellich type inequality in $L^p(\mathbb{G})$). *Let $\mathbb{G}$ be a homogeneous group of homogeneous dimension Q and let $|\cdot|$ be a homogeneous quasi-norm on $\mathbb{G}$. Let $1 < p < Q$ and $\alpha \in \mathbb{R}$. Then for all complex-valued functions $f \in C_0^\infty(\mathbb{G}\backslash\{0\})$ we have*

$$\left\| \frac{1}{|x|^\alpha} \left(\mathcal{R}f + \frac{Q-1}{|x|} f \right) \right\|_{L^p(\mathbb{G})} \geq \frac{|Q+p'\alpha|}{p'} \left\| \frac{f}{|x|^{1+\alpha}} \right\|_{L^p(\mathbb{G})},$$

where the constant $\frac{4}{Q(Q-4)}$ is sharp and it is attained if and only if $f = 0$.

As a consequence of Theorem 3.1.5 we have

Corollary 3.1.7. *Let $\mathbb{G}$ be a homogeneous group of homogeneous dimension Q. Let $|\cdot|$ be any homogeneous quasi-norm on $\mathbb{G}$ and $1 < p < Q/2$. Then for any complex-valued $f \in C_0^\infty(\mathbb{G}\backslash\{0\})$, we have*

$$\int_{\mathbb{G}} \frac{|\tilde{\mathcal{R}}f|^p}{|x|^{p\alpha}} dx = \frac{|Q+p'\alpha|^p}{(p')^p} \int_{\mathbb{G}} \frac{|\mathcal{R}f|^p}{|x|^{p(1+\alpha)}} dx + p \int_{\mathbb{G}} \frac{1}{|x|^{p\alpha}} R_p \left(\frac{Q+p'\alpha}{p'} \frac{\mathcal{R}f}{|x|}, \tilde{\mathcal{R}}f \right) dx \tag{3.23}$$

for any $\alpha \in \mathbb{R}$. Here R_p is as in (2.22). As a consequence, we obtain the following weighted L^p-Rellich type inequality

$$\frac{|Q+p'\alpha|}{p'} \left\| \frac{\mathcal{R}f}{|x|^{(1+\alpha)}} \right\|_{L^p(\mathbb{G})} \leq \left\| \frac{\tilde{\mathcal{R}}f}{|x|^\alpha} \right\|_{L^p(\mathbb{G})} \tag{3.24}$$

for any $\alpha \in \mathbb{R}$ and any complex-valued function $f \in C_0^\infty(\mathbb{G}\backslash\{0\})$. Moreover, the inequality (3.24) is sharp and equality holds if and only if $f = 0$.

Proof of Corollary 3.1.7. The equality (3.23) is exactly (3.22) with f being replaced by $\mathcal{R}f$. The inequality (3.24) follows immediately from (3.23) by dropping the non-negative remainder term on the right-hand side of (3.23). The sharpness of (3.24) is proved by using approximations of the function $r^{-(Q'p(1+\alpha))/p}$. If the equality occurs in (3.24) for some function f, then by (3.23) we must have

$$\tilde{\mathcal{R}}f = \frac{Q+p'\alpha}{p'|x|} \mathcal{R}f,$$

which is equivalent to

$$\mathcal{R}^2 f + \frac{Q-p(1+\alpha)}{p|x|} \mathcal{R}f = 0.$$

This can be also expressed as

$$\mathbb{E}(\mathcal{R}f) = -\frac{Q-p(1+\alpha)}{p} \mathcal{R}f.$$

Hence $\mathcal{R}f$ is positively homogeneous of degree $-(Q'p(1+\alpha))/p$, which forces $\mathcal{R}f = 0$ since $\mathcal{R}f/|x|^{1+\alpha}$ is in $L^p(\mathbb{G})$. Thus, we get $f = 0$. □

Theorem 3.1.8 (Another weighted L^p-Rellich inequality). *Let $\mathbb{G}$ be a homogeneous group of homogeneous dimension Q. Let $1 < p < Q/2$. Let $|\cdot|$ be a homogeneous quasi-norm on $\mathbb{G}$. Then we have the following properties.*

(1) *For all $\alpha \in \mathbb{R}$ and for all complex-valued functions $f \in C_0^\infty(\mathbb{G}\backslash\{0\})$ we have the identity*

$$\begin{aligned}\left\|\frac{\tilde{\mathcal{R}}f}{|x|^\alpha}\right\|^p_{L^p(\mathbb{G})} &= |C_{p,\alpha}|^p \int_{\mathbb{G}} \frac{|f|^p}{|x|^{p(2+\alpha)}}dx + p\int_{\mathbb{G}} \frac{1}{|x|^{p\alpha}} R_p\left(C_{p,\alpha}\frac{f}{|x|^2}, -\tilde{\mathcal{R}}f\right)dx \\ &+ p|C_{p,\alpha}|^{p-2}C_{p,\alpha}(p-1)\int_{\mathbb{G}} \frac{|f|^{p-2}}{|x|^{p(2+\alpha)-2}}\left|\mathcal{R}|f| + \frac{Q-p(2+\alpha)}{p|x|}|f|\right|^2 dx \\ &+ p|C_{p,\alpha}|^{p-2}C_{p,\alpha}\int_{\mathbb{G}} \frac{|f|^{p-4}(\mathrm{Im}(f\overline{\mathcal{R}f}))^2}{|x|^{p(2+\alpha)-2}}dx. \end{aligned} \tag{3.25}$$

Here R_p is as in (2.22) and

$$C_{p,\alpha} = \frac{(Q-2p-p\alpha)(Q+p'\alpha)}{pp'}, \quad \frac{1}{p}+\frac{1}{p'} = 1. \tag{3.26}$$

(2) *For any $\alpha \in (-(p-1)Q/p, (Q-2p)/p)$ and all complex-valued functions $f \in C_0^\infty(\mathbb{G}\backslash\{0\})$ we have*

$$C_{p,\alpha}\left\|\frac{f}{|x|^{2+\alpha}}\right\|_{L^p(\mathbb{G})} \le \left\|\frac{\tilde{\mathcal{R}}f}{|x|^\alpha}\right\|_{L^p(\mathbb{G})}. \tag{3.27}$$

Moreover, the constant $C_{p,\alpha} > 0$ as in (3.26) is sharp and equality in (3.27) holds if and only if $f = 0$.

Proof of Theorem 3.1.8. Inequality (3.27) in Part (2) follows from Part (1) in view of the positivity of the constant $C_{p,\alpha} > 0$ under the corresponding conditions on α, so we prove Part (1) now. For the argument below we may assume that $C_{p,\alpha} \neq 0$, and that the constant on the left-hand side in the following estimate is non-zero. Then a direct calculation using Proposition 1.2.10 and integration by parts gives

$$\begin{aligned}&\frac{(Q-p(2+\alpha))(Q-p(2+\alpha)+1)}{p}\int_{\mathbb{G}} \frac{|f(x)|^p}{|x|^{p(2+\alpha)}}dx \\ &= \frac{(Q-p(2+\alpha))(Q-p(2+\alpha)+1)}{p}\int_0^\infty r^{Q-p(2+\alpha)-1}\int_{\wp}|f(ry)|^p d\sigma(y)dr \\ &= \frac{(Q-p(2+\alpha)+1)}{p}\int_0^\infty (r^{Q-p(2+\alpha)})'\int_{\wp}|f(ry)|^p d\sigma(y)dr \\ &= -(Q-p(2+\alpha)+1)\mathrm{Re}\int_0^\infty r^{Q-p(2+\alpha)}\int_{\wp}|f(ry)|^{p-2}f(ry)\overline{\mathcal{R}f(ry)}d\sigma(y)dr\end{aligned}$$

$$
\begin{aligned}
&= -\mathrm{Re}\int_0^\infty (r^{Q-p(2+\alpha)+1})' \int_{\wp} |f(ry)|^{p-2} f(ry)\overline{\mathcal{R}f(ry)}d\sigma(y)dr \\
&= \mathrm{Re}\int_0^\infty r^{Q-p(2+\alpha)+1} \int_{\wp} (p-2)|f(ry)|^{p-4}(\mathrm{Re}(f(ry)\overline{\mathcal{R}f(ry)}))^2 \\
&\quad + |f(ry)|^{p-2}|\mathcal{R}f(ry)|^2 + |f(ry)|^{p-2}f(ry)\overline{\mathcal{R}^2 f(ry)}d\sigma(y)dr \\
&= \mathrm{Re}\int_{\mathbb{G}} \frac{1}{|x|^{p(2+\alpha)-2}}\left((p-2)|f|^{p-4}(\mathrm{Re}(f\overline{\mathcal{R}f}))^2 + |f|^{p-2}|\mathcal{R}f|^2 + |f|^{p-2}f\overline{\mathcal{R}^2 f}\right)dx \\
&= \mathrm{Re}\int_{\mathbb{G}} \frac{|f|^{p-2}f}{|x|^{(p-1)(2+\alpha)}}\frac{\overline{\tilde{\mathcal{R}}f}}{|x|^\alpha}dx - (Q-1)\mathrm{Re}\int_{\mathbb{G}}\frac{|f|^{p-2}f\overline{\mathcal{R}f}}{|x|^{p(2+\alpha)-1}}dx \\
&\quad + (p-1)\int_{\mathbb{G}}\frac{|f|^{p-4}(\mathrm{Re}(f\overline{\mathcal{R}f}))^2}{|x|^{p(2+\alpha)-2}}dx + \int_{\mathbb{G}}\frac{|f|^{p-4}(\mathrm{Im}(f\overline{\mathcal{R}f}))^2}{|x|^{p(2+\alpha)-2}}dx \\
&= \mathrm{Re}\int_{\mathbb{G}} \frac{|f|^{p-2}f}{|x|^{(p-1)(2+\alpha)}}\frac{\overline{\tilde{\mathcal{R}}f}}{|x|^\alpha}dx + \frac{(Q-1)(Q-p(2+\alpha))}{p}\int_{\mathbb{G}}\frac{|f|^p}{|x|^{p(2+\alpha)}}dx \\
&\quad + \frac{4(p-1)}{p^2}\int_{\mathbb{G}}\frac{(\mathcal{R}(|f|^{p/2}))^2}{|x|^{p(2+\alpha)-2}}dx + \int_{\mathbb{G}}\frac{|f|^{p-4}(\mathrm{Im}(f\overline{\mathcal{R}f}))^2}{|x|^{p(2+\alpha)-2}}dx. \qquad (3.28)
\end{aligned}
$$

By using Theorem 3.1.1 for $|f|^{\frac{p}{2}}$, we obtain

$$
\begin{aligned}
\int_{\mathbb{G}}\frac{(\mathcal{R}(|f|^{p/2}))^2}{|x|^{p(2+\alpha)-2}}dx &= \frac{(Q-p(2+\alpha))^2}{4}\int_{\mathbb{G}}\frac{|f|^p}{|x|^{p(2+\alpha)}}dx \\
&\quad + \int_{\mathbb{G}}\frac{1}{|x|^{p(2+\alpha)-2}}\left|\mathcal{R}|f|^{p/2} + \frac{Q-p(2+\alpha)}{2|x|}|f|^{p/2}\right|^2 dx.
\end{aligned}
$$

Plugging this in (3.28) we get

$$
\begin{aligned}
&\int_{\mathbb{G}}\frac{|f(x)|^p}{|x|^{p(2+\alpha)}}dx \\
&= -\frac{1}{C_{p,\alpha}}\mathrm{Re}\int_{\mathbb{G}}\frac{|f|^{p-2}f}{|x|^{(p-1)(2+\alpha)}}\frac{\overline{\tilde{\mathcal{R}}f}}{|x|^\alpha}dx - \frac{1}{C_{p,\alpha}}\int_{\mathbb{G}}\frac{|f|^{p-4}(\mathrm{Im}(f\overline{\mathcal{R}f}))^2}{|x|^{p(2+\alpha)-2}}dx \\
&\quad - \frac{1}{C_{p,\alpha}}\frac{4(p-1)}{p^2}\int_{\mathbb{G}}\frac{1}{|x|^{p(2+\alpha)-2}}\left|\mathcal{R}|f|^{\frac{p}{2}} + \frac{Q-p(2+\alpha)}{2|x|}|f|^{\frac{p}{2}}\right|^2 dx \\
&= \frac{p-1}{p}\int_{\mathbb{G}}\frac{|f|^p}{|x|^{p(2+\alpha)}}dx + \frac{1}{p}\frac{1}{|C_{p,\alpha}|^p}\int_{\mathbb{G}}\frac{|\tilde{\mathcal{R}}f|^p}{|x|^{p\alpha}}dx - \int_{\mathbb{G}}R_p\left(\frac{f}{|x|^{2+\alpha}}, -\frac{1}{C_{p,\alpha}}\frac{\tilde{\mathcal{R}}f}{|x|^\alpha}\right)dx \\
&\quad - \frac{4(p-1)}{p^2 C_{p,\alpha}}\int_{\mathbb{G}}\frac{1}{|x|^{p(2+\alpha)-2}}\left|\mathcal{R}|f|^{\frac{p}{2}} + \frac{Q-p(2+\alpha)}{2|x|}|f|^{\frac{p}{2}}\right|^2 dx \\
&\quad - \frac{1}{C_{p,\alpha}}\int_{\mathbb{G}}\frac{|f|^{p-4}(\mathrm{Im}(f\overline{\mathcal{R}f}))^2}{|x|^{p(2+\alpha)-2}}dx. \qquad (3.29)
\end{aligned}
$$

The equality (3.25) now follows from (3.29) and the equality

$$\mathcal{R}(|f|^{\frac{p}{2}}) = \frac{p}{2}|f|^{\frac{p}{2}-1}\mathcal{R}(|f|).$$

Part (2). Clearly, inequality (3.27) follows from equality (3.25). As in the proof of Theorem 3.1.4, the sharpness of (3.27) follows by considering appropriate approximations of the function $r^{-(Q-p(2+\alpha))/p}$. Moreover, if we have equality in (3.27) for some function f, then in view of Part (1) we must have

$$\mathcal{R}(|f|) + \frac{Q-p(2+\alpha)}{p|x|}|f| = 0.$$

This means that

$$\mathbb{E}(|f|) = -\frac{Q-p(2+\alpha)}{p}|f|.$$

By Proposition 1.3.1 the function $|f|$ must be positively homogeneous of degree $-(Q-p(2+\alpha))/p$. Since $|f|/|x|^{2+\alpha}$ is in $L^p(\mathbb{G})$ we must have $f=0$. □

3.1.3 Stability of Rellich type inequalities

The method used in the previous section also allows one to obtain the following stability property for Rellich type inequalities. We present such a result following [RS18].

Theorem 3.1.9 (Stability of Rellich type inequalities). *Let $\mathbb{G}$ be a homogeneous group of homogeneous dimension Q. Let $|\cdot|$ be a homogeneous quasi-norm on $\mathbb{G}$ and let $p \geq 1$. Let $k \geq 2$, $k \in \mathbb{N}$, be such that $kp < Q$. Then for all real-valued radial functions $u \in C_0^\infty(\mathbb{G})$ we have*

$$\begin{aligned}\int_{\mathbb{G}} \frac{|\tilde{\mathcal{R}}u|^p}{|x|^{(k-2)p}}dx - K_{k,p}^p \int_{\mathbb{G}} \frac{|u|^p}{|x|^{kp}}dx \\ \geq C \sup_{R>0} \int_{\mathbb{G}} \frac{\left||u(x)|^{\frac{p-2}{2}}u(x) - R^{\frac{Q-kp}{2}}|u(R)|^{\frac{p-2}{2}}u(R)|x|^{-\frac{Q-kp}{2}}\right|^2}{|x|^{kp}\left|\log\frac{R}{|x|}\right|^2}dx,\end{aligned} \tag{3.30}$$

where

$$\tilde{\mathcal{R}}f = \mathcal{R}^2 f + \frac{Q-1}{|x|}\mathcal{R}f$$

and

$$K_{k,p} = \frac{(Q-kp)[(k-2)p+(p-1)Q]}{p^2}. \tag{3.31}$$

Proof of Theorem 3.1.9. For $k \geq 2$, $k \in \mathbb{N}$, and $kp < Q$, as in the assumptions of the theorem, let us denote

$$v(r) := r^{\frac{Q-kp}{p}} u(r), \quad \text{where} \quad r \in [0, \infty). \tag{3.32}$$

In particular, we have $v(0) = 0$ and $v(\infty) = 0$. We then calculate as follows:

$$\begin{aligned}
-\tilde{\mathcal{R}}u = & -\mathcal{R}^2\left(r^{\frac{kp-Q}{p}}v(r)\right) - \frac{Q-1}{r}\mathcal{R}\left(r^{\frac{kp-Q}{p}}v(r)\right) \\
= & -\mathcal{R}\left(\frac{kp-Q}{p}r^{\frac{kp-Q}{p}-1}v(r) + r^{\frac{kp-Q}{p}}\mathcal{R}v(r)\right) \\
& - \frac{Q-1}{r}\frac{kp-Q}{p}r^{\frac{kp-Q}{p}-1}v(r) - \frac{Q-1}{r}r^{\frac{kp-Q}{p}}\mathcal{R}v(r) \\
= & -\frac{kp-Q}{p}\left(\frac{kp-Q}{p}-1\right)r^{\frac{kp-Q}{p}-2}v(r) - \frac{kp-Q}{p}r^{\frac{kp-Q}{p}-1}\mathcal{R}v(r) \\
& - \frac{kp-Q}{p}r^{\frac{kp-Q}{p}-1}\mathcal{R}v(r) - r^{\frac{kp-Q}{p}}\mathcal{R}^2v(r) \\
& - \frac{Q-1}{r}\frac{kp-Q}{p}r^{\frac{kp-Q}{p}-1}v(r) - \frac{Q-1}{r}r^{\frac{kp-Q}{p}}\mathcal{R}v(r) \\
= & -r^{\frac{kp-Q}{p}-2}\left(\frac{(kp-Q)(kp-Q-p)}{p^2} + \frac{(Q-1)(kp-Q)}{p}\right)v(r) \\
& - r^{\frac{kp-Q}{p}-2}r^2\left(\mathcal{R}^2v(r) + \frac{1}{r}\left(\frac{2(kp-Q)}{p} + (Q-1)\right)\mathcal{R}v(r)\right) \\
= & \, r^{k-2-\frac{Q}{p}}(K_{k,p}v(r) - r^2\tilde{\mathcal{R}}_k v(r)),
\end{aligned}$$

where $K_{k,p}$ is as in (3.31), and where we denote

$$\tilde{\mathcal{R}}_k f := \mathcal{R}^2 f + \frac{2k + \frac{Q(p-2)}{p} - 1}{r}\mathcal{R}f.$$

By using the first inequality in Lemma 2.4.2 with $a = K_{k,p}v(r)$ and $b = r^2\tilde{\mathcal{R}}_k v(r)$, and the fact that $\int_0^\infty |v|^{p-2}vv'dr = 0$ in view of $v(0) = 0$ and $v(\infty) = 0$, we obtain

$$\begin{aligned}
J := & \int_{\mathbb{G}} \frac{|\tilde{\mathcal{R}}u|^p}{|x|^{(k-2)p}}dx - K_{k,p}^p \int_{\mathbb{G}} \frac{|u|^p}{|x|^{kp}}dx \\
= & |\wp| \int_0^\infty |-\tilde{\mathcal{R}}u(r)|^p r^{Q-1-(k-2)p}dr - K_{k,p}^p|\wp| \int_0^\infty |u(r)|^p r^{Q-kp-1}dr \\
= & |\wp| \int_0^\infty \left(|K_{k,p}v(r) - r^2\tilde{\mathcal{R}}_k v(r)|^p - (K_{k,p}v(r))^p\right) r^{-1}dr \\
\geq & -p|\wp|K_{k,p}^{p-1} \int_0^\infty |v|^{p-2}v\tilde{\mathcal{R}}_k v r dr
\end{aligned}$$

$$= -p|\wp|K_{k,p}^{p-1}\int_0^\infty |v|^{p-2}v\left(v'' + \frac{2k+\frac{Q(p-2)}{p}-1}{r}v'\right)rdr$$

$$= -p|\wp|K_{k,p}^{p-1}\int_0^\infty |v|^{p-2}vv''rdr.$$

On the other hand, we have

$$-\int_0^\infty |v|^{p-2}vv''rdr = (p-1)\int_0^\infty |v|^{p-2}(v')^2rdr + \int_0^\infty |v|^{p-2}vv'dr$$

$$= (p-1)\int_0^\infty |v|^{p-2}(v')^2rdr$$

$$= \frac{4(p-1)}{p^2}\int_0^\infty \left(\frac{p-2}{2}\right)^2 |v|^{p-2}(v')^2dr$$

$$+ \frac{4(p-1)}{p^2}\int_0^\infty (p-2)|v|^{p-2}(v')^2 + |v|^{p-2}(v')^2rdr$$

$$= \frac{4(p-1)}{p^2}\int_0^\infty \left(\left(|v|^{\frac{p-2}{2}}\right)' v + |v|^{\frac{p-2}{2}}v'\right)^2 rdr$$

$$= \frac{4(p-1)}{p^2}\int_0^\infty |(|v|^{\frac{p-2}{2}}v)'|^2rdr$$

$$= \frac{4(p-1)}{|\wp_2|p^2}\int_{\mathbb{G}_2} \left|\mathcal{R}(|v|^{\frac{p-2}{2}}v)\right|^2 dx,$$

where $\mathbb{G}_2$ is a homogeneous group of homogeneous degree 2 and $|\wp_2|$ is the measure of the corresponding unit 2-quasi-ball. By using Remark 2.2.9 for $|v|^{\frac{p-2}{2}}v \in C_0^\infty(\mathbb{G}_2\backslash\{0\})$ in $p = Q = 2$ case, and combining the above equalities, we obtain

$$J \geq C_1\int_{\mathbb{G}_2} \frac{\left||v(x)|^{\frac{p-2}{2}}v(x) - |v(R\frac{x}{|x|})|^{\frac{p-2}{2}}v(R\frac{x}{|x|})\right|^2}{|x|^2\left|\log\frac{R}{|x|}\right|^2}dx$$

$$= C_1\int_0^\infty \frac{\left||v(r)|^{\frac{p-2}{2}}v(r) - |v(R)|^{\frac{p-2}{2}}v(R)\right|^2}{r\left|\log\frac{R}{r}\right|^2}dr$$

$$= C_1\int_0^\infty \frac{\left||u(r)|^{\frac{p-2}{2}}u(r) - R^{\frac{Q-kp}{2}}|u(R)|^{\frac{p-2}{2}}u(R)r^{-\frac{Q-kp}{2}}\right|^2}{r^{1-Q+kp}\left|\log\frac{R}{r}\right|^2}dr$$

for any $R > 0$. That is, we have

$$J \geq C\sup_{R>0}\int_{\mathbb{G}} \frac{\left||u(x)|^{\frac{p-2}{2}}u(x) - R^{\frac{Q-kp}{2}}|u(R)|^{\frac{p-2}{2}}u(R)|x|^{-\frac{Q-kp}{2}}\right|^2}{|x|^{kp}\left|\log\frac{R}{|x|}\right|^2}dx.$$

The proof is complete. □

3.1.4 Higher-order Hardy–Rellich inequalities

In this section we show that by iterating the already established weighted Hardy inequalities we get inequalities of higher order. An interesting feature is that we also obtain the exact formula for the remainder which yields the sharpness of the constants as well. That is, when $p = 2$, one can iterate the exact representation formulae of the remainder that we obtained in Theorem 3.1.1. It implies higher-order remainder equalities that can be then also used to argue the sharpness of the constant.

Theorem 3.1.10 (Higher-order Hardy–Rellich identities and inequalities). *Let $\mathbb{G}$ be a homogeneous group of homogeneous dimension $Q \geq 3$. Let $\alpha \in \mathbb{R}$ and $k \in \mathbb{N}$ be such that*

$$\prod_{j=0}^{k-1} \left| \frac{Q-2}{2} - (\alpha + j) \right| \neq 0.$$

Then for all complex-valued functions $f \in C_0^\infty(\mathbb{G} \backslash \{0\})$ we have

$$\left\| \frac{f}{|x|^{k+\alpha}} \right\|_{L^2(\mathbb{G})} \leq \left[\prod_{j=0}^{k-1} \left| \frac{Q-2}{2} - (\alpha + j) \right| \right]^{-1} \left\| \frac{1}{|x|^\alpha} \mathcal{R}^k f \right\|_{L^2(\mathbb{G})}, \tag{3.33}$$

where the constant above is sharp, and is attained if and only if $f = 0$.

Moreover, for all $k \in \mathbb{N}$ and $\alpha \in \mathbb{R}$, the following identity holds:

$$\begin{aligned} \left\| \frac{1}{|x|^\alpha} \mathcal{R}^k f \right\|_{L^2(\mathbb{G})}^2 &= \left[\prod_{j=0}^{k-1} \left(\frac{Q-2}{2} - (\alpha + j) \right)^2 \right] \left\| \frac{f}{|x|^{k+\alpha}} \right\|_{L^2(\mathbb{G})}^2 \\ &+ \sum_{l=1}^{k-1} \left[\prod_{j=0}^{l-1} \left(\frac{Q-2}{2} - (\alpha + j) \right)^2 \right] \\ &\times \left\| \frac{1}{|x|^{l+\alpha}} \mathcal{R}^{k-l} f + \frac{Q - 2(l+1+\alpha)}{2|x|^{l+1+\alpha}} \mathcal{R}^{k-l-1} f \right\|_{L^2(\mathbb{G})}^2 \\ &+ \left\| \frac{1}{|x|^\alpha} \mathcal{R}^k f + \frac{Q-2-2\alpha}{2|x|^{1+\alpha}} \mathcal{R}^{k-1} f \right\|_{L^2(\mathbb{G})}^2. \end{aligned} \tag{3.34}$$

Remark 3.1.11. Let us point out some special cases of inequality (3.33).

1. For $k = 1$ inequality (3.33) gives the weighted L^2-Hardy inequalities from Corollary 2.1.6.
2. In particular, for $k = 1$ and $\alpha = 0$, inequality (3.33) gives the L^2-Hardy inequality, i.e., (2.2) in the case of $p = 2$.
3. For $k = 2$, inequality (3.33) can be thought of as a (weighted) Hardy–Rellich type inequality, while for larger k this corresponds to higher-order (weighted) Rellich inequalities.

Proof of Theorem 3.1.10. For any $\alpha \in \mathbb{R}$ and $Q \geq 3$ let us iterate the identity

$$\begin{aligned} \left\| \frac{1}{|x|^{\alpha}} \mathcal{R} f \right\|^2_{L^2(\mathbb{G})} &= \left(\frac{Q-2}{2} - \alpha \right)^2 \left\| \frac{f}{|x|^{\alpha+1}} \right\|^2_{L^2(\mathbb{G})} \\ &\quad + \left\| \frac{1}{|x|^{\alpha}} \mathcal{R} f + \frac{Q-2(\alpha+1)}{2|x|^{\alpha+1}} f \right\|^2_{L^2(\mathbb{G})}, \end{aligned} \tag{3.35}$$

given in (2.16), as follows. First, replacing f in (3.35) by $\mathcal{R}f$ we have

$$\begin{aligned} \left\| \frac{1}{|x|^{\alpha}} \mathcal{R}^2 f \right\|^2_{L^2(\mathbb{G})} &= \left(\frac{Q-2}{2} - \alpha \right)^2 \left\| \frac{\mathcal{R} f}{|x|^{\alpha+1}} \right\|^2_{L^2(\mathbb{G})} \\ &\quad + \left\| \frac{1}{|x|^{\alpha}} \mathcal{R}^2 f + \frac{Q-2(\alpha+1)}{2|x|^{\alpha+1}} \mathcal{R} f \right\|^2_{L^2(\mathbb{G})}. \end{aligned} \tag{3.36}$$

Furthermore, replacing α by $\alpha+1$, (3.35) implies that

$$\begin{aligned} \left\| \frac{1}{|x|^{\alpha+1}} \mathcal{R} f \right\|^2_{L^2(\mathbb{G})} &= \left(\frac{Q-2}{2} - (\alpha+1) \right)^2 \left\| \frac{f}{|x|^{\alpha+2}} \right\|^2_{L^2(\mathbb{G})} \\ &\quad + \left\| \frac{1}{|x|^{\alpha+1}} \mathcal{R} f + \frac{Q-2(\alpha+2)}{2|x|^{\alpha+2}} f \right\|^2_{L^2(\mathbb{G})}. \end{aligned}$$

Combination of this with (3.36) gives

$$\begin{aligned} \left\| \frac{1}{|x|^{\alpha}} \mathcal{R}^2 f \right\|^2_{L^2(\mathbb{G})} &= \left(\frac{Q-2}{2} - \alpha \right)^2 \left(\frac{Q-2}{2} - (\alpha+1) \right)^2 \left\| \frac{f}{|x|^{\alpha+2}} \right\|^2_{L^2(\mathbb{G})} \\ &\quad + \left(\frac{Q-2}{2} - \alpha \right)^2 \left\| \frac{1}{|x|^{\alpha+1}} \mathcal{R} f + \frac{Q-2(\alpha+2)}{2|x|^{\alpha+2}} f \right\|^2_{L^2(\mathbb{G})} \\ &\quad + \left\| \frac{1}{|x|^{\alpha}} \mathcal{R}^2 f + \frac{Q-2-2\alpha}{2|x|^{\alpha+1}} \mathcal{R} f \right\|^2_{L^2(\mathbb{G})}. \end{aligned}$$

By using this iteration process further, we eventually arrive at the family of identities

$$\begin{aligned} \left\| \frac{1}{|x|^{\alpha}} \mathcal{R}^k f \right\|^2_{L^2(\mathbb{G})} &= \left[\prod_{j=0}^{k-1} \left(\frac{Q-2}{2} - (\alpha+j) \right)^2 \right] \left\| \frac{f}{|x|^{k+\alpha}} \right\|^2_{L^2(\mathbb{G})} \\ &\quad + \sum_{l=1}^{k-1} \left[\prod_{j=0}^{l-1} \left(\frac{Q-2}{2} - (\alpha+j) \right)^2 \right] \\ &\quad \times \left\| \frac{1}{|x|^{l+\alpha}} \mathcal{R}^{k-l} f + \frac{Q-2(l+1+\alpha)}{2|x|^{l+1+\alpha}} \mathcal{R}^{k-l-1} f \right\|^2_{L^2(\mathbb{G})} \end{aligned}$$

$$+\left\|\frac{1}{|x|^{\alpha}}\mathcal{R}^{k}f+\frac{Q-2-2\alpha}{2|x|^{1+\alpha}}\mathcal{R}^{k-1}f\right\|^{2}_{L^{2}(\mathbb{G})},\quad k=1,2,\ldots,$$

which give (3.34).

Now, by dropping positive terms, these identities imply that

$$\left\|\frac{1}{|x|^{\alpha}}\mathcal{R}^{k}f\right\|^{2}_{L^{2}(\mathbb{G})}\geq C_{k,Q}\left\|\frac{f}{|x|^{k+\alpha}}\right\|^{2}_{L^{2}(\mathbb{G})},$$

where

$$C_{k,Q}=\prod_{j=0}^{k-1}\left(\frac{Q-2}{2}-(\alpha+j)\right)^{2}.$$

If $C_{k,Q}\neq 0$, this can be written as

$$\left\|\frac{f}{|x|^{k+\alpha}}\right\|^{2}_{L^{2}(\mathbb{G})}\leq\frac{1}{C_{k,Q}}\left\|\frac{1}{|x|^{\alpha}}\mathcal{R}^{k}f\right\|^{2}_{L^{2}(\mathbb{G})},$$

which gives (3.33).

Now it remains to show the sharpness of the constant and the equality in (3.33). To do this, we rewrite the equality

$$\frac{\mathcal{R}^{k-l}f}{|x|^{l+\alpha}}+\frac{Q-2(l+1+\alpha)}{2|x|^{l+1+\alpha}}\mathcal{R}^{k-l-1}f=0$$

as

$$|x|\mathcal{R}(\mathcal{R}^{k-l-1}f)+\frac{Q-2(l+1+\alpha)}{2}(\mathcal{R}^{k-l-1}f)=0,$$

and by Proposition 1.3.1, Part (i), this means that $\mathcal{R}^{k-l-1}f$ is positively homogeneous of degree $-\frac{Q}{2}+l+1+\alpha$. So, all the remainder terms vanish if f is positively homogeneous of degree $k-\frac{Q}{2}+\alpha$. As this can be approximated by functions in $C_0^{\infty}(\mathbb{G}\backslash\{0\})$, the constant $C_{k,Q}$ is sharp.

If this constant was attained, it would be on functions f which are positively homogeneous of degree $k-\frac{Q}{2}+\alpha$, in which case $\frac{f}{|x|^{k+\alpha}}$ would be positively homogeneous of degree $-\frac{Q}{2}$. These are in L^2 if and only if they are zero. □

Theorem 3.1.4 and Theorem 3.1.10 were proved in [RS17b]. They can be extended further to derivatives of higher order. Such results have been recently obtained in [Ngu17] and in the rest of this subsection we describe such extensions.

Theorem 3.1.12 (Further higher-order Hardy–Rellich identities and inequalities). *Let $\mathbb{G}$ be a homogeneous group of homogeneous dimension Q and let $|\cdot|$ be a homogeneous quasi-norm on $\mathbb{G}$. We denote*

$$\tilde{\mathcal{R}}f:=\mathcal{R}^{2}f+\frac{Q-1}{|x|}\mathcal{R}f,\quad C_{\beta}:=\frac{(Q+2\beta)(Q-4-2\beta)}{4},\tag{3.37}$$

for $\beta \in \mathbb{R}$, for the constants appearing below. Let $k \in \mathbb{N}$ be a positive integer and let $\alpha \in \mathbb{R}$. Then for all complex-valued functions $f \in C_0^\infty(\mathbb{G}\backslash\{0\})$ we have the following identities:

(1) *If $Q \geq 4k+1$, then we have*

$$
\begin{aligned}
&\left(\prod_{i=0}^{k-1} C_{2i+\alpha}\right)^2 \left\|\frac{f}{|x|^{2k+\alpha}}\right\|^2_{L^2(\mathbb{G})} \\
&= \left\|\frac{\tilde{\mathcal{R}}^k f}{|x|^\alpha}\right\|^2_{L^2(\mathbb{G})} - \left\|\frac{1}{|x|^\alpha}\left|\tilde{\mathcal{R}}^k f + C_\alpha \frac{\tilde{\mathcal{R}}^{k-1} f}{|x|^2}\right|\right\|^2_{L^2(\mathbb{G})} \\
&\quad - \sum_{j=1}^{k-1}\left(\prod_{i=0}^{k-1} C_{2i+\alpha}\right)^2 \left\|\frac{1}{|x|^{2j+\alpha}}\left|\tilde{\mathcal{R}}^{k-j} f + C_{2j+\alpha}\frac{\tilde{\mathcal{R}}^{k-j-1} f}{|x|^2}\right|\right\|^2_{L^2(\mathbb{G})} \\
&\quad - 2C_\alpha \left\|\frac{1}{|x|^{1+\alpha}}\left|\mathcal{R}(\tilde{\mathcal{R}}^{k-1} f) + \frac{Q-4-2\alpha}{2|x|}\tilde{\mathcal{R}}^{k-1} f\right|\right\|^2_{L^2(\mathbb{G})} \\
&\quad - 2\sum_{j=1}^{k-1}\left(\prod_{i=0}^{k-1} C_{2i+\alpha}\right)^2 C_{2j+\alpha} \qquad (3.38)\\
&\qquad \times \left\|\frac{1}{|x|^{1+\alpha+2j}}\left|\mathcal{R}(\tilde{\mathcal{R}}^{k-j-1} f) + \frac{Q-4-2\alpha-4j}{2|x|}\tilde{\mathcal{R}}^{k-j-1} f\right|\right\|^2_{L^2(\mathbb{G})}.
\end{aligned}
$$

(2) *If $Q \geq 4k+3$, then we have*

$$
\begin{aligned}
&\left(\frac{Q-2-2\alpha}{2}\prod_{i=0}^{k-1} C_{2i+1+\alpha}\right)^2 \left\|\frac{f}{|x|^{2k+1+\alpha}}\right\|^2_{L^2(\mathbb{G})} \\
&= \left\|\frac{\mathcal{R}(\tilde{\mathcal{R}}^k f)}{|x|^\alpha}\right\|^2_{L^2(\mathbb{G})} - \left\|\frac{1}{|x|^\alpha}\left|\mathcal{R}(\tilde{\mathcal{R}}^k f) + \frac{Q-2-2\alpha}{2|x|}\tilde{\mathcal{R}}^k f\right|\right\|^2_{L^2(\mathbb{G})} \\
&\quad - \frac{(Q-2-2\alpha)^2}{4}\left\|\frac{1}{|x|^\alpha}\left|\tilde{\mathcal{R}}^k f + C_{1+\alpha}\frac{\tilde{\mathcal{R}}^{k-1} f}{|x|^2}\right|\right\|^2_{L^2(\mathbb{G})} \\
&\quad - \frac{(Q-2-2\alpha)^2}{4}\sum_{j=1}^{k-1}\left(\prod_{i=0}^{k-1} C_{2i+\alpha}\right)^2 \\
&\qquad \times \left\|\frac{1}{|x|^{1+2j+\alpha}}\left|\tilde{\mathcal{R}}^{k-j} f + C_{2j+1+\alpha}\frac{\tilde{\mathcal{R}}^{k-j-1} f}{|x|^2}\right|\right\|^2_{L^2(\mathbb{G})} \\
&\quad - \frac{(Q-2-2\alpha)^2}{4} C_{\alpha+1}\left\|\frac{1}{|x|^{2+\alpha}}\left|\mathcal{R}(\tilde{\mathcal{R}}^{k-1} f) + \frac{Q-6-2\alpha}{2|x|}\tilde{\mathcal{R}}^{k-1} f\right|\right\|^2_{L^2(\mathbb{G})}
\end{aligned}
$$

$$-\frac{(Q-2-2\alpha)^2}{4}\sum_{j=1}^{k-1}\left(\prod_{i=0}^{k-1}C_{2i+1+\alpha}\right)^2 C_{2j+1+\alpha} \tag{3.39}$$
$$\times\left\|\frac{1}{|x|^{2+\alpha+2j}}\left|\mathcal{R}(\tilde{\mathcal{R}}^{k-j-1}f)+\frac{Q-6-2\alpha-4j}{2|x|}\tilde{\mathcal{R}}^{k-j-1}f\right|\right\|^2_{L^2(\mathbb{G})}.$$

As consequences, for all complex-valued function $f \in C_0^\infty(\mathbb{G}\backslash\{0\})$ we have the following estimates:

(3) *If $Q \geq 4k+1$ and $\alpha \in (-Q/2,(Q-4k)/2)$, then we have*

$$\left(\prod_{i=0}^{k-1}C_{2i+\alpha}\right)\left\|\frac{f}{|x|^{2k+\alpha}}\right\|_{L^2(\mathbb{G})} \leq \left\|\frac{\tilde{\mathcal{R}}^k f}{|x|^\alpha}\right\|_{L^2(\mathbb{G})}. \tag{3.40}$$

(4) *If $Q \geq 4k+3$ and $\alpha \in (-(Q+2)/2,(Q-4k-2)/2)$, then we have*

$$\left(\frac{Q-2-2\alpha}{2}\prod_{i=0}^{k-1}C_{2i+1+\alpha}\right)\left\|\frac{f}{|x|^{2k+1+\alpha}}\right\|_{L^2(\mathbb{G})} \leq \left\|\frac{\mathcal{R}(\tilde{\mathcal{R}}^k f)}{|x|^\alpha}\right\|_{L^2(\mathbb{G})}. \tag{3.41}$$

Moreover, these inequalities in Parts (3) *and* (4) *are sharp and the equalities hold if only if $f=0$.*

Proof of Theorem 3.1.12. The equality (3.38) follows by induction using Theorem 3.1.4, which gives the case of $k=1$. Furthermore, we have

$$\int_{\mathbb{G}}\frac{|\mathcal{R}(\tilde{\mathcal{R}}^k f)|^2}{|x|^{2\alpha}}dx = \frac{(Q-2-2\alpha)^2}{4}\int_{\mathbb{G}}\frac{|\tilde{\mathcal{R}}^k f|^2}{|x|^{2+2\alpha}}dx$$
$$+\int_{\mathbb{G}}\frac{1}{|x|^{2\alpha}}\left|\mathcal{R}(\tilde{\mathcal{R}}^k f)+\frac{Q-2-2\alpha}{2|x|}\tilde{\mathcal{R}}^k f\right|^2 dx. \tag{3.42}$$

Thus, (3.39) follows from (3.38) and (3.42). The inequalities (3.40) and (3.41) follow directly from these equalities since $C_{\alpha+2i}>0, C_{\alpha+2i+1}>0$ for any $i=0,\ldots,k-1$ corresponding to each case. To check the sharpness of constants, we use the arguments similar to those in the proof of Theorem 3.1.4, considering approximations of the function $r^{-(Q-4k-2\alpha)/2}$. Indeed, one has

$$\int_{\mathbb{G}}\frac{|f_\epsilon|^2}{|x|^{4k+2\alpha}}dx = (-\ln\epsilon)|\wp|+O(1),$$

and

$$\int_{\mathbb{G}}\frac{|\tilde{\mathcal{R}}^k f|^2}{|x|^{2\alpha}}dx = \left(\prod_{i=0}^{k-1}C_{2i+\alpha}\right)^2(-\ln\epsilon)|\wp|+O(1).$$

This proves the sharpness of (3.40). For (3.41), we use the approximation of the function $r^{-(Q-2-4k-2\alpha)/2}$ and similar calculations.

Finally, suppose that for some function f we have an equality in (3.40). Then by (3.38), we must have

$$\mathcal{R}f + \frac{Q-4k-2\alpha}{2|x|} f = 0.$$

This is equivalent to

$$\mathbb{E}f = -\frac{Q-4k-2\alpha}{2} f.$$

By Proposition 1.3.1, Part (i), it follows that f is positively homogeneous of order $-(Q-2\alpha-4k)/2$. Since $f/|x|^{\alpha+2k} \in L^2(\mathbb{G})$, this forces to have $f=0$. The sharpness of (3.41) is proved in a similar way. □

Theorem 3.1.12 can be used to derive further weighted Rellich type inequalities. First we start with one iteration, continuing using the notation $\tilde{\mathcal{R}}$ as in (3.37).

Theorem 3.1.13 (Iterated weighted Rellich inequality). *Let $\mathbb{G}$ be a homogeneous group of homogeneous dimension $Q \geq 5$, and let $|\cdot|$ be a homogeneous quasi-norm on $\mathbb{G}$. Then for any $\alpha \in \mathbb{R}$ and for all complex-valued functions $f \in C_0^\infty(\mathbb{G}\backslash\{0\})$ we have the identity*

$$\begin{aligned}\left\|\frac{\tilde{\mathcal{R}}f}{|x|^\alpha}\right\|^2_{L^2(\mathbb{G})} &= \frac{(Q+2\alpha)^2}{4}\left\|\frac{\mathcal{R}f}{|x|^{1+\alpha}}\right\|^2_{L^2(\mathbb{G})} \\ &\quad + \left\|\frac{1}{|x|^\alpha}\left|\mathcal{R}^2 f + \frac{Q-2-2\alpha}{2|x|}\mathcal{R}f\right|\right\|^2_{L^2(\mathbb{G})}.\end{aligned} \tag{3.43}$$

As a consequence, for any $\alpha \in \mathbb{R}$ we get the inequality

$$\frac{|Q+2\alpha|}{2}\left\|\frac{\mathcal{R}f}{|x|^{1+\alpha}}\right\|_{L^2(\mathbb{G})} \leq \left\|\frac{\tilde{\mathcal{R}}f}{|x|^\alpha}\right\|_{L^2(\mathbb{G})}, \tag{3.44}$$

for all complex-valued functions $f \in C_0^\infty(\mathbb{G}\backslash\{0\})$. For $Q+2\alpha \neq 0$ the inequality (3.44) is sharp and the equality holds if and only if $f=0$.

Proof of Theorem 3.1.13. By definition (3.37) we have

$$\int_{\mathbb{G}} \frac{|\tilde{\mathcal{R}}f|^2}{|x|^{2\alpha}}dx = \int_{\mathbb{G}} \frac{|\mathcal{R}(\mathcal{R}f)|^2}{|x|^{2\alpha}}dx + 2(Q-1)\mathrm{Re}\int_{\mathbb{G}} \frac{\mathcal{R}(\mathcal{R}f)\overline{\mathcal{R}f}}{|x|^{2\alpha+1}}dx + (Q-1)^2\int_{\mathbb{G}} \frac{|\mathcal{R}f|^2}{|x|^{2+2\alpha}}dx.$$

Moreover, we have

$$2\mathrm{Re}\int_{\mathbb{G}} \frac{\mathcal{R}(\mathcal{R}f)\overline{\mathcal{R}f}}{|x|^{2\alpha+1}}dx = -(Q-2-2\alpha)\int_{\mathbb{G}} \frac{|\mathcal{R}f|^2}{|x|^{2+2\alpha}}dx.$$

On the other hand, by Theorem 3.1.1 we have the identity

$$\int_{\mathbb{G}} \frac{|\mathcal{R}(\mathcal{R}f)|^2}{|x|^{2\alpha}}dx = \frac{(Q-2-2\alpha)^2}{4}\int_{\mathbb{G}} \frac{|\mathcal{R}f|^2}{|x|^{2+2\alpha}}dx + \int_{\mathbb{G}} \frac{1}{|x|^{2\alpha}}\left|\mathcal{R}^2 f + \frac{Q-2-2\alpha}{2|x|}\mathcal{R}f\right|^2 dx.$$

Combining these inequalities we obtain (3.43).

Identity (3.43) then implies inequality (3.44). The sharpness of inequality (3.44) can be verified by using approximations of the function $r^{-(Q-2\alpha-4k)/2}$. If the equality in (3.44) holds for some function f, then we must have

$$\mathcal{R}^2 f + \frac{Q-2-2\alpha}{2|x|}\mathcal{R}f = 0.$$

This is equivalent to

$$\mathbb{E}(\mathcal{R}f) = -\frac{Q-2-2\alpha}{2}\mathcal{R}f$$

which implies, by Proposition 1.3.1, Part (i), that $\mathcal{R}f$ is positively homogeneous of degree $-(Q-2\alpha-2k)/2$. Since $\mathcal{R}f/|x|^{1+\alpha}$ is in $L^2(\mathbb{G})$, this implies that $\mathcal{R}f=0$. Consequently, we must also have $f=0$. □

Theorem 3.1.13 implies further identities and inequalities.

Theorem 3.1.14 (Further higher-order Rellich type identities and inequalities). *Let $\mathbb{G}$ be a homogeneous group of homogeneous dimension Q with a homogeneous quasi-norm $|\cdot|$. Let $k,l\in\mathbb{N}$ be such that $Q\geq 4k+1$ and $k\geq l+1$. Then for all complex-valued functions $f\in C_0^\infty(\mathbb{G}\backslash\{0\})$ we have*

$$\left\|\frac{\tilde{\mathcal{R}}^k f}{|x|^{\alpha}}\right\|^2_{L^2(\mathbb{G})} = \frac{4}{(Q-2\alpha)^2}\left(\prod_{i=0}^{k-l-1}\frac{Q^2-4(2i+\alpha)^2}{4}\right)^2\left\|\frac{\mathcal{R}\tilde{\mathcal{R}}^l f}{|x|^{2(k-l)-1+\alpha}}\right\|^2_{L^2(\mathbb{G})}$$
$$+\left(\prod_{i=0}^{k-l-2}C_{2i+\alpha}\right)^2\left\|\frac{\left|\mathcal{R}^2\tilde{\mathcal{R}}^l f + \frac{Q+2-4(k-l)-2\alpha}{2|x|}\mathcal{R}\tilde{\mathcal{R}}^l f\right|}{|x|^{2(k-l-1)+\alpha}}\right\|^2_{L^2(\mathbb{G})}$$
$$+\left\|\frac{\left|\tilde{\mathcal{R}}^k f + C_\alpha\frac{\tilde{\mathcal{R}}^{k-1}f}{|x|^2}\right|}{|x|^{\alpha}}\right\|^2_{L^2(\mathbb{G})} + \sum_{j=1}^{k-l-2}\left(\prod_{i=0}^{j-1}C_{2i+\alpha}\right)^2\left\|\frac{\left|\tilde{\mathcal{R}}^{k-j}f + C_{2j+\alpha}\frac{\tilde{\mathcal{R}}^{k-j-1}f}{|x|^2}\right|}{|x|^{2j+\alpha}}\right\|^2_{L^2(\mathbb{G})}$$
$$+2C_\alpha\left\|\frac{\left|\mathcal{R}(\tilde{\mathcal{R}}^{k-1}f) + \frac{Q-4-2\alpha}{2|x|}\tilde{\mathcal{R}}^{k-1}f\right|}{|x|^{1+\alpha}}\right\|^2_{L^2(\mathbb{G})}$$

$$+2\sum_{j=1}^{k-l-2}\left(\prod_{i=0}^{j-1}C_{2i+\alpha}\right)^{2}C_{2j+\alpha}\left\|\frac{\left|\mathcal{R}(\tilde{\mathcal{R}}^{k-j-1}f)+\frac{Q-4-2\alpha-4j}{2|x|}\tilde{\mathcal{R}}^{k-j-1}f\right|}{|x|^{1+2j+\alpha}}\right\|^{2}_{L^{2}(\mathbb{G})} \tag{3.45}$$

and

$$\left\|\frac{\mathcal{R}\tilde{\mathcal{R}}^{k}f}{|x|^{\alpha}}\right\|^{2}_{L^{2}(\mathbb{G})}=\left(\prod_{i=0}^{k-l-1}\frac{Q^{2}-4(1+2i+\alpha)^{2}}{4}\right)^{2}\left\|\frac{\mathcal{R}\tilde{\mathcal{R}}^{l}f}{|x|^{2(k-l)+\alpha}}\right\|^{2}_{L^{2}(\mathbb{G})}$$
$$+\frac{(Q-2-2\alpha)^{2}}{4}\left(\prod_{i=0}^{k-l-2}C_{2i+1+\alpha}\right)^{2}\left\|\frac{\left|\mathcal{R}^{2}\tilde{\mathcal{R}}^{l}f+\frac{Q+2-4(k-l)-2\alpha}{2|x|}\mathcal{R}\tilde{\mathcal{R}}^{l}f\right|}{|x|^{2(k-l-1)+1+\alpha}}\right\|^{2}_{L^{2}(\mathbb{G})}$$
$$+\frac{(Q-2-2\alpha)^{2}}{4}\left\|\frac{\left|\tilde{\mathcal{R}}^{k}f+C_{\alpha}\frac{\tilde{\mathcal{R}}^{k-1}f}{|x|^{2}}\right|}{|x|^{\alpha}}\right\|^{2}_{L^{2}(\mathbb{G})}$$
$$+\frac{(Q-2-2\alpha)^{2}}{4}\sum_{j=1}^{k-l-2}\left(\prod_{i=0}^{j-1}C_{2i+1+\alpha}\right)^{2}\left\|\frac{\left|\tilde{\mathcal{R}}^{k-j}f+C_{2j+1+\alpha}\frac{\tilde{\mathcal{R}}^{k-j-1}f}{|x|^{2}}\right|}{|x|^{2j+\alpha+1}}\right\|^{2}_{L^{2}(\mathbb{G})}$$
$$+2C_{1+\alpha}\frac{(Q-2-2\alpha)^{2}}{4}\left\|\frac{\left|\mathcal{R}(\tilde{\mathcal{R}}^{k-1}f)+\frac{Q-6-2\alpha}{2|x|}\tilde{\mathcal{R}}^{k-1}f\right|}{|x|^{2+\alpha}}\right\|^{2}_{L^{2}(\mathbb{G})}$$
$$+2\frac{(Q-2-2\alpha)^{2}}{4}\sum_{j=1}^{k-l-2}\left(\prod_{i=0}^{j-1}C_{2i+1+\alpha}\right)^{2}C_{2j+1+\alpha}$$
$$\times\left\|\frac{\left|\mathcal{R}(\tilde{\mathcal{R}}^{k-j-1}f)+\frac{Q-6-2\alpha-4j}{2|x|}\tilde{\mathcal{R}}^{k-j-1}f\right|}{|x|^{1+2j+\alpha}}\right\|^{2}_{L^{2}(\mathbb{G})}$$
$$+\left\|\frac{\left|\mathcal{R}\tilde{\mathcal{R}}^{k}f+\frac{Q-2-2\alpha}{2}\frac{\tilde{\mathcal{R}}^{k}f}{|x|}\right|}{|x|^{\alpha}}\right\|^{2}_{L^{2}(\mathbb{G})}. \tag{3.46}$$

As consequences, we have the following weighted Rellich type inequalities for all complex-valued functions $f\in C_{0}^{\infty}(\mathbb{G}\backslash\{0\})$*:*

(1) *For any* $\alpha\in(-Q/2,(Q-4(k-l-1))/2)$ *if* $k\geq l+2$ *and* $\alpha\in\mathbb{R}$ *if* $k=l+1$*, we have*

$$\frac{4}{(Q-2\alpha)^{2}}\left(\prod_{i=0}^{k-l-1}\frac{Q^{2}-4(2i+\alpha)^{2}}{4}\right)^{2}\left\|\frac{\mathcal{R}\tilde{\mathcal{R}}^{l}f}{|x|^{2(k-l)-1+\alpha}}\right\|^{2}_{L^{2}(\mathbb{G})}\leq\left\|\frac{\tilde{\mathcal{R}}^{k}f}{|x|^{\alpha}}\right\|^{2}_{L^{2}(\mathbb{G})}. \tag{3.47}$$

(2) *For any* $\alpha \in (-(Q+2)/2, (Q-4(k-l)+2)/2)$ *if* $k \geq l+2$ *and* $\alpha \in \mathbb{R}$ *if* $k = l+1$, *we have*

$$\left(\prod_{i=0}^{k-l-1} \frac{Q^2-4(1+2i+\alpha)^2}{4}\right)^2 \left\| \frac{\mathcal{R}\tilde{\mathcal{R}}^l f}{|x|^{2(k-l)+\alpha}} \right\|^2_{L^2(\mathbb{G})} \leq \left\| \frac{\mathcal{R}\tilde{\mathcal{R}}^k f}{|x|^{\alpha}} \right\|^2_{L^2(\mathbb{G})}. \tag{3.48}$$

Moreover, inequalities (3.47) *and* (3.48) *are sharp and there is an equality in each of them if and only if* $f = 0$.

Proof of Theorem 3.1.14. Let $g := \tilde{\mathcal{R}}^{k+1} f$. Applying (3.38) to $\tilde{\mathcal{R}}^{k-l-1} g$ and then Theorem 3.1.13 to $\tilde{\mathcal{R}}\tilde{\mathcal{R}}^l f$, one obtains equality (3.45). Note that

$$\frac{Q+4(k-l-1)+2\alpha}{2} \prod_{i=0}^{k-l-2} C_{2i+\alpha} = \frac{2}{Q-2\alpha} \prod_{i=0}^{k-l-1} \frac{Q^2-4(2i+\alpha)^2}{4}.$$

By applying Theorem 3.1.1 to $\mathcal{R}\tilde{\mathcal{R}}^k f$, and then using (3.45) for $\tilde{\mathcal{R}}^k f$ with weights $|x|^{2(1+\alpha)}$, together with the equality

$$\frac{Q-2-2\alpha}{2} \frac{(Q+4(k-l-1)+2+2\alpha)^2}{4} \prod_{i=0}^{k-l-2} C_{2i+\alpha} = \prod_{i=0}^{k-l-1} \frac{Q^2-4(1+\alpha+2i)^2}{4},$$

we also obtain (3.46).

Consequently, inequalities (3.47) and (3.48) follow from respective identities (3.45) and (3.46) by dropping non-negative remainder terms on the right-hand side. The sharpness of (3.47) and (3.48) follows by considering, respectively, as in the previous theorems, approximations of the function $r^{-(Q-2\alpha-4k)/2}$ and of the function $r^{-(Q-2\alpha-2-4k)/2}$.

If for some function f we have the equality in (3.47), then (3.45) implies that we have

$$\mathcal{R}(\tilde{\mathcal{R}}^{k-1} f) + \frac{Q-4-2\alpha}{2|x|} \tilde{\mathcal{R}}^{k-1} f = 0.$$

This means that

$$\mathbb{E}(\tilde{\mathcal{R}}^{k-1} f) = -\frac{Q-4-2\alpha}{2} \tilde{\mathcal{R}}^{k-1} f.$$

By Proposition 1.3.1, Part (i), it follows that $\tilde{\mathcal{R}}^{k-1} f$ is positively homogeneous of order $-(Q-4-2\alpha)/2$. Since $\tilde{\mathcal{R}}^{k-1} f/|x|^{2\alpha} \in L^2(\mathbb{G})$, we then must have $\tilde{\mathcal{R}}^{k-1} f = 0$, which in turn implies $f = 0$. The same arguments also work for (3.48). $\square$

Using the established theorems, as a corollary one gets the following weighted L^p-Hardy–Rellich type identities which will, in turn, imply the corresponding inequalities.

Theorem 3.1.15 (Higher-order weighted L^p-Hardy–Rellich type identities). *Let $\mathbb{G}$ be a homogeneous group of homogeneous dimension Q, with a homogeneous quasi-norm $|\cdot|$ on $\mathbb{G}$. Let $k \in \mathbb{N}$ be a positive integer, let $1 < p < Q/k$, and let $\alpha \in \mathbb{R}$. Then for all complex-valued functions $f \in C_0^\infty(\mathbb{G}\backslash\{0\})$ we have the identities:*

(1) *If $k = 2l$, $l \geq 2$, then*

$$\begin{aligned}
&\left\| \frac{\tilde{\mathcal{R}}^l f}{|x|^\alpha} \right\|_{L^p(\mathbb{G})}^p \\
&= \left| \prod_{i=0}^{l-1} C_{p,2i+\alpha} \right|^p \int_{\mathbb{G}} \frac{|f|^p}{|x|^{p(k+\alpha)}} dx + p \int_{\mathbb{G}} \frac{1}{|x|^{p\alpha}} R_p \left(C_{p,\alpha} \frac{\tilde{\mathcal{R}}^{l-1} f}{|x|^2}, -\tilde{\mathcal{R}}^l f \right) dx \\
&+ p \sum_{j=1}^{l-1} \left| \prod_{i=0}^{j-1} C_{p,2i+\alpha} \right|^p \int_{\mathbb{G}} \frac{1}{|x|^{p(2j+\alpha)}} R_p \left(C_{p,2j+\alpha} \frac{\tilde{\mathcal{R}}^{l-j-1} f}{|x|^2}, -\tilde{\mathcal{R}}^{l-j} f \right) dx \\
&+ B_{p,\alpha}(p-1) \int_{\mathbb{G}} \frac{|\tilde{\mathcal{R}}^{l-1} f|^{p-2}}{|x|^{p(2j+\alpha)-2}} \left| \mathcal{R}|\tilde{\mathcal{R}}^{l-1} f| + \frac{Q - p(2+\alpha)}{p|x|} |\tilde{\mathcal{R}}^{l-1} f| \right|^2 dx \\
&+ B_{p,\alpha} \int_{\mathbb{G}} \frac{|\tilde{\mathcal{R}}^{l-1} f|^{p-4} (\mathrm{Im}(\tilde{\mathcal{R}}^{l-1} f \overline{\mathcal{R}\tilde{\mathcal{R}}^{l-1} f}))^2}{|x|^{p(2+\alpha)-2}} dx \qquad (3.49) \\
&+ \sum_{j=1}^{l-1} \left| \prod_{i=0}^{j-1} C_{p,2i+\alpha} \right|^p B_{p,\alpha+2j} \int_{\mathbb{G}} \frac{|\tilde{\mathcal{R}}^{l-j-1} f|^{p-4} (\mathrm{Im}(\tilde{\mathcal{R}}^{l-j-1} f \overline{\mathcal{R}\tilde{\mathcal{R}}^{l-j-1} f}))^2}{|x|^{p(2(j+1)+\alpha)-2}} dx \\
&+ (p-1) \int_{\mathbb{G}} \frac{|\tilde{\mathcal{R}}^{l-j-1} f|^{p-2}}{|x|^{p(2(j+1)+\alpha)-2}} \left| \mathcal{R}|\tilde{\mathcal{R}}^{l-j-1} f| + \frac{Q - p(2(j+1)+\alpha)}{p|x|} |\tilde{\mathcal{R}}^{l-j-1} f| \right|^2 dx,
\end{aligned}$$

where R_p is as in (2.22), *i.e.,*

$$R_p(\xi, \eta) := \frac{1}{p}|\eta|^p + \frac{p-1}{p}|\xi|^p - \mathrm{Re}(|\xi|^{p-2}\xi\overline{\eta}),$$

as well as

$$B_{p,\alpha+2j} = p|C_{p,\alpha+2j}|^{p-2} C_{p,\alpha+2j},$$

and

$$C_{p,\alpha} = \frac{(Q-2p-p\alpha)(Q+p'\alpha)}{pp'}, \quad p' = \frac{p}{p-1}.$$

(2) *If $k = 2l+1$, $l \geq 1$, then*

$$\begin{aligned}
\left\| \frac{\mathcal{R}(\tilde{\mathcal{R}}^l f)}{|x|^\alpha} \right\|_{L^p(\mathbb{G})}^p &= \left| \frac{Q - p(1+\alpha)}{p} \right|^p \left| \prod_{i=0}^{l-1} C_{p,2i+1+\alpha} \right|^p \int_{\mathbb{G}} \frac{|f|^p}{|x|^{p(k+\alpha)}} dx \\
&+ p \int_{\mathbb{G}} \frac{1}{|x|^{p\alpha}} R_p \left(-\frac{Q-p(1+\alpha)}{p} \frac{\tilde{\mathcal{R}}^l f}{|x|}, \mathcal{R}\tilde{\mathcal{R}}^l f \right) dx \\
&+ A_{p,\alpha} \int_{\mathbb{G}} \frac{1}{|x|^{p(1+\alpha)}} R_p \left(C_{p,1+\alpha} \frac{\tilde{\mathcal{R}}^{l-1} f}{|x|^2}, -\tilde{\mathcal{R}}^l f \right) dx
\end{aligned}$$

$$
\begin{aligned}
&+ A_{p,\alpha} \sum_{j=1}^{l-1} \left| \prod_{i=0}^{j-1} C_{p,2i+1+\alpha} \right|^p \int_{\mathbb{G}} \frac{1}{|x|^{p(2j+1+\alpha)}} R_p \left(C_{p,2j+1+\alpha} \frac{\tilde{\mathcal{R}}^{l-j-1} f}{|x|^2}, -\tilde{\mathcal{R}}^{l-j} f \right) dx \\
&+ A_{p,\alpha} \frac{B_{p,\alpha+1}}{p} \left[(p-1) \int_{\mathbb{G}} \frac{|\tilde{\mathcal{R}}^{l-1} f|^{p-2}}{|x|^{p(3+\alpha)-2}} \left| \mathcal{R} |\tilde{\mathcal{R}}^{l-1} f| + \frac{Q-p(3+\alpha)}{p|x|} |\tilde{\mathcal{R}}^{l-1} f| \right|^2 dx \right] \\
&+ A_{p,\alpha} \frac{B_{p,\alpha+1}}{p} \left[\int_{\mathbb{G}} \frac{|\tilde{\mathcal{R}}^{l-1} f|^{p-4} (\mathrm{Im}(\tilde{\mathcal{R}}^{l-1} f \overline{\mathcal{R} \tilde{\mathcal{R}}^{l-1} f}))^2}{|x|^{p(3+\alpha)-2}} dx \right] \qquad (3.50) \\
&+ A_{p,\alpha} \sum_{j=1}^{l-1} \left| \prod_{i=0}^{j-1} C_{p,2i+1+\alpha} \right|^p \frac{B_{p,\alpha+2j+1}}{p} \left[\int_{\mathbb{G}} \frac{|\tilde{\mathcal{R}}^{l-j-1} f|^{p-4} (\mathrm{Im}(\tilde{\mathcal{R}}^{l-j-1} f \overline{\mathcal{R} \tilde{\mathcal{R}}^{l-j-1} f}))^2}{|x|^{p(2j+3+\alpha)-2}} dx \right. \\
&\left. + (p-1) \int_{\mathbb{G}} \frac{|\tilde{\mathcal{R}}^{l-j-1} f|^{p-2}}{|x|^{p(2j+3+\alpha)-2}} \left| \mathcal{R} |\tilde{\mathcal{R}}^{l-j-1} f| + \frac{Q-p(2j+3+\alpha)}{p|x|} |\tilde{\mathcal{R}}^{l-j-1} f| \right|^2 dx \right],
\end{aligned}
$$

where $A_{p,\alpha} = p \left| \frac{Q-p(1+\alpha)}{p} \right|^p$.

Proof of Theorem 3.1.15. The equality (3.49) follows from (3.25). The equality (3.50) is consequence of (3.49) and (3.32). Note that in (3.50), if $k = 2l+1, l \geq 1$, then the terms concerning the sum from 1 to $l-1$ do not appear if $l = 1$. □

By dropping the non-negative remainder terms in (3.49) and (3.50), we obtain the following higher-order weighted L^p-Hardy–Rellich type inequalities.

Corollary 3.1.16 (Higher-order weighted L^p-Hardy–Rellich type inequalities). *Let $\mathbb{G}$ be a homogeneous group of homogeneous dimension Q and let $|\cdot|$ be a homogeneous quasi-norm on $\mathbb{G}$. Then for any $\alpha \in \mathbb{R}$ and all complex-valued functions $f \in C_0^\infty(\mathbb{G}\backslash\{0\})$ we have:*

(1) *if $1 < p < Q/2k$ and $\alpha \in (-Q(p-1))/p, (Q-2pk)/p)$, then*

$$
\left(\prod_{i=0}^{k-1} C_{p,2i+\alpha} \right) \left\| \frac{f}{|x|^{(2k+\alpha)}} \right\|_{L^p(\mathbb{G})} \leq \left\| \frac{\tilde{\mathcal{R}}^k f}{|x|^\alpha} \right\|_{L^p(\mathbb{G})}. \qquad (3.51)
$$

(2) *if $1 < p < Q/(2k+1)$ and $\alpha \in (-(Q+p')/p', (Q-p(2k+1))/p)$, then*

$$
\frac{Q-p(1+\alpha)}{p} \left(\prod_{i=0}^{k-1} C_{p,2i+1+\alpha} \right) \left\| \frac{f}{|x|^{(2k+1+\alpha)}} \right\|_{L^p(\mathbb{G})} \leq \left\| \frac{\mathcal{R}\tilde{\mathcal{R}}^k f}{|x|^\alpha} \right\|_{L^p(\mathbb{G})}. \qquad (3.52)
$$

In the above inequalities

$$
C_{p,\alpha} = \frac{(Q-2p-p\alpha)(Q+p'\alpha)}{pp'}, \quad p' = \frac{p}{p-1}.
$$

Inequalities (3.51) and (3.52) are sharp and equalities hold in each of them if and only if $f = 0$.

Proof of Corollary 3.1.16. Inequalities (3.51) and (3.52) are immediate consequences of equalities (3.49) and (3.50), respectively, since $C_{p,2i+\alpha+1} \geq 0$ for $0 \leq i \leq k-1$ under the respective assumptions on indices. The sharpness of constants in (3.51) and (3.52) can be checked by approximating the function $r^{-(Q-p(2k+\alpha))/p}$ and the function $r^{-(Q-p(2k+1+\alpha))/p}$, respectively, by smooth compactly supported functions. Moreover, if for some function f an equality holds in any of these inequalities, then it follows from Theorem 3.1.15 that $|f|$ is positively homogeneous of degree $-(Q-p(2k+\alpha))/p$ and $-(Q-p(2k+1+\alpha))/p$, respectively, which implies that $f=0$ in view of the condition of L^p integrability. ☐

As usual, Hardy inequalities imply the corresponding uncertainty principles.

Corollary 3.1.17 (Higher-order Hardy–Rellich uncertainty principles). *Let $\mathbb{G}$ be a homogeneous group of homogeneous dimension Q and let $|\cdot|$ be a homogeneous quasi-norm on $\mathbb{G}$. Let $k \in \mathbb{N}$ be a positive integer and let $p>1$. Then for all complex-valued functions $f \in C_0^\infty(\mathbb{G}\backslash\{0\})$ we have:*

(1) *if $k=2l$, $l \geq 1$, and $\alpha \in (-Q(p-1)/p),(Q-2pl)/p)$, then*

$$\left(\prod_{i=0}^{l-1} C_{p,2i+\alpha}\right) \int_{\mathbb{G}} |f|^2 dx \leq \left(\int_{\mathbb{G}} \frac{|\tilde{\mathcal{R}}^l f|^p}{|x|^{p\alpha}} dx\right)^{1/p} \left(\int_{\mathbb{G}} |f|^{p'} |x|^{p'(2l+\alpha)} dx\right)^{1/p'}. \tag{3.53}$$

(2) *if $k=2l+1$, $l \geq 0$, and $\alpha \in (-(Q+p')/p',(Q-p(2l+1))/p)$ if $l \geq 1$ and $\alpha \in \mathbb{R}$ if $l=0$, then*

$$\begin{aligned} &\left|\frac{\frac{Q-p(1+\alpha)}{p}}{C_{p,2l+1+\alpha}} \prod_{i=0}^{l} C_{p,2i+1+\alpha}\right| \int_{\mathbb{G}} |f|^2 dx \\ &\leq \left(\int_{\mathbb{G}} \frac{|\mathcal{R}\tilde{\mathcal{R}}^l f|^p}{|x|^{p\alpha}} dx\right)^{1/p} \left(\int_{\mathbb{G}} |f|^{p'} |x|^{p'(2l+1+\alpha)} dx\right)^{1/p'}. \end{aligned} \tag{3.54}$$

Proof of Corollary 3.1.17. By the Hölder inequality, we have

$$\int_{\mathbb{G}} |f|^2 dx = \int_{\mathbb{G}} \frac{|f|}{|x|^{2l+\alpha}} |f||x|^{2l+\alpha} dx \leq \left(\int_{\mathbb{G}} \frac{|f|^p}{|x|^{p(2l+\alpha)}} dx\right)^{\frac{1}{p}} \left(\int_{\mathbb{G}} |f|^{p'} |x|^{p'(2l+\alpha)} dx\right)^{\frac{1}{p'}}.$$

Further, applying inequality (3.51) to this, we get (3.53) with $C_{2i+\alpha}>0$ for $0 \leq i \leq l-1$ and $\alpha \in (-Q(p-1)/p,(Q-2pl)/p)$. Inequality (3.54) is proved in the same way. ☐

Combining Corollary 3.28 and Theorem 2.24, we obtain the weighted L^p-Rellich type inequality below which is an L^p-analogue of Theorem 3.1.14.

Theorem 3.1.18 (Higher-order L^p-weighted Rellich identities and inequalities). *Let $\mathbb{G}$ be a homogeneous group of homogeneous dimension Q and let $|\cdot|$ be any* *homogeneous quasi-norm on* $\mathbb{G}$. *Let* $k, l \in \mathbb{N}$ *be non-negative integers such that* $k \geq l+1$ *and let* $p > 1$. *Then for any* $\alpha \in \mathbb{R}$ *and for all complex-valued functions* $f \in C_0^\infty(\mathbb{G}\backslash\{0\})$ *we have*

$$\begin{aligned}
&\left\|\frac{\tilde{\mathcal{R}}^k f}{|x|^\alpha}\right\|^p_{L^p(\mathbb{G})} \\
&= \frac{p^p}{|Q-p\alpha|^p}\left|\prod_{i=0}^{k-l-1}\frac{(Q-p(2i+\alpha))(Q+p'(2i+\alpha))}{pp'}\right|^p \int_{\mathbb{G}} \frac{|\mathcal{R}\tilde{\mathcal{R}}^l f|^p}{|x|^{p(2(k-l)-1+\alpha)}}dx \\
&\quad + p\left|\prod_{i=0}^{k-l-2} C_{p,2i+\alpha}\right|^p \int_{\mathbb{G}} \frac{R_p\left(\frac{Q+p'(2(k-l-1)+\alpha)}{p'}\frac{\mathcal{R}\tilde{\mathcal{R}}^l f}{|x|}, \tilde{\mathcal{R}}^{l+1}f\right)}{|x|^{p(2(k-l-1)+\alpha)}}dx \\
&\quad + p\int_{\mathbb{G}} \frac{R_p\left(-C_{p,\alpha}\frac{\tilde{\mathcal{R}}^{k-1}f}{|x|^2}, \tilde{\mathcal{R}}^k f\right)}{|x|^{p\alpha}}dx \\
&\quad + p\sum_{j=1}^{k-l-2}\left|\prod_{i=0}^{j-1} C_{p,2i+\alpha}\right|^p \int_{\mathbb{G}} \frac{R_p\left(-C_{p,2j+\alpha}\frac{\tilde{\mathcal{R}}^{k-j-1}f}{|x|^2}, \tilde{\mathcal{R}}^{k-j}f\right)}{|x|^{p(2j+\alpha)}}dx \\
&\quad + B_{p,\alpha}(p-1)\int_{\mathbb{G}} \frac{|\tilde{\mathcal{R}}^{k-1}f|^{p-2}}{|x|^{p(2+\alpha)-2}}\left|\mathcal{R}|\tilde{\mathcal{R}}^{k-1}f| + \frac{Q-p(2+\alpha)}{p|x|}|\tilde{\mathcal{R}}^{k-1}f|\right|^2 dx \\
&\quad + B_{p,\alpha}\int_{\mathbb{G}} \frac{|\tilde{\mathcal{R}}^{k-1}f|^{p-4}(\mathrm{Im}(\tilde{\mathcal{R}}^{l-1}f\overline{\mathcal{R}\tilde{\mathcal{R}}^{k-1}f}))^2}{|x|^{p(2+\alpha)-2}}dx \\
&\quad + \sum_{j=1}^{k-l-2}\left|\prod_{i=0}^{j-1} C_{p,2i+\alpha}\right|^p B_{p,\alpha+2j}\int_{\mathbb{G}} \frac{|\tilde{\mathcal{R}}^{k-j-1}f|^{p-4}(\mathrm{Im}(\tilde{\mathcal{R}}^{k-j-1}f\overline{\mathcal{R}\tilde{\mathcal{R}}^{k-j-1}f}))^2}{|x|^{p(2(j+1)+\alpha)-2}}dx \\
&\quad + (p-1)\int_{\mathbb{G}} \frac{|\tilde{\mathcal{R}}^{k-j-1}f|^{p-2}}{|x|^{p(2(j+1)+\alpha)-2}}\Big|\mathcal{R}|\tilde{\mathcal{R}}^{k-j-1}f| \\
&\quad + \frac{Q-p(2(j+1)+\alpha)}{p|x|}|\tilde{\mathcal{R}}^{k-j-1}f|\Big|^2 dx,
\end{aligned} \tag{3.55}$$

where R_p *is as in* (2.22):

$$R_p(\xi,\eta) := \frac{1}{p}|\eta|^p + \frac{p-1}{p}|\xi|^p - \mathrm{Re}(|\xi|^{p-2}\xi\overline{\eta}),$$

$$B_{p,\alpha+2j} = p|C_{p,\alpha+2j}|^{p-2}C_{p,\alpha+2j},$$

and

$$C_{p,\alpha} = \frac{(Q-2p-p\alpha)(Q+p'\alpha)}{pp'}, \quad p' = \frac{p}{p-1}.$$

We also have

$$
\begin{aligned}
&\left\|\frac{\mathcal{R}\tilde{\mathcal{R}}^k f}{|x|^\alpha}\right\|^p_{L^p(\mathbb{G})} \\
&= \left|\prod_{i=0}^{k-l-1}\frac{(Q-p(2i+1+\alpha))(Q+p'(2i+1+\alpha))}{pp'}\right|^p \int_{\mathbb{G}} \frac{|\mathcal{R}\tilde{\mathcal{R}}^l f|^p}{|x|^{p(2(k-l)-1+\alpha)}}dx \\
&\quad + A_{p,\alpha}\left|\prod_{i=0}^{k-l-2} C_{p,2i+1+\alpha}\right|^p \int_{\mathbb{G}} \frac{R_p\left(\frac{Q+p'(2(k-l)-1+\alpha)}{p'}\frac{\mathcal{R}\tilde{\mathcal{R}}^l f}{|x|}, \tilde{\mathcal{R}}^{l+1}f\right)}{|x|^{p(2(k-l)-1+\alpha)}}dx \\
&\quad + A_{p,\alpha}\int_{\mathbb{G}} \frac{R_p\left(-C_{p,1+\alpha}\frac{\tilde{\mathcal{R}}^{k-j-1}f}{|x|^2}, \tilde{\mathcal{R}}^{k-j}f\right)}{|x|^{p(1+\alpha)}}dx \\
&\quad + A_{p,\alpha}\sum_{j=1}^{k-l-2}\left|\prod_{i=0}^{j-1} C_{p,2i+1+\alpha}\right|^p \int_{\mathbb{G}} \frac{R_p\left(-C_{p,2j+1+\alpha}\frac{\tilde{\mathcal{R}}^{k-j-1}f}{|x|^2}, \tilde{\mathcal{R}}^{k-j}f\right)}{|x|^{p(2j+1+\alpha)}}dx \\
&\quad + A_{p,\alpha}|\frac{B_{p,\alpha+1}}{p}(p-1)\int_{\mathbb{G}} \frac{|\tilde{\mathcal{R}}^{k-1}f|^{p-2}}{|x|^{p(3+\alpha)-2}}\left|\mathcal{R}|\tilde{\mathcal{R}}^{k-1}f| + \frac{Q-p(3+\alpha)}{p|x|}|\tilde{\mathcal{R}}^{k-1}f|\right|^2 dx \\
&\quad + A_{p,\alpha}\frac{B_{p,\alpha+1}}{p}\int_{\mathbb{G}} \frac{|\tilde{\mathcal{R}}^{k-1}f|^{p-4}(\operatorname{Im}(\tilde{\mathcal{R}}^{k-1}f\overline{\mathcal{R}\tilde{\mathcal{R}}^{k-1}f}))^2}{|x|^{p(3+\alpha)-2}}dx \\
&\quad + A_{p,\alpha}\sum_{j=1}^{k-l-2}\left|\prod_{i=0}^{j-1} C_{p,2i+1+\alpha}\right|^p \frac{B_{p,2j+\alpha+1}}{p} \\
&\quad \times \int_{\mathbb{G}} \frac{|\tilde{\mathcal{R}}^{k-j-1}f|^{p-4}(\operatorname{Im}(\tilde{\mathcal{R}}^{k-j-1}f\overline{\mathcal{R}\tilde{\mathcal{R}}^{k-j-1}f}))^2}{|x|^{p(2(j+1)+1+\alpha)-2}}dx \\
&\quad + (p-1)\int_{\mathbb{G}} \frac{|\tilde{\mathcal{R}}^{k-j-1}f|^{p-2}}{|x|^{p(2(j+1)+1+\alpha)-2}} \\
&\quad \times \left|\mathcal{R}|\tilde{\mathcal{R}}^{k-j-1}f| + \frac{Q-p(2(j+1)+\alpha)}{p|x|}|\tilde{\mathcal{R}}^{k-j-1}f|\right|^2 dx \\
&\quad + p\int_{\mathbb{G}} \frac{1}{|x|^{p\alpha}}R_p\left(-\frac{Q-p-p\alpha}{p}\frac{\tilde{\mathcal{R}}^k f}{|x|}, \mathcal{R}\tilde{\mathcal{R}}^k f\right)dx, \qquad (3.56)
\end{aligned}
$$

where

$$A_{p,\alpha} = p\left|\frac{Q-p(1+\alpha)}{p}\right|^p.$$

Consequently, we obtain the following inequalities for all functions $f \in C_0^\infty(\mathbb{G}\backslash\{0\})$: for any $\alpha \in (-Q(p-1)/p, (Q-2p(k-l-1))/2)$ if $k-l \geq 2$ and for any $\alpha \in \mathbb{R}$

if $k = l+1$, we have

$$\frac{p}{|Q-p\alpha|}\left|\prod_{i=0}^{k-l-1}\frac{(Q-p(2i+\alpha))(Q+p'(2i+\alpha))}{pp'}\right|\left\|\frac{\mathcal{R}\tilde{\mathcal{R}}^{l}f}{|x|^{(2(k-l)-1+\alpha)}}\right\|_{L^p(\mathbb{G})} \leq \left\|\frac{\tilde{\mathcal{R}}^{k}f}{|x|^{\alpha}}\right\|_{L^p(\mathbb{G})}; \tag{3.57}$$

and for any $\alpha \in (-(Q+p')/p', (Q-p(2(k-l-1)+1))/p)$ if $k-l \geq 2$ and for any $\alpha \in \mathbb{R}$ if $k = l+1$, we have

$$\left|\prod_{i=0}^{k-l-1}\frac{(Q-p(2i+1+\alpha))(Q+p'(2i+1+\alpha))}{pp'}\right|\left\|\frac{\mathcal{R}\tilde{\mathcal{R}}^{l}f}{|x|^{(2(k-l)-1+\alpha)}}\right\|_{L^p(\mathbb{G})} \leq \left\|\frac{\mathcal{R}\tilde{\mathcal{R}}^{k}f}{|x|^{\alpha}}\right\|_{L^p(\mathbb{G})}. \tag{3.58}$$

Moreover, these inequalities (3.57) and (3.58) are sharp and equality in any of them holds if and only if $f = 0$.

Now let us present an extension of the critical Hardy inequality to the higher-order derivatives. To do this, still following [Ngu17] until the end of this section, we present a critical Rellich inequality for $\tilde{\mathcal{R}}$ as follows.

Theorem 3.1.19 (Critical Rellich identity and inequality for $\tilde{\mathcal{R}}$). *Let $\mathbb{G}$ be a homogeneous group of homogeneous dimension $Q \geq 3$ and let $|\cdot|$ be a homogeneous quasi-norm on $\mathbb{G}$. Let $f \in C_0^{\infty}(\mathbb{G}\backslash\{0\})$ be any complex-valued function and denote*

$$f_R(x) := f(Rx/|x|)$$

for any $x \in \mathbb{G}$ and $R > 0$. Then we have

$$\begin{aligned}\left\|\frac{\tilde{\mathcal{R}}f}{|x|^{\frac{Q}{p}-2}}\right\|^p_{L^p(\mathbb{G})} &= \left(\frac{(p-1)(Q-2)}{p}\right)^p \int_{\mathbb{G}} \frac{|f-f_R|^p}{|x|^Q\left|\ln\frac{R}{|x|}\right|^p}dx \\ &+p\int_{\mathbb{G}}\frac{1}{|x|^{Q-2p}}R_p\left((Q-2)\frac{\mathcal{R}f}{|x|}, \tilde{\mathcal{R}}f\right)dx \\ &+p(Q-2)^p\int_{\mathbb{G}}\frac{1}{|x|^{Q-p}}R_p\left(-\frac{p-1}{p}\frac{f-f_R}{|x|\ln\frac{R}{|x|}}, \mathcal{R}f\right)dx,\end{aligned} \tag{3.59}$$

for any $1 < p < \infty$ and any $R > 0$. Here R_p is as in (2.22). As a consequence, for all complex-valued functions $f \in C_0^{\infty}(\mathbb{G}\backslash\{0\})$ we have

$$\frac{(p-1)(Q-2)}{p}\sup_{R>0}\left\|\frac{f-f_R}{|x|^{\frac{Q}{p}}\left|\ln\frac{R}{|x|}\right|}\right\|_{L^p(\mathbb{G})} \leq \left\|\frac{\tilde{\mathcal{R}}f}{|x|^{\frac{Q}{p}-2}}\right\|_{L^p(\mathbb{G})}, \quad 1<p<\infty, \tag{3.60}$$

with the constant sharp $(p-1)(Q-2)/p$.

Proof of Theorem 3.1.19. Let $1 < p < \infty$. Let us first restate equality (2.48) in the form

$$\begin{aligned}\int_{\mathbb{G}} \frac{|\mathcal{R}f|^p}{|x|^{Q-p}}dx =& \left(\frac{p-1}{p}\right)^p \int_{\mathbb{G}} \frac{|f-f_R|^p}{|x|^Q|\ln\frac{R}{|x|}|^p}dx \\ &+p\int_{\mathbb{G}} \frac{1}{|x|^{Q-p}}R_p\left(-\frac{p-1}{p}\frac{f-f_R}{|x|\ln\frac{R}{|x|}}, \mathcal{R}f\right)dx,\end{aligned} \tag{3.61}$$

for any $R > 0$. It follows from Theorem 3.1.7 for $\alpha = (Q-2p)/p$ that

$$\begin{aligned}\int_{\mathbb{G}} \frac{|\tilde{\mathcal{R}}f|^p}{|x|^{Q-2p}}dx =& (Q-2)^p \int_{\mathbb{G}} \frac{|\mathcal{R}f|^p}{|x|^{Q-p}}dx \\ &+p\int_{\mathbb{G}} \frac{1}{|x|^{Q-2p}}R_p\left((Q-2)\frac{\mathcal{R}f}{|x|}, \tilde{\mathcal{R}}f\right)dx.\end{aligned} \tag{3.62}$$

The combination of (3.61) and (3.62) gives (3.59), which in turn implies (3.60).

Let us now check the sharpness of (3.60). For small enough $\epsilon, \delta > 0$ and for $R > 2$, let us define the function

$$f_\delta(x) := \left(\ln\frac{R}{|x|}\right)^{1-\frac{1}{p}-\delta} g(|x|),$$

where g is the function as in the proof of Theorem 3.1.4. A straightforward calculation gives

$$\mathcal{R}f_\delta(r) = -\left(1-\frac{1}{p}-\delta\right)\frac{1}{r}\left(\ln\frac{R}{r}\right)^{-\frac{1}{p}-\delta} g(r) + \left(\ln\frac{R}{r}\right)^{1-\frac{1}{p}-\delta} g'(r)$$

and

$$\begin{aligned}\mathcal{R}^2 f_\delta(r) =& -2\left(1-\frac{1}{p}-\delta\right)\frac{1}{r}\left(\ln\frac{R}{r}\right)^{-\frac{1}{p}-\delta} g'(r) \\ &-\left(1-\frac{1}{p}-\delta\right)\left(\frac{1}{p}+\delta\right)\frac{1}{r^2}\left(\ln\frac{R}{r}\right)^{-1-\frac{1}{p}-\delta} g(r) \\ &+\left(1-\frac{1}{p}-\delta\right)\frac{1}{r^2}\left(\ln\frac{R}{r}\right)^{-\frac{1}{p}-\delta} g(r) + \left(\ln\frac{R}{r}\right)^{1-\frac{1}{p}-\delta} g''(r).\end{aligned}$$

Therefore, we have

$$\begin{aligned}\tilde{\mathcal{R}}f_\delta(r) =& -2\left(1-\frac{1}{p}-\delta\right)\frac{1}{r}\left(\ln\frac{R}{r}\right)^{-\frac{1}{p}-\delta} g'(r) \\ &-(Q-2)\left(1-\frac{1}{p}-\delta\right)\frac{1}{r^2}\left(\ln\frac{R}{r}\right)^{-\frac{1}{p}-\delta} g(r) \\ &-\left(1-\frac{1}{p}-\delta\right)\left(\frac{1}{p}+\delta\right)\frac{1}{r^2}\left(\ln\frac{R}{r}\right)^{-1-\frac{1}{p}-\delta} g(r) + \left(\ln\frac{R}{r}\right)^{1-\frac{1}{p}-\delta} \tilde{\mathcal{R}}g(r).\end{aligned}$$

Since $(f_\delta)_R = 0$ we obtain

$$\int_{\mathbb{G}} \frac{|f_\delta - (f_\delta)_R|^p}{|x|^Q |\ln \frac{R}{|x|}|^p} dx = |\wp| \int_0^2 \frac{1}{r}(\ln R - \ln r)^{-1-\delta p} g(r)^p dr$$
$$\geq |\wp| \int_0^1 \frac{1}{r}(\ln R - \ln r)^{-1-\delta p} dr$$
$$= \frac{1}{\delta p}(\ln R)^{p\sigma} |\wp|.$$

Thus,

$$\lim_{\delta \to 0} \int_{\mathbb{G}} \frac{|f_\delta - (f_\delta)_R|^p}{|x|^Q |\ln \frac{R}{|x|}|^p} dx = \infty.$$

At the same time, direct calculations also give

$$\int_{\mathbb{G}} \frac{1}{|x|^{Q-2p}} \left| \frac{1}{|x|} \left(\ln \frac{R}{|x|}\right)^{-\frac{1}{p}-\delta} g'(|x|) \right|^p dx = O(1),$$

$$\int_{\mathbb{G}} \frac{1}{|x|^{Q-2p}} \left| \frac{1}{|x|^2} \left(\ln \frac{R}{|x|}\right)^{-1-\frac{1}{p}-\delta} g(|x|) \right|^p dx = O(1),$$

$$\int_{\mathbb{G}} \frac{1}{|x|^{Q-2p}} \left| \left(\ln \frac{R}{|x|}\right)^{1-\frac{1}{p}-\delta} \tilde{\mathcal{R}} g(|x|) \right|^p dx = O(1),$$

and

$$\int_{\mathbb{G}} \frac{1}{|x|^{Q-2p}} \left| \frac{1}{|x|} \left(\ln \frac{R}{|x|}\right)^{-\frac{1}{p}-\delta} g'(|x|) \right|^p dx = |\wp| \int_0^2 \frac{1}{r}(\ln R - \ln r)^{-1-\delta p} g(r)^p dr$$
$$= \int_{\mathbb{G}} \frac{|f_\delta - (f_\delta)_R|^p}{|x|^Q |\ln \frac{R}{|x|}|^p} dx$$

Consequently, we obtain

$$\lim_{\delta \to 0} \frac{\int_{\mathbb{G}} \frac{|\tilde{\mathcal{R}} f_\delta|^p}{|x|^{Q-2p}} dx}{\int_{\mathbb{G}} \frac{|f_\delta - (f_\delta)_R|^p}{|x|^Q |\ln \frac{R}{|x|}|^p} dx} = \left(\frac{(p-1)(Q-2)}{p} \right)^p.$$

This proves the sharpness of (3.60). □

Consequently, the following identities hold true.

Theorem 3.1.20 (Higher-order critical Rellich identities for $\tilde{\mathcal{R}}$). *Let $\mathbb{G}$ be a homogeneous group of homogeneous dimension $Q \geq 3$ and let $|\cdot|$ be a homogeneous quasi-norm on $\mathbb{G}$. Let $f \in C_0^\infty(\mathbb{G}\backslash\{0\})$ be any complex-valued function and denote*

$$f_R(x) := f(Rx/|x|)$$

for $x \in \mathbb{G}$ and $R > 0$.

Then for $2 \leq k < Q/2$ we have

$$\left\| \frac{\tilde{\mathcal{R}}^k f}{|x|^{\frac{Q}{p}-2k}} \right\|_{L^p(\mathbb{G})}^p = \left(\frac{Q-2}{p'}\right)^p \left(\prod_{i=1}^{k-1} a_{i,Q}\right)^p \int_{\mathbb{G}} \frac{|f-f_R|^p}{|x|^Q \left|\ln \frac{R}{|x|}\right|^p} dx \tag{3.63}$$

$$+ p \left(\prod_{i=1}^{k-1} a_{i,Q}\right)^p \int_{\mathbb{G}} \frac{1}{|x|^{Q-2p}} R_p\left((Q-2)\frac{\mathcal{R}f}{|x|}, \tilde{\mathcal{R}}f\right) dx$$

$$+ p(Q-2)^p \left(\prod_{i=1}^{k-1} a_{i,Q}\right)^p \int_{\mathbb{G}} \frac{1}{|x|^{Q-p}} R_p\left(-\frac{p-1}{p}\frac{f-f_R}{|x|\ln\frac{R}{|x|}}, \mathcal{R}f\right) dx$$

$$+ p \int_{\mathbb{G}} \frac{1}{|x|^{Q-2kp}} R_p\left(-a_{k-1,Q}\frac{\tilde{\mathcal{R}}^{k-1}f}{|x|^2}, \tilde{\mathcal{R}}^k f\right) dx$$

$$+ p \sum_{j=1}^{k-2} \left(\prod_{i=1}^{k-1} a_{i,Q}\right)^p \int_{\mathbb{G}} \frac{1}{|x|^{Q-2(k-j)p}} R_p\left(-a_{k-j-1,Q}\frac{\tilde{\mathcal{R}}^{k-j-1}f}{|x|^2}, \tilde{\mathcal{R}}^{k-j} f\right) dx$$

$$+ p a_{k-1,Q}^{p-1}(p-1) \int_{\mathbb{G}} \frac{|\tilde{\mathcal{R}}^{k-1}f|^{p-2}}{|x|^{Q-2(k-1)p-2}} \left|\mathcal{R}|\tilde{\mathcal{R}}^{k-1}f| + \frac{2(k-1)}{|x|}|\tilde{\mathcal{R}}^{k-1}f|\right|^2 dx$$

$$+ p a_{k-1,Q}^{p-1} \int_{\mathbb{G}} \frac{|\tilde{\mathcal{R}}^{k-1}f|^{p-4}(\mathrm{Im}(\tilde{\mathcal{R}}^{k-1}f\overline{\mathcal{R}\tilde{\mathcal{R}}^{k-1}f}))^2}{|x|^{Q-2(k-1)p-2}} dx$$

$$+ p \sum_{j=1}^{l-1} \left(\prod_{i=k-j}^{k-1} a_{k-i-1,Q}\right)^p a_{k-i-1,Q}^{p-1}$$

$$\times \int_{\mathbb{G}} \frac{|\tilde{\mathcal{R}}^{k-j-1}f|^{p-4}(\mathrm{Im}(\tilde{\mathcal{R}}^{k-j-1}f\overline{\mathcal{R}\tilde{\mathcal{R}}^{k-j-1}f}))^2}{|x|^{Q-2(k-j-1)p-2}}$$

$$+ p \sum_{j=1}^{l-1} \left(\prod_{i=k-j}^{k-1} a_{k-i-1,Q}\right)^p a_{k-i-1,Q}^{p-1}(p-1)$$

$$\times \int_{\mathbb{G}} \frac{|\tilde{\mathcal{R}}^{k-j-1}f|^{p-2}}{|x|^{Q-2(k-j-1)p-2}} \left|\mathcal{R}|\tilde{\mathcal{R}}^{k-j-1}f| + \frac{Q-2p(2(j+1)+\alpha)}{p|x|}|\tilde{\mathcal{R}}^{k-j-1}f|\right|^2 dx,$$

where

$$a_{j,Q} = 2j(Q-2j-2)$$

and R_p is as in (2.22). For $1 \leq k < (Q-1)/2$ we also have

$$\left\| \frac{\mathcal{R}\tilde{\mathcal{R}}^k f}{|x|^{\frac{Q}{p}-(2k+1)}} \right\|_{L^p(\mathbb{G})}^p = (2k)^p \left(\frac{(p-1)(Q-2)}{p}\right)^p \left(\prod_{i=1}^{k-1} a_{i,Q}\right)^p \int_{\mathbb{G}} \frac{|f-f_R|^p}{|x|^Q \left|\ln \frac{R}{|x|}\right|^p} dx$$

$$+ p(2k)^p \left(\prod_{i=1}^{k-1} a_{i,Q}\right)^p \int_{\mathbb{G}} \frac{1}{|x|^{Q-2p}} R_p\left((Q-2)\frac{\mathcal{R}f}{|x|}, \tilde{\mathcal{R}}f\right) dx$$

$$
\begin{aligned}
&+p(2k)^p(Q-2)^p\left(\prod_{i=1}^{k-1}a_{i,Q}\right)^p\int_{\mathbb{G}}\frac{1}{|x|^{Q-p}}R_p\left(-\frac{p-1}{p}\frac{f-f_R}{|x|\ln\frac{R}{|x|}},\mathcal{R}f\right)dx\\
&+p(2k)^p\int_{\mathbb{G}}\frac{1}{|x|^{Q-2kp}}R_p\left(-a_{k-1,Q}\frac{\tilde{\mathcal{R}}^{k-1}f}{|x|^2},\tilde{\mathcal{R}}^kf\right)dx\\
&+p(2k)^p\sum_{j=1}^{k-2}\left(\prod_{i=k-j}^{k-1}a_{i,Q}\right)^p\int_{\mathbb{G}}\frac{1}{|x|^{Q-2(k-j)p}}R_p\left(-a_{k-j-1,Q}\frac{\tilde{\mathcal{R}}^{k-j-1}f}{|x|^2},\tilde{\mathcal{R}}^{k-j}f\right)dx\\
&+p(2k)^pa_{k-1,Q}^{p-1}(p-1)\int_{\mathbb{G}}\frac{|\tilde{\mathcal{R}}^{k-1}f|^{p-2}}{|x|^{Q-2(k-1)p-2}}\left|\mathcal{R}|\tilde{\mathcal{R}}^{k-1}f|+\frac{2(k-1)}{|x|}|\tilde{\mathcal{R}}^{k-1}f|\right|^2dx\\
&+p(2k)^pa_{k-1,Q}^{p-1}\int_{\mathbb{G}}\frac{|\tilde{\mathcal{R}}^{k-1}f|^{p-4}(\mathrm{Im}(\tilde{\mathcal{R}}^{k-1}f\overline{\mathcal{R}\tilde{\mathcal{R}}^{k-1}f}))^2}{|x|^{Q-2(k-1)p-2}}dx\\
&+p(2k)^p\sum_{j=1}^{k-2}\left(\prod_{i=1}^{k-1}a_{k-i-1,Q}\right)^pa_{k-i-1,Q}^{p-1}\\
&\times\int_{\mathbb{G}}\frac{|\tilde{\mathcal{R}}^{k-j-1}f|^{p-4}(\mathrm{Im}(\tilde{\mathcal{R}}^{k-j-1}f\overline{\mathcal{R}\tilde{\mathcal{R}}^{k-j-1}f}))^2}{|x|^{Q-2(k-j-1)p-2}}dx\\
&+p(2k)^p\sum_{j=1}^{k-2}\left(\prod_{i=1}^{k-1}a_{k-i-1,Q}\right)^pa_{k-i-1,Q}^{p-1}(p-1)\\
&\times\int_{\mathbb{G}}\frac{|\tilde{\mathcal{R}}^{k-j-1}f|^{p-2}}{|x|^{Q-2(k-j-1)p-2}}\left|\mathcal{R}|\tilde{\mathcal{R}}^{k-j-1}f|+\frac{Q-2p(2(j+1)+\alpha)}{p|x|}|\tilde{\mathcal{R}}^{k-j-1}f|\right|^2dx\\
&+p\int_{\mathbb{G}}\frac{1}{|x|^{Q-(2k+1)p}}R_p\left(-2k\frac{\tilde{\mathcal{R}}^kf}{|x|},\mathcal{R}\tilde{\mathcal{R}}^kf\right)dx. \qquad (3.64)
\end{aligned}
$$

Proof of Theorem 3.1.20. Denoting $g=\tilde{\mathcal{R}}f$ and applying (3.49) to the function g with $l=k-1$, and then using (3.59), we arrive at (3.63). To prove (3.64) we use Theorem 2.1.8 with $\alpha=(Q-2(k+1)p)/p$ which gives

$$
\begin{aligned}
\left\|\frac{\mathcal{R}\tilde{\mathcal{R}}^kf}{|x|^{\frac{Q}{p}-(2k+1)}}\right\|_{L^p(\mathbb{G})}^p&=(2k)^p\int_{\mathbb{G}}\frac{|\tilde{\mathcal{R}}^kf|^p}{|x|^{Q-2kp}}dx\\
&+p\int_{\mathbb{G}}\frac{1}{|x|^{Q-(2k+1)p}}R_p\left(-2k\frac{\tilde{\mathcal{R}}^kf}{|x|},\mathcal{R}\tilde{\mathcal{R}}^kf\right)dx. \qquad (3.65)
\end{aligned}
$$

Now, the combination of (3.65) and (3.63) implies (3.64). □

As a consequence of Theorem 3.1.20 we have the corresponding inequalities.

Corollary 3.1.21 (Higher-order critical Rellich inequalities for $\tilde{\mathcal{R}}$)**.** *Let $\mathbb{G}$ be a homogeneous group of homogeneous dimension $Q\geq 3$ and let $|\cdot|$ be a homogeneous* *quasi-norm on $\mathbb{G}$. Let $f \in C_0^\infty(\mathbb{G}\backslash\{0\})$ be any complex-valued function and denote*

$$f_R(x) := f(Rx/|x|)$$

with $x \in \mathbb{G}$ and $R > 0$. Let $1 < p < \infty$. Then for any $2 \leq k < Q/2$ we have

$$\frac{2^{k-1}(k-1)!(p-1)}{p} \prod_{i=0}^{k-1}(Q-2i-2) \sup_{R>0} \left\| \frac{f - f_R}{|x|^{\frac{Q}{p}} \left|\ln \frac{R}{|x|}\right|} \right\|_{L^p(\mathbb{G})} \leq \left\| \frac{\tilde{\mathcal{R}} f}{|x|^{\frac{Q}{p}-2k}} \right\|_{L^p(\mathbb{G})}. \tag{3.66}$$

Furthermore, for any $1 \leq k \leq (Q-1)/2$ we have

$$\frac{2^k k!(p-1)}{p} \prod_{i=0}^{k-1}(Q-2i-2) \sup_{R>0} \left\| \frac{f - f_R}{|x|^{\frac{Q}{p}} \left|\ln \frac{R}{|x|}\right|} \right\|_{L^p(\mathbb{G})} \leq \left\| \frac{\mathcal{R}\tilde{\mathcal{R}}^k f}{|x|^{\frac{Q}{p}-2(k+1)}} \right\|_{L^p(\mathbb{G})}. \tag{3.67}$$

Moreover, the constants in inequalities (3.66) and (3.67) are sharp.

Proof of Corollary 3.1.21. Inequalities (3.66) and (3.67) are immediate consequences of identities (3.63) and (3.64), respectively, since we have $a_{i,Q} \geq 0$ for $1 \leq i \leq k-1$, as well as the equalities

$$\frac{(Q-2)(p-1)}{p} \prod_{i=1}^{k-1} a_{i,Q} = \frac{2^{k-1}(k-1)!}{p'} \prod_{i=0}^{k-1}(Q-2i-2),$$

and

$$\frac{2k(Q-2)(p-1)}{p} \prod_{i=1}^{k-1} a_{i,Q} = \frac{2^k k!(p-1)}{p} \prod_{i=0}^{k-1}(Q-2i-2).$$

To show the sharpness of the constants in (3.66) and (3.67) we can use the same argument as in the proof of Theorem 3.1.19 with the test function

$$f_\delta(x) := \left(\ln \frac{R}{|x|}\right)^{1-\frac{1}{p}-\delta} g(|x|).$$

This completes the proof. □

3.2 Sobolev type inequalities

In this section we restate some of Hardy inequalities from the previous chapter in terms of the Euler operator $\mathbb{E}$ from Section 1.3.2, relating them to Sobolev type inequalities. We also briefly recast some of their proofs for the convenience of the reader since fixing the notation in terms of the Euler and radial operators will be useful in further arguments, especially for the analysis in Chapter 10.

The classical *Sobolev inequality* in $\mathbb{R}^n$ has the form

$$\|g\|_{L^p(\mathbb{R}^n)} \leq C(p)\|\nabla g\|_{L^{p^*}(\mathbb{R}^n)}, \quad 1 < p, p^* < \infty, \tag{3.68}$$

where ∇ is the standard gradient in $\mathbb{R}^n$ and

$$\frac{1}{p} = \frac{1}{p^*} - \frac{1}{n}.$$

The aim of this section is to discuss another version of the Sobolev inequality (we call it *Sobolev type inequality*) with respect to the operator $x \cdot \nabla$ instead of ∇, that is, the inequality

$$\|g\|_{L^p(\mathbb{R}^n)} \leq C'(p)\|x \cdot \nabla g\|_{L^q(\mathbb{R}^n)}. \tag{3.69}$$

For any $\lambda > 0$, by setting $g(x) = h(\lambda x)$ in (3.69), it is straightforward to see that $p = q$ is a necessary condition to have inequality (3.69). So, one may concentrate on the case $p = q$. In the case of $\mathbb{R}^n$ this inequality was analysed by Ozawa and Sasaki [OS09], now we concentrate on the setting of general homogeneous groups.

We also note that the classical Sobolev inequality (3.68) can be extended to nilpotent Lie groups: for stratified groups see Folland [Fol75], for graded groups see [FR17], with a general summary presented also in [FR16], and for another version on general homogeneous groups see Section 4.3.

3.2.1 Hardy and Sobolev type inequalities

The inequality (3.70) below is such an extension of (3.69) formulated in terms of the Euler operator $\mathbb{E}$. Moreover, it turns out to be possible to derive a formula for the remainder in this inequality. For indices $1 < p < Q$ this Sobolev type inequality implies the Hardy inequality and for $p = 2$ they are equivalent. All this is the subject of the following statement, an analogue of Theorem 2.1.1.

In this section, unless stated otherwise, $\mathbb{G}$ is a homogeneous group of homogeneous dimension $Q \geq 1$ and $|\cdot|$ is a homogeneous quasi-norm on $\mathbb{G}$.

Proposition 3.2.1 (Sobolev type and Hardy inequalities). *We have the following properties.*

(i) *For all complex-valued functions $f \in C_0^\infty(\mathbb{G}\backslash\{0\})$,*

$$\|f\|_{L^p(\mathbb{G})} \leq \frac{p}{Q}\|\mathbb{E}f\|_{L^p(\mathbb{G})}, \quad 1 < p < \infty, \tag{3.70}$$

where the constant $\frac{p}{Q}$ is sharp and the equality is attained if and only if $f = 0$.

(ii) *Let*

$$u := u(x) = -\frac{p}{Q}\mathbb{E}f(x), \ v := v(x) = f(x).$$

Then we have the following expression for the remainder:

$$\|u\|^p_{L^p(\mathbb{G})} - \|v\|^p_{L^p(\mathbb{G})} = p\int_{\mathbb{G}} I_p(v,u)|v-u|^2 dx, \quad 1 < p < \infty, \tag{3.71}$$

for all real-valued functions $f \in C_0^\infty(\mathbb{G}\backslash\{0\})$, *where*

$$I_p(h,g) = (p-1)\int_0^1 |\xi h + (1-\xi)g|^{p-2}\xi d\xi.$$

(iii) *For all complex-valued functions* $f \in C_0^\infty(\mathbb{G}\backslash\{0\})$ *the identity* (3.71) *with* $p = 2$ *holds and can be written in the form*

$$\|\mathbb{E}f\|^2_{L^2(\mathbb{G})} = \left(\frac{Q}{2}\right)^2 \|f\|^2_{L^2(\mathbb{G})} + \left\|\mathbb{E}f + \frac{Q}{2}f\right\|^2_{L^2(\mathbb{G})}, \quad Q \geq 1. \tag{3.72}$$

(iv) *In the case* $p = 2$ *and* $Q \geq 3$ *the inequality* (3.70) *is equivalent to Hardy's inequality, i.e., for any* $g \in C_0^\infty(\mathbb{G}\backslash\{0\})$ *the inequality*

$$\left\|\frac{g}{|x|}\right\|_{L^2(\mathbb{G})} \leq \frac{2}{Q-2}\|\mathcal{R}g\|_{L^2(\mathbb{G})} \equiv \frac{2}{Q-2}\left\|\frac{1}{|x|}\mathbb{E}g\right\|_{L^2(\mathbb{G})}. \tag{3.73}$$

(v) *In the case* $1 < p < Q$ *the inequality* (3.70) *yields Hardy's inequality for any* $f \in C_0^\infty(\mathbb{G}\backslash\{0\})$, *i.e., the inequality*

$$\left\|\frac{f}{|x|}\right\|_{L^p(\mathbb{G})} \leq \frac{p}{Q-p}\|\mathcal{R}f\|_{L^p(\mathbb{G})}. \tag{3.74}$$

Remark 3.2.2.

1. As mentioned above in the Euclidian case, for any $\lambda > 0$, substituting $g(x) = h(\lambda x)$ into the Sobolev type inequality

$$\|g\|_{L^p(\mathbb{G})} \leq C(p)\|\mathbb{E}g\|_{L^q(\mathbb{G})}, \quad 1 < p, q < \infty, \tag{3.75}$$

and using the fact that the Euler operator is a homogeneous operator of order zero, we obtain that $p = q$ is a necessary condition for having inequality (3.75).

2. In the Euclidean case $\mathbb{G} = \mathbb{R}^n$ the inequality (3.70) was observed in [BEHL08]: for any $n \geq 1$ and $1 \leq p < \infty$, for all $f \in C_0^\infty(\mathbb{R}^n)$ we have

$$\|f\|_{L^p(\mathbb{R}^n)} \leq \frac{p}{n}\|x \cdot \nabla f\|_{L^p(\mathbb{R}^n)}.$$

Indeed, this is a consequence of a simple integration by parts:

$$\begin{aligned} n\int_{\mathbb{R}^n}|f(x)|^p dx &= \int_{\mathbb{R}^n}\operatorname{div}(x)|f(x)|^p dx \\ &= -p\operatorname{Re}\int_{\mathbb{R}^n} x\cdot\nabla f(x)|f(x)|^{p-2}\overline{f(x)}dx \\ &\leq p\left(\int_{\mathbb{R}^n}|x\cdot\nabla f(x)|^p dx\right)^{1/p}\left(\int_{\mathbb{R}^n}|f(x)|^p dx\right)^{\frac{p-1}{p}}, \end{aligned}$$

using Hölder's inequality in the last line.

There is a weighted version of the above inequality given in Theorem 6.6.1, see also Remark 6.6.2, Part 3.

3. The analysis in this section is based on [RSY18d].

Proof of Proposition 3.2.1. Let us first show that Part (ii) implies Part (i). By dropping non-negative term in the right-hand side of (3.71), we get

$$\|f\|_{L^p(\mathbb{G})} \leq \frac{p}{Q}\|\mathbb{E}f\|_{L^p(\mathbb{G})}, \quad 1<p<\infty,\ \ Q\geq 1, \tag{3.76}$$

for all real-valued functions $f\in C_0^\infty(\mathbb{G}\backslash\{0\})$. Consequently, this inequality is valid for all complex-valued functions if we use the identity

$$\forall z\in\mathbb{C}:\ |z|^p = \left(\int_{-\pi}^{\pi}|\cos\theta|^p d\theta\right)^{-1}\int_{-\pi}^{\pi}|\operatorname{Re}(z)\cos\theta+\operatorname{Im}(z)\sin\theta|^p\,d\theta, \tag{3.77}$$

see (2.8).

So, the inequality (3.70) holds true and the expression for the remainder implies that the constant $\frac{p}{Q}$ is sharp.

Let us show that this constant is attained only for $f=0$. In view of the identity (3.77), it is enough to check this only for real-valued functions f. If the right-hand side of (3.71) is zero, then we must have the equality

$$-\frac{p}{Q}\mathbb{E}f(x)=f(x),$$

which yields that

$$\mathbb{E}f=-\frac{Q}{p}f.$$

By the property of the Euler operator in Lemma 1.3.1 this means that f is positively homogeneous function of order $-\frac{Q}{p}$, i.e., there exists a function $h:\wp\to\mathbb{C}$, where $\wp$ is defined by (1.12), such that

$$f(x)=|x|^{-\frac{Q}{p}}h\left(\frac{x}{|x|}\right). \tag{3.78}$$

In particular, (3.78) means that f cannot be compactly supported unless it is identically equal to zero. Therefore, we have shown that Part (ii), namely (3.71) implies Part (i) of Theorem 3.2.1.

Nevertheless, let us also give another direct proof of (3.70) for complex-valued functions without using formula (3.77) and without using the remainder formula in Part (ii).

Proof of Part (i). Introducing polar coordinates $(r,y)=(|x|,\frac{x}{|x|})\in(0,\infty)\times\wp$ on $\mathbb{G}$, where the sphere $\wp$ is defined in (1.12), we now apply the polar decomposition formula (1.13). This and integrating by parts yield

$$\begin{aligned}\int_{\mathbb{G}}|f(x)|^p dx &= \int_0^\infty\int_\wp |f(ry)|^p r^{Q-1}d\sigma(y)dr\\ &= -\frac{p}{Q}\int_0^\infty r^Q\operatorname{Re}\int_\wp |f(ry)|^{p-2}f(ry)\overline{\frac{df(ry)}{dr}}d\sigma(y)dr \qquad (3.79)\\ &= -\frac{p}{Q}\operatorname{Re}\int_{\mathbb{G}}|f(x)|^{p-2}f(x)\overline{\mathbb{E}f(x)}dx.\end{aligned}$$

By using the Hölder inequality with an index q such that $\frac{1}{p}+\frac{1}{q}=1$ we obtain

$$\begin{aligned}\int_{\mathbb{G}}|f(x)|^p dx &= -\frac{p}{Q}\operatorname{Re}\int_{\mathbb{G}}|f(x)|^{p-2}f(x)\overline{\mathbb{E}f(x)}dx\\ &\leq \frac{p}{Q}\left(\int_{\mathbb{G}}\left||f(x)|^{p-2}f(x)\right|^q dx\right)^{\frac{1}{q}}\left(\int_{\mathbb{G}}|\mathbb{E}f(x)|^p dx\right)^{\frac{1}{p}}\\ &= \frac{p}{Q}\left(\int_{\mathbb{G}}|f(x)|^p dx\right)^{1-\frac{1}{p}}\|\mathbb{E}f\|_{L^p(\mathbb{G})},\end{aligned}$$

which gives inequality (3.70) in Part (i).

Proof of Part (ii). With the notation

$$u:=u(x)=-\frac{p}{Q}\mathbb{E}f \quad\text{and}\quad v:=v(x)=f(x),$$

the formula (3.79) can be reformulated as

$$\|v\|^p_{L^p(\mathbb{G})}=\operatorname{Re}\int_{\mathbb{G}}|v|^{p-2}v\overline{u}dx. \qquad (3.80)$$

For any real-valued function f formula (3.79) becomes

$$\int_{\mathbb{G}}|f(x)|^p dx=-\frac{p}{Q}\int_{\mathbb{G}}|f(x)|^{p-2}f(x)\mathbb{E}f(x)dx,$$

and (3.80) becomes

$$\|v\|^p_{L^p(\mathbb{G})}=\int_{\mathbb{G}}|v|^{p-2}vudx. \qquad (3.81)$$

Moreover, for all L^p-integrable real-valued functions u and v the following equalities hold

$$\begin{aligned}\|u\|^p_{L^p(\mathbb{G})} - \|v\|^p_{L^p(\mathbb{G})} &+ p\int_{\mathbb{G}}(|v|^p - |v|^{p-2}vu)dx \\ &= \int_{\mathbb{G}}(|u|^p + (p-1)|v|^p - p|v|^{p-2}vu)dx \\ &= p\int_{\mathbb{G}} I_p(v,u)|v-u|^2dx, \quad 1<p<\infty,\end{aligned}$$

where

$$I_p(v,u) = (p-1)\int_0^1 |\xi v + (1-\xi)u|^{p-2}\xi d\xi.$$

Combining this with (3.81) we arrive at

$$\|u\|^p_{L^p(\mathbb{G})} - \|v\|^p_{L^p(\mathbb{G})} = p\int_{\mathbb{G}} I_p(v,u)|v-u|^2dx,$$

which proves the equality (3.71).

Now we prove Part (iii). If $p=2$, then the identity (3.80) can be rewritten as

$$\|v\|^2_{L^2(\mathbb{G})} = \mathrm{Re}\int_{\mathbb{G}} v\overline{u}dx.$$

Thus, we have

$$\begin{aligned}\|u\|^2_{L^2(\mathbb{G})} - \|v\|^2_{L^2(\mathbb{G})} &= \|u\|^2_{L^2(\mathbb{G})} - \|v\|^2_{L^2(\mathbb{G})} + 2\int_{\mathbb{G}}(|v|^2 - \mathrm{Re}\, v\overline{u})dx \\ &= \int_{\mathbb{G}}(|u|^2 + |v|^2 - 2\mathrm{Re}\, v\overline{u})dx \\ &= \int_{\mathbb{G}} |u-v|^2dx,\end{aligned}$$

which gives (3.72).

To show Part (iv) first we verify that the inequality (3.70) implies (3.73). A direct calculation shows

$$\begin{aligned}\|\mathbb{E}f\|^2_{L^2(\mathbb{G})} &= \left\|\mathbb{E}\frac{g}{|x|}\right\|^2_{L^2(\mathbb{G})} \\ &= \int_0^\infty\int_{\wp}\left|\left(r\frac{d}{dr}\right)\frac{g(ry)}{r}\right|^2 r^{Q-1}d\sigma(y)dr \\ &= \left\|-\frac{g}{|x|} + \frac{d}{d|x|}g\right\|^2_{L^2(\mathbb{G})} \\ &= \left\|\frac{g}{|x|}\right\|^2_{L^2(\mathbb{G})} - 2\mathrm{Re}\int_{\mathbb{G}}\frac{g}{|x|}\overline{\frac{d}{d|x|}g}dx + \left\|\frac{d}{d|x|}g\right\|^2_{L^2(\mathbb{G})},\end{aligned}$$

where $g = |x|f$. From

$$\begin{aligned}
-2\mathrm{Re}\int_{\mathbb{G}} \frac{g}{|x|}\overline{\frac{d}{d|x|}g}dx &= -2\mathrm{Re}\int_0^\infty\int_\wp \frac{g(ry)}{r}\overline{\frac{d}{dr}g(ry)}r^{Q-1}d\sigma(y)dr \\
&= -\mathrm{Re}\int_0^\infty\int_\wp \frac{d}{dr}(|\overline{g}|^2)r^{Q-2}d\sigma(y)dr \\
&= (Q-2)\mathrm{Re}\int_0^\infty\int_\wp |\overline{g}|^2 r^{Q-3}d\sigma(y)dr \\
&= (Q-2)\left\|\frac{g}{|x|}\right\|^2_{L^2(\mathbb{G})},
\end{aligned}$$

it follows that

$$\|\mathbb{E}f\|^2_{L^2(\mathbb{G})} = (Q-1)\left\|\frac{g}{|x|}\right\|^2_{L^2(\mathbb{G})} + \left\|\frac{d}{d|x|}g\right\|^2_{L^2(\mathbb{G})}. \tag{3.82}$$

Combining (3.70) and (3.82) we obtain

$$\left\|\frac{g}{|x|}\right\|^2_{L^2(\mathbb{G})} \leq \frac{4}{Q^2}\left((Q-1)\left\|\frac{g}{|x|}\right\|^2_{L^2(\mathbb{G})} + \left\|\frac{d}{d|x|}g\right\|^2_{L^2(\mathbb{G})}\right),$$

which gives (3.73).

Conversely, let us assume that (3.73) is valid. Then with the notation $f = g/|x|$ we get

$$\begin{aligned}
\left\|\frac{d}{d|x|}(|x|f)\right\|^2_{L^2(\mathbb{G})} &= \|f + \mathbb{E}f\|^2_{L^2(\mathbb{G})} \\
&= \|f\|^2_{L^2(\mathbb{G})} + 2\mathrm{Re}\int_{\mathbb{G}} f(x)\overline{\mathbb{E}f(x)}dx + \|\mathbb{E}f\|^2_{L^2(\mathbb{G})}.
\end{aligned}$$

Hence by (1.42) and (3.73) it follows that

$$\|f\|^2_{L^2(\mathbb{G})} \leq \frac{4}{(Q-2)^2}\left(\|\mathbb{E}f\|^2_{L^2(\mathbb{G})} - (Q-1)\|f\|^2_{L^2(\mathbb{G})}\right),$$

which gives

$$\|f\|_{L^2(\mathbb{G})} \leq \frac{2}{Q}\|\mathbb{E}f\|_{L^2(\mathbb{G})}.$$

Now it remains to prove Part (v). We will show that the inequality (3.70) gives (3.74). We have

$$\|\mathcal{R}(|x|f)\|_{L^p(\mathbb{G})} = \|\mathbb{E}f + f\|_{L^p(\mathbb{G})} \geq \|\mathbb{E}f\|_{L^p(\mathbb{G})} - \|f\|_{L^p(\mathbb{G})}.$$

Finally, by using the inequality (3.70) we establish

$$\|\mathcal{R}(|x|f)\|_{L^p(\mathbb{G})} \geq \frac{Q-p}{p}\|f\|_{L^p(\mathbb{G})},$$

which implies the Hardy inequality (3.74). □

3.2.2 Weighted L^p-Sobolev type inequalities

Now we establish weighted L^p-Sobolev type inequalities on the homogeneous group $\mathbb{G}$ of homogeneous dimension $Q \geq 1$.

Theorem 3.2.3 (Weighted L^p-Sobolev type inequalities). *For all complex-valued functions $f \in C_0^\infty(\mathbb{G}\backslash\{0\})$, $1 < p < \infty$, and any homogeneous quasi-norm $|\cdot|$ on $\mathbb{G}$ for $\alpha p \neq Q$ we have*

$$\left\| \frac{f}{|x|^\alpha} \right\|_{L^p(\mathbb{G})} \leq \left| \frac{p}{Q-\alpha p} \right| \left\| \frac{1}{|x|^\alpha} \mathbb{E} f \right\|_{L^p(\mathbb{G})} \quad \textit{for all } \alpha \in \mathbb{R}. \tag{3.83}$$

If $\alpha p \neq Q$ then the constant $\left|\frac{p}{Q-\alpha p}\right|$ is sharp.

For $\alpha p = Q$ we have

$$\left\| \frac{f}{|x|^{\frac{Q}{p}}} \right\|_{L^p(\mathbb{G})} \leq p \left\| \frac{\log |x|}{|x|^{\frac{Q}{p}}} \mathbb{E} f \right\|_{L^p(\mathbb{G})}, \tag{3.84}$$

where the constant p is sharp.

Proof of Theorem 3.2.3. Using the integration by parts formula from Proposition 1.2.10, for $\alpha p \neq Q$ we obtain

$$\begin{aligned} \int_{\mathbb{G}} \frac{|f(x)|^p}{|x|^{\alpha p}} dx &= \int_0^\infty \int_{\wp} |f(ry)|^p r^{Q-1-\alpha p} d\sigma(y) dr \\ &= -\frac{p}{Q-\alpha p} \int_0^\infty r^{Q-\alpha p} \operatorname{Re} \int_{\wp} |f(ry)|^{p-2} f(ry) \overline{\frac{df(ry)}{dr}} d\sigma(y) dr \\ &\leq \left| \frac{p}{Q-\alpha p} \right| \int_{\mathbb{G}} \frac{|\mathbb{E} f(x)||f(x)|^{p-1}}{|x|^{\alpha p}} dx = \left| \frac{p}{Q-\alpha p} \right| \int_{\mathbb{G}} \frac{|\mathbb{E} f(x)||f(x)|^{p-1}}{|x|^{\alpha+\alpha(p-1)}} dx. \end{aligned}$$

By Hölder's inequality, it follows that

$$\int_{\mathbb{G}} \frac{|f(x)|^p}{|x|^{\alpha p}} dx \leq \left| \frac{p}{Q-\alpha p} \right| \left(\int_{\mathbb{G}} \frac{|\mathbb{E} f(x)|^p}{|x|^{\alpha p}} dx \right)^{\frac{1}{p}} \left(\int_{\mathbb{G}} \frac{|f(x)|^p}{|x|^{\alpha p}} dx \right)^{\frac{p-1}{p}},$$

which gives (3.83).

Now we show the sharpness of the constant. We need to check the equality condition in the above Hölder inequality. Let us consider the function

$$g(x) = \frac{1}{|x|^C},$$

where $C \in \mathbb{R}, C \neq 0$ and $\alpha p \neq Q$. Then by a direct calculation we obtain

$$\left| \frac{1}{C} \right|^p \left(\frac{|\mathbb{E} g(x)|}{|x|^\alpha} \right)^p = \left(\frac{|g(x)|^{p-1}}{|x|^{\alpha(p-1)}} \right)^{\frac{p}{p-1}},$$

which satisfies the equality condition in Hölder's inequality. This gives the sharpness of the constant $\left|\frac{p}{Q-\alpha p}\right|$ in (3.83).

Now let us prove (3.84). Using integration by parts, we have

$$\int_{\mathbb{G}} \frac{|f(x)|^p}{|x|^Q} dx = \int_0^\infty \int_{\wp} |f(ry)|^p r^{Q-1-Q} d\sigma(y) dr$$

$$= -p \int_0^\infty \log r \mathrm{Re} \int_{\wp} |f(ry)|^{p-2} f(ry) \overline{\frac{df(ry)}{dr}} d\sigma(y) dr$$

$$\leq p \int_{\mathbb{G}} \frac{|\mathbb{E}f(x)||f(x)|^{p-1}}{|x|^Q} |\log|x|| dx = p \int_{\mathbb{G}} \frac{|\mathbb{E}f(x)| \log|x||}{|x|^{\frac{Q}{p}}} \frac{|f(x)|^{p-1}}{|x|^{\frac{Q(p-1)}{p}}} dx.$$

By Hölder's inequality, it follows that

$$\int_{\mathbb{G}} \frac{|f(x)|^p}{|x|^Q} dx \leq p \left(\int_{\mathbb{G}} \frac{|\mathbb{E}f(x)|^p |\log|x||^p}{|x|^Q} dx \right)^{1/p} \left(\int_{\mathbb{G}} \frac{|f(x)|^p}{|x|^Q} dx \right)^{(p-1)/p},$$

which gives (3.84).

Now we show the sharpness of the constant. We need to check the equality condition in the above Hölder inequality. Let us consider the function

$$h(x) = (\log|x|)^C,$$

where $C \in \mathbb{R}$ and $C \neq 0$. Then by a direct calculation we obtain

$$\left|\frac{1}{C}\right|^p \left(\frac{|\mathbb{E}h(x)||\log|x||}{|x|^{\frac{Q}{p}}} \right)^p = \left(\frac{|h(x)|^{p-1}}{|x|^{\frac{Q(p-1)}{p}}} \right)^{p/(p-1)},$$

which satisfies the equality condition in Hölder's inequality. This gives the sharpness of the constant p in (3.84). □

Let us consider separately the case $p = 2$, that is, let us restate Theorem 2.1.5 in terms of the operator $\mathbb{E}$:

Proposition 3.2.4 (An identity for Euler operator). *For every complex-valued function $f \in C_0^\infty(\mathbb{G}\backslash\{0\})$ we have*

$$\left\| \frac{1}{|x|^\alpha} \mathbb{E}f \right\|_{L^2(\mathbb{G})}^2 = \left(\frac{Q}{2} - \alpha \right)^2 \left\| \frac{f}{|x|^\alpha} \right\|_{L^2(\mathbb{G})}^2 + \left\| \frac{1}{|x|^\alpha} \mathbb{E}f + \frac{Q-2\alpha}{2|x|^\alpha} f \right\|_{L^2(\mathbb{G})}^2, \quad (3.85)$$

for any $\alpha \in \mathbb{R}$.

Remark 3.2.5.

1. By dropping the non-negative last term in (3.85) we immediately get the following inequality for $\alpha \in \mathbb{R}$ with $Q - 2\alpha \neq 0$:

$$\left\| \frac{f}{|x|^\alpha} \right\|_{L^2(\mathbb{G})} \leq \frac{2}{|Q-2\alpha|} \left\| \frac{1}{|x|^\alpha} \mathbb{E}f \right\|_{L^2(\mathbb{G})}, \quad (3.86)$$

for all complex-valued functions $f \in C_0^\infty(\mathbb{G}\backslash\{0\})$, where the constant in (3.86) is sharp and the equality is attained if and only if $f = 0$. This statement on the constant and the equality follows by the same argument as that in Remark 2.1.7.

2. By iterating the established weighted Sobolev inequality (3.83) one obtains inequalities of higher order. Thus, for $1 < p < \infty$, $k \in \mathbb{N}$ and $\alpha \in \mathbb{R}$ with $Q \neq \alpha p$ we have
$$\left\|\frac{f}{|x|^\alpha}\right\|_{L^p(\mathbb{G})} \leq \left|\frac{p}{Q-\alpha p}\right|^k \left\|\frac{\mathbb{E}^k f}{|x|^\alpha}\right\|_{L^p(\mathbb{G})} \tag{3.87}$$
for any complex-valued function $f \in C_0^\infty(\mathbb{G}\backslash\{0\})$.
3. For $k = 1$ (3.87) implies the weighted Sobolev inequality and for $k = 1$ and $\alpha = 0$ this gives the Sobolev inequality. In the case $k = 2$ this can be thought of as a (weighted) Sobolev–Rellich type inequality.

3.2.3 Stubbe type remainder estimates

The remainders in Hardy inequalities may be described in different ways: there may be equalities or estimates of different forms. These are discussed in some detail at various spaces of this book. Here, we give a remainder estimate in the most basic case of L^2. Such a type of inequalities have been analysed on $\mathbb{R}^n$ by Stubbe [Stu90], and here we give its general version on homogeneous groups.

In the proof of the following statement we will use the useful feature that some estimates involving radial derivatives of the Euler operator can be proved first for radial functions, and then extended to non-radial ones by a more abstract argument, see Section 1.3.3.

Theorem 3.2.6 (Stubbe type remainder estimate). *Let $\mathbb{G}$ be a homogeneous group of homogeneous dimension $Q \geq 3$. Let $|\cdot|$ be a homogeneous quasi-norm. Then we have for all $f \in C_0^\infty(\mathbb{G})$ and $0 \leq \delta < \frac{Q^2}{4}$, the inequality*
$$\int_{\mathbb{G}} |\mathbb{E}f(x)|^2 dx - \delta \int_{\mathbb{G}} |f(x)|^2 dx \geq \frac{\left(\frac{Q^2}{4} - \delta\right)^{\frac{Q-1}{Q}}}{\left(\frac{(Q-2)^2}{4}\right)^{\frac{Q-1}{Q}}} S_Q \left(\int_{\mathbb{G}} |x|^{2^*} |g(|x|)|^{2^*} dx\right)^{2/2^*} \tag{3.88}$$
with sharp constant, where
$$g(|x|) = \mathcal{M}(f)(|x|) := \frac{1}{|\wp|}\int_{\wp} f(|x|, y) d\sigma(y)$$
and
$$S_Q := |\wp|^{\frac{2}{Q}} Q^{\frac{Q-2}{Q}} (Q-2) \left(\frac{\Gamma(Q/2)\Gamma(1+Q/2)}{\Gamma(Q)}\right)^{2/Q}. \tag{3.89}$$

Remark 3.2.7.

1. In the Abelian case $\mathbb{G} = (\mathbb{R}^n, +)$, we have $Q = n$, and inequality (3.88) becomes

$$\int_{\mathbb{R}^n} |(x\cdot\nabla)f(x)|^2 dx - \delta \int_{\mathbb{R}^n} |f(x)|^2 dx \geq \frac{\left(\frac{n^2}{4}-\delta\right)^{\frac{n-1}{n}}}{\left(\frac{(n-2)^2}{4}\right)^{\frac{n-1}{n}}} S_n \left(\int_{\mathbb{G}} |x|^{2^*} |g(|x|)|^{2^*} dx\right)^{\frac{2}{2^*}}. \tag{3.90}$$

2. An interesting observation is that the constant in the above inequality on $\mathbb{R}^n$ is sharp for any quasi-norm $|\cdot|$, that is, it does not depend on the quasi-norm $|\cdot|$. Therefore, this inequality is new already in the Euclidean setting of $\mathbb{R}^n$. When $|x| = \sqrt{x_1^2 + x_2^2 + \cdots + x_n^2}$ is the Euclidean distance, the inequality (3.90) was investigated in $\mathbb{R}^n$ in [BEHL08, Corollary 4.4] and in [Xia11, Theorem 1.1].
3. The following result, proved in [Bli30], will be useful in the proof: Let f be a non-negative function. Then for $s \geq 0$ and $q > p > 1$ we have

$$\left(\int_0^\infty \left|\int_0^s f(r)dr\right|^q r^{q/p-q-1} dr\right)^{p/q} \leq C_{p,q} \int_0^\infty |f(r)|^p dr, \tag{3.91}$$

where

$$C_{p,q} = (q - q/p)^{-p/q} \left(\frac{(q/p-1)\Gamma\left(\frac{pq}{q-p}\right)}{\Gamma\left(\frac{p}{q-p}\right)\Gamma\left(\frac{p(q-1)}{q-p}\right)}\right)^{(q-p)/q}$$

is sharp. Moreover, the equality in (3.91) is attained for functions of the form

$$f(r) = c_1(c_2 r^{q/p-1} + 1)^{q/(p-q)}, \; c_1 > 0, \; c_2 > 0. \tag{3.92}$$

Proof of Theorem 3.2.6. First we prove inequality (3.88) for $|\cdot|$-radial functions $f(x) = \widetilde{f}(|x|)$. Then we have $g(r) = \widetilde{f}(r)$ since

$$g(|x|) = \mathcal{M}(f)(|x|) := \frac{1}{|\wp|}\int_{\wp} f(|x|, y) d\sigma(y),$$

and we also have $\mathbb{E}f(x) = |x|\widetilde{f}'(|x|)$. We calculate

$$\int_{\mathbb{G}} |\mathbb{E}\widetilde{f}(|x|)|^2 dx - \beta(Q-\beta)\int_{\mathbb{G}} |\widetilde{f}(|x|)|^2 dx = |\wp|\left(\int_0^\infty |\widetilde{f}'(r)|^2 r^{Q+1} dr - \beta(Q-\beta)\int_0^\infty |\widetilde{f}(r)|^2 r^{Q-1} dr\right), \tag{3.93}$$

where $0 \leq \beta < Q/2$. Using the notation $h(|x|) := |x|^{\beta}\widetilde{f}(|x|)$, and integrating by parts we obtain

$$
\begin{aligned}
&\int_0^{\infty} |h'(r)|^2 r^{Q+1-2\beta} dr \\
&= \int_0^{\infty} |\beta r^{\beta-1}\widetilde{f}(r) + r^{\beta}\widetilde{f}'(r)|^2 r^{Q+1-2\beta} dr \\
&= \beta^2 \int_0^{\infty} |\widetilde{f}(r)|^2 r^{Q-1} dr + \int_0^{\infty} |\widetilde{f}'(r)|^2 r^{Q+1} dr + \beta \int_0^{\infty} \frac{d}{dr}|\widetilde{f}(r)|^2 r^{Q} dr \\
&= \int_0^{\infty} |\widetilde{f}'(r)|^2 r^{Q+1} dr - \beta(Q-\beta) \int_0^{\infty} |\widetilde{f}(r)|^2 r^{Q-1} dr.
\end{aligned}
\tag{3.94}
$$

Moreover, by changing the variables $s = r^{Q-2\beta}$ we get

$$
\begin{aligned}
\int_0^{\infty} |h'(r)|^2 r^{Q+1-2\beta} dr &= \int_0^{\infty} |(Q-2\beta) s^{\frac{Q-2\beta-1}{Q-2\beta}} h'(s)|^2 s^{\frac{Q-2\beta+1}{Q-2\beta}} \frac{s^{\frac{1}{Q-2\beta}-1} ds}{Q-2\beta} \\
&= (Q-2\beta) \int_0^{\infty} s^2 |h'(s)|^2 ds.
\end{aligned}
\tag{3.95}
$$

Combining (3.94) and (3.95), we restate (3.93) as

$$
\int_{\mathbb{G}} |\mathbb{E}\widetilde{f}(|x|)|^2 dx - \beta(Q-\beta) \int_{\mathbb{G}} |\widetilde{f}(|x|)|^2 dx = |\wp|(Q-2\beta) \left(\int_0^{\infty} s^2 |h'(s)|^2 ds \right). \tag{3.96}
$$

Now setting $\phi(s) := h'(s)$ and $\psi(s) := s^{-2}\phi(s^{-1})$, and using (3.91) with $p = 2$ and $q = 2^*$, we obtain

$$
\begin{aligned}
&\int_0^{\infty} s^2 |h'(s)|^2 ds = \int_0^{\infty} s^2 |\phi(s)|^2 ds = \int_0^{\infty} |\psi(s)|^2 ds \\
&\geq \left(\frac{Q}{Q-2}\right)^{\frac{Q-2}{Q}} \left(\frac{\Gamma(Q/2)\Gamma(1+Q/2)}{\Gamma(Q)}\right)^{\frac{2}{Q}} \left(\int_0^{\infty} \left| \int_0^{s} |\psi(t)| dt \right|^{2^*} s^{\frac{2-2Q}{Q-2}} ds \right)^{\frac{2}{2^*}} \\
&= \left(\frac{Q}{Q-2}\right)^{\frac{Q-2}{Q}} \left(\frac{\Gamma(Q/2)\Gamma(1+Q/2)}{\Gamma(Q)}\right)^{\frac{2}{Q}} \left(\int_0^{\infty} \left| \int_{s^{-1}}^{\infty} |\phi(t)| dt \right|^{2^*} s^{\frac{2-2Q}{Q-2}} ds \right)^{\frac{2}{2^*}} \\
&\geq \left(\frac{Q}{Q-2}\right)^{\frac{Q-2}{Q}} \left(\frac{\Gamma(Q/2)\Gamma(1+Q/2)}{\Gamma(Q)}\right)^{\frac{2}{Q}} \left(\int_0^{\infty} |h(s^{-1})|^{2^*} s^{\frac{2-2Q}{Q-2}} ds \right)^{\frac{2}{2^*}} \\
&= \left(\frac{Q}{Q-2}\right)^{\frac{Q-2}{Q}} \left(\frac{\Gamma(Q/2)\Gamma(1+Q/2)}{\Gamma(Q)}\right)^{\frac{2}{Q}} \left(\int_0^{\infty} |h(s)|^{2^*} s^{\frac{2}{Q-2}} ds \right)^{\frac{2}{2^*}} \\
&= (Q-2\beta)^{\frac{2}{2^*}} \left(\frac{Q}{Q-2}\right)^{\frac{Q-2}{Q}} \left(\frac{\Gamma(Q/2)\Gamma(1+Q/2)}{\Gamma(Q)}\right)^{\frac{2}{Q}} \left(\int_0^{\infty} |r\widetilde{f}(r)|^{2^*} r^{Q-1} dr \right)^{\frac{2}{2^*}},
\end{aligned}
$$

where we have used $s = r^{Q-2\beta}$ and $h(r) = r^{\beta}\widetilde{f}(r)$ in the last line. Thus, now (3.96) implies that

$$\begin{aligned}
&\int_{\mathbb{G}} |\mathbb{E}\widetilde{f}(|x|)|^2 dx - \beta(Q-\beta)\int_{\mathbb{G}} |\widetilde{f}(|x|)|^2 dx \\
&\geq |\wp|(Q-2\beta)^{\frac{2Q-2}{Q}} \left(\frac{Q}{Q-2}\right)^{\frac{Q-2}{Q}} \left(\frac{\Gamma(Q/2)\Gamma(1+Q/2)}{\Gamma(Q)}\right)^{\frac{2}{Q}} \left(\int_0^{\infty} |r\widetilde{f}(r)|^{2^*} r^{Q-1} dr\right)^{\frac{2}{2^*}} \\
&= |\wp|^{\frac{2}{Q}} (Q-2\beta)^{\frac{2Q-2}{Q}} \left(\frac{Q}{Q-2}\right)^{\frac{Q-2}{Q}} \left(\frac{\Gamma(Q/2)\Gamma(1+Q/2)}{\Gamma(Q)}\right)^{\frac{2}{Q}} \left(\int_{\mathbb{G}} |x|^{2^*} |\widetilde{f}(|x|)|^{2^*} dx\right)^{\frac{2}{2^*}}.
\end{aligned} \tag{3.97}$$

Here, denoting $\beta = (Q - \sqrt{Q^2 - 4\delta})/2$ for $0 \leq \delta < Q^2/4$ and recalling that $g(|x|) = \widetilde{f}(|x|)$, we see that (3.97) (with (3.89)) yields

$$\begin{aligned}
&\int_{\mathbb{G}} |\mathbb{E}\widetilde{f}(|x|)|^2 dx - \delta\int_{\mathbb{G}} |\widetilde{f}(|x|)|^2 dx \\
&\geq |\wp|^{\frac{2}{Q}} (Q^2-4\delta)^{\frac{Q-1}{Q}} \left(\frac{Q}{Q-2}\right)^{\frac{Q-2}{Q}} \left(\frac{\Gamma(Q/2)\Gamma(1+Q/2)}{\Gamma(Q)}\right)^{\frac{2}{Q}} \left(\int_{\mathbb{G}} |x|^{2^*} |g(|x|)|^{2^*} dx\right)^{\frac{2}{2^*}} \\
&= \left(\frac{Q^2-4\delta}{(Q-2)^2}\right)^{\frac{Q-1}{Q}} S_Q \left(\int_{\mathbb{G}} |x|^{2^*} |g(|x|)|^{2^*} dx\right)^{\frac{2}{2^*}}.
\end{aligned} \tag{3.98}$$

That is, we obtain (3.88) with sharp constant for all $|\cdot|$-radial functions $f \in C_0^{\infty}(\mathbb{G})$.

Finally, using Proposition 1.3.3, and in (3.98) with $g(|x|) = \widetilde{f}(|x|)$, we get (3.88) for non-radial functions. Clearly, the constant in (3.88) is sharp, since this constant is sharp for radial functions by Remark 3.2.7, Part (3). □

3.3 Caffarelli–Kohn–Nirenberg inequalities

This section is devoted to deriving the Caffarelli–Kohn–Nirenberg inequalities in the setting of homogeneous groups. Here we will be working with the radial operators and general quasi-norms. The case of stratified groups with the horizontal gradient and weights will be discussed in Section 6.7.

First, we recall the classical Caffarelli–Kohn–Nirenberg inequalities on $\mathbb{R}^n$ due to Caffarelli, Kohn and Nirenberg [CKN84], with $|\cdot|$ denoting the usual Euclidean distance:

Theorem 3.3.1 (Classical Caffarelli–Kohn–Nirenberg inequality). *Let $n \in \mathbb{N}$ and let p, q, r, a, b, d, $\delta \in \mathbb{R}$ be such that $p, q \geq 1$, $r > 0$, $0 \leq \delta \leq 1$, and*

$$\frac{1}{p} + \frac{a}{n}, \; \frac{1}{q} + \frac{b}{n}, \; \frac{1}{r} + \frac{c}{n} > 0, \tag{3.99}$$

where

$$c = \delta d + (1-\delta)b.$$

Then there exists a positive constant C such that

$$\||x|^c f\|_{L^r(\mathbb{R}^n)} \leq C \||x|^a |\nabla f|\|^{\delta}_{L^p(\mathbb{R}^n)} \||x|^b f\|^{1-\delta}_{L^q(\mathbb{R}^n)} \tag{3.100}$$

holds for all $f \in C_0^\infty(\mathbb{R}^n)$, if and only if the following conditions hold:

$$\frac{1}{r} + \frac{c}{n} = \delta\left(\frac{1}{p} + \frac{a-1}{n}\right) + (1-\delta)\left(\frac{1}{q} + \frac{b}{n}\right), \tag{3.101}$$

$$a - d \geq 0 \quad \textit{if} \quad \delta > 0, \tag{3.102}$$

$$a - d \leq 1 \quad \text{if} \quad \delta > 0 \quad \textit{and} \quad \frac{1}{r} + \frac{c}{n} = \frac{1}{p} + \frac{a-1}{n}. \tag{3.103}$$

Thus, in this section we are interested in inequalities of this type, and we show that some Caffarelli–Kohn–Nirenberg inequalities continue to hold in the setting of homogeneous groups, in particular, including the cases of anisotropic structures on $\mathbb{R}^n$, i.e., the quasi-norm $|\cdot|$ does not need to be the Euclidean norm $|\cdot|_E$ on $\mathbb{R}^n$.

Some inequalities will be obtained as a consequence of the weighted Hardy inequalities. As a particular case of such weighted inequalities we can think of the inequalities

$$\left(\int_{\mathbb{R}^n} |x|_E^{-p\beta} |f|^p dx\right)^{\frac{2}{p}} \leq C_{\alpha,\beta} \int_{\mathbb{R}^n} |x|_E^{-2\alpha} |\nabla f|^2 dx, \tag{3.104}$$

for $f \in C_0^\infty(\mathbb{R}^n)$, where for $n \geq 3$:

$$-\infty < \alpha < \frac{n-2}{2}, \ \alpha \leq \beta \leq \alpha + 1, \text{ and } p = \frac{2n}{n-2+2(\beta-\alpha)},$$

and for $n = 2$:

$$-\infty < \alpha < 0, \ \alpha < \beta \leq \alpha + 1, \text{ and } p = \frac{2}{\beta - \alpha}.$$

Here

$$|x|_E = \sqrt{x_1^2 + \cdots + x_n^2}$$

is the standard Euclidean norm. Moreover, we are interested in replacing the Euclidean norm by a general quasi-norm as well as extending such inequalities to general homogeneous groups.

As a special case we can highlight the case of $p = 2$ that was also studied by [WW03] in the Euclidean setting. Here, for all $f \in C_0^\infty(\mathbb{R}^n)$ we have

$$\int_{\mathbb{R}^n} |x|_E^{-2(\alpha+1)} |f|^2 dx \leq \widetilde{C}_\alpha \int_{\mathbb{R}^n} |x|_E^{-2\alpha} |\nabla f|^2 dx,$$

with any $n \geq 2$ and $-\infty < \alpha < 0$, which in turn can be written for any $f \in C_0^\infty(\mathbb{R}^n\backslash\{0\})$ as

$$\left\| \frac{1}{|x|_E^{\alpha+1}} |f| \right\|_{L^2(\mathbb{R}^n)} \leq C_\alpha \left\| \frac{1}{|x|_E^{\alpha}} |\nabla f| \right\|_{L^2(\mathbb{R}^n)}, \tag{3.105}$$

for all $\alpha \in \mathbb{R}$.

A homogeneous group version of the inequality (3.105) was obtained in Corollary 2.1.6, that is, it was proved that if $\mathbb{G}$ is a homogeneous group of homogeneous dimension Q, then for any homogeneous quasi-norm $|\cdot|$ on $\mathbb{G}$ and for every $f \in C_0^\infty(\mathbb{G}\backslash\{0\})$ we have

$$\frac{|Q-2-2\alpha|}{2} \left\| \frac{f}{|x|^{\alpha+1}} \right\|_{L^2(\mathbb{G})} \leq \left\| \frac{1}{|x|^{\alpha}} \mathcal{R} f \right\|_{L^2(\mathbb{G})} \quad \text{for all } \alpha \in \mathbb{R}, \tag{3.106}$$

where $\mathcal{R}$ is the radial derivative operator with respect to the norm $|\cdot|$. Note that if $\alpha \neq \frac{Q-2}{2}$, then the constant in (3.106) was shown to be sharp for any homogeneous quasi-norm $|\cdot|$ on $\mathbb{G}$.

Remark 3.3.2.

1. An alternative formulation of Theorem 3.3.1 emphasizing the appearing indices was given by D'Ancona and Luca [DL12].
2. The improved versions of the Caffarelli–Kohn–Nirenberg inequality for radially symmetric functions with respect to the range of parameters were investigated in [NDD12]. In [ZHD15] and [HZ11], weighted Hardy type inequalities were obtained for the generalized Baouendi–Grushin vector fields: for $\gamma = 0$ it gives the standard gradient in $\mathbb{R}^n$. We also refer to [HNZ11], [Han15] for weighted Hardy inequalities on the Heisenberg group, to [HZD11] and [ZHD14] on the H-type groups, and a recent paper [Yac18] on Lie groups of polynomial growth.
3. The analysis in this section is based on [ORS18] as well as on [RSY17b] and [RSY18b].

3.3.1 L^p-Caffarelli–Kohn–Nirenberg inequalities

In this section we generalize inequality (3.106) to L^p-cases for all $1 < p < \infty$. Since all the inequalities are of similar type we will keep calling them the Caffarelli–Kohn–Nirenberg inequalities.

Theorem 3.3.3 (Caffarelli–Kohn–Nirenberg inequality for L^p-norms). *Let $\mathbb{G}$ be a homogeneous group of homogeneous dimension $Q \geq 2$ and let $|\cdot|$ be a homogeneous quasi-norm on $\mathbb{G}$. Then we have*

$$\frac{|Q-\gamma|}{p} \left\| \frac{f}{|x|^{\frac{\gamma}{p}}} \right\|_{L^p(\mathbb{G})}^p \leq \left\| \frac{1}{|x|^{\alpha}} \mathcal{R} f \right\|_{L^p(\mathbb{G})} \left\| \frac{f}{|x|^{\frac{\beta}{p-1}}} \right\|_{L^p(\mathbb{G})}^{p-1}, \tag{3.107}$$

for all complex-valued functions $f \in C_0^\infty(\mathbb{G}\backslash\{0\})$, $1 < p < \infty$, *and all* $\alpha, \beta \in \mathbb{R}$ *with*

$$\gamma = \alpha + \beta + 1.$$

If $\gamma \neq Q$ *then the constant* $\frac{|Q-\gamma|}{p}$ *is sharp.*

Before proving this theorem let us point out some of its implications.

Remark 3.3.4.

1. In the Euclidean case $\mathbb{G} = (\mathbb{R}^n, +)$, Theorem 3.3.3 gives the inequality

$$\frac{|n-\gamma|}{p}\left\|\frac{f}{|x|^{\frac{\gamma}{p}}}\right\|^p_{L^p(\mathbb{R}^n)} \leq \left\|\frac{1}{|x|^\alpha}\frac{df}{d|x|}\right\|_{L^p(\mathbb{R}^n)}\left\|\frac{f}{|x|^{\frac{\beta}{p-1}}}\right\|^{p-1}_{L^p(\mathbb{R}^n)},$$

with the optimal constant. In particular, for the standard Euclidean distance $|x|_E = \sqrt{x_1^2 + \cdots + x_n^2}$, by using the Cauchy–Schwarz inequality, it follows that

$$\frac{|n-\gamma|}{p}\left\|\frac{f}{|x|_E^{\frac{\gamma}{p}}}\right\|^p_{L^p(\mathbb{R}^n)} \leq \left\|\frac{1}{|x|_E^\alpha}\nabla f\right\|_{L^p(\mathbb{R}^n)}\left\|\frac{f}{|x|_E^{\frac{\beta}{p-1}}}\right\|^{p-1}_{L^p(\mathbb{R}^n)},$$

for all $f \in C_0^\infty(\mathbb{R}^n\backslash\{0\})$, with the sharp constant.

2. In the case $\alpha = 0$, $\beta = p-1$, and $1 < p < Q$, the inequality (3.107) implies the homogeneous group version of the L^p-Hardy inequality

$$\left\|\frac{1}{|x|}f\right\|_{L^p(\mathbb{G})} \leq \frac{p}{Q-p}\|\mathcal{R}f\|_{L^p(\mathbb{G})}, \tag{3.108}$$

again with $\frac{p}{Q-p}$ being the best constant, see Section 2.1.1.

3. For $\mathbb{G} = (\mathbb{R}^n, +)$, $n \geq 3$, inequality (3.108) gives

$$\left\|\frac{f}{|x|}\right\|_{L^p(\mathbb{R}^n)} \leq \frac{p}{n-p}\left\|\frac{df}{d|x|}\right\|_{L^p(\mathbb{R}^n)}. \tag{3.109}$$

For the Euclidean distance, by the Cauchy–Schwarz inequality, it implies the classical Hardy inequality:

$$\left\|\frac{f}{|x|_E}\right\|_{L^p(\mathbb{R}^n)} \leq \frac{p}{n-p}\|\nabla f\|_{L^p(\mathbb{R}^n)},$$

for all $f \in C_0^\infty(\mathbb{R}^n\backslash\{0\})$.

4. For the Euclidean distance, the exact formulae of the difference between the right-hand side and the left-hand side of inequality (3.109) were investigated by Ioku, Ishiwata and Ozawa [IIO16b], see also Machihara, Ozawa and Wadade [MOW17a] as well as [IIO16a].

5. One can obtain a number of Heisenberg–Pauli–Weyl type uncertainty inequities directly from the inequality (3.107) which have various consequences and applications. For instance, if $\alpha p = \alpha + \beta + 1$, inequality (3.107) gives

$$\frac{|Q-\alpha p|}{p}\left\|\frac{f}{|x|^{\alpha}}\right\|^{p}_{L^p(\mathbb{G})} \leq \left\|\frac{\mathcal{R}f}{|x|^{\alpha}}\right\|_{L^p(\mathbb{G})}\left\||x|^{\frac{1}{p-1}}\frac{f}{|x|^{\alpha}}\right\|^{p-1}_{L^p(\mathbb{G})}. \tag{3.110}$$

For $\alpha + \beta + 1 = 0$ and $\alpha = -p$, inequality (3.107) gives

$$\frac{Q}{p}\|f\|^{p}_{L^p(\mathbb{G})} \leq \||x|^{p}\mathcal{R}f\|_{L^p(\mathbb{G})}\left\|\frac{f}{|x|}\right\|^{p-1}_{L^p(\mathbb{G})}, \tag{3.111}$$

all with sharp constants.

The Hardy inequality immediately implies a version of the Heisenberg–Pauli–Weyl uncertainty principle.

Corollary 3.3.5 (Heisenberg–Pauli–Weyl type uncertainty principle). *Let $\mathbb{G}$ be a homogeneous group of homogeneous dimension $Q \geq 2$ and let $|\cdot|$ be a homogeneous quasi-norm on $\mathbb{G}$. Then we have*

$$\|f\|^{2}_{L^2(\mathbb{G})} \leq \frac{p}{Q-p}\|\mathcal{R}f\|_{L^p(\mathbb{G})}\||x|f\|_{L^{\frac{p}{p-1}}(\mathbb{G})}, \quad 1<p<Q, \tag{3.112}$$

for all $f \in C_0^{\infty}(\mathbb{G}\backslash\{0\})$.

Proof of Corollary 3.3.5. By applying the Hölder inequality to (3.108) with $1 < p < Q$, we get

$$\|f\|^{2}_{L^2(\mathbb{G})} \leq \left\|\frac{1}{|x|}f\right\|_{L^p(\mathbb{G})}\||x|f\|_{L^{\frac{p}{p-1}}(\mathbb{G})} \leq \frac{p}{Q-p}\|\mathcal{R}f\|_{L^p(\mathbb{G})}\||x|f\|_{L^{\frac{p}{p-1}}(\mathbb{G})}, \tag{3.113}$$

giving (3.112). ☐

Remark 3.3.6.

1. For the Euclidean space $\mathbb{G} = (\mathbb{R}^n, +)$ and $p = 2$, inequality (3.112) implies the uncertainty principle for any homogeneous quasi-norm $|x|$ on $\mathbb{R}^n$:

$$\left(\int_{\mathbb{R}^n}|f(x)|^2dx\right)^2 \leq \left(\frac{2}{n-2}\right)^2\int_{\mathbb{R}^n}\left|\frac{df(x)}{d|x|}\right|^2dx\int_{\mathbb{R}^n}|x|^2|f(x)|^2dx. \tag{3.114}$$

In turn this gives the classical uncertainty principle on $\mathbb{R}^n$ with the standard Euclidean distance:

$$\left(\int_{\mathbb{R}^n}|f(x)|^2dx\right)^2 \leq \left(\frac{2}{n-2}\right)^2\int_{\mathbb{R}^n}|\nabla f(x)|^2dx\int_{\mathbb{R}^n}|x|_E^2|f(x)|^2dx,$$

which is the Heisenberg–Pauli–Weyl uncertainty principle on the Euclidean spaces $\mathbb{R}^n$, $n > 2$.

2. The stratified groups version of Corollary 3.3.5 will be discussed in Section 4.7.

Proof of Theorem 3.3.3. In the case $\gamma = Q$ the inequality is trivial, so we may assume that $\gamma \neq Q$. To use the polar decomposition in Proposition 1.2.10 we denote $(r, y) = (|x|, \frac{x}{|x|}) \in (0, \infty) \times \wp$ on $\mathbb{G}$, where $\wp$ is the unit quasi-sphere. Then a direct calculation gives that

$$\begin{aligned}
\int_{\mathbb{G}} \frac{|f(x)|^p}{|x|^\gamma} dx &= \int_0^\infty \int_\wp \frac{|f(ry)|^p}{r^\gamma} r^{Q-1} d\sigma(y) dr \\
&= \frac{1}{Q-\gamma} \int_0^\infty \int_\wp |f(ry)|^p \frac{d\, r^{Q-\gamma}}{dr} d\sigma(y) dr \\
&= -\frac{1}{Q-\gamma} \mathrm{Re} \int_0^\infty \int_\wp p f(ry) |f(ry)|^{p-2} \left(\overline{\frac{df(ry)}{dr}} \right) \frac{1}{r^{\gamma-1}} r^{Q-1} d\sigma(y) dr \\
&= -\frac{p}{Q-\gamma} \mathrm{Re} \int_{\mathbb{G}} f(x) \frac{|f(x)|^{p-2}}{|x|^{\gamma-1}} \left(\overline{\frac{d}{d|x|} f(x)} \right) dx \\
&\leq \left| \frac{p}{Q-\gamma} \right| \int_{\mathbb{G}} \frac{|f(x)|^{p-1}}{|x|^{\gamma-1}} \left| \frac{d}{d|x|} f(x) \right| dx \\
&= \left| \frac{p}{Q-\gamma} \right| \int_{\mathbb{G}} \frac{|f(x)|^{p-1}}{|x|^{\alpha+\beta}} |\mathcal{R} f(x)|\, dx \\
&\leq \left| \frac{p}{Q-\gamma} \right| \left(\int_{\mathbb{G}} \frac{|\mathcal{R} f(x)|^p}{|x|^{\alpha p}} dx \right)^{1/p} \left(\int_{\mathbb{G}} \frac{|f(x)|^p}{|x|^{\frac{\beta p}{p-1}}} dx \right)^{(p-1)/p},
\end{aligned}$$

using the Hölder inequality in the last line. Thus, we obtain

$$\left| \frac{Q-\gamma}{p} \right| \int_{\mathbb{G}} \frac{|f(x)|^p}{|x|^\gamma} dx \leq \left(\int_{\mathbb{G}} \frac{|\mathcal{R} f(x)|^p}{|x|^{\alpha p}} dx \right)^{1/p} \left(\int_{\mathbb{G}} \frac{|f(x)|^p}{|x|^{\frac{\beta p}{p-1}}} dx \right)^{(p-1)/p},$$

yielding (3.107).

Now it remains to show the sharpness of the constant. To do it we have to examine the equality condition in the Hölder inequality. Let us consider the following function

$$g(x) = \begin{cases} e^{-\frac{C}{\lambda}|x|^\lambda}, & \lambda := \alpha - \frac{\beta}{p-1} + 1 \neq 0, \\ \frac{1}{|x|^C}, & \alpha - \frac{\beta}{p-1} + 1 = 0, \end{cases}$$

where $C = \left| \frac{Q-\gamma}{p} \right|$ and $\gamma \neq Q$. Then one can readily check that

$$\left| \frac{p}{Q-\gamma} \right|^p \frac{|\mathcal{R} g(x)|^p}{|x|^{\alpha p}} = \frac{|g(x)|^p}{|x|^{\frac{\beta p}{p-1}}},$$

satisfying the equality condition in the Hölder inequality. This means that the constant $|(Q-\gamma)/p|$ is sharp. □

3.3.2 Higher-order L^p-Caffarelli–Kohn–Nirenberg inequalities

By iterating the L^p-Caffarelli–Kohn–Nirenberg type inequalities from Theorem 3.3.3 we can establish higher-order inequalities.

Theorem 3.3.7 (Higher-order L^p-Caffarelli–Kohn–Nirenberg inequalities). *Let $\mathbb{G}$ be a homogeneous group of homogeneous dimension $Q \geq 2$ and let $|\cdot|$ be a homogeneous quasi-norm on $\mathbb{G}$. Then for all $k, m \in \mathbb{N}$ and all $1 < p < \infty$ we have*

$$\frac{|Q-\gamma|}{p}\left\|\frac{f}{|x|^{\frac{\gamma}{p}}}\right\|^p_{L^p(\mathbb{G})} \leq \widetilde{A}_{\alpha,m}\widetilde{A}_{\beta,k}\left\|\frac{1}{|x|^{\alpha-m}}\mathcal{R}^{m+1}f\right\|_{L^p(\mathbb{G})}\left\|\frac{1}{|x|^{\frac{\beta}{p-1}-k}}\mathcal{R}^k f\right\|^{p-1}_{L^p(\mathbb{G})}, \tag{3.115}$$

for all complex-valued function $f \in C_0^\infty(\mathbb{G}\backslash\{0\})$,

$$\gamma = \alpha + \beta + 1,$$

and $\alpha \in \mathbb{R}$ such that $\prod_{j=0}^{m-1}|Q-p(\alpha-j)| \neq 0$, and

$$\widetilde{A}_{\alpha,m} := p^m\left[\prod_{j=0}^{m-1}|Q-p(\alpha-j)|\right]^{-1},$$

as well as $\beta \in \mathbb{R}$ such that $\prod_{j=0}^{k-1}\left|Q-p(\frac{\beta}{p-1}-j)\right| \neq 0$, and

$$\widetilde{A}_{\beta,k} := p^{k(p-1)}\left[\prod_{j=0}^{k-1}\left|Q-p\left(\frac{\beta}{p-1}-j\right)\right|\right]^{-(p-1)}.$$

For $p=2$ the above constants are sharp.

Proof of Theorem 3.3.7. First, let us consider in (3.107) the case

$$\beta = \gamma\left(1-\frac{1}{p}\right).$$

In this case we have $\beta = (\alpha+1)(p-1)$ and $\gamma = p(\alpha+1)$, so that inequality (3.107) becomes

$$\left\|\frac{f}{|x|^{\alpha+1}}\right\|_{L^p(\mathbb{G})} \leq \frac{p}{|Q-p(\alpha+1)|}\left\|\frac{1}{|x|^{\alpha}}\mathcal{R}f\right\|_{L^p(\mathbb{G})}, \quad 1<p<\infty, \tag{3.116}$$

for all $f \in C_0^\infty(\mathbb{G}\backslash\{0\})$ and every $\alpha \in \mathbb{R}$ with $\alpha \neq \frac{Q}{p}-1$.

Now taking $\mathcal{R}f$ instead of f and $\alpha-1$ instead of α in (3.116) we consequently obtain

$$\left\|\frac{\mathcal{R}f}{|x|^{\alpha}}\right\|_{L^p(\mathbb{G})} \leq \frac{p}{|Q-p\alpha|}\left\|\frac{1}{|x|^{\alpha-1}}\mathcal{R}^2 f\right\|_{L^p(\mathbb{G})},$$

for $\alpha \neq \frac{Q}{p}$. Combining it with (3.116) we get

$$\left\|\frac{f}{|x|^{\alpha+1}}\right\|_{L^p(\mathbb{G})} \leq \frac{p}{|Q-p(\alpha+1)|}\frac{p}{|Q-p\alpha|}\left\|\frac{1}{|x|^{\alpha-1}}\mathcal{R}^2 f\right\|_{L^p(\mathbb{G})},$$

for every $\alpha \in \mathbb{R}$ such that $\alpha \neq \frac{Q}{p}-1$ and $\alpha \neq \frac{Q}{p}$. This iteration process yields

$$\left\|\frac{f}{|x|^{\theta+1}}\right\|_{L^p(\mathbb{G})} \leq A_{\theta,k}\left\|\frac{1}{|x|^{\theta+1-k}}\mathcal{R}^k f\right\|_{L^p(\mathbb{G})}, \quad 1<p<\infty, \tag{3.117}$$

for all $f \in C_0^\infty(\mathbb{G}\backslash\{0\})$ and all $\theta \in \mathbb{R}$ such that $\prod_{j=0}^{k-1}|Q-p(\theta+1-j)| \neq 0$, and

$$A_{\theta,k} := p^k\left[\prod_{j=0}^{k-1}|Q-p(\theta+1-j)|\right]^{-1}.$$

Similarly, we get

$$\left\|\frac{\mathcal{R}f}{|x|^{\vartheta+1}}\right\|_{L^p(\mathbb{G})} \leq A_{\vartheta,m}\left\|\frac{1}{|x|^{\vartheta+1-m}}\mathcal{R}^{m+1} f\right\|_{L^p(\mathbb{G})}, \quad 1<p<\infty, \tag{3.118}$$

for all $f \in C_0^\infty(\mathbb{G}\backslash\{0\})$ and all $\vartheta \in \mathbb{R}$ such that $\prod_{j=0}^{m-1}|Q-p(\vartheta+1-j)| \neq 0$, and

$$A_{\vartheta,m} := p^m\left[\prod_{j=0}^{m-1}|Q-p(\vartheta+1-j)|\right]^{-1}.$$

Now putting $\vartheta+1=\alpha$ and $\theta+1=\frac{\beta}{p-1}$ into (3.118) and (3.117), respectively, we arrive at (3.115).

Let us now show the sharpness of the constants in the case $p=2$. This will follow from having an exact form of the remainder in these inequalities. Recall Theorem 3.1.10 saying that if $Q \geq 3$, $\alpha \in \mathbb{R}$, $k \in \mathbb{N}$ and $\prod_{j=0}^{k-1}\left|\frac{Q-2}{2}-(\alpha+j)\right| \neq 0$, then for all complex-valued functions $f \in C_0^\infty(\mathbb{G}\backslash\{0\})$ we have

$$\left\|\frac{f}{|x|^{k+\alpha}}\right\|_{L^2(\mathbb{G})} \leq \left[\prod_{j=0}^{k-1}\left|\frac{Q-2}{2}-(\alpha+j)\right|\right]^{-1}\left\|\frac{1}{|x|^{\alpha}}\mathcal{R}^k f\right\|_{L^2(\mathbb{G})}, \tag{3.119}$$

where the constant is sharp. In addition, from (3.34) we have the identity

$$\begin{aligned}\left\|\frac{1}{|x|^{\alpha}}\mathcal{R}^{k}f\right\|^{2}_{L^{2}(\mathbb{G})} &= \left[\prod_{j=0}^{k-1}\left(\frac{Q-2}{2}-(\alpha+j)\right)^{2}\right]\left\|\frac{f}{|x|^{k+\alpha}}\right\|^{2}_{L^{2}(\mathbb{G})} \\ &+\sum_{l=1}^{k-1}\left[\prod_{j=0}^{l-1}\left(\frac{Q-2}{2}-(\alpha+j)\right)^{2}\right] \\ &\times\left\|\frac{1}{|x|^{l+\alpha}}\mathcal{R}^{k-l}f+\frac{Q-2(l+1+\alpha)}{2|x|^{l+1+\alpha}}\mathcal{R}^{k-l-1}f\right\|^{2}_{L^{2}(\mathbb{G})} \\ &+\left\|\frac{1}{|x|^{\alpha}}\mathcal{R}^{k}f+\frac{Q-2-2\alpha}{2|x|^{1+\alpha}}\mathcal{R}^{k-1}f\right\|^{2}_{L^{2}(\mathbb{G})}, \end{aligned} \tag{3.120}$$

for all $k\in\mathbb{N}$ and $\alpha\in\mathbb{R}$. When $p=2$, Theorem 3.3.3 can be restated that for each $f\in C_0^{\infty}(\mathbb{G}\backslash\{0\})$ we have

$$\frac{|Q-\gamma|}{2}\left\|\frac{f}{|x|^{\frac{\gamma}{2}}}\right\|^{2}_{L^{2}(\mathbb{G})}\leq\left\|\frac{1}{|x|^{\alpha}}\mathcal{R}f\right\|_{L^{2}(\mathbb{G})}\left\|\frac{f}{|x|^{\beta}}\right\|_{L^{2}(\mathbb{G})},\quad\forall\alpha,\,\beta\in\mathbb{R}, \tag{3.121}$$

where $\gamma=\alpha+\beta+1$. The sharpness then follows from the following remark. □

Remark 3.3.8. Combining (3.121) with (3.119) (or with (3.120)), one can obtain a number of inequalities with sharp constants, for example:

$$\frac{|Q-\gamma|}{2}\left\|\frac{f}{|x|^{\frac{\gamma}{2}}}\right\|^{2}_{L^{2}(\mathbb{G})}\leq C_j(\beta,k)\left\|\frac{1}{|x|^{\alpha}}\mathcal{R}f\right\|_{L^{2}(\mathbb{G})}\left\|\frac{1}{|x|^{\beta-k}}\mathcal{R}^{k}f\right\|_{L^{2}(\mathbb{G})}, \tag{3.122}$$

for $\gamma=\alpha+\beta+1$ and all $\alpha,\,\beta\in\mathbb{R}$ and $k\in\mathbb{N}$, such that,

$$C_j(\beta,k):=\left[\prod_{j=0}^{k-1}\left|\frac{Q-2}{2}-(\beta-k+j)\right|\right]^{-1}\neq 0,$$

as well as

$$\frac{|Q-\gamma|}{2}\left\|\frac{f}{|x|^{\frac{\gamma}{2}}}\right\|^{2}_{L^{2}(\mathbb{G})}\leq C_j(\alpha,k)\left\|\frac{1}{|x|^{\alpha-k}}\mathcal{R}^{k+1}f\right\|_{L^{2}(\mathbb{G})}\left\|\frac{f}{|x|^{\beta}}\right\|_{L^{2}(\mathbb{G})}, \tag{3.123}$$

for $\gamma=\alpha+\beta+1$ and all $\alpha,\,\beta\in\mathbb{R}$ and $k\in\mathbb{N}$, such that,

$$C_j(\alpha,k):=\left[\prod_{j=0}^{k-1}\left|\frac{Q-2}{2}-(\alpha+k+j)\right|\right]^{-1}\neq 0.$$

It follows from (3.120) that these constants $C_j(\beta,k)$ and $C_j(\alpha,k)$ in (3.122) and (3.123) are sharp.

3.3.3 New type of L^p-Caffarelli–Kohn–Nirenberg inequalities

In this section, we introduce new Caffarelli–Kohn–Nirenberg type inequalities on homogeneous groups.

Theorem 3.3.9 (New types of L^p-Caffarelli–Kohn–Nirenberg inequalities). *Let $\mathbb{G}$ be a homogeneous group of homogeneous dimension $Q \geq 2$. Let $|\cdot|$ be a homogeneous quasi-norm on $\mathbb{G}$. Let $0 < \delta < 1$. Then for all $f \in C_0^\infty(\mathbb{G}\backslash\{0\})$ we have*

$$\left\| \frac{f}{|x|^{\frac{Q-p}{p}+\delta}} \right\|_{L^p(\mathbb{G})} \leq p^\delta \left\| \frac{\log|x|}{|x|^{\frac{Q-p}{p}}} \mathcal{R}f \right\|_{L^p(\mathbb{G})}^{\delta} \left\| \frac{f}{|x|^{\frac{Q-p}{p}}} \right\|_{L^p(\mathbb{G})}^{1-\delta}, \quad 1 < p < \infty. \tag{3.124}$$

Moreover, we have

$$\left\| \frac{f}{|x|^{\frac{Q-p}{p}+\delta}} \right\|_{L^p(\mathbb{G})} \leq \left\| \frac{\mathcal{R}f}{|x|^{\frac{Q-2p}{p}}} \right\|_{L^p(\mathbb{G})}^{1-\delta} \left\| \frac{f}{|x|^{\frac{Q}{p}}} \right\|_{L^p(\mathbb{G})}^{\delta}, \quad 1 < p < \infty. \tag{3.125}$$

Remark 3.3.10. In the Abelian case $\mathbb{G} = (\mathbb{R}^n, +)$ and $Q = n$, (3.124) implies a new type of the Caffarelli–Kohn–Nirenberg inequality for any quasi-norm on $\mathbb{R}^n$: For any function $f \in C_0^\infty(\mathbb{R}^n\backslash\{0\})$ and any $1 < p < \infty$ we have

$$\left\| \frac{f}{|x|^{\frac{n-p}{p}+\delta}} \right\|_{L^p(\mathbb{R}^n)} \leq p^\delta \left\| \frac{\log|x|}{|x|^{\frac{n-p}{p}}} \left(\frac{x}{|x|} \cdot \nabla f \right) \right\|_{L^p(\mathbb{R}^n)}^{\delta} \left\| \frac{f}{|x|^{\frac{n-p}{p}}} \right\|_{L^p(\mathbb{R}^n)}^{1-\delta}. \tag{3.126}$$

By the Schwarz inequality with the standard Euclidean distance given by $|x|_E = \sqrt{x_1^2 + x_2^2 + \cdots + x_n^2}$, we obtain the Euclidean form of the Caffarelli–Kohn–Nirenberg type inequality for any quasi-norm on $\mathbb{R}^n$, for $1 < p < \infty$, and for all functions $f \in C_0^\infty(\mathbb{R}^n\backslash\{0\})$:

$$\left\| \frac{f}{|x|_E^{\frac{n-p}{p}+\delta}} \right\|_{L^p(\mathbb{R}^n)} \leq p^\delta \left\| \frac{\log|x|_E}{|x|_E^{\frac{n-p}{p}}} \nabla f \right\|_{L^p(\mathbb{R}^n)}^{\delta} \left\| \frac{f}{|x|_E^{\frac{n-p}{p}}} \right\|_{L^p(\mathbb{R}^n)}^{1-\delta}, \tag{3.127}$$

where ∇ is the standard gradient in $\mathbb{R}^n$. Similarly, we can write the inequality (3.125) in the Euclidean case as

$$\left\| \frac{f}{|x|_E^{\frac{n-p}{p}+\delta}} \right\|_{L^p(\mathbb{R}^n)} \leq \left\| \frac{\nabla f}{|x|_E^{\frac{n-2p}{p}}} \right\|_{L^p(\mathbb{R}^n)}^{1-\delta} \left\| \frac{f}{|x|_E^{\frac{n}{p}}} \right\|_{L^p(\mathbb{R}^n)}^{\delta}, \quad 1 < p < \infty. \tag{3.128}$$

Note that since $\frac{1}{p} + (-\frac{n}{p})\frac{1}{n} = 0$, the inequality (3.128) does not follow from the Caffarelli–Kohn–Nirenberg inequality in Theorem 3.3.3, thus providing an extension of (3.107) in terms of indices but also in terms of a possibility of choosing any homogeneous quasi-norm on $\mathbb{R}^n$.

Proof of Theorem 3.3.9. A direct calculation shows that we have

$$\left\|\frac{f}{|x|^{\frac{Q-p}{p}+\delta}}\right\|^p_{L^p(\mathbb{G})} = \int_{\mathbb{G}} \frac{|f(x)|^p}{|x|^{Q-p+\delta p}}dx = \int_{\mathbb{G}} \frac{|f(x)|^{\delta p}}{|x|^{\delta Q}} \cdot \frac{|f(x)|^{p(1-\delta)}}{|x|^{(Q-p)(1-\delta)}}dx.$$

Using Hölder's inequality it follows that

$$\left\|\frac{f}{|x|^{\frac{Q-p}{p}+\delta}}\right\|^p_{L^p(\mathbb{G})} \leq \left(\int_{\mathbb{G}} \frac{|f(x)|^p}{|x|^Q}dx\right)^\delta \left(\int_{\mathbb{G}} \frac{|f(x)|^p}{|x|^{Q-p}}dx\right)^{1-\delta}. \tag{3.129}$$

By Theorem 3.2.3, we have

$$\int_{\mathbb{G}} \frac{|f(x)|^p}{|x|^Q}dx \leq p^p \int_{\mathbb{G}} \frac{(\log|x|)^p}{|x|^Q}|\mathbb{E}f(x)|^p dx, \quad 1 < p < \infty,$$

where $\mathbb{E} = |x|\mathcal{R}$ is the Euler operator. It implies that

$$\int_{\mathbb{G}} \frac{|f(x)|^p}{|x|^Q}dx \leq p^p \int_{\mathbb{G}} \frac{(\log|x|)^p}{|x|^{Q-p}}|\mathcal{R}f(x)|^p dx, \quad 1 < p < \infty.$$

Using this in (3.129), one obtains

$$\left\|\frac{f}{|x|^{\frac{Q-p}{p}+\delta}}\right\|^p_{L^p(\mathbb{G})} \leq p^{p\delta} \left(\int_{\mathbb{G}} \frac{(\log|x|)^p}{|x|^{Q-p}}|\mathcal{R}f(x)|^p dx\right)^\delta \left(\int_{\mathbb{G}} \frac{|f(x)|^p}{|x|^{Q-p}}dx\right)^{1-\delta},$$

which implies (3.124).

Now let us prove (3.125). Using Theorem 3.2.3, one has

$$\int_{\mathbb{G}} \frac{|f(x)|^p}{|x|^{Q-p}}dx \leq \int_{\mathbb{G}} \frac{|\mathbb{E}f(x)|^p}{|x|^{Q-p}}dx, \quad 1 < p < \infty.$$

Then, using this in (3.130), we obtain

$$\begin{aligned}\left\|\frac{f}{|x|^{\frac{Q-p}{p}+\delta}}\right\|^p_{L^p(\mathbb{G})} &\leq \left(\int_{\mathbb{G}} \frac{|f(x)|^p}{|x|^Q}dx\right)^\delta \left(\int_{\mathbb{G}} \frac{|\mathbb{E}f(x)|^p}{|x|^{Q-p}}dx\right)^{1-\delta} \\ &= \left(\int_{\mathbb{G}} \frac{|f(x)|^p}{|x|^Q}dx\right)^\delta \left(\int_{\mathbb{G}} \frac{|\mathcal{R}f(x)|^p}{|x|^{Q-2p}}dx\right)^{1-\delta},\end{aligned}$$

which gives (3.125), completing the proof. □

3.3.4 Extended Caffarelli–Kohn–Nirenberg inequalities

In this section, we extend the range of indices for Theorem 3.3.1. Again, we work in the setting of general homogeneous groups: $\mathbb{G}$ is a homogeneous group of homogeneous dimension $Q \geq 1$ and $|\cdot|$ is a homogeneous quasi-norm on $\mathbb{G}$.

Theorem 3.3.11 (Extended Caffarelli–Kohn–Nirenberg inequalities). *Let $1 < p, q < \infty$, $0 < r < \infty$, with $p + q \geq r$. Let $\delta \in [0,1] \cap \left[\frac{r-q}{r}, \frac{p}{r}\right]$ and a, b, $c \in \mathbb{R}$. Assume that*

$$\frac{\delta r}{p} + \frac{(1-\delta)r}{q} = 1 \quad \text{and} \quad c = \delta(a-1) + b(1-\delta).$$

Then for all $f \in C_0^\infty(\mathbb{G}\backslash\{0\})$ we have the following inequalities:

(i) *If $Q \neq p(1-a)$, then we have*

$$\||x|^c f\|_{L^r(\mathbb{G})} \leq \left|\frac{p}{Q - p(1-a)}\right|^\delta \||x|^a \mathcal{R} f\|^\delta_{L^p(\mathbb{G})} \||x|^b f\|^{1-\delta}_{L^q(\mathbb{G})}. \tag{3.130}$$

(ii) *If $Q = p(1-a)$, then we have*

$$\||x|^c f\|_{L^r(\mathbb{G})} \leq p^\delta \||x|^a \log|x| \mathcal{R} f\|^\delta_{L^p(\mathbb{G})} \||x|^b f\|^{1-\delta}_{L^q(\mathbb{G})}. \tag{3.131}$$

The constant in the inequality (3.130) is sharp for $p = q$ with $a - b = 1$ or $p \neq q$ with $p(1-a) + bq \neq 0$. Moreover, the constants in (3.130) and (3.131) are sharp for $\delta = 0$ or $\delta = 1$.

To compare these inequalities with those in Theorem 3.3.1 let us first formulate the isotropic version of Theorem 3.3.11 in the usual setting of $\mathbb{R}^n$, and its further implication in the case of the Euclidean norm.

Corollary 3.3.12. *Let $|\cdot|$ be a homogeneous quasi-norm on $\mathbb{R}^n$, $n \in \mathbb{N}$. Let $1 < p, q < \infty$, $0 < r < \infty$, with $p + q \geq r$, $\delta \in [0,1] \cap \left[\frac{r-q}{r}, \frac{p}{r}\right]$ and a, b, $c \in \mathbb{R}$. Assume that*

$$\frac{\delta r}{p} + \frac{(1-\delta)r}{q} = 1 \quad \text{and} \quad c = \delta(a-1) + b(1-\delta).$$

Then we have the following estimates:

(i) *If $n \neq p(1-a)$, then for any function $f \in C_0^\infty(\mathbb{R}^n\backslash\{0\})$ we have*

$$\||x|^c f\|_{L^r(\mathbb{R}^n)} \leq \left|\frac{p}{n - p(1-a)}\right|^\delta \left\||x|^a \left(\frac{x}{|x|} \cdot \nabla f\right)\right\|^\delta_{L^p(\mathbb{R}^n)} \||x|^b f\|^{1-\delta}_{L^q(\mathbb{R}^n)}. \tag{3.132}$$

(ii) *In the critical case $n = p(1-a)$ for any function $f \in C_0^\infty(\mathbb{R}^n\backslash\{0\})$ we have*

$$\||x|^c f\|_{L^r(\mathbb{R}^n)} \leq p^\delta \left\||x|^a \log|x| \left(\frac{x}{|x|} \cdot \nabla f\right)\right\|^\delta_{L^p(\mathbb{R}^n)} \||x|^b f\|^{1-\delta}_{L^q(\mathbb{R}^n)}. \tag{3.133}$$

(iii) *If $|\cdot|_E$ is the Euclidean norm on $\mathbb{R}^n$, inequalities (3.132) and (3.133) imply, respectively,*

$$\||x|_E^c f\|_{L^r(\mathbb{R}^n)} \leq \left|\frac{p}{n - p(1-a)}\right|^\delta \||x|_E^a \nabla f\|^\delta_{L^p(\mathbb{R}^n)} \||x|_E^b f\|^{1-\delta}_{L^q(\mathbb{R}^n)} \tag{3.134}$$

for $n \neq p(1-a)$, and

$$\left\| |x|_E^c f \right\|_{L^r(\mathbb{R}^n)} \leq p^\delta \left\| |x|_E^a \log|x| \nabla f \right\|^\delta_{L^p(\mathbb{R}^n)} \left\| |x|_E^b f \right\|^{1-\delta}_{L^q(\mathbb{R}^n)}, \tag{3.135}$$

for $n = p(1-a)$.

The inequality (3.132) holds for any homogeneous quasi-norm $|\cdot|$, and the constant $\left|\frac{p}{n-p(1-a)}\right|^\delta$ is sharp for $p = q$ with $a - b = 1$, or for $p \neq q$ with $p(1-a) + bq \neq 0$. Furthermore, the constants $\left|\frac{p}{n-p(1-a)}\right|^\delta$ and p^δ are sharp for $\delta = 0, 1$.

Remark 3.3.13.

1. If the conditions (3.99) on the parameters hold, then the inequality (3.134) is contained in the inequalities in Theorem 3.3.1. However, already in this case, if we require $p = q$ with $a - b = 1$ or $p \neq q$ with $p(1-a) + bq \neq 0$, then (3.134) yields the inequality (3.100) with sharp constant. Moreover, the constants $\left|\frac{p}{n-p(1-a)}\right|^\delta$ and p^δ are sharp for $\delta = 0$ or $\delta = 1$. The conditions

$$\frac{\delta r}{p} + \frac{(1-\delta)r}{q} = 1 \quad \text{and} \quad c = \delta(a-1) + b(1-\delta)$$

imply the condition (3.101) of Theorem 3.3.1, as well as conditions (3.102)–(3.103) which are all necessary for having estimates of this type, at least under the conditions (3.99).

2. If the conditions (3.99) are not satisfied, then the inequality (3.134) is not covered by Theorem 3.3.1. So, this gives an extension of Theorem 3.3.1 with respect to the range of parameters. Indeed, let us take, for example,

$$1 < p = q = r < \infty, \; a = -\frac{n-2p}{p}, \; b = -\frac{n}{p}, \; c = -\frac{n-\delta p}{p}.$$

Then by (3.134), for all $f \in C_0^\infty(\mathbb{R}^n \backslash \{0\})$ we have the inequalities

$$\left\| \frac{f}{|x|_E^{\frac{n-\delta p}{p}}} \right\|_{L^p(\mathbb{R}^n)} \leq \left\| \frac{\nabla f}{|x|_E^{\frac{n-2p}{p}}} \right\|^\delta_{L^p(\mathbb{R}^n)} \left\| \frac{f}{|x|_E^{\frac{n}{p}}} \right\|^{1-\delta}_{L^p(\mathbb{R}^n)}, \tag{3.136}$$

for all $1 < p < \infty$ and $0 \leq \delta \leq 1$, where ∇ is the standard gradient in $\mathbb{R}^n$. Since we have

$$\frac{1}{q} + \frac{b}{n} = \frac{1}{p} + \frac{1}{n}\left(-\frac{n}{p}\right) = 0,$$

we see that conditions (3.99) fail, so that the inequality (3.136) is not covered by Theorem 3.3.1.

Proof of Theorem 3.3.11. Case $\delta = 0$. In this case we have $q = r$ and $b = c$ by $\frac{\delta r}{p} + \frac{(1-\delta)r}{q} = 1$ and $c = \delta(a-1) + b(1-\delta)$, respectively. Then, the inequalities (3.130) and (3.131) are equivalent to the trivial estimate

$$\||x|^b f\|_{L^q(\mathbb{G})} \leq \left\||x|^b f\right\|_{L^q(\mathbb{G})},$$

with clearly a sharp constant.

Case $\delta = 1$. In this case we have $p = r$ and $a - 1 = c$. By Theorem 3.2.3, we have for $Q + pc = Q + p(a-1) \neq 0$ the inequality

$$\||x|^c f\|_{L^r(\mathbb{G})} \leq \left|\frac{p}{Q+pc}\right| \||x|^c \mathbb{E} f\|_{L^r(\mathbb{G})},$$

where $\mathbb{E} = |x|\mathcal{R}$ is the Euler operator. Using this estimate we get

$$\||x|^c f\|_{L^r(\mathbb{G})} \leq \left|\frac{p}{Q+pc}\right| \||x|^{c+1} \mathcal{R} f\|_{L^r(\mathbb{G})} = \left|\frac{p}{Q-p(1-a)}\right| \||x|^a \mathcal{R} f\|_{L^p(\mathbb{G})},$$

which implies (3.130). For $Q + pc = Q + p(a-1) = 0$ by Theorem 3.2.3 we obtain

$$\begin{aligned} \||x|^c f\|_{L^r(\mathbb{G})} &\leq p\||x|^c \log|x| \mathbb{E} f\|_{L^r(\mathbb{G})} \\ &= p\||x|^{c+1} \log|x| \mathcal{R} f\|_{L^r(\mathbb{G})} \\ &= p\||x|^a \log|x| \mathcal{R} f\|_{L^p(\mathbb{G})}, \end{aligned}$$

which gives (3.131). In this case, the constants in (3.130) and (3.131) are sharp, since the constants in Theorem 3.2.3 are sharp.

Case $\delta \in (0,1) \cap \left[\frac{r-q}{r}, \frac{p}{r}\right]$. Using $c = \delta(a-1) + b(1-\delta)$, a direct calculation gives

$$\||x|^c f\|_{L^r(\mathbb{G})} = \left(\int_{\mathbb{G}} |x|^{cr} |f(x)|^r dx\right)^{1/r} = \left(\int_{\mathbb{G}} \frac{|f(x)|^{\delta r}}{|x|^{\delta r(1-a)}} \frac{|f(x)|^{(1-\delta)r}}{|x|^{-br(1-\delta)}} dx\right)^{1/r}.$$

Since we have $\delta \in (0,1) \cap \left[\frac{r-q}{r}, \frac{p}{r}\right]$ and $p + q \geq r$, then by using Hölder's inequality for $\frac{\delta r}{p} + \frac{(1-\delta)r}{q} = 1$, we obtain

$$\begin{aligned} \||x|^c f\|_{L^r(\mathbb{G})} &\leq \left(\int_{\mathbb{G}} \frac{|f(x)|^p}{|x|^{p(1-a)}} dx\right)^{\delta/p} \left(\int_{\mathbb{G}} \frac{|f(x)|^q}{|x|^{-bq}} dx\right)^{(1-\delta)/q} \\ &= \left\|\frac{f}{|x|^{1-a}}\right\|^{\delta}_{L^p(\mathbb{G})} \left\|\frac{f}{|x|^{-b}}\right\|^{1-\delta}_{L^q(\mathbb{G})}. \end{aligned} \tag{3.137}$$

Here we note that when $p = q$ and $a - b = 1$, the equality in Hölder's inequality holds for any function. We also note that in the case $p \neq q$ the function

$$h(x) = |x|^{\frac{1}{(p-q)}(p(1-a)+bq)} \tag{3.138}$$

satisfies Hölder's equality condition

$$\frac{|h|^p}{|x|^{p(1-a)}} = \frac{|h|^q}{|x|^{-bq}}.$$

If $Q \neq p(1-a)$, then by Theorem 3.2.3 we have

$$\begin{aligned}\left\|\frac{f}{|x|^{1-a}}\right\|^{\delta}_{L^p(\mathbb{G})} &\leq \left|\frac{p}{Q-p(1-a)}\right|^{\delta}\left\|\frac{\mathbb{E}f}{|x|^{1-a}}\right\|^{\delta}_{L^p(\mathbb{G})} \\ &= \left|\frac{p}{Q-p(1-a)}\right|^{\delta}\left\|\frac{\mathcal{R}f}{|x|^{-a}}\right\|^{\delta}_{L^p(\mathbb{G})}, \quad 1<p<\infty.\end{aligned} \tag{3.139}$$

Putting this in (3.125), one has

$$\||x|^c f\|_{L^r(\mathbb{G})} \leq \left|\frac{p}{Q-p(1-a)}\right|^{\delta}\left\|\frac{\mathcal{R}f}{|x|^{-a}}\right\|^{\delta}_{L^p(\mathbb{G})}\left\|\frac{f}{|x|^{-b}}\right\|^{1-\delta}_{L^q(\mathbb{G})}.$$

We note that in the case of $p=q$ and $a-b=1$, Hölder's equality condition of the inequalities (3.137) and (3.139) holds true for functions $\frac{1}{|x|^c}$, $c \in \mathbb{R}\backslash\{0\}$. Moreover, in the case of $p \neq q$ and $p(1-a)+bq \neq 0$, Hölder's equality condition of the inequalities (3.137) and (3.139) holds true for the function $h(x)$ in (3.138). Therefore, the constant in (3.130) is sharp when $p=q$ and $a-b=1$, or when $p \neq q$ and $p(1-a)+bq \neq 0$.

Now let us consider the case $Q=p(1-a)$. Using Theorem 3.2.3, one has

$$\left\|\frac{f}{|x|^{1-a}}\right\|^{\delta}_{L^p(\mathbb{G})} \leq p^{\delta}\left\|\frac{\log|x|}{|x|^{1-a}}\mathbb{E}f\right\|^{\delta}_{L^p(\mathbb{G})}, \quad 1<p<\infty.$$

Then, putting this in (3.137), we obtain

$$\begin{aligned}\||x|^c f\|_{L^r(\mathbb{G})} &\leq p^{\delta}\left\|\frac{\log|x|}{|x|^{1-a}}\mathbb{E}f\right\|^{\delta}_{L^p(\mathbb{G})}\left\|\frac{f}{|x|^{-b}}\right\|^{1-\delta}_{L^q(\mathbb{G})} \\ &= p^{\delta}\left\|\frac{\log|x|}{|x|^{-a}}\mathcal{R}f\right\|^{\delta}_{L^p(\mathbb{G})}\left\|\frac{f}{|x|^{-b}}\right\|^{1-\delta}_{L^q(\mathbb{G})},\end{aligned}$$

completing the proof. ☐

Chapter 4

Fractional Hardy Inequalities

In this chapter we present results concerning fractional forms of Hardy inequalities. Such a topic is well investigated in the Abelian Euclidean setting and we will be providing relevant references in the sequel. For a general survey of fractional Laplacians in the Euclidean setting see, e.g., [Gar17]. However, as usual, the general approach based on homogeneous groups allows one to get insights also in the Abelian case, for example, from the point of view of the possibility of choosing an arbitrary quasi-norm. Moreover, another application of the setting of homogeneous groups is that the results can be equally applied to both elliptic and subelliptic problems.

We start by discussing fractional Sobolev and Hardy inequalities on the homogeneous groups. As a consequence of these inequalities, we derive a Lyapunov type inequality for the fractional p-sub-Laplacian, which also implies an estimate of the first eigenvalue in a quasi-ball for the Dirichlet fractional p-sub-Laplacian. We also extend this analysis to systems of fractional p-sub-Laplacians and to Riesz potential operators.

4.1 Gagliardo seminorms and fractional p-sub-Laplacians

Throughout this chapter $\mathbb{G}$ will be a homogeneous group of homogeneous dimension Q. Let $|\cdot|$ be a homogeneous quasi-norm on $\mathbb{G}$. We start with the definition of the fractional p-sub-Laplacian.

Definition 4.1.1 (Fractional p-sub-Laplacian). Let $p>1$ and let $s\in(0,1)$. For a measurable and compactly supported function u the *fractional p-sub-Laplacian* $(-\Delta_{p,|\cdot|})^s=(-\Delta_p)^s$ on $\mathbb{G}$ is defined by the formula

$$(-\Delta_p)^s u(x) := 2\lim_{\delta\searrow 0}\int_{\mathbb{G}\setminus B(x,\delta)}\frac{|u(x)-u(y)|^{p-2}(u(x)-u(y))}{|y^{-1}x|^{Q+sp}}dy,\quad x\in\mathbb{G},\qquad (4.1)$$

M. Ruzhansky, D. Suragan, *Hardy Inequalities on Homogeneous Groups*, Progress in Mathematics 327, https://doi.org/10.1007/978-3-030-02895-4_5

where $B(x,\delta) = B_{|\cdot|}(x,\delta)$ is a quasi-ball with respect to the quasi-norm $|\cdot|$, with radius δ centred at $x \in \mathbb{G}$.

Definition 4.1.2 (Gagliardo seminorm and fractional Sobolev spaces)**.** For a measurable function $u : \mathbb{G} \to \mathbb{R}$, its *Gagliardo seminorm* is defined as

$$[u]_{s,p,|\cdot|} = [u]_{s,p} := \left(\int_{\mathbb{G}} \int_{\mathbb{G}} \frac{|u(x) - u(y)|^p}{|y^{-1}x|^{Q+sp}} dxdy \right)^{1/p}. \tag{4.2}$$

For $p \geq 1$ and $s \in (0,1)$, the functional space

$$W^{s,p}(\mathbb{G}) = \{u \in L^p(\mathbb{G}) : u \text{ is measurable and } [u]_{s,p} < +\infty\}, \tag{4.3}$$

endowed with the norm

$$\|u\|_{W^{s,p}(\mathbb{G})} := (\|u\|^p_{L^p(\mathbb{G})} + [u]^p_{s,p})^{1/p}, \quad u \in W^{s,p}(\mathbb{G}), \tag{4.4}$$

is called the *fractional Sobolev space on* $\mathbb{G}$. Sometimes, to emphasize the dependence on a particular quasi-norm, we may write $[u]_{s,p,|\cdot|}$ and $W^{s,p,|\cdot|}$ but we note that the space $W^{s,p,|\cdot|}$ is independent of a particular choice of a quasi-norm due to their equivalence, see Proposition 1.2.3.

Similarly, if $\Omega \subset \mathbb{G}$ is a Haar measurable set, we define the *fractional Sobolev space* $W^{s,p}(\Omega)$ on Ω by

$$W^{s,p}(\Omega) = \Bigg\{ u \in L^p(\Omega) : u \text{ is measurable} \\ \text{and } \left(\int_{\Omega} \int_{\Omega} \frac{|u(x) - u(y)|^p}{|y^{-1}x|^{Q+sp}} dxdy \right)^{1/p} < +\infty \Bigg\},$$

endowed with the norm

$$\|u\|_{W^{s,p}(\Omega)} := \left(\|u\|^p_{L^p(\Omega)} + \int_{\Omega} \int_{\Omega} \frac{|u(x) - u(y)|^p}{|y^{-1}x|^{Q+sp}} dxdy \right)^{1/p}. \tag{4.5}$$

Moreover, the Sobolev space $W^{s,p}_0(\Omega)$ is defined as the completion of $C^\infty_0(\Omega)$ with respect to the norm $\|u\|_{W^{s,p}(\Omega)}$.

4.2 Fractional Hardy inequalities on homogeneous groups

In the present section we establish fractional Hardy inequalities on homogeneous groups.

Theorem 4.2.1 (Fractional Hardy inequality). *Let $\mathbb{G}$ be a homogeneous group of homogeneous dimension Q with a homogeneous quasi-norm $|\cdot|$. Let $p > 1$, $s \in (0,1)$ and $Q > sp$. Then for all $u \in C_0^\infty(\mathbb{G})$ we have*

$$2\mu(\gamma)\int_{\mathbb{G}} \frac{|u(x)|^p}{|x|^{ps}}dx \leq [u]^p_{s,p,|\cdot|}, \tag{4.6}$$

where $\mu(\gamma)$ is defined in (4.10).

Remark 4.2.2. In [AB17] the authors studied the weighted fractional p-Laplacian and established the following weighted fractional L^p-Hardy inequality:

$$C\int_{\mathbb{R}^N} \frac{|u(x)|^p}{|x|_E^{ps+2\beta}}dx \leq \int_{\mathbb{R}^N}\int_{\mathbb{R}^N} \frac{|u(x)-u(y)|^p}{|x-y|_E^{N+ps}|x|_E^\beta |y|_E^\beta}dxdy, \tag{4.7}$$

where $\beta < \frac{N-ps}{2}$, $u \in C_0^\infty(\mathbb{R}^N)$, $C > 0$ is a positive constant, and $|\cdot|_E$ is the Euclidean distance in $\mathbb{R}^N$.

Before we prove Theorem 4.2.1, let us establish the following two lemmas that will be instrumental in the proof.

Lemma 4.2.3. *We fix a homogeneous quasi-norm $|\cdot|$ on $\mathbb{G}$. Let $\omega \in W_0^{s,p}(\Omega)$ and assume that $w > 0$ in $\Omega \subset \mathbb{G}$. Assume that $(-\Delta_p)^s\omega = \nu > 0$ with $\nu \in L^1_{\rm loc}(\Omega)$. Then for all $u \in C_0^\infty(\Omega)$, we have*

$$\frac{1}{2}\int_\Omega\int_\Omega \frac{|u(x)-u(y)|^p}{|y^{-1}x|^{Q+ps}}dxdy \geq \left\langle (-\Delta_p)^s\omega, \frac{|u|^p}{\omega^{p-1}}\right\rangle.$$

Proof of Lemma 4.2.3. Using the notations

$$v := \frac{|u|^p}{|\omega|^{p-1}} \text{ and } k(x,y) := \frac{1}{|y^{-1}x|^{Q+ps}},$$

we get

$$\begin{aligned}\langle(-\Delta_p)^s\omega(x), v(x)\rangle &= \int_\Omega v(x)dx\int_\Omega |\omega(x)-\omega(y)|^{p-2}(\omega(x)-\omega(y))k(x,y)dy \\ &= \int_\Omega \frac{|u(x)|^p}{|\omega(x)|^{p-1}}dx\int_\Omega |\omega(x)-\omega(y)|^{p-2}(\omega(x)-\omega(y))k(x,y)dy.\end{aligned}$$

Let us show that $k(x,y)$ is symmetric, that is, $k(x,y)=k(y,x)$ for all $x,y\in\mathbb{G}$. This readily follows, since by the definition of the quasi-norm we have $|x^{-1}|=|x|$ for all $x\in\mathbb{G}$. So, since $k(x,y)$ is symmetric, we obtain

$$\begin{aligned}&\langle(-\Delta_p)^s\omega(x),v(x)\rangle\\&=\frac{1}{2}\int_\Omega\int_\Omega\left(\frac{|u(x)|^p}{|\omega(x)|^{p-1}}-\frac{|u(y)|^p}{|\omega(y)|^{p-1}}\right)|\omega(x)-\omega(y)|^{p-2}(\omega(x)-\omega(y))k(x,y)dydx.\end{aligned}$$

Let $g:=\frac{u}{\omega}$ and

$$R(x,y):=|u(x)-u(y)|^p-(|g(x)|^p\omega(x)-|g(y)|^p\omega(y))|\omega(x)-\omega(y)|^{p-2}(\omega(x)-\omega(y)).$$

Then we have

$$\langle(-\Delta_p)^s\omega,v\rangle+\frac{1}{2}\int_\Omega\int_\Omega R(x,y)k(x,y)dydx=\frac{1}{2}\int_\Omega\int_\Omega|u(x)-u(y)|^pk(x,y)dydx.$$

By the symmetry argument, we can assume that $\omega(x)\geq\omega(y)$. By using the inequality (see, e.g., [FS08, Lemma 2.6])

$$|a-t|^p\geq(1-t)^{p-1}(|a|^p-t),\ p>1,\ t\in[0,1],\ a\in\mathbb{C}, \tag{4.8}$$

with $t=\frac{\omega(y)}{\omega(x)}$ and $a=\frac{g(x)}{g(y)}$, we see that $R(x,y)\geq 0$. Therefore, we have proved the inequality

$$\langle(-\Delta_p)^s\omega,v\rangle\leq\frac{1}{2}\int_\Omega\int_\Omega\frac{|u(x)-u(y)|^p}{|y^{-1}x|^{Q+sp}}dydx,$$

completing the proof of Lemma 4.2.3. □

Lemma 4.2.4. *Let $p>1$ and $\gamma\in\left(0,\frac{Q-ps}{p-1}\right)$. Then there exists a positive constant $\mu(\gamma)>0$ such that*

$$(-\Delta_p)^s(|x|^{-\gamma})=\mu(\gamma)\frac{1}{|x|^{ps+\gamma(p-1)}}\quad a.e.\ in\ \mathbb{G}\setminus\{0\}. \tag{4.9}$$

Proof of Lemma 4.2.4. Let us denote $\omega(x):=|x|^{-\gamma}$. We set $r=|x|$ and $\rho=|y|$ with $x=rx'$ and $y=\rho y'$ where $|x'|=|y'|=1$. Then we have

$$\begin{aligned}(-\Delta_p)^s\omega&=\int_0^{+\infty}||x|^{-\gamma}-|y|^{-\gamma}|^{p-2}(|x|^{-\gamma}-|y|^{-\gamma})|y|^{Q-1}\left(\int_{|y'|=1}\frac{d\sigma(y)}{|y^{-1}x|^{Q+ps}}\right)d|y|\\&=\frac{1}{|x|^{ps+\gamma(p-1)}}\int_0^{+\infty}\left|1-\frac{|y|^{-\gamma}}{|x|^{-\gamma}}\right|^{p-2}\\&\quad\times\left(1-\frac{|y|^{-\gamma}}{|x|^{-\gamma}}\right)\frac{|y|^{Q-1}}{|x|^{Q-1}}\left(\int_{|y'|=1}\frac{d\sigma(y)}{\left|\left(\frac{|y|}{|x|}y'\right)^{-1}\circ x'\right|^{Q+ps}}\right)d|y|.\end{aligned}$$

Let $\rho = \frac{|y|}{|x|}$ and $L(\rho) = \int_{|y'|=1} \frac{d\sigma(y)}{|(\rho y')^{-1} \circ x'|^{Q+ps}}$. Then we have

$$(-\Delta_p)^s \omega = \frac{1}{|x|^{ps+\gamma(p-1)}} \int_0^{+\infty} |1-\rho^{-\gamma}|^{p-2}(1-\rho^{-\gamma})L(\rho)\rho^{Q-1}d\rho.$$

We then have (4.9) with

$$\mu(\gamma) := \int_0^{+\infty} \phi(\rho)d\rho \tag{4.10}$$

for

$$\phi(\rho) = |1-\rho^{-\gamma}|^{p-2}(1-\rho^{-\gamma})L(\rho)\rho^{Q-1}.$$

Now it remains to show that $\mu(\gamma)$ is positive and bounded. Firstly, let us show that $\mu(\gamma)$ is bounded. We have

$$\mu(\gamma) = \int_0^1 \phi(\rho)d\rho + \int_1^{+\infty} \phi(\rho)d\rho = I_1 + I_2. \tag{4.11}$$

Using the new variable $\zeta = \frac{1}{\rho}$ we have $L(\rho) = L\left(\frac{1}{\zeta}\right) = \zeta^{Q+ps}L(\zeta)$ for any $\zeta > 0$. Thus, we get that

$$\mu(\gamma) = \int_1^{+\infty} (\rho^{-\gamma}-1)^{p-1}(\rho^{Q-1-\gamma(p-1)} - \rho^{ps-1})L(\rho)d\rho. \tag{4.12}$$

For $\rho \to 1$ we have

$$(\rho^{-\gamma}-1)^{p-1}(\rho^{Q-1-\gamma(p-1)} - \rho^{ps-1})L(\rho) \simeq (\rho-1)^{-1-ps+p} \in L^1(1,2). \tag{4.13}$$

Similarly, for $\rho \to \infty$ we get

$$(\rho^{-\gamma}-1)^{p-1}(\rho^{Q-1-\gamma(p-1)} - \rho^{ps-1})L(\rho) \simeq \rho^{-1-ps} \in L^1(2,\infty). \tag{4.14}$$

These properties show that $\mu(\gamma)$ is bounded. On the other hand, by (4.12) with $\gamma \in \left(0, \frac{Q-ps}{p-1}\right)$, we see that $\mu(\gamma)$ is positive.

Lemma 4.2.4 is proved. □

Proof of Theorem 4.2.1. Let $u \in C_0^\infty(\mathbb{G})$ and $\gamma < \frac{Q-ps}{p-1}$. By Lemma 4.2.4 and Lemma 4.2.3 we readily obtain that

$$\begin{aligned}\frac{1}{2}[u]_{s,p}^p &= \frac{1}{2}\int_{\mathbb{G}}\int_{\mathbb{G}} \frac{|u(x)-u(y)|^p}{|y^{-1}x|^{Q+ps}}dxdy \\ &\geq \left\langle (-\Delta_p)^s(|x|^{-\gamma}), \frac{|u(x)|^p}{|x|^{-\gamma(p-1)}} \right\rangle = \mu(\gamma)\int_{\mathbb{G}} \frac{|u(x)|^p}{|x|^{ps}}dx.\end{aligned}$$

This completes the proof of Theorem 4.2.1. □

4.3 Fractional Sobolev inequalities on homogeneous groups

In this section we establish fractional Sobolev inequalities on homogeneous groups.

Theorem 4.3.1 (Fractional Sobolev inequality). *Let $\mathbb{G}$ be a homogeneous group of homogeneous dimension Q. Let us fix a homogeneous quasi-norm $|\cdot|$ on $\mathbb{G}$. Let $p>1$, $s\in(0,1)$ and $Q>sp$, and let us set $p^*:=\frac{Qp}{Q-sp}$. Then there is a positive constant $C=C(Q,p,s,|\cdot|)$ such that we have*

$$\|u\|_{L^{p^*}(\mathbb{G})}\leq C[u]_{s,p,|\cdot|}, \tag{4.15}$$

for all measurable compactly supported functions $u:\mathbb{G}\to\mathbb{R}$.

Remark 4.3.2. In [DNPV12] the authors obtained the fractional Sobolev inequality in the case $N>sp$, $1<p<\infty$, and $s\in(0,1)$. Namely, for all measurable and compactly supported functions u one has

$$\|u\|_{L^{p^*}(\mathbb{R}^N)}\leq C[u]_{s,p}, \tag{4.16}$$

where $p^*=\frac{Np}{N-sp}$, $C=C(N,p,s)>0$ is a suitable constant independent of u, and

$$[u]^p_{s,p}=\int_{\mathbb{R}^N}\int_{\mathbb{R}^N}\frac{|u(x)-u(y)|^p}{|x-y|_E^{N+sp}}dxdy,$$

with $|\cdot|_E$ being the Euclidean distance in $\mathbb{R}^N$.

To prove the above analogue of the fractional Sobolev inequality, first we present the following two lemmas.

Lemma 4.3.3. *Let $p>1$, $s\in(0,1)$, and let $K\subset\mathbb{G}$ be a Haar measurable set. Fix $x\in\mathbb{G}$ and a quasi-norm $|\cdot|$ on $\mathbb{G}$. Then we have*

$$\int_{K^c}\frac{dy}{|y^{-1}x|^{Q+sp}}\geq C|K|^{-sp/Q}, \tag{4.17}$$

where $C=C(Q,s,p,|\cdot|)$ is a positive constant, $K^c:=\mathbb{G}\setminus K$, and $|K|$ is the Haar measure of K.

Proof of Lemma 4.3.3. Let $\delta:=\left(\frac{|K|}{\omega_Q}\right)^{1/Q}$, where ω_Q is a surface measure of the unit quasi-ball on $\mathbb{G}$. For the corresponding quasi-ball $B(x,\delta)=B_{|\cdot|}(x,\delta)$ centred at x with radius δ, we have

$$\begin{aligned}|K^c\cap B(x,\delta)|&=|B(x,\delta)|-|K\cap B(x,\delta)|\\&=|K|-|K\cap B(x,\delta)|=|K\cap B^c(x,\delta)|,\end{aligned} \tag{4.18}$$

where $|\cdot|$ (by abuse of notation, only in this proof) is the Haar measure on $\mathbb{G}$. Then,

$$\begin{aligned}\int_{K^c}\frac{dy}{|y^{-1}x|^{Q+sp}} &= \int_{K^c\cap B(x,\delta)}\frac{dy}{|y^{-1}x|^{Q+sp}}+\int_{K^c\cap B^c(x,\delta)}\frac{dy}{|y^{-1}x|^{Q+sp}}\\ &\geq \int_{K^c\cap B(x,\delta)}\frac{dy}{\delta^{Q+sp}}+\int_{K^c\cap B^c(x,\delta)}\frac{dy}{|y^{-1}x|^{Q+sp}}\\ &= \frac{|K^c\cap B(x,\delta)|}{\delta^{Q+sp}}+\int_{K^c\cap B^c(x,\delta)}\frac{dy}{|y^{-1}x|^{Q+sp}}.\end{aligned}$$

By using (4.18) we obtain

$$\begin{aligned}\int_{K^c}\frac{dy}{|y^{-1}x|^{Q+sp}} &\geq \frac{|K^c\cap B(x,\delta)|}{\delta^{Q+sp}}+\int_{K^c\cap B^c(x,\delta)}\frac{dy}{|y^{-1}x|^{Q+sp}}\\ &= \frac{|K\cap B^c(x,\delta)|}{\delta^{Q+sp}}+\int_{K^c\cap B^c(x,\delta)}\frac{dy}{|y^{-1}x|^{Q+sp}}\\ &\geq \int_{K\cap B^c(x,\delta)}\frac{dy}{|y^{-1}x|^{Q+sp}}+\int_{K^c\cap B^c(x,\delta)}\frac{dy}{|y^{-1}x|^{Q+sp}}\\ &= \int_{B^c(x,\delta)}\frac{dy}{|y^{-1}x|^{Q+sp}}.\end{aligned}$$

Now using the polar decomposition formula in Proposition 1.2.10 we obtain that

$$\int_{K^c}\frac{dy}{|y^{-1}x|^{Q+sp}}\geq C|K|^{-sp/Q}, \tag{4.19}$$

completing the proof. □

We now establish a useful technical estimate for the Gagliardo seminorm $[u]_{s,p}$ defined in (4.2).

Lemma 4.3.4. *Let $p>1$, $s\in(0,1)$ and $Q>sp$. Let $u\in L^{\infty}(\mathbb{G})$ be compactly supported and denote $a_k:=|\{|u|>2^k\}|$ for any $k\in\mathbb{Z}$. Then we have*

$$C\sum_{k\in\mathbb{Z},\,a_k\neq 0}a_{k+1}a_k^{-sp/Q}2^{kp}\leq [u]^p_{s,p}, \tag{4.20}$$

where $C=C(Q,p,s,|\cdot|)$ is a positive constant.

Proof of Lemma 4.3.4. We define

$$A_k:=\{|u|>2^k\},\ k\in\mathbb{Z}, \tag{4.21}$$

and

$$D_k:=A_k\setminus A_{k+1}=\{2^k<|u|\leq 2^{k+1}\}\ \text{and}\ d_k:=|D_k|, \tag{4.22}$$

the Haar measure of D_k. Since $A_{k+1} \subseteq A_k$, it follows that

$$a_{k+1} \leq a_k. \tag{4.23}$$

By the assumption $u \in L^{\infty}(\mathbb{G})$ is compactly supported, a_k and d_k are bounded and vanish when k is large enough. Also, we notice that the D_k's are disjoint, therefore,

$$\bigcup_{l\in\mathbb{Z},\ l\leq k} D_l = A_{k+1}^c \tag{4.24}$$

and

$$\bigcup_{l\in\mathbb{Z},\ l\geq k} D_l = A_k. \tag{4.25}$$

From (4.25) it follows that

$$\sum_{l\in\mathbb{Z},\ l\geq k} d_l = a_k \tag{4.26}$$

and

$$d_k = a_k - \sum_{l\in\mathbb{Z},\ l\geq k+1} d_l. \tag{4.27}$$

Since a_k and d_k are bounded and vanish when k is large enough, (4.26) and (4.27) are convergent. We define the convergent series

$$S := \sum_{l\in\mathbb{Z},\ a_{l-1}\neq 0} 2^{lp} a_{l-1}^{-sp/Q} d_l. \tag{4.28}$$

We have that $D_k \subseteq A_k \subseteq A_{k-1}$, therefore, $a_{i-1}^{-sp/Q} d_l \leq a_{i-1}^{-sp/Q} a_{l-1}$. Thus,

$$\{(i,l) \in \mathbb{Z} \text{ s.t. } a_{i-1} \neq 0 \text{ and } a_{i-1}^{-sp/Q} d_l \neq 0\} \subseteq \{(i,l) \in \mathbb{Z} \text{ s.t. } a_{l-1} \neq 0\}. \tag{4.29}$$

By using (4.29) and (4.23), we can estimate

$$\begin{aligned}
\sum_{\substack{i\in\mathbb{Z},\\ a_{i-1}\neq 0}} \sum_{\substack{l\in\mathbb{Z},\\ l\geq i+1}} 2^{ip} a_{i-1}^{-sp/Q} d_l &= \sum_{\substack{i\in\mathbb{Z},\\ a_{i-1}\neq 0}} \sum_{\substack{l\in\mathbb{Z},\\ l\geq i+1,\\ a^{sp/Q} d_l\neq 0}} 2^{ip} a_{i-1}^{-sp/Q} d_l \\
&\leq \sum_{i\in\mathbb{Z}} \sum_{\substack{l\in\mathbb{Z},\\ l\geq i+1,\\ a_{l-1}\neq 0}} 2^{ip} a_{i-1}^{-sp/Q} d_l = \sum_{\substack{l\in\mathbb{Z},\\ a_{l-1}\neq 0}} \sum_{\substack{i\in\mathbb{Z},\\ i\leq l-1}} 2^{ip} a_{i-1}^{-sp/Q} d_l \\
&\leq \sum_{\substack{l\in\mathbb{Z},\\ a_{l-1}\neq 0}} \sum_{\substack{i\in\mathbb{Z},\\ i\leq l-1}} 2^{ip} a_{l-1}^{-sp/Q} d_l = \sum_{\substack{l\in\mathbb{Z},\\ a_{l-1}\neq 0}} \sum_{k=0}^{+\infty} 2^{p(l-1-k)} a_{l-1}^{-sp/Q} d_l \leq S.
\end{aligned} \tag{4.30}$$

Notice that

$$||u(x)| - |u(y)|| \leq |u(x) - u(y)|,$$

for any $x, y \in \mathbb{G}$. If we fix $i \in \mathbb{Z}$ and $x \in D_i$, then for any $j \in \mathbb{Z}$ with $j \leq i-2$, for any $y \in D_j$ using the above inequality, we obtain that

$$|u(x) - u(y)| \geq 2^i - 2^{j+1} \geq 2^i - 2^{i-1} \geq 2^{i-1}.$$

Then, using (4.24), we have

$$\begin{aligned}\sum_{j\in\mathbb{Z},\, j\leq i-2} \int_{D_j} \frac{|u(x)-u(y)|^p}{|y^{-1}x|^{Q+sp}} dy &\geq 2^{(i-1)p} \sum_{j\in\mathbb{Z},\, j\leq i-2} \int_{D_j} \frac{dy}{|y^{-1}x|^{Q+sp}} \\ &= 2^{(i-1)p} \int_{A_{i-1}^c} \frac{dy}{|y^{-1}x|^{Q+sp}}.\end{aligned} \tag{4.31}$$

Now using (4.31) and Lemma 4.3.3, we obtain that

$$\sum_{j\in\mathbb{Z},\, j\leq i-2} \int_{D_j} \frac{|u(x)-u(y)|^p}{|y^{-1}x|^{Q+sp}} dy \geq C 2^{ip} a_{i-1}^{-sp/Q},$$

with some positive constant C. That is, for any $i \in \mathbb{Z}$, we have

$$\sum_{j\in\mathbb{Z},\ j\leq i-2} \int_{D_i} \int_{D_j} \frac{|u(x)-u(y)|^p}{|y^{-1}x|^{Q+sp}} dxdy \geq C 2^{ip} a_{i-1}^{-sp/Q} d_i. \tag{4.32}$$

From (4.32) and (4.27) we get

$$\begin{aligned}&\sum_{j\in\mathbb{Z},\ j\leq i-2} \int_{D_i} \int_{D_j} \frac{|u(x)-u(y)|^p}{|y^{-1}x|^{Q+sp}} dxdy \\ &\geq C \left(2^{ip} a_{i-1}^{-sp/Q} a_i - \sum_{l\in\mathbb{Z},\, l\geq i+1} 2^{ip} a_{i-1}^{-sp/Q} d_l \right).\end{aligned} \tag{4.33}$$

By (4.32) and (4.28) it follows that

$$\begin{aligned}&\sum_{i\in\mathbb{Z}, a_{i-1}\neq 0} \sum_{j\in\mathbb{Z}, j\leq i-2} \int_{D_i} \int_{D_j} \frac{|u(x)-u(y)|^p}{|y^{-1}x|^{Q+sp}} dxdy \\ &\geq C \sum_{i\in\mathbb{Z}, a_{i-1}\neq 0} 2^{ip} a_{i-1}^{-sp/Q} d_i \geq C\, S.\end{aligned} \tag{4.34}$$

Then, by using (4.30), (4.33) and (4.34), we obtain that

$$\begin{aligned}&\sum_{i\in\mathbb{Z}, a_{i-1}\neq 0} \sum_{j\in\mathbb{Z}, j\leq i-2} \int_{D_i} \int_{D_j} \frac{|u(x)-u(y)|^p}{|y^{-1}x|^{Q+sp}} dxdy \\ &\geq C \sum_{i\in\mathbb{Z},\, a_{i-1}\neq 0} 2^{ip} a_{i-1}^{-sp/Q} a_i - C \sum_{i\in\mathbb{Z}, a_{i-1}\neq 0} \sum_{l\in\mathbb{Z}, l\geq i+1} 2^{ip} a_{i-1}^{-sp/Q} d_l\end{aligned}$$

$$\geq C \sum_{i\in\mathbb{Z},\, a_{i-1}\neq 0} 2^{ip} a_{i-1}^{-sp/Q} a_i - C\, S$$

$$\geq C \sum_{i\in\mathbb{Z}, a_{i-1}\neq 0} 2^{ip} a_{i-1}^{-sp/Q} a_i - \sum_{i\in\mathbb{Z}, a_{i-1}\neq 0} \sum_{j\in\mathbb{Z}, j\leq i-2} \int_{D_i}\int_{D_j} \frac{|u(x)-u(y)|^p}{|y^{-1}x|^{Q+sp}} dxdy.$$

This means that

$$\sum_{i\in\mathbb{Z}, a_{i-1}\neq 0} \sum_{j\in\mathbb{Z}, j\leq i-2} \int_{D_i}\int_{D_j} \frac{|u(x)-u(y)|^p}{|y^{-1}x|^{Q+sp}} dxdy \geq \frac{C}{2} \sum_{i\in\mathbb{Z}, a_{i-1}\neq 0} 2^{ip} a_{i-1}^{-sp/Q} a_i, \tag{4.35}$$

for some constant $C > 0$. By symmetry and using (4.35), we arrive at

$$\begin{aligned}
[u]^p_{s,p,|\cdot|} &= \int_{\mathbb{G}}\int_{\mathbb{G}} \frac{|u(x)-u(y)|^p}{|y^{-1}x|^{Q+sp}} dxdy = \sum_{i,j\in\mathbb{Z}} \int_{D_i}\int_{D_j} \frac{|u(x)-u(y)|^p}{|y^{-1}x|^{Q+sp}} dxdy \\
&\geq 2 \sum_{i,j\in\mathbb{Z},\, j<i} \int_{D_i}\int_{D_j} \frac{|u(x)-u(y)|^p}{|y^{-1}x|^{Q+sp}} dxdy \\
&\geq 2 \sum_{i\in\mathbb{Z}, a_{i-1}\neq 0} \sum_{j\in\mathbb{Z}, j\leq i-2} \int_{Di}\int_{D_j} \frac{|u(x)-u(y)|^p}{|y^{-1}x|^{Q+sp}} dxdy \\
&\geq C \sum_{i\in\mathbb{Z},\, a_{i-1}\neq 0} 2^{ip} a_{i-1}^{-sp/Q} a_i,
\end{aligned}$$

completing the proof of Lemma 4.3.4. □

Proof of Theorem 4.3.1. Assume that Gagliardo's seminorm $[u]_{s,p}$ is bounded, i.e., that

$$[u]^p_{s,p} = \int_{\mathbb{G}}\int_{\mathbb{G}} \frac{|u(x)-u(y)|^p}{|y^{-1}x|^{Q+sp}} dxdy < +\infty. \tag{4.36}$$

Suppose also that $u \in L^\infty(\mathbb{G})$.

If (4.36) is satisfied for bounded functions, it holds also for the function u_n, obtained from u by cutting at levels $-n$ and n, that is, for

$$u_n := \max\{\min\{u(x), n\}, -n\},$$

for any $n \in \mathbb{R}$ and $x \in \mathbb{G}$. Thus, using the fact that

$$\lim_{n\to+\infty} \|u_n\|_{L^p(\mathbb{G})} = \|u\|_{L^p(\mathbb{G})},$$

$1 < p < \infty$, and by using (4.36) with the dominated convergence theorem, we obtain that

$$\begin{aligned}\lim_{n\to+\infty}[u_n]^p_{s,p} &= \lim_{n\to+\infty}\int_{\mathbb{G}}\int_{\mathbb{G}}\frac{|u_n(x)-u_n(y)|^p}{|y^{-1}x|^{Q+sp}}dxdy \\ &= \int_{\mathbb{G}}\int_{\mathbb{G}}\frac{|u(x)-u(y)|^p}{|y^{-1}x|^{Q+sp}}dxdy = [u]^p_{s,p}.\end{aligned} \tag{4.37}$$

Defining a_k and A_k as in Lemma 4.3.4, we have

$$\begin{aligned}\|u\|_{L^{p^*}(\mathbb{G})} &= \left(\sum_{k\in\mathbb{Z}}\int_{A_k\setminus A_{k+1}}|u(x)|^{p^*}dx\right)^{1/p^*} \\ &\leq \left(\sum_{k\in\mathbb{Z}}\int_{A_k\setminus A_{k+1}}2^{(k+1)p^*}dx\right)^{1/p^*} \\ &\leq \left(\sum_{k\in\mathbb{Z}}2^{(k+1)p^*}a_k\right)^{1/p^*}.\end{aligned}$$

Recall the following fact from [DNPV12, Lemma 6.2]: let $T, p > 1$ and $s \in (0,1)$ be such that $Q > sp$, $m \in \mathbb{Z}$, and assume that a_k is a bounded, decreasing, non-negative sequence with $a_k = 0$ for any $k \geq m$; then we have

$$\sum_{k\in\mathbb{Z}} a_k^{(Q-sp)/Q}T^k \leq C \sum_{k\in\mathbb{Z},\, a_k\neq 0} a_{k+1}a_k^{-sp/Q}T^k, \tag{4.38}$$

for some positive constant $C = C(Q, s, p, T)$.

Then, with $p/p^* = 1 - sp/Q < 1$ and $T = 2^p$, this fact yields

$$\begin{aligned}\|u\|^p_{L^{p^*}(\mathbb{G})} &\leq 2^p\left(\sum_{k\in\mathbb{Z}}2^{kp^*}a_k\right)^{p/p^*} \leq 2^p\sum_{k\in\mathbb{Z}}2^{kp}a_k^{(Q-sp)/Q} \\ &\leq C\sum_{k\in\mathbb{Z},\, a_k\neq 0}2^{kp}a_k^{-sp/Q}a_{k+1}\end{aligned} \tag{4.39}$$

for a positive constant $C = C(Q, p, s, |\cdot|)$. Finally, using Lemma 4.3.4 we arrive at

$$\begin{aligned}\|u\|^p_{L^{p^*}(\mathbb{G})} &\leq C\sum_{k\in\mathbb{Z},\, a_k\neq 0}2^{kp}a_k^{-sp/Q}a_{k+1} \\ &\leq C\int_{\mathbb{G}}\int_{\mathbb{G}}\frac{|u(x)-u(y)|^p}{|y^{-1}x|^{Q+sp}}dxdy = C[u]^p_{s,p,|\cdot|}.\end{aligned}$$

Theorem 4.3.1 is proved. □

4.4 Fractional Gagliardo–Nirenberg inequalities

In this section we discuss an analogue of the fractional Gagliardo–Nirenberg inequality on homogeneous groups. As it can be partly expected, in its proof we will use an already established version of a fractional Sobolev inequality on the homogeneous groups.

Theorem 4.4.1 (Fractional Gagliargo–Nirenberg inequality). *Let $\mathbb{G}$ be a homogeneous group of homogeneous dimension Q with a quasi-norm $|\cdot|$. Assume that $Q \geq 2$, $s \in (0,1)$, $p > 1$, $\alpha \geq 1$, $\tau > 0$, $a \in (0,1]$, $Q > sp$ and*

$$\frac{1}{\tau} = a\left(\frac{1}{p} - \frac{s}{Q}\right) + \frac{1-a}{\alpha}.$$

Then there exists $C = C(s,p,Q,a,\alpha) > 0$ such that

$$\|u\|_{L^\tau(\mathbb{G})} \leq C[u]^a_{s,p,|\cdot|}\|u\|^{1-a}_{L^\alpha(\mathbb{G})} \tag{4.40}$$

holds for all $u \in C^1_c(\mathbb{G})$.

Remark 4.4.2. Theorem 4.4.1 was proved in [KRS18a]. In the Abelian case $(\mathbb{R}^N, +)$ with the standard Euclidean distance instead of the quasi-norm, Theorem 4.4.1 covers the fractional Gagliardo–Nirenberg inequality which was proved in [NS18a].

Proof of Theorem 4.4.1. By using Hölder's inequality, for every $\frac{1}{\tau} = a\left(\frac{1}{p} - \frac{s}{Q}\right) + \frac{1-a}{\alpha}$ we get

$$\|u\|^\tau_{L^\tau(\mathbb{G})} = \int_{\mathbb{G}} |u|^\tau dx = \int_{\mathbb{G}} |u|^{a\tau}|u|^{(1-a)\tau} dx \leq \|u\|^{a\tau}_{L^{p^*}(\mathbb{G})}\|u\|^{(1-a)\tau}_{L^\alpha(\mathbb{G})}, \tag{4.41}$$

where $p^* = \frac{Qp}{Q-sp}$. From (4.41), by using the fractional Sobolev inequality (Theorem 4.3.1), we obtain

$$\|u\|^\tau_{L^\tau(\mathbb{G})} \leq \|u\|^{a\tau}_{L^{p^*}(\mathbb{G})}\|u\|^{(1-a)\tau}_{L^\alpha(\mathbb{G})} \leq C[u]^{a\tau}_{s,p,|\cdot|}\|u\|^{(1-a)\tau}_{L^\alpha(\mathbb{G})},$$

that is,

$$\|u\|_{L^\tau(\mathbb{G})} \leq C[u]^a_{s,p,|\cdot|}\|u\|^{1-a}_{L^\alpha(\mathbb{G})}, \tag{4.42}$$

where C is a positive constant independent of u. Theorem 4.4.1 is proved. □

4.5 Fractional Caffarelli–Kohn–Nirenberg inequalities

In this section we discuss the weighted fractional Caffarelli–Kohn–Nirenberg inequalities on homogeneous groups. First, let us define a weighted version of fractional Sobolev spaces from Definition 4.1.2.

Definition 4.5.1 (Weighted fractional Sobolev spaces)**.** We define the *weighted fractional Sobolev space* on a homogeneous group $\mathbb{G}$ with homogeneous dimension Q and homogeneous quasi-norm $|\cdot|$ by

$$W^{s,p,\beta}(\mathbb{G}) = \Big\{ u \in L^p(\mathbb{G}) : u \text{ is measurable}, \tag{4.43}$$

$$[u]_{s,p,\beta,|\cdot|} := \left(\int_{\mathbb{G}} \int_{\mathbb{G}} \frac{|x|^{\beta_1 p}|y|^{\beta_2 p}|u(x)-u(y)|^p}{|y^{-1}x|^{Q+sp}} dxdy \right)^{\frac{1}{p}} < +\infty \Big\},$$

where $\beta_1,\ \beta_2 \in \mathbb{R}$ with $\beta = \beta_1 + \beta_2$. We note that the space $W^{s,p,\beta}(\mathbb{G})$ depends on β_1 and β_2. At the same time, it is independent of a particular choice of a quasi-norm due to their equivalence, see Proposition 1.2.3.

For a Haar measurable set $\Omega \subset \mathbb{G}$, $p \geq 1$, $s \in (0,1)$ and $\beta_1,\ \beta_2 \in \mathbb{R}$ with $\beta = \beta_1 + \beta_2$, we define the weighted fractional Sobolev space on Ω by

$$W^{s,p,\beta}(\Omega) = \Big\{ u \in L^p(\Omega) : u \text{ is measurable}, \tag{4.44}$$

$$[u]_{s,p,\beta,|\cdot|,\Omega} := \left(\int_{\Omega} \int_{\Omega} \frac{|x|^{\beta_1 p}|y|^{\beta_2 p}|u(x)-u(y)|^p}{|y^{-1}x|^{Q+sp}} dxdy \right)^{\frac{1}{p}} < +\infty \Big\}.$$

Theorem 4.5.2 (Fractional Caffarelli–Kohn–Nirenberg inequality)**.** *Let $\mathbb{G}$ be a homogeneous group of homogeneous dimension Q. Let $Q \geq 2$, $s \in (0,1)$, $p > 1$, $\alpha \geq 1$, $\tau > 0$, $a \in (0,1]$, β_1, β_2, β, μ, $\gamma \in \mathbb{R}$, $\beta_1 + \beta_2 = \beta$ and*

$$\frac{1}{\tau} + \frac{\gamma}{Q} = a\left(\frac{1}{p} + \frac{\beta - s}{Q}\right) + (1-a)\left(\frac{1}{\alpha} + \frac{\mu}{Q}\right). \tag{4.45}$$

In addition, assume that, $0 \leq \beta - \sigma$ with $\gamma = a\sigma + (1-a)\mu$. We also assume

$$\beta - \sigma \leq s \ \text{ only if } \ \frac{1}{\tau} + \frac{\gamma}{Q} = \frac{1}{p} + \frac{\beta - s}{Q}. \tag{4.46}$$

Then when $\frac{1}{\tau} + \frac{\gamma}{Q} > 0$ we have

$$\||x|^{\gamma}u\|_{L^{\tau}(\mathbb{G})} \leq C[u]^{a}_{s,p,\beta,|\cdot|}\||x|^{\mu}u\|^{1-a}_{L^{\alpha}(\mathbb{G})}, \tag{4.47}$$

for all $u \in C^1_c(\mathbb{G})$, and when $\frac{1}{\tau} + \frac{\gamma}{Q} < 0$ we have

$$\||x|^{\gamma}u\|_{L^{\tau}(\mathbb{G})} \leq C[u]^{a}_{s,p,\beta,|\cdot|}\||x|^{\mu}u\|^{1-a}_{L^{\alpha}(\mathbb{G})}, \tag{4.48}$$

for all $u \in C^1_c(\mathbb{G} \setminus \{0\})$.

Remark 4.5.3.

1. The critical case $\frac{1}{\tau}+\frac{\gamma}{Q}=0$ will be considered in Theorem 4.5.5.
2. In the Abelian (Euclidean) case $(\mathbb{R}^N,+)$ with the usual Euclidean distance instead of the quasi-norm in Theorem 4.5.2, we get the (Abelian) fractional Caffarelli–Kohn–Nirenberg inequality (see, e.g., [NS18a], Theorem 1.1). In (4.48) by setting $a=1$, $\tau=p$, $\beta_1=\beta_2=0$, and $\gamma=-s$, one has an analogue of the fractional Hardy inequality on homogeneous groups.
3. In the Abelian (Euclidean) case $(\mathbb{R}^N,+)$ again with the usual Euclidian distance instead of the quasi-norm and by taking in (4.48) the values $a=1$, $\tau=p$, $\beta_1=\beta_2=0$, and $\gamma=-s$, we get the fractional Hardy inequality (see [FS74, Theorem 1.1]).
4. The results of this section were obtained in [KRS18a] and we follow the presentation there in our proof.

To prove the fractional weighted Caffarelli–Kohn–Nirenberg inequality on $\mathbb{G}$ we will use Theorem 4.4.1 in the proof of the following lemma.

Lemma 4.5.4. *Let $Q\geq 2$, $s\in(0,1)$, $p>1$, $\alpha\geq 1$, $\tau>0$, $a\in(0,1]$ and*

$$\frac{1}{\tau}\geq a\left(\frac{1}{p}-\frac{s}{Q}\right)+\frac{1-a}{\alpha}.$$

Let $\lambda>0$ and $0<r<R$ and set

$$\Omega=\{x\in\mathbb{G}:\ \lambda r<|x|<\lambda R\}.$$

Then, for all $u\in C^1(\overline{\Omega})$, we have

$$\left(\fint_{\Omega}|u-u_{\Omega}|^{\tau}dx\right)^{\frac{1}{\tau}}\leq C_{r,R}\lambda^{\frac{a(sp-Q)}{p}}[u]^{a}_{s,p,|\cdot|,\Omega}\left(\fint_{\Omega}|u|^{\alpha}dx\right)^{\frac{1-a}{\alpha}},\tag{4.49}$$

where $C_{r,R}$ is a positive constant independent of u and λ.

Proof of Lemma 4.5.4. Without loss of generality, we assume that $0<s'\leq s$ and $\tau'\geq\tau$ are such that

$$\frac{1}{\tau'}=a\left(\frac{1}{p}-\frac{s'}{Q}\right)+\frac{1-a}{\alpha},$$

and $\lambda=1$ with

$$\Omega_1:=\{x\in\mathbb{G}:\ r<|x|<R\}.$$

By using Theorem 4.4.1, Jensen's inequality and $[u]_{s',p,|\cdot|,\Omega}\leq C[u]_{s,p,|\cdot|,\Omega}$, we compute

$$\left(\fint_{\Omega_1}|u-u_{\Omega_1}|^{\tau}dx\right)^{1/\tau}=\frac{1}{|\Omega_1|^{\frac{1}{\tau}}}\|u-u_{\Omega_1}\|_{\tau}\leq C_{r,R}\|u-u_{\Omega_1}\|_{L^{\tau'}(\Omega_1)}$$

$$
\begin{aligned}
&\le C_{r,R}[u-u_{\Omega_1}]^a_{s',p,|\cdot|,\Omega_1}\|u\|^{1-a}_{L^\alpha(\Omega_1)}\\
&\le C_{r,R}\left(\int_{\Omega_1}\int_{\Omega_1}\frac{|u(x)-u_{\Omega_1}-u(y)+u_{\Omega_1}|^p}{|y^{-1}x|^{Q+s'p}}dxdy\right)^{a/p}\|u\|^{1-a}_{L^\alpha(\Omega_1)}\\
&\le C_{r,R}[u]^a_{s,p,|\cdot|,\Omega_1}\|u\|^{1-a}_{L^\alpha(\Omega_1)}\le C_{r,R}[u]^a_{s,p,|\cdot|,\Omega_1}\left(⨍_{\Omega_1}|u|^\alpha dx\right)^{(1-a)/\alpha},
\end{aligned}
$$

where $C_{r,R}>0$. Let us apply the above inequality to $u(\lambda x)$ instead of $u(x)$. This yields

$$
\begin{aligned}
&\left(⨍_{\Omega_1}\left|u(\lambda x)-⨍_{\Omega_1}u(\lambda x)dx\right|^\tau dx\right)^{1/\tau}\\
&\le C_{r,R}\left(\int_{\Omega_1}\int_{\Omega_1}\frac{|u(\lambda x)-u(\lambda y)|^p}{|y^{-1}x|^{Q+sp}}dxdy\right)^{a/p}\left(\frac{1}{|\Omega_1|}\int_{\Omega_1}|u(\lambda x)|^\alpha dx\right)^{(1-a)/\alpha}.
\end{aligned}
$$

Therefore, we have

$$
\begin{aligned}
&\left(⨍_{\Omega}\left|u(x)-⨍_{\Omega}u(x)dx\right|^\tau dx\right)^{\frac{1}{\tau}}=\left(\frac{1}{|\Omega|}\int_{\Omega}\left|u(x)-\frac{1}{|\Omega|}\int_{\Omega}u(x)dx\right|^\tau dx\right)^{\frac{1}{\tau}}\\
&=\left(\frac{1}{|\Omega|}\int_{\Omega}\left|u(\lambda y)-\frac{1}{|\Omega|}\int_{\Omega}u(\lambda y)d(\lambda y)\right|^\tau d(\lambda y)\right)^{\frac{1}{\tau}}\\
&=\left(\frac{1}{|\Omega_1|}\int_{\Omega_1}\frac{\lambda^Q}{\lambda^Q}\left|u(\lambda y)-\frac{\lambda^Q}{\lambda^Q|\Omega_1|}\int_{\Omega_1}u(\lambda y)dy\right|^\tau dy\right)^{\frac{1}{\tau}}\\
&=\left(\frac{1}{|\Omega_1|}\int_{\Omega_1}\left|u(\lambda y)-\frac{1}{|\Omega_1|}\int_{\Omega_1}u(\lambda y)dy\right|^\tau dy\right)^{\frac{1}{\tau}}\\
&\le C_{r,R}\left(\int_{\Omega_1}\int_{\Omega_1}\frac{|u(\lambda x)-u(\lambda y)|^p}{|y^{-1}x|^{Q+sp}}dxdy\right)^{\frac{a}{p}}\left(\frac{1}{|\Omega_1|}\int_{\Omega_1}|u(\lambda x)|^\alpha dx\right)^{\frac{1-a}{\alpha}}\\
&=C_{r,R}\left(\int_{\Omega_1}\int_{\Omega_1}\frac{\lambda^{2Q}\lambda^{Q+sp}|u(\lambda x)-u(\lambda y)|^p}{\lambda^{2Q}\lambda^{Q+sp}|y^{-1}x|^{Q+sp}}dxdy\right)^{\frac{a}{p}}\left(\frac{1}{|\Omega_1|}\int_{\Omega_1}\frac{\lambda^Q}{\lambda^Q}|u(\lambda x)|^\alpha dx\right)^{\frac{1-a}{\alpha}}\\
&=C_{r,R}\left(\int_{\Omega}\int_{\Omega}\frac{\lambda^{sp-Q}|u(\lambda x)-u(\lambda y)|^p}{|(\lambda y)^{-1}\lambda x|^{Q+sp}}d(\lambda x)d(\lambda y)\right)^{\frac{a}{p}}\left(\frac{1}{|\Omega|}\int_{\Omega}|u(\lambda x)|^\alpha d(\lambda x)\right)^{\frac{1-a}{\alpha}}\\
&=C_{r,R}\left(\int_{\Omega}\int_{\Omega}\frac{\lambda^{sp-Q}|u(x)-u(y)|^p}{|y^{-1}x|^{Q+sp}}dxdy\right)^{\frac{a}{p}}\left(\frac{1}{|\Omega|}\int_{\Omega}|u(x)|^\alpha dx\right)^{\frac{1-a}{\alpha}}\\
&=C_{r,R}\lambda^{\frac{a(sp-Q)}{p}}[u]^a_{s,p,|\cdot|,\Omega}\left(\frac{1}{|\Omega|}\int_{\Omega}|u(x)|^\alpha dx\right)^{\frac{1-a}{\alpha}}.
\end{aligned}
$$

This completes the proof. □

Proof of Theorem 4.5.2. First let us consider the case (4.46), that is, $\beta - \sigma \leq s$ and $\frac{1}{\tau} + \frac{\gamma}{Q} = \frac{1}{p} + \frac{\beta - s}{Q}$. By using Lemma 4.5.4 with $\lambda = 2^k$, $r = 1$, $R = 2$ and $\Omega = A_k$, we get

$$\left(\fint_{A_k} |u - u_{A_k}|^\tau dx \right)^{1/\tau} \leq C 2^{\frac{ak(sp-Q)}{p}} [u]^a_{s,p,|\cdot|,A_k} \left(\fint_{A_k} |u|^\alpha dx \right)^{(1-a)/\alpha}. \quad (4.50)$$

Here and below A_k is the quasi-annulus defined by

$$A_k := \{x \in \mathbb{G} : \; 2^k \leq |x| < 2^{k+1}\},$$

for $k \in \mathbb{Z}$. Now by using (4.50) we obtain

$$\begin{aligned} \int_{A_k} |u|^\tau dx &= \int_{A_k} |u - u_{A_k} + u_{A_k}|^\tau dx \\ &\leq C \left(\int_{A_k} |u_{A_k}|^\tau dx + \int_{A_k} |u - u_{A_k}|^\tau dx \right) \\ &= C \left(\int_{A_k} |u_{A_k}|^\tau dx + \frac{|A_k|}{|A_k|} \int_{A_k} |u - u_{A_k}|^\tau dx \right) \\ &= C \left(|A_k||u_{A_k}|^\tau + |A_k| \fint_{A_k} |u - u_{A_k}|^\tau dx \right) \qquad (4.51) \\ &\leq C \left(|A_k||u_{A_k}|^\tau + 2^{\frac{ak(sp-Q)\tau}{p}} |A_k| [u]^{a\tau}_{s,p,|\cdot|,A_k} \left(\frac{1}{|A_k|} \int_{A_k} |u|^\alpha dx \right)^{\frac{(1-a)\tau}{\alpha}} \right) \\ &\leq C \left(2^{Qk} |u_{A_k}|^\tau + 2^{\frac{ak(sp-Q)\tau}{p}} 2^{kQ} 2^{-\frac{Q(1-a)\tau k}{\alpha}} [u]^{a\tau}_{s,p,|\cdot|,A_k} \|u\|^{(1-a)\tau}_{L^\alpha(A_k)} \right). \end{aligned}$$

Then, from (4.51) we get

$$\begin{aligned} &\int_{A_k} |x|^{\gamma\tau} |u|^\tau dx \leq 2^{(k+1)\gamma\tau} \int_{A_k} |u|^\tau dx \\ &\leq C 2^{(Q+\gamma\tau)k} |u_{A_k}|^\tau + C 2^{\gamma\tau k} 2^{kQ} 2^{\frac{ak(sp-Q)\tau}{p}} 2^{-\frac{Q(1-a)\tau k}{\alpha}} [u]^{a\tau}_{s,p,|\cdot|,A_k} \|u\|^{(1-a)\tau}_{L^\alpha(A_k)} \\ &= C 2^{(Q+\gamma\tau)k} |u_{A_k}|^\tau \\ &\quad + C 2^{\left(\gamma\tau + Q + \frac{a(sp-Q)\tau}{p} - \frac{Q(1-a)\tau}{\alpha}\right)k} \left(\int_{A_k} \int_{A_k} \frac{2^{kp\beta_1} 2^{kp\beta_2} |u(x) - u(y)|^p}{2^{kp\beta} |y^{-1}x|^{Q+sp}} dx dy \right)^{\frac{a\tau}{p}} \\ &\quad \times \left(\int_{A_k} \frac{2^{k\alpha\mu}}{2^{k\alpha\mu}} |u(x)|^\alpha dx \right)^{\frac{(1-a)\tau}{\alpha}} \\ &\leq C 2^{(Q+\gamma\tau)k} |u_{A_k}|^\tau \\ &\quad + C 2^{\left(\gamma\tau + Q + \frac{a(sp-Q)\tau}{p} - \frac{Q(1-a)\tau}{\alpha} - a\beta\tau - \mu\tau(1-a)\right)k} \\ &\quad \times \left(\int_{A_k} \int_{A_k} \frac{|x|^{p\beta_1} |y|^{p\beta_2} |u(x) - u(y)|^p}{|y^{-1}x|^{Q+sp}} dx dy \right)^{\frac{a\tau}{p}} \end{aligned}$$

$$\times \left(\int_{A_k} |x|^{\alpha\mu} |u(x)|^{\alpha} dx \right)^{\frac{(1-a)\tau}{\alpha}}$$
$$\le C 2^{(Q+\gamma\tau)k} |u_{A_k}|^{\tau} \tag{4.52}$$
$$+ C 2^{\left(\gamma\tau + Q + \frac{a(sp-Q)\tau}{p} - \frac{Q(1-a)\tau}{\alpha} - a\beta\tau - \mu\tau(1-a)\right)k} [u]^{a\tau}_{s,p,\beta,|\cdot|,A_k} \| |x|^{\mu} u \|^{(1-a)\tau}_{L^{\alpha}(A_k)}.$$

Here by (4.45), we have

$$\begin{aligned} &\gamma\tau + Q + \frac{a(sp-Q)\tau}{p} - \frac{Q(1-a)\tau}{\alpha} - a\beta\tau - \mu\tau(1-a) \\ &= Q\tau \left(\frac{\gamma}{Q} + \frac{1}{\tau} + \frac{a(sp-Q)}{Qp} - \frac{(1-a)}{\alpha} - \frac{a\beta}{Q} - \frac{\mu(1-a)}{Q} \right) \\ &= Q\tau \left(a \left(\frac{1}{p} + \frac{\beta - s}{Q} \right) + (1-a) \left(\frac{1}{\alpha} + \frac{\mu}{Q} \right) \right) \\ &\quad + Q\tau \left(\frac{a(sp-Q)}{Qp} - \frac{(1-a)}{\alpha} - \frac{a\beta}{Q} - \frac{\mu(1-a)}{Q} \right) = 0. \end{aligned} \tag{4.53}$$

Thus, we obtain

$$\int_{A_k} |x|^{\gamma\tau} |u|^{\tau} dx \le C 2^{(\gamma\tau+Q)k} |u_{A_k}|^{\tau} + C [u]^{a\tau}_{s,p,\beta,|\cdot|,A_k} \| |x|^{\mu} u \|^{(1-a)\tau}_{L^{\alpha}(A_k)}, \tag{4.54}$$

and by summing over k from m to n, we get

$$\begin{aligned} \int_{\cup_{k=m}^{n} A_k} |x|^{\gamma\tau} |u|^{\tau} dx &= \int_{\{2^m < |x| < 2^{n+1}\}} |x|^{\gamma\tau} |u|^{\tau} dx \\ &\le C \sum_{k=m}^{n} 2^{(\gamma\tau+Q)k} |u_{A_k}|^{\tau} + C \sum_{k=m}^{n} [u]^{a\tau}_{s,p,\beta,|\cdot|,A_k} \| |x|^{\mu} u \|^{(1-a)\tau}_{L^{\alpha}(A_k)}, \end{aligned} \tag{4.55}$$

where $k, m, n \in \mathbb{Z}$ and $m \le n-2$.

To prove (4.47) let us choose n such that

$$\operatorname{supp} u \subset B_{2^n}, \tag{4.56}$$

where B_{2^n} is a quasi-ball of $\mathbb{G}$ with the radius 2^n.

Let us consider the following integral

$$\begin{aligned} &⨍_{A_{k+1}\cup A_k} \left| u - ⨍_{A_{k+1}\cup A_k} u \right|^{\tau} dx = \frac{1}{|A_{k+1}| + |A_k|} \int_{A_{k+1}\cup A_k} \left| u - ⨍_{A_{k+1}\cup A_k} u \right|^{\tau} dx \\ &= \frac{1}{|A_{k+1}| + |A_k|} \left(\int_{A_{k+1}} \left| u - ⨍_{A_{k+1}\cup A_k} u \right|^{\tau} dx + \int_{A_k} \left| u - ⨍_{A_{k+1}\cup A_k} u \right|^{\tau} dx \right). \end{aligned}$$

On the other hand, a direct calculation gives

$$
\begin{aligned}
&\fint_{A_{k+1}\cup A_k}\left|u-\fint_{A_{k+1}\cup A_k}u\right|^{\tau}dx\\
&=\frac{1}{|A_{k+1}|+|A_k|}\left(\int_{A_{k+1}}\left|u-\fint_{A_{k+1}\cup A_k}u\right|^{\tau}dx+\int_{A_k}\left|u-\fint_{A_{k+1}\cup A_k}u\right|^{\tau}dx\right)\\
&\geq\frac{1}{|A_{k+1}|+|A_k|}\int_{A_k}\left|u-\fint_{A_{k+1}\cup A_k}u\right|^{\tau}dx\\
&\geq\frac{1}{|A_{k+1}|+|A_k|}\left|\int_{A_k}\left(u-\fint_{A_{k+1}\cup A_k}u\right)dx\right|^{\tau}\\
&=\frac{1}{|A_{k+1}|+|A_k|}\left|\int_{A_k}udx-\frac{|A_k|}{|A_{k+1}|+|A_k|}\int_{A_k}udx-\frac{|A_k|}{|A_{k+1}|+|A_k|}\int_{A_{k+1}}udx\right|^{\tau}\\
&=\frac{1}{|A_{k+1}|+|A_k|}\left|\frac{|A_{k+1}|}{|A_{k+1}|+|A_k|}\int_{A_k}udx-\frac{|A_k|}{|A_{k+1}|+|A_k|}\int_{A_{k+1}}udx\right|^{\tau}\\
&=\frac{1}{(|A_{k+1}|+|A_k|)^{\tau+1}}\left||A_{k+1}|\int_{A_k}udx-|A_k|\int_{A_{k+1}}udx\right|^{\tau}\\
&=\frac{|A_{k+1}|^{\tau}|A_k|^{\tau}}{(|A_{k+1}|+|A_k|)^{\tau+1}}\left|\frac{1}{|A_k|}\int_{A_k}udx-\frac{1}{|A_{k+1}|}\int_{A_{k+1}}udx\right|^{\tau}\\
&=\frac{|A_{k+1}|^{\tau}|A_k|^{\tau}}{(|A_{k+1}|+|A_k|)^{\tau+1}}|u_{A_{k+1}}-u_{A_k}|^{\tau}\\
&\geq C|u_{A_{k+1}}-u_{A_k}|^{\tau}.
\end{aligned}\tag{4.57}
$$

From (4.57) and Lemma 4.5.4, we obtain

$$
\begin{aligned}
|u_{A_{k+1}}-u_{A_k}|^{\tau}&\leq C\fint_{A_{k+1}\cup A_k}\left|u-\fint_{A_{k+1}\cup A_k}u\right|^{\tau}dx\\
&\leq C2^{\frac{ak(sp-Q)}{p}}[u]^{\tau a}_{s,p,|\cdot|,A_{k+1}\cup A_k}\left(\fint_{A_{k+1}\cup A_k}|u|^{\alpha}dx\right)^{\frac{(1-a)\tau}{\alpha}}.
\end{aligned}\tag{4.58}
$$

By using this fact, taking $\tau=1$ we have

$$
\begin{aligned}
|u_{A_k}|&\leq|u_{A_{k+1}}-u_{A_k}|+|u_{A_{k+1}}|\\
&\leq|u_{A_{k+1}}|+C2^{\frac{ak(sp-Q)}{p}}[u]^{a}_{s,p,|\cdot|,A_{k+1}\cup A_k}\left(\fint_{A_{k+1}\cup A_k}|u|^{\alpha}dx\right)^{\frac{(1-a)}{\alpha}}.
\end{aligned}\tag{4.59}
$$

On the other hand (see, e.g., [NS18b, Lemma 2.2]), there exists a positive constant C depending $\xi > 1$ and $\eta > 1$ such that $1 < \zeta < \xi$,

$$(|a| + |b|)^\eta \le \zeta |a|^\eta + \frac{C}{(\zeta - 1)^{\eta-1}} |b|^\eta, \quad \forall\, a, b \in \mathbb{R}. \tag{4.60}$$

Thus, by using with $\eta = \tau$, $\zeta = 2^{\gamma\tau+Q}c$, where $c = \frac{2}{1+2^{\gamma\tau+Q}} < 1$, since $\gamma\tau + Q > 0$, from the previous inequality we have

$$2^{(\gamma\tau+Q)k}|u_{A_k}|^\tau \le c2^{(k+1)(\gamma\tau+Q)}|u_{A_{k+1}}|^\tau + C[u]^{\tau a}_{s,p,\beta,|\cdot|,A_{k+1}\cup A_k} \||x|^\mu u\|^{(1-a)\tau}_{L^\alpha(A_{k+1}\cup A_k)}.$$

By summing over k from m to n and by using (4.56) we have

$$\begin{aligned}\sum_{k=m}^{n} 2^{(\gamma\tau+Q)k}|u_{A_k}|^\tau &\le \sum_{k=m}^{n} c2^{(k+1)(\gamma\tau+Q)}|u_{A_{k+1}}|^\tau \\ &\quad + C\sum_{k=m}^{n} [u]^{\tau a}_{s,p,\beta,|\cdot|,A_{k+1}\cup A_k} \||x|^\mu u\|^{(1-a)\tau}_{L^\alpha(A_{k+1}\cup A_k)}.\end{aligned} \tag{4.61}$$

By using (4.61), we compute

$$\begin{aligned}(1-c)\sum_{k=m}^{n} 2^{(\gamma\tau+Q)k}|u_{A_k}|^\tau &\le 2^{(\gamma\tau+Q)m}|u_{A_m}|^\tau + (1-c)\sum_{k=m+1}^{n} 2^{(\gamma\tau+Q)k}|u_{A_k}|^\tau \\ &\le C\sum_{k=m}^{n} [u]^{\tau a}_{s,p,\beta,|\cdot|,A_{k+1}\cup A_k} \||x|^\mu u\|^{(1-a)\tau}_{L^\alpha(A_{k+1}\cup A_k)}.\end{aligned}$$

This yields

$$\sum_{k=m}^{n} 2^{(\gamma\tau+Q)k}|u_{A_k}|^\tau \le C\sum_{k=m}^{n} [u]^{\tau a}_{s,p,\beta,|\cdot|,A_{k+1}\cup A_k} \||x|^\mu u\|^{(1-a)\tau}_{L^\alpha(A_{k+1}\cup A_k)}. \tag{4.62}$$

From (4.55) and (4.62), we have

$$\int_{\{2^m<|x|<2^{n+1}\}} |x|^{\gamma\tau}|u|^\tau dx \le C\sum_{k=m}^{n} [u]^{\tau a}_{s,p,\beta,|\cdot|,A_{k+1}\cup A_k} \||x|^\mu u\|^{(1-a)\tau}_{L^\alpha(A_{k+1}\cup A_k)}. \tag{4.63}$$

Let $s, t \ge 0$ be such that $s + t \ge 1$. Then for any $x_k, y_k \ge 0$, we have

$$\sum_{k=m}^{n} x_k^s y_k^t \le \left(\sum_{k=m}^{n} x_k\right)^s \left(\sum_{k=m}^{n} y_k\right)^t. \tag{4.64}$$

By using this inequality in (4.63) with $s = \frac{\tau a}{p}$, $t = \frac{(1-a)\tau}{\alpha}$, $\frac{a}{p} + \frac{1-a}{\alpha} \ge \frac{1}{\tau}$ and $s \ge \beta - \sigma$, we obtain

$$\int_{\{|x|>2^m\}} |x|^{\gamma\tau}|u|^\tau dx \le C[u]^{a\tau}_{s,p,\beta,|\cdot|,\cup_{k=m}^{\infty}A_k} \||x|^\mu u\|^{(1-a)\tau}_{L^\alpha(\cup_{k=m}^{\infty}A_k)}.$$

Inequality (4.47) is proved.

Let us prove (4.48). The strategy of the proof is similar to the previous case. Choose m such that

$$\operatorname{supp} u \cap B_{2^m} = \emptyset. \tag{4.65}$$

From Lemma 4.5.4 we have

$$|u_{A_{k+1}} - u_{A_k}|^\tau \le C 2^{\frac{a\tau k(sp-Q)}{p}} [u]^{\tau a}_{s,p,|\cdot|,A_{k+1}\cup A_k} \left(\fint_{A_{k+1}\cup A_k} |u|^\alpha dx \right)^{\frac{(1-a)\tau}{\alpha}}.$$

By (4.60) and choosing $c = \frac{1+2^{\gamma\tau+Q}}{2} < 1$, since $\gamma\tau + Q < 0$, we have

$$2^{(\gamma\tau+Q)(k+1)} |u_{A_{k+1}}|^\tau \le c 2^{k(\gamma\tau+Q)} |u_{A_k}|^\tau + C[u]^{\tau a}_{s,p,\beta,|\cdot|,A_{k+1}\cup A_k} \| |x|^\mu u \|^{(1-a)\tau}_{L^\alpha(A_{k+1}\cup A_k)},$$

and by summing over k from m to n and by using (4.65) we obtain

$$\sum_{k=m}^{n} 2^{(\gamma\tau+Q)k} |u_{A_k}|^\tau \le C \sum_{k=m-1}^{n-1} [u]^{\tau a}_{s,p,\beta,|\cdot|,A_{k+1}\cup A_k} \| |x|^\mu u \|^{(1-a)\tau}_{L^\alpha(A_{k+1}\cup A_k)}. \tag{4.66}$$

From (4.55) and (4.66), we establish that

$$\int_{\{2^m<|x|<2^{n+1}\}} |x|^{\gamma\tau} |u|^\tau dx \le C \sum_{k=m-1}^{n-1} [u]^{\tau a}_{s,p,\beta,|\cdot|,A_{k+1}\cup A_k} \| |x|^\mu u \|^{(1-a)\tau}_{L^\alpha(A_{k+1}\cup A_k)}.$$

Now by using (4.64) we get

$$\int_{\{|x|<2^{n+1}\}} |x|^{\gamma\tau} |u|^\tau dx \le C[u]^{\tau a}_{s,p,\beta,|\cdot|,\cup_{k=-\infty}^{n} A_k} \| |x|^\mu u \|^{(1-a)\tau}_{L^\alpha(\cup_{k=-\infty}^{n} A_k)}.$$

The proof of the case $s \ge \beta - \sigma$ is complete.

Let us prove the case of $\beta - \sigma > s$. Without loss of generality, we assume that

$$[u]_{s,p,\beta,|\cdot|} = \|u\|_{L^\alpha(\mathbb{G})} = 1,$$

where

$$\frac{1}{p} + \frac{\beta - s}{Q} \ne \frac{1}{\alpha} + \frac{\mu}{Q}.$$

We also assume that $a_1 > 0$, $1 > a_2$ and τ_1, $\tau_2 > 0$ with

$$\frac{1}{\tau_2} = \frac{a_2}{p} + \frac{1-a_2}{\alpha},$$

and

$$\begin{aligned} &\text{if } \frac{a}{p} + \frac{1-a}{\alpha} - \frac{as}{Q} > 0, \quad \text{then } \frac{1}{\tau_1} = \frac{a_1}{p} + \frac{1-a_1}{\alpha} - \frac{a_1 s}{Q}, \\ &\text{if } \frac{a}{p} + \frac{1-a}{\alpha} - \frac{as}{Q} \le 0, \quad \text{then } \frac{1}{\tau} > \frac{1}{\tau_1} \ge \frac{a_1}{p} + \frac{1-a_1}{\alpha} - \frac{a_1 s}{Q}. \end{aligned} \tag{4.67}$$

Taking $\gamma_1 = a_1\beta + (1-a_1)\mu$ and $\gamma_2 = a_2(\beta - s) + (1-a_2)\mu$, we obtain

$$\frac{1}{\tau_1} + \frac{\gamma_1}{Q} \geq a_1\left(\frac{1}{p} + \frac{\beta - s}{Q}\right) + (1-a_1)\left(\frac{1}{\alpha} + \frac{\mu}{Q}\right) \tag{4.68}$$

and

$$\frac{1}{\tau_2} + \frac{\gamma_2}{Q} = a_2\left(\frac{1}{p} + \frac{\beta - s}{Q}\right) + (1-a_2)\left(\frac{1}{\alpha} + \frac{\mu}{Q}\right). \tag{4.69}$$

Let a_1 and a_2 be such that

$$|a - a_1| \text{ and } |a - a_2| \text{ are small enough,} \tag{4.70}$$

$$\begin{aligned} a_2 < a < a_1, \ \text{if } \frac{1}{p} + \frac{\beta - s}{Q} > \frac{1}{\alpha} + \frac{\mu}{Q}, \\ a_1 < a < a_2, \ \text{if } \frac{1}{p} + \frac{\beta - s}{Q} < \frac{1}{\alpha} + \frac{\mu}{Q}. \end{aligned} \tag{4.71}$$

By using (4.70)–(4.71) in (4.68), (4.69) and (4.45), we establish

$$\frac{1}{\tau_1} + \frac{\gamma_1}{Q} > \frac{1}{\tau} + \frac{\gamma}{Q} > \frac{1}{\tau_2} + \frac{\gamma_2}{Q} > 0.$$

From (4.67) in the case $\frac{a}{p} + \frac{1-a}{\alpha} - \frac{as}{Q} > 0$ with $a > 0$, $\beta - \sigma > s$ and (4.70), we get

$$\frac{1}{\tau} - \frac{1}{\tau_1} = (a - a_1)\left(\frac{1}{p} - \frac{s}{Q} - \frac{1}{\alpha}\right) + \frac{a}{Q}(\beta - \sigma) > 0, \tag{4.72}$$

and

$$\frac{1}{\tau} - \frac{1}{\tau_2} = (a - a_2)\left(\frac{1}{p} - \frac{1}{\alpha}\right) + \frac{a}{Q}(\beta - \sigma - s) > 0. \tag{4.73}$$

From (4.67), (4.72) and (4.73), we have

$$\tau_1 > \tau, \quad \tau_2 > \tau.$$

Thus, using this, (4.70) and Hölder's inequality, we obtain

$$\||x|^{\gamma}u\|_{L^{\tau}(\mathbb{G}\setminus B_1)} \leq C\||x|^{\gamma_1}u\|_{L^{\tau_1}(\mathbb{G})},$$

and

$$\||x|^{\gamma}u\|_{L^{\tau}(B_1)} \leq C\||x|^{\gamma_2}u\|_{L^{\tau_2}(\mathbb{G})},$$

where B_1 is the unit quasi-ball. By using the previous case, we establish

$$\||x|^{\gamma_1}u\|_{L^{\tau_1}(\mathbb{G})} \leq C[u]^{a_1}_{s,p,\beta,|\cdot|}\||x|^{\mu}u\|^{1-a_1}_{L^{\alpha}(\mathbb{G})} \leq C,$$

and

$$\||x|^{\gamma_2}u\|_{L^{\tau_2}(\mathbb{G})} \leq C[u]^{a_2}_{s,p,\beta,|\cdot|}\||x|^{\mu}u\|^{1-a_2}_{L^{\alpha}(\mathbb{G})} \leq C.$$

The proof of Theorem 4.5.2 is complete. □

Now we consider the critical case $\frac{1}{\tau} + \frac{\gamma}{Q} = 0$.

Theorem 4.5.5 (Fractional critical Caffarelli–Kohn–Nirenberg inequality). *Let $\mathbb{G}$ be a homogeneous group of homogeneous dimension Q. Let $Q \geq 2$, $s \in (0,1)$, $p > 1$, $\alpha \geq 1$, $\tau > 1$, $a \in (0,1]$, β_1, β_2, β, μ, $\gamma \in \mathbb{R}$, $\beta_1 + \beta_2 = \beta$,*

$$\frac{1}{\tau} + \frac{\gamma}{Q} = a\left(\frac{1}{p} + \frac{\beta - s}{Q}\right) + (1-a)\left(\frac{1}{\alpha} + \frac{\mu}{Q}\right). \tag{4.74}$$

In addition, assume that $0 \leq \beta - \sigma \leq s$ with $\gamma = a\sigma + (1-a)\mu$.

If $\frac{1}{\tau} + \frac{\gamma}{Q} = 0$ and $\operatorname{supp} u \subset B_R = \{x \in \mathbb{G} : |x| < R\}$, then we have

$$\left\| \frac{|x|^{\gamma}}{\ln \frac{2R}{|x|}} u \right\|_{L^{\tau}(\mathbb{G})} \leq C[u]^a_{s,p,\beta,|\cdot|} \| |x|^{\mu} u \|^{1-a}_{L^{\alpha}(\mathbb{G})}, \tag{4.75}$$

for all $u \in C^1_c(\mathbb{G})$.

Proof of Theorem 4.5.5. The proof is similar to the proof of Theorem 4.5.2. In (4.54), summing over k from m to n and fixing $\varepsilon > 0$, we have

$$\begin{aligned} \int_{\{|x|>2^m\}} \frac{|x|^{\gamma\tau}}{\ln^{1+\varepsilon}\left(\frac{2R}{|x|}\right)} |u|^{\tau} dx &\leq C \sum_{k=m}^{n} \frac{1}{(n+1-k)^{1+\varepsilon}} |u_{A_k}|^{\tau} \\ &\quad + C \sum_{k=m}^{n} [u]^{a\tau}_{s,p,\beta,|\cdot|,A_k} \| |x|^{\mu} u \|^{(1-a)\tau}_{L^{\alpha}(A_k)}. \end{aligned} \tag{4.76}$$

From Lemma 4.5.4, we have

$$|u_{A_{k+1}} - u_{A_k}| \leq C 2^{\frac{ak(sp-Q)}{p}} [u]^a_{s,p,|\cdot|,A_{k+1}\cup A_k} \left(\int_{A_{k+1}\cup A_k} |u|^{\alpha} dx \right)^{\frac{1-a}{\alpha}}.$$

By using (4.60) with $\zeta = \frac{(n+1-k)^{\varepsilon}}{(n+\frac{1}{2}-k)^{\varepsilon}}$ we get

$$\begin{aligned} \frac{|u_{A_k}|^{\tau}}{(n+1-k)^{\varepsilon}} &\leq \frac{|u_{A_{k+1}}|^{\tau}}{(n+\frac{1}{2}-k)^{\varepsilon}} \\ &\quad + C(n+1-k)^{\tau-1-\varepsilon} [u]^{a\tau}_{s,p,\beta,|\cdot|,A_{k+1}\cup A_k} \| |x|^{\mu} u \|^{(1-a)\tau}_{L^{\alpha}(A_{k+1}\cup A_k)}. \end{aligned} \tag{4.77}$$

For $\varepsilon > 0$ and $n \geq k$, we have

$$\frac{1}{(n-k+1)^{\varepsilon}} - \frac{1}{(n-k+\frac{3}{2})^{\varepsilon}} \sim \frac{1}{(n-k+1)^{1+\varepsilon}}. \tag{4.78}$$

By using this fact, (4.77), (4.78) and $\varepsilon = \tau - 1$, we obtain

$$\sum_{k=m}^{n} \frac{|u_{A_k}|^{\tau}}{(n+1-k)^{\tau}} \leq C \sum_{k=m}^{n} [u]^{a\tau}_{s,p,\beta,|\cdot|,A_{k+1}\cup A_k} \| q^{\mu}(x) u \|^{(1-a)\tau}_{L^{\alpha}(A_{k+1}\cup A_k)}. \tag{4.79}$$

From (4.76) and (4.79), we establish

$$\int_{\{|x|>2^m\}} \frac{|x|^{\gamma\tau}}{\ln^{\tau}\frac{2R}{|x|}}|u|^{\tau}dx \leq C\sum_{k=m}^{n}[u]^{a\tau}_{s,p,\beta,|\cdot|,A_{k+1}\cup A_k}\||x|^{\mu}u\|^{(1-a)\tau}_{L^{\alpha}(A_{k+1}\cup A_k)}.$$

By using (4.64) with (4.74) and $0\leq\beta-\sigma\leq s$, where $s=\frac{\tau a}{p}$, $t=\frac{(1-a)\tau}{\alpha}$, we have $s+t\geq 1$ and we arrive at

$$\int_{\{|x|>2^m\}} \frac{|x|^{\gamma\tau}}{\ln^{\tau}\frac{2R}{|x|}}|u|^{\tau}dx \leq C\sum_{k=m}^{n}[u]^{a\tau}_{s,p,\beta,|\cdot|,\cup_{k=m}^{\infty}A_k}\||x|^{\mu}u\|^{(1-a)\tau}_{L^{\alpha}(\cup_{k=m}^{\infty}A_k)}.$$

Theorem 4.5.5 is proved. □

4.6 Lyapunov inequalities on homogeneous groups

In this section we give an application of the preceding results to derive a Lyapunov type inequality for the fractional p-sub-Laplacian with homogeneous Dirichlet boundary condition on homogeneous groups. First, we summarize the basic results concerning the classical Lyapunov inequality.

Remark 4.6.1 (Euclidean Lyapunov inequalities).

1. In [Lya07], Lyapunov obtained the following result for the one-dimensional homogeneous Dirichlet boundary value problem. Consider the second-order ordinary differential equation

$$\begin{cases} u''(x)+\omega(x)u(x)=0, \ x\in(a,b), \\ u(a)=u(b)=0. \end{cases} \tag{4.80}$$

   Then, if (4.80) has a non-trivial solution u, and $\omega=\omega(x)$ is a real-valued and continuous function on $[a,b]$, then we must have

$$\int_a^b |\omega(x)|dx > \frac{4}{b-a}. \tag{4.81}$$

   Inequality (4.81) is called a (classical) *Lyapunov inequality.*

2. Nowadays, there are many extensions of the Lyapunov inequality (4.81). For example, in [Elb81] the Lyapunov inequality for the one-dimensional Dirichlet p-Laplacian was obtained: if

$$\begin{cases} (|u'(x)|^{p-2}u'(x))'+\omega(x)u(x)=0, \ x\in(a,b), \ 1<p<\infty, \\ u(a)=u(b)=0, \end{cases} \tag{4.82}$$

has a non-trivial solution u for $\omega \in L^1(a,b)$, then

$$\int_a^b |\omega(x)|dx > \frac{2^p}{(b-a)^{p-1}}, \quad 1 < p < \infty. \tag{4.83}$$

Obviously, taking $p = 2$ in (4.83), we recover (4.81).

3. In [JKS17], the following Lyapunov inequality was obtained for the multi-dimensional fractional p-Laplacian $(-\Delta_p)^s$, $1 < p < \infty$, $s \in (0,1)$, with a homogeneous Dirichlet boundary condition, that is, for the equation

$$\begin{cases} (-\Delta_p)^s u = \omega(x)|u|^{p-2}u, \ x \in \Omega, \\ u(x) = 0, \ x \in \mathbb{R}^N \setminus \Omega, \end{cases} \tag{4.84}$$

where $\Omega \subset \mathbb{R}^N$ is a measurable set, $1 < p < \infty$, and $s \in (0,1)$. More precisely, let $\omega \in L^\theta(\Omega)$ with $N > sp$, $\frac{N}{sp} < \theta < \infty$, be a non-negative weight. Suppose that the problem (4.84) has a non-trivial weak solution $u \in W_0^{s,p}(\Omega)$. Then we have

$$\left(\int_\Omega \omega^\theta(x)\, dx \right)^{1/\theta} > \frac{C}{r_\Omega^{sp-\frac{N}{\theta}}}, \tag{4.85}$$

where $C > 0$ is a universal constant and r_Ω is the inner radius of Ω.

The appearance of the inner radius in (4.85) motivates one to define its analogue also in the setting of homogeneous groups.

Definition 4.6.2 (Inner quasi-radius). Let $p > 1$ and $s \in (0,1)$ be such that $Q > sp$. Let $\Omega \subset \mathbb{G}$ be a Haar measurable set. We denote by $r_{\Omega,q}$ the *inner quasi-radius* of Ω, that is,

$$r_\Omega = r_{\Omega,|\cdot|} := \sup\{|x| : \ x \in \Omega\}. \tag{4.86}$$

Clearly, the exact values depend on the choice of a homogeneous quasi-norm $|\cdot|$. As before, if the quasi-norm is fixed, we can omit it from the notation.

4.6.1 Lyapunov type inequality for fractional p-sub-Laplacians

We now turn our attention to the Lyapunov inequalities for the fractional p-sub-Laplacian $(-\Delta_p)^s$ from Definition 4.1.1. Let us consider the boundary value problem

$$\begin{cases} (-\Delta_p)^s u(x) = \omega(x)|u(x)|^{p-2}u(x), \ x \in \Omega, \\ u(x) = 0, \quad x \in \mathbb{G} \setminus \Omega, \end{cases} \tag{4.87}$$

where $\omega \in L^\infty(\Omega)$. A function $u \in W_0^{s,p}(\Omega)$ is called a *weak solution of the problem* (4.87) if we have

$$\int_\Omega \int_\Omega \frac{|u(x)-u(y)|^{p-2}(u(x)-u(y))(v(x)-v(y))}{|y^{-1}x|^{Q+sp}} dxdy$$
$$= \int_\Omega \omega(x)|u(x)|^{p-2}u(x)v(x)dx \quad \text{for all } v \in W_0^{s,p}(\Omega).$$

Theorem 4.6.3 (Lyapunov inequality for fractional p-sub-Laplacian). *Let $\mathbb{G}$ be a homogeneous group of homogeneous dimension Q. Let $\Omega \subset \mathbb{G}$ be a Haar measurable set. Let $\omega \in L^{\theta}(\Omega)$ be a non-negative function with $\frac{Q}{sp} < \theta < \infty$, and with $Q > ps$. Suppose that the boundary value problem (4.87) has a non-trivial weak solution $u \in W_0^{s,p}(\Omega)$. Then we have*

$$\|\omega\|_{L^{\theta}(\Omega)} \geq C \Big/ r_{\Omega,|\cdot|}^{sp-Q/\theta}, \tag{4.88}$$

for some $C = C(Q, p, s, |\cdot|) > 0$.

Proof of Theorem 4.6.3. We fix a homogeneous quasi-norm $|\cdot|$ thus also eliminating it from the notation. Let $\alpha := \frac{\theta - \theta/sp}{\theta - 1} \in (0,1)$ and let p^* be the Sobolev conjugate exponent as in Theorem 4.3.1. Let us define

$$\beta := \alpha p + (1-\alpha)p^*.$$

Let $\beta = p\theta'$ with $1/\theta + 1/\theta' = 1$. Then we have

$$\int_{\Omega} \frac{|u(x)|^{\beta}}{r_{\Omega}^{\alpha sp}} dx \leq \int_{\Omega} \frac{|u(x)|^{\beta}}{|x|^{\alpha sp}} dx. \tag{4.89}$$

On the other hand, the Hölder inequality with exponents $\nu = \alpha^{-1}$ and its conjugate $1/\nu + 1/\nu' = 1$ implies

$$\begin{aligned}\int_{\Omega} \frac{|u(x)|^{\beta}}{|x|^{\alpha sp}} dx &\leq \int_{\Omega} \frac{|u(x)|^{\alpha p}|u(x)|^{(1-\alpha)p^*}}{|x|^{\alpha sp}} dx \\ &\leq \left(\int_{\Omega} \frac{|u(x)|^{p}}{|x|^{sp} dx}\right)^{\alpha} \left(\int_{\Omega} |u(x)|^{p^*} dx\right)^{1-\alpha}.\end{aligned}$$

Further, by using Theorem 4.3.1 and Theorem 4.2.1, we get

$$\begin{aligned}\int_{\Omega} \frac{|u(x)|^{\beta}}{|x|^{\alpha sp}} dx &\leq C_1^{\alpha} \left(\int_{\Omega}\int_{\Omega} \frac{|u(x)-u(y)|^p}{|y^{-1}x|^{Q+sp}} dx dy\right)^{\alpha/p} C_2^{(1-\alpha)p^*/p} [u]_{s,p}^{(1-\alpha)p^*/p} \\ &\leq C_1^{\alpha} [u]_{s,p}^{\alpha} C_2^{(1-\alpha)p^*/p} [u]_{s,p}^{(1-\alpha)p^*/p} \\ &= C\left([u]_{s,p}^{p}\right)^{(\alpha p + (1-\alpha)p^*)/p} = C\left(\int_{\Omega} \omega(x)|u(x)|^p dx\right)^{\theta'} \\ &\leq C\left(\int_{\Omega} \omega^{\theta}(x) dx\right)^{\theta'/\theta} \int_{\Omega} |u(x)|^{p\theta'} dx = C\|\omega\|_{L^{\theta}(\Omega)}^{\theta'} \int_{\Omega} |u(x)|^{\beta} dx.\end{aligned}$$

That is, we obtain

$$\int_{\Omega} \frac{|u(x)|^{\beta}}{|x|^{\alpha sp}} dx \leq C\|\omega\|_{L^{\theta}(\Omega)}^{\theta'} \int_{\Omega} |u(x)|^{\beta} dx.$$

Thus, from (4.89) we have

$$\frac{1}{r_{\Omega}^{\alpha sp}} \int_{\Omega} |u(x)|^{\beta} dx \leq \int_{\Omega} \frac{|u(x)|^{\beta}}{|x|^{\alpha sp}} dx \leq C\|\omega\|_{L^{\theta}(\Omega)}^{\theta'} \int_{\Omega} |u(x)|^{\beta} dx.$$

Finally, we arrive at

$$\frac{C}{r_{\Omega}^{sp-Q/\theta}} \leq \|\omega\|_{L^{\theta}(\Omega)}.$$

Theorem 4.6.3 is proved. □

As an application of the Lyapunov inequality, let us consider the spectral problem for the (nonlinear) fractional p-sub-Laplacian $(-\Delta_p)^s$, $1 < p < \infty$, $s \in (0,1)$, with the Dirichlet boundary condition:

$$\begin{cases} (-\Delta_p)^s u = \lambda |u|^{p-2}u, \ x \in \Omega, \\ u(x) = 0, \ x \in \mathbb{G} \setminus \Omega. \end{cases} \tag{4.90}$$

We define the corresponding *Rayleigh quotient* by

$$\lambda_1 := \inf_{u \in W_0^{s,p}(\Omega),\ u \neq 0} \frac{[u]_{s,p}^p}{\|u\|_{L^p(\mathbb{G})}^p}. \tag{4.91}$$

Clearly, its precise value may depend on the choice of the homogeneous quasi-norm $|\cdot|$. As a consequence of Theorem 4.6.3 we have

Theorem 4.6.4 (First eigenvalue for the fractional p-sub-Laplacian). *Let $\mathbb{G}$ be a homogeneous group of homogeneous dimension Q. Let λ_1 be the first eigenvalue of problem (4.90) given by (4.91). Let $Q > sp$, $s \in (0,1)$ and $1 < p < \infty$. Then we have*

$$\lambda_1 \geq \sup_{\frac{Q}{sp} < \theta < \infty} \frac{C}{|\Omega|^{\frac{1}{\theta}} r_{\Omega,|\cdot|}^{sp-Q/\theta}}, \tag{4.92}$$

where C is a positive constant given in Theorem 4.6.3 and $|\Omega|$ is the Haar measure of Ω.

Proof of Theorem 4.6.4. In Theorem 4.6.3, taking $\omega = \lambda \in L^{\theta}(\Omega)$ and using Lyapunov type inequality (4.88), we get that

$$\|\omega\|_{L^{\theta}(\Omega)} = \|\lambda\|_{L^{\theta}(\Omega)} = \left(\int_{\Omega} \lambda^{\theta} dx\right)^{1/\theta} \geq \frac{C}{r_{\Omega,|\cdot|}^{sp-Q/\theta}}.$$

For every $\theta > \frac{Q}{sp}$, we have

$$\lambda_1 \geq \frac{C}{|\Omega|^{\frac{1}{\theta}} r_{\Omega,|\cdot|}^{sp-Q/\theta}}.$$

Thus, we obtain

$$\lambda_1 \geq \sup_{\frac{Q}{sp} < \theta < \infty} \frac{C}{|\Omega|^{\frac{1}{\theta}} r_{\Omega,|\cdot|}^{sp-Q/\theta}},$$

for all $\frac{Q}{sp} < \theta < \infty$. Theorem 4.6.4 is proved. □

4.6.2 Lyapunov type inequality for systems

In the previous section we have presented the Lyapunov type inequality for the fractional p-sub-Laplacian with the homogeneous Dirichlet condition. Now we discuss the Lyapunov type inequality for the fractional p-sub-Laplacian system for the homogeneous Dirichlet problem.

Let $\Omega \subset \mathbb{G}$ be a Haar measurable set, let $\omega_i \in L^1(\Omega)$, $\omega_i \geq 0$, $s_i \in (0,1)$, $p_i \in (1,\infty)$. As before in Definition 4.1.1, for each p_i we denote by $(-\Delta_{p_i})^{s_i}$ the fractional p-sub-Laplacian on $\mathbb{G}$ defined by

$$(-\Delta_{p_i})^{s_i} u_i(x) = 2 \lim_{\delta \searrow 0} \int_{\mathbb{G}\setminus B(x,\delta)} \frac{|u_i(x)-u_i(y)|^{p_i-2}(u_i(x)-u_i(y))}{|y^{-1}x|^{Q+s_ip_i}} dy, \quad x \in \mathbb{G}, \quad i = 1,\ldots,n.$$

Here $B(x,\delta)$ is a quasi-ball with respect to a fixed quasi-norm $|\cdot|$, with radius δ, centred at $x \in \mathbb{G}$. Let α_i be positive parameters such that

$$\sum_{i=1}^{n} \frac{\alpha_i}{p_i} = 1. \tag{4.93}$$

Then, we consider the following system of the fractional p-sub-Laplacians:

$$\begin{cases} (-\Delta_{p_1})^{s_1} u_1(x) = \omega_1(x)|u_1(x)|^{\alpha_1-2}u_1(x)|u_2(x)|^{\alpha_2}\cdots|u_n(x)|^{\alpha_n}, \ x \in \Omega, \\ (-\Delta_{p_2})^{s_2} u_2(x) = \omega_2(x)|u_1(x)|^{\alpha_1}|u_2(x)|^{\alpha_2-2}u_2(x)\cdots|u_n(x)|^{\alpha_n}, \ x \in \Omega, \\ \quad\cdots \\ (-\Delta_{p_n})^{s_n} u_n(x) = \omega_n(x)|u_1(x)|^{\alpha_1}|u_2(x)|^{\alpha_2}\cdots|u_n(x)|^{\alpha_n-2}u_n(x), \ x \in \Omega, \end{cases} \tag{4.94}$$

with homogeneous Dirichlet conditions

$$u_i(x) = 0, \quad x \in \mathbb{G}\setminus\Omega, \quad i = 1,\ldots,n. \tag{4.95}$$

We denote by r_Ω the inner quasi-radius of Ω, that is,

$$r_\Omega = r_{\Omega,|\cdot|} := \max\{|x| : x \in \Omega\}.$$

Definition 4.6.5 (Weak solutions of the p-sub-Laplacian system)**.** We will say that $(u_1,\ldots,u_n) \in \prod_{i=1}^n W_0^{s_i,p_i}(\Omega)$ is a *weak solution* of (4.94)–(4.95) if for all $(v_1,\ldots,v_n) \in \prod_{i=1}^n W_0^{s_i,p_i}(\Omega)$, we have

$$\int_{\mathbb{G}}\int_{\mathbb{G}} \frac{|u_i(x)-u_i(y)|^{p_i-2}(u_i(x)-u_i(y))(v_i(x)-v_i(y))}{|y^{-1}x|^{Q+s_ip_i}} dxdy \tag{4.96}$$
$$= \int_\Omega \omega_i(x)\left(\prod_{j=1}^{i-1}|u_j(x)|^{\alpha_j}\right)\left(\prod_{j=i+1}^{n}|u_j(x)|^{\alpha_j}\right)|u_i(x)|^{\alpha_i-2}u_i(x)v_i(x)dx,$$

for every $i = 1,\ldots,n$.

Now we present the following analogue of the Lyapunov type inequality for the fractional p-sub-Laplacian system on $\mathbb{G}$.

Theorem 4.6.6 (Lyapunov inequality for fractional p-sub-Laplacian system). *Let $\mathbb{G}$ be a homogeneous group of homogeneous dimension Q. Let $s_i \in (0,1)$ and $p_i \in (1,\infty)$ be such that $Q > s_i p_i$ with $i = 1, \ldots, n$. Let $\omega_i \in L^\theta(\Omega)$ be a non-negative weight and assume that*

$$1 < \max_{i=1,\ldots,n} \left\{ \frac{Q}{s_i p_i} \right\} < \theta < \infty.$$

If (4.94)–(4.95) admits a non-trivial weak solution, then we have

$$\prod_{i=1}^{n} \|\omega_i\|_{L^\theta(\Omega)}^{\frac{\theta\alpha_i}{p_i}} \geq C r_{\Omega,q}^{Q-\theta\sum_{j=1}^{n} s_j\alpha_j}, \tag{4.97}$$

where C is a positive constant.

Proof of Theorem 4.6.6. Set

$$\xi_i := \gamma_i p_i + (1-\gamma_i)p_i^*, \quad i = 1, \ldots, n,$$

and

$$\gamma_i := \frac{\theta - \frac{Q}{s_i p_i}}{\theta - 1}, \tag{4.98}$$

where $p_i^* = \frac{Q}{Q - s_i p_i}$ is the Sobolev conjugate exponent as in Theorem 4.3.1. Notice that for all $i = 1, \ldots, n$ we have $\gamma_i \in (0,1)$ and $\xi_i = p_i\theta'$, where $\theta' = \frac{\theta}{\theta-1}$. Then for every i we have

$$\int_\Omega \frac{|u_i(x)|^{\xi_i}}{r_{\Omega,q}^{\gamma_i s_i p_i}} dx \leq \int_\Omega \frac{|u_i(x)|^{\xi_i}}{|x|^{\gamma_i s_i p_i}} dx,$$

and by using the Hölder inequality with $\nu_i = \frac{1}{\gamma_i}$ and $\frac{1}{\nu_i} + \frac{1}{\nu_i'} = 1$, we get

$$\begin{aligned} \int_\Omega \frac{|u_i(x)|^{\xi_i}}{|x|^{\gamma_i s_i p_i}} dx &= \int_\Omega \frac{|u_i(x)|^{\gamma_i p_i} |u_i(x)|^{(1-\gamma_i)p_i^*}}{|x|^{\gamma_i s_i p_i}} dx \\ &\leq \left(\int_\Omega \frac{|u_i(x)|^{p_i}}{|x|^{s_i p_i}} dx \right)^{\gamma_i} \left(\int_\Omega |u_i(x)|^{p_i^*} dx \right)^{1-\gamma_i}. \end{aligned} \tag{4.99}$$

On the other hand, from Theorem 4.3.1, we obtain

$$\left(\int_\Omega |u_i(x)|^{p_i^*} dx \right)^{1-\gamma_i} \leq C[u_i]_{s_i,p_i,|\cdot|}^{p_i^*(1-\gamma_i)},$$

and from Theorem 4.2.1, we have

$$\left(\int_\Omega \frac{|u_i(x)|^{p_i}}{|x|^{s_i p_i}} dx \right)^{\gamma_i} \leq C[u_i]_{s_i,p_i,|\cdot|}^{p_i\gamma_i}.$$

Thus, from (4.99) and by setting $u_i(x) = v_i(x)$ in (4.96), we get

$$\int_\Omega \frac{|u_i(x)|^{\xi_i}}{|x|^{\gamma_i s_i p_i}} dx \le C([u_i]^{p_i}_{s_i,p_i,|\cdot|,\Omega})^{\frac{\xi_i}{p_i}} \le C([u_i]^{p_i}_{s_i,p_i,|\cdot|})^{\frac{\xi_i}{p_i}}$$
$$= C\left(\int_\Omega \omega_i(x) \prod_{j=1}^n |u_j|^{\alpha_j} dx\right)^{\frac{\xi_i}{p_i}} = C\left(\int_\Omega \omega_i(x) \prod_{j=1}^n |u_j|^{\alpha_j} dx\right)^{\theta'},$$

for every $i = 1, \ldots, n$. Therefore, by using the Hölder inequality with exponents θ and θ', we obtain

$$\int_\Omega \frac{|u_i(x)|^{\xi_i}}{|x|^{\gamma_i s_i p_i}} dx \le C\|\omega_i\|^{\frac{\theta}{\theta-1}}_{L^\theta(\Omega)} \int_\Omega \prod_{j=1}^n |u_j(x)|^{\alpha_j \theta'} dx.$$

Again by using the Hölder inequality and (4.93), we get

$$\int_\Omega \prod_{j=1}^n |u_j(x)|^{\alpha_j\theta'} dx \le \prod_{j=1}^n \left(\int_\Omega |u_j|^{\theta' p_j} dx\right)^{\frac{\alpha_j}{p_j}}.$$

It yields that

$$\int_\Omega \frac{|u_i(x)|^{\xi_i}}{|x|^{\gamma_i s_i p_i}} dx \le C\|\omega_i\|^{\frac{\theta}{\theta-1}}_{L^\theta(\Omega)} \prod_{j=1}^n \left(\int_\Omega |u_j|^{\theta' p_j} dx\right)^{\frac{\alpha_j}{p_j}}.$$

That is, we have

$$\int_\Omega \frac{|u_i(x)|^{\xi_i}}{r_\Omega^{\gamma_i s_i p_i}} dx \le \int_\Omega \frac{|u_i(x)|^{\xi_i}}{|x|^{\gamma_i s_i p_i}} dx \le C\|\omega_i\|^{\frac{\theta}{\theta-1}}_{L^\theta(\Omega)} \prod_{j=1}^n \left(\int_\Omega |u_j|^{\theta' p_j} dx\right)^{\frac{\alpha_j}{p_j}}.$$

Thus, for every $e_i > 0$ we have

$$\left(\int_\Omega \frac{|u_i(x)|^{\xi_i}}{r_\Omega^{\gamma_i s_i p_i}} dx\right)^{e_i} = \frac{1}{r_\Omega^{e_i \gamma_i s_i p_i}} \left(\int_\Omega |u_i(x)|^{\xi_i} dx\right)^{e_i}$$
$$\le C\|\omega_i\|^{\frac{e_i\theta}{\theta-1}}_{L^\theta(\Omega)} \prod_{j=1}^n \left(\int_\Omega |u_j|^{\theta' p_j} dx\right)^{\frac{e_i\alpha_j}{p_j}},$$

so that

$$\frac{1}{r_\Omega^{\sum_{j=1}^n \gamma_j s_j p_j e_j}} \prod_{i=1}^n \left(\int_\Omega |u_i(x)|^{\theta' p_i} dx\right)^{e_i}$$
$$\le C\left(\prod_{i=1}^n \|\omega_i\|^{\frac{e_i\theta}{\theta-1}}_{L^\theta(\Omega)}\right)\left(\prod_{i=1}^n \left(\int_\Omega |u_i(x)|^{\theta' p_i} dx\right)^{\frac{\alpha_i \sum_{j=1}^n e_j}{p_i}}\right).$$

This implies

$$\frac{1}{r_\Omega^{\sum_{j=1}^n \gamma_j s_j p_j e_j}} \leq C\left(\prod_{i=1}^n \|\omega_i\|_{L^\theta(\Omega)}^{\frac{e_i\theta}{\theta-1}}\right)\left(\prod_{i=1}^n \left(|u_i(x)|^{\theta' p_i}dx\right)^{\frac{\alpha_i\sum_{j=1}^n e_j}{p_i}-e_i}\right), \quad (4.100)$$

where C is a positive constant. Then, we choose e_i, $i = 1,\dots,n$, such that $\frac{\alpha_i\sum_{j=1}^n e_j}{p_i} - e_i = 0$. Consequently, from (4.93) we have the solution of this system

$$e_i = \frac{\alpha_i}{p_i}, \quad i = 1,\dots,n. \quad (4.101)$$

Combining (4.100), (4.98) and (4.101) we arrive at

$$\prod_{i=1}^n \|\omega_i\|_{L^\theta(\Omega)}^{\frac{\theta\alpha_i}{p_i}} \geq C r_\Omega^{Q-\theta\sum_{j=1}^n s_j\alpha_j}. \quad (4.102)$$

Theorem 4.6.6 is proved. □

In order to discuss an application of the Lyapunov type inequality for the fractional p-sub-Laplacian system on $\mathbb{G}$, we consider the spectral problem for the system of fractional p-sub-Laplacians:

$$\begin{cases} (-\Delta_{p_1,q})^{s_1}u_1(x) = \lambda_1\alpha_1\varphi(x)|u_1(x)|^{\alpha_1-2}u_1(x)|u_2(x)|^{\alpha_2}\cdots|u_n(x)|^{\alpha_n}, & x\in\Omega,\\ (-\Delta_{p_2,q})^{s_2}u_2(x) = \lambda_2\alpha_2\varphi(x)|u_1(x)|^{\alpha_1}|u_2(x)|^{\alpha_2-2}u_2(x)\cdots|u_n(x)|^{\alpha_n}, & x\in\Omega,\\ \cdots \\ (-\Delta_{p_n,q})^{s_n}u_n(x) = \lambda_n\alpha_n\varphi(x)|u_1(x)|^{\alpha_1}|u_2(x)|^{\alpha_2}\cdots|u_n(x)|^{\alpha_n-2}u_n(x), & x\in\Omega, \end{cases} \quad (4.103)$$

with

$$u_i(x) = 0, \quad x\in\mathbb{G}\setminus\Omega, \quad i=1,\dots,n, \quad (4.104)$$

where $\Omega\subset\mathbb{G}$ is a Haar measurable set, $\varphi\in L^1(\Omega)$, $\varphi\geq 0$ and $s_i\in(0,1)$, $p_i\in(1,\infty)$, $i=1,\dots,n$.

Definition 4.6.7 (Eigenvalues of p-sub-Laplacian system (4.103)–(4.104))**.** We say that $\lambda = (\lambda_1,\dots,\lambda_n)$ is an eigenvalue if the problem (4.103)–(4.104) admits at least one non-trivial weak solution $(u_1,\dots,u_n)\in\prod_{i=1}^n W_0^{s_i,p_i}(\Omega)$.

Theorem 4.6.8 (Eigenvalue estimates for p-sub-Laplacian system)**.** *Let $s_i\in(0,1)$ and $p_i\in(1,\infty)$ be such that $Q > s_ip_i$, for all $i=1,\dots,n$, and*

$$1 < \max_{i=1,\dots,n}\left\{\frac{Q}{s_ip_i}\right\} < \theta < \infty.$$

Let $\varphi\in L^\theta(\Omega)$ with $\|\varphi\|_{L^\theta(\Omega)}\neq 0$. Then we have

$$\lambda_k \geq \frac{C}{\alpha_k}\left(\frac{1}{\prod_{i=1,i\neq k}^n \lambda_i^{\frac{\alpha_i}{p_i}}}\right)^{\frac{p_k}{\alpha_k}}\left(\frac{1}{r_\Omega^{\theta\sum_{i=1}^n\alpha_is_i-Q}\prod_{i=1,i\neq k}^n\alpha_i^{\frac{\theta\alpha_i}{p_i}}\int_\Omega\varphi^\theta(x)dx}\right)^{\frac{p_k}{\theta\alpha_k}},$$

where C is a positive constant and $k=1,\dots,n$.

Proof of Theorem 4.6.8. In Theorem 4.6.6 by setting $\omega_k = \lambda_k\alpha_k\varphi(x)$, $k = 1, \dots, n$, we get

$$\alpha_k^{\frac{\theta\alpha_k}{p_k}}\lambda_k^{\frac{\theta\alpha_k}{p_k}}\prod_{i=1,i\neq k}^{n}(\alpha_i\lambda_i)^{\frac{\theta\alpha_i}{p_i}}\prod_{i=1}^{n}\|\varphi\|_{L^\theta(\Omega)}^{\frac{\theta\alpha_i}{p_i}} \geq Cr_{\Omega,q}^{Q-\theta\sum_{j=1}^{n}s_j\alpha_j}.$$

So from (4.93) we have

$$\alpha_k^{\frac{\theta\alpha_k}{p_k}}\lambda_k^{\frac{\theta\alpha_k}{p_k}}\prod_{i=1,i\neq k}^{n}(\alpha_i\lambda_i)^{\frac{\theta\alpha_i}{p_i}}\int_\Omega\varphi^\theta(x)dx \geq Cr_{\Omega,q}^{Q-\theta\sum_{j=1}^{n}s_j\alpha_j}.$$

This yields

$$\lambda_k^{\frac{\theta\alpha_k}{p_k}} \geq \frac{C}{\alpha_k^{\frac{\theta\alpha_k}{p_k}}r_{\Omega,q}^{\theta\sum_{j=1}^{n}s_j\alpha_j-Q}\prod_{i=1,i\neq k}^{n}(\alpha_i\lambda_i)^{\frac{\theta\alpha_i}{p_i}}\int_\Omega\varphi^\theta(x)dx}, \quad k = 1, \dots, n.$$

Therefore, we have

$$\lambda_k \geq \frac{C}{\alpha_k}\left(\frac{1}{\prod_{i=1,i\neq k}^{n}\lambda_i^{\frac{\alpha_i}{p_i}}}\right)^{\frac{p_k}{\alpha_k}}\left(\frac{1}{r_\Omega^{\theta\sum_{i=1}^{n}\alpha_i s_i-Q}\prod_{i=1,i\neq k}^{n}\alpha_i^{\frac{\theta\alpha_i}{p_i}}\int_\Omega\varphi^\theta(x)dx}\right)^{\frac{p_k}{\theta\alpha_k}}, \quad k = 1, \dots, n,$$

completing the proof. □

4.6.3 Lyapunov type inequality for Riesz potentials

In this section we discuss the Lyapunov type inequality for the Riesz potential operators on homogeneous groups. As an application, we discuss a two-sided estimate for the first eigenvalue of the Riesz potential.

Definition 4.6.9 (Riesz potentials). Let $\mathbb{G}$ be a homogeneous group of homogeneous dimension Q with a quasi-norm $|\cdot|$. The *Riesz potential* on a Haar measurable set $\Omega \subset \mathbb{G}$ is the operator given by the formula

$$\mathfrak{R}u(x) = \int_\Omega \frac{u(y)}{|y^{-1}x|^{Q-2s}}dy, \quad 0 < 2s < Q. \tag{4.105}$$

The (*weighted*) *Riesz potential* is defined by

$$\mathfrak{R}(\omega u)(x) = \int_\Omega \frac{\omega(y)u(y)}{|y^{-1}x|^{Q-2s}}dy, \quad 0 < 2s < Q. \tag{4.106}$$

A Lyapunov type inequality for the weighted Riesz potentials can be formulated as follows.

Theorem 4.6.10 (Lyapunov type inequality for weighted Riesz potentials). *Let $\mathbb{G}$ be a homogeneous group of homogeneous dimension Q with a quasi-norm $|\cdot|$. Let $\Omega \subset \mathbb{G}$ be a Haar measurable set, let $Q \geq 2 > 2s > 0$ and $1 < p < 2$. Assume that $\omega \in L^{\frac{p}{2-p}}(\Omega)$, $\frac{1}{|y^{-1}x|^{Q-2s}} \in L^{\frac{p}{p-1}}(\Omega \times \Omega)$ and*

$$C_0 = \left\| \frac{1}{|y^{-1}x|^{Q-2s}} \right\|_{L^{\frac{p}{p-1}}(\Omega\times\Omega)} < \infty.$$

Let $u \in L^{\frac{p}{p-1}}(\Omega)$, $u \neq 0$, satisfy

$$\mathfrak{R}(\omega u)(x) = \int_{\Omega} \frac{\omega(y)u(y)}{|y^{-1}x|^{Q-2s}} dy = u(x), \text{ for a.e. } x \in \Omega. \tag{4.107}$$

Then

$$\|\omega\|_{L^{\frac{p}{2-p}}(\Omega)} \geq \frac{1}{C_0}. \tag{4.108}$$

Proof of Theorem 4.6.10. In (4.107), by using Hölder's inequality for $p, \theta > 1$ with $\frac{1}{p} + \frac{1}{p'} = 1$ and $\frac{1}{\theta} + \frac{1}{\theta'} = 1$, we get

$$\begin{aligned}
|u(x)| &= \left| \int_{\Omega} \frac{\omega(y)u(y)}{|y^{-1}x|^{Q-2s}} dy \right| \\
&\leq \left(\int_{\Omega} |\omega(y)u(y)|^p dy \right)^{\frac{1}{p}} \left(\int_{\Omega} \left| \frac{1}{|y^{-1}x|^{Q-2s}} \right|^{p'} dy \right)^{\frac{1}{p'}} \\
&\leq \left(\int_{\Omega} |\omega(y)|^{p\theta} dy \right)^{\frac{1}{p\theta}} \left(\int_{\Omega} |u(y)|^{\theta' p} dy \right)^{\frac{1}{\theta' p}} \left(\int_{\Omega} \left| \frac{1}{|y^{-1}x|^{Q-2s}} \right|^{p'} dy \right)^{\frac{1}{p'}} \\
&= \|\omega\|_{L^{p\theta}(\Omega)} \|u\|_{L^{p\theta'}(\Omega)} \left(\int_{\Omega} \left| \frac{1}{|y^{-1}x|^{Q-2s}} \right|^{p'} dy \right)^{\frac{1}{p'}}.
\end{aligned}$$

Let p' be such that $p' = p\theta'$, so that $\theta = \frac{1}{2-p}$. Then we have

$$|u(x)| \leq \|\omega\|_{L^{\frac{p}{2-p}}(\Omega)} \|u\|_{L^{\frac{p}{p-1}}(\Omega)} \left(\int_{\Omega} \left| \frac{1}{|y^{-1}x|^{Q-2s}} \right|^{\frac{p}{p-1}} dy \right)^{\frac{p-1}{p}}. \tag{4.109}$$

From (4.109) we calculate

$$\begin{aligned}
\|u\|_{L^{\frac{p}{p-1}}(\Omega)} &\leq \|\omega\|_{L^{\frac{p}{2-p}}(\Omega)} \|u\|_{L^{\frac{p}{p-1}}(\Omega)} \left(\int_{\Omega} \int_{\Omega} \left| \frac{1}{|y^{-1}x|^{Q-2s}} \right|^{\frac{p}{p-1}} dxdy \right)^{\frac{p-1}{p}} \\
&= C_0 \|\omega\|_{L^{\frac{p}{2-p}}(\Omega)} \|u\|_{L^{\frac{p}{p-1}}(\Omega)}.
\end{aligned}$$

Finally, since $u \neq 0$, this implies

$$\|\omega\|_{L^{\frac{p}{2-p}}(\Omega)} \geq \frac{1}{C_0},$$

completing the proof. □

Let us now consider the following spectral problem for the Riesz potential:

$$\mathfrak{R}u(x) = \int_\Omega \frac{u(y)}{|y^{-1}x|^{Q-2s}} dy = \lambda u(x), \quad x \in \Omega, \quad 0 < 2s < Q. \tag{4.110}$$

We recall the Rayleigh quotient for the Riesz potential:

$$\lambda_1(\Omega) = \sup_{u \neq 0} \frac{\int_\Omega \int_\Omega \frac{u(x)u(y)}{|y^{-1}x|^{Q-2s}} dxdy}{\|u\|^2_{L^2(\Omega)}}, \tag{4.111}$$

where $\lambda_1(\Omega)$ is the first eigenvalue of the Riesz potential. A direct consequence of Theorem 4.6.10 is the following estimate for this first eigenvalue.

Theorem 4.6.11 (First eigenvalue of the Riesz potential)**.** *Let $\mathbb{G}$ be a homogeneous group of homogeneous dimension Q. Let $\Omega \subset \mathbb{G}$ be a Haar measurable set, let $Q \geq 2 > 2s > 0$ and $1 < p < 2$ with $\frac{1}{|y^{-1}x|^{Q-2s}} \in L^{\frac{p}{p-1}}(\Omega \times \Omega)$ and set*

$$C_0 := \left\| \frac{1}{|y^{-1}x|^{Q-2s}} \right\|_{L^{\frac{p}{p-1}}(\Omega\times\Omega)}.$$

Then for the spectral problem (4.110), we have

$$\lambda_1(\Omega) \leq C_0 |\Omega|^{\frac{2-p}{p}}. \tag{4.112}$$

Proof of Theorem 4.6.11. By using (4.111), Theorem 4.6.10 and $\omega = \frac{1}{\lambda_1(\Omega)}$, we obtain

$$\lambda_1(\Omega) \leq C_0 |\Omega|^{\frac{2-p}{p}}. \tag{4.113}$$

Theorem 4.6.11 is proved. □

Euclidean case. Let us now record several applications of the above constructions in the case of the Abelian group $(\mathbb{R}^N, +)$. With the Euclidean distance $|\cdot|_E$, the Riesz potential is given by

$$\mathfrak{R}u(x) = \int_\Omega \frac{u(y)}{|x-y|_E^{N-2s}} dy, \quad 0 < 2s < N, \quad \Omega \subset \mathbb{R}^N, \tag{4.114}$$

and the weighted Riesz potential is

$$\mathfrak{R}(\omega u)(x) = \int_\Omega \frac{\omega(y)u(y)}{|x-y|_E^{N-2s}} dy, \quad 0 < 2s < N, \quad \Omega \subset \mathbb{R}^N. \tag{4.115}$$

Then, in Theorem 4.6.10, setting $\mathbb{G} = (\mathbb{R}^N, +)$ and taking the standard Euclidean distance instead of the quasi-norm, we obtain

Theorem 4.6.12 (Euclidean Lyapunov inequality for the Riesz potential). *Let $\Omega \subset \mathbb{R}^N$ be a measurable set with $|\Omega| < \infty$, $1 < p < 2$ and let $N \geq 2 > 2s > 0$. Let $\omega \in L^{\frac{p}{2-p}}(\Omega)$, $\frac{1}{|x-y|_E^{N-2s}} \in L^{\frac{p}{p-1}}(\Omega \times \Omega)$ and set*

$$S := \left\| \frac{1}{|x-y|_E^{N-2s}} \right\|_{L^{\frac{p}{p-1}}(\Omega\times\Omega)}.$$

In addition, assume that $u \in L^{\frac{p}{p-1}}(\Omega)$, $u \neq 0$, satisfies

$$\mathfrak{R}(\omega u)(x) = u(x), \quad x \in \Omega.$$

Then we have

$$\|\omega\|_{L^{\frac{p}{2-p}}(\Omega)} \geq \frac{1}{S}.$$

Now let us consider the spectral problem for the Euclidean Riesz potential from (4.114):

$$\mathfrak{R}u(x) = \int_\Omega \frac{u(y)}{|x-y|_E^{N-2s}} dy = \lambda u(x), \quad 0 < 2s < N. \tag{4.116}$$

Theorem 4.6.13 (Isoperimetric inequality for the Euclidean Riesz potential). *Let $\Omega \subset \mathbb{R}^N$ be a set with $|\Omega| < \infty$, $1 < p < 2$ and $N \geq 2 > 2s > 0$ and $1 < p < 2$. Assume that $\omega \in L^{\frac{p}{2-p}}(\Omega)$, $\frac{1}{|x-y|_E^{N-2s}} \in L^{\frac{p}{p-1}}(\Omega \times \Omega)$ and set*

$$S := \left\| \frac{1}{|x-y|_E^{N-2s}} \right\|_{L^{\frac{p}{p-1}}(\Omega\times\Omega)}.$$

Then, for the spectral problem (4.116), we have,

$$\lambda_1(\Omega) \leq \lambda_1(B) \leq S|B|^{\frac{2-p}{p}},$$

where $B \subset \mathbb{R}^N$ is an open ball, $\lambda_1(\Omega)$ is the first eigenvalue of the spectral problem (4.116), for all sets Ω with $|\Omega| = |B|$.

Proof of Theorem 4.6.13. The proof of $\lambda_1(B) \leq S|B|^{\frac{2-p}{p}}$ is the same as the proof of Theorem 4.6.11. From [RRS16] we have

$$\lambda_1(B) \geq \lambda_1(\Omega),$$

which completes the proof of Theorem 4.6.13. □

4.7 Hardy inequalities for fractional sub-Laplacians on stratified groups

In this section we discuss the Hardy inequalities involving fractional powers of the sub-Laplacian from a different point of view. First, we observe another way of writing the well-known one-dimensional L^p-Hardy inequality

$$\int_0^\infty \left|\frac{f(x)}{x}\right|^p dx \leq \left(\frac{p}{p-1}\right)^p \int_0^\infty \left|\frac{df(x)}{dx}\right|^p dx, \quad p>1.$$

Consequently, one can replace $f(x)$ by $xf(x)$ and restate this inequality in terms of the boundedness of the operator

$$Tf(x) = \frac{d(xf(x))}{dx},$$

or of its dual operator

$$T^*f(x) = x\frac{df(x)}{dx}.$$

In this section, we discuss this point of view and its extension to the subelliptic setting. For fractional powers of the sub-Laplacian on stratified groups this has been analysed in [CCR15], with subsequent extensions to more general hypoelliptic operators and more general graded groups in [RY18a]. We discuss this matter in the spirit of the former.

4.7.1 Riesz kernels on stratified Lie groups

In this section we briefly recapture some properties of the Riesz kernels of the sub-Laplacians on stratified Lie groups. Historically, these have been consistently developed in [Fol75]. For a detailed unifying description containing also higher-order hypoelliptic operators on general graded groups, their fractional powers, and the corresponding Riesz and Bessel kernels we refer to the exposition in [FR16, Section 4.3] where one can find proofs of most of the properties described in this section.

The analysis of second-order hypoelliptic operators, including the sub-Laplacian, is significantly simpler than that of higher-order operators. This difference is based on the Hunt theorem [Hun56] which asserts that the semigroup $\{e^{-t\mathcal{L}}\}_{t>0}$ generated by the sub-Laplacian $\mathcal{L}$ (or more precisely, by its unique self-adjoint extension) consists of convolutions with probability measures. Moreover, the hypoellipticity of $\mathcal{L}$ implies that these measures are absolutely continuous with respect to the Haar measure and have densities in the Schwartz space [FS82], which are strictly positive by the Bony maximum principle [Bon69]. Therefore, for all $t\in\mathbb{R}^+$, there exists the heat kernel of $\mathcal{L}$, that is, a function $p_t\in\mathcal{S}(\mathbb{G})$ such that

$$e^{-t\mathcal{L}}f(x) = f*p_t(x) = \int_{\mathbb{G}} f(xy^{-1})p_t(y)dy$$

for all $x \in \mathbb{G}$ and all $f \in L^2(\mathbb{G})$. Since $\mathcal{L}$ is homogeneous of order two with respect to the dilations D_r of a stratified group $\mathbb{G}$, we also have

$$p_t(x) = t^{-Q/2} P(D_{t^{-1/2}} x)$$

for all $x \in \mathbb{G}$, where $P = p_1$, which this is a strictly positive function in the Schwartz class $\mathcal{S}(\mathbb{G})$ and

$$\int_{\mathbb{G}} P(x) dx = 1.$$

Let us define the following fractional integral kernel F_α on $\mathbb{G} \backslash \{0\}$ by

$$F_\alpha(x) := \int_0^\infty t^{\alpha/2} p_t(x) \frac{dt}{t} = \int_0^\infty t^{(\alpha-Q)/2} P(D_{t^{-1/2}} x) \frac{dt}{t}, \quad \operatorname{Re} \alpha < Q, \tag{4.117}$$

which converges absolutely and uniformly on compact subsets of $\mathbb{G} \backslash \{0\}$ to a smooth function, homogeneous of degree $\alpha - Q$. Since P is positive F_α is positive when α is real. Thus, the function $|\cdot|_\alpha$, given by the formula

$$|x|_\alpha := \begin{cases} (F_0(x))^{-1/(Q-\alpha)} & \text{if } x \in \mathbb{G} \backslash \{0\}, \\ 0 & \text{if } x = 0, \end{cases} \tag{4.118}$$

is non-negative and homogeneous of degree 1, and vanishes only at the origin. So $|\cdot|_\alpha$ is a homogeneous norm, and for further discussions it will be convenient to use the norm $|\cdot|_0$, which is a limit of the norms $|\cdot|_\alpha$.

The fractional integral $F_\alpha(x)$ in (4.117) is holomorphic with respect to α in $\mathcal{H}_Q := \{\alpha \in \mathbb{C} : \operatorname{Re} \alpha < Q\}$, and its derivatives

$$F_\alpha^{(k)}(x) = \frac{1}{2^k} \int_0^\infty t^{\alpha/2} (\log t)^k p_t(x) \frac{dt}{t}$$

are the absolutely convergent integrals. The functions F_α are smooth and homogeneous of degree $\alpha - Q$ and locally integrable on $\mathbb{G}$ when $0 < \operatorname{Re} \alpha < Q$. Thus, the associated distributions are defined by

$$\begin{aligned} \langle F_\alpha, \phi \rangle := \int_{\mathbb{G}} F_\alpha(x) \phi(x) dx &= \int_{\mathbb{G}} \int_0^\infty \phi(x) t^{(\alpha-Q)/2-1} P(D_{t^{-1/2}} x) dt dx \\ &= 2 \int_0^\infty \int_{\mathbb{G}} s^{\alpha-1} \phi(D_s x) P(x) dx ds \end{aligned} \tag{4.119}$$

for all $\phi \in C_0^\infty(\mathbb{G})$. Now it is clear that

$$I_\alpha := \Gamma(\alpha/2)^{-1} F_\alpha$$

is the *Riesz kernel of order* α, i.e., the convolution kernel of $\mathcal{L}^{-\alpha/2}$, that is,

$$\mathcal{L}^{-\alpha/2} f = f * I_\alpha, \tag{4.120}$$

with

$$I_\alpha|_{\alpha=0} = \delta_0. \tag{4.121}$$

The family of Riesz kernels can be analytically continued as distributions to the half-plane $\mathcal{H}_Q$ by the identity

$$I_\alpha = \mathcal{L}(I_{\alpha+2}).$$

The analytic continuation to the strip $\{\alpha \in \mathbb{C} : -1 < \operatorname{Re}\alpha < Q\}$ will be enough for our purpose. An explicit expression is obtained by rewriting (4.119) in the form

$$\begin{aligned} \langle I_\alpha, \phi\rangle = \frac{2}{\Gamma\left(\frac{\alpha}{2}\right)} \int_{\mathbb{G}} \int_0^1 s^{\alpha-1}(\phi(D_s x) - \phi(0)) ds P(x) dx + \frac{2\phi(0)}{\alpha\Gamma(\frac{\alpha}{2})} \\ + \frac{2}{\Gamma(\frac{\alpha}{2})} \int_{\mathbb{G}} \int_1^\infty s^{\alpha-1}\phi(D_s x) ds P(x) dx. \end{aligned} \tag{4.122}$$

Since $\Gamma'(1) = -\gamma$ and hence

$$\frac{1}{\Gamma(t)} = \frac{t}{\Gamma(t+1)} = t - \gamma t^2 + O(t^3) \text{ as } t \to 0, \tag{4.123}$$

and

$$s^t = 1 + \log s \int_0^t s^u du,$$

the equality (4.122) implies that for α near 0, for $\phi \in C_0^\infty(\mathbb{G})$ we have

$$c\langle I_\alpha, \phi\rangle = \phi(0) + \langle \Lambda, \phi\rangle\alpha + O(\alpha^2), \tag{4.124}$$

where Λ is the distribution defined by

$$\begin{aligned} \langle \Lambda, \phi\rangle := \int_{\mathbb{G}} \int_0^1 \frac{1}{s}(\phi(D_s x) - \phi(0)) ds P(x) dx + \frac{\gamma\phi(0)}{2} \\ + \int_{\mathbb{G}} \int_1^\infty \frac{1}{s}\phi(D_s x) ds P(x) dx. \end{aligned} \tag{4.125}$$

Thus, (4.124) is the Taylor expansion of I_α around 0. By the analytic continuation of (4.122) one obtains the following representation of the distribution Λ in terms of the homogeneous norm $|\cdot|_0$ defined in (4.118): There exists $a > 0$ such that

$$\langle \Lambda, \phi\rangle = \frac{1}{2}\left(\int_{\{x\in\mathbb{G}:|x|_0<a\}} (\phi(x) - \phi(0))|x|_0^{-Q} dx + \int_{\{x\in\mathbb{G}:|x|_0>a\}} \phi(x)|x|_0^{-Q} dx\right), \tag{4.126}$$

for all $\phi \in C_0^\infty$.

Let $0 < \operatorname{Re}\alpha$, $0 < \operatorname{Re}\beta$ and $\operatorname{Re}(\alpha+\beta) < Q$. Then the convolution of the Riesz potentials of orders α and β is defined pointwise by absolutely convergent integrals as well as

$$I_\alpha * I_\beta = I_{\alpha+\beta} \tag{4.127}$$

is satisfied pointwise and in the sense of distributions. This identity is equivalent to the functional equality $\mathcal{L}^{-\alpha/2}\mathcal{L}^{-\beta/2} = \mathcal{L}^{-(\alpha+\beta)/2}$ and can be obtained from (4.117) by using the properties of the heat kernel. This fact can be also extended by using the analytical continuation. That is,

$$(\phi * I_\alpha) * I_\beta = \phi * I_{\alpha+\beta}$$

holds if $\operatorname{Re}\alpha > -1$, $\operatorname{Re}\beta > -1$ and $\operatorname{Re}(\alpha+\beta) < Q$. These distributions will be useful to present a family of Hardy type inequalities in the following sections. We refer to [FR16, Section 4.3] for the detailed discussion of these and other properties of the Riesz kernels and fractional powers of left invariant hypoelliptic operators.

4.7.2 Hardy inequalities for fractional powers of sub-Laplacians

As in this whole section, let $\mathcal{L}$ be the sub-Laplacian on the stratified Lie group $\mathbb{G}$ of homogeneous dimension Q. Let us define the operator T_α on $C_0^\infty(\mathbb{G})$ by the formula

$$T_\alpha f := |\cdot|^{-\alpha}\mathcal{L}^{-\alpha/2} f = |\cdot|^{-\alpha}(f * I_\alpha), \quad -1 < \operatorname{Re}\alpha < Q.$$

Consequently, the operator

$$T_\alpha^* g = (|\cdot|^{-\overline{\alpha}} g) * I_{\overline{\alpha}}$$

satisfies

$$\langle f, T_\alpha^* g\rangle = \langle T_\alpha f, g\rangle \tag{4.128}$$

for all $f, g \in C_0^\infty(\mathbb{G})$. It turns out that the operator T_α is bounded on $L^p(\mathbb{G})$ when $1 < p < \infty$ and $0 < \alpha < Q/p$ which, in turn, can be formulated as a version of a Hardy inequality:

Theorem 4.7.1 (Hardy inequality for fractional powers of sub-Laplacian). *Let $\mathbb{G}$ be a stratified Lie group of homogeneous dimension Q. Let $1 < p < \infty$ and $0 < \alpha < Q/p$. Then the operator T_α extends uniquely to a bounded operator on $L^p(\mathbb{G})$.*

Applying this to $\mathcal{L}^{\alpha/2} f$ rather than f we get that for all $f \in C_0^\infty(\mathbb{G})$, we have

$$\left\|\frac{f}{|x|^\alpha}\right\|_{L^p(\mathbb{G})} \le \|T_\alpha\|_{L^p(\mathbb{G})\to L^p(\mathbb{G})} \|\mathcal{L}^{\alpha/2} f\|_{L^p(\mathbb{G})}, \quad 1 < p < \infty. \tag{4.129}$$

Remark 4.7.2 (Hardy–Sobolev inequalities).

1. Theorem 4.7.1 was established in [CCR15]. It is easy to see that combined with the Sobolev inequality, it implies the following *Hardy–Sobolev inequality*:

   Let $0 \le b < Q$ and $\frac{a}{Q} = \frac{1}{p} - \frac{1}{q} + \frac{b}{qQ}$. Then there exists a positive constant $C > 0$ such that we have

$$\left\|\frac{f}{|x|^{\frac{b}{q}}}\right\|_{L^q(\mathbb{G})} \le C\|(-\mathcal{L})^{\frac{a}{2}} f\|_{L^p(\mathbb{G})}. \tag{4.130}$$

Such an inequality can be interpreted as a weighted Sobolev embedding of the homogeneous Sobolev space $\dot{L}^p_a(\mathbb{G})$ over L^p of order a, based on the sub-Laplacian $\mathcal{L}$. The theory of such spaces has been extensively developed by Folland [Fol75]. We refer to [RY18a] for the inequality (4.130).

2. The asymptotic behaviour of $\|T_\alpha\|_{L^p(\mathbb{G})\to L^p(\mathbb{G})}$ with respect to α will be given in Theorem 4.7.3; in its proof we will follow the original proof in [CCR15], as well as for Theorem 4.7.4.
3. (Critical global Hardy inequality for $a = Q/p$) Let $1 < p < r < \infty$ and $p < q < (r-1)p'$, where $1/p + 1/p' = 1$. Then there exists a positive constant $C = C(p, q, r, Q) > 0$ such that we have

$$\left\| \frac{f}{\left(\log\left(e + \frac{1}{|x|}\right)\right)^{\frac{r}{q}} |x|^{\frac{Q}{q}}} \right\|_{L^q(\mathbb{G})} \leq C(\|f\|_{L^p(\mathbb{G})} + \|(-\mathcal{L})^{\frac{Q}{2p}} f\|_{L^p(\mathbb{G})}). \quad (4.131)$$

4. (Critical local Hardy inequality for $a = Q/p$) Let $1 < p < \infty$ and $\beta \in [0, Q)$. Let $r > 0$ be given and let x_0 be any point of $\mathbb{G}$. Then for any $p \leq q < \infty$ there exists a positive constant $C = C(p, Q, \beta, r, q)$ such that we have

$$\left\| \frac{f}{|x|^{\frac{\beta}{q}}} \right\|_{L^q(B(x_0,r))} \leq C q^{1-1/p} (\|f\|_{L^p(B(x_0,r))} + \|(-\mathcal{L})^{\frac{Q}{2p}} f\|_{L^p(B(x_0,r))}), \quad (4.132)$$

and such that

$$\lim_{q\to\infty} \sup C(p, Q, \beta, r, q) < \infty.$$

Inequalities (4.131) and (4.132) have been obtained in [RY18a], to which we refer for their proofs as well as for the expressions for the asymptotically sharp constants in these inequalities.

5. (Trudinger–Moser inequality on stratified groups) Let $Q \geq 3$ and let $|\cdot|$ be a homogeneous quasi-norm on $\mathbb{G}$. Then there exists a constant $\alpha_Q > 0$ such that we have

$$\sup_{\|f\|_{L^Q_1(\mathbb{G})}\leq 1} \int_{\mathbb{G}} \frac{1}{|x|^\beta} \left(\exp(\alpha |f(x)|^{Q'}) - \sum_{k=0}^{Q-2} \frac{\alpha^k |f(x)|^{kQ'}}{k!} \right) dx < \infty \quad (4.133)$$

for any $\beta \in [0, Q)$ and $\alpha \in (0, \alpha_Q(1 - \beta/Q))$, where $Q' = Q/(Q-1)$. When $\alpha > \alpha_Q(1 - \beta/Q)$, the integral in (4.133) is still finite for any $f \in L^Q_1(\mathbb{G})$, but the supremum is infinite. The space $L^Q_1(\mathbb{G})$ is the Sobolev space with the norm $\|f\|_{L^Q_1(\mathbb{G})} = \|f\|_{L^Q(\mathbb{G})} + \|\nabla_H f\|_{L^Q(\mathbb{G})}$.

The inequality (4.133) was obtained in [RY18a], also with a rather explicit expression for the constant α_Q. We refer to the above paper also for an extensive history of this subject.

6. (Hardy–Sobolev inequalities on graded groups) In fact, Hardy–Sobolev inequalities (4.130), (4.131) and (4.132), Trudinger–Moser inequalities (4.133) and their local versions, have been obtained in [RY18a] for general homogeneous left invariant hypoelliptic differential operators (the so-called Rockland operators) on **general graded Lie groups**. The methods of the proof, however, are rather different, and somewhat more involved, than the proofs for the sub-Laplacians presented in this section. Therefore, here we do not present them in full generality: these results, their proofs, and expressions for (asymptotically) best constants can be found in [RY18a].

Proof of Theorem 4.7.1. As usual here we denote the conjugate number to p by p'. Since the integral kernel of the operator T_α is positive, its L^p-boundedness follows from the following *Schur test* (see [FR75]):

Suppose that there exist a positive function g and constants $A_{\alpha,p}$ and $B_{\alpha,p}$ such that we have

$$T_\alpha(g^{p'})(x) \leq A_{\alpha,p} g^{p'}(x) \text{ and } T^*_\alpha(g^p)(x) \leq B_{\alpha,p} u^p(x) \tag{4.134}$$

for almost all $x \in \mathbb{G}$. Then the estimate

$$\|T_\alpha f\|_{L^p(\mathbb{G})} \leq A^{1/p'}_{\alpha,p} B^{1/p}_{\alpha,p} \|f\|_{L^p(\mathbb{G})} \tag{4.135}$$

holds for all $f \in L^p(\mathbb{G})$.

In order to produce g as in (4.134), let us consider the family of functions $g_\gamma := |\cdot|^{\gamma-Q}$, $\gamma > 0$, and consider the convolutions $g^{p'}_\gamma * I_\alpha$ and $(|\cdot|^{-\alpha} g^p_\gamma) * I_\alpha$ related to the computations of $T_\alpha(g^{p'}_\gamma)$ and $T^*_\alpha(g^p_\gamma)$. Let

$$\beta' = Q + (\gamma - Q)p' \text{ and } \beta = Q - \alpha + (\gamma - Q)p, \tag{4.136}$$

so that $g^{p'}_\gamma$ and $|\cdot|^{-\alpha} g^p_\gamma$ have the same homogeneity as $I_{\beta'}$ and I_β, respectively. As in the case of (4.127) these convolution integrals converge absolutely in $\mathbb{G}\backslash\{0\}$ if and only if $0 < \beta < Q - \alpha$ and $0 < \beta' < Q - \alpha$, i.e., for γ such that

$$\max\left(\frac{Q}{p}, \frac{\alpha}{p} + \frac{Q}{p'}\right) < \gamma < Q - \frac{\alpha}{p'}. \tag{4.137}$$

Thus, if this condition is satisfied, then both $T_\alpha(g^{p'}_\gamma)$ and $T^*_\alpha(g^p_\gamma)$ are positive functions, continuous away from the origin, and of the same homogeneity as $I_{\beta'}$ and I_β, respectively. The pointwise estimates (4.134) follow directly from the homogeneity. Finally, for γ the condition (4.137) is nontrivial if and only if $0 < \alpha < Q/p$, completing the proof of Theorem 4.7.1. □

By Theorem 4.7.1, the operator T_α is L^p-bounded and the constant in the inequality (4.129) depends on the choice of a homogeneous quasi-norm. While the inequality itself holds true for any homogeneous quasi-norm (due to their equivalence), in the case of particular homogeneous norm $|\cdot|_0$ from the family (4.118) one can give the following estimate for it.

Theorem 4.7.3 (Estimate for L^p-operator norm). *Let $\mathbb{G}$ be a stratified Lie group of homogeneous dimension Q. Let $1 < p < \infty$ and $0 < \alpha < Q/p$. For the particular homogeneous norm $|\cdot|_0$, the operator norm $\|T_\alpha\|_{L^p(\mathbb{G})\to L^p(\mathbb{G})}$ satisfies*

$$\|T_\alpha\|_{L^p(\mathbb{G})\to L^p(\mathbb{G})} \le 1 + C\alpha + O(\alpha^2).$$

For the rest of this subsection, we assume that $|\cdot| = |\cdot|_0$ and that the operator T_α is defined using $|\cdot|_0$.

Proof of Theorem 4.7.3. The proof of Theorem 4.7.1, and especially (4.135), show that, if

$$0 < \alpha < \frac{Q}{p} \quad \text{and} \quad \frac{\alpha}{p} + \frac{Q}{p'} < \gamma < Q - \frac{\alpha}{p'}, \tag{4.138}$$

then

$$\|T_\alpha\|_{L^p(\mathbb{G})\to L^p(\mathbb{G})} \le A^{1/p'}_{\alpha,\gamma,p} B^{1/p}_{\alpha,\gamma,p}, \tag{4.139}$$

where

$$\begin{aligned} A_{\alpha,\gamma,p} &:= \sup_{|y|=1} \left(|\cdot|_0^{(\gamma-Q)p'} * I_\alpha\right)(y) &&= \sup_{|y|=1} \left(|\cdot|_0^{\beta'-Q} * I_\alpha\right)(y), \\ B_{\alpha,\gamma,p} &:= \sup_{|y|=1} \left(|\cdot|_0^{(\gamma-Q)p-\alpha} * I_\alpha\right)(y) &&= \sup_{|y|=1} \left(|\cdot|_0^{\beta-Q} * I_\alpha\right)(y). \end{aligned} \tag{4.140}$$

First let us show that there is a constant C_p such that

$$\|T_\alpha f\|_{L^p(\mathbb{G})\to L^p(\mathbb{G})} \le 1 + C_p\alpha + O(\alpha^2), \quad 1 < p < \infty, \tag{4.141}$$

holds for all sufficiently small positive α. Assume that $Q/p' < \gamma < Q$. Then (4.138) is valid for α in a neighborhood of 0^+. Moreover, with β and β' as in (4.136), the constants $A_{\alpha,\gamma,p}$ and $B_{\alpha,\gamma,p}$ in (4.140) are bounded by $1 + L_{\beta'}\alpha + O(\alpha^2)$ and $1 + L_\beta\alpha + O(\alpha^2)$, respectively. By (4.121) we obtain

$$\|T_\alpha\|_{L^p(\mathbb{G})\to L^p(\mathbb{G})} \le 1 + \left(\frac{L_{\beta'}}{p'} + \frac{L_\beta}{p}\right)\alpha + O(\alpha^2),$$

which confirms (4.141). Combining this with Theorem 4.7.1 completes the proof of Theorem 4.7.3. □

Theorem 4.7.4 (A logarithmic version of uncertainty principle). *Let $\mathbb{G}$ be a stratified Lie group and let $1 < p < \infty$. Then we have*

$$\int_{\mathbb{G}} (\log|x|)|f(x)|^p dx + \int_{\mathbb{G}} \mathrm{Re}((\log \mathcal{L}^{1/2} f)(x)\overline{f(x)})|f(x)|^{p-2} dx \ge -C_p\|f\|^p_{L^p(\mathbb{G})}.$$

Proof of Theorem 4.7.4. Now let us recall the distribution Λ introduced in (4.125). We also define the operator

$$T_0' f := \left(\frac{d}{d\alpha} T_\alpha f\right)\Big|_{\alpha=0} = -\log \mathcal{L}^{1/2} f - (\log|\cdot|_0) f = -f * \Lambda - (\log|\cdot|_0) f. \tag{4.142}$$

Setting $\phi = |y \cdot|_0^{\beta-Q}$ in (4.126), the integrals converge absolutely for any $|y| = 1$, and so (4.124) also extends to

$$|\cdot|_0^{\beta-Q} * I_\alpha(y) = 1 + \alpha |\cdot|_0^{\beta-Q} * \Lambda(y) + O(\alpha^2);$$

the term $O(\alpha^2)$ is uniform in $|y| = 1$. Taking the supremum over y yields the logarithmic uncertainty inequality for T_0', that is, this shows that if α is small and $0 < \alpha < Q - \beta$, then

$$\sup_{|y|=1} \left(|\cdot|_0^{\beta-Q} * I_\alpha \right)(y) \leq 1 + L_\beta \alpha + O(\alpha^2), \quad 0 < \beta < Q,$$

where

$$L_\beta = \sup_{|y|=1} \left(|\cdot|_0^{\beta-Q} * \Lambda(y) \right). \tag{4.143}$$

Let $1 < p < \infty$. Taking $f \in C_0^\infty(\mathbb{G})$ with $\|f\|_{L^p(\mathbb{G})} = 1$ and restricting ourselves to positive values of α for which (4.141) holds and $1 + C_p\alpha > 0$, we consider the function

$$\Phi_\epsilon(\alpha) := (1 + (C_p + \epsilon)\alpha)^p - \|T_\alpha f\|^p_{L^p(\mathbb{G})},$$

where $\epsilon \geq 0$. The expression $\|T_\alpha f\|^p_{L^p(\mathbb{G})}$ can be differentiated in α at 0, and

$$\frac{d}{d\alpha} \left(\|T_\alpha f\|^p_{L^p(\mathbb{G})} \right) \bigg|_{\alpha=0} = p \int_{\mathbb{G}} \operatorname{Re}\left(T_0' f(x) \overline{f(x)} \right) |f(x)|^{p-2} dx.$$

Hence Φ_ϵ is differentiable at 0. Now $\Phi_\epsilon(0) = 0$ and $\Phi_\epsilon(\alpha) \geq 0$ for all sufficiently small α, by (4.141), and so $\Phi_\epsilon'(0) \geq 0$. Now we let ϵ tend to 0, and deduce that $\Phi_\epsilon(0) \geq 0$. Thus, the statement of Theorem 4.7.4 follows from (4.142). □

4.7.3 Landau–Kolmogorov inequalities on stratified groups

In this section we discuss the stratified groups version of the Landau–Kolmogorov inequality, and some of its consequences. We start with a related simpler version of such an inequality.

Proposition 4.7.5 (Two weighted inequalities).

(1) *Let $\mathbb{G}$ be a homogeneous group. If $1 \leq p, q, r \leq \infty$, and*

$$\frac{1}{p} = \frac{\theta}{q} + \frac{1-\theta}{r}, \quad 0 \leq \theta \leq 1,$$

then

$$\| |\cdot|^\alpha f \|_{L^p(\mathbb{G})} \leq \| |\cdot|^{\alpha/\theta} f \|^\theta_{L^q(\mathbb{G})} \|f\|^{1-\theta}_{L^r(\mathbb{G})}, \quad \alpha \geq 0,$$

holds for all $f \in C_0^\infty(\mathbb{G})$.

(2) *Let $\mathbb{G}$ be a stratified group and let $\mathcal{L}$ be a sub-Laplacian on $\mathbb{G}$. Let $\beta > 0$, $\gamma > 0$, $p > 1$, $q \geq 1$, $r > 1$ be such that*

$$\gamma < \frac{Q}{r} \quad \text{and} \quad \frac{\beta+\gamma}{p} = \frac{\gamma}{q} + \frac{\beta}{r}.$$

Then

$$\|f\|_{L^p(\mathbb{G})} \leq C \| |\cdot|^\beta f \|_{L^q(\mathbb{G})}^{\gamma/(\beta+\gamma)} \|\mathcal{L}^{\gamma/2} f\|_{L^r(\mathbb{G})}^{\beta/(\beta+\gamma)} \tag{4.144}$$

holds for all $f \in C_0^\infty(\mathbb{G})$.

Proof. Part (1). Writing

$$(|\cdot|^\alpha |f|)^p = \left(|\cdot|^\alpha |f|^\theta\right)^p \left(|f|^{1-\theta}\right)^p,$$

then applying the Hölder inequality (with index s) we have

$$\left(\int_{\mathbb{G}} (|x|^\alpha |f(x)|^p) dx\right)^{\frac{1}{p}} \leq \left(\int_{\mathbb{G}} \left(|x|^\alpha |f(x)|^\theta\right)^{ps} dx\right)^{\frac{1}{ps}} \left(\int_{\mathbb{G}} (|f(x)|^{1-\theta})^{ps'} dx\right)^{\frac{1}{ps'}}.$$

If the index s is chosen so that $q = \theta ps$, then $r = (1-\theta)ps'$, finishing the proof.

Part (2). By using the Hölder inequality and (4.129) with (4.141) we have

$$\begin{aligned}\|f\|_{L^p(\mathbb{G})} &\leq \| |\cdot|^\beta f\|_{L^q(\mathbb{G})}^{\gamma/(\beta+\gamma)} \| |\cdot|^{-\gamma} f\|_{L^r(\mathbb{G})}^{\beta/(\beta+\gamma)} \\ &\leq C \| |\cdot|^\beta f\|_{L^q(\mathbb{G})}^{\gamma/(\beta+\gamma)} \|\mathcal{L}^{\gamma/2} f\|_{L^r(\mathbb{G})}^{\beta/(\beta+\gamma)},\end{aligned}$$

yielding (4.144). □

Now we present the following Landau–Kolmogorov type inequality.

Theorem 4.7.6 (Landau–Kolmogorov type inequality). *Let $\mathbb{G}$ be a stratified group and let $\mathcal{L}$ be a sub-Laplacian on $\mathbb{G}$. If $1 < p, q, r < \infty$, and*

$$\frac{1}{p} = \frac{\theta}{q} + \frac{1-\theta}{r}, \quad 0 \leq \theta \leq 1,$$

then we have

$$\|\mathcal{L}^{\alpha/2} f\|_{L^p(\mathbb{G})} \leq C \|\mathcal{L}^{\alpha/2\theta} f\|_{L^q(\mathbb{G})}^\theta \|f\|_{L^r(\mathbb{G})}^{1-\theta}, \quad \alpha \geq 0, \tag{4.145}$$

for all $f \in C_0^\infty(\mathbb{G})$.

Proof of Theorem 4.7.6. To prove this theorem we will use the well-known complex interpolation methods (see, e.g., [BL76] or [Ste56]). First, we have to see that the operator $\mathcal{L}^{iy}$, where $y \in \mathbb{R}$, is bounded on $L^s(\mathbb{G})$ with $1 < s < \infty$, and that

$$\|\mathcal{L}^{iy/2}\|_{L^s \to L^s} \leq C(s)\phi(y) \quad \text{with} \quad \phi(y) = e^{\gamma|y|}.$$

In the case of the stratified groups this follows from the Mihlin–Hörmander multiplier theorem (see, e.g., [Chr91], [MM90]).

Let $g \in L^{p'}(\mathbb{G})$ be a compactly supported simple function of norm one. For z in the strip $S := \{z \in \mathbb{C} : 0 \leq \operatorname{Re} z \leq \frac{\alpha}{\theta}\}$, let us define

$$g_z(x) := \|g\|^{az+b}_{L^{p'}(\mathbb{G})} g(x)|g(x)|^{cx+d}$$

for all $x \in \mathbb{G}$, where

$$a = \frac{\theta p'}{\alpha}\left(\frac{1}{q} - \frac{1}{p}\right), \; b = -\frac{p'}{r'}, \; c = \frac{\theta p'}{\alpha}\left(\frac{1}{q} - \frac{1}{p}\right), \; d = \frac{p'}{r'} - 1.$$

For $\operatorname{Re} z = \eta$ and

$$\frac{1}{s} = \frac{1}{z} - \frac{\theta\eta}{\alpha}\left(\frac{1}{q} - \frac{1}{p}\right),$$

it follows that

$$\|g_z\|^{s'}_{L^{s'}(\mathbb{G})} = \int_{\mathbb{G}} |g_z(x)|^{s'} dx = \|g\|^{s'(a\eta+b)+p'}_{L^{p'}(\mathbb{G})} = 1.$$

Further, let us fix $f \in C_0^\infty(\mathbb{G})$ and set

$$h(z) := e^{z^2} \int_{\mathbb{G}} \mathcal{L}^{z/2} f(x) g_z(x) dx.$$

By the assumptions on f, for each $z \in S$, the function $\mathcal{L}^{z/2} f$ on $\mathbb{G}$ is smooth, while g_z is a simple function with compact support, and so $h(z)$ is well defined. In addition, if $z = \eta + iy$, then

$$\begin{aligned}|h(z)| &\leq e^{\eta^2 - y^2} \|\mathcal{L}^{z/2} f\|_{L^s(\mathbb{G})} \|g_z\|_{L^{s'}(\mathbb{G})} \\ &= e^{\eta^2 - y^2} \|\mathcal{L}^{iy/2} \mathcal{L}^{\eta/2} f\|_{L^s(\mathbb{G})} \leq C e^{-y^2} \phi(y) \|\mathcal{L}^{\eta/2} f\|_{L^s(\mathbb{G})},\end{aligned}$$

hence

$$|h(\eta + iy)| \leq C \|\mathcal{L}^{\eta/2} f\|_{L^s(\mathbb{G})}.$$

Moreover, if $\operatorname{Re} z = 0$, then $|h(z)| \leq C\|f\|_{L^r(\mathbb{G})}$, while if $\operatorname{Re} z = \alpha/\theta$, then $|h(z)| \leq C\|\mathcal{L}^{\alpha/2\theta} f\|_{L^q(\mathbb{G})}$. On the other hand, the Phragmen–Lindelölf theorem implies that

$$|h(\alpha)| \leq C \|\mathcal{L}^{\alpha/2\theta} f\|^{\theta}_{L^q(\mathbb{G})} \|f\|^{1-\theta}_{L^r(\mathbb{G})}.$$

Since

$$h(\alpha) = \int_{\mathbb{G}} \mathcal{L}^{\alpha/2} f(x) g_\alpha(x) dx,$$

and g_α is an arbitrary simple function on $\mathbb{G}$ with compact support and $L^{p'}(\mathbb{G})$-norm equal to 1, this completes the proof. □

Let us present the following consequence of the Landau–Kolmogorov inequality.

Corollary 4.7.7 (Consequence of the Landau–Kolmogorov inequality)**.** *Let* $\beta > 0$, $\delta > 0$, $p > 1$, $s \geq 1$, $r > 1$ *be such that*

$$\frac{\beta+\delta}{p} = \frac{\delta}{s} + \frac{\beta}{r}.$$

Then we have

$$\|f\|_{L^p(\mathbb{G})} \leq C\||\cdot|^{\beta} f\|_{L^s(\mathbb{G})}^{\delta/(\beta+\delta)} \|\mathcal{L}^{\delta/2} f\|_{L^r(\mathbb{G})}^{\beta/(\beta+\delta)}$$

for all $f \in C_0^\infty(\mathbb{G})$.

Proof of Corollary 4.7.7. Obviously, if $\delta < Q/r$, there is nothing to prove. Otherwise, we use Proposition 4.7.5, Part (2), and the version of the Landau–Kolmogorov inequality (4.145). Let $0 < \theta < Q/(r\delta)$ and $\gamma = \theta\delta$. Then Proposition 4.7.5, Part (2), gives that

$$\|f\|_{L^p(\mathbb{G})} \leq C\||\cdot|^{\beta} f\|_{L^q(\mathbb{G})}^{\gamma/(\beta+\gamma)} \|\mathcal{L}^{\gamma/2} f\|_{L^r(\mathbb{G})}^{\beta/(\beta+\gamma)}$$

for all $f \in C_0^\infty(\mathbb{G})$. By Theorem 4.7.6, with r, s and p instead of $p, q,$ and r, we have

$$\|\mathcal{L}^{\gamma/2} f\|_{L^r(\mathbb{G})} \leq C\|\mathcal{L}^{\delta/2} f\|_{L^s(\mathbb{G})}^{\theta} \|f\|_{L^p(\mathbb{G})}^{1-\theta},$$

that is,

$$\|f\|_{L^p(\mathbb{G})} \leq C\||\cdot|_0^{\beta} f\|_{L^q(\mathbb{G})}^{\gamma/(\beta+\gamma)} \|\mathcal{L}^{\gamma/2} f\|_{L^s(\mathbb{G})}^{\theta\beta/(\beta+\gamma)} \|f\|_{L^p(\mathbb{G})}^{(1-\theta)\beta/(\beta+\gamma)}.$$

This completes the proof. ◻

Chapter 5

Integral Hardy Inequalities on Homogeneous Groups

In this chapter we discuss the integral form of Hardy inequalities where instead of estimating a function by its gradient, we estimate the integral by the function itself. This stems from the original version of Hardy's inequality [Har20]:

$$\int_b^\infty \left(\frac{\int_b^x f(t)dt}{x} \right)^p dx \leq \left(\frac{p}{p-1} \right)^p \int_b^\infty f(x)^p dx, \tag{5.1}$$

where $p > 1$, $b > 0$, and $f \geq 0$ is a non-negative function. We analyse the weighted versions of such inequalities in the setting of general homogeneous groups. Most of the results of this chapter have been obtained in [RY18a] and here we follow the presentation of this paper.

5.1 Two-weight integral Hardy inequalities

Here we discuss the weighted Hardy inequalities in the integral form extending that in (5.1). It turns out that one can actually derive the necessary and sufficient conditions on weights for these inequalities to hold.

Theorem 5.1.1 (Integral Hardy inequalities for $p \leq q$). *Let $\mathbb{G}$ be a homogeneous group of homogeneous dimension Q and let $1 < p \leq q < \infty$. Let $\phi_1 > 0$, $\phi_2 > 0$ be positive functions on $\mathbb{G}$. Then we have the following properties:*

(1) *The inequality*

$$\left(\int_{\mathbb{G}} \left(\int_{B(0,|x|)} f(z)dz \right)^q \phi_1(x)dx \right)^{1/q} \leq C_1 \left(\int_{\mathbb{G}} (f(x))^p \psi_1(x)dx \right)^{1/p} \tag{5.2}$$

M. Ruzhansky, D. Suragan, *Hardy Inequalities on Homogeneous Groups*, Progress in Mathematics 327, https://doi.org/10.1007/978-3-030-02895-4_6

holds for all $f \geq 0$ a.e. on $\mathbb{G}$ if and only if

$$A_1 := \sup_{R>0} \left(\int_{\{|x| \geq R\}} \phi_1(x) dx \right)^{1/q} \left(\int_{\{|x| \leq R\}} (\psi_1(x))^{-(p'-1)} dx \right)^{1/p'} < \infty. \tag{5.3}$$

(2) *The inequality*

$$\left(\int_{\mathbb{G}} \left(\int_{\mathbb{G} \setminus B(0,|x|)} f(z) dz \right)^q \phi_2(x) dx \right)^{1/q} \leq C_2 \left(\int_{\mathbb{G}} (f(x))^p \psi_2(x) dx \right)^{1/p} \tag{5.4}$$

holds for all $f \geq 0$ a.e. on $\mathbb{G}$ if and only if

$$A_2 := \sup_{R>0} \left(\int_{\{|x| \leq R\}} \phi_2(x) dx \right)^{1/q} \left(\int_{\{|x| \geq R\}} (\psi_2(x))^{-(p'-1)} dx \right)^{1/p'} < \infty. \tag{5.5}$$

(3) *If $\{C_i\}_{i=1}^2$ are the smallest constants for which (5.2) and (5.4) hold, then*

$$A_i \leq C_i \leq (p')^{\frac{1}{p'}} p^{\frac{1}{q}} A_i, \quad i = 1, 2. \tag{5.6}$$

Before we prove this theorem let us give a few comments.

Remark 5.1.2.

1. In the Abelian case $\mathbb{G} = (\mathbb{R}^n, +)$ and $Q = n$, if we take $p = q > 1$, and

   $$\phi_1(x) = |B(0, |x|)|^{-p} \text{ and } \psi_1(x) = 1$$

   in (5.2), then we have $A_1 = (p-1)^{-1/p}$ and

   $$\left(\int_{\mathbb{R}^n} \left| \frac{1}{|B(0,|x|)|} \int_{B(0,|x|)} f(z) dz \right|^p dx \right)^{1/p} \leq \frac{p}{p-1} \left(\int_{\mathbb{R}^n} |f(x)|^p dx \right)^{1/p}, \tag{5.7}$$

   where $|B(0, |x|)|$ is the volume of the ball $B(0, |x|)$. The inequality (5.7) was obtained in [CG95].

2. Theorem 5.1.1 was obtained in [RY18a] and here we follow the proof from that paper. However, due to the fact that the formulations do not make use of the differential structure, the statement can be actually extended to general metric measure spaces with polar decomposition. More specifically, consider a metric space $\mathbb{X}$ with a Borel measure dx allowing for the following *polar decomposition* at $a \in \mathbb{X}$: we assume that there is a locally integrable function $\lambda \in L^1_{\text{loc}}$ such that for all $f \in L^1(\mathbb{X})$ we have

   $$\int_{\mathbb{X}} f(x) dx = \int_0^\infty \int_\Sigma f(r, \omega) \lambda(r, \omega) d\omega dr, \tag{5.8}$$

for some set $\Sigma \subset \mathbb{X}$ with a measure on it denoted by $d\omega$, and $(r,\omega) \to a$ as $r \to 0$. In the case of homogeneous groups such a polar decomposition is given in Proposition 1.2.10.

Let us denote by $B(a,r)$ the ball in $\mathbb{X}$ with centre a and radius r, i.e.,

$$B(a,r) := \{x \in \mathbb{X} : d(x,a) < r\},$$

where d is the metric on $\mathbb{X}$. Once and for all we will fix some point $a \in \mathbb{X}$, and we will write

$$|x|_a := d(a,x).$$

Then the following result was obtained in [RV18] which we record here without proof.

Theorem 5.1.3 (Integral Hardy inequality in metric measure spaces). *Let $1 < p \le q < \infty$ and let $s > 0$. Let $\mathbb{X}$ be a metric measure space with a polar decomposition (5.8) at a. Let $u, v > 0$ be measurable functions positive a.e. in $\mathbb{X}$ such that $u \in L^1(\mathbb{X}\backslash\{a\})$ and $v^{1-p'} \in L^1_{\mathrm{loc}}(\mathbb{X})$. Denote*

$$U(x) := \int_{\mathbb{X}\backslash B(a,|x|_a)} u(y)dy \qquad \text{and} \qquad V(x) := \int_{B(a,|x|_a)} v^{1-p'}(y)dy$$

Then the inequality

$$\left(\int_{\mathbb{X}}\left(\int_{B(a,|x|_a)}|f(y)|dy\right)^q u(x)dx\right)^{1/q} \le C\left\{\int_{\mathbb{X}}|f(x)|^p v(x)dx\right\}^{1/p} \tag{5.9}$$

holds for all measurable functions $f : \mathbb{X} \to \mathbb{C}$ if and only if any of the following equivalent conditions hold:

1. $\mathcal{D}_1 := \sup_{x\neq a}\left\{U^{\frac{1}{q}}(x)V^{\frac{1}{p'}}(x)\right\} < \infty.$

2. $\mathcal{D}_2 := \sup_{x\neq a}\left\{\int_{\mathbb{X}\backslash B(a,|x|_a)} u(y)V^{q(\frac{1}{p'}-s)}(y)dy\right\}^{1/q} V^s(x) < \infty.$

3. $\mathcal{D}_3 := \sup_{x\neq a}\left\{\int_{B(a,|x|_a)} u(y)V^{q(\frac{1}{p'}+s)}(y)dy\right\}^{1/q} V^{-s}(x) < \infty,$ *provided that $u, v^{1-p'} \in L^1(\mathbb{X})$.*

4. $\mathcal{D}_4 := \sup_{x\neq a}\left\{\int_{B(a,|x|_a)} v^{1-p'}(y)U^{p'(\frac{1}{q}-s)}(y)dy\right\}^{1/p'} U^s(x) < \infty.$

5. $\mathcal{D}_5 := \sup_{x\neq a}\left\{\int_{\mathbb{X}\backslash B(a,|x|_a)} v^{1-p'}(y)U^{p'(\frac{1}{q}+s)}(y)dy\right\}^{1/p'} U^{-s}(x) < \infty,$ *provided that $u, v^{1-p'} \in L^1(\mathbb{X})$.*

Moreover, the constant C for which (5.9) holds and quantities $\mathcal{D}_1$–$\mathcal{D}_5$ are related by

$$\mathcal{D}_1 \le C \le \mathcal{D}_1 (p')^{\frac{1}{p'}} p^{\frac{1}{q}}, \tag{5.10}$$

and

$$\mathcal{D}_1 \le (\max(1, p's))^{\frac{1}{q}} \mathcal{D}_2,\ \mathcal{D}_2 \le \left(\max(1, \frac{1}{p's})\right)^{1/q} \mathcal{D}_1,$$

$$\left(\frac{sp'}{1+p's}\right)^{1/q} \mathcal{D}_3 \le \mathcal{D}_1 \le (1+sp')^{\frac{1}{q}} \mathcal{D}_3,$$

$$\mathcal{D}_1 \le (\max(1, qs))^{\frac{1}{p'}} \mathcal{D}_4,\ \mathcal{D}_4 \le \left(\max(1, \frac{1}{qs})\right)^{1/p'} \mathcal{D}_1,$$

$$\left(\frac{sq}{1+qs}\right)^{1/p'} \mathcal{D}_5 \le \mathcal{D}_1 \le (1+sq)^{\frac{1}{p'}} \mathcal{D}_5.$$

3. As such, Theorem 5.1.3 is an extension of (5.1) to the setting of metric measures spaces $\mathbb{X}$ with the polar decomposition (5.8): in particular, for $p = q$ and real-valued non-negative measurable $f \ge 0$, inequality (5.9) becomes

$$\int_{\mathbb{X}} \left(\int_{B(a,|x|_a)} f(y) dy \right)^p u(x) dx \le C \int_{\mathbb{X}} f(x)^p v(x) dx,$$

as an extension of (5.1). Indeed, in this case we can take $u(x) = \frac{1}{x^p}$, $v(x) = 1$, $\mathbb{X} = [b, \infty)$, $a = b$, so that Theorem 5.1.3 implies (5.1).

4. Let us give an application of Theorem 5.1.3 in the setting of homogeneous groups, recovering a two-weighted result obtained in [RV18]:

Corollary 5.1.4 (Characterization for homogeneous weights). *Let $\mathbb{G}$ be a homogeneous group of homogeneous dimension Q, equipped with a homogeneous quasi-norm $|\cdot|$. Let $1 < p \le q < \infty$ and let $\alpha, \beta \in \mathbb{R}$. Then the inequality*

$$\left(\int_{\mathbb{G}} \left(\int_{B(0,|x|)} |f(y)| dy \right)^q |x|^\alpha dx \right)^{\frac{1}{q}} \le C \left(\int_{\mathbb{G}} |f(x)|^p |x|^\beta dx \right)^{\frac{1}{p}} \tag{5.11}$$

holds for all measurable functions $f : \mathbb{G} \to \mathbb{C}$ if and only if $\alpha + Q < 0$, $\beta(1-p') + Q > 0$ and $\frac{\alpha+Q}{q} + \frac{\beta(1-p')+Q}{p'} = 0$. Moreover, the best constant C for (5.11) satisfies

$$\frac{\sigma^{\frac{1}{q}+\frac{1}{p'}}}{|\alpha+Q|^{\frac{1}{q}} (\beta(1-p')+Q)^{\frac{1}{p'}}} \le C \le (p')^{\frac{1}{p'}} p^{\frac{1}{q}} \frac{\sigma^{\frac{1}{q}+\frac{1}{p'}}}{|\alpha+Q|^{\frac{1}{q}} (\beta(1-p')+Q)^{\frac{1}{p'}}},$$

where σ is the area of the unit sphere in $\mathbb{G}$ with respect to the quasi-norm $|\cdot|$.

Let us show how Theorem 5.1.3 implies Corollary 5.1.4. If we take $a = 0$, and the power weights

$$u(x) = |x|^{\alpha} \qquad \text{and} \qquad v(x) = |x|^{\beta},$$

then the inequality (5.9) holds for $1 < p \leq q < \infty$ if and only if

$$\mathcal{D}_1 = \sup_{r>0} \left(\sigma \int_r^{\infty} \rho^{\alpha} \rho^{Q-1} d\rho \right)^{1/q} \left(\sigma \int_0^r \rho^{\beta(1-p')} \rho^{Q-1} d\rho \right)^{1/p'} < \infty,$$

where σ is the area of the unit sphere in $\mathbb{G}$ with respect to the quasi-norm $|\cdot|$. For this supremum to be well defined we need to have $\alpha + Q < 0$ and $\beta(1-p') + Q > 0$. Consequently, we can calculate

$$\begin{aligned}
\mathcal{D}_1 &= \sigma^{\left(\frac{1}{q}+\frac{1}{p'}\right)} \sup_{r>0} \left(\int_r^{\infty} \rho^{\alpha+Q-1} d\rho \right)^{1/q} \left(\int_0^r \rho^{\beta(1-p')+Q-1} d\rho \right)^{1/p'} \\
&= \sigma^{\left(\frac{1}{q}+\frac{1}{p'}\right)} \sup_{r>0} \frac{r^{\frac{\alpha+Q}{q}}}{|\alpha+Q|^{\frac{1}{q}}} \frac{r^{\frac{\beta(1-p')+Q}{p'}}}{(\beta(1-p')+Q)^{\frac{1}{p'}}},
\end{aligned}$$

which is finite if and only if the power of r is zero. Consequently, Corollary 5.1.4 follows from Theorem 5.1.3.

Proof of Theorem 5.1.1. We will prove Part (1) of the theorem since the proof of Part (2) is similar. The obtained estimates will also show the corresponding part of the statement in Part (3).

Thus, let us first show that (5.3) implies (5.2). Using the polar decomposition in Proposition 1.2.10 and denoting $r = |x|$, we write

$$\begin{aligned}
&\int_{\mathbb{G}} \phi_1(x) \left[\int_{B(0,r)} f(z) dz \right]^q dx \\
&= \int_0^{\infty} \int_{\wp} r^{Q-1} \phi_1(ry) \left[\int_0^r \int_{\wp} s^{Q-1} f(sy) d\sigma(y) ds \right]^q d\sigma(y) dr.
\end{aligned} \tag{5.12}$$

Denoting

$$g(r) := \left\{ \int_{\wp} \int_0^r s^{Q-1} (\psi_1(sy))^{1-p'} ds d\sigma(y) \right\}^{1/(pp')}, \tag{5.13}$$

and using Hölder's inequality, we can estimate

$$\begin{aligned}
&\int_0^r \int_{\wp} s^{Q-1} f(sy) d\sigma(y) ds \\
&= \int_{\wp} \int_0^r s^{(Q-1)/p} f(sy) (\psi_1(sy))^{1/p} g(s) s^{(Q-1)/p'} \\
&\quad \times \left((\psi_1(sy))^{1/p} g(s) \right)^{-1} ds d\sigma(y)
\end{aligned}$$

$$\leq \left(\int_{\wp}\int_0^r s^{Q-1}\left[f(sy)(\psi_1(sy))^{1/p}g(s)\right]^p dsd\sigma(y)\right)^{1/p}$$
$$\times\left(\int_{\wp}\int_0^r s^{Q-1}\left[(\psi_1(sy))^{1/p}g(s)\right]^{-p'} dsd\sigma(y)\right)^{1/p'}. \tag{5.14}$$

Let us introduce the following notations:

$$U(s) := \int_{\wp} s^{Q-1}\left(f(sy)(\psi_1(sy))^{1/p}g(s)\right)^p d\sigma(y), \tag{5.15}$$

$$V(r) := \int_0^r\int_{\wp} s^{Q-1}\left((\psi_1(sy))^{1/p}g(s)\right)^{-p'} d\sigma(y)ds, \tag{5.16}$$

$$W_1(r) := \int_{\wp} r^{Q-1}\phi_1(ry)d\sigma(y), \tag{5.17}$$

for $s, r > 0$. Plugging (5.14) into (5.12) we obtain

$$\begin{aligned}&\int_{\mathbb{G}}\phi_1(x)\left(\int_{B(0,r)} f(z)dz\right)^q dx\\ &\leq \int_0^\infty W_1(r)\left(\int_0^r U(s)ds\right)^{q/p}(V(r))^{q/p'}dr.\end{aligned} \tag{5.18}$$

We now recall the following continuous version of the *Minkowski inequality* (see, e.g., [DHK97, Formula 2.1]):

Let $\theta \geq 1$. Then for all $f_1(x), f_2(x) \geq 0$ on $(0,\infty)$, we have

$$\int_0^\infty f_1(x)\left(\int_0^x f_2(z)dz\right)^\theta dx \leq \left(\int_0^\infty f_2(z)\left(\int_z^\infty f_1(x)dx\right)^{1/\theta} dz\right)^\theta. \tag{5.19}$$

Using this inequality with $\theta = q/p \geq 1$ in the right-hand side of (5.18), we can estimate

$$\int_{\mathbb{G}}\phi_1(x)\left(\int_{B(0,r)} f(z)dz\right)^q dx \leq \left(\int_0^\infty U(s)\left(\int_s^\infty W_1(r)(V(r))^{q/p'}dr\right)^{p/q} ds\right)^{q/p}. \tag{5.20}$$

Let us introduce one more temporary notation

$$T(s) := \int_{\wp} s^{Q-1}(\psi_1(sy))^{1-p'}d\sigma(y).$$

Using (5.13), (5.16), the integration by parts, (5.3) and (5.17), we compute

$$
\begin{aligned}
V(r) &= \int_{\wp}\int_0^r s^{Q-1}(\psi_1(sy))^{1-p'}\left(\int_0^s\int_{\wp} t^{Q-1}(\psi_1(tw))^{1-p'}d\sigma(w)dt\right)^{-1/p} dsd\sigma(y)\\
&= \int_0^r T(s)\left(\int_0^s T(t)dt\right)^{-1/p} ds = p'\int_0^r \frac{d}{ds}\left(\int_0^s T(t)dt\right)^{1/p'} ds\\
&= p'\left(\int_0^r T(s)ds\right)^{1/p'} = p'\left(\int_0^r\int_{\wp} s^{Q-1}(\psi_1(sy))^{1-p'}d\sigma(y)ds\right)^{1/p'}\\
&\le p'A_1\left(\int_r^\infty s^{Q-1}\int_{\wp}\phi_1(sw)d\sigma(w)ds\right)^{-1/q} = p'A_1\left(\int_r^\infty W_1(s)ds\right)^{-1/q}.
\end{aligned}
$$

Similarly, applying the integration by parts and (5.3), this implies

$$
\begin{aligned}
\int_s^\infty W_1(r)(V(r))^{q/p'}dr &= (p'A_1)^{q/p'}\int_s^\infty W_1(r)\left(\int_r^\infty W_1(s)ds\right)^{-1/p'} dr\\
&= (p'A_1)^{q/p'}p\left(\int_s^\infty W_1(r)dr\right)^{1/p}\\
&= (p'A_1)^{q/p'}p\left(\int_s^\infty\int_{\wp} r^{Q-1}\phi_1(ry)d\sigma(y)dr\right)^{1/p}\\
&\le (p'A_1)^{q/p'}pA_1^{q/p}\left(\int_0^s r^{Q-1}\int_{\wp}(\psi_1(ry))^{1-p'}d\sigma(y)dr\right)^{-q/(p'p)}\\
&= A_1^q(p')^{q/p'}p(g(s))^{-q},
\end{aligned}
\tag{5.21}
$$

where we have used (5.13) in the last line. Putting (5.21) in (5.20) and recalling (5.15), we obtain

$$
\begin{aligned}
\int_{\mathbb{G}}\phi_1(x)\left(\int_{B(0,r)} f(z)dz\right)^q dx &\le \left(\int_0^\infty U(s)A_1^p(p')^{p-1}p^{p/q}(g(s))^{-p}ds\right)^{q/p}\\
&= A_1^q(p')^{q/p'}p\left(\int_0^\infty U(s)(g(s))^{-p}ds\right)^{q/p}\\
&= A_1^q(p')^{q/p'}p\left(\int_0^\infty\int_{\wp} s^{Q-1}(f(sy))^p\psi_1(sy)d\sigma(y)ds\right)^{q/p}\\
&= A_1^q(p')^{q/p'}p\left(\int_{\mathbb{G}}\psi_1(x)(f(x))^p dx\right)^{q/p},
\end{aligned}
\tag{5.22}
$$

yielding (5.2) with $C_1 = A_1(p')^{1/p'}p^{1/q}$.

We now show the converse, namely, that (5.2) implies (5.3). For that, we set

$$f(x) := (\psi_1(x))^{1-p'}\chi_{(0,R)}(|x|),$$

with $R > 0$. For this f we observe the equality

$$\begin{aligned}&\left(\int_{\mathbb{G}}\psi_1(x)(f(x))^p dx\right)^{1/p}\left(\int_{|x|\le R}(\psi_1(x))^{1-p'}dx\right)^{-1/p}\\&=\left(\int_{|x|\le R}(\psi_1(x))^{1-p'}dx\right)^{1/p}\left(\int_{|x|\le R}(\psi_1(x))^{1-p'}dx\right)^{-1/p}=1.\end{aligned}\tag{5.23}$$

Consequently, by (5.2) we have

$$\begin{aligned}C&=C\left(\int_{\mathbb{G}}\psi_1(x)(f(x))^p dx\right)^{1/p}\left(\int_{|x|\le R}(\psi_1(x))^{1-p'}dx\right)^{-1/p}\\&\ge\left(\int_{\mathbb{G}}\phi_1(x)\left(\int_{|z|\le|x|}f(z)dz\right)^q dx\right)^{1/q}\left(\int_{|x|\le R}(\psi_1(x))^{1-p'}dx\right)^{-1/p}\\&\ge\left(\int_{|x|\ge R}\phi_1(x)\left(\int_{|z|\le|x|}f(z)dz\right)^q dx\right)^{1/q}\left(\int_{|x|\le R}(\psi_1(x))^{1-p'}dx\right)^{-1/p}\\&=\left(\int_{|x|\ge R}\phi_1(x)dx\right)^{1/q}\left(\int_{|z|\le R}(\psi_1(z))^{1-p'}dz\right)^{1/p'}.\end{aligned}\tag{5.24}$$

Combining (5.23) and (5.24), we obtain (5.3) with $C \ge A_1$. □

The next case is the version of Theorem 5.1.1 for the indices $p > q$:

Theorem 5.1.5 (Integral Hardy inequalities for $p > q$). *Let $\mathbb{G}$ be a homogeneous group of homogeneous dimension Q and let $1 < q < p < \infty$ and $1/\delta = 1/q - 1/p$. Let ϕ_3 and ϕ_4 be positive functions on $\mathbb{G}$. Then we have the following properties:*

(1) *The inequality*

$$\left(\int_{\mathbb{G}}\left(\int_{B(0,|x|)}f(z)dz\right)^q\phi_3(x)dx\right)^{1/q}\le C_1\left(\int_{\mathbb{G}}(f(x))^p\psi_3(x)dx\right)^{1/p}\tag{5.25}$$

holds for all $f \ge 0$ if and only if

$$\int_{\mathbb{G}}\left(\int_{\mathbb{G}\backslash B(0,|x|)}\phi_3(z)dz\right)^{\delta/q}\left(\int_{B(0,|x|)}(\psi_3(z))^{1-p'}dz\right)^{\delta/q'}(\psi_3(x))^{1-p'}dx<\infty.\tag{5.26}$$

(2) *The inequality*

$$\left(\int_{\mathbb{G}}\left(\int_{\mathbb{G}\backslash B(0,|x|)} f(z)dz\right)^q \phi_4(x)dx\right)^{1/q} \leq C_2\left(\int_{\mathbb{G}}(f(x))^p\psi_4(x)dx\right)^{1/p} \tag{5.27}$$

holds for all $f \geq 0$ if and only if

$$\int_{\mathbb{G}}\left(\int_{B(0,|x|)} \phi_4(z)dz\right)^{\delta/q}\left(\int_{\mathbb{G}\backslash B(0,|x|)} (\psi_4(z))^{1-p'}dz\right)^{\delta/q'} (\psi_4(x))^{1-p'}dx < \infty. \tag{5.28}$$

Proof of Theorem 5.1.5. We will prove Part (1) of the theorem since Part (2) is similar. We will denote

$$A_3 := \int_{\mathbb{G}}\left(\int_{\mathbb{G}\backslash B(0,|x|)} \phi_3(z)dz\right)^{\delta/q}\left(\int_{B(0,|x|)} (\psi_3(z))^{1-p'}dz\right)^{\delta/q'} (\psi_3(x))^{1-p'}dx.$$

First we prove that if $A_3 < \infty$, then we have inequality (5.25). Denote

$$W_2(r) := \int_{\wp} r^{Q-1}\phi_3(ry)d\sigma(y) \tag{5.29}$$

and

$$G(s) := \int_{\wp} s^{Q-1}h(sy)(\psi_3(sy))^{1-p'}d\sigma(y) \tag{5.30}$$

for $h \geq 0$ on $\mathbb{G}$ to be chosen later. Using the polar decomposition in Proposition 1.2.10 we have the following equalities:

$$\begin{aligned}
&\int_{\mathbb{G}}\phi_3(x)\left(\int_{B(0,|x|)} h(z)(\psi_3(z))^{1-p'}dz\right)^q dx\\
&= \int_0^\infty\int_{\wp} r^{Q-1}\phi_3(rw)d\sigma(w)\left(\int_0^r\int_{\wp} s^{Q-1}h(sy)(\psi_3(sy))^{1-p'}d\sigma(y)ds\right)^q dr\\
&= \int_0^\infty W_2(r)\left(\int_0^r G(s)ds\right)^q dr\\
&= q\int_0^\infty G(s)\left(\int_0^s G(r)dr\right)^{q-1}\left(\int_s^\infty W_2(r)dr\right)ds\\
&= q\int_{\wp}\int_0^\infty s^{Q-1}h(sy)(\psi_3(sy))^{1-p'}\left(\int_0^s\int_{\wp} r^{Q-1}h(rw)(\psi_3(rw))^{1-p'}d\sigma(w)dr\right)^{q-1}\\
&\quad\times\left(\int_s^\infty W_2(r)dr\right)dsd\sigma(y)
\end{aligned}$$

$$
\begin{aligned}
&= q\int_{\wp}\int_0^{\infty} s^{Q-1}h(sy)(\psi_3(sy))^{(1-p')(\frac{1}{p}+\frac{q-1}{p}+\frac{p-q}{p})}\\
&\times\left(\frac{\int_{\wp}\int_0^s r^{Q-1}h(rw)(\psi_3(rw))^{1-p'}drd\sigma(w)}{\int_{\wp}\int_0^s r^{Q-1}(\psi_3(rw))^{1-p'}drd\sigma(w)}\right)^{q-1}\\
&\times\left(\left(\int_{\wp}\int_0^s r^{Q-1}(\psi_3(rw))^{1-p'}drd\sigma(w)\right)^{q-1}\left(\int_s^{\infty}W_2(r)dr\right)\right)dsd\sigma(y).
\end{aligned}
$$

Here, using Hölder's inequality with three factors with indices $\frac{1}{p}+\frac{q-1}{p}+\frac{p-q}{p}=1$, we can estimate

$$
\int_{\mathbb{G}}\phi_3(x)\left(\int_{B(0,|x|)}h(z)(\psi_3(z))^{1-p'}dz\right)^q dx \le qK_1K_2K_3, \tag{5.31}
$$

where

$$
\begin{aligned}
K_1 &= \left(\int_{\wp}\int_0^{\infty} s^{Q-1}(h(sy))^p(\psi_3(sy))^{1-p'}dsd\sigma(y)\right)^{1/p}\\
&= \left(\int_{\mathbb{G}}(h(x))^p(\psi_3(x))^{1-p'}dx\right)^{1/p},
\end{aligned} \tag{5.32}
$$

$$
\begin{aligned}
K_2 = &\left(\int_{\wp}\int_0^{\infty} s^{Q-1}(\psi_3(sy))^{1-p'}\right.\\
&\times\left.\left(\frac{\int_{\wp}\int_0^s r^{Q-1}h(rw)(\psi_3(rw))^{1-p'}drd\sigma(w)}{\int_{\wp}\int_0^s r^{Q-1}(\psi_3(rw))^{1-p'}drd\sigma(w)}\right)^p dsd\sigma(y)\right)^{\frac{q-1}{p}}
\end{aligned} \tag{5.33}
$$

and

$$
\begin{aligned}
K_3 = &\left(\int_{\wp}\int_0^{\infty} s^{Q-1}(\psi_3(sy))^{1-p'}\left(\int_{\wp}\int_0^s r^{Q-1}(\psi_3(rw))^{1-p'}drd\sigma(w)\right)^{\frac{(q-1)p}{p-q}}\right.\\
&\times\left.\left(\int_s^{\infty}W_2(r)dr\right)^{\frac{p}{p-q}}dsd\sigma(y)\right)^{\frac{p-q}{p}}.
\end{aligned} \tag{5.34}
$$

Leaving K_1 as it is, we will estimate K_2 and K_3. We rewrite K_2 as

$$
K_2 = \left(\int_{\mathbb{G}}\frac{(\psi_3(x))^{1-p'}}{(\int_{B(0,|x|)}(\psi_3(z))^{1-p'}dz)^p}\left(\int_{B(0,|x|)}(\psi_3(z))^{1-p'}h(z)dz\right)^p dx\right)^{(q-1)/p}.
$$

We want to apply (5.2) to K_2 with indices $p=q$, and with functions $f(x)=(\psi_3(x))^{1-p'}h(x)$ and

$$
\phi_1(x) = \frac{(\psi_3(x))^{1-p'}}{(\int_{B(0,|x|)}(\psi_3(z))^{1-p'}dz)^p}, \quad \psi_1(x) = (\psi_3(x))^{(1-p')(1-p)}.
$$

For that, we will check the condition that

$$\begin{aligned} A_1(R) = &\left(\int_{|x| \geq R} (\psi_3(x))^{1-p'} \left(\int_{B(0,|x|)} (\psi_3(z))^{1-p'} dz \right)^{-p} dx \right)^{1/p} \\ &\times \left(\int_{|x| \leq R} (\psi_3(x))^{1-p'} dx \right)^{1/p'} < \infty \end{aligned} \tag{5.35}$$

holds uniformly for all $R > 0$. Assuming (5.35) uniformly in $R > 0$ for a moment, the inequality (5.2) would imply that

$$\begin{aligned} K_2 &\leq C \left(\int_{\mathbb{G}} (\psi_3(x))^{(1-p')(1-p+p)} (h(x))^p dx \right)^{(q-1)/p} \\ &= C \left(\int_{\mathbb{G}} (h(x))^p (\psi_3(x))^{1-p'} dx \right)^{(q-1)/p}, \end{aligned} \tag{5.36}$$

which is something we will use later. So, let us check (5.35). For this, we denote

$$S(s) := \int_{\wp} s^{Q-1} (\psi_3(sw))^{1-p'} d\sigma(w).$$

Using integration by parts we have

$$\begin{aligned} A_1(R) &= \left(\int_{\wp} \int_R^{\infty} r^{Q-1} (\psi_3(rw))^{1-p'} \left(\int_0^r S(s) ds \right)^{-p} dr d\sigma(w) \right)^{\frac{1}{p}} \left(\int_0^R S(s) ds \right)^{\frac{1}{p'}} \\ &= \left(\int_R^{\infty} \left(\int_0^r S(s) ds \right)^{-p} S(r) dr \right)^{1/p} \left(\int_0^R S(s) ds \right)^{1/p'} \\ &\leq \left(\frac{1}{p-1} \left(\int_0^R S(s) ds \right)^{1-p} \right)^{1/p} \left(\int_0^R S(s) ds \right)^{1/p'} = (p-1)^{-1/p} < \infty, \end{aligned}$$

so that (5.36) is confirmed. Next, for K_3, taking into account

$$\frac{1}{\delta} = \frac{1}{q} - \frac{1}{p} = \frac{p-q}{pq}$$

and using (5.26), we have

$$\begin{aligned} K_3 = &\left(\int_0^{\infty} \int_{\wp} \left(\int_s^{\infty} W_2(r) dr \right)^{\delta/q} \left(\int_{\wp} \int_0^r r^{Q-1} (\psi_3(rw))^{1-p'} dr d\sigma(w) \right)^{\delta/q'} \right. \\ &\left. \times s^{Q-1} (\psi_3(sy))^{1-p'} d\sigma(y) ds \right)^{(p-q)/p} \end{aligned}$$

$$
= \left(\int_{\mathbb{G}} \left(\int_{\mathbb{G}\setminus B(0,|x|)} \phi_3(z)dz \right)^{\frac{\delta}{q}} \left(\int_{B(0,|x|)} (\psi_3(z))^{1-p'} dz \right)^{\frac{\delta}{q'}} (\psi_3(x))^{1-p'} dx \right)^{\frac{p-q}{p}}
$$

$$
= A_3^{\frac{p-q}{p}} < \infty. \tag{5.37}
$$

Now, plugging (5.32), (5.36) and (5.37) into (5.31), we obtain

$$
\begin{aligned}
&\int_{\mathbb{G}} \phi_3(x) \left(\int_{B(0,|x|)} h(z)(\psi_3(z))^{1-p'} dz \right)^q dx \\
&\leq C A_3^{\frac{p-q}{p}} \left(\int_{\mathbb{G}} (h(x))^p (\psi_3(x))^{1-p'} dx \right)^{\frac{1}{p}+\frac{q-1}{p}},
\end{aligned}
$$

which implies (5.25) after taking $h := f\psi_3^{p'-1}$.

Let us now show the converse, namely, that (5.25) implies (5.26). For this, we consider a sequence of functions

$$
\begin{aligned}
f_k(x) := &\left(\int_{|y|\geq|x|} \phi_3(z)dz \right)^{\delta/(pq)} \left(\int_{\alpha_k\leq|z|\leq|x|} (\psi_3(z))^{1-p'} dz \right)^{\delta/(pq')} \\
&\times (\psi_3(x))^{1-p'} \chi_{(\alpha_k,\beta_k)}(|x|), \; k = 1, 2, \ldots .
\end{aligned}
$$

Inserting these functions in the place of $f(x)$ in (5.25), we obtain (5.26), if we take $0 < \alpha_k < \beta_k$ with $\alpha_k \searrow 0$ and $\beta_k \nearrow \infty$ for $k \to \infty$. □

5.2 Convolution Hardy inequalities

In this section we discuss integral Hardy inequalities in the convolution form. Such inequalities are particularly useful if we make particular choices of the convolution kernels. For example, by taking the Riesz kernels of hypoelliptic differential operators on graded groups, such inequalities can be used to derive a number of hypoelliptic versions of Hardy inequalities. While this topic falls outside the scope of this book, we refer to [RY18a] for such applications. The inequalities that we will present here have been established in [RY18a] and we follow the proofs there in our exposition.

Theorem 5.2.1 (Convolution Hardy inequality). *Let $\mathbb{G}$ be a homogeneous Lie group of homogeneous dimension Q and with a homogeneous quasi-norm $|\cdot|$. Let $1 < p \leq q < \infty$, $0 < a < Q/p$, $0 \leq b < Q$ and $\frac{a}{Q} = \frac{1}{p} - \frac{1}{q} + \frac{b}{qQ}$. Assume that there is $C_2 = C_2(a, Q) > 0$ such that*

$$
|T_a^{(1)}(x)| \leq C_2 |x|^{a-Q} \tag{5.38}
$$

holds for all $x \neq 0$. Then there exists a positive constant $C_1 = C_1(p, q, a, b) > 0$ such that

$$\left\| \frac{f * T_a^{(1)}}{|x|^{\frac{b}{q}}} \right\|_{L^q(\mathbb{G})} \leq C_1 \|f\|_{L^p(\mathbb{G})} \tag{5.39}$$

holds for all $f \in L^p(\mathbb{G})$.

The critical case $b = Q$ of Theorem 5.2.1 will be shown in Theorem 5.2.5.

Remark 5.2.2 (Riesz kernels). Let us briefly describe a typical situation when condition (5.38) is satisfied. Without going much into detail, let us assume that $\mathscr{R}$ is a positive homogeneous left invariant hypoelliptic differential operator on $\mathbb{G}$ of homogeneous degree ν. The existence of such an operator implies that the group $\mathbb{G}$ is graded, see [FR16, Section 4.1]. The operators $\mathscr{R}$ satisfying the above properties are called *Rockland operators.*

Let h_t denote the heat kernel associated to the operator $\mathscr{R}$, see [FR16, Section 4.3.4] for a thorough treatment of this and for the proof of the following notes. This heat kernel satisfies the following properties, see [FR16, Theorem 4.2.7 and Lemma 4.3.8]:

Theorem 5.2.3 (Heat kernels). *Let $\mathscr{R}$ be homogeneous left invariant hypoelliptic differential operator on $\mathbb{G}$ of homogeneous degree ν and let h_t be the associated heat kernel. Then each h_t is Schwartz and we have*

$$\forall s, t > 0 \;\; h_t * h_s = h_{t+s}, \tag{5.40}$$

$$\forall x \in \mathbb{G}, r, t > 0 \;\; h_{r^\nu t}(rx) = r^{-Q} h_t(x), \tag{5.41}$$

$$\forall x \in \mathbb{G} \;\; h_t(x) = \overline{h_t(x^{-1})}, \tag{5.42}$$

$$\int_{\mathbb{G}} h_t(x) dx = 1. \tag{5.43}$$

Moreover, we have

$$\exists C = C_{\alpha,N,\ell} > 0 \;\; \forall t \in (0,1] \;\; \sup_{|x|=1} |\partial_t^\ell X^\alpha h_t(x)| \leq C_{\alpha,N} t^N \tag{5.44}$$

for any $N \in \mathbb{N}_0$, $\alpha \in \mathbb{N}_0^n$ and $\ell \in \mathbb{N}_0$.

Furthermore, for any multi-index $\alpha \in \mathbb{N}_0^n$ and any real number a with $0 < a < (Q + [\alpha])/\nu$ there exists a positive constant $C > 0$ such that

$$\int_0^\infty t^{a-1} |X^\alpha h_t(x)| dt \leq C |x|^{-Q-[\alpha]+\nu a}. \tag{5.45}$$

The fractional powers $\mathscr{R}^{-a/\nu}$ for $\{a \in \mathbb{R},\ 0 < a < Q\}$ and $(I + \mathscr{R})^{-a/\nu}$ for $a \in \mathbb{R}_+$ are called *Riesz and Bessel potentials*, respectively, and they are well defined, see [FR16, Chapter 4.2]. Let us denote their respective kernels by $\mathcal{I}_a$ and $\mathcal{B}_a$. Then we have the relations

$$\mathcal{I}_a(x) = \frac{1}{\Gamma\left(\frac{a}{\nu}\right)} \int_0^\infty t^{\frac{a}{\nu}-1} h_t(x) dt \tag{5.46}$$

for $0 < a < Q$ with $a \in \mathbb{R}$, and

$$\mathcal{B}_a(x) = \frac{1}{\Gamma\left(\frac{a}{\nu}\right)} \int_0^\infty t^{\frac{a}{\nu}-1} e^{-t} h_t(x) dt \tag{5.47}$$

for $a > 0$, where Γ denotes the Gamma function. Consequently, it can be shown (see [FR16, Section 4.3.4]) that for any $0 < a < Q$ there exists a positive constant $C = C(Q, a)$ such that

$$|\mathcal{I}_a(x)| \leq C|x|^{-(Q-a)} \tag{5.48}$$

holds for all $x \neq 0$. Therefore, the Riesz kernel $\mathcal{I}_a$ gives a typical example of an operator satisfying condition (5.38).

Proof of Theorem 5.2.1. We split the integral in the left-hand side of (5.39) into three parts:

$$\int_{\mathbb{G}} |(f * T_a^{(1)})(x)|^q \frac{dx}{|x|^b} \leq 3^q (M_1 + M_2 + M_3), \tag{5.49}$$

with

$$M_1 := \int_{\mathbb{G}} \left(\int_{\{2|y|<|x|\}} |T_a^{(1)}(y^{-1}x) f(y)| dy \right)^q \frac{dx}{|x|^b},$$

$$M_2 := \int_{\mathbb{G}} \left(\int_{\{|x|\leq 2|y|<4|x|\}} |T_a^{(1)}(y^{-1}x) f(y)| dy \right)^q \frac{dx}{|x|^b}$$

and

$$M_3 := \int_{\mathbb{G}} \left(\int_{\{|y|>2|x|\}} |T_a^{(1)}(y^{-1}x) f(y)| dy \right)^q \frac{dx}{|x|^b}.$$

First, let us *estimate* M_1. We can assume without loss of generality that $|\cdot|$ is a norm (such a norm always exists, see Proposition 1.2.4, Part (2)) since replacing the seminorm by an equivalent one only changes the appearing constants.

Observe that by the reverse triangle inequality and the assumption $2|y| < |x|$ we have

$$|y^{-1}x| \geq |x| - |y| > |x| - \frac{|x|}{2} = \frac{|x|}{2}, \tag{5.50}$$

which is $|x| < 2|y^{-1}x|$. Taking into account this and that $T_a^{(1)}(x)$ is bounded by a radial function which is non-increasing with respect to $|x|$, we can estimate

$$\begin{aligned} M_1 &\leq \int_{\mathbb{G}} \left(\int_{\{2|y|<|x|\}} |f(y)|dy \right)^q \left(\sup_{\{|x|<2|z|\}} |T_a^{(1)}(z)| \right)^q \frac{dx}{|x|^b} \\ &\leq C \int_{\mathbb{G}} \left(\int_{\{2|y|<|x|\}} |f(y)|dy \right)^q \left(\frac{|x|}{2} \right)^{(a-Q)q} \frac{dx}{|x|^b}. \end{aligned} \tag{5.51}$$

We will now apply Theorem 5.1.1, Part (1), to estimate M_1. For this we need to check condition (5.3), that is, that

$$\sup_{R>0} \left(\int_{\{2R<|x|\}} \left(\frac{|x|}{2} \right)^{(a-Q)q} \frac{dx}{|x|^b} \right)^{\frac{1}{q}} \left(\int_{\{|x|<R\}} dx \right)^{\frac{1}{p'}} < \infty. \tag{5.52}$$

To check this, we consider two cases: $R \geq 1$ and $0 < R < 1$. For $R \geq 1$, we can estimate

$$\begin{aligned} &\left(\int_{\{2R<|x|\}} \left(\frac{|x|}{2} \right)^{(a-Q)q} \frac{dx}{|x|^b} \right)^{\frac{1}{q}} \left(\int_{\{|x|<R\}} dx \right)^{\frac{1}{p'}} \\ &\qquad \leq CR^{\frac{Q}{p'}} \left(\int_{\{2R<|x|\}} \left(\frac{|x|}{2} \right)^{(a-Q)q} \frac{dx}{|x|^b} \right)^{\frac{1}{q}} \\ &\qquad \leq CR^{\frac{Q}{p'}} \left(\int_{\{2R<|x|\}} |x|^{(a-Q)q-b} dx \right)^{\frac{1}{q}} \\ &\qquad \leq CR^{\frac{Q}{p'}} R^{\frac{(a-Q)q-b+Q}{q}} \\ &\qquad \leq C, \end{aligned} \tag{5.53}$$

which is uniformly bounded since $\frac{a}{Q} = \frac{1}{p} - \frac{1}{q} + \frac{b}{qQ}$ and $(a-Q)q-b+Q = -\frac{Qq}{p'} \neq 0$. Now let us check the condition (5.52) for $0 < R < 1$. Here, taking into account that $(a-Q)q - b + Q = -\frac{Qq}{p'} \neq 0$ we have

$$\int_{\{2R<|x|\}} \left(\frac{|x|}{2} \right)^{(a-Q)q} \frac{dx}{|x|^b} \leq CR^{(a-Q)q-b+Q}. \tag{5.54}$$

It follows with $\frac{a}{Q} = \frac{1}{p} - \frac{1}{q} + \frac{b}{qQ}$ that

$$\left(\int_{\{2R<|x|\}} \left(\frac{|x|}{2} \right)^{(a-Q)q} \frac{dx}{|x|^b} \right)^{\frac{1}{q}} \left(\int_{\{|x|<R\}} dx \right)^{\frac{1}{p'}} \leq CR^{a-Q-\frac{b}{q}+\frac{Q}{q}} R^{Q/p'} \leq C \tag{5.55}$$

holds for any $0 < R < 1$. Thus, we have checked (5.52). Applying Theorem 5.1.1, Part (1), we obtain

$$M_1^{\frac{1}{q}} \leq (p')^{\frac{1}{p'}} p^{\frac{1}{q}} A_1 \|f\|_{L^p(\mathbb{G})}. \tag{5.56}$$

Let us now *estimate* M_2. For this, we decompose M_2 as

$$M_2 = \sum_{k\in\mathbb{Z}} \int_{\{2^k \leqslant |x| < 2^{k+1}\}} \left(\int_{\{|x|\leqslant 2|y|\leqslant 4|x|\}} |T_a^{(1)}(y^{-1}x) f(y)| dy \right)^q \frac{dx}{|x|^b}.$$

Since $|x| \leqslant 2|y| \leqslant 4|x|$ and $2^k \leqslant |x| < 2^{k+1}$, we have $2^{k-1} \leqslant |y| < 2^{k+2}$. As in (5.50), assuming $|\cdot|$ is the norm and using the triangle inequality, we have

$$3|x| = |x| + 2|x| \geq |x| + |y| \geq |y^{-1}x|, \tag{5.57}$$

which implies $0 \leq |y^{-1}x| \leq 3|x| < 3 \cdot 2^{k+1}$. If we denote

$$\widetilde{I}_a(x) := C_2 |x|^{a-Q},$$

then by the assumption we have

$$|T_a^{(1)}(x)| \leq \widetilde{I}_a(x).$$

Taking into account these observations and applying Young's inequality in Proposition 1.2.13 with $1 + \frac{1}{q} = \frac{1}{r} + \frac{1}{p}$, $r \in [1, \infty]$, we can estimate M_2 by

$$\begin{aligned}
M_2 &\leq \sum_{k\in\mathbb{Z}} 2^{-kb} \int_{\mathbb{G}} (([f \cdot \chi_{\{2^{k-1}\leqslant|\cdot|<2^{k+2}\}}] * \widetilde{I}_a)(x))^q dx \\
&= \sum_{k\in\mathbb{Z}} 2^{-kb} \|[f \cdot \chi_{\{2^{k-1}\leqslant|\cdot|<2^{k+2}\}}] * \widetilde{I}_a\|^q_{L^q(\mathbb{G})} \\
&\leq \sum_{k\in\mathbb{Z}} 2^{-kb} \|\widetilde{I}_a \cdot \chi_{\{0\leqslant|\cdot|<3\cdot 2^{k+1}\}}\|^q_{L^r(\mathbb{G})} \|f \cdot \chi_{\{2^{k-1}\leqslant|\cdot|<2^{k+2}\}}\|^q_{L^p(\mathbb{G})} \\
&= C_2 \sum_{k\in\mathbb{Z}} 2^{-kb} \left(\int_{|x|<3\cdot 2^{k+1}} |x|^{(a-Q)r} dx \right)^{\frac{q}{r}} \|f \cdot \chi_{\{2^{k-1}\leqslant|x|<2^{k+2}\}}\|^q_{L^p(\mathbb{G})} \qquad (5.58) \\
&\leq C \sum_{k\in\mathbb{Z}} 2^{-kb} (3 \cdot 2^{k+1})^{\left(\frac{(a-Q)pq}{pq+p-q} + Q\right)\frac{pq+p-q}{p}} \|f \cdot \chi_{\{2^{k-1}\leqslant|x|<2^{k+2}\}}\|^q_{L^p(\mathbb{G})} \\
&= C \sum_{k\in\mathbb{Z}} 2^{-kb} (3 \cdot 2^{k+1})^b \|f \cdot \chi_{\{2^{k-1}\leqslant|x|<2^{k+2}\}}\|^q_{L^p(\mathbb{G})} \\
&\leq C \sum_{k\in\mathbb{Z}} \|f \cdot \chi_{\{2^{k-1}\leqslant|x|<2^{k+2}\}}\|^q_{L^p(\mathbb{G})} \\
&\leq C \|f\|^q_{L^p(\mathbb{G})},
\end{aligned}$$

since $\frac{(a-Q)pq}{pq+p-q} + Q = \frac{bp}{pq+p-q} > 0$ and $q \geq p$.

Now let us *estimate* M_3. Without loss of generality, we may assume again that $|\cdot|$ is a norm. Then, similarly to (5.50) we note that $2|x| < |y|$ implies $|y| < 2|y^{-1}x|$. Consequently we can estimate M_3 as

$$M_3 \leq \int_{\mathbb{G}} \left(\int_{\{|y|>2|x|\}} \left(\frac{|y|}{2} \right)^{(a-Q)} |f(y)| dy \right)^q \frac{dx}{|x|^b}.$$

We will apply Theorem 5.1.1, Part (2), to estimate M_3. For this, we need to check that

$$\sup_{R>0} \left(\int_{\{|x|<R\}} \frac{dx}{|x|^b} \right)^{\frac{1}{q}} \left(\int_{\{2R<|x|\}} \left(\frac{|x|}{2} \right)^{(a-Q)p'} dx \right)^{\frac{1}{p'}} < \infty. \tag{5.59}$$

To verify this, we consider two cases: $R \geq 1$ and $0 < R < 1$. First, for $R \geq 1$, using the assumption $|T_a^{(1)}(x)| \leq C|x|^{a-Q}$ and that $Q \neq ap$, one gets

$$\left(\int_{\{2R<|x|\}} \left(\frac{|x|}{2} \right)^{(a-Q)p'} dx \right)^{\frac{1}{p'}} \leq C \left(\int_{\{2R<|x|\}} |x|^{(a-Q)p'} dx \right)^{\frac{1}{p'}} \leq CR^{a-\frac{Q}{p}}. \tag{5.60}$$

Since $R \geq 1$ we can estimate

$$\left(\int_{\{|x|<R\}} \frac{dx}{|x|^b} \right)^{\frac{1}{q}} \left(\int_{\{2R<|x|\}} \left(\frac{|x|}{2} \right)^{(a-Q)p'} dx \right)^{\frac{1}{p'}} \leq CR^{a-\frac{Q}{p}+\frac{Q-b}{q}} \leq C,$$

since $b < Q$ and $\frac{a}{Q} = \frac{1}{p} - \frac{1}{q} + \frac{b}{qQ}$. Now let us check the condition (5.59) for the range $0 < R < 1$. In this case, noting that $ap - Q < 0$ we have

$$\int_{\{2R<|x|\}} \left(\frac{|x|}{2} \right)^{(a-Q)p'} dx \leq C \int_{\{2R<|x|\}} |x|^{(a-Q)p'} dx \leq CR^{(a-Q)p'+Q}. \tag{5.61}$$

Since

$$\int_{\{|x|<R\}} \frac{dx}{|x|^b} \leqslant CR^{Q-b},$$

we have

$$\left(\int_{\{|x|<R\}} \frac{dx}{|x|^b} \right)^{\frac{1}{q}} \left(\int_{\{2R<|x|\}} \left(\frac{|x|}{2} \right)^{(a-Q)p'} dx \right)^{\frac{1}{p'}} \leq CR^{\frac{Q-b}{q}} R^{\frac{(a-Q)p'+Q}{p'}} \leq C, \tag{5.62}$$

since $Q > b$ and $\frac{a}{Q} = \frac{1}{p} - \frac{1}{q} + \frac{b}{qQ}$. Thus, we have checked (5.59). Consequently, the application of Theorem 5.1.1, Part (2), to M_3 yields

$$M_3^{\frac{1}{q}} \leq (p')^{\frac{1}{p'}} p^{\frac{1}{q}} A_2 \|f\|_{L^p(\mathbb{G})}. \tag{5.63}$$

Thus, (5.56), (5.63) and (5.58) complete the proof of Theorem 5.2.1. □

Remark 5.2.4 (Schur test argument). In the case $p = q$, we can also prove Theorem 5.2.1 by using Schur's test ([FR75]) as in the proof of Theorem 4.7.1. For the Riesz kernels of the sub-Laplacian on stratified groups such an argument was used in [CCR15], and the argument below was given in [RY18a].

For $p = q$, the condition $\frac{a}{Q} = \frac{1}{p} - \frac{1}{q} + \frac{b}{qQ}$ in Theorem 5.2.1 implies that $b = ap$, so we are interested in the L^p-boundedness of the operator

$$S_a f := |x|^{-b/p}(f * |x|^{a-Q}) = |x|^{-a}(f * |x|^{a-Q}).$$

The adjoint, defined by $(f, S_a^* g) = (S_a f, g)$, is given by $S_a^* g := (|x|^{-a}g) * |x|^{a-Q}$.

We now recall again the *Schur test*:

Assume that the integral operator S has a positive integral kernel, and that there exist a positive function h and constants A_p and B_p such that

$$S(h^{p'})(x) \leq A_p(h(x))^{p'} \quad \text{and} \quad S^*(h^p)(x) \leq B_p(h(x))^p$$

hold for almost all $x \in \mathbb{G}$. Then we have

$$\|S_a f\|_{L^p(\mathbb{G})} \leq A_p^{1/p'} B_p^{1/p} \|f\|_{L^p(\mathbb{G})}$$

for all $f \in L^p(\mathbb{G})$.

Let us now take $h_c(x) := |x|^{c-Q}$ with $c > 0$ and consider the convolution integrals

$$h_c^{p'} * |x|^{a-Q} \quad \text{and} \quad (|x|^{-b/p} h_c^p) * |x|^{a-Q},$$

which arise in the computation of $S_a(h_c^{p'})$ and $S_a^*(h_c^p)$. We see that the homogeneity orders of $h_c^{p'}$ and $|x|^{-b/p}h_c^p$ are $(c-Q)p'$ and $(c-Q)p - b/p$, respectively. Then, the homogeneity orders of $h_c^{p'} * |x|^{a-Q}$ and $(|x|^{-b/p}h_c^p) * |x|^{a-Q}$ are $a - Q + (c-Q)p'$ and $a - Q + (c-Q)p - b/p$, respectively. Therefore, these convolution integrals converge absolutely in $\mathbb{G}\backslash\{0\}$ if and only if $0 < (c-Q)p' + Q < Q - a$ and $0 < (c-Q)p - b/p + Q < Q - a$, that is, if

$$\max\left(\frac{Q}{p}, \frac{a}{p} + \frac{Q}{p'}\right) < c < Q - \frac{a}{p'}$$

since $b = ap$. This condition is true if $0 < a < Q/p$.

Thus, it follows from Schur's test that

$$\||x|^{-a}(f * |x|^{a-Q})\|_{L^p(\mathbb{G})} \leq A_{a,p}^{1/p'} B_{a,p}^{1/p} \|f\|_{L^p(\mathbb{G})},$$

where $0 < a < Q/p$, $1 < p < \infty$, and $f \in L^p(\mathbb{G})$.

Taking into account this and $|T_a^{(1)}(x)| \leq C|x|^{a-Q}$, we obtain

$$\left\|\frac{f * T_a^{(1)}}{|x|^{\frac{b}{p}}}\right\|_{L^p(\mathbb{G})} \leq C\left\|\frac{|f| * |T_a^{(1)}|}{|x|^{\frac{b}{p}}}\right\|_{L^p(\mathbb{G})} \leq C\||x|^{-a}(|f| * |x|^{a-Q})\|_{L^p(\mathbb{G})} \leq C\|f\|_{L^p(\mathbb{G})}, \tag{5.64}$$

which proves Theorem 5.2.1 in the case $p = q$.

Let us now show the critical case $b = Q$ of Theorem 5.2.1.

Theorem 5.2.5 (Critical convolution Hardy inequality). *Let $\mathbb{G}$ be a homogeneous Lie group of homogeneous dimension Q with a homogeneous quasi-norm $|\cdot|$. Let $1 < p < r < \infty$ and $p < q < (r-1)p'$, where $1/p + 1/p' = 1$. Assume that for $a = Q/p$ we have*

$$|T_a^{(2)}(x)| \leq C_2 \begin{cases} |x|^{a-Q}, & \text{for } x \in \mathbb{G}\backslash\{0\}, \\ |x|^{-Q}, & \text{for } x \in \mathbb{G} \text{ with } |x| \geq 1, \end{cases} \tag{5.65}$$

for some positive $C_2 = C_2(a, Q)$. Then there exists a positive constant $C_1 = C_1(p, q, r, Q) > 0$ such that

$$\left\| \frac{f * T_{Q/p}^{(2)}}{\left(\log\left(e + \frac{1}{|x|}\right)\right)^{\frac{r}{q}} |x|^{\frac{Q}{q}}} \right\|_{L^q(\mathbb{G})} \leq C_1 \|f\|_{L^p(\mathbb{G})} \tag{5.66}$$

holds for all $f \in L^p(\mathbb{G})$.

Remark 5.2.6. We note that compared to the condition (5.38), the decay assumption in (5.65) for large x is stronger. Continuing with the notation of Remark 5.2.2, we observe that the Bessel kernel (5.47) of the operator $(I + \mathscr{R})^{-a/\nu}$ for $0 < a < Q$ satisfies (5.65): there exists a positive constant $C = C(Q, a) > 0$ such that we have, in particular,

$$|\mathcal{B}_a(x)| \leq \begin{cases} C|x|^{-(Q-a)}, & \text{for } x \in \mathbb{G}\backslash\{0\}, \\ C|x|^{-Q}, & \text{for } x \in \mathbb{G} \text{ with } |x| \geq 1. \end{cases} \tag{5.67}$$

We refer to [RY18a] for further details, as well as to the original proof of Theorem 5.2.5 that we follow here.

Proof of Theorem 5.2.5. Let us split the integral in the left-hand side of (5.66) into three parts,

$$\int_{\mathbb{G}} |(f * T_{Q/p}^{(2)})(x)|^q \frac{dx}{\left|\log\left(e + \frac{1}{|x|}\right)\right|^r |x|^Q} \leq 3^q (N_1 + N_2 + N_3), \tag{5.68}$$

where

$$N_1 := \int_{\mathbb{G}} \left(\int_{\{2|y| < |x|\}} |T_{Q/p}^{(2)}(y^{-1}x) f(y)| dy \right)^q \frac{dx}{\left|\log\left(e + \frac{1}{|x|}\right)\right|^r |x|^Q},$$

$$N_2 := \int_{\mathbb{G}} \left(\int_{\{|x| \leq 2|y| < 4|x|\}} |T_{Q/p}^{(2)}(y^{-1}x) f(y)| dy \right)^q \frac{dx}{\left|\log\left(e + \frac{1}{|x|}\right)\right|^r |x|^Q}$$

and

$$N_3 := \int_{\mathbb{G}} \left(\int_{\{|y|>2|x|\}} |T^{(2)}_{Q/p}(y^{-1}x) f(y)| dy \right)^q \frac{dx}{\left|\log\left(e + \frac{1}{|x|}\right)\right|^r |x|^Q}.$$

We begin by *estimating* N_1. Similar to the argument in (5.50), in the region $2|y| < |x|$ we have

$$|y^{-1}x| \geq |x| - |y| > |x| - \frac{|x|}{2} = \frac{|x|}{2}, \tag{5.69}$$

which is $|x| < 2|y^{-1}x|$. Denote

$$|T^{(2)}_a(x)| \leq \widetilde{B}_a(x) := C_2 \begin{cases} |x|^{a-Q}, & \text{for } x \in \mathbb{G}\backslash\{0\}, \\ |x|^{-Q}, & \text{for } x \in \mathbb{G} \text{ with } |x| \geq 1. \end{cases} \tag{5.70}$$

Since $T^{(2)}_{Q/p}(x)$ is bounded by $\widetilde{B}_{Q/p}(x)$ which is non-increasing with respect to $|x|$, then using (5.69) we get

$$\begin{aligned} N_1 &\leq \int_{\mathbb{G}} \left(\int_{\{2|y|<|x|\}} |f(y)| dy \right)^q \left(\sup_{\{|x|<2|z|\}} |T^{(2)}_{Q/p}(z)| \right)^q \frac{dx}{\left|\log\left(e + \frac{1}{|x|}\right)\right|^r |x|^Q} \\ &\leq \int_{\mathbb{G}} \left(\int_{\{2|y|<|x|\}} |f(y)| dy \right)^q \left(\widetilde{B}_{Q/p}\left(\frac{x}{2}\right) \right)^q \frac{dx}{\left|\log\left(e + \frac{1}{|x|}\right)\right|^r |x|^Q}. \end{aligned}$$

We will now apply Theorem 5.1.1, Part (1), to estimate N_1. For this we have to check the condition (5.3), that is, that

$$\left(\int_{\{2R<|x|\}} \left(\widetilde{B}_{Q/p}\left(\frac{x}{2}\right) \right)^q \frac{dx}{\left|\log\left(e + \frac{1}{|x|}\right)\right|^r |x|^Q} \right)^{\frac{1}{q}} \left(\int_{\{|x|<R\}} dx \right)^{\frac{1}{p'}} \leq A_1 \tag{5.71}$$

holds uniformly for all $R > 0$. To verify this uniform boundedness, we consider two cases: $R \geq 1$ and $0 < R < 1$. First, for $R \geq 1$, using the second equality in (5.70), we can estimate

$$\begin{aligned} &\left(\int_{\{2R<|x|\}} \left(\widetilde{B}_{Q/p}\left(\frac{x}{2}\right) \right)^q \frac{dx}{\left|\log\left(e + \frac{1}{|x|}\right)\right|^r |x|^Q} \right)^{\frac{1}{q}} \left(\int_{\{|x|<R\}} dx \right)^{\frac{1}{p'}} \\ &\leq CR^{\frac{Q}{p'}} \left(\int_{\{2R<|x|\}} \left(\widetilde{B}_{Q/p}\left(\frac{x}{2}\right) \right)^q \frac{dx}{|x|^Q} \right)^{\frac{1}{q}} = CR^{\frac{Q}{p'}} \left(\int_{\{2R<|x|\}} |x|^{-Qq-Q} dx \right)^{\frac{1}{q}} \\ &\leq CR^{-Q} R^{\frac{Q}{p'}} \leq C. \end{aligned} \tag{5.72}$$

Next, let us check (5.71) for $0 < R < 1$. We split the integral into two terms,

$$\begin{aligned}\int_{\{2R<|x|\}} \left(\widetilde{B}_{Q/p}\left(\frac{x}{2}\right)\right)^q \frac{dx}{\left|\log\left(e+\frac{1}{|x|}\right)\right|^r |x|^Q} &= \int_{\{2R<|x|<2\}} \left(\widetilde{B}_{Q/p}\left(\frac{x}{2}\right)\right)^q \frac{dx}{\left|\log\left(e+\frac{1}{|x|}\right)\right|^r |x|^Q} \\ &+ \int_{\{|x|\geqslant 2\}} \left(\widetilde{B}_{Q/p}\left(\frac{x}{2}\right)\right)^q \frac{dx}{\left|\log\left(e+\frac{1}{|x|}\right)\right|^r |x|^Q}. \end{aligned} \tag{5.73}$$

We note that the second integral in the right-hand side of (5.73) is finite by the second equality in (5.70). For the first integral, using the first equality in (5.70), we can estimate

$$\begin{aligned}\int_{\{2R<|x|<2\}} \left|\widetilde{B}_{Q/p}\left(\frac{x}{2}\right)\right|^q \frac{dx}{\left|\log\left(e+\frac{1}{|x|}\right)\right|^r |x|^Q} &\leq \int_{\{2R<|x|<2\}} \left|\widetilde{B}_{Q/p}\left(\frac{x}{2}\right)\right|^q \frac{dx}{|x|^Q} \\ &\leq C\int_{\{2R<|x|<2\}} |x|^{-Qq/p'-Q}dx \\ &\leq CR^{-Qq/p'}.\end{aligned}$$

Combining this with (5.73), we obtain

$$\begin{aligned}\left(\int_{\{2R<|x|\}} \left|\widetilde{B}_{Q/p}\left(\frac{x}{2}\right)\right|^q \frac{dx}{\left|\log\left(e+\frac{1}{|x|}\right)\right|^r |x|^Q}\right)^{\frac{1}{q}} \left(\int_{\{|x|<R\}} dx\right)^{\frac{1}{p'}} \\ \leq C(R^{-Q/p'}+1)R^{Q/p'} \leq C\end{aligned}$$

uniformly for all $0 < R < 1$. Thus, we have verified (5.71), so that applying Theorem 5.1.1, Part (1), to N_1 we obtain

$$N_1^{\frac{1}{q}} \leq (p')^{\frac{1}{p'}} p^{\frac{1}{q}} A_1 \|f\|_{L^p(\mathbb{G})}. \tag{5.74}$$

Now let us *estimate* N_2. We decompose it into the sum

$$\begin{aligned}N_2 = \sum_{k\in\mathbb{Z}} \int_{\{2^k\leqslant|x|<2^{k+1}\}} \left(\int_{\{|x|\leqslant 2|y|\leqslant 4|x|\}} |T^{(2)}_{Q/p}(y^{-1}x)f(y)|dy\right)^q \\ \times \frac{dx}{\left|\log\left(e+\frac{1}{|x|}\right)\right|^r |x|^Q}.\end{aligned}$$

Since the function $\left(\log\left(\frac{1}{|x|}\right)\right)^r |x|^Q$ is non-decreasing with respect to $|x|$ near the origin, there exists an integer $k_0 \in \mathbb{Z}$ with $k_0 \leqslant -3$ such that this function is non-decreasing in $|x| \in (0, 2^{k_0+1})$. Fixing this k_0, we decompose N_2 further as

$$N_2 = N_{21} + N_{22}, \tag{5.75}$$

where

$$N_{21} := \sum_{k=-\infty}^{k_0} \int_{\{2^k \leqslant |x| < 2^{k+1}\}} \left(\int_{\{|x|\leqslant 2|y| \leqslant 4|x|\}} |T^{(2)}_{Q/p}(y^{-1}x) f(y)| dy \right)^q \times \frac{dx}{\left|\log\left(e + \frac{1}{|x|}\right)\right|^r |x|^Q}$$

and

$$N_{22} := \sum_{k=k_0+1}^{\infty} \int_{\{2^k \leqslant |x| < 2^{k+1}\}} \left(\int_{\{|x|\leqslant 2|y| \leqslant 4|x|\}} |T^{(2)}_{Q/p}(y^{-1}x) f(y)| dy \right)^q \times \frac{dx}{\left|\log\left(e + \frac{1}{|x|}\right)\right|^r |x|^Q}.$$

Let us first estimate N_{22}. Since $|x| \leqslant 2|y| \leqslant 4|x|$ and $2^k \leqslant |x| < 2^{k+1}$, we must also have $2^{k-1} \leqslant |y| < 2^{k+2}$. Before starting to estimate N_{22}, using (5.65) and $q > p$, let us show that

$$\begin{aligned} \int_{\mathbb{G}} |T^{(2)}_{Q/p}(x)|^{\tilde{r}} dx &= \int_{|x|<1} |T^{(2)}_{Q/p}(x)|^{\tilde{r}} dx + \int_{|x|\geq 1} |T^{(2)}_{Q/p}(x)|^{\tilde{r}} dx \\ &\leq C_2 \left(\int_{|x|<1} |x|^{-\frac{Qq(p-1)}{pq+p-q}} dx + \int_{|x|\geq 1} |x|^{-\frac{Qpq}{pq+p-q}} dx \right) < \infty, \end{aligned} \tag{5.76}$$

where $\tilde{r} \in [1, \infty]$ is such that $1 + \frac{1}{q} = \frac{1}{\tilde{r}} + \frac{1}{p}$.

Then, (5.76) and Young's inequality in Proposition 1.2.13 with $1 + \frac{1}{q} = \frac{1}{\tilde{r}} + \frac{1}{p}$ and $\tilde{r} \in [1, \infty]$ imply that

$$\begin{aligned} N_{22} &\leqslant C \sum_{k=k_0+1}^{\infty} \int_{\{2^k \leqslant |x| < 2^{k+1}\}} \left(\int_{\{|x|\leqslant 2|y| \leqslant 4|x|\}} |T^{(2)}_{Q/p}(y^{-1}x) f(y)| dy \right)^q dx \\ &\leqslant C \| [f \cdot \chi_{\{2^{k-1} \leqslant |\cdot| < 2^{k+2}\}}] * T^{(2)}_{Q/p} \|^q_{L^q(\mathbb{G})} \\ &\leqslant C \|T^{(2)}_{Q/p}\|^q_{L^{\tilde{r}}(\mathbb{G})} \sum_{k=k_0+1}^{\infty} \|f \cdot \chi_{\{2^{k-1} \leqslant |\cdot| < 2^{k+2}\}}\|^q_{L^p(\mathbb{G})} \\ &= C \sum_{k=k_0+1}^{\infty} \left(\int_{\{2^k \leqslant |x| < 2^{k+1}\}} |f(x)|^p dx \right)^{q/p} \end{aligned}$$

$$\leqslant C\left(\sum_{k\in\mathbb{Z}}\int_{\{2^k\leqslant|x|<2^{k+1}\}}|f(x)|^p dx\right)^{q/p}$$
$$= C\|f\|^q_{L^p(\mathbb{G})}. \tag{5.77}$$

Next, let us estimate N_{21}. As in (5.69), assuming $|\cdot|$ is the norm and using the triangle inequality and $|y|\leqslant 2|x|$, we can estimate

$$3|x| = |x| + 2|x| \geq |x| + |y| \geq |y^{-1}x|. \tag{5.78}$$

Since $\left(\log\left(\frac{1}{|x|}\right)\right)^r |x|^Q$ is non-decreasing in $|x|\in(0,2^{k_0+1})$ and $3|x|\geqslant|y^{-1}x|$, we have

$$\left(\log\left(\frac{1}{|x|}\right)\right)^r |x|^Q \geq \left(\log\left(\frac{1}{|y^{-1}x/3|}\right)\right)^r \left|\frac{y^{-1}x}{3}\right|^Q.$$

Consequently, this and (5.65) yield

$$N_{21} \leq C\sum_{k=-\infty}^{k_0}\int_{\{2^k\leq|x|<2^{k+1}\}}\left(\int_{\{|x|\leq 2|y|\leq 4|x|\}}|y^{-1}x|^{-\frac{Q}{p'}}|f(y)|dy\right)^q$$
$$\times\frac{dx}{\left(\log\left(\frac{1}{|x|}\right)\right)^r|x|^Q}$$
$$= C\sum_{k=-\infty}^{k_0}\int_{\{2^k\leq|x|<2^{k+1}\}}\left(\int_{\{|x|\leq 2|y|\leq 4|x|\}}\frac{|y^{-1}x|^{-\frac{Q}{p'}}|f(y)|}{\left(\left(\log\left(\frac{1}{|x|}\right)\right)^r|x|^Q\right)^{\frac{1}{q}}}dy\right)^q dx$$

$$\leq C\sum_{k=-\infty}^{k_0}\int_{\{2^k\leq|x|<2^{k+1}\}}\left(\int_{\{|x|\leq 2|y|\leq 4|x|\}}\frac{|y^{-1}x|^{-\frac{Q}{p'}}|f(y)|}{\left(\left(\log\left(\frac{1}{|(y^{-1}x)/3|}\right)\right)^r|(y^{-1}x)/3|^Q\right)^{\frac{1}{q}}}dy\right)^q dx.$$

Since $|x|\leq 2|y|\leq 4|x|$ and $2^k\leq|x|<2^{k+1}$ with $k\leq k_0$, we must also have $2^{k-1}\leq|y|<2^{k+2}$ and $|y^{-1}x|\leq 3|x|<3\cdot 2^{k_0+1}\leq 3/4$, using (5.78) and $k_0\leq -3$. Taking into account these and setting

$$g(x) := \frac{\chi_{B_{\frac{3}{4}}(0)}(x)}{\left(\log\left(\frac{1}{|x|}\right)\right)^{\frac{r}{q}}|x|^{\frac{Q}{q}+\frac{Q}{p'}}},$$

we have for N_{21} that

$$N_{21}\leq C\sum_{k=-\infty}^{k_0}\int_{\{2^k\leq|x|<2^{k+1}\}}\left(\int_{\{|x|\leq 2|y|\leq 4|x|\}}\frac{|f(y)|}{\left(\log\left(\frac{1}{|y^{-1}x|}\right)\right)^{\frac{r}{q}}|y^{-1}x|^{\frac{Q}{q}+\frac{Q}{p'}}}dy\right)^q dx$$
$$\leq C\sum_{k=-\infty}^{k_0}\|[f\cdot\chi_{\{2^{k-1}\leq|\cdot|<2^{k+2}\}}]*g\|^q_{L^q(\mathbb{G})}.$$

Since $p < q < (r-1)p'$, we use Young's inequality in Proposition 1.2.13 with $1 + \frac{1}{q} = \frac{1}{\tilde{r}} + \frac{1}{p}$ and $\tilde{r} \in [1, \infty)$, to get

$$N_{21} \leq C\|g\|^q_{L^{\tilde{r}}(\mathbb{G})} \sum_{k=-\infty}^{k_0} \|f \cdot \chi_{\{2^{k-1} \leq |\cdot| < 2^{k+2}\}}\|^q_{L^p(\mathbb{G})} \leq C\|f\|^q_{L^p(\mathbb{G})}, \tag{5.79}$$

provided that $g \in L^{\tilde{r}}(\mathbb{G})$. Since $\left(\frac{Q}{q} + \frac{Q}{p'}\right)\tilde{r} = Q$, $\frac{r\tilde{r}}{q} = \frac{rp'}{p'+q}$ and $q < (r-1)p'$, then changing variables, we obtain

$$\|g\|^{\tilde{r}}_{L^{\tilde{r}}(\mathbb{G})} = \int_{B(0,3/4)} \frac{dx}{\left(\log\left(\frac{1}{x}\right)\right)^{\frac{rp'}{p'+q}} |x|^Q} = C\int_{\log\left(\frac{4}{3}\right)}^{\infty} \frac{dt}{t^{\frac{rp'}{p'+q}}} < \infty.$$

Let us *estimate* N_3 now. Without loss of generality, we may assume again that $|\cdot|$ is the norm. Similarly to (5.69) we obtain $|y| < 2|y^{-1}x|$ from $2|x| < |y|$. Then, we have for N_3 that

$$N_3 \leq \int_{\mathbb{G}} \left(\int_{\{|y|>2|x|\}} \left|\widetilde{B}_{Q/p}\left(\frac{y}{2}\right)\right| |f(y)| dy\right)^q \frac{dx}{\left|\log\left(e + \frac{1}{|x|}\right)\right|^r |x|^Q}.$$

We will apply Theorem 5.1.1, Part (2), for the required estimate of N_3. For this we have to check the following condition:

$$\left(\int_{\{|x|<R\}} \frac{dx}{\left|\log\left(e + \frac{1}{|x|}\right)\right|^r |x|^Q}\right)^{\frac{1}{q}} \left(\int_{\{2R<|x|\}} \left|\widetilde{B}_{Q/p}\left(\frac{x}{2}\right)\right|^{p'} dx\right)^{\frac{1}{p'}} \leq A_2. \tag{5.80}$$

To check this, let us consider the cases: $R \geq 1$ and $0 < R < 1$. Then, for $R \geq 1$ by the second equality in (5.70), we get

$$\left(\int_{\{2R<|x|\}} \left|\widetilde{B}_{Q/p}\left(\frac{x}{2}\right)\right|^{p'} dx\right)^{\frac{1}{p'}} \leq C\left(\int_{\{2R<|x|\}} |x|^{-Qp'} dx\right)^{\frac{1}{p'}} \leq CR^{-\frac{Q}{p}}. \tag{5.81}$$

Moreover, we have

$$\int_{\{|x|<R\}} \frac{dx}{\left|\log\left(e + \frac{1}{|x|}\right)\right|^r |x|^Q} = \int_{\{|x|<\frac{1}{2}\}} \frac{dx}{\left|\log\left(e + \frac{1}{|x|}\right)\right|^r |x|^Q} + \int_{\{\frac{1}{2}\leq|x|<R\}} \frac{dx}{\left|\log\left(e + \frac{1}{|x|}\right)\right|^r |x|^Q},$$

and we note that the first summand in the right-hand side of above is finite since $r > 1$. For the second term, we get

$$\int_{\{\frac{1}{2}\leq|x|<R\}} \frac{dx}{\left|\log\left(e + \frac{1}{|x|}\right)\right|^r |x|^Q} \leq \int_{\{\frac{1}{2}\leq|x|<R\}} \frac{dx}{|x|^Q} \leq C(1 + \log R). \tag{5.82}$$

Combining (5.81) and (5.82), we have for $R \geq 1$ that

$$\left(\int_{\{|x|<R\}} \frac{dx}{\left|\log\left(e+\frac{1}{|x|}\right)\right|^r |x|^Q}\right)^{\frac{1}{q}} \left(\int_{\{2R<|x|\}} \left|\widetilde{B}_{Q/p}\left(\frac{x}{2}\right)\right|^{p'} dx\right)^{\frac{1}{p'}} \leq CR^{-\frac{Q}{p}}(1+\log R)^{\frac{1}{q}} \leq C.$$

Now let us check the condition (5.80) for $0 < R < 1$. We split the integral into two terms:

$$\begin{aligned}\int_{\{2R<|x|\}} &\left|\widetilde{B}_{Q/p}\left(\frac{x}{2}\right)\right|^{p'} dx \\ &= \int_{\{2R<|x|<2\}} \left|\widetilde{B}_{Q/p}\left(\frac{x}{2}\right)\right|^{p'} dx + \int_{\{|x|\geqslant 2\}} \left|\widetilde{B}_{Q/p}\left(\frac{x}{2}\right)\right|^{p'} dx.\end{aligned} \tag{5.83}$$

We note that the second integral in the right-hand side of above is finite by the second equality in (5.70). Then, using the first equality in (5.70) we get for the first integral that

$$\int_{\{2R<|x|<2\}} \left|\widetilde{B}_{Q/p}\left(\frac{x}{2}\right)\right|^{p'} dx \leq C \int_{\{2R<|x|<2\}} |x|^{-Q} dx \leq C \log\left(\frac{1}{R}\right).$$

Combined with (5.83), it follows that

$$\int_{\{2R<|x|\}} \left|\widetilde{B}_{Q/p}\left(\frac{x}{2}\right)\right|^{p'} dx \leq C\left(1+\log\left(\frac{1}{R}\right)\right). \tag{5.84}$$

Since

$$\int_{\{|x|<R\}} \frac{dx}{\left|\log\left(e+\frac{1}{|x|}\right)\right|^r |x|^Q} \leqslant C\left(\log\left(e+\frac{1}{R}\right)\right)^{-(r-1)},$$

and (5.84), and taking into account $r > 1$ and $q < (r-1)p'$ we obtain that

$$\begin{aligned}&\left(\int_{\{|x|<R\}} \frac{dx}{\left|\log\left(e+\frac{1}{|x|}\right)\right|^r |x|^Q}\right)^{\frac{1}{q}} \left(\int_{\{2R<|x|\}} \left|\widetilde{B}_{Q/p}\left(\frac{x}{2}\right)\right|^{p'} dx\right)^{\frac{1}{p'}} \\ &\leqslant C\left(\log\left(e+\frac{1}{R}\right)\right)^{-\frac{r-1}{q}} \left(1+\left(\log\left(\frac{1}{R}\right)\right)^{\frac{1}{p'}}\right) \leqslant C.\end{aligned} \tag{5.85}$$

Thus, we have checked (5.80). Consequently, applying Theorem 5.1.1, Part (2), for the term N_3, we obtain

$$N_3^{\frac{1}{q}} \leq (p')^{\frac{1}{p'}} p^{\frac{1}{q}} A_2 \|f\|_{L^p(\mathbb{G})}. \tag{5.86}$$

Finally, a combination of (5.74), (5.86), (5.75), (5.77), (5.79) and (5.68) completes the proof of Theorem 5.2.5. □

5.3 Hardy–Littlewood–Sobolev inequalities on homogeneous groups

In this section we discuss the Hardy–Littlewood–Sobolev inequality on homogeneous groups. We show that it can be obtained as a simple consequence of the convolution Hardy inequality in Theorem 5.2.1. In fact, the argument implies a little more.

Theorem 5.3.1 (Hardy–Littlewood–Sobolev inequality). *Let $\mathbb{G}$ be a homogeneous Lie group of homogeneous dimension Q with a homogeneous quasi-norm $|\cdot|$. Let $0<\lambda<Q$ and $1<p,q<\infty$ be such that*

$$1/p+1/q+(\alpha+\lambda)/Q=2$$

with $0\leq\alpha<Q/p'$ and $\alpha+\lambda\leq Q$, where $1/p+1/p'=1$. Then there exists a positive constant $C=C(Q,\lambda,p,\alpha)>0$ such that

$$\left|\int_{\mathbb{G}}\int_{\mathbb{G}}\frac{\overline{f(x)}g(y)}{|x|^{\alpha}|y^{-1}x|^{\lambda}}dxdy\right|\leq C\|f\|_{L^p(\mathbb{G})}\|g\|_{L^q(\mathbb{G})} \tag{5.87}$$

holds for all $f\in L^p(\mathbb{G})$ and $g\in L^q(\mathbb{G})$.

Remark 5.3.2 (Stein–Weiss inequality).

1. The original Hardy–Littlewood–Sobolev inequality goes back to the work of Hardy–Littlewood [HL27], [HL30] and Sobolev [Sob38]. More specifically, in [HL27], Hardy and Littlewood considered the one-dimensional fractional operator on $(0,\infty)$, given by

$$T_{\lambda}f(x)=\int_0^{\infty}\frac{f(y)}{|x-y|^{\lambda}}dy,\quad 0<\lambda<1, \tag{5.88}$$

and proved that if $1<p<q<\infty$ and $\frac{1}{q}=\frac{1}{p}+\lambda-1$, then there is $C>0$ such that

$$\|T_{\lambda}f\|_{L^q(0,\infty)}\leq C\|f\|_{L^p(0,\infty)},$$

holds for all $f\in L^p(0,\infty)$. The N-dimensional analogue of (5.88) can be written by the formula

$$I_{\lambda}f(x)=\int_{\mathbb{R}^N}\frac{f(y)}{|x-y|^{\lambda}}dy,\quad 0<\lambda<N. \tag{5.89}$$

Consequently, if was shown by Sobolev in [Sob38] that if $1<p<q<\infty$ and $\frac{1}{q}=\frac{1}{p}+\frac{\lambda}{N}-1$, then there is $C>0$ such that

$$\|I_{\lambda}f\|_{L^q(\mathbb{R}^N)}\leq C\|f\|_{L^p(\mathbb{R}^N)},$$

holds for all $f\in L^p(\mathbb{R}^N)$. In [SW58], Stein and Weiss obtained the following two-weight extension of the Hardy–Littlewood–Sobolev inequality, which is nowadays called the *Stein–Weiss inequality.* More specifically, they have shown that if $0<\lambda<N$, $1<p<\infty$, $\alpha<\frac{N(p-1)}{p}$, $\beta<\frac{N}{q}$, $\alpha+\beta\geq 0$,

$\frac{1}{q} = \frac{1}{p} + \frac{\lambda+\alpha+\beta}{N} - 1$, and $1 < p \leq q < \infty$, then there is $C > 0$ such that we have

$$\| |x|^{-\beta} I_\lambda f \|_{L^q(\mathbb{R}^N)} \leq C \| |x|^{\alpha} f \|_{L^p(\mathbb{R}^N)}. \tag{5.90}$$

2. An extension of the Hardy–Littlewood–Sobolev inequality to the Heisenberg groups was considered in [FS74]. The sharp constants in the Hardy–Littlewood–Sobolev inequality in the cases of $\mathbb{R}^N$ and the Heisenberg group were obtained in [Lie83] and [FL12], respectively.
3. In [GMS10] the analogues of the Stein–Weiss inequality were obtained on Carnot groups. Note that in [HLZ12] the authors also proved an analogue of the Stein–Weiss inequality on the Heisenberg groups.
4. On general homogeneous groups the statement of Theorem 5.3.1 will be here obtained as a consequence of the integral Hardy inequalities. The estimate (5.87) contains the Hardy–Littlewood–Sobolev inequality and half of the Stein–Weiss inequality. The full Stein–Weiss inequality on homogeneous groups was obtained in [KRS18b]: *Let*

$$I_\lambda u(x) := \int_{\mathbb{G}} \frac{u(y)}{|y^{-1}x|^\lambda} dy, \quad 0 < \lambda < Q. \tag{5.91}$$

Let $0 < \lambda < Q$, $1 < p < \infty$, $\alpha < \frac{Q}{p'}$, $\beta < \frac{Q}{q}$, $\alpha + \beta \geq 0$, $\frac{1}{q} = \frac{1}{p} + \frac{\alpha+\beta+\lambda}{Q} - 1$, *where* $\frac{1}{p} + \frac{1}{p'} = 1$ *and* $\frac{1}{q} + \frac{1}{q'} = 1$. *Then for* $1 < p \leq q < \infty$ *we have*

$$\| |x|^{-\beta} I_\lambda u \|_{L^q(\mathbb{G})} \leq C \| |x|^{\alpha} u \|_{L^p(\mathbb{G})}. \tag{5.92}$$

5. **(Differential Stein–Weiss inequality on graded groups)**. Continuing with the notation of Remark 5.2.2, let $\dot{L}^p_a(\mathbb{G})$ be the homogeneous Sobolev space over L^p of order a associated to a Rockland operator. Such spaces are well defined on graded groups and do not depend on a particular choice of a Rockland operator, we refer to [FR17] or to [FR16, Section 4.4] for the extensive analysis and exposition of their properties. The following differential version of the Stein–Weiss inequality was obtained in [RY18a]:

   Theorem 5.3.3 (Differential Stein–Weiss inequality). *Let* $\mathbb{G}$ *be a graded group of homogeneous dimension* Q *and let* $|\cdot|$ *be a quasi-norm on* $\mathbb{G}$. *Let* $1 < p, q < \infty$, $0 \leq a < Q/p$ *and* $0 \leq b < Q/q$. *Let* $0 < \lambda < Q$, $0 \leq \alpha < a + Q/p'$ *and* $0 \leq \beta \leq b$ *be such that* $(Q - ap)/(pQ) + (Q - q(b - \beta))/(qQ) + (\alpha + \lambda)/Q = 2$ *and* $\alpha + \lambda \leq Q$, *where* $1/p + 1/p' = 1$. *Then there exists a positive constant* $C = C(Q, \lambda, p, \alpha, \beta, a, b)$ *such that*

$$\left| \int_{\mathbb{G}} \int_{\mathbb{G}} \frac{\overline{f(x)} g(y)}{|x|^\alpha |y^{-1}x|^\lambda |y|^\beta} dx dy \right| \leq C \|f\|_{\dot{L}^p_a(\mathbb{G})} \|g\|_{\dot{L}^q_b(\mathbb{G})} \tag{5.93}$$

   holds for all $f \in \dot{L}^p_a(\mathbb{G})$ *and* $g \in \dot{L}^q_b(\mathbb{G})$.

   While the setting of graded groups falls outside the scope of this book, we follow [RY18a] in the proof of Theorem 5.3.1 below.

Proof of Theorem 5.3.1. Let $T_a(x) := |x|^{a-Q}$ with $0 < a < Q/r$ for some $1 < r < \infty$. Then, using Hölder's inequality we have

$$\begin{aligned}\left|\int_{\mathbb{G}}\int_{\mathbb{G}} \frac{\overline{f(x)}g(y)}{|x|^{\alpha}|y^{-1}x|^{\lambda}}dxdy\right| &= \left|\int_{\mathbb{G}} \overline{f(x)}\frac{(g*T_{Q-\lambda})(x)}{|x|^{\alpha}}dx\right| \\ &\leq \|f\|_{L^p(\mathbb{G})}\left\|\frac{g*T_{Q-\lambda}}{|x|^{\alpha}}\right\|_{L^{p'}(\mathbb{G})}.\end{aligned} \tag{5.94}$$

Note that the conditions $\alpha+\lambda \leq Q$ and $1/p+1/q+(\alpha+\lambda)/Q = 2$ imply $q \leq p'$, while $0<\lambda<Q$, $\alpha < Q/p'$ and $1/p+1/q+(\alpha+\lambda)/Q = 2$ give

$$0 < Q-\lambda = Q - Q\left(2-\frac{1}{p}-\frac{1}{q}\right)+\alpha < Q - Q\left(2-\frac{1}{p}-\frac{1}{q}\right)+\frac{Q}{p'} = Q/q.$$

Since we have $1 < q \leq p' < \infty$, $0 \leq \alpha p' < Q$, $0 < Q-\lambda < Q/q$ and $(Q-\lambda)/Q = 1/q - 1/p' + \alpha/Q$, using Theorem 5.2.1 in (5.94) we obtain (5.87). □

Remark 5.3.4 (Reversed Hardy–Littlewood–Sobolev inequality)**.** Let us make some remarks concerning the reversed Hardy–Littlewood–Sobolev inequality on homogeneous groups. Namely, consider the inequality

$$\int_{\mathbb{G}}\int_{\mathbb{G}} f(x)|y^{-1}x|^{\lambda}f(y)dxdy \geq C_{Q,\lambda,p}\|f\|^{\theta}_{L^1(\mathbb{G})}\|f\|^{2-\theta}_{L^p(\mathbb{G})} \tag{5.95}$$

for any $0 \leq f \in L^1 \cap L^p(\mathbb{G})$ with $f \not\equiv 0$ and $0 < p < 1$, where $\lambda > 0$ and $\theta := (2Q - p(2Q+\lambda))/(Q(1-p))$.

In the Euclidean case $\mathbb{G} = (\mathbb{R}^n, +)$, i.e., with $Q = n$, the case $p = 2n/(2n+\lambda)$ was investigated in [DZ15] and [NN17], while the case $p > n/(n+\lambda)$ was studied in [DFH18].

Following [RY18a], let us briefly recapture the argument in the setting of general homogeneous groups. Namely, let us show that in the case $0 < p \leq Q/(Q+\lambda)$ the inequality (5.95) is not valid, namely, (5.95) fails for any $C_{Q,\lambda,p} > 0$. In the Euclidean case this was shown in [CDP18] when $p < n/(n+\lambda)$ and in [DFH18] when $p \leq n/(n+\lambda)$.

Let f be a non-negative function with compact support, and let h be a non-negative smooth function such that $\int_{\mathbb{G}} h(x)dx = 1$. Then, for some $A > 0$, consider the function

$$f_\varepsilon(x) := f(x) + A\varepsilon^{-Q}h(x/\varepsilon).$$

Suppose now that (5.95) holds for some $C_{Q,\lambda,p} > 0$. Putting this f_ε in the inequality (5.95), we obtain

$$\begin{aligned}C_{Q,\lambda,p} &\leq \frac{\int_{\mathbb{G}}\int_{\mathbb{G}} f_\varepsilon(x)|y^{-1}x|^{\lambda}f_\varepsilon(y)dxdy}{\|f_\varepsilon\|^{\theta}_{L^1(\mathbb{G})}\|f_\varepsilon\|^{2-\theta}_{L^p(\mathbb{G})}} \\ &\to \frac{\int_{\mathbb{G}}\int_{\mathbb{G}} f(x)|y^{-1}x|^{\lambda}f(y)dxdy + 2A\int_{\mathbb{G}}|x|^{\lambda}f(x)dx}{(\int_{\mathbb{G}} f(x)dx + A)^{\theta}(\int_{\mathbb{G}}(f(x))^p dx)^{(2-\theta)/p}}\end{aligned} \tag{5.96}$$

as $\varepsilon \to 0_+$, where we have used that

- $\int_{\mathbb{G}} f_\varepsilon(x)dx = \int_{\mathbb{G}} f(x)dx + A$;
- when $\varepsilon \to 0_+$, we have
$$\int_{\mathbb{G}} (f_\varepsilon(x))^p dx \to \int_{\mathbb{G}} (f(x))^p dx;$$
- when $\varepsilon \to 0_+$, we have
$$\begin{aligned}
&\int_{\mathbb{G}}\int_{\mathbb{G}} f_\varepsilon(x)|y^{-1}x|^\lambda f_\varepsilon(y)dxdy \\
&= \int_{\mathbb{G}}\int_{\mathbb{G}} f(x)|y^{-1}x|^\lambda f(y)dxdy + 2A\int_{\mathbb{G}}\int_{\mathbb{G}} f(x)|(\varepsilon^{-1}y)^{-1}x|^\lambda h(y)dxdy \\
&+ A^2\varepsilon^{-2Q}\int_{\mathbb{G}}\int_{\mathbb{G}} h\left(\frac{x}{\varepsilon}\right) h\left(\frac{y}{\varepsilon}\right) dxdy \\
&\to \int_{\mathbb{G}}\int_{\mathbb{G}} f(x)|y^{-1}x|^\lambda f(y)dxdy + 2A\int_{\mathbb{G}} |x|^\lambda f(x)dx,
\end{aligned}$$
since $\int_{\mathbb{G}} h(x)dx = 1$.

Note that in (5.96) we can take also the limit as $A \to +\infty$ since it is valid for all $A > 0$. Then, when $\theta > 1$, that is, for $p < Q/(Q+\lambda)$, taking $A \to +\infty$ in (5.96) we see that $C_{Q,\lambda,p} = 0$. In the case $\theta = 1$, that is, for $p = Q/(Q+\lambda)$, taking the limit as $A \to +\infty$ in (5.96) we get

$$C_{Q,\lambda,p} \le \frac{2\int_{\mathbb{G}} |x|^\lambda f(x)dx}{\left(\int_{\mathbb{G}} (f(x))^p dx\right)^{1/p}}. \tag{5.97}$$

Finally, we show that the right-hand side of (5.97) goes to zero as $R \to \infty$ if we insert the function

$$f_R(x) = \begin{cases} |x|^{-(Q+\lambda)}, & \text{for } 1 \le |x| \le R, \\ 0, & \text{otherwise,} \end{cases} \tag{5.98}$$

for any $R > 1$. Indeed, in this case $p = Q/(Q+\lambda)$, and from (5.97) we obtain that

$$C_{Q,\lambda,p} \le \frac{2\int_{\mathbb{G}} |x|^\lambda f_R(x)dx}{\left(\int_{\mathbb{G}} (f_R(x))^p dx\right)^{1/p}} = 2(|\wp| \log R)^{-\lambda/Q} \to 0 \tag{5.99}$$

as $R \to \infty$, where $|\wp|$ is a $Q-1$-dimensional surface measure of the unit quasi-sphere in $\mathbb{G}$.

Summarizing, we conclude that for $0 < p \le Q/(Q+\lambda)$ the reversed Hardy–Littlewood–Sobolev inequality (5.95) is not valid with any constant $C_{Q,\lambda,p} > 0$.

5.4 Maximal weighted integral Hardy inequality

Here we present a maximal Hardy inequality in the integral form, involving the maximal function

$$(\mathcal{M}f)(x) := \frac{1}{|B(0,|x|)|}\int_{B(0,|x|)} f(z)dz.$$

Theorem 5.4.1 (Maximal integral weighted Hardy inequality). *Let $\mathbb{G}$ be a homogeneous group of homogeneous dimension Q with a homogeneous quasi-norm $|\cdot|$. Let ϕ and ψ be positive functions defined on $\mathbb{G}$. Then there exists a constant $C>0$ such that*

$$\int_{\mathbb{G}} \phi(x)\exp(\mathcal{M}\log f)(x)dx \leq C\int_{\mathbb{G}} \psi(x)f(x)dx \tag{5.100}$$

holds for all positive $f \geq 0$ if and only if

$$A := \sup_{R>0} R^Q \int_{|x|\geq R} \frac{\phi(x)\exp\left(\mathcal{M}\log\frac{1}{\psi}\right)(x)}{|x|^{2Q}}dx < \infty. \tag{5.101}$$

Remark 5.4.2. Inequalities of the type of those in Theorem 5.4.1 in the Abelian case $\mathbb{G} = (\mathbb{R}^n, +)$ were studied in [HKK01] for the one-dimensional case $n = 1$, and in [DHK97] for the multidimensional case $n \geq 1$. Theorem 5.4.1 was proved in [RSY18a] and we follow the presentation there.

Proof of Theorem 5.4.1. Let us first show (5.101) implies (5.100) for all $f \geq 0$. Denoting

$$W_3(x) := \phi(x)\exp\left(\mathcal{M}\log\frac{1}{\psi}\right)(x), \tag{5.102}$$

as well as $u(x) := f(x)\psi(x)$, and $z = |x|\xi$, we have

$$\begin{aligned}
&\int_{\mathbb{G}} \phi(x)\exp(\mathcal{M}\log f)(x)dx \\
&= \int_{\mathbb{G}} \phi(x)\exp\left(\frac{1}{|B(0,|x|)|}\left(\int_{B(0,|x|)} \log\left(\frac{1}{\phi}\right)(z)dz + \int_{|z|\leq|x|} \log(\phi f)(z)dz\right)\right)dx \\
&= \int_{\mathbb{G}} \phi(x)\exp\left(\mathcal{M}\log\frac{1}{\phi}\right)(x)\exp\left(\frac{1}{|B(0,|x|)|}\int_{|z|\leq|x|}\log(\phi(z)f(z))dz\right)dx \\
&= \int_{\mathbb{G}} W_3(x)\exp\left(\frac{1}{|B(0,|x|)|}\int_{|z|\leq|x|}\log(u(z))dz\right)dx \\
&= \int_{\mathbb{G}} W_3(x)\exp\left(\frac{1}{|x|^Q|B(0,1)|}\int_{B(0,1)}\log(u(|x|\xi))|x|^Q d\xi\right)dx.
\end{aligned} \tag{5.103}$$

Since

$$\int_{B(0,1)} \log(|\xi|^Q)d\xi = Q\int_{\wp}\int_0^1 r^{Q-1}\log r dr d\sigma(y) = -|B(0,1)|,$$

and by using Jensen's inequality, we obtain

$$
\begin{aligned}
&\int_{\mathbb{G}} \phi(x)\exp(\mathcal{M}\log f)(x)dx \\
&= \int_{\mathbb{G}} W_3(x)\exp\left(\frac{1}{|B(0,1)|}\left(\int_{B(0,1)}\log(|\xi|^Q u(|x|\xi))d\xi - \int_{B(0,1)}\log(|\xi|^Q)d\xi\right)\right)dx \\
&= \int_{\mathbb{G}} W_3(x)\exp\left(\frac{1}{|B(0,1)|}\int_{B(0,1)}\log(|\xi|^Q u(|x|\xi))d\xi + 1\right)dx \\
&= e\int_{\mathbb{G}} W_3(x)\exp\left(\frac{1}{|B(0,1)|}\int_{B(0,1)}\log(|\xi|^Q u(|x|\xi))d\xi\right)dx \\
&\le \frac{e}{|B(0,1)|}\int_{\mathbb{G}} W_3(x)\int_{B(0,1)}|\xi|^Q u(|x|\xi)d\xi dx \\
&= \frac{e}{|B(0,1)|}\int_{\wp}\int_{\wp}\int_0^\infty r^{Q-1}W_3(rw)\int_0^1 s^{2Q-1}u(rsy)dsdrd\sigma(y)d\sigma(w),
\end{aligned}
$$

where $|\xi| = s$ and $|x| = r$. Furthermore, with $t = rs$ we get

$$
\begin{aligned}
&\int_{\mathbb{G}} \phi(x)\exp(\mathcal{M}\log f)(x)dx \\
&\le \frac{e}{|B(0,1)|}\int_{\wp}\int_{\wp}\int_0^\infty r^{Q-1}W_3(rw)\int_0^1 s^{2Q-1}u(rsy)dsdrd\sigma(y)d\sigma(w) \\
&= \frac{e}{|B(0,1)|}\int_{\wp}\int_{\wp}\int_0^1 s^{2Q-1}\int_0^\infty W_3\left(\frac{t}{s}w\right)\left(\frac{t}{s}\right)^{Q-1}u(ty)\frac{dt}{s}dsd\sigma(y)d\sigma(w) \\
&= \frac{e}{|B(0,1)|}\int_{\wp}\int_{\wp}\int_0^1 s^{Q-1}\int_0^\infty W_3\left(\frac{t}{s}w\right)t^{Q-1}u(ty)dtdsd\sigma(y)d\sigma(w) \\
&= \frac{e}{|B(0,1)|}\int_{\wp}\int_{\wp}\int_0^\infty t^{Q-1}u(ty)\left(\int_0^1 s^{Q-1}W_3\left(\frac{tw}{s}\right)ds\right)dtd\sigma(y)d\sigma(w) \\
&= \frac{e}{|B(0,1)|}\int_{\wp}\int_{\wp}\int_0^\infty t^{Q-1}u(ty)\left(\int_t^\infty \left(\frac{t}{r}\right)^{Q-1}W_3(rw)t\frac{dr}{r^2}\right)dtd\sigma(y)d\sigma(w) \\
&= \frac{e}{|B(0,1)|}\int_{\wp}\int_{\wp}\int_0^\infty t^{2Q-1}u(ty)\left(\int_t^\infty r^{-Q-1}W_3(rw)dr\right)dtd\sigma(y)d\sigma(w) \\
&\le \frac{e}{|B(0,1)|}\int_{\wp}\int_{\wp}\int_0^\infty t^{Q-1}u(ty)t^Q\left(\int_t^\infty \frac{r^{Q-1}W_3(rw)dr}{r^{2Q}}\right)dtd\sigma(w)d\sigma(y) \\
&= \frac{e}{|B(0,1)|}\int_{\mathbb{G}} u(x)\left(|x|^Q\int_{|z|\ge|x|}\frac{W_3(z)}{|z|^{2Q}}dz\right)dx \le A\frac{e}{|B(0,1)|}\int_{\mathbb{G}} u(x)dx,
\end{aligned}
$$

yielding (5.100), where we have used (5.101) in the last line.

We now show that (5.100) implies (5.101). From (5.103) we notice that (5.100) is equivalent to

$$\int_{\mathbb{G}} W_3(x)\exp\left(\frac{1}{|B(0,|x|)|}\int_{|z|\le|x|}\log(u(z))dz\right)dx \le C\int_{\mathbb{G}} u(x)dx. \tag{5.104}$$

Furthermore, for a function

$$u(x) = R^{-Q}\chi_{(0,R)}(|x|) + e^{-2Q}|x|^{-2Q}R^{Q}\chi_{(R,\infty)}(|x|),\ x\in\mathbb{G},\ R>0, \tag{5.105}$$

we have

$$\begin{aligned}
&\int_{\mathbb{G}} W_3(x)\exp\left(\frac{1}{|B(0,|x|)|}\int_{|z|\le|x|}\log(u(z))dz\right)dx\\
&\quad\le C\int_{\mathbb{G}} u(x)dx = C\int_{\wp}\int_0^\infty s^{Q-1}u(s)dsd\sigma(y)\\
&\quad= C|\wp|\left(\int_0^R s^{Q-1}R^{-Q}ds + \int_R^\infty e^{-2Q}s^{Q-1}R^{Q}s^{-2Q}ds\right)\\
&\quad= C|\wp|\left(\frac{1}{Q}+\frac{e^{-2Q}}{Q}\right) =: C(Q) < \infty,
\end{aligned}$$

since χ is the cut-off function. Thus, from this, by plugging (5.105) into the left-hand side of (5.104) we calculate

$$\begin{aligned}
\infty > C(Q) &\ge \int_{\mathbb{G}} W_3(x)\exp\left(\frac{1}{|B(0,|x|)|}\int_{|z|\le|x|}\log(u(z))dz\right)dx\\
&= \int_{\wp}\int_0^\infty s^{Q-1}W_3(sy)\exp\left(\frac{1}{|B(0,s)|}\int_{\wp}\int_0^s r^{Q-1}\log(u(r))drd\sigma(w)\right)dsd\sigma(y)\\
&= \int_{\wp}\int_0^\infty s^{Q-1}W_3(sy)\exp\left(\frac{|\wp|}{s^Q|B(0,1)|}\int_0^s r^{Q-1}\log(u(r))dr\right)dsd\sigma(y)\\
&= \int_{\wp}\left(\int_0^R s^{Q-1}W_3(sy)\exp\left(\frac{|\wp|}{s^Q|B(0,1)|}\int_0^s r^{Q-1}\log(u(r))dr\right)ds\right)d\sigma(y)\\
&\quad+\int_{\wp}\left(\int_R^\infty s^{Q-1}W_3(sy)\exp\left(\frac{|\wp|}{s^Q|B(0,1)|}\left(\int_0^R r^{Q-1}\log(R^{-Q})dr\right.\right.\right.\\
&\quad\quad\left.\left.\left.+\int_R^s r^{Q-1}\log(e^{-2Q}r^{-2Q}R^{Q})dr\right)\right)ds\right)d\sigma(y)\\
&\ge \int_{\wp}\left(\int_R^\infty s^{Q-1}W_3(sy)\exp\left(\frac{|\wp|}{s^Q|B(0,1)|}\left(\int_0^R r^{Q-1}\log(R^{-Q})dr\right.\right.\right.\\
&\quad\left.\left.\left.+\int_R^s r^{Q-1}\log(e^{-2Q}r^{-2Q}R^{Q})dr\right)\right)ds\right)d\sigma(y)
\end{aligned}$$

$$
\begin{aligned}
&= \int_{\wp} \left(\int_R^\infty s^{Q-1} W_3(sy) \exp\left(\frac{|\wp|}{s^Q |B(0,1)|} \left(\int_0^R r^{Q-1} \log(R^{-Q}) dr \right.\right.\right. \\
&\quad + \int_R^s r^{Q-1} \log(e^{-2Q}) dr - 2Q \int_R^s r^{Q-1} (\log r) dr \\
&\quad \left.\left.\left. + \int_R^s r^{Q-1} \log(R^Q) dr \right)\right) ds \right) d\sigma(y) \\
&= \int_{\wp} \left(\int_R^\infty s^{Q-1} W_3(sy) \exp\left(\frac{|\wp|}{s^Q |B(0,1)|} \left(\frac{R^Q \log(R^{-Q})}{Q} - 2Q \frac{s^Q - R^Q}{Q} \right.\right.\right. \\
&\quad \left.\left.\left. -2Q \left[\frac{r^Q \log r}{Q} - \frac{r^Q}{Q^2} \right]_R^s + \frac{s^Q - R^Q}{Q} \log(R^Q) \right)\right) ds \right) d\sigma(y) \\
&= \int_{\wp} \int_R^\infty s^{Q-1} W_3(sy) \\
&\qquad \exp\left(\frac{|\wp|}{|B(0,1)|} \left(-2 + \frac{2R^Q}{s^Q} - 2 \log s + \frac{2}{Q} - \frac{2R^Q}{Qs^Q} + \log R \right)\right) ds d\sigma(y) \\
&\geq e^{(2-2Q)} \int_{\wp} \int_R^\infty s^{Q-1} W_3(sy) \frac{R^{\frac{|\wp|}{|B(0,1)|}}}{s^{\frac{2|\wp|}{|B(0,1)|}}} ds d\sigma(y) \\
&= e^{2-2Q} R^Q \int_{|x| \geq R} \frac{W_3(x)}{|x|^{2Q}} dx = e^{2-2Q} R^Q \int_{|x| \geq R} \frac{\phi(x) \exp\left(\mathcal{M} \log \frac{1}{\psi} \right)(x)}{|x|^{2Q}} dx,
\end{aligned}
$$

which implies (5.101), where we have used $\frac{|\wp|}{|B(0,1)|} = Q$, $\frac{2R^Q}{s^Q} - \frac{2R^Q}{Qs^Q} > 0$, and (5.102) in the last two lines. □

Chapter 6

Horizontal Inequalities on Stratified Groups

In this chapter we discuss versions of some of the inequalities from the previous chapters in the setting of stratified groups. Because of the stratified structure here we can use the horizontal gradient in the estimates.

As already outlined in the introduction there are three versions of estimates on stratified groups available in the literature:

(A) Using the homogeneous semi-norm, sometimes called the $\mathcal{L}$-gauge, given by the appropriate power of the fundamental solution of the sub-Laplacian $\mathcal{L}$. Thus, if $d(x)$ is the $\mathcal{L}$-gauge, then $d(x)^{2-Q}$ is a constant multiple of Folland's [Fol75] fundamental solution of the sub-Laplacian $\mathcal{L}$, with Q being the homogeneous dimension of the stratified group $\mathbb{G}$; these will be discussed in Chapter 7.

(B) Using the Carnot–Carathéodory distance, i.e., the control distance associated to the sub-Laplacian.

(C) Using the Euclidean distance on the first stratum of the group.

The constants in the corresponding inequalities may depend on the quasi-norm that one is using. There is an extensive literature on Hardy type inequalities of stratified Lie groups, see, e.g., [DGP11], [GL90], [GK08], [Gri03], [JS11], [KS16], [KÖ13], [Lia13], [NZW01]). For example, in the case (A) the Hardy inequality takes the form

$$\left\|\frac{f}{d(x)}\right\|_{L^p(\mathbb{G})} \leq \frac{p}{Q-p}\|\nabla_H f\|_{L^p(\mathbb{G})}, \quad Q\geq 3,\ 1<p<Q, \tag{6.1}$$

where Q is the homogeneous dimension of the stratified group $\mathbb{G}$, ∇_H is the horizontal gradient, and $d(x)$ is the $\mathcal{L}$-gauge from (A). The analysis in the case (A) in terms of the fundamental solution of the sub-Laplacian will be the subject of Chapter 11. The results on Hardy and other inequalities for the case (B) are less extensive, mostly devoted to the case of the Heisenberg group.

In this chapter we concentrate on the case (C). Thus, here throughout we adopt the notations from Section 1.4.8 concerning the stratified groups. In this

M. Ruzhansky, D. Suragan, *Hardy Inequalities on Homogeneous Groups*, Progress in Mathematics 327, https://doi.org/10.1007/978-3-030-02895-4_7

case there are several additional properties available assisting the analysis, such as formulae (1.72) and (1.73), making certain calculations possible. It is also worth noting that generally, for the same types of inequalities, the optimal constants in case (C) are different than the optimal constants in case (A).

Some analysis of inequalities of the case (C) is known in the literature, see, e.g., [BT02a] and [D'A04b]. However, in this chapter we aim at developing an independent point of view based on the divergence relations (1.72) and (1.73).

Notation. As already mentioned, throughout this chapter we adopt the notations from Section 1.4.8 concerning the stratified groups. In particular, $\mathbb{G}$ will always be a stratified group of homogeneous dimension Q, with N being the dimension of the first stratum. Also, x' will denote the variables in the first stratum of $\mathbb{G}$, and ∇_H will denote the horizontal gradient. To simplify the notation, we denote simply by

$$|x'| = \sqrt{x_1'^2 + \cdots + x_N'^2}$$

the Euclidean norm on the first stratum of $\mathbb{G}$, which can be identified with $\mathbb{R}^N$.

6.1 Horizontal L^p-Caffarelli–Kohn–Nirenberg type inequalities

In this section we establish the horizontal version on stratified groups of the L^p-Caffarelli–Kohn–Nirenberg type inequalities from Section 3.3.1. In particular, this would imply the horizontal version, as in the case (C) above, of the L^p-Hardy inequality with the sharp constant:

$$\left\| \frac{f}{|x'|} \right\|_{L^p(\mathbb{G})} \leq \frac{p}{N-p} \|\nabla_H f\|_{L^p(\mathbb{G})}, \quad 1 < p < N. \tag{6.2}$$

We obtain it as a special case of the following more general inequality, see Remark 6.1.2, Part 3.

Theorem 6.1.1 (Horizontal L^p-Caffarelli–Kohn–Nirenberg inequalities). *For any $\alpha, \beta \in \mathbb{R}$ and every complex-valued function $f \in C_0^\infty(\mathbb{G}\backslash\{x' = 0\})$, we have*

$$\frac{|N-\gamma|}{p} \left\| \frac{f}{|x'|^{\frac{\gamma}{p}}} \right\|^p_{L^p(\mathbb{G})} \leq \left\| \frac{1}{|x'|^\alpha} \nabla_H f \right\|_{L^p(\mathbb{G})} \left\| \frac{f}{|x'|^{\frac{\beta}{p-1}}} \right\|^{p-1}_{L^p(\mathbb{G})}, \quad 1 < p < \infty, \tag{6.3}$$

where $\gamma = \alpha + \beta + 1$. If $\gamma \neq N$ then the constant $\frac{|N-\gamma|}{p}$ is sharp.

Before proving Theorem 6.1.1, let us point out some of its consequences.

Remark 6.1.2.

1. In the Abelian case $\mathbb{G} = (\mathbb{R}^n, +)$, we have $N = n$, $\nabla_H = \nabla = (\partial_{x_1}, \ldots, \partial_{x_n})$, so (6.3) gives the L^p-Caffarelli–Kohn–Nirenberg type inequality for $\mathbb{R}^n$ with the sharp constant:
$$\frac{|n-\gamma|}{p} \left\| \frac{f}{|x|_E^{\frac{\gamma}{p}}} \right\|_{L^p(\mathbb{R}^n)}^p \leq \left\| \frac{1}{|x|_E^{\alpha}} \nabla f \right\|_{L^p(\mathbb{R}^n)} \left\| \frac{f}{|x|_E^{\frac{\beta}{p-1}}} \right\|_{L^p(\mathbb{R}^n)}^{p-1}, \tag{6.4}$$
for all $f \in C_0^\infty(\mathbb{R}^n \backslash \{0\})$, and $|x|_E = \sqrt{x_1^2 + \cdots + x_n^2}$. In this case it becomes a special case of Theorem 3.3.3 because a particular (Euclidean) norm is used. In this case the inequality of this type has been analysed in, e.g., [Cos08] and [DJSJ13].
2. (Horizontal weighted L^p-Hardy inequality) In the case
$$\beta = \gamma \left(1 - \frac{1}{p}\right),$$
i.e., with $\beta = (\alpha+1)(p-1)$ and $\gamma = p(\alpha+1)$, inequality (6.3) implies the *horizontal weighted L^p-Hardy type inequality*
$$\frac{|N - p(\alpha+1)|}{p} \left\| \frac{f}{|x'|^{\alpha+1}} \right\|_{L^p(\mathbb{G})} \leq \left\| \frac{1}{|x'|^{\alpha}} \nabla_H f \right\|_{L^p(\mathbb{G})}, \quad 1 < p < \infty, \tag{6.5}$$
for any $f \in C_0^\infty(\mathbb{G} \backslash \{x' = 0\})$ and all $\alpha \in \mathbb{R}$, with sharp constant in (6.5) for $p(\alpha+1) \neq N$.
3. (Horizontal L^p-Hardy inequality) In particular, in the case of $\alpha = 0$, the inequality (6.5) implies the following stratified group version of *horizontal L^p-Hardy inequality* with the sharp constant:
$$\left\| \frac{f}{|x'|} \right\|_{L^p(\mathbb{G})} \leq \frac{p}{N-p} \|\nabla_H f\|_{L^p(\mathbb{G})}, \quad 1 < p < N. \tag{6.6}$$
Such a type of inequalities was also considered in [D'A04b], and in [Yen16] in the case of the Heisenberg group.

   In the case $p = 2$ this inequality can be in turn sharpened to the following inequality: If $N \geq 3$ and $\alpha \in \mathbb{R}$, then for all complex-valued functions $f \in C_0^\infty(\mathbb{G} \backslash \{x' = 0\})$ we have
$$\left\| \frac{f}{|x'|} \right\|_{L^2(\mathbb{G})} \leq \frac{2}{N-2} \left\| \frac{x' \cdot \nabla_H f}{|x'|} \right\|_{L^2(\mathbb{G})}, \tag{6.7}$$
where the constant $\frac{2}{N-2}$ is sharp. This refinement will be shown in Theorem 6.4.4 by using the factorization method.

4. Clearly, when $\mathbb{G} = (\mathbb{R}^n, +)$, $n \geq 3$, (6.6) implies the classical Hardy inequality for $\mathbb{R}^n$:
$$\left\| \frac{f}{|x|_E} \right\|_{L^p(\mathbb{R}^n)} \leq \frac{p}{n-p} \|\nabla f\|_{L^p(\mathbb{R}^n)},$$
for all $f \in C_0^\infty(\mathbb{R}^n \backslash \{0\})$, and $|x|_E = \sqrt{x_1^2 + \cdots + x_n^2}$.

Similar to Corollary 3.3.5, Theorem 6.1.1, and even the corresponding Hardy inequality, immediately implies a version of the Heisenberg–Pauli–Weyl uncertainty principle.

Corollary 6.1.3 (Horizontal Heisenberg–Pauli–Weyl uncertainty principle). *For all $f \in C_0^\infty(\mathbb{G}\backslash\{x' = 0\})$ we have*
$$\|f\|^2_{L^2(\mathbb{G})} \leq \frac{p}{N-p} \|\nabla_H f\|_{L^p(\mathbb{G})} \||x'|f\|_{L^{\frac{p}{p-1}}(\mathbb{G})}, \quad 1 < p < N. \tag{6.8}$$

Proof of Corollary 6.1.3. Using (6.6) the Hölder inequality we immediately obtain
$$\begin{aligned} \|f\|^2_{L^2(\mathbb{G})} &\leq \left\| \frac{1}{|x'|} f \right\|_{L^p(\mathbb{G})} \||x'|f\|_{L^{\frac{p}{p-1}}(\mathbb{G})} \\ &\leq \frac{p}{N-p} \|\nabla_H f\|_{L^p(\mathbb{G})} \||x'|f\|_{L^{\frac{p}{p-1}}(\mathbb{G})}, \quad 1 < p < N, \end{aligned} \tag{6.9}$$
giving (6.8). □

Remark 6.1.4.

1. In the Abelian case $\mathbb{G} = (\mathbb{R}^n, +)$, taking $N = n$, we get that (6.8) with $p = 2$ implies the classical uncertainty principle on $\mathbb{R}^n$, namely,
$$\left(\int_{\mathbb{R}^n} |f(x)|^2 dx \right)^2 \leq \left(\frac{2}{n-2} \right)^2 \int_{\mathbb{R}^n} |\nabla f(x)|^2 dx \int_{\mathbb{R}^n} |x|_E^2 |f(x)|^2 dx, \tag{6.10}$$
for all $f \in C_0^\infty(\mathbb{R}^n \backslash \{0\})$. This is the Heisenberg–Pauli–Weyl uncertainty principle on $\mathbb{R}^n$. We note that we can also obtain (6.10) already as a consequence of the radial Heisenberg–Pauli–Weyl uncertainty principle on homogeneous groups, see Remark 3.3.6, Part 1. However, since the proofs of Theorem 3.3.3 and Theorem 6.1.1 are different, they give two different proofs of (6.10).
2. We can point out some inequalities with sharp constants as special cases of (6.3). For example, for $\alpha p = \alpha + \beta + 1$ we get
$$\frac{|N - \alpha p|}{p} \left\| \frac{f}{|x'|^\alpha} \right\|^p_{L^p(\mathbb{G})} \leq \left\| \frac{\nabla_H f}{|x'|^\alpha} \right\|_{L^p(\mathbb{G})} \left\| |x'|^{\frac{1}{p-1} - \alpha} f \right\|^{p-1}_{L^p(\mathbb{G})}. \tag{6.11}$$
Also, if $0 = \alpha + \beta + 1$ and $\alpha = -p$, then
$$\frac{N}{p} \|f\|^p_{L^p(\mathbb{G})} \leq \||x'|^p \nabla_H f\|_{L^p(\mathbb{G})} \left\| \frac{f}{|x'|} \right\|^{p-1}_{L^p(\mathbb{G})}, \tag{6.12}$$
with constants in both of these inequalities being sharp.

Proof of Theorem 6.1.1. We may assume that $\gamma \neq N$ since for $\gamma = N$ the inequality (6.3) is trivial. By using the identity (1.73), the divergence theorem, and Schwarz' inequality, one calculates

$$\begin{aligned}\int_{\mathbb{G}} \frac{|f(x)|^p}{|x'|^\gamma} dx &= \frac{1}{N-\gamma} \int_{\mathbb{G}} |f(x)|^p \mathrm{div}_H \left(\frac{x'}{|x'|^\gamma}\right) dx \\ &= -\frac{1}{N-\gamma} \mathrm{Re} \int_{\mathbb{G}} p f(x) |f(x)|^{p-2} \frac{\overline{x' \cdot \nabla_H f}}{|x'|^\gamma} dx \\ &\leq \left|\frac{p}{N-\gamma}\right| \int_{\mathbb{G}} \frac{|f(x)|^{p-1}}{|x'|^\gamma} |x' \cdot \nabla_H f| \, dx \\ &\leq \left|\frac{p}{N-\gamma}\right| \int_{\mathbb{G}} \frac{|f(x)|^{p-1}}{|x'|^{\alpha+\beta}} |\nabla_H f(x)| \, dx \\ &\leq \left|\frac{p}{N-\gamma}\right| \left(\int_{\mathbb{G}} \frac{|\nabla_H f(x)|^p}{|x'|^{\alpha p}} dx\right)^{\frac{1}{p}} \left(\int_{\mathbb{G}} \frac{|f(x)|^p}{|x'|^{\frac{\beta p}{p-1}}} dx\right)^{\frac{p-1}{p}}.\end{aligned}$$

Here in the last line we used Hölder's inequality. This gives

$$\left|\frac{N-\gamma}{p}\right| \int_{\mathbb{G}} \frac{|f(x)|^p}{|x'|^\gamma} dx \leq \left(\int_{\mathbb{G}} \frac{|\nabla_H f(x)|^p}{|x'|^{\alpha p}} dx\right)^{\frac{1}{p}} \left(\int_{\mathbb{G}} \frac{|f(x)|^p}{|x'|^{\frac{\beta p}{p-1}}} dx\right)^{\frac{p-1}{p}},$$

proving (6.3). Let us now show the sharpness of the constant. For this, we look at the equality condition in Hölder's inequality. Let us consider the function

$$g(x) = \begin{cases} e^{-\frac{C}{\lambda}|x'|^\lambda}, & \lambda := \alpha - \frac{\beta}{p-1} + 1 \neq 0, \\ \frac{1}{|x'|^C}, & \alpha - \frac{\beta}{p-1} + 1 = 0, \end{cases}$$

where $C = \left|\frac{N-\gamma}{p}\right|$ and $\gamma \neq N$. Then it can be checked that

$$\left|\frac{p}{N-\gamma}\right|^p \frac{|\nabla_H g(x)|^p}{|x'|^{\alpha p}} = \frac{|g(x)|^p}{|x'|^{\frac{\beta p}{p-1}}}.$$

Finally, approximating this function by functions in $C_0^\infty(\mathbb{G}\backslash\{x'=0\})$ completes the proof. ∎

6.1.1 Badiale–Tarantello conjecture

The idea of the proof of Theorem 6.1.1 implies the following similar fact in $\mathbb{R}^n$ which we may split into two factors as $\mathbb{R}^n = \mathbb{R}^N \times \mathbb{R}^{n-N}$. The best constant in the Hardy inequality of the type of inequality (6.13) was conjectured by Badiale and Tarantello in [BT02a, Remark 2.3]. Although it was subsequently established in [SSW03], here we follow [RS17e] to present an independent proof of a more general result.

Proposition 6.1.5 (L^p-Caffarelli–Kohn–Nirenberg inequality for Euclidean decomposition). *Let $x = (x', x'') \in \mathbb{R}^N \times \mathbb{R}^{n-N}$, $1 \le N \le n$, and $\alpha, \beta \in \mathbb{R}$. Then for any $f \in C_0^\infty(\mathbb{R}^n \backslash \{x' = 0\})$, and all $1 < p < \infty$, we have*

$$\frac{|N-\gamma|}{p} \left\| \frac{f}{|x'|^{\frac{\gamma}{p}}} \right\|^p_{L^p(\mathbb{R}^n)} \le \left\| \frac{1}{|x'|^\alpha} \nabla f \right\|_{L^p(\mathbb{R}^n)} \left\| \frac{f}{|x'|^{\frac{\beta}{p-1}}} \right\|^{p-1}_{L^p(\mathbb{R}^n)}, \tag{6.13}$$

where $\gamma = \alpha + \beta + 1$ and $|x'|$ is the Euclidean norm on $\mathbb{R}^N$. If $\gamma \neq N$ then the constant $\frac{|N-\gamma|}{p}$ is sharp.

Proof of Proposition 6.1.5. The proof is a modification of the proof of Theorem 6.1.1. For $\gamma = N$ the inequality (6.13) is trivial, so let us assume $\gamma \neq N$. Thus, by using the identity

$$\operatorname{div}_N \frac{x'}{|x'|^\gamma} = \frac{N-\gamma}{|x'|^\gamma},$$

for all $\gamma \in \mathbb{R}$ and $x' \in \mathbb{R}^N$ with $|x'| \neq 0$, where div_N is the standard divergence on $\mathbb{R}^N$, and applying the divergence theorem and Schwarz' inequality one can calculate

$$\begin{aligned}
\int_{\mathbb{R}^n} \frac{|f(x)|^p}{|x'|^\gamma} dx &= \frac{1}{N-\gamma} \int_{\mathbb{R}^n} |f(x)|^p \operatorname{div}_N \left(\frac{x'}{|x'|^\gamma} \right) dx \\
&= -\frac{1}{N-\gamma} \operatorname{Re} \int_{\mathbb{R}^n} p f(x) |f(x)|^{p-2} \frac{\overline{x' \cdot \nabla_N f}}{|x'|^\gamma} dx \\
&\le \left| \frac{p}{N-\gamma} \right| \int_{\mathbb{R}^n} \frac{|f(x)|^{p-1}}{|x'|^\gamma} |x' \cdot \nabla_N f| \, dx \\
&= \left| \frac{p}{N-\gamma} \right| \int_{\mathbb{R}^n} \frac{|f(x)|^{p-1}}{|x'|^\gamma} |x'_0 \cdot \nabla f| \, dx \\
&\le \left| \frac{p}{N-\gamma} \right| \int_{\mathbb{R}^n} \frac{|f(x)|^{p-1}}{|x'|^{\alpha+\beta}} |\nabla f(x)| \, dx \\
&\le \left| \frac{p}{N-\gamma} \right| \left(\int_{\mathbb{R}^n} \frac{|\nabla f(x)|^p}{|x'|^{\alpha p}} dx \right)^{\frac{1}{p}} \left(\int_{\mathbb{R}^n} \frac{|f(x)|^p}{|x'|^{\frac{\beta p}{p-1}}} dx \right)^{\frac{p-1}{p}},
\end{aligned}$$

where $x'_0 = (x', 0) \in \mathbb{R}^n$, that is $|x'_0| = |x'|$, ∇_N is the standard gradient on $\mathbb{R}^N$, and ∇ is the gradient on $\mathbb{R}^n$. Here we have used Hölder's inequality in the last line. This gives

$$\left| \frac{N-\gamma}{p} \right| \int_{\mathbb{R}^n} \frac{|f(x)|^p}{|x'|^\gamma} dx \le \left(\int_{\mathbb{R}^n} \frac{|\nabla f(x)|^p}{|x'|^{\alpha p}} dx \right)^{\frac{1}{p}} \left(\int_{\mathbb{R}^n} \frac{|f(x)|^p}{|x'|^{\frac{\beta p}{p-1}}} dx \right)^{\frac{p-1}{p}},$$

which proves (6.13). Again as in the proof of Theorem 6.1.1 let us examine the equality condition in the above Hölder inequality. Thus, we consider

$$g(x) = \begin{cases} e^{-\frac{C}{\lambda}|x'|^{\lambda}}, & \lambda := \alpha - \frac{\beta}{p-1} + 1 \neq 0, \\ \frac{1}{|x'|^C}, & \alpha - \frac{\beta}{p-1} + 1 = 0, \end{cases}$$

where $C = \left|\frac{N-\gamma}{p}\right|$ and $\gamma \neq N$. Then it can be directly checked that

$$\left|\frac{p}{N-\gamma}\right|^p \frac{|\nabla g(x)|^p}{|x'|^{\alpha p}} = \left|\frac{p}{N-\gamma}\right|^p \frac{|\nabla_N g(x)|^p}{|x'|^{\alpha p}} = \frac{|g(x)|^p}{|x'|^{\frac{\beta p}{p-1}}},$$

which satisfies the equality condition in Hölder's inequality. Approximating this function by functions in $C_0^\infty(\mathbb{R}^n \backslash \{x' = 0\})$ shows that the constant $\left|\frac{N-\gamma}{p}\right|$ is sharp. □

Remark 6.1.6. For $\beta = (\alpha+1)(p-1)$ and $\gamma = p(\alpha+1)$ the inequality (6.13) gives that

$$\frac{|N - p(\alpha+1)|}{p} \left\| \frac{f}{|x'|^{\alpha+1}} \right\|_{L^p(\mathbb{R}^n)} \leq \left\| \frac{1}{|x'|^{\alpha}} \nabla f \right\|_{L^p(\mathbb{R}^n)}, \quad 1 < p < \infty, \tag{6.14}$$

for all $f \in C_0^\infty(\mathbb{R}^n \backslash \{x' = 0\})$ and for all $\alpha \in \mathbb{R}$, with the sharp constant. For $\alpha = 0$ and $1 < p < N$, $2 \leq N \leq n$, the inequality (6.14) implies that

$$\left\| \frac{f}{|x'|} \right\|_{L^p(\mathbb{R}^n)} \leq \frac{p}{N-p} \|\nabla f\|_{L^p(\mathbb{R}^n)}, \tag{6.15}$$

again with $\frac{p}{N-p}$ being the best constant.

6.1.2 Horizontal higher-order versions

We can iterate the L^p-Caffarelli–Kohn–Nirenberg type inequalities from Theorem 6.1.1 to obtain higher-order inequalities. Let us denote inductively

$$\nabla_H^2 f := \nabla_H |\nabla_H f| \text{ and } \nabla_H^m f := \nabla_H |\nabla_H^{m-1} f|, \quad m \in \mathbb{N}.$$

Then as a consequence of Theorem 6.1.1 we obtain

Corollary 6.1.7 (Higher-order horizontal L^p-Caffarelli–Kohn–Nirenberg type inequalities). *For any $k, m \in \mathbb{N}$ and $1 < p < \infty$ we have*

$$\frac{|N-\gamma|}{p} \left\| \frac{f}{|x'|^{\frac{\gamma}{p}}} \right\|_{L^p(\mathbb{G})}^p \leq \widetilde{A}_{\alpha,m} \widetilde{A}_{\beta,k} \left\| \frac{1}{|x'|^{\alpha-m}} \nabla_H^{m+1} f \right\|_{L^p(\mathbb{G})} \left\| \frac{1}{|x'|^{\frac{\beta}{p-1}-k}} \nabla_H^k f \right\|_{L^p(\mathbb{G})}^{p-1}, \tag{6.16}$$

for all $f \in C_0^\infty(\mathbb{G}\backslash\{x'=0\})$, *where* $\gamma = \alpha+\beta+1$, *and all* $\alpha \in \mathbb{R}$ *such that* $\prod_{j=0}^{m-1}|N-p(\alpha-j)| \neq 0$, *and*

$$\widetilde{A}_{\alpha,m} := p^m \left[\prod_{j=0}^{m-1}|N-p(\alpha-j)|\right]^{-1},$$

as well as all $\beta \in \mathbb{R}$ *such that* $\prod_{j=0}^{k-1}\left|N-p\left(\frac{\beta}{p-1}-j\right)\right| \neq 0$, *and*

$$\widetilde{A}_{\beta,k} := p^{k(p-1)} \left[\prod_{j=0}^{k-1}\left|N-p\left(\frac{\beta}{p-1}-j\right)\right|\right]^{-(p-1)}.$$

Proof of Corollary 6.1.7. Taking $|\nabla_H f|$ instead of f and $\alpha-1$ instead of α in (6.5) we consequently get

$$\left\|\frac{\nabla_H f}{|x'|^\alpha}\right\|_{L^p(\mathbb{G})} \leq \frac{p}{|N-p\alpha|}\left\|\frac{1}{|x'|^{\alpha-1}}\nabla_H^2 f\right\|_{L^p(\mathbb{G})},$$

for $\alpha \neq \frac{N}{p}$. Combining it with (6.5) we obtain

$$\left\|\frac{f}{|x'|^{\alpha+1}}\right\|_{L^p(\mathbb{G})} \leq \frac{p}{|N-p(\alpha+1)|}\frac{p}{|N-p\alpha|}\left\|\frac{1}{|x'|^{\alpha-1}}\nabla_H^2 f\right\|_{L^p(\mathbb{G})},$$

for each $\alpha \in \mathbb{R}$ such that $\alpha \neq \frac{N}{p}-1$ and $\alpha \neq \frac{N}{p}$. This iteration process gives

$$\left\|\frac{f}{|x'|^{\theta+1}}\right\|_{L^p(\mathbb{G})} \leq A_{\theta,k}\left\|\frac{1}{|x'|^{\theta+1-k}}\nabla_H^k f\right\|_{L^p(\mathbb{G})}, \quad 1<p<\infty, \tag{6.17}$$

for all $f \in C_0^\infty(\mathbb{G}\backslash\{x'=0\})$ and all $\theta \in \mathbb{R}$ such that $\prod_{j=0}^{k-1}|N-p(\theta+1-j)| \neq 0$, and

$$A_{\theta,k} := p^k \left[\prod_{j=0}^{k-1}|N-p(\theta+1-j)|\right]^{-1}.$$

Similarly, we have

$$\left\|\frac{\nabla_H f}{|x'|^{\vartheta+1}}\right\|_{L^p(\mathbb{G})} \leq A_{\vartheta,m}\left\|\frac{1}{|x'|^{\vartheta+1-m}}\nabla_H^{m+1} f\right\|_{L^p(\mathbb{G})}, \quad 1<p<\infty, \tag{6.18}$$

for all $f \in C_0^\infty(\mathbb{G}\backslash\{x'=0\})$ and all $\vartheta \in \mathbb{R}$ such that $\prod_{j=0}^{m-1}|N-p(\vartheta+1-j)| \neq 0$, and

$$A_{\vartheta,m} := p^m \left[\prod_{j=0}^{m-1}|N-p(\vartheta+1-j)|\right]^{-1}.$$

Now putting $\vartheta+1=\alpha$ and $\theta+1=\frac{\beta}{p-1}$ into (6.18) and (6.17), respectively, from (6.3) we obtain (6.16). □

6.2 Horizontal Hardy and Rellich inequalities

First of all let us record the horizontal Hardy inequalities discussed in Remark 6.1.2:

Corollary 6.2.1 (Horizontal L^p Hardy inequalities). *Let $\mathbb{G}$ be a stratified group with N being the dimension of the first stratum. Then for any $1 < p < \infty$, $\alpha \in \mathbb{R}$, and for all $f \in C_0^\infty(\mathbb{G}\backslash\{x' = 0\})$ we have*

$$\left\| \frac{1}{|x'|^\alpha} \nabla_H f \right\|_{L^p(\mathbb{G})} \geq \frac{|N - p(\alpha+1)|}{p} \left\| \frac{f}{|x'|^{\alpha+1}} \right\|_{L^p(\mathbb{G})}. \tag{6.19}$$

If $p(\alpha + 1) \neq N$ then the constant $\frac{|N-p(\alpha+1)|}{p}$ is sharp.

From this we move on to Rellich inequalities.

Theorem 6.2.2 (Horizontal Rellich inequalities). *Let $\mathbb{G}$ be a stratified group with $N \geq 3$ being the dimension of the first stratum. Let $\delta \in \mathbb{R}$ with $-N/2 \leq \delta \leq -1$. Then for all functions $f \in C_0^\infty(\mathbb{G}\backslash\{x' = 0\})$ we have*

$$\left\| \frac{\mathcal{L}f}{|x'|^\delta} \right\|_{L^2(\mathbb{G})} \geq \left| \frac{(N-2\delta-4)(N+2\delta)}{4} \right| \left\| \frac{f}{|x'|^{\delta+2}} \right\|_{L^2(\mathbb{G})}. \tag{6.20}$$

If $N + 2\delta \neq 0$, then the constant in (6.20) is sharp.

We note that another version of the Rellich inequality will be given in Corollary 6.5.2.

Proof of Theorem 6.2.2. In the case $p = 2$ and $\alpha = \delta + 1$, Corollary 6.2.1 implies

$$\left\| \frac{\nabla_H f}{|x'|^{\delta+1}} \right\|_{L^2(\mathbb{G})} \geq \left| \frac{N-2\delta-4}{2} \right| \left\| \frac{f}{|x'|^{\delta+2}} \right\|_{L^2(\mathbb{G})}. \tag{6.21}$$

It also follows that the constant $\left|\frac{N-2\delta-4}{2}\right|$ is sharp when $N - 2\delta - 4 \neq 0$. On the other hand, Corollary 6.5.2 gives (see also Theorem 6.8.1 with $p = 2$, $\gamma = 2\beta$ and $\alpha = \beta - 1$)

$$\left\| \frac{\mathcal{L}f}{|x'|^{\beta-1}} \right\|_{L^2(\mathbb{G})} \geq \left| \frac{N+2\beta-2}{2} \right| \left\| \frac{\nabla_H f}{|x'|^{\beta}} \right\|_{L^2(\mathbb{G})} \tag{6.22}$$

for $2 - N \leq 2\beta \leq 0$ and $N \geq 3$.

Putting $\delta + 1$ instead of β this gives

$$\left\| \frac{\mathcal{L}f}{|x'|^{\delta}} \right\|_{L^2(\mathbb{G})} \geq \left| \frac{N+2\delta}{2} \right| \left\| \frac{\nabla_H f}{|x'|^{\delta+1}} \right\|_{L^2(\mathbb{G})} \tag{6.23}$$

for $2 - N \leq 2\delta + 2 \leq 0$ and $N \geq 3$. Combining (6.21) and (6.23), we obtain (6.20).

Now, to show the sharpness of the constant in (6.20), we observe first that in Theorem 6.2.2 the sharpness of the constant is reduced to that in Theorem

6.1.1 which in turn is obtained by checking the equality condition in Hölder's inequality. Namely, the function $|x'|^{C_1}$ satisfies this equality condition for any real number $C_1 \neq 0$. Similarly, in Theorem 6.8.1, the sharpness of the constant will be obtained again from the equality condition in Hölder's inequality, so that we see that the same function $|x'|^{C_1}$ satisfies the equality condition. Therefore, the constant in (6.23) is sharp when $N + 2\delta \neq 0$, so, the constant in (6.20) is sharp for $N + 2\delta \neq 0$. □

6.3 Critical horizontal Hardy type inequality

For $p = N$ the inequality (6.2) fails, and in this section we consider its critical versions.

Theorem 6.3.1 (Critical horizontal Hardy inequality). *For a bounded domain $\Omega \subset \mathbb{G}$ with $0 \in \Omega$ and for all $f \in C_0^\infty(\Omega\backslash\{x' = 0\})$ we have*

$$\left\| \frac{f}{|x'| \log \frac{R}{|x'|}} \right\|_{L^N(\Omega)} \leq \frac{N}{N-1} \left\| \frac{x'}{|x'|} \cdot \nabla_H f \right\|_{L^N(\Omega)}, \quad 1 < N < \infty, \tag{6.24}$$

where $R = \sup_{x\in\Omega} |x'|$.

To show Theorem 6.3.1 we will first prove the following more abstract theorem, and then the proof of Theorem 6.3.1 will follows directly from this. Moreover, it will imply a number of other estimates, for example the critical L^N-Poincaré inequality.

Theorem 6.3.2. *Let $0 \in \Omega \subset \mathbb{G}$ be a bounded domain. Let $g : (1, \infty) \to \mathbb{R}$ be a C^2-function such that*

$$g'(t) < 0, \quad g''(t) > 0, \tag{6.25}$$

for all $t > 1$, and such that

$$\frac{(-g'(t))^{2(N-1)}}{(g''(t))^{N-1}} \leq C < \infty, \quad \text{for all } t > 1. \tag{6.26}$$

Then we have

$$\begin{aligned} \left(\frac{N-1}{N}\right)^N \int_\Omega \frac{|f(x)|^N}{|x'|^N} \left(-g'\left(\log\frac{Re}{|x'|}\right)\right)^{N-2} g''\left(\log\frac{Re}{|x'|}\right) dx \\ \leq \int_\Omega \frac{\left(-g'\left(\log\frac{Re}{|x'|}\right)\right)^{2(N-1)}}{\left(g''\left(\log\frac{Re}{|x'|}\right)\right)^{N-1}} \left|\frac{x'}{|x'|} \cdot \nabla_H f(x)\right|^N dx, \end{aligned} \tag{6.27}$$

for all $f \in C_0^\infty(\Omega\backslash\{x' = 0\})$, with $R = \sup_{x\in\Omega} |x'|$.

Proof of Theorem 6.3.2. For $\epsilon > 0$ a direct calculation shows

$$|\nabla_H G_\epsilon(x)|^{N-2}\nabla_H G_\epsilon(x) = (-g'(F_\epsilon(x)))^{N-1}\left(\frac{|x'|^{N-2}x'}{(|x'|^2+\epsilon^2)^{N-1}}\right),$$

where

$$F_\epsilon(x) := \log\frac{R_\epsilon e}{\sqrt{|x'|^2+\epsilon^2}}, \quad R_\epsilon := \sup_{x\in\Omega}\sqrt{|x'|^2+2\epsilon^2},$$

and

$$G_\epsilon(x) = g(F_\epsilon(x)).$$

Since $g'(t) < 0$, with $\mathcal{L}_N$ as in (1.71), we have

$$\begin{aligned}\mathcal{L}_N G_\epsilon(x) &= \operatorname{div}_H(|\nabla_H G_\epsilon(x)|^{N-2}\nabla_H G_\epsilon(x)) \\ &= (N-1)(-g'(F_\epsilon(x)))^{N-2} g''(F_\epsilon(x))\frac{|x'|^N}{(|x'|^2+\epsilon^2)^N} \\ &\quad + (N-1)(-g'(F_\epsilon(x)))^{N-1}\frac{2\epsilon^2|x'|^{N-2}}{(|x'|^2+\epsilon^2)^N}.\end{aligned}$$

The divergence theorem gives

$$\begin{aligned}\int_\Omega |f|^N \mathcal{L}_N G_\epsilon(x)dx &= \int_\Omega |f|^N \operatorname{div}_H(|\nabla_H G_\epsilon(x)|^{N-2}\nabla_H G_\epsilon(x))dx \\ &= -\int_\Omega \nabla_H|f|^N\cdot(|\nabla_H G_\epsilon(x)|^{N-2}\nabla_H G_\epsilon(x))dx.\end{aligned} \tag{6.28}$$

We have

$$\begin{aligned}&\int_\Omega |f|^N \mathcal{L}_N G_\epsilon(x)dx \\ &= (N-1)\int_\Omega |f|^N (-g'(F_\epsilon(x)))^{N-2} g''(F_\epsilon(x))\frac{|x'|^N}{(|x'|^2+\epsilon^2)^N}dx \\ &\quad + (N-1)\int_\Omega |f|^N (-g'(F_\epsilon(x)))^{N-1}\frac{2\epsilon^2|x'|^{N-2}}{(|x'|^2+\epsilon^2)^N}dx \\ &\geq (N-1)\int_\Omega |f|^N (-g'(F_\epsilon(x)))^{N-2} g''(F_\epsilon(x))\frac{|x'|^N}{(|x'|^2+\epsilon^2)^N}dx.\end{aligned} \tag{6.29}$$

Moreover,

$$\begin{aligned}&\left|-\int_\Omega \nabla_H|f(x)|^N\cdot(|\nabla_H G_\epsilon(x)|^{N-2}\nabla_H G_\epsilon(x))dx\right| \\ &= \left|N\int_\Omega |f(x)|^{N-2}f(x)(-g'(F_\epsilon(x)))^{N-1}\left(\frac{|x'|^{N-2}x'\cdot\nabla_H f}{(|x'|^2+\epsilon^2)^{N-1}}\right)dx\right|\end{aligned}$$

$$
\begin{aligned}
&= N \int_{\Omega} |f(x)|^{N-1} \left(-g'(F_\epsilon(x))\right)^{N-1} \left(\frac{|x'|^{N-2}\,|x' \cdot \nabla_H f|}{(|x'|^2+\epsilon^2)^{N-1}} \right) dx \\
&\leq N \left(\int_{\Omega} \left(-g'(F_\epsilon(x))\right)^{N-2} g''(F_\epsilon(x)) \frac{|x'|^N |f(x)|^N}{(|x'|^2+\epsilon^2)^N} dx \right)^{\frac{N-1}{N}} \\
&\quad \times \left(\int_{\Omega} \left(-g'(F_\epsilon(x))\right)^{2(N-1)} \left(g''(F_\epsilon(x))\right)^{-(N-1)} \left| \frac{x'}{|x'|} \cdot \nabla_H f \right|^N dx \right)^{\frac{1}{N}}. \qquad (6.30)
\end{aligned}
$$

Combining (6.28), (6.29) and (6.30) we obtain

$$
\begin{aligned}
&(N-1) \int_{\Omega} |f|^N \left(-g'(F_\epsilon(x))\right)^{N-2} g''(F_\epsilon(x)) \frac{|x'|^N}{(|x'|^2+\epsilon^2)^N} dx \\
&\quad \leq N \left(\int_{\Omega} \left(-g'(F_\epsilon(x))\right)^{N-2} g''(F_\epsilon(x)) \frac{|x'|^N |f(x)|^N}{(|x'|^2+\epsilon^2)^N} dx \right)^{\frac{N-1}{N}} \\
&\quad \times \left(\int_{\Omega} \left(-g'(F_\epsilon(x))\right)^{2(N-1)} \left(g''(F_\epsilon(x))\right)^{-(N-1)} \left| \frac{x'}{|x'|} \cdot \nabla_H f \right|^N dx \right)^{\frac{1}{N}},
\end{aligned}
$$

which means

$$
\begin{aligned}
&\left(\frac{N-1}{N}\right)^N \int_{\Omega} |f|^N \left(-g'(F_\epsilon(x))\right)^{N-2} g''(F_\epsilon(x)) \frac{|x'|^N}{(|x'|^2+\epsilon^2)^N} dx \\
&\quad \leq \int_{\Omega} \left(-g'(F_\epsilon(x))\right)^{2(N-1)} \left(g''(F_\epsilon(x))\right)^{-(N-1)} \left| \frac{x'}{|x'|} \cdot \nabla_H f \right|^N dx.
\end{aligned}
$$

Now letting $\epsilon \to 0$ we obtain (6.27). □

Proof of Theorem 6.3.1. If we take

$$
g(t) = -\log(t-1),
$$

for $t > 1$, then we see that this function satisfies all assumptions of Theorem 6.3.2. That is,

$$
g'(t) = -\frac{1}{t-1} < 0, \quad g''(t) = \frac{1}{(t-1)^2} > 0,
$$

and

$$
\frac{(-g'(t))^{2(N-1)}}{(g''(t))^{N-1}} = 1, \quad \text{for all } \ t > 1.
$$

Therefore, putting

$$
g'\left(\log \frac{Re}{|x'|}\right) = -\frac{1}{\log \frac{R}{|x'|}} \quad \text{and} \quad g''\left(\log \frac{Re}{|x'|}\right) = \frac{1}{\left(\log \frac{R}{|x'|}\right)^2}
$$

in (6.27) we obtain (6.24). □

One can obtain a number of inequalities from Theorem 6.3.2 by choosing different functions $g(t)$. For example, we get the following analogue of the L^N-Poincaré inequality for the horizontal gradient.

Corollary 6.3.3 (Horizontal critical Poincaré inequality). *Let* $R := \sup_{x\in\Omega}|x'|$. *Then for all* $f \in C_0^\infty(\Omega\backslash\{x'=0\})$ *we have*

$$\|f\|_{L^N(\Omega)} \leq R\,\|\nabla_H f\|_{L^N(\Omega)}\,. \tag{6.31}$$

Proof. Let us take

$$g(t) = e^{\frac{Nt}{1-N}}, \quad t > 1,$$

in (6.27). Then we have

$$\|f\|_{L^N(\Omega)} \leq \left\| |x'| \frac{x'}{|x'|} \cdot \nabla_H f \right\|_{L^N(\Omega)}.$$

For $R = \sup_{x\in\Omega}|x'|$, the Cauchy–Schwarz inequality implies (6.31). □

Remark 6.3.4. In the Euclidean case the idea of proving Theorem 6.3.1 using Theorem 6.3.2 was realized in [Tak15]. In this section our presentation followed [RS17e].

6.4 Two-parameter Hardy–Rellich inequalities by factorization

In this section we apply the factorization method, similar to the ideas explained in Section 2.1.5, but now in the setting of stratified groups. As a result we obtain two-parameter inequalities analogous to the Gesztesy–Littlejohn type inequalities described in Example 2.1.13. The presentation of this section follows [RY17].

Theorem 6.4.1 (Two-parameter Hardy–Rellich inequalities). *Let* $\mathbb{G}$ *be a stratified group with* $N \geq 2$ *being the dimension of the first stratum, and let* $\alpha, \beta \in \mathbb{R}$. *Then for all complex-valued functions* $f \in C_0^\infty(\mathbb{G}\backslash\{x'=0\})$ *we have*

$$\begin{aligned}\|\mathcal{L}f\|^2_{L^2(\mathbb{G})} \geq{}& (\alpha(N-2)-2\beta)\left\|\frac{\nabla_H f}{|x'|}\right\|^2_{L^2(\mathbb{G})} \\ &- \alpha^2\left\|\frac{x'\cdot\nabla_H f}{|x'|^2}\right\|^2_{L^2(\mathbb{G})} + C_{N,\alpha,\beta}\left\|\frac{f}{|x'|^2}\right\|^2_{L^2(\mathbb{G})},\end{aligned} \tag{6.32}$$

where

$$C_{N,\alpha,\beta} = \alpha(N-4)(N-2) - \alpha^2(N-2) + 2\beta(4-N) - \beta^2 + \alpha\beta(N-2).$$

Remark 6.4.2.

1. Using the Cauchy–Schwarz inequality
$$\int_{\mathbb{G}} \frac{|x'\cdot(\nabla_H f)(x)|^2}{|x'|^4}dx \le \int_{\mathbb{G}} \frac{|(\nabla_H f)(x)|^2}{|x'|^2}dx,$$
inequality (6.32) implies the inequality
$$\|\mathcal{L}f\|^2_{L^2(\mathbb{G})} \ge (\alpha(N-2)-2\beta-\alpha^2)\left\|\frac{\nabla_H f}{|x'|}\right\|^2_{L^2(\mathbb{G})} + C_{N,\alpha,\beta}\left\|\frac{f}{|x'|^2}\right\|^2_{L^2(\mathbb{G})}. \tag{6.33}$$

2. In the Abelian case $\mathbb{G}=(\mathbb{R}^n,+)$, we have $N=n$, $\nabla_H=\nabla=(\partial_{x_1},\ldots,\partial_{x_n})$ is the usual (full) gradient, so (6.32) implies for $\alpha,\beta\in\mathbb{R}$ and for any $f\in C_0^\infty(\mathbb{R}^n\backslash\{0\})$ with $n\ge 2$ the inequality
$$\begin{aligned}\|\Delta f\|^2_{L^2(\mathbb{R}^n)} \ge (\alpha(n-2)-2\beta)&\left\|\frac{\nabla f}{|x|}\right\|^2_{L^2(\mathbb{R}^n)} \\ &- \alpha^2\left\|\frac{x\cdot\nabla f}{|x|^2}\right\|^2_{L^2(\mathbb{R}^n)} + C_{n,\alpha,\beta}\left\|\frac{f}{|x|^2}\right\|^2_{L^2(\mathbb{R}^n)}.\end{aligned} \tag{6.34}$$

   This can be also compared with inequality (2.34).

Proof of Theorem 6.4.1. For two parameters $\alpha,\beta\in\mathbb{R}$, let us define
$$T_{\alpha,\beta} := -\mathcal{L} + \alpha\frac{x'\cdot\nabla_H}{|x'|^2} + \frac{\beta}{|x'|^2}.$$
One can readily check that its formal adjoint is given by
$$T^+_{\alpha,\beta} := -\mathcal{L} - \alpha\frac{x'\cdot\nabla_H}{|x'|^2} - \frac{\alpha(N-2)-\beta}{|x'|^2},$$
for $x'\neq 0$. Then, by a direct calculation for any function $f\in C_0^\infty(\mathbb{G}\backslash\{x'=0\})$ we have
$$\begin{aligned}&(T^+_{\alpha,\beta}T_{\alpha,\beta}f)(x)\\ &= \left(-\mathcal{L} - \alpha\frac{x'\cdot\nabla_H}{|x'|^2} - \frac{\alpha(N-2)-\beta}{|x'|^2}\right)\left(-(\mathcal{L}f)(x) + \alpha\frac{x'\cdot(\nabla_H f)}{|x'|^2} + \frac{\beta f(x)}{|x'|^2}\right)\\ &= (\mathcal{L}^2 f)(x) + \alpha\left(-\mathcal{L}\left(\frac{x'\cdot(\nabla_H f)}{|x'|^2}\right)(x) + \frac{x'\cdot\nabla_H}{|x'|^2}(\mathcal{L}f)(x) + \frac{N-2}{|x'|^2}(\mathcal{L}f)(x)\right)\\ &\quad + \beta\left(-\mathcal{L}\left(\frac{f}{|x'|^2}\right)(x) - \frac{(\mathcal{L}f)(x)}{|x'|^2}\right)\\ &\quad + \alpha\beta\left(-\frac{x'\cdot\nabla_H}{|x'|^2}\left(\frac{f}{|x'|^2}\right)(x) + \frac{x'\cdot(\nabla_H f)(x)}{|x'|^4} - \frac{(N-2)f(x)}{|x'|^4}\right)\end{aligned}$$

$$+ \alpha^2 \left(- \left(\frac{x' \cdot \nabla_H}{|x'|^2} \right) \left(\frac{x' \cdot (\nabla_H f)}{|x'|^2} \right) (x) - (N-2) \frac{x' \cdot (\nabla_H f)(x)}{|x'|^4} \right) + \beta^2 \frac{f(x)}{|x'|^4}.$$

Now we calculate in the other direction,

$$\begin{aligned}
&(T_{\alpha,\beta} T^{+}_{\alpha,\beta} f)(x) \\
&= \left(-\mathcal{L} + \alpha \frac{x' \cdot \nabla_H}{|x'|^2} + \frac{\beta}{|x'|^2} \right) \\
&\quad \times \left(-(\mathcal{L}f)(x) - \alpha \frac{x' \cdot (\nabla_H f)(x)}{|x'|^2} - \frac{(\alpha(N-2) - \beta) f(x)}{|x'|^2} \right) \\
&= (\mathcal{L}^2 f)(x) + \alpha \left(\mathcal{L} \left(\frac{x' \cdot (\nabla_H f)}{|x'|^2} \right)(x) - \frac{x' \cdot \nabla_H}{|x'|^2} (\mathcal{L}f)(x) + (N-2) \mathcal{L} \left(\frac{f}{|x'|^2} \right)(x) \right) \\
&\quad + \beta \left(-\mathcal{L} \left(\frac{f}{|x'|^2} \right)(x) - \frac{(\mathcal{L}f)(x)}{|x'|^2} \right) \\
&\quad + \alpha\beta \left(\frac{x' \cdot \nabla_H}{|x'|^2} \left(\frac{f}{|x'|^2} \right)(x) - \frac{x' \cdot (\nabla_H f)(x)}{|x'|^4} - \frac{(N-2) f(x)}{|x'|^4} \right) \\
&\quad + \alpha^2 \left(- \left(\frac{x' \cdot \nabla_H}{|x'|^2} \right) \left(\frac{x' \cdot (\nabla_H f)}{|x'|^2} \right)(x) - (N-2) \frac{x' \cdot \nabla_H}{|x'|^2} \left(\frac{f}{|x'|^2} \right)(x) \right) \\
&\quad + \beta^2 \frac{f(x)}{|x'|^4}.
\end{aligned} \tag{6.35}$$

Using that

$$\begin{aligned}
\mathcal{L} \left(\frac{f}{|x'|^2} \right)(x) &= \sum_{j=1}^{N} X_j^2 \left(\frac{f}{|x'|^2} \right)(x) \\
&= \sum_{j=1}^{N} X_j \left(\frac{X_j f}{|x'|^2} - \frac{2 x'_j f}{|x'|^4} \right)(x) \\
&= \sum_{j=1}^{N} \left(\frac{(X_j^2 f)(x)}{|x'|^2} - \frac{4 x'_j (X_j f)(x)}{|x'|^4} + \frac{8 (x'_j)^2 f(x)}{|x'|^6} - \frac{2 f(x)}{|x'|^4} \right) \\
&= \frac{(\mathcal{L}f)(x)}{|x'|^2} - \frac{4 x' \cdot (\nabla_H f)(x)}{|x'|^4} - (2N - 8) \frac{f(x)}{|x'|^4}
\end{aligned}$$

and

$$\begin{aligned}
\frac{x' \cdot \nabla_H}{|x'|^2} \left(\frac{f}{|x'|^2} \right)(x) &= \frac{f(x)}{|x'|^2} \sum_{j=1}^{N} x'_j X_j (|x'|^{-2}) + \frac{x' \cdot (\nabla_H f)(x)}{|x'|^4} \\
&= -2 \sum_{j=1}^{N} \frac{(x'_j)^2 f(x)}{|x'|^6} + \frac{x' \cdot (\nabla_H f)(x)}{|x'|^4} \\
&= -2 \frac{f(x)}{|x'|^4} + \frac{x' \cdot (\nabla_H f)(x)}{|x'|^4},
\end{aligned}$$

in (6.35), we can write the sum $(T^+_{\alpha,\beta}T_{\alpha,\beta}f)(x)+(T_{\alpha,\beta}T^+_{\alpha,\beta}f)(x)$ as

$$\begin{aligned}
&(T^+_{\alpha,\beta}T_{\alpha,\beta}f)(x)+(T_{\alpha,\beta}T^+_{\alpha,\beta}f)(x)\\
&\quad=2(\mathcal{L}^2f)(x)+2\alpha(N-2)\left(\frac{(\mathcal{L}f)(x)}{|x'|^2}-2\frac{x'\cdot(\nabla_Hf)(x)}{|x'|^4}+(4-N)\frac{f(x)}{|x'|^4}\right)\\
&\qquad+2\beta\left(-\frac{2(\mathcal{L}f)(x)}{|x'|^2}+\frac{4x'\cdot(\nabla_Hf)(x)}{|x'|^4}+(2N-8)\frac{f(x)}{|x'|^4}\right)-2\alpha\beta(N-2)\frac{f(x)}{|x'|^4}\\
&\qquad+2\alpha^2\Bigg(-\left(\frac{x'\cdot\nabla_H}{|x'|^2}\right)\left(\frac{x'\cdot(\nabla_Hf)}{|x'|^2}\right)(x)\\
&\qquad\qquad-(N-2)\frac{x'\cdot(\nabla_Hf)(x)}{|x'|^4}+(N-2)\frac{f(x)}{|x'|^4}\Bigg)\\
&\qquad+2\beta^2\frac{f(x)}{|x'|^4}.
\end{aligned}\tag{6.36}$$

In order to simplify this, let us rewrite the following expression:

$$\begin{aligned}
&-2\left(\frac{x'\cdot\nabla_H}{|x'|^2}\right)\left(\frac{x'\cdot(\nabla_Hf)}{|x'|^2}\right)(x)-2(N-2)\frac{x'\cdot(\nabla_Hf)(x)}{|x'|^4}+2(N-2)\frac{f(x)}{|x'|^4}\\
&\quad=-\frac{2\sum_{j,k=1}^N(x'_jX_j)(x'_k(X_kf))(x)}{|x'|^4}-2\sum_{j,k=1}^N x'_j(-2)|x'|^{-3}X_j|x'|\frac{x'_k(X_kf)(x)}{|x'|^2}\\
&\qquad-2(N-2)\frac{x'\cdot(\nabla_Hf)(x)}{|x'|^4}+2(N-2)\frac{f(x)}{|x'|^4}\\
&\quad=-\frac{2\sum_{k=1}^N x'_k(X_kf)(x)}{|x'|^4}-\frac{2\sum_{j,k=1}^N x'_jx'_kX_j(X_kf)(x)}{|x'|^4}+\frac{4\sum_{k=1}^N x'_k(X_kf)(x)}{|x'|^4}\\
&\qquad-2(N-2)\frac{x'\cdot(\nabla_Hf)(x)}{|x'|^4}+2(N-2)\frac{f(x)}{|x'|^4}\\
&\quad=-2(N-3)\frac{x'\cdot(\nabla_Hf)(x)}{|x'|^4}-\frac{2\sum_{j,k=1}^N x'_jx'_k(X_jX_kf)(x)}{|x'|^4}+2(N-2)\frac{f(x)}{|x'|^4}.
\end{aligned}$$

Now putting this in (6.36), we obtain

$$\begin{aligned}
&(T^+_{\alpha,\beta}T_{\alpha,\beta}f)(x)+(T_{\alpha,\beta}T^+_{\alpha,\beta}f)(x)\\
&\quad=2(\mathcal{L}^2f)(x)+(2\alpha(N-2)-4\beta)\frac{(\mathcal{L}f)(x)}{|x'|^2}\\
&\qquad+(-4\alpha(N-2)-2\alpha^2(N-3)+8\beta)\frac{x'\cdot(\nabla_Hf)(x)}{|x'|^4}\\
&\qquad+(2\alpha(N-2)(4-N)+2\alpha^2(N-2)-2\alpha\beta(N-2)\\
&\qquad+(4N-16)\beta+2\beta^2)\frac{f(x)}{|x'|^4}-2\alpha^2\frac{\sum_{j,k=1}^N x'_jx'_k(X_jX_kf)(x)}{|x'|^4}.
\end{aligned}\tag{6.37}$$

In general, the non-negativity of $T^+_{\alpha,\beta}T_{\alpha,\beta} + T_{\alpha,\beta}T^+_{\alpha,\beta}$ and integration by parts imply

$$\int_{\mathbb{G}} |(T_{\alpha,\beta}f)(x)|^2 dx + \int_{\mathbb{G}} |(T^+_{\alpha,\beta}f)(x)|^2 dx = \int_{\mathbb{G}} \overline{f(x)}((T^+_{\alpha,\beta}T_{\alpha,\beta} + T_{\alpha,\beta}T^+_{\alpha,\beta})f)(x)dx \geq 0.$$

Putting (6.37) into this inequality, one calculates

$$\begin{aligned} 2\int_{\mathbb{G}} |(\mathcal{L}f)(x)|^2 dx &+ (2\alpha(N-2) - 4\beta) \int_{\mathbb{G}} \frac{\overline{f(x)}(\mathcal{L}f)(x)}{|x'|^2} dx \\ &+ (-4\alpha(N-2) - 2\alpha^2(N-3) + 8\beta) \int_{\mathbb{G}} \frac{\overline{f(x)}(x' \cdot (\nabla_H f)(x))}{|x'|^4} dx \\ &+ (2\alpha(N-2)(4-N) + 2\alpha^2(N-2) - 2\alpha\beta(N-2) \\ &+ (4N-16)\beta + 2\beta^2) \int_{\mathbb{G}} \frac{|f(x)|^2}{|x'|^4} dx \\ &- 2\alpha^2 \sum_{j,k=1}^{N} \int_{\mathbb{G}} \frac{\overline{f(x)} x'_j x'_k (X_j X_k f)(x)}{|x'|^4} dx \geq 0. \end{aligned} \tag{6.38}$$

Using the identities

$$\begin{aligned} \int_{\mathbb{G}} \frac{\overline{f(x)}(\mathcal{L}f)(x)}{|x'|^2} dx &= -\sum_{j=1}^{N} \int_{\mathbb{G}} \frac{\overline{(X_j f)(x)}(X_j f)(x)}{|x'|^2} \\ &\quad - \sum_{j=1}^{N} \int_{\mathbb{G}} \overline{f(x)}(-2)|x'|^{-3} X_j |x'| (X_j f)(x) dx \\ &= 2\int_{\mathbb{G}} \frac{\overline{f(x)}(x' \cdot (\nabla_H f)(x))}{|x'|^4} dx - \int_{\mathbb{G}} \frac{|(\nabla_H f)(x)|^2}{|x'|^2} dx \end{aligned}$$

and

$$\begin{aligned} &\sum_{j,k=1}^{N} \int_{\mathbb{G}} \frac{\overline{f(x)} x'_j x'_k (X_j X_k f)(x)}{|x'|^4} dx \\ &= -(N-1) \sum_{k=1}^{N} \int_{\mathbb{G}} \frac{\overline{f(x)} x'_k (X_k f)(x)}{|x'|^4} dx - 2\sum_{k=1}^{N} \int_{\mathbb{G}} \frac{\overline{f(x)} x'_k (X_k f)(x)}{|x'|^4} dx \\ &\quad - \sum_{j,k=1}^{N} \int_{\mathbb{G}} \frac{x'_j x'_k \overline{(X_j f)(x)}(X_k f)(x)}{|x'|^4} dx \\ &\quad - \sum_{j,k=1}^{N} \int_{\mathbb{G}} \overline{f(x)} x'_j x'_k (X_k f)(x)(-4)|x'|^{-5} X_j |x'| dx \end{aligned}$$

$$
\begin{aligned}
&= -(N+1)\sum_{k=1}^{N}\int_{\mathbb{G}} \frac{\overline{f(x)}x'_k(X_k f)(x)}{|x'|^4}dx + 4\sum_{j,k=1}^{N}\int_{\mathbb{G}}\frac{\overline{f(x)}(x'_j)^2 x'_k(X_k f)(x)}{|x'|^6}dx \\
&\quad -\int_{\mathbb{G}}\frac{|x'\cdot(\nabla_H f)(x)|^2}{|x'|^4}dx \\
&= -(N-3)\int_{\mathbb{G}}\frac{\overline{f(x)}(x'\cdot(\nabla_H f)(x))}{|x'|^4}dx - \int_{\mathbb{G}}\frac{|x'\cdot(\nabla_H f)(x)|^2}{|x'|^4}dx
\end{aligned}
$$

in (6.38), we obtain

$$
\begin{aligned}
&2\int_{\mathbb{G}}|(\mathcal{L}f)(x)|^2dx + (4\alpha(N-2)-8\beta-4\alpha(N-2)+8\beta-2\alpha^2(N-3)+2\alpha^2(N-3)) \\
&\quad\times\int_{\mathbb{G}}\frac{\overline{f(x)}(x'\cdot(\nabla_H f)(x))}{|x'|^4}dx \\
&\quad+(2\alpha(N-2)(4-N)+2\alpha^2(N-2)-2\alpha\beta(N-2)+(4N-16)\beta+2\beta^2) \\
&\quad\times\int_{\mathbb{G}}\frac{|f(x)|^2}{|x'|^4}dx-(2\alpha(N-2)-4\beta)\int_{\mathbb{G}}\frac{|(\nabla_H f)(x)|^2}{|x'|^2}dx \\
&\quad+2\alpha^2\int_{\mathbb{G}}\frac{|x'\cdot(\nabla_H f)(x)|^2}{|x'|^4}dx \geq 0,
\end{aligned}
$$

which implies (6.32). □

The factorization method can be used to give an elementary proof of the horizontal L^2-weighted inequality given in Remark 6.1.2, Part 3.

Proposition 6.4.3 (Horizontal L^2-Hardy inequality). *Let $\mathbb{G}$ be a stratified group with $N \geq 3$ being the dimension of the first stratum. Let $\alpha \in \mathbb{R}$. Then for all complex-valued functions $f \in C_0^\infty(\mathbb{G}\backslash\{x'=0\})$ we have*

$$
\|\nabla_H f\|_{L^2(\mathbb{G})} \geq \frac{N-2}{2}\left\|\frac{f}{|x'|}\right\|_{L^2(\mathbb{G})}, \tag{6.39}
$$

where the constant $\frac{N-2}{2}$ is sharp.

Proof of Proposition 6.4.3. Let

$$
\widetilde{T}_\alpha := \nabla_H + \alpha\frac{x'}{|x'|^2}.
$$

One can readily check that its formal adjoint is given by

$$
\widetilde{T}_\alpha^+ = -\mathrm{div}_H + \alpha\frac{x'}{|x'|^2},
$$

where $x' \neq 0$.

Using (1.73) we have

$$\begin{aligned}\widetilde{T}_\alpha^+\widetilde{T}_\alpha f &= -(\mathcal{L}f)(x) - \alpha\mathrm{div}_H\left(\frac{x'}{|x'|^2}f\right)(x) + \alpha\frac{x'\cdot(\nabla_H f)(x)}{|x'|^2} + \alpha^2\frac{f(x)}{|x'|^2}\\ &= -(\mathcal{L}f)(x) - \alpha\mathrm{div}_H\left(\frac{x'}{|x'|^2}\right)f(x) + \alpha^2\frac{f(x)}{|x'|^2}\\ &= -(\mathcal{L}f)(x) + \frac{\alpha(\alpha+2-N)}{|x'|^2}f(x).\end{aligned}$$

By integrating by parts and using the non-negativity of $\widetilde{T}_\alpha^+\widetilde{T}_\alpha$ we have

$$\begin{aligned}0 &\le \int_{\mathbb{G}} |\widetilde{T}_\alpha f|^2 dx\\ &= \int_{\mathbb{G}} \overline{f(x)}(\widetilde{T}_\alpha^+\widetilde{T}_\alpha f)(x)dx\\ &= \int_{\mathbb{G}} |\nabla_H f|^2 dx + \alpha(\alpha+2-N)\int_{\mathbb{G}} \frac{|f(x)|^2}{|x'|^2}dx.\end{aligned}$$

It follows from this that

$$\int_{\mathbb{G}} |\nabla_H f(x)|^2 dx \ge \alpha(N-2-\alpha)\int_{\mathbb{G}} \frac{|f(x)|^2}{|x'|^2}dx.$$

By maximizing the constant with respect to α we obtain (6.39). The sharpness of the constant follows from Remark 6.1.2, Part 3. □

By modifying the differential expression $\widetilde{T}_\alpha$ in the proof of Proposition 6.4.3 we can also show the following refinement of the L^2-Hardy inequality (6.39). The fact that it is indeed a refinement, that is, that (6.40) implies (6.39) follows by the Cauchy–Schwarz inequality. Consequently, the sharpness of the constant in (6.40) also follows from the sharpness of the constant in (6.39).

Theorem 6.4.4 (Refined horizontal L^2-Hardy inequality). *Let $\mathbb{G}$ be a stratified group with $N \ge 3$ being the dimension of the first stratum. Let $\alpha \in \mathbb{R}$. Then for all complex-valued functions $f \in C_0^\infty(\mathbb{G}\backslash\{x'=0\})$ we have*

$$\left\|\frac{x'\cdot\nabla_H f}{|x'|}\right\|_{L^2(\mathbb{G})} \ge \frac{N-2}{2}\left\|\frac{f}{|x'|}\right\|_{L^2(\mathbb{G})}, \tag{6.40}$$

where the constant $\frac{N-2}{2}$ is sharp.

Proof of Theorem 6.4.4. Let us define

$$\widehat{T}_\alpha := \frac{x'\cdot\nabla_H}{|x'|} + \frac{\alpha}{|x'|}.$$

One can readily check that its formal adjoint is given by

$$\widehat{T}_\alpha^+ := -\frac{x' \cdot \nabla_H}{|x'|} + \frac{\alpha - N + 1}{|x'|},$$

where $x' \neq 0$. Using (1.73) we get

$$\begin{aligned}
&\left(\frac{x' \cdot \nabla_H}{|x'|}\right)\left(\frac{x' \cdot (\nabla_H f)(x)}{|x'|}\right) \\
&= \frac{\sum_{j,k=1}^N (x'_j X_j)(x'_k (X_k f)(x))}{|x'|^2} + \sum_{j,k=1}^N x'_j (-1)|x'|^{-2} X_j |x'| \frac{x'_k (X_k f)(x)}{|x'|} \\
&= \frac{\sum_{k=1}^N x'_k (X_k f)(x)}{|x'|^2} + \frac{\sum_{j,k=1}^N x'_j x'_k (X_j X_k f)(x)}{|x'|^2} - \frac{\sum_{k=1}^N x'_k (X_k f)(x)}{|x'|^2} \\
&= \frac{\sum_{j,k=1}^N x'_j x'_k (X_j X_k f)(x)}{|x'|^2}
\end{aligned}$$

and

$$\begin{aligned}
(x' \cdot \nabla_H)\left(\frac{f}{|x'|}\right) &= \frac{\sum_{k=1}^N x'_k (X_k f)(x)}{|x'|} + \sum_{k=1}^N x'_k (-1)|x'|^{-2} X_k |x'| f(x) \\
&= \frac{x' \cdot (\nabla_H f)(x)}{|x'|} - \frac{f(x)}{|x'|}.
\end{aligned}$$

From these identities we get

$$\begin{aligned}
\widehat{T}_\alpha^+ \widehat{T}_\alpha f(x) = &- \frac{\sum_{j,k=1}^N x'_j x'_k (X_j X_k f)(x)}{|x'|^2} \\
&- (N-1)\frac{x' \cdot (\nabla_H f)(x)}{|x'|^2} + \frac{\alpha(\alpha + 2 - N)}{|x'|^2} f(x).
\end{aligned}$$

Using (1.73) again we have

$$\begin{aligned}
&\sum_{j,k=1}^N \int_{\mathbb{G}} \frac{\overline{f(x)} x'_j x'_k (X_j X_k f)(x)}{|x'|^2} dx \\
&= -(N-1) \sum_{k=1}^N \int_{\mathbb{G}} \frac{\overline{f(x)} x'_k (X_k f)(x)}{|x'|^2} dx - 2 \sum_{k=1}^N \int_{\mathbb{G}} \frac{\overline{f(x)} x'_k (X_k f)(x)}{|x'|^2} dx \\
&\quad - \sum_{j,k=1}^N \int_{\mathbb{G}} \frac{x'_j x'_k \overline{(X_j f)(x)} (X_k f)(x)}{|x'|^2} dx \\
&\quad - \sum_{j,k=1}^N \int_{\mathbb{G}} \overline{f(x)} x'_j x'_k (X_k f)(x) (-2) |x'|^{-3} X_j |x'| dx
\end{aligned}$$

$$
\begin{aligned}
&= -(N+1)\sum_{k=1}^{N}\int_{\mathbb{G}}\frac{\overline{f(x)}x'_k(X_kf)(x)}{|x'|^2}dx + 2\sum_{j,k=1}^{N}\int_{\mathbb{G}}\frac{\overline{f(x)}(x'_j)^2x'_k(X_kf)(x)}{|x'|^4}dx\\
&\quad -\int_{\mathbb{G}}\frac{|x'\cdot(\nabla_H f)(x)|^2}{|x'|^2}dx\\
&= -(N-1)\int_{\mathbb{G}}\frac{\overline{f(x)}(x'\cdot(\nabla_H f)(x))}{|x'|^2}dx - \int_{\mathbb{G}}\frac{|x'\cdot(\nabla_H f)(x)|^2}{|x'|^2}dx. \qquad (6.41)
\end{aligned}
$$

Taking into account this, integrating by parts, and using the non-negativity of the operator $\widehat{T}_\alpha^+\widehat{T}_\alpha$, we get

$$
\begin{aligned}
0 &\le \int_{\mathbb{G}}|\widehat{T}_\alpha f|^2dx = \int_{\mathbb{G}}\overline{f(x)}(\widehat{T}_\alpha^+\widehat{T}_\alpha f)(x)dx\\
&= -\int_{\mathbb{G}}\left(\frac{\sum_{j,k=1}^{N}x'_jx'_k\overline{f(x)}(X_jX_kf)(x)}{|x'|^2} + \frac{(N-1)\overline{f(x)}(x'\cdot(\nabla_H f)(x))}{|x'|^2}\right)dx\\
&\quad +\alpha(\alpha-N+2)\int_{\mathbb{G}}\frac{|f(x)|^2}{|x'|^2}dx.
\end{aligned}
$$

Consequently, using (6.41) we obtain

$$
\int_{\mathbb{G}}\left(\frac{|x'\cdot(\nabla_H f)(x)|^2}{|x'|^2} + \alpha(\alpha-N+2)\frac{|f(x)|^2}{|x'|^2}\right)dx \ge 0.
$$

It now follows that

$$
\int_{\mathbb{G}}\frac{|x'\cdot(\nabla_H f)(x)|^2}{|x'|^2}dx \ge \alpha((N-2)-\alpha)\int_{\mathbb{G}}\frac{|f(x)|^2}{|x'|^2}dx.
$$

By maximizing $\alpha((N-2)-\alpha)$ with respect to α we obtain (6.40). □

Further two-parameter inequalities by factorization method are possible in the setting of the Heisenberg group, for which we can refer the reader to [RY17].

6.5 Hardy–Rellich type inequalities and embedding results

We now discuss several refinements of the Hardy–Rellich inequalities with respect to the variables in the first stratum. We recall that N stands for the dimension of the first stratum of a stratified Lie group $\mathbb{G}$ here. As a consequence, we formulate several corollaries for the embeddings of the appearing function spaces.

Theorem 6.5.1 (Horizontal L^2-Hardy–Rellich type inequalities). *Let α, $\beta \in \mathbb{R}$. Let $N \geq 2$ be the dimension of the first stratum of a stratified Lie group $\mathbb{G}$, and let $|\cdot|$ be the Euclidean norm on $\mathbb{R}^N$. Then for all $f \in C_0^\infty(\mathbb{G}\backslash\{x'=0\})$ we have*

$$\left(\frac{N-(\alpha+\beta+3)}{2}\int_{\mathbb{G}}\frac{|\nabla_H f|^2}{|x'|^{\alpha+\beta+1}}dx+(\alpha+\beta+1)\int_{\mathbb{G}}\frac{(x'\cdot\nabla_H f)^2}{|x'|^{\alpha+\beta+3}}dx\right)^2 \leq \int_{\mathbb{G}}\frac{|\mathcal{L}f|^2}{|x'|^{2\beta}}dx\int_{\mathbb{G}}\frac{|\nabla_H f|^2}{|x'|^{2\alpha}}dx. \tag{6.42}$$

Moreover, if $\alpha+\beta+1\leq 0$ then we have

$$\frac{N+\alpha+\beta-1}{2}\int_{\mathbb{G}}\frac{|\nabla_H f|^2}{|x'|^{\alpha+\beta+1}}dx \leq \left(\int_{\mathbb{G}}\frac{|\mathcal{L}f|^2}{|x'|^{2\beta}}dx\right)^{1/2}\left(\int_{\mathbb{G}}\frac{|\nabla_H f|^2}{|x'|^{2\alpha}}dx\right)^{1/2}. \tag{6.43}$$

The inequality (6.43) can be considered as a special case ($p=2$) of Theorem 6.8.1. In particular, taking $\alpha=\beta+1$, we obtain the following Rellich type inequality:

Corollary 6.5.2 (Horizontal L^2-Rellich type inequality). *Let N be the dimension of the first stratum of a stratified Lie group $\mathbb{G}$ and let $\alpha\leq 0$. Then for all $f\in C_0^\infty(\mathbb{G}\backslash\{x'=0\})$ we have*

$$\frac{(N+2\alpha-2)^2}{4}\int_{\mathbb{G}}\frac{|\nabla_H f|^2}{|x'|^{2\alpha}}dx \leq \int_{\mathbb{G}}\frac{|\mathcal{L}f|^2}{|x'|^{2\alpha-2}}dx. \tag{6.44}$$

Furthermore, we have

$$\frac{(N+2\alpha-2)^2(N-2\alpha-2)^2}{16}\int_{\mathbb{G}}\frac{|f(x)|^2}{|x'|^{2\alpha+2}}dx \leq \int_{\mathbb{G}}\frac{|\mathcal{L}f|^2}{|x'|^{2\alpha-2}}dx. \tag{6.45}$$

We can compare it with another version given in Theorem 6.2.2.

Proof of Corollary 6.5.2. Inequality (6.44) follows from Theorem 6.5.1 by taking $\alpha=\beta+1$. Inequality (6.45) follows from (6.44) and Corollary 6.2.1 with $p=2$ which says that

$$\left\|\frac{1}{|x'|^\alpha}\nabla_H f\right\|_{L^2(\mathbb{G})} \geq \frac{|N-2(\alpha+1)|}{2}\left\|\frac{f}{|x'|^{\alpha+1}}\right\|_{L^2(\mathbb{G})}.$$

Note that the sharpness of the constant follows from the fact that in both inequalities the best constants are attained when there are equalities in the corresponding Hölder inequalities in their proofs, and these are attained on powers of $|x'|$. □

We note that another version of the horizontal Rellich inequality is given in Theorem 6.2.2.

Note that when $\mathbb{G} = (\mathbb{R}^n, +)$, that is, $N = n$, $\nabla_H = \nabla = (\partial_{x_1}, \ldots, \partial_{x_n})$, then (6.42) implies the following Hardy–Rellich type inequality for all $f \in C_0^\infty(\mathbb{R}^n \backslash \{0\})$:

$$\left(\frac{n-(\alpha+\beta+3)}{2}\int_{\mathbb{R}^n}\frac{|\nabla f|^2}{|x|_E^{\alpha+\beta+1}}dx + (\alpha+\beta+1)\int_{\mathbb{R}^n}\frac{(x\cdot\nabla f)^2}{|x|_E^{\alpha+\beta+3}}dx\right)^2 \leq \int_{\mathbb{R}^n}\frac{|\Delta f|^2}{|x|_E^{2\beta}}dx\int_{\mathbb{R}^n}\frac{|\nabla f|^2}{|x|_E^{2\alpha}}dx, \tag{6.46}$$

where $|x|_E = \sqrt{x_1^2+\cdots+x_n^2}$. This inequality was also discussed in [Cos08] and [DJSJ13].

Proof of Theorem 6.5.1. First we note that for all $s \in \mathbb{R}$ we have

$$\int_{\mathbb{G}}\left|\frac{\nabla_H f}{|x'|^\alpha} + s\frac{x'}{|x'|^{\beta+1}}\mathcal{L}f\right|^2 dx \geq 0,$$

that is,

$$\int_{\mathbb{G}}\frac{|\nabla_H f|^2}{|x'|^{2\alpha}}dx + 2s\int_{\mathbb{G}}\frac{x'\cdot\nabla_H f}{|x'|^{\alpha+\beta+1}}\mathcal{L}f dx + s^2\int_{\mathbb{G}}\frac{|\mathcal{L}f|^2}{|x'|^{2\beta}}dx \geq 0. \tag{6.47}$$

Since

$$\int_{\mathbb{G}}\frac{x'\cdot\nabla_H f}{|x'|^{\alpha+\beta+1}}\mathcal{L}f dx = \int_{\mathbb{G}}\mathrm{div}_H(\nabla_H f)\left(\frac{x'\cdot\nabla_H f}{|x'|^{\alpha+\beta+1}}\right)dx$$

by using the divergence theorem (Theorem 1.4.5) and (1.73) we obtain

$$\int_{\mathbb{G}}\mathrm{div}_H(\nabla_H f)\left(\frac{x'\cdot\nabla_H f}{|x'|^{\alpha+\beta+1}}\right)dx = -\frac{1}{2}\int_{\mathbb{G}}\frac{x'}{|x'|^{\alpha+\beta+1}}\cdot\nabla_H(|\nabla_H f|^2)dx - \int_{\mathbb{G}}\frac{|\nabla_H f|^2}{|x'|^{\alpha+\beta+1}}dx + (\alpha+\beta+1)\int_{\mathbb{G}}\frac{(x'\cdot\nabla_H f)^2}{|x'|^{\alpha+\beta+3}}dx.$$

Again by Theorem 1.4.5 and (1.73) we have the equality

$$-\frac{1}{2}\int_{\mathbb{G}}\frac{x'}{|x'|^{\alpha+\beta+1}}\cdot\nabla_H(|\nabla_H f|^2)dx = \frac{N-(\alpha+\beta+1)}{2}\int_{\mathbb{G}}\frac{|\nabla_H f|^2}{|x'|^{\alpha+\beta+1}}dx.$$

Thus,

$$\int_{\mathbb{G}}\frac{x'\cdot\nabla_H f}{|x'|^{\alpha+\beta+1}}\mathcal{L}f dx = \frac{N-(\alpha+\beta+3)}{2}\int_{\mathbb{G}}\frac{|\nabla_H f|^2}{|x'|^{\alpha+\beta+1}}dx + (\alpha+\beta+1)\int_{\mathbb{G}}\frac{(x'\cdot\nabla_H f)^2}{|x'|^{\alpha+\beta+3}}dx. \tag{6.48}$$

Therefore, the inequality (6.47) can be rewritten as

$$s^2\int_{\mathbb{G}}\frac{|\mathcal{L}f|^2}{|x'|^{2\beta}}dx+2s\left(\frac{N-(\alpha+\beta+3)}{2}\int_{\mathbb{G}}\frac{|\nabla_H f|^2}{|x'|^{\alpha+\beta+1}}dx\right.$$
$$\left.+(\alpha+\beta+1)\int_{\mathbb{G}}\frac{(x'\cdot\nabla_H f)^2}{|x'|^{\alpha+\beta+3}}dx\right)+\int_{\mathbb{G}}\frac{|\nabla_H f|^2}{|x'|^{2\alpha}}dx\geq 0.$$

Denoting

$$a:=\int_{\mathbb{G}}\frac{|\mathcal{L}f|^2}{|x'|^{2\beta}}dx,$$

$$b:=\frac{N-(\alpha+\beta+3)}{2}\int_{\mathbb{G}}\frac{|\nabla_H f|^2}{|x'|^{\alpha+\beta+1}}dx+(\alpha+\beta+1)\int_{\mathbb{G}}\frac{(x'\cdot\nabla_H f)^2}{|x'|^{\alpha+\beta+3}}dx,$$

and

$$c:=\int_{\mathbb{G}}\frac{|\nabla_H f|^2}{|x'|^{2\alpha}}dx$$

we arrive at

$$as^2+2bs+c\geq 0,$$

which is equivalent to $b^2-ac\leq 0$. Thus, we have

$$\left(\frac{N-(\alpha+\beta+3)}{2}\int_{\mathbb{G}}\frac{|\nabla_H f|^2}{|x'|^{\alpha+\beta+1}}dx+(\alpha+\beta+1)\int_{\mathbb{G}}\frac{(x'\cdot\nabla_H f)^2}{|x'|^{\alpha+\beta+3}}dx\right)^2$$
$$\leq\int_{\mathbb{G}}\frac{|\mathcal{L}f|^2}{|x'|^{2\beta}}dx\int_{\mathbb{G}}\frac{|\nabla_H f|^2}{|x'|^{2\alpha}}dx.$$

This shows the inequality (6.42). Now let us show the inequality (6.43). By using Schwarz' and Hölder's inequality we obtain

$$\int_{\mathbb{G}}\frac{x'\cdot\nabla_H f}{|x'|^{\alpha+\beta+1}}\mathcal{L}f dx\leq\int_{\mathbb{G}}\frac{|\nabla_H f|}{|x'|^{\alpha+\beta}}\mathcal{L}f dx\leq\left(\int_{\mathbb{G}}\frac{|\mathcal{L}f|^2}{|x'|^{2\beta}}dx\right)^{1/2}\left(\int_{\mathbb{G}}\frac{|\nabla_H f|^2}{|x'|^{2\alpha}}dx\right)^{1/2}.$$

On the other hand, since $\alpha+\beta+1\leq 0$ by Schwarz' inequality we have

$$\frac{N-(\alpha+\beta+3)}{2}\int_{\mathbb{G}}\frac{|\nabla_H f|^2}{|x'|^{\alpha+\beta+1}}dx+(\alpha+\beta+1)\int_{\mathbb{G}}\frac{(x'\cdot\nabla_H f)^2}{|x'|^{\alpha+\beta+3}}dx$$
$$\geq\frac{N-(\alpha+\beta+3)}{2}\int_{\mathbb{G}}\frac{|\nabla_H f|^2}{|x'|^{\alpha+\beta+1}}dx+(\alpha+\beta+1)\int_{\mathbb{G}}\frac{|\nabla_H f|^2}{|x'|^{\alpha+\beta+1}}dx$$
$$=\frac{N+\alpha+\beta-1}{2}\int_{\mathbb{G}}\frac{|\nabla_H f|^2}{|x'|^{\alpha+\beta+1}}dx.$$

Combining the above inequalities with (6.48) we obtain (6.42). □

The special case of Theorem 6.1.1 with $p = 2$ can be also shown using the divergence formula techniques in the proof of Theorem 6.5.1:

Corollary 6.5.3 (Horizontal L^2-Caffarelli–Kohn–Nirenberg inequalities). *Let $\mathbb{G}$ be a homogeneous stratified group with N being the dimension of the first stratum. Let $\alpha, \beta \in \mathbb{R}$. Then for all $f \in C_0^\infty(\mathbb{G}\backslash\{x' = 0\})$ we have*

$$\frac{|N-\gamma|}{2}\left\|\frac{f}{|x'|^{\frac{\gamma}{2}}}\right\|^2_{L^2(\mathbb{G})} \leq \left\|\frac{\nabla_H f}{|x'|^{\alpha}}\right\|_{L^2(\mathbb{G})}\left\|\frac{f}{|x'|^{\beta}}\right\|_{L^2(\mathbb{G})}, \tag{6.49}$$

where $\gamma = \alpha + \beta + 1$, and the constant $\frac{|N-\gamma|}{2}$ is sharp.

Proof. For all $f \in C_0^\infty(\mathbb{G}\backslash\{x' = 0\})$, $\alpha, \beta \in \mathbb{R}$ and $s \in \mathbb{R}$ we have

$$\int_{\mathbb{G}}\left|\frac{\nabla_H f}{|x'|^{\beta}} + s\frac{x'}{|x'|^{\alpha+1}}f\right|^2 dx \geqslant 0.$$

This can be written as

$$\int_{\mathbb{G}}\frac{|\nabla_H f|^2}{|x'|^{2\beta}}dx + s^2\int_{\mathbb{G}}\frac{|f|^2}{|x'|^{2\alpha}}dx + 2s\int_{\mathbb{G}} f\frac{x'\cdot\nabla_H f}{|x'|^{\gamma}}dx \geqslant 0.$$

By the divergence theorem (Theorem 1.4.5) we have

$$\int_{\mathbb{G}} f\frac{x'\cdot\nabla_H f}{|x'|^{\gamma}}dx = -\frac{N-\gamma}{2}\int_{\mathbb{G}}\frac{|f|^2}{|x'|^{\gamma}}dx.$$

Denoting

$$a := \int_{\mathbb{G}}\frac{|f|^2}{|x'|^{2\alpha}}dx, \quad b := |N-\gamma|\int_{\mathbb{G}}\frac{|f|^2}{|x'|^{\gamma}}, \quad c := \int_{\mathbb{G}}\frac{|\nabla_H f|^2}{|x'|^{2\beta}}dx,$$

this means that

$$as^2 - bs + c \geqslant 0,$$

which is equivalent to $b^2 - 4ac \leqslant 0$, that is,

$$|N-\gamma|^2\left(\int_{\mathbb{G}}\frac{|f|^2}{|x'|^{\gamma}}\right)^2 \leqslant 4\left(\int_{\mathbb{G}}\frac{|f|^2}{|x'|^{2\alpha}}dx\right)\left(\int_{\mathbb{G}}\frac{|\nabla_H f|^2}{|x'|^{2\beta}}dx\right),$$

which gives (6.49). □

The appearance of the horizontal weights in Theorem 6.5.1 prompts one to define the following weighted Sobolev type spaces on the stratified Lie group $\mathbb{G}$ (in Chapter 10 we will be discussing analogous spaces but there on general homogeneous groups).

Definition 6.5.4 (Sobolev types spaces with horizontal weights). Let us define the following spaces:

(1) Let $L^2_\alpha(\mathbb{G})$ be the completion of $C_0^\infty(\mathbb{G}\backslash\{x'=0\})$ with respect to the norm

$$\|f\|_{L^2_\alpha} := \left(\int_{\mathbb{G}} \frac{|f|^2}{|x'|^{2\alpha}} dx\right)^{1/2}.$$

(2) Let $D^{1,2}_\gamma(\mathbb{G})$ be the completion of $C_0^\infty(\mathbb{G}\backslash\{x'=0\})$ with respect to the norm

$$\|f\|_{D^{1,2}_\gamma(\mathbb{G})} := \left(\int_{\mathbb{G}} \frac{|\nabla_H f|^2}{|x'|^{2\gamma}} dx\right)^{1/2}.$$

(3) Let $D^{2,2}_\gamma(\mathbb{G})$ be the completion of $C_0^\infty(\mathbb{G}\backslash\{x'=0\})$ with respect to the norm

$$\|f\|_{D^{2,2}_\gamma(\mathbb{G})} := \left(\int_{\mathbb{G}} \frac{|\mathcal{L}f|^2}{|x'|^{2\gamma}} dx\right)^{1/2}.$$

(4) Let $H^1_{\alpha,\beta}(\mathbb{G})$ be the completion of $C_0^\infty(\mathbb{G}\backslash\{x'=0\})$ with respect to the norm

$$\|f\|_{H^1_{\alpha,\beta}} := \left(\int_{\mathbb{G}} \left[\frac{|f|^2}{|x'|^{2\alpha}} + \frac{|\nabla_H f|^2}{|x'|^{2\beta}}\right] dx\right)^{1/2}.$$

(5) Let $H^2_{\alpha,\beta}(\mathbb{G})$ be the completion of $C_0^\infty(\mathbb{G}\backslash\{x'=0\})$ with respect to the norm

$$\|f\|_{H^2_{\alpha,\beta}(\mathbb{G})} := \left(\int_{\mathbb{G}} \frac{|\nabla_H f|^2}{|x'|^{2\alpha}} + \frac{|\mathcal{L}f|^2}{|x'|^{2\beta}} dx\right)^{1/2}.$$

Theorem 6.5.5 (Several horizontal embeddings). *Let α, $\beta \in \mathbb{R}$. We have the following continuous embeddings*

(i) $H^2_{\alpha,\beta}(\mathbb{G}) \subset D^{2,2}_{\frac{\alpha+\beta+1}{2}}(\mathbb{G})$ *for* $\alpha+\beta-1 \neq N$.

(ii) $D^{2,2}_\alpha(\mathbb{G}) \subset D^{1,2}_{\alpha+1}(\mathbb{G})$ *for* $\alpha \leq \frac{N}{2}-2$.

(iii) $H^1_{\alpha,\beta}(\mathbb{G}) \subset L^2_{\gamma/2}(\mathbb{G})$ *and* $H^1_{\beta,\alpha}(\mathbb{G}) \subset L^2_{\gamma/2}(\mathbb{G})$ *for* $\gamma = \alpha+\beta+1$, *provided that* $\gamma \neq N$.

Proof of Theorem 6.5.5. Since $N \neq \alpha+\beta-1$, from (6.43) we obtain

$$\begin{aligned}\int_{\mathbb{G}} \frac{|\nabla_H f|^2}{|x'|^{2\frac{(\alpha+\beta+1)}{2}}} dx &\leq \frac{2}{|N+\alpha+\beta-1|}\left(\int_{\mathbb{G}} \frac{|\mathcal{L}f|^2}{|x'|^{2\beta}} dx\right)^{1/2}\left(\int_{\mathbb{G}} \frac{|\nabla_H f|^2}{|x'|^{2\alpha}} dx\right)^{1/2} \\ &\leq \frac{2}{|N+\alpha+\beta-1|}\left(\int_{\mathbb{G}} \frac{|\mathcal{L}f|^2}{|x'|^{2\beta}} dx + \int_{\mathbb{G}} \frac{|\nabla_H f|^2}{|x'|^{2\alpha}} dx\right),\end{aligned}$$

for all $f \in C_0^\infty(\mathbb{G}\backslash\{x'=0\})$. This proves Part (i).

Part (ii) follows from the inequality (6.43), namely assuming $\alpha+\beta+3 \leq N$ and letting $\beta = \alpha+1$, $\alpha \neq \frac{N}{2}$.

The first inequality in Part (iii) follows from inequality (6.49). Since the spaces are symmetric with respect to the parameters α, β we also have the second embedding. □

Using inequality (6.43) and choosing different values of α and β we can obtain a number of Heisenberg–Pauli–Weyl type uncertainty inequalities. Let us list some interesting cases.

Corollary 6.5.6 (Horizontal Heisenberg–Pauli–Weyl type uncertainty inequalities). *We have the following inequalities:*

(1) *For* $\alpha \leq \frac{N}{2} - 2$ *and any* $f \in H^2_{\alpha,\alpha+1}(\mathbb{G})$,

$$\frac{|N+2\alpha|}{2} \int_{\mathbb{G}} \frac{|\nabla_H f|^2}{|x'|^{2(\alpha+1)}} dx \leq \left(\int_{\mathbb{G}} \frac{|\mathcal{L} f|^2}{|x'|^{2(\alpha+1)}} dx \right)^{1/2} \left(\int_{\mathbb{G}} \frac{|\nabla_H f|^2}{|x'|^{2\alpha}} dx \right)^{1/2}.$$

(2) *For* $N \geq 3$ *and any* $f \in D^{1,2}_0(\mathbb{G})$,

$$\frac{|N-2|}{2} \int_{\mathbb{G}} |\nabla_H f|^2 dx \leq \left(\int_{\mathbb{G}} |x'|^{2(\alpha+1)} |\nabla_H f|^2 dx \right)^{1/2} \left(\int_{\mathbb{G}} \frac{|\mathcal{L} f|^2}{|x'|^{2\alpha}} dx \right)^{1/2}.$$

(3) *For any* $f \in D^{1,2}_1(\mathbb{G})$,

$$\frac{N}{2} \int_{\mathbb{G}} \frac{|\nabla_H f|^2}{|x'|^2} dx \leq \left(\int_{\mathbb{G}} |\nabla_H f|^2 dx \right)^{1/2} \left(\int_{\mathbb{G}} \frac{|\mathcal{L} f|^2}{|x'|^2} dx \right)^{1/2}.$$

(4) *For* $N \geq 2$ *and any* $f \in D^{1,2}_{1/2}(\mathbb{G})$,

$$\frac{N-1}{2} \int_{\mathbb{G}} \frac{|\nabla_H f|^2}{|x'|} dx \leq \left(\int_{\mathbb{G}} |x'|^2 |\nabla_H f|^2 dx \right)^{1/2} \left(\int_{\mathbb{G}} \frac{|\mathcal{L} f|^2}{|x'|^2} dx \right)^{1/2}.$$

(5) *For* $N \geq 2$ *and any* $f \in D^{1,2}_{1/2}(\mathbb{G})$,

$$\frac{N-1}{2} \int_{\mathbb{G}} \frac{|\nabla_H f|^2}{|x'|} dx \leq \left(\int_{\mathbb{G}} |\nabla_H f|^2 dx \right)^{1/2} \left(\int_{\mathbb{G}} |\mathcal{L} f|^2 dx \right)^{1/2}.$$

Moreover, the following inequalities hold true with sharp constants:

(6) *For any* $f \in D^{1,2}(\mathbb{G})$, *taking* $\alpha = 1, \beta = 0$,

$$\left(\frac{N-2}{2} \right)^2 \int_{\mathbb{G}} \frac{|f|^2}{|x'|^2} dx \leq \int_{\mathbb{G}} |\nabla_H f|^2 dx.$$

(7) *For any* $f \in H^1_{\beta+1,\beta}(\mathbb{G})$, *taking* $\alpha = \beta + 1$,

$$\left(\frac{N-2(\beta+1)}{2} \right)^2 \int_{\mathbb{G}} \frac{|f|^2}{|x'|^{2(\beta+1)}} dx \leq \int_{\mathbb{G}} \frac{|\nabla_H f|^2}{|x'|^{2\beta}} dx.$$

(8) *For any* $f \in H^1_{\alpha,\alpha+1}(\mathbb{G})$, *taking* $\beta = \alpha + 1$,

$$\left(\frac{N-2(\alpha+1)}{2}\right)^2 \int_{\mathbb{G}} \frac{|f|^2}{|x'|^{2(\alpha+1)}} dx \leq \left(\int_{\mathbb{G}} \frac{|f|^2}{|x'|^{2\alpha}} dx\right)^{1/2} \left(\int_{\mathbb{G}} \frac{|\nabla_H f|^2}{|x'|^{2(\alpha+1)}} dx\right)^{1/2}.$$

(9) *For any* $f \in H^1_{-(\beta+1),\beta}(\mathbb{G})$, *taking* $\alpha = -(\beta+1)$, *then* $f \in L^2(\mathbb{G})$ *and*

$$\left(\frac{N}{2}\right) \int_{\mathbb{G}} |u|^2 dx \leq \left(\int_{\mathbb{G}} |x'|^{2(\beta+1)} |f|^2 dx\right)^{1/2} \left(\int_{\mathbb{G}} \frac{|\nabla_H f|^2}{|x'|^{2\beta}} dx\right)^{1/2}.$$

(10) *For any* $f \in H^1_{0,1}(\mathbb{G})$, *taking* $\alpha = 0, \beta = 1$, *then* $f \in L^2_1(\mathbb{G})$ *and*

$$\left|\frac{N-2}{2}\right| \int_{\mathbb{G}} \frac{|u|^2}{|x'|^2} dx \leq \left(\int_{\mathbb{G}} |f|^2 dx\right)^{1/2} \left(\int_{\mathbb{G}} \frac{|\nabla_H f|^2}{|x'|^2} dx\right)^{1/2}.$$

(11) *For any* $f \in H^1_{-1,1}(\mathbb{G})$, $N > 1$, *taking* $\alpha = -1, \beta = 1$, *then* $f \in L^2_{1/2}(\mathbb{G})$ *and*

$$\left(\frac{N-1}{2}\right) \int_{\mathbb{G}} \frac{|u|^2}{|x'|^2} dx \leq \left(\int_{\mathbb{G}} |x'|^2 |f|^2 dx\right)^{1/2} \left(\int_{\mathbb{G}} \frac{|\nabla_H f|^2}{|x'|^2} dx\right)^{1/2}.$$

(12) *For any* $f \in H^1(\mathbb{G}) = H^1_{0,0}(\mathbb{G})$, $N > 1$, *taking* $\alpha = 0, \beta = 0$, *then* $f \in L^2_{1/2}(\mathbb{G})$ *and*

$$\left(\frac{N-1}{2}\right) \int_{\mathbb{G}} \frac{|u|^2}{|x'|^2} dx \leq \left(\int_{\mathbb{G}} |f|^2 dx\right)^{1/2} \left(\int_{\mathbb{G}} |\nabla_H f|^2 dx\right)^{1/2}.$$

6.6 Horizontal Sobolev type inequalities

In this section, first, we are interested in Sobolev inequalities, so let us repeat them briefly again for the sake of the reader comparing to the full homogeneous group version discussed in Section 3.2.2. The (Euclidean) Sobolev inequality in its simplest form has the form

$$\|g\|_{L^p(\mathbb{R}^n)} \leq C(p) \|\nabla g\|_{L^{p^*}(\mathbb{R}^n)},$$

for all $1 < p, p^* < \infty$ with

$$\frac{1}{p} = \frac{1}{p^*} - \frac{1}{n}.$$

Here ∇ is the usual gradient in $\mathbb{R}^n$. The following version of a Sobolev type inequality with respect to the operator $x \cdot \nabla$ instead of the standard gradient ∇ was considered in [BEHL08, OS09]:

$$\|g\|_{L^p(\mathbb{R}^n)} \leq C'(p) \|x \cdot \nabla g\|_{L^q(\mathbb{R}^n)}. \tag{6.50}$$

By putting $g(x) = h(\lambda x)$, $\lambda > 0$, into this inequality, we see that $p = q$ is a necessary condition to have (6.50).

We can notice that in formula (6.50) the operator $x \cdot \nabla$ can be interpreted as the homogeneous Euler operator on general homogeneous groups (see Section 1.3.2) as well as an operator on stratified groups by substituting x and ∇ with the corresponding *horizontal* operations x' and ∇_H related to the first stratum of the group.

The homogeneous groups version of such inequalities was discussed in Section 3.2.2. Thus, we will now concentrate on the horizontal interpretation presenting a range of Caffarelli–Kohn–Nirenberg and weighted L^p-Sobolev type inequalities on stratified Lie groups. All the inequalities can be obtained with sharp constants.

The presentation of the following results follows [RSY17a].

We start with an L^p-weighted Sobolev type inequality.

Theorem 6.6.1 (Horizontal weighted L^p-Sobolev type inequality). *Let $\mathbb{G}$ be a stratified group with N being the dimension of the first stratum. For any $f \in C_0^\infty(\mathbb{G}\backslash\{x'=0\})$, and all $\alpha \in \mathbb{R}$, we have*

$$\frac{|N-\alpha p|}{p}\left\|\frac{f}{|x'|^\alpha}\right\|_{L^p(\mathbb{G})} \leq \left\|\frac{x'\cdot\nabla_H f}{|x'|^\alpha}\right\|_{L^p(\mathbb{G})}, \quad 1<p<\infty, \tag{6.51}$$

where $|\cdot|$ is the Euclidean norm on $\mathbb{R}^N$. The constant $\frac{|N-\alpha p|}{p}$ is sharp when $N \neq \alpha p$.

Remark 6.6.2.

1. In the Abelian case $\mathbb{G}=(\mathbb{R}^n,+)$, that is, $N=n$ and $\nabla_H=\nabla=(\partial_{x_1},\ldots,\partial_{x_n})$, the inequality (6.51) yields the L^p-weighted Sobolev type inequality for $\mathbb{G}=\mathbb{R}^n$ with the sharp constant:

$$\frac{|n-\alpha p|}{p}\left\|\frac{f}{|x|_E^\alpha}\right\|_{L^p(\mathbb{R}^n)} \leq \left\|\frac{x\cdot\nabla f}{|x|_E^\alpha}\right\|_{L^p(\mathbb{R}^n)}, \tag{6.52}$$

for all $f \in C_0^\infty(\mathbb{R}^n\backslash\{0\})$, and $|x|_E = \sqrt{x_1^2+\cdots+x_n^2}$. This Euclidean inequality was shown in [OS09].

2. Using Schwarz' inequality in the right-hand side of (6.51) we see that (6.51) is a refinement of the L^p-weighted Hardy inequality on stratified groups: For any $f \in C_0^\infty(\mathbb{G}\backslash\{x'=0\})$, and all $\alpha \in \mathbb{R}$, we have

$$\frac{|N-\alpha p|}{p}\left\|\frac{f}{|x'|^\alpha}\right\|_{L^p(\mathbb{G})} \leq \left\|\frac{\nabla_H f}{|x'|^{\alpha-1}}\right\|_{L^p(\mathbb{G})}, \quad 1<p<\infty, \tag{6.53}$$

where $|\cdot|$ is the Euclidean norm on $\mathbb{R}^N$. If $N \neq \alpha p$ then the constant $\frac{|N-\alpha p|}{p}$ is sharp. Thus, (6.51) can be regarded as a refinement of (6.53). In the case of $p=2$ they are actually equivalent, see Theorem 6.6.3. These results have been obtained in [RSY17a].

3. For $\alpha = 0$, inequality (6.51) gives the weighted version of the homogeneous groups inequality in Proposition 3.2.1, Part (i). In particular, in the Euclidean case of $\mathbb{R}^n$ we have $N = n$, and this gives the weighted version of the inequality in Remark 3.2.2, Part 2.

Proof of Theorem 6.6.1. Let us assume $\alpha p \neq N$ since when $\alpha p = N$ there is nothing to prove. By using the identity (1.73) and the divergence theorem we obtain

$$\begin{aligned}
\int_{\mathbb{G}} \frac{|f(x)|^p}{|x'|^{\alpha p}} &= \frac{1}{N-\alpha p}\int_{\mathbb{G}} |f(x)|^p \operatorname{div}_H\left(\frac{x'}{|x'|^{\alpha p}}\right) dx \\
&= -\frac{p}{N-\alpha p}\operatorname{Re}\int_{\mathbb{G}} pf(x)|f(x)|^{p-2}\frac{\overline{x'\cdot \nabla_H f}}{|x'|^{\alpha p}} dx \\
&\leq \left|\frac{p}{N-\alpha p}\right| \int_{\mathbb{G}} \frac{|f(x)|^{p-1}}{|x'|^{\alpha p}} |x'\cdot \nabla_H f| dx \\
&\leq \left|\frac{p}{N-\alpha p}\right| \int_{\mathbb{G}} \frac{|f(x)|^{p-1}}{|x'|^{\alpha(p-1)}} \frac{|x'\cdot \nabla_H f|}{|x'|^{\alpha}} dx \\
&\leq \left|\frac{p}{N-\alpha p}\right| \left(\frac{|f(x)|^p}{|x'|^{\alpha p}} dx\right)^{(p-1)/p} \left(\frac{|x'\cdot \nabla_H f|^p}{|x'|^{\alpha p}} dx\right)^{1/p},
\end{aligned}$$

which implies (6.51). Here in the last line the Hölder inequality has been used. Now it remains to show the sharpness of the constant. Observe that the function

$$h_1(x) = \frac{1}{|x'|^{\frac{|N-\alpha p|}{p}}}, \quad N \neq \alpha p,$$

satisfies the equality condition in the Hölder inequality

$$\left|\frac{p}{N-\alpha p}\right|^p \frac{|x'\cdot \nabla_H h_1(x)|^p}{|x'|^{\alpha p}} = \frac{|h_1(x)|^p}{|x'|^{\alpha p}}.$$

This means that the constant $\frac{|N-\alpha p|}{p}$ is sharp. □

In the case of L^2 the horizontal Sobolev type inequality is actually equivalent to the Hardy inequality:

Theorem 6.6.3 (Equivalence of Sobolev type and Hardy inequalities in L^2). *Let $\mathbb{G}$ be a stratified group with N being the dimension of the first stratum with $N \geq 3$. Then the following two statements are equivalent:*

(a) *For any $f \in C_0^\infty(\mathbb{G}\backslash\{x'=0\})$, we have*

$$\|f\|_{L^2(\mathbb{G})} \leq \frac{2}{N}\|x'\cdot \nabla_H f\|_{L^2(\mathbb{G})}. \tag{6.54}$$

(b) *For any* $g \in C_0^\infty(\mathbb{G}\backslash\{x'=0\})$, *we have*

$$\left\|\frac{g}{|x'|}\right\|_{L^2(\mathbb{G})} \leq \frac{2}{N-2}\left\|\frac{x'}{|x'|}\cdot\nabla_H g\right\|_{L^2(\mathbb{G})}. \tag{6.55}$$

Proof of Theorem 6.6.3. Setting $g = |x'|f$ we obtain that

$$\|x'\cdot\nabla_H f\|^2_{L^2(\mathbb{G})} = \left\|-\frac{g}{|x'|}+\frac{x'}{|x'|}\cdot\nabla_H g\right\|^2_{L^2(\mathbb{G})} \tag{6.56}$$

$$= \left\|\frac{g}{|x'|}\right\|^2_{L^2(\mathbb{G})} - 2\mathrm{Re}\int_{\mathbb{G}}\frac{\overline{g(x)}}{|x'|}\frac{x'}{|x'|}\cdot\nabla_H g(x)dx + \left\|\frac{x'}{|x'|}\cdot\nabla_H g\right\|^2_{L^2(\mathbb{G})}.$$

By (1.73), one calculates

$$\begin{aligned}-2\mathrm{Re}\int_{\mathbb{G}}\frac{\overline{g(x)}}{|x'|}\frac{x'}{|x'|}\cdot\nabla_H g(x)dx &= -\int_{\mathbb{G}}\frac{x'}{|x'|^2}\nabla_H|g(x)|^2dx\\ &= \int_{\mathbb{G}}\mathrm{div}_H\left(\frac{x'}{|x'|^2}\right)|g(x)|^2dx\\ &= (N-2)\int_{\mathbb{G}}\frac{|g(x)|^2}{|x'|^2}dx.\end{aligned}$$

We obtain from the statement (a) and (6.56) that

$$\left\|\frac{g}{|x'|}\right\|^2_{L^2(\mathbb{G})} \leq \frac{4}{N^2}\left((N-1)\left\|\frac{g}{|x'|}\right\|^2_{L^2(\mathbb{G})} + \left\|\frac{x'}{|x'|}\cdot\nabla_H g\right\|^2_{L^2(\mathbb{G})}\right),$$

which implies (6.55). This shows that the statement (a) gives (b).

Conversely, assume that (b) holds. Put $f = g/|x'|$. Then we obtain

$$\begin{aligned}\left\|\frac{x'}{|x'|}\cdot\nabla_H(|x'|f)\right\|^2_{L^2(\mathbb{G})} &= \|f + x'\cdot\nabla_H f\|^2_{L^2(\mathbb{G})}\\ &= \|f\|^2_{L^2(\mathbb{G})} + 2\mathrm{Re}\int_{\mathbb{G}} x'f(x)\overline{\nabla_H f}dx + \|x'\cdot\nabla_H f\|^2_{L^2(\mathbb{G})}.\end{aligned}$$

Using (1.73), we have

$$2\mathrm{Re}\int_{\mathbb{G}} x'f(x)\overline{\nabla_H f}dx = -N\|f\|^2_{L^2(\mathbb{G})}.$$

It follows from the statement (b) that

$$\|f\|^2_{L^2(\mathbb{G})} \leq \frac{4}{(N-2)^2}(\|x'\cdot\nabla_H f\|^2_{L^2(\mathbb{G})} - (N-1)\|f\|^2_{L^2(\mathbb{G})}),$$

which implies (6.54). □

6.7 Horizontal extended Caffarelli–Kohn–Nirenberg inequalities

We now present horizontal extended Caffarelli–Kohn–Nirenberg inequalities in the setting of stratified groups. We recall that another version of such inequalities on general homogeneous groups involving the radial derivative was discussed in Section 3.3.

Theorem 6.7.1 (Horizontal Caffarelli–Kohn–Nirenberg type inequalities). *Let $1 < p, q < \infty$, $0 < r < \infty$ with $p + q \geq r$, $\delta \in [0, 1] \cap \left[\frac{r-q}{r}, \frac{p}{r}\right]$ and a, b, $c \in \mathbb{R}$. In addition, assume that*

$$\frac{\delta r}{p} + \frac{(1-\delta)r}{q} = 1 \quad \text{and} \quad c = \delta(a-1) + b(1-\delta).$$

Let $\mathbb{G}$ be a stratified group with N being the dimension of the first stratum with $N \neq p(1-a)$. Then the following inequality holds:

$$\||x'|^c f\|_{L^r(\mathbb{G})} \leq \left|\frac{p}{N + p(a-1)}\right|^{\delta} \||x'|^a \nabla_H f\|^{\delta}_{L^p(\mathbb{G})} \||x'|^b f\|^{1-\delta}_{L^q(\mathbb{G})} \tag{6.57}$$

for all $f \in C_0^\infty(\mathbb{G}\backslash\{0\})$. The constant in the inequality (6.57) is sharp for $p = q$ with $a - b = 1$ or $p \neq q$ with $p(1-a) + bq \neq 0$, or for $\delta = 0, 1$.

Remark 6.7.2.

1. In the Abelian case $\mathbb{G} = (\mathbb{R}^n, +)$, we have $N = n$ and $\nabla_H = \nabla = (\partial_{x_1}, \ldots, \partial_{x_n})$, so (6.57) implies the following Caffarelli–Kohn–Nirenberg type inequality for $\mathbb{G} = \mathbb{R}^n$: Let $1 < p, q < \infty$, $0 < r < \infty$ with $p + q \geq r$ and $\delta \in [0, 1] \cap \left[\frac{r-q}{r}, \frac{p}{r}\right]$ and a, b, $c \in \mathbb{R}$. Assume that $\frac{\delta r}{p} + \frac{(1-\delta)r}{q} = 1$ and $c = \delta(a-1) + b(1-\delta)$. Then we have

$$\||x|^c f\|_{L^r(\mathbb{R}^n)} \leq \left|\frac{p}{n + p(a-1)}\right|^{\delta} \||x|^a \nabla f\|^{\delta}_{L^p(\mathbb{R}^n)} \||x|^b f\|^{1-\delta}_{L^q(\mathbb{R}^n)}, \tag{6.58}$$

   for all $f \in C_0^\infty(\mathbb{R}^n\backslash\{0\})$, $|x| = \sqrt{x_1^2 + \cdots + x_n^2}$, and $n \neq p(1-a)$. The constant in the inequality (6.58) is sharp for $p = q$ with $a - b = 1$ or $p \neq q$ with $p(1-a) + bq \neq 0$, or for $\delta = 0, 1$.

2. The inequalities (6.58) give an extension of the Caffarelli–Kohn–Nirenberg in equalities Theorem 3.3.3 with respect to the range of indices. For example, let us take $1 < p = q = r < \infty$, $a = -\frac{n-2p}{p}$, $b = -\frac{n}{p}$ and $c = -\frac{n-\delta p}{p}$. Then by (6.58), for all $f \in C_0^\infty(\mathbb{R}^n\backslash\{0\})$ and all $1 < p < \infty$, $0 \leq \delta \leq 1$, we have the inequality

$$\left\|\frac{f}{|x|^{\frac{n-\delta p}{p}}}\right\|_{L^p(\mathbb{R}^n)} \leq \left\|\frac{\nabla f}{|x|^{\frac{n-2p}{p}}}\right\|^{\delta}_{L^p(\mathbb{R}^n)} \left\|\frac{f}{|x|^{\frac{n}{p}}}\right\|^{1-\delta}_{L^p(\mathbb{R}^n)}, \tag{6.59}$$

where ∇ is the standard gradient in $\mathbb{R}^n$. Since we have

$$\frac{1}{q}+\frac{b}{n}=\frac{1}{p}+\frac{1}{n}\left(-\frac{n}{p}\right)=0,$$

we see that (3.99) fails, so that the inequality (6.59) is not covered by Theorem 3.3.1.

Proof of Theorem 6.7.1. Case $\delta=0$. Notice that in this case we have $q=r$ and $b=c$ by $\frac{\delta r}{p}+\frac{(1-\delta)r}{q}=1$ and $c=\delta(a-1)+b(1-\delta)$, respectively. Then, the inequality (6.57) is reduced to

$$\||x'|^b f\|_{L^q(\mathbb{G})}\leq \left\||x'|^b f\right\|_{L^q(\mathbb{G})},$$

which is trivial.

Case $\delta=1$. In this case we have $p=r$ and $a-1=c$. By (6.53), for $N+cp\neq 0$ we obtain

$$\||x'|^c f\|_{L^r(\mathbb{G})}\leq \left|\frac{p}{N+cp}\right| \||x'|^{c+1}\nabla_H f\|_{L^r(\mathbb{G})}.$$

The constants in (6.53) is sharp, therefore, in this case the constant in (6.57) is sharp.

Case $\delta\in(0,1)\cap\left[\frac{r-q}{r},\frac{p}{r}\right]$. By using $c=\delta(a-1)+b(1-\delta)$, a direct calculation gives

$$\||x'|^c f\|_{L^r(\mathbb{G})}=\left(\int_{\mathbb{G}}|x'|^{cr}|f(x)|^r dx\right)^{1/r}=\left(\int_{\mathbb{G}}\frac{|f(x)|^{\delta r}}{|x'|^{\delta r(1-a)}}\frac{|f(x)|^{(1-\delta)r}}{|x'|^{-br(1-\delta)}}dx\right)^{1/r}.$$

Since we have $\delta\in(0,1)\cap\left[\frac{r-q}{r},\frac{p}{r}\right]$ and $p+q\geq r$, then by using Hölder's inequality for $\frac{\delta r}{p}+\frac{(1-\delta)r}{q}=1$, we obtain

$$\begin{aligned}\||x'|^c f\|_{L^r(\mathbb{G})}&\leq\left(\int_{\mathbb{G}}\frac{|f(x)|^p}{|x'|^{p(1-a)}}dx\right)^{\delta/p}\left(\int_{\mathbb{G}}\frac{|f(x)|^q}{|x'|^{-bq}}dx\right)^{(1-\delta)/q}\\&=\left\|\frac{f}{|x'|^{1-a}}\right\|^{\delta}_{L^p(\mathbb{G})}\left\|\frac{f}{|x'|^{-b}}\right\|^{1-\delta}_{L^q(\mathbb{G})}.\end{aligned}\tag{6.60}$$

When $p=q$ and $a-b=1$, the Hölder equality condition is satisfied for all compactly supported smooth functions. We also note that in the case $p\neq q$ the function

$$h_2(x)=|x'|^{\frac{1}{(p-q)}(p(1-a)+bq)}\tag{6.61}$$

satisfies the Hölder equality condition:

$$\frac{|h_2(x)|^p}{|x'|^{p(1-a)}}=\frac{|h_2(x)|^q}{|x'|^{-bq}}.$$

If $N \neq p(1-a)$, then by (6.53), we have

$$\left\| \frac{f}{|x'|^{1-a}} \right\|^{\delta}_{L^p(\mathbb{G})} \leq \left| \frac{p}{N+p(a-1)} \right|^{\delta} \left\| \frac{\nabla_H f}{|x'|^{-a}} \right\|^{\delta}_{L^p(\mathbb{G})}. \tag{6.62}$$

Combining this with (6.60), we get

$$\| |x'|^c f \|_{L^r(\mathbb{G})} \leq \left| \frac{p}{N+p(a-1)} \right|^{\delta} \left\| \frac{\nabla_H f}{|x'|^{-a}} \right\|^{\delta}_{L^p(\mathbb{G})} \left\| \frac{f}{|x'|^{-b}} \right\|^{1-\delta}_{L^q(\mathbb{G})}.$$

When we prove (6.62), in the same way as in the proof of Theorem 6.6.1, we note that

$$h_3(x) = |x'|^C, \quad C \neq 0, \tag{6.63}$$

satisfies the Hölder equality condition. Therefore, in the case $p = q$, $a - b = 1$ the Hölder equality condition of the inequalities (6.60) and (6.62) holds true for $h_3(x)$ in (6.63). Moreover, in the case $p \neq q$ and $p(1-a) + bq \neq 0$ the Hölder equality condition of the inequalities (6.60) and (6.62) holds true for $h_2(x)$ in (6.61). Therefore, the constant in (6.57) is sharp when $p = q$, $a - b = 1$ or $p \neq q$, $p(1-a) + bq \neq 0$. □

6.8 Horizontal Hardy–Rellich type inequalities for p-sub-Laplacians

We prove the following Hardy–Rellich type inequalities for p-sub-Laplacians on the stratified group $\mathbb{G}$. As usual, N is the dimension of the first stratum and $|\cdot|$ is the Euclidean norm on it, identified with $\mathbb{R}^N$.

Theorem 6.8.1 (Horizontal Hardy–Rellich inequalities for p-sub-Laplacian). *Let $1 < p < N$ with $\frac{1}{p} + \frac{1}{q} = 1$ and $\alpha, \beta \in \mathbb{R}$ be such that*

$$\frac{p-N}{p-1} \leq \gamma := \alpha + \beta + 1 \leq 0.$$

Then for all $f \in C_0^\infty(\mathbb{G} \backslash \{x' = 0\})$ we have

$$\frac{N + \gamma(p-1) - p}{p} \left\| \frac{\nabla_H f}{|x'|^{\frac{\gamma}{p}}} \right\|^p_{L^p(\mathbb{G})} \leq \left\| \frac{1}{|x'|^\alpha} \mathcal{L}_p f \right\|_{L^p(\mathbb{G})} \left\| \frac{\nabla_H f}{|x'|^\beta} \right\|_{L^q(\mathbb{G})}, \tag{6.64}$$

where $\mathcal{L}_p$ is the p-sub-Laplacian operator defined by

$$\mathcal{L}_p f := \operatorname{div}_H(|\nabla_H f|^{p-2} \nabla_H f). \tag{6.65}$$

Remark 6.8.2.

1. For $\beta = 0$, $\alpha = -1$ and $q = \frac{p}{p-1}$, the inequality (6.64) gives a stratified group Rellich type inequality for the p-sub-Laplacian $\mathcal{L}_p$:

$$\|\nabla_H f\|^p_{L^p(\mathbb{G})} \leq \frac{p}{N-p} \||x'|\mathcal{L}_p f\|_{L^p(\mathbb{G})} \|\nabla_H f\|_{L^{\frac{p}{p-1}}(\mathbb{G})}, \quad 1 < p < N, \tag{6.66}$$

for all $f \in C_0^\infty(\mathbb{G}\backslash\{x' = 0\})$.

2. For $\alpha = 0$, $\beta = -1$, the inequality (6.64) implies the following Heisenberg–Pauli–Weyl type uncertainty principle for the p-sub-Laplacian $\mathcal{L}_p$: for $1 < p < N$ and for all $f \in C_0^\infty(\mathbb{G}\backslash\{x' = 0\})$ we have

$$\|\nabla_H f\|^p_{L^p(\mathbb{G})} \leq \frac{p}{N-p} \|\mathcal{L}_p f\|_{L^p(\mathbb{G})} \||x'|\nabla_H f\|_{L^q(\mathbb{G})}, \quad \frac{1}{p} + \frac{1}{q} = 1. \tag{6.67}$$

Proof of Theorem 6.8.1. As in the proof of Theorem 6.1.1 we have

$$\begin{aligned}\int_{\mathbb{G}} \frac{|\nabla_H f(x)|^p}{|x'|^\gamma} dx &= \frac{1}{N-\gamma} \int_{\mathbb{G}} |\nabla_H f(x)|^p \mathrm{div}_H \left(\frac{x'}{|x'|^\gamma}\right) dx \\ &= -\frac{1}{N-\gamma} \int_{\mathbb{G}} \frac{p}{2} |\nabla_H f(x)|^{p-2} \frac{x' \cdot \nabla_H |\nabla_H f(x)|^2}{|x'|^\gamma} dx \\ &= \frac{p}{2(\gamma - N)} \int_{\mathbb{G}} |\nabla_H f(x)|^{p-2} \frac{x' \cdot \nabla_H |\nabla_H f(x)|^2}{|x'|^\gamma} dx.\end{aligned} \tag{6.68}$$

Moreover, we have

$$\begin{aligned}\int_{\mathbb{G}} \frac{\mathcal{L}_p f}{|x'|^\gamma} x' \cdot \nabla_H f(x) dx &= \int_{\mathbb{G}} \frac{\mathrm{div}_H(|\nabla_H f(x)|^{p-2} \nabla_H f(x))}{|x'|^\gamma} x' \cdot \nabla_H f(x) dx \\ &= -\int_{\mathbb{G}} |\nabla_H f(x)|^{p-2} \nabla_H f(x) \cdot \nabla_H \left(\frac{x' \cdot \nabla_H f(x)}{|x'|^\gamma}\right) dx\end{aligned}$$

$$= -\int_{\mathbb{G}} |\nabla_H f(x)|^{p-2} \left(\frac{|\nabla_H f(x)|^2}{|x'|^\gamma} + \frac{x' \cdot \nabla_H |\nabla_H f(x)|^2}{2|x'|^\gamma} - \frac{\gamma |x' \cdot \nabla_H f(x)|^2}{|x'|^{\gamma+2}}\right) dx,$$

that is,

$$\begin{aligned}&\int_{\mathbb{G}} \frac{|\nabla_H f(x)|^{p-2}}{|x'|^\gamma} x' \cdot \nabla_H |\nabla_H f(x)|^2 dx \\ &= 2\gamma \int_{\mathbb{G}} |\nabla_H f(x)|^{p-2} \frac{|x' \cdot \nabla_H f(x)|^2}{|x'|^{\gamma+2}} dx - 2 \int_{\mathbb{G}} \frac{|\nabla_H f(x)|^p}{|x'|^\gamma} dx \\ &\quad - 2 \int_{\mathbb{G}} \frac{\mathcal{L}_p f}{|x'|^\gamma} x' \cdot \nabla_H f(x) dx.\end{aligned}$$

Putting this in the right-hand side of (6.68) we obtain

$$\begin{aligned}\int_{\mathbb{G}} \frac{|\nabla_H f(x)|^p}{|x'|^\gamma} dx = & \frac{p\gamma}{\gamma - N} \int_{\mathbb{G}} |\nabla_H f(x)|^{p-2} \frac{|x' \cdot \nabla_H f(x)|^2}{|x'|^{\gamma+2}} dx \\ & - \frac{p}{\gamma - N} \int_{\mathbb{G}} \frac{|\nabla_H f(x)|^p}{|x'|^\gamma} dx - \frac{p}{\gamma - N} \int_{\mathbb{G}} \frac{\mathcal{L}_p f}{|x'|^\gamma} x' \cdot \nabla_H f(x) dx.\end{aligned}$$

Thus,

$$\begin{aligned}\int_{\mathbb{G}} \frac{\mathcal{L}_p f}{|x'|^\gamma} x' \cdot \nabla_H f(x) dx = & \frac{N - p - \gamma}{p} \int_{\mathbb{G}} \frac{|\nabla_H f(x)|^p}{|x'|^\gamma} dx \\ & + \gamma \int_{\mathbb{G}} |\nabla_H f(x)|^{p-2} \frac{|x' \cdot \nabla_H f(x)|^2}{|x'|^{\gamma+2}} dx.\end{aligned}$$

Since $\gamma \leq 0$, applying the Cauchy–Schwarz inequality to the last integrants we get

$$\begin{aligned}& \int_{\mathbb{G}} \frac{\mathcal{L}_p f}{|x'|^\gamma} x' \cdot \nabla_H f(x) dx \\ & = \frac{N - p - \gamma}{p} \int_{\mathbb{G}} \frac{|\nabla_H f(x)|^p}{|x'|^\gamma} dx + \gamma \int_{\mathbb{G}} |\nabla_H f(x)|^{p-2} \frac{|x' \cdot \nabla_H f(x)|^2}{|x'|^{\gamma+2}} dx \\ & \geq \frac{N - p - \gamma}{p} \int_{\mathbb{G}} \frac{|\nabla_H f(x)|^p}{|x'|^\gamma} dx + \gamma \int_{\mathbb{G}} \frac{|\nabla_H f(x)|^p}{|x'|^\gamma} dx \\ & = \frac{N + \gamma(p-1) - p}{p} \int_{\mathbb{G}} \frac{|\nabla_H f(x)|^p}{|x'|^\gamma} dx. \end{aligned} \tag{6.69}$$

Moreover, again applying the Cauchy–Schwarz inequality and the Hölder inequality we obtain

$$\begin{aligned}\int_{\mathbb{G}} \frac{\mathcal{L}_p f}{|x'|^\gamma} x' \cdot \nabla_H f(x) dx & \leq \int_{\mathbb{G}} \frac{\mathcal{L}_p f}{|x'|^{\gamma-1}} |\nabla_H f(x)| \, dx \\ & \leq \left(\int_{\mathbb{G}} \left| \frac{\mathcal{L}_p f}{|x'|^\alpha} \right|^p dx \right)^{1/p} \left(\int_{\mathbb{G}} \left| \frac{\nabla_H f}{|x'|^\beta} \right|^q dx \right)^{1/q}.\end{aligned}$$

Combining it with (6.69), the proof of Theorem 6.8.1 is complete. □

6.8.1 Inequalities for weighted p-sub-Laplacians

In this section, for a non-negative function $0 \leq \rho \in C^1(\mathbb{G})$ we consider the corresponding weighted p-sub-Laplacian

$$\mathcal{L}_{p,\rho} f = \operatorname{div}_H \left(\rho(x) |\nabla_H f|^{p-2} \nabla_H f \right), \quad 1 < p < \infty. \tag{6.70}$$

Depending on the function ρ, it satisfies the following inequalities.

Theorem 6.8.3 (Inequalities for weighted p-sub-Laplacian). *Let $0 < F \in C^\infty(\mathbb{G})$ and $0 \le \eta \in L^1_{\rm loc}(\mathbb{G})$ be such that*

$$\eta F^{p-1} \le -\mathcal{L}_{p,\rho} F \tag{6.71}$$

holds almost everywhere in $\mathbb{G}$. Then for each $2 \le p < \infty$ there is a positive constant $C_p > 0$ such that we have

$$\|\eta^{\frac{1}{p}} f\|^p_{L^p(\mathbb{G})} + C_p \left\| \rho^{\frac{1}{p}} F \nabla_H \frac{f}{F} \right\|^p_{L^p(\mathbb{G})} \le \|\rho^{\frac{1}{p}} \nabla_H f\|^p_{L^p(\mathbb{G})}, \tag{6.72}$$

for all real-valued functions $f \in C_0^\infty(\mathbb{G})$.

Proof of Theorem 6.8.3. We observe first that for all $x, y \in \mathbb{R}^n$ there exists a positive number C_p such that

$$|x|^p + C_p |y|^p + p|x|^{p-2} x \cdot y \le |x+y|^p, \quad 2 \le p < \infty. \tag{6.73}$$

Therefore, we have the estimate

$$\begin{aligned} |g|^p |\nabla_H F|^p + C_p F^p |\nabla_H g|^p + F|\nabla_H F|^{p-2} \nabla_H F \cdot \nabla_H |g|^p \\ \le |g \nabla_H F + F \nabla_H g|^p = |\nabla_H f|^p, \end{aligned}$$

with $g = \frac{f}{F}$. This implies that

$$\begin{aligned} \int_{\mathbb{G}} \rho(x) |\nabla_H f(x)|^p dx \ge & \int_{\mathbb{G}} \rho(x) |\nabla_H F(x)|^p |g(x)|^p dx \\ & + C_p \int_{\mathbb{G}} \rho(x) |\nabla_H g(x)|^p |F(x)|^p dx \\ & - \int_{\mathbb{G}} {\rm div}_H(\rho(x) F(x) |\nabla_H F(x)|^{p-2} \nabla_H F(x)) |g(x)|^p dx \\ \ge & C_p \int_{\mathbb{G}} \rho(x) |\nabla_H g(x)|^p |F(x)|^p dx \\ & + \int_{\mathbb{G}} -{\rm div}_H(\rho(x) |\nabla_H F(x)|^{p-2} \nabla_H F(x)) F(x) |g(x)|^p dx. \end{aligned}$$

Using the assumption (6.71) it follows that

$$\int_{\mathbb{G}} \eta(x) |g(x)|^p |F(x)|^p dx + C_p \int_{\mathbb{G}} \rho(x) |\nabla_H g(x)|^p |F(x)|^p dx \le \int_{\mathbb{G}} \rho(x) |\nabla_H f(x)|^p dx.$$

Since $g = \frac{f}{F}$ we obtain

$$\|\eta^{\frac{1}{p}} f\|^p_{L^p(\mathbb{G})} + C_p \left\| \rho^{\frac{1}{p}} F \nabla_H \left(\frac{f}{F}\right) \right\|^p_{L^p(\mathbb{G})} \le \|\rho^{\frac{1}{p}} \nabla_H f\|^p_{L^p(\mathbb{G})},$$

proving (6.72). □

Remark 6.8.4.

1. For $p = 2$, the inequality (6.73) becomes an equality with $C_2 = 1$. Therefore, the proof yields a remainder formula for $p = 2$ in the form

$$\left\| \rho^{\frac{1}{2}} F \nabla_H \frac{f}{F} \right\|^2_{L^2(\mathbb{G})} = \|\rho^{\frac{1}{2}} \nabla_H f\|^2_{L^2(\mathbb{G})} - \|\eta^{\frac{1}{2}} f\|^2_{L^2(\mathbb{G})}. \tag{6.74}$$

2. In the case of $1 < p < 2$ the inequality (6.73) can be also stated in the form that for all $x, y \in \mathbb{R}^n$ there exists a positive constant $C_p > 0$ such that

$$|x|^p + C_p \frac{|y|^p}{(|x| + |y|)^{2-p}} + p|x|^{p-2} x \cdot y \leq |x + y|^p, \quad 1 < p < 2, \tag{6.75}$$

see, e.g., [Lin90, Lemma 4.2]. Thus, from the proof it then follows that we have

$$\|\eta^{\frac{1}{p}} f\|^p_{L^p(\mathbb{G})} + C_p \left\| \rho^{\frac{1}{2}} \left(\left| \frac{f}{F} \nabla_H F \right| + F \left| \nabla_H \left(\frac{f}{F} \right) \right| \right)^{\frac{p-2}{2}} |F| \nabla_H \left(\frac{f}{F} \right) \right\|^2_{L^2(\mathbb{G})} \leq \|\rho^{\frac{1}{p}} \nabla_H f\|^p_{L^p(\mathbb{G})}, \tag{6.76}$$

for all real-valued functions $f \in C_0^\infty(\mathbb{G})$.

As a special case, we can apply Theorem 6.8.3 to the usual p-sub-Laplacian by taking the function $\rho \equiv 1$. In turn, this gives another proof of the L^p-Hardy inequality (6.6):

Corollary 6.8.5 (Horizontal L^p-Hardy inequality). *For $f \in C_0^\infty(\mathbb{G}\backslash\{0\})$ we have*

$$\left\| \frac{f}{|x'|} \right\|_{L^p} \leq \frac{p}{N - p} \|\nabla_H f\|_{L^p}, \quad 1 < p < N. \tag{6.77}$$

Proof of Corollary 6.8.5. In Theorem 6.8.3 setting $\rho = 1$ and

$$F_\epsilon = |x'_\epsilon|^{-\frac{\theta-p-2}{p}} = \left((x'_1 + \epsilon)^2 + \cdots + (x'_n + \epsilon)^2\right)^{-\frac{\theta-p-2}{2p}},$$

for a given $\epsilon > 0$, using the identity (1.72) we obtain

$$\begin{aligned}
-\mathcal{L}_{p,1} F_\epsilon &= -\mathrm{div}_H \left(|\nabla_H F_\epsilon|^{p-2} \nabla_H F_\epsilon \right) \\
&= -\mathrm{div}_H \left(|\nabla_H |x'_\epsilon|^{-\frac{\theta-p-2}{p}}|^{p-2} \nabla_H |x'_\epsilon|^{-\frac{\theta-p-2}{p}} \right) \\
&= \frac{\theta - p - 2}{p} \left| \frac{\theta - p - 2}{p} \right|^{p-2} \left(\frac{\theta - p - 2}{p} - \theta + 2 + N \right) |x'_\epsilon|^{-\frac{(\theta-p-2)(p-1)}{p} - p}
\end{aligned}$$

$$= \left(\left| \frac{\theta - p - 2}{p} \right|^p + \frac{\theta - p - 2}{p} \left| \frac{\theta - p - 2}{p} \right|^{p-2} (-\theta + 2 + N) \right) |x'_\epsilon|^{-\frac{(\theta-p-2)(p-1)}{p} - p}. \tag{6.78}$$

If $1 < p < \theta - 2$ and $\theta \leq 2 + N$, then (6.78) gives

$$-\mathcal{L}_{p,1} F_\epsilon \geq \left| \frac{\theta - p - 2}{p} \right|^p \frac{1}{|x'_\epsilon|^p} F_\epsilon^{p-1},$$

that is, according to the assumption in Theorem 6.8.3, we can set

$$\eta(x) = \left| \frac{\theta - p - 2}{p} \right|^p \frac{1}{|x'_\epsilon|^p}.$$

It follows that (6.72) (and also (6.76)) implies

$$\left\| \frac{f}{|x'|} \right\|_{L^p} \leq \frac{p}{\theta - p - 2} \|\nabla_H f\|_{L^p}, \quad 1 < p < \theta - 2,\ \theta \leq 2 + N,\ \theta \in \mathbb{R}.$$

Optimizing with respect to θ we obtain (6.77). □

Remark 6.8.6.

1 A version of Theorem 6.8.3 in the Euclidean case was shown in [Yen16]. In the presentation of this section we followed [RS17e].

2. The Heisenberg group version of (6.77) was shown in [D'A04b]. Here it is worth to recall that on the Heisenberg group we have $Q = N + 2$.

3. We have included Corollary 6.8.5 as a consequence of Theorem 6.8.3 to demonstrate that this method actually also yields best constants in some inequalities, as this constant in the L^p-Hardy inequality (6.6) was sharp.

6.9 Horizontal Rellich inequalities for sub-Laplacians with drift

In this section, we discuss (weighted) Rellich inequalities for sub-Laplacians with drift. For this, we assume all the notation of Section 1.4.6 where sub-Laplacians with drift have been discussed.

In this section we will discuss the horizontal versions, that is, with the weights being the powers of $|x'|$. In Section 7.4, we will discuss a version with weights in terms of the $\mathcal{L}$-gauge but that analysis is currently available only in the setting of polarizable Carnot groups. In the presentation of this section as well as of Section 7.4 we follow [RY18b].

The following result shows that the drift allows one to improve over the Rellich inequality without drift, given in Theorem 6.2.2.

Theorem 6.9.1 (Horizontal Rellich inequalities for sub-Laplacians with drift). *Let $\mathbb{G}$ be a stratified group with $N \geq 3$ being the dimension of the first stratum. Let $\delta \in \mathbb{R}$ with $-N/2 \leq \delta \leq -1$. Then for all functions $f \in C_0^\infty(\mathbb{G}\backslash\{x'=0\})$ we have*

$$\begin{aligned}\left\|\frac{\mathcal{L}_X f}{|x'|^{\delta}}\right\|^2_{L^2(\mathbb{G},\mu_X)} \geq & \left(\frac{(N-2\delta-4)(N+2\delta)}{4}\right)^2 \left\|\frac{f}{|x'|^{\delta+2}}\right\|^2_{L^2(\mathbb{G},\mu_X)} \\ & + \gamma^2 b_X^2 \frac{(N-2\delta-2)(N+2\delta-2)}{2} \left\|\frac{f}{|x'|^{\delta+1}}\right\|^2_{L^2(\mathbb{G},\mu_X)} \\ & + \gamma^4 b_X^4 \left\|\frac{f}{|x'|^{\delta}}\right\|^2_{L^2(\mathbb{G},\mu_X)},\end{aligned} \tag{6.79}$$

where $\mathcal{L}_X$ and b_X are defined in (1.93) and (1.95), respectively. If $(N+2\delta)(N+2\delta-2) \neq 0$, then the constants in (6.79) are sharp. Moreover, when $\delta = 0$ and $N > 4$, for all functions $f \in C_0^\infty(\mathbb{G}\backslash\{x'=0\})$ we have

$$\begin{aligned}\|\mathcal{L}_X f\|^2_{L^2(\mathbb{G},\mu_X)} \geq & \left(\frac{N(N-4)}{4}\right)^2 \left\|\frac{f}{|x'|^2}\right\|^2_{L^2(\mathbb{G},\mu_X)} \\ & + \gamma^2 b_X^2 \frac{(N-2)^2}{2} \left\|\frac{f}{|x'|}\right\|^2_{L^2(\mathbb{G},\mu_X)} + \gamma^4 b_X^4 \|f\|^2_{L^2(\mathbb{G},\mu_X)},\end{aligned} \tag{6.80}$$

with sharp constants. The constants in (6.79) and (6.80) are sharp in the sense that there is a sequence of functions such that the equalities in (6.79) and (6.80) are attained in the limit of this sequence of functions, respectively.

Remark 6.9.2.

1. The improvement in Rellich inequalities with drift compared to the standard ones as in Theorem 6.2.2 can be seen since for $(N-2\delta-2)(N+2\delta-2) \geq 0$, by dropping positive terms in (6.79) we get the following 'standard' Rellich type inequality for all functions $f \in C_0^\infty(\mathbb{G}\backslash\{x'=0\})$

$$\left\|\frac{\mathcal{L}_X f}{|x'|^{\delta}}\right\|^2_{L^2(\mathbb{G},\mu_X)} \geq \left(\frac{(N-2\delta-4)(N+2\delta)}{4}\right)^2 \left\|\frac{f}{|x'|^{\delta+2}}\right\|^2_{L^2(\mathbb{G},\mu_X)}, \tag{6.81}$$

where $\delta \in \mathbb{R}$ with $-N/2 \leq \delta \leq -1$ and $N \geq 3$.

Similarly, from (6.80) we obtain for $N > 4$ and for all functions $f \in C_0^\infty(\mathbb{G}\backslash\{x'=0\})$ the inequality

$$\|\mathcal{L}_X f\|_{L^2(\mathbb{G},\mu_X)} \geq \frac{N(N-4)}{4} \left\|\frac{f}{|x'|^2}\right\|_{L^2(\mathbb{G},\mu_X)}, \tag{6.82}$$

which can be compared to the Rellich inequality in Corollary 6.5.2.

2. In the Euclidean case $\mathbb{G} = (\mathbb{R}^n, +)$, we have $N = n$, $\nabla_H = \nabla = (\partial_{x_1}, \ldots, \partial_{x_n})$ is the usual full gradient, and setting

$$X = \sum_{i=1}^{n} a_i \partial_{x_i}$$

for $a_i \in \mathbb{R}$ for $i = 1, \ldots, n$, and $\delta = -\alpha$, $\gamma \in \mathbb{R}$, (6.79) implies, for $\alpha \geq 1$ and $n \geq \max\{3, 2\alpha\}$, $n + 2\alpha - 4 > 0$, that for all functions $f \in C_0^\infty(\mathbb{R}^n \backslash \{0\})$ we have

$$\begin{aligned} &\left\| |x|^\alpha \left(\Delta + \gamma \sum_{i=1}^{n} a_i \partial_{x_i} \right) f \right\|^2_{L^2(\mathbb{R}^n, \mu_X)} \\ &\geq \frac{(n + 2\alpha - 4)^2 (n - 2\alpha)^2}{16} \left\| |x|_E^{\alpha-2} f \right\|^2_{L^2(\mathbb{R}^n, \mu_X)} \\ &+ \gamma^2 b_X^2 \frac{(n + 2\alpha - 2)(n - 2\alpha - 2)}{2} \left\| |x|_E^{\alpha-1} f \right\|^2_{L^2(\mathbb{R}^n, \mu_X)} \\ &+ \gamma^4 b_X^4 \left\| |x|_E^\alpha f \right\|^2_{L^2(\mathbb{R}^n, \mu_X)}, \end{aligned} \tag{6.83}$$

with the measure μ_X on $\mathbb{R}^n$ given by

$$d\mu_X = e^{-\gamma \sum_{i=1}^n a_i x_i} dx,$$

where dx is the Lebesgue measure, and

$$b_X = \frac{1}{2} \left(\sum_{j=1}^{n} a_j^2 \right)^{1/2}.$$

If $(n - 2\alpha)(n - 2\alpha - 2) \neq 0$ with $\alpha \geq 1$ and $n \geq 2\alpha$, then the constants in (6.83) are sharp, in the sense that there is a sequence of functions such that the equality in (6.83) is attained in the limit of this sequence of functions.

In particular, for $\alpha = 0$, in the Euclidean setting of $\mathbb{R}^n$ with $n \geq 5$, for all $a_i \in \mathbb{R}$ for $i = 1, \ldots, n$ and $\gamma \in \mathbb{R}$, and all $f \in C_0^\infty(\mathbb{R}^n \backslash \{0\})$ we have a family of inequalities

$$\begin{aligned} &\left\| \left(\Delta + \gamma \sum_{i=1}^{n} a_i \partial_{x_i} \right) f \right\|^2_{L^2(\mathbb{R}^n, \mu_X)} \\ &\geq \frac{n^2 (n-4)^2}{16} \left\| \frac{f}{|x|^2} \right\|^2_{L^2(\mathbb{R}^n, \mu_X)} + \gamma^4 b_X^4 \| f \|^2_{L^2(\mathbb{R}^n, \mu_X)} \\ &+ \gamma^2 b_X^2 \frac{(n-2)^2}{2} \left\| \frac{f}{|x|} \right\|^2_{L^2(\mathbb{R}^n, \mu_X)}. \end{aligned} \tag{6.84}$$

All the constants in (6.84) are sharp in the sense that there is a sequence of functions such that the equality in (6.84) is attained in the limit of this sequence of functions.

Proof of Theorem 6.9.1. We denote by χ the positive character on $\mathbb{G}$ that appeared in Proposition 1.4.14. Let $g = g(x) \in C_0^\infty(\mathbb{G}\backslash\{x'=0\})$ be such that $f = \chi^{-1/2}g$. Since the mapping (1.101) is an isomorphism we have

$$\left\|\frac{\mathcal{L}_X f}{|x'|^\delta}\right\|_{L^2(\mathbb{G},\mu_X)} = \left\|\frac{\chi^{1/2}\mathcal{L}_X f}{|x'|^\delta}\right\|_{L^2(\mathbb{G},\mu)} = \left\|\frac{\chi^{1/2}\mathcal{L}_X(\chi^{-1/2}g)}{|x'|^\delta}\right\|_{L^2(\mathbb{G},\mu)}.$$

By this, (1.100) and integration by parts, we have the equalities

$$\begin{aligned}
\left\|\frac{\mathcal{L}_X f}{|x'|^\delta}\right\|^2_{L^2(\mathbb{G},\mu_X)} &= \left\|\frac{(\mathcal{L}_0+\gamma^2 b_X^2)g}{|x'|^\delta}\right\|^2_{L^2(\mathbb{G},\mu)}\\
&= \left\|\frac{\mathcal{L}_0 g}{|x'|^\delta}\right\|^2_{L^2(\mathbb{G},\mu)} + 2\gamma^2 b_X^2 \mathrm{Re}\int_{\mathbb{G}} \frac{\mathcal{L}_0 g(x)\overline{g(x)}}{|x'|^{2\delta}}dx + \gamma^4 b_X^4 \left\|\frac{g}{|x'|^\delta}\right\|^2_{L^2(\mathbb{G},\mu)}\\
&= \left\|\frac{\mathcal{L}_0 g}{|x'|^\delta}\right\|^2_{L^2(\mathbb{G},\mu)} - 2\gamma^2 b_X^2 \mathrm{Re}\sum_{j=1}^N\int_{\mathbb{G}} \frac{X_j^2 g(x)\overline{g(x)}}{|x'|^{2\delta}}dx + \gamma^4 b_X^4 \left\|\frac{g}{|x'|^\delta}\right\|^2_{L^2(\mathbb{G},\mu)}\\
&= \left\|\frac{\mathcal{L}_0 g}{|x'|^\delta}\right\|^2_{L^2(\mathbb{G},\mu)} + 2\gamma^2 b_X^2 \int_{\mathbb{G}} \frac{|\nabla_H g(x)|^2}{|x'|^{2\delta}}\\
&\quad - 4\delta\gamma^2 b_X^2 \mathrm{Re}\sum_{j=1}^N\int_{\mathbb{G}} \frac{x'_j X_j g(x)\overline{g(x)}}{|x'|^{2\delta+2}}dx + \gamma^4 b_X^4 \left\|\frac{g}{|x'|^\delta}\right\|^2_{L^2(\mathbb{G},\mu)}.
\end{aligned} \tag{6.85}$$

Since we also have the equality

$$\begin{aligned}
&\mathrm{Re}\sum_{j=1}^N\int_{\mathbb{G}} \frac{x'_j X_j g(x)\overline{g(x)}}{|x'|^{2\delta+2}}dx\\
&\quad = (2\delta+2-N)\int_{\mathbb{G}}\frac{|g(x)|^2}{|x'|^{2\delta+2}}dx - \mathrm{Re}\sum_{j=1}^N\int_{\mathbb{G}} \frac{x'_j g(x)\overline{X_j g(x)}}{|x'|^{2\delta+2}}dx,
\end{aligned}$$

we obtain

$$\mathrm{Re}\sum_{j=1}^N\int_{\mathbb{G}} \frac{x'_j X_j g(x)\overline{g(x)}}{|x'|^{2\delta+2}}dx = \frac{2\delta+2-N}{2}\int_{\mathbb{G}}\frac{|g(x)|^2}{|x'|^{2\delta+2}}dx.$$

If we plug this into (6.85) we get

$$\begin{aligned}
\left\|\frac{\mathcal{L}_X f}{|x'|^\delta}\right\|^2_{L^2(\mathbb{G},\mu_X)} &= \left\|\frac{\mathcal{L}_0 g}{|x'|^\delta}\right\|^2_{L^2(\mathbb{G},\mu)} + 2\gamma^2 b_X^2\left\|\frac{\nabla_H g}{|x'|^\delta}\right\|^2_{L^2(\mathbb{G},\mu)}\\
&\quad + 2\delta(N-2\delta-2)\gamma^2 b_X^2\left\|\frac{g}{|x'|^{\delta+1}}\right\|^2_{L^2(\mathbb{G},\mu)} + \gamma^4 b_X^4\left\|\frac{g}{|x'|^\delta}\right\|^2_{L^2(\mathbb{G},\mu)}.
\end{aligned} \tag{6.86}$$

Using the Rellich (6.20) and Hardy (6.21) inequalities, we get from (6.86) that

$$\begin{aligned}\left\|\frac{\mathcal{L}_X f}{|x'|^\delta}\right\|^2_{L^2(\mathbb{G},\mu_X)} \geq & \left(\frac{(N-2\delta-4)(N+2\delta)}{4}\right)^2 \left\|\frac{g}{|x'|^{\delta+2}}\right\|^2_{L^2(\mathbb{G},\mu)} \\ & + \gamma^4 b_X^4 \left\|\frac{g}{|x'|^\delta}\right\|^2_{L^2(\mathbb{G},\mu)} + 2\gamma^2 b_X^2 \left(\frac{N-2\delta-2}{2}\right)^2 \left\|\frac{g}{|x'|^{\delta+1}}\right\|^2_{L^2(\mathbb{G},\mu)} \\ & + 2\delta(N-2\delta-2)\gamma^2 b_X^2 \left\|\frac{g}{|x'|^{\delta+1}}\right\|^2_{L^2(\mathbb{G},\mu)}.\end{aligned}$$

It follows then that

$$\begin{aligned}\left\|\frac{\mathcal{L}_X f}{|x'|^\delta}\right\|^2_{L^2(\mathbb{G},\mu_X)} \geq & \left(\frac{(N-2\delta-4)(N+2\delta)}{4}\right)^2 \left\|\frac{f}{|x'|^{\delta+2}}\right\|^2_{L^2(\mathbb{G},\mu_X)} \\ & + \gamma^4 b_X^4 \left\|\frac{f}{|x'|^\delta}\right\|^2_{L^2(\mathbb{G},\mu_X)} \\ & + \gamma^2 b_X^2 \frac{(N-2\delta-2)(N+2\delta-2)}{2} \left\|\frac{f}{|x'|^{\delta+1}}\right\|^2_{L^2(\mathbb{G},\mu_X)}.\end{aligned}$$

As we have discussed in the proof of Theorem 6.2.2, since the same function satisfies the equality conditions in Hölder's inequalities, the constants in (6.79) are sharp.

To obtain (6.80), that is the unweighted case $\delta = 0$, we use the inequality (6.21) and (6.43) in Corollary 6.5.2 that gives the inequality

$$\|\mathcal{L}f\|_{L^2(\mathbb{G})} \geq \frac{N(N-4)}{4} \left\|\frac{f}{|x'|^2}\right\|_{L^2(\mathbb{G})}, \quad N \geq 5, \tag{6.87}$$

for $f \in C_0^\infty(\mathbb{G}\backslash\{x'=0\})$. Since it is known from Corollary 6.5.2 that the constant $\frac{N(N-4)}{4}$ is sharp in (6.87), using the same argument as for the constants in (6.79), we obtain the sharpness of the constants in (6.80). $\square$

6.10 Horizontal anisotropic Hardy and Rellich inequalities

In this section we discuss the anisotropic versions of horizontal Hardy and Rellich inequalities. These inequalities appear in the analysis of anisotropic p-sub-Laplacians. The presentation of this section follows [RSS18a]. To put the notions in perspective, we start by recalling the Euclidean counterparts of the appearing objects.

The anisotropic Laplacian (on $\mathbb{R}^N$) is defined by

$$\sum_{i=1}^{N} \frac{\partial}{\partial x_i}\left(\left|\frac{\partial u}{\partial x_i}\right|^{p_i-2}\frac{\partial u}{\partial x_i}\right), \tag{6.88}$$

for $p_i > 1$, with $i = 1, \ldots, N$. Note that choosing $p_i = 2$ or $p_i = p$ for all i in (6.88) we get the Laplacian and the pseudo-p-Laplacian, respectively.

A subelliptic analogue of the operator in (6.88) is the anisotropic p-sub-Laplacian on stratified groups which is the operator of the form

$$\mathcal{L}_p f := \sum_{i=1}^{N} X_i\left(|X_i f|^{p_i-2}X_i f\right), \quad 1 < p_i < \infty,$$

where X_i, $i = 1, \ldots, N$, are the generators of the first stratum of a stratified Lie group.

Following the classical scheme for the analysis of such operators, first, we present the horizontal versions of the so-called Picone type identities. As a consequence, Hardy and Rellich type inequalities for anisotropic sub-Laplacians can be obtained.

6.10.1 Horizontal Picone identities

First, we discuss the horizontal Picone type identity on a stratified group $\mathbb{G}$.

Lemma 6.10.1 (Horizontal Picone identity). *Let $\Omega \subset \mathbb{G}$ be an open set of a stratified group $\mathbb{G}$, and let N be the dimension of the first stratum of $\mathbb{G}$. Let u, v be differentiable a.e. in Ω, $v > 0$ a.e. in Ω and $u \geq 0$. Denote*

$$R(u,v) := \sum_{i=1}^{N} |X_i u|^{p_i} - \sum_{i=1}^{N} X_i\left(\frac{u^{p_i}}{v^{p_i-1}}\right)|X_i v|^{p_i-2} X_i v, \tag{6.89}$$

and

$$L(u,v) := \sum_{i=1}^{N} |X_i u|^{p_i} - \sum_{i=1}^{N} p_i \frac{u^{p_i-1}}{v^{p_i-1}}|X_i v|^{p_i-2} X_i v X_i u + \sum_{i=1}^{N}(p_i-1)\frac{u^{p_i}}{v^{p_i}}|X_i v|^{p_i}, \tag{6.90}$$

where $p_i > 1$, $i = 1, \ldots, N$. *Then we have*

$$L(u, v) = R(u, v) \geq 0. \tag{6.91}$$

In addition, we have $L(u, v) = 0$ *a.e. in* Ω *if and only if* $u = cv$ *a.e. in* Ω *with a positive constant* c.

Remark 6.10.2.

1. The Euclidean case of Lemma 6.10.1 was obtained by Feng and Cui [FC17].
2. Our proof of Lemma 6.10.1 follows [RSS18a] and is based on the method of Allegretto and Huang [AH98] for the (Euclidean) p-Laplacian, see also [NZW01].

Proof of Lemma 6.10.1. A direct computation gives

$$\begin{aligned}
R(u, v) &= \sum_{i=1}^{N} |X_i u|^{p_i} - \sum_{i=1}^{N} X_i \left(\frac{u^{p_i}}{v^{p_i - 1}} \right) |X_i v|^{p_i - 2} X_i v \\
&= \sum_{i=1}^{N} |X_i u|^{p_i} - \sum_{i=1}^{N} \frac{p_i u^{p_i - 1} X_i u v^{p_i - 1} - u^{p_i} (p_i - 1) v^{p_i - 2} X_i v}{(v^{p_i - 1})^2} |X_i v|^{p_i - 2} X_i v \\
&= \sum_{i=1}^{N} |X_i u|^{p_i} - \sum_{i=1}^{N} p_i \frac{u^{p_i - 1}}{v^{p_i - 1}} |X_i v|^{p_i - 2} X_i v X_i u + \sum_{i=1}^{N} (p_i - 1) \frac{u^{p_i}}{v^{p_i}} |X_i v|^{p_i} \\
&= L(u, v).
\end{aligned}$$

This proves the equality in (6.91). Now we rewrite $L(u, v)$ to see that $L(u, v) \geq 0$, that is, we write

$$\begin{aligned}
L(u, v) = \sum_{i=1}^{N} |X_i u|^{p_i} &- \sum_{i=1}^{N} p_i \frac{u^{p_i - 1}}{v^{p_i - 1}} |X_i v|^{p_i - 1} |X_i u| + \sum_{i=1}^{N} (p_i - 1) \frac{u^{p_i}}{v^{p_i}} |X_i v|^{p_i} \\
&+ \sum_{i=1}^{N} p_i \frac{u^{p_i - 1}}{v^{p_i - 1}} |X_i v|^{p_i - 2} \left(|X_i v| |X_i u| - X_i v X_i u \right) = S_1 + S_2,
\end{aligned}$$

where we denote

$$\begin{aligned}
S_1 := &\sum_{i=1}^{N} p_i \left[\frac{1}{p_i} |X_i u|^{p_i} + \frac{p_i - 1}{p_i} \left(\left(\frac{u}{v} |X_i v| \right)^{p_i - 1} \right)^{\frac{p_i}{p_i - 1}} \right] \\
&- \sum_{i=1}^{N} p_i \frac{u^{p_i - 1}}{v^{p_i - 1}} |X_i v|^{p_i - 1} |X_i u|,
\end{aligned}$$

and

$$S_2 := \sum_{i=1}^{N} p_i \frac{u^{p_i - 1}}{v^{p_i - 1}} |X_i v|^{p_i - 2} \left(|X_i v| |X_i u| - X_i v X_i u \right).$$

We can see that $S_2 \geq 0$ due to $|X_i v||X_i u| \geq X_i v X_i u$. To check that we also have $S_1 \geq 0$, we will use Young's inequality for $a \geq 0$ and $b \geq 0$:

$$ab \leq \frac{a^{p_i}}{p_i} + \frac{b^{q_i}}{q_i}, \tag{6.92}$$

for $p_i > 1$, $q_i > 1$ and $\frac{1}{p_i} + \frac{1}{q_i} = 1$, for all $i = 1, \ldots, N$. The equality in (6.92) holds if and only if $a^{p_i} = b^{q_i}$, that is, if $a = b^{\frac{1}{p_i - 1}}$.

Let us now take $a = |X_i u|$ and $b = \left(\frac{u}{v}|X_i v|\right)^{p_i - 1}$ and apply (6.92) to get

$$p_i |X_i u| \left(\frac{u}{v}|X_i v|\right)^{p_i - 1} \leq p_i \left[\frac{1}{p_i}|X_i u|^{p_i} + \frac{p_i - 1}{p_i}\left(\left(\frac{u}{v}|X_i v|\right)^{p_i - 1}\right)^{\frac{p_i}{p_i - 1}}\right]. \tag{6.93}$$

From this we see that $S_1 \geq 0$ which proves that $L(u, v) = S_1 + S_2 \geq 0$.

It is easy to see that $u = cv$ implies $R(u, v) = 0$. Now let us prove that $L(u, v) = 0$ implies $u = cv$. Due to $u(x) \geq 0$ and since $L(u, v)(x_0) = 0$, $x_0 \in \Omega$, we can consider two cases $u(x_0) > 0$ and $u(x_0) = 0$.

(a) For the case $u(x_0) > 0$ we conclude from $L(u, v)(x_0) = 0$ that $S_1 = 0$ and $S_2 = 0$. Then $S_1 = 0$ implies

$$|X_i u| = \frac{u}{v}|X_i v|, \quad i = 1, \ldots, N, \tag{6.94}$$

and $S_2 = 0$ implies

$$|X_i v||X_i u| - X_i v X_i u = 0, \quad i = 1, \ldots, N. \tag{6.95}$$

The combination of (6.94) and (6.95) gives

$$\frac{X_i u}{X_i v} = \frac{u}{v} = c, \quad \text{with} \quad c \neq 0, \quad i = 1, \ldots, N. \tag{6.96}$$

(b) Let us denote

$$\Omega^* := \{x \in \Omega : u(x) = 0\}.$$

If $\Omega^* \neq \Omega$, then suppose that $x_0 \in \partial\Omega^*$. Then there exists a sequence $x_k \notin \Omega^*$ such that $x_k \to x_0$. In particular, $u(x_k) \neq 0$, and hence by Case (a) we have $u(x_k) = cv(x_k)$. Passing to the limit we get $u(x_0) = cv(x_0)$. Since $u(x_0) = 0$ and $v(x_0) \neq 0$, we get that $c = 0$. But then by Case (a) again, since $u = cv$ and $u \neq 0$ in $\Omega \backslash \Omega^*$, it is impossible to have $c = 0$. This contradiction implies that $\Omega^* = \Omega$.

This completes the proof of Lemma 6.10.1. □

The following consequence of Lemma 6.10.1 will be instrumental in the proof of the horizontal anisotropic Hardy inequality in Theorem 6.10.5.

Lemma 6.10.3. *Let $\Omega \subset \mathbb{G}$ be an open set of a stratified group $\mathbb{G}$, and let N be the dimension of the first stratum of $\mathbb{G}$. Let constants $K_i > 0$ and functions $H_i(x)$ with $i = 1, \ldots, N$, be such that for an a.e. differentiable function v, such that $v > 0$ a.e. in Ω, we have*

$$- X_i(|X_i v|^{p_i-2} X_i v) \geq K_i H_i(x) v^{p_i-1}, \quad i = 1, \ldots, N. \tag{6.97}$$

Then, for all non-negative functions $u \in C^1(\Omega)$ we have

$$\sum_{i=1}^{N} \int_\Omega |X_i u|^{p_i} dx \geq \sum_{i=1}^{N} K_i \int_\Omega H_i(x) u^{p_i} dx. \tag{6.98}$$

Proof of Lemma 6.10.3. In view of (6.91) and (6.97) we have

$$\begin{aligned} 0 \leq \int_\Omega L(u,v)dx &= \int_\Omega R(u,v)dx \\ &= \sum_{i=1}^{N} \int_\Omega |X_i u|^{p_i} dx - \sum_{i=1}^{N} \int_\Omega X_i \left(\frac{u^{p_i}}{v^{p_i-1}}\right) |X_i v|^{p_i-2} X_i v dx \\ &= \sum_{i=1}^{N} \int_\Omega |X_i u|^{p_i} dx + \sum_{i=1}^{N} \int_\Omega \frac{u^{p_i}}{v^{p_i-1}} X_i \left(|X_i v|^{p_i-2} X_i v\right) dx \\ &\leq \sum_{i=1}^{N} \int_\Omega |X_i u|^{p_i} dx - \sum_{i=1}^{N} K_i \int_\Omega H_i(x) u^{p_i} dx, \end{aligned}$$

proving the statement. □

We now present the second-order horizontal Picone type identity that will be instrumental in the proof of Theorem 6.10.6 giving the Rellich type inequality for the anisotropic sub-Laplacians.

Lemma 6.10.4 (Second-order horizontal Picone identity). *Let $\Omega \subset \mathbb{G}$ be an open set of a stratified group $\mathbb{G}$, and let N be the dimension of the first stratum of $\mathbb{G}$. Let u, v be twice differentiable a.e. in Ω and satisfying the following conditions: $u \geq 0$, $v > 0$, $X_i^2 v < 0$ a.e. in Ω for $p_i > 1$, $i = 1, \ldots, N$. Then we have*

$$L_1(u,v) = R_1(u,v) \geq 0, \tag{6.99}$$

where

$$R_1(u,v) := \sum_{i=1}^{N} |X_i^2 u|^{p_i} - \sum_{i=1}^{N} X_i^2 \left(\frac{u^{p_i}}{v^{p_i-1}}\right) |X_i^2 v|^{p_i-2} X_i^2 v,$$

and

$$L_1(u,v) := \sum_{i=1}^{N} |X_i^2 u|^{p_i} - \sum_{i=1}^{N} p_i \left(\frac{u}{v}\right)^{p_i-1} X_i^2 u X_i^2 v |X_i^2 v|^{p_i-2}$$

$$+\sum_{i=1}^{N}(p_i-1)\left(\frac{u}{v}\right)^{p_i}|X_i^2v|^{p_i}$$
$$-\sum_{i=1}^{N}p_i(p_i-1)\frac{u^{p_i-2}}{v^{p_i-1}}|X_i^2v|^{p_i-2}X_i^2v\left(X_iu-\frac{u}{v}X_iv\right)^2.$$

Proof of Lemma 6.10.4. A direct computation gives

$$\begin{aligned}X_i^2\left(\frac{u^{p_i}}{v^{p_i-1}}\right)&=X_i\left(p_i\frac{u^{p_i-1}}{v^{p_i-1}}X_iu-(p_i-1)\frac{u^{p_i}}{v^{p_i}}X_iv\right)\\&=p_i(p_i-1)\frac{u^{p_i-2}}{v^{p_i-2}}\left(\frac{(X_iu)v-u(X_iv)}{v^2}\right)X_iu+p_i\frac{u^{p_i-1}}{v^{p_i-1}}X_i^2u\\&\quad-p_i(p_i-1)\frac{u^{p_i-1}}{v^{p_i-1}}\left(\frac{(X_iu)v-u(X_iv)}{v^2}\right)X_iv-(p_i-1)\frac{u^{p_i}}{v^{p_i}}X_i^2v\\&=p_i(p_i-1)\left(\frac{u^{p_i-2}}{v^{p_i-1}}|X_iu|^2-2\frac{u^{p_i-1}}{v^{p_i}}X_ivX_iu+\frac{u^{p_i}}{v^{p_i+1}}|X_iv|^2\right)\\&\quad+p_i\frac{u^{p_i-1}}{v^{p_i-1}}X_i^2u-(p_i-1)\frac{u^{p_i}}{v^{p_i}}X_i^2v\\&=p_i(p_i-1)\frac{u^{p_i-2}}{v^{p_i-1}}\left(X_iu-\frac{u}{v}X_iv\right)^2+p_i\frac{u^{p_i-1}}{v^{p_i-1}}X_i^2u-(p_i-1)\frac{u^{p_i}}{v^{p_i}}X_i^2v,\end{aligned}$$

which yields (6.99). By Young's inequality (6.92) we have

$$\frac{u^{p_i-1}}{v^{p_i-1}}X_i^2uX_i^2v|X_i^2v|^{p_i-2}\le\frac{|X_i^2u|^{p_i}}{p_i}+\frac{1}{q_i}\frac{u^{p_i}}{v^{p_i}}|X_i^2v|^{p_i},\quad i=1,\dots,N,$$

where $p_i>1$, $q_i>1$, $\frac{1}{p_i}+\frac{1}{q_i}=1$. Since $X_i^2v<0$ we arrive at

$$\begin{aligned}L_1(u,v)&\ge\sum_{i=1}^{N}|X_i^2u|^{p_i}+\sum_{i=1}^{N}(p_i-1)\frac{u^{p_i}}{v^{p_i}}|X_i^2v|^{p_i}-\sum_{i=1}^{N}p_i\left(\frac{|X_i^2u|^{p_i}}{p_i}+\frac{1}{q_i}\frac{u^{p_i}}{v^{p_i}}|X_i^2v|^{p_i}\right)\\&\quad-\sum_{i=1}^{N}p_i(p_i-1)\frac{u^{p_i-2}}{v^{p_i-1}}|X_i^2v|^{p_i-2}X_i^2v\left|X_iu-\frac{u}{v}X_iv\right|^2\\&=\sum_{i=1}^{N}\left(p_i-1-\frac{p_i}{q_i}\right)\frac{u^{p_i}}{v^{p_i}}|X_i^2v|^{p_i}\\&\quad-\sum_{i=1}^{N}p_i(p_i-1)\frac{u^{p_i-2}}{v^{p_i-1}}|X_i^2v|^{p_i-2}X_i^2v\left|X_iu-\frac{u}{v}X_iv\right|^2\ge0.\end{aligned}$$

This completes the proof of Lemma 6.10.4. □

6.10.2 Horizontal anisotropic Hardy type inequality

As a consequence of the horizontal Picone type identity in Lemma 6.10.1 we can obtain the Hardy type inequality for the anisotropic sub-Laplacian on stratified Lie groups. We recall that for $x \in \mathbb{G}$ we write customarily

$$x = (x', x''),$$

with coordinates x' corresponding to the first stratum of $\mathbb{G}$.

Theorem 6.10.5 (Horizontal anisotropic Hardy type inequality). *Let $\mathbb{G}$ be a stratified group with N being the dimension of its first stratum, and let $\Omega \subset \mathbb{G}\backslash\{x' = 0\}$ be an open set. Let $1 < p_i < N$ for all $i = 1, \ldots, N$. Then we have*

$$\sum_{i=1}^{N} \int_{\Omega} |X_i u|^{p_i} dx \geq \sum_{i=1}^{N} \left(\frac{p_i - 1}{p_i}\right)^{p_i} \int_{\Omega} \frac{|u|^{p_i}}{|x_i'|^{p_i}} dx, \tag{6.100}$$

for all $u \in C^1(\Omega)$.

Proof of Theorem 6.10.5. The proof is based on the application of Lemma 6.10.3. For this, we introduce the auxiliary function

$$v := \prod_{j=1}^{N} |x_j'|^{\alpha_j} = |x_i'|^{\alpha_i} V_i, \tag{6.101}$$

where $V_i = \prod_{j=1, j\neq i}^{N} |x_j'|^{\alpha_j}$ and $\alpha_j = \frac{p_j - 1}{p_j}$. Then we have

$$\begin{aligned} X_i v &= \alpha_i V_i |x_i'|^{\alpha_i - 2} x_i', \\ |X_i v|^{p_i - 2} &= \alpha_i^{p_i - 2} V_i^{p_i - 2} |x_i'|^{\alpha_i p_i - 2\alpha_i - p_i + 2}, \\ |X_i v|^{p_i - 2} X_i v &= \alpha_i^{p_i - 1} V_i^{p_i - 1} |x_i'|^{\alpha_i p_i - \alpha_i - p_i} x_i'. \end{aligned}$$

Consequently, we also have

$$- X_i(|X_i v|^{p_i - 2} X_i v) = \left(\frac{p_i - 1}{p_i}\right)^{p_i} \frac{v^{p_i - 1}}{|x_i'|^{p_i}}. \tag{6.102}$$

To complete the proof of Theorem 6.10.5, we choose $K_i = \left(\frac{p_i - 1}{p_i}\right)^{p_i}$ and $H_i(x) = \frac{1}{|x_i'|^{p_i}}$, and use Lemma 6.10.3. □

6.10.3 Horizontal anisotropic Rellich type inequality

Now we present the horizontal anisotropic Rellich type inequality on stratified Lie groups.

Theorem 6.10.6 (Horizontal anisotropic Rellich type inequality). *Let $\mathbb{G}$ be a stratified group with N being the dimension of its first stratum, and let $\Omega \subset \mathbb{G}\backslash\{x'=0\}$ be an open set. Then for a function $u \geq 0$, $u \in C^2(\Omega)$, and $2 < \alpha_i < N-2$ we have the following inequality*

$$\sum_{i=1}^{N} \int_{\Omega} |X_i^2 u|^{p_i} dx \geq \sum_{i=1}^{N} C_i(\alpha_i, p_i) \int_{\Omega} \frac{|u|^{p_i}}{|x_i'|^{2p_i}} dx, \tag{6.103}$$

where $1 < p_i < N$ for $i = 1, \ldots, N$, and

$$C_i(\alpha_i, p_i) = (\alpha_i(\alpha_i - 1))^{p_i - 1}(\alpha_i p_i - 2p_i - \alpha_i + 2)(\alpha_i p_i - 2p_i - \alpha_i + 1).$$

Proof of Theorem 6.10.6. We introduce the auxiliary function

$$v := \prod_{j=1}^{N} |x_j'|^{\alpha_j} = |x_i'|^{\alpha_i} V_i,$$

we choose α_j later, and where $V_i := \prod_{j=1, j\neq i}^{N} |x_j'|^{\alpha_j}$. Then we have

$$\begin{aligned} X_i^2 v &= X_i(\alpha_i V_i |x_i'|^{\alpha_i - 2} x_i') = \alpha_i(\alpha_i - 1) V_i |x_i'|^{\alpha_i - 2}, \\ |X_i^2 v|^{p_i - 2} &= (\alpha_i(\alpha_i - 1))^{p_i - 2} V_i^{p_i - 2} |x_i'|^{\alpha_i p_i - 2p_i - 2\alpha_i + 4}, \\ |X_i^2 v|^{p_i - 2} X_i^2 v &= (\alpha_i(\alpha_i - 1))^{p_i - 1} V_i^{p_i - 1} |x_i'|^{\alpha_i p_i - 2p_i - \alpha_i + 2}. \end{aligned}$$

Consequently, we obtain

$$\begin{aligned} &X_i^2(|X_i^2 v|^{p_i - 2} X_i^2 v) \\ &= (\alpha_i(\alpha_i - 1))^{p_i - 1} V_i^{p_i - 1} X_i^2(|x_i'|^{\alpha_i p_i - 2p_i - \alpha_i + 2}) \\ &= (\alpha_i(\alpha_i - 1))^{p_i - 1}(\alpha_i p_i - 2p_i - \alpha_i + 2) V_i^{p_i - 1} X_i \left(|x_i'|^{\alpha_i p_i - 2p_i - \alpha_i} x_i'\right) \\ &= (\alpha_i(\alpha_i - 1))^{p_i - 1}(\alpha_i p_i - 2p_i - \alpha_i + 2)(\alpha_i p_i - 2p_i - \alpha_i + 1) \\ &\quad \times V_i^{p_i - 1} |x_i'|^{\alpha_i(p_i - 1) - 2p_i}. \end{aligned}$$

Thus, for a twice differentiable function $v > 0$ a.e. in Ω with $X_i^2 v < 0$, we have

$$X_i^2(|X_i^2|^{p_i - 2} X_i^2 v) = C_i(\alpha_i, p_i) \frac{v^{p_i - 1}}{|x_i'|^{2p_i}} \tag{6.104}$$

a.e. in Ω. Using (6.104) we compute

$$\begin{aligned} 0 &\leq \int_{\Omega} L_1(u, v) dx = \int_{\Omega} R_1(u, v) dx \\ &= \sum_{i=1}^{N} \int_{\Omega} |X_i^2 u|^{p_i} dx - \sum_{i=1}^{N} \int_{\Omega} X_i^2 \left(\frac{u^{p_i}}{v^{p_i - 1}}\right) |X_i^2 v|^{p_i - 2} X_i^2 v dx \end{aligned}$$

$$= \sum_{i=1}^{N} \int_{\Omega} |X_i^2 u|^{p_i} dx - \sum_{i=1}^{N} \int_{\Omega} \frac{u^{p_i}}{v^{p_i-1}} X_i^2 \left(|X_i^2 v|^{p_i-2} X_i^2 v\right) dx$$
$$= \sum_{i=1}^{N} \int_{\Omega} |X_i^2 u|^{p_i} dx - \sum_{i=1}^{N} C_i(\alpha_i, p_i) \int_{\Omega} \frac{|u|^{p_i}}{|x_i'|^{2p_i}} dx.$$

The proof of Theorem 6.10.6 is complete. □

6.11 Horizontal Hardy inequalities with multiple singularities

In this section we obtain the analogue of the Hardy inequality with multiple singularities on stratified Lie groups. The singularities will be represented by a family of points $\{a_k\}_{k=1}^m \in \mathbb{G}$. We will be using the usual notation $a_k = (a_k', a_k'')$, with a_k' corresponding to the first stratum of $\mathbb{G}$. In turn, we can also write $a_k' = (a_{k1}', \dots, a_{kN}')$. From (1.17) it follows that

$$(xa_k^{-1})' = x' - a_k'.$$

We denote by $(xa_k^{-1})_j' = x_j' - a_{kj}'$ the jth component of xa_k^{-1}.

Theorem 6.11.1 (Horizontal Hardy inequality with multiple singularities). *Let $\mathbb{G}$ be a stratified group with N being the dimension of its first stratum, and let $\Omega \subset \mathbb{G}$ be an open set. Let $N \geq 3$, $x = (x', x'') \in \mathbb{G}$ with $x' = (x_1', \dots, x_N')$ being in the first stratum of $\mathbb{G}$, and let $a_k \in \mathbb{G}, k = 1, \dots, m$, be the singularities. Then we have*

$$\int_{\Omega} |\nabla_H u|^2 dx \geq \left(\frac{N-2}{2}\right)^2 \int_{\Omega} \frac{\sum_{j=1}^{N} \left| \sum_{k=1}^{m} \frac{(xa_k^{-1})_j'}{|(xa_k^{-1})'|^N} \right|^2}{\left(\sum_{k=1}^{m} \frac{1}{|(xa_k^{-1})'|^{N-2}}\right)^2} |u|^2 dx, \tag{6.105}$$

for all $u \in C_0^\infty(\Omega)$.

Remark 6.11.2. The Euclidean case of the inequality (6.105) was obtained by Kapitanski and Laptev [KL16]. Theorem 6.11.1 was obtained in [RSS18a] and our presentation here follows the arguments there.

Proof of Theorem 6.11.1. Let us fix a vector-valued function

$$\mathcal{A}(x) = (\mathcal{A}_1(x), \dots, \mathcal{A}_N(x))$$

to be specified later. Also let λ be a real parameter. We start with the inequality

$$0 \leq \int_{\Omega} \sum_{j=1}^{N} (|X_j u - \lambda \mathcal{A}_j u|^2) dx$$
$$= \int_{\Omega} \left(|\nabla_H u|^2 - 2\lambda \mathrm{Re} \sum_{j=1}^{N} \overline{\mathcal{A}_j u} X_j u + \lambda^2 \sum_{j=1}^{N} |\mathcal{A}_j|^2 |u|^2 \right) dx.$$

By using the integration by parts we get

$$-\int_{\Omega}\left(\lambda^2\sum_{j=1}^{N}|\mathcal{A}_j|^2+\lambda\mathrm{div}_H\mathcal{A}\right)|u|^2dx\le\int_{\Omega}|\nabla_H u|^2dx. \tag{6.106}$$

We differentiate the integral on the left-hand side with respect to λ to optimize it, yielding

$$2\lambda|\mathcal{A}|^2+\mathrm{div}_H\mathcal{A}=0,$$

for all $x\in\Omega$. This is the condition that we impose on $\mathcal{A}(x)$, that is, the quotient $\frac{\mathrm{div}_H\mathcal{A}(x)}{|\mathcal{A}(x)|^2}$ must be constant. For $\lambda=\frac{1}{2}$ we get

$$\mathrm{div}_H\mathcal{A}(x)=-|\mathcal{A}(x)|^2. \tag{6.107}$$

Then putting (6.107) in (6.106) we have the following Hardy inequality

$$\frac{1}{4}\int_{\Omega}\sum_{j=1}^{N}|\mathcal{A}_j(x)|^2|u|^2dx\le\int_{\Omega}|\nabla_H u|^2dx. \tag{6.108}$$

Now if we assume that $\mathcal{A}=\nabla_H\phi$ for some function ϕ, then (6.107) becomes

$$\mathcal{L}\phi+|\nabla_H\phi|^2=0.$$

It follows that the function

$$w=e^{\phi}\ge 0$$

is harmonic with respect to the sub-Laplacian $\mathcal{L}$. Thus, w is a constant >0 or it has a singularity. Let us now take

$$w(x):=\sum_{k=1}^{m}\frac{1}{|(xa_k^{-1})'|^{N-2}},$$

and then also

$$\phi(x):=\ln(w(x)).$$

Therefore

$$\begin{aligned}\mathcal{A}(x)=\nabla_H(\ln w)&=\frac{1}{w}\nabla_H\left(\sum_{k=1}^{m}|(xa_k^{-1})'|^{2-N}\right)\\&=\frac{1}{w}\sum_{k=1}^{m}\nabla_H\left(\sum_{j=1}^{N}((xa_k^{-1})'_j)^2\right)^{\frac{2-N}{2}}\\&=-\frac{N-2}{w}\left(\sum_{k=1}^{m}\frac{(xa_k^{-1})'}{|(xa_k^{-1})'|^N}\right),\end{aligned}$$

and

$$|\mathcal{A}(x)|^2 = \sum_{j=1}^{N} |\mathcal{A}_j(x)|^2 = \left(\frac{N-2}{w}\right)^2 \sum_{j=1}^{N} \left| \sum_{k=1}^{m} \frac{(xa_k^{-1})'_j}{|(xa_k^{-1})'|^N} \right|^2 .$$

The inequality (6.105) now follows from (6.108), completing the proof of Theorem 6.11.1. ◻

We then also obtain the corresponding uncertainty principle.

Corollary 6.11.3 (Uncertainty principle with multiple singularities). *Let $\mathbb{G}$ be a stratified group with N being the dimension of its first stratum, and let $\Omega \subset \mathbb{G}$ be an open set. Let $N \geq 3$, $x = (x', x'') \in \mathbb{G}$ with $x' = (x'_1, \ldots, x'_N)$ corresponding to the first stratum of $\mathbb{G}$. Let $a_k \in \mathbb{G}$, $k = 1, \ldots, m$, be the singularities, and let $1 < p_i < N$ for $i = 1, \ldots, N$. Then we have*

$$\frac{N-2}{2} \int_\Omega |u|^2 dx \leq \left(\int_\Omega |\nabla_H u|^2 dx\right)^{1/2} \left(\int_\Omega \frac{\left(\sum_{k=1}^m \frac{1}{|(xa_k^{-1})'|^{N-2}}\right)^2}{\sum_{j=1}^N \left|\sum_{k=1}^m \frac{(xa_k^{-1})'_j}{|(xa_k^{-1})'|^N}\right|^2} |u|^2 dx \right)^{1/2},$$

for all $u \in C_0^\infty(\Omega)$.

Proof of Corollary 6.11.3. By (6.105) and the Cauchy–Schwarz inequality we get

$$\begin{aligned}
&\int_\Omega |\nabla_H u|^2 dx \int_\Omega \frac{\left(\sum_{k=1}^m \frac{1}{|(xa_k^{-1})'|^{N-2}}\right)^2}{\sum_{j=1}^N \left|\sum_{k=1}^m \frac{(xa_k^{-1})'_j}{|(xa_k^{-1})'|^N}\right|^2} |u|^2 dx \\
&\geq \left(\frac{N-2}{2}\right)^2 \int_\Omega \frac{\sum_{j=1}^N \left|\sum_{k=1}^m \frac{(xa_k^{-1})'_j}{|(xa_k^{-1})'|^N}\right|^2}{\left(\sum_{k=1}^m \frac{1}{|(xa_k^{-1})'|^{N-2}}\right)^2} |u|^2 dx \\
&\quad \times \int_\Omega \frac{\left(\sum_{k=1}^m \frac{1}{|(xa_k^{-1})'|^{N-2}}\right)^2}{\sum_{j=1}^N \left|\sum_{k=1}^m \frac{(xa_k^{-1})'_j}{|(xa_k^{-1})'|^N}\right|^2} |u|^2 dx \\
&\geq \left(\frac{N-2}{2}\right)^2 \left(\int_\Omega |u|^2 dx\right)^2 .
\end{aligned}$$

The proof is complete. ◻

6.12 Horizontal many-particle Hardy inequality

In this section we discuss Hardy inequalities for $n \geq 1$ particles on stratified Lie groups. We denote by $\mathbb{G}^n$ the product

$$\mathbb{G}^n := \overbrace{\mathbb{G} \times \cdots \times \mathbb{G}}^{n}.$$

We consider the points $x = (x_1, \ldots, x_n) \in \mathbb{G}^n$, with $x_j \in \mathbb{G}$. The horizontal component of $x \in \mathbb{G}^n$ will be denoted by $x' = (x'_1, \ldots, x'_n)$, with $x'_i = (x'_{i1}, \ldots, x'_{iN})$ being the coordinates corresponding to the first stratum of $\mathbb{G}$ for $i = 1, \ldots, n$. The (horizontal) distance between particles $x_i, x_j \in \mathbb{G}$ can be defined by

$$r_{ij} := |(x_i x_j^{-1})'| = |x'_i - x'_j| = \sqrt{\sum_{k=1}^{N} (x'_{ik} - x'_{jk})^2}.$$

We will also use the notation

$$\nabla_{H_i} = (X_{i1}, \ldots, X_{iN})$$

for the horizontal gradient associated to the ith particle. We denote

$$\nabla_{H^n} := (\nabla_{H_1}, \ldots, \nabla_{H_n}), \qquad \text{and} \qquad \mathcal{L}_i := \sum_{k=1}^{N} X_{ik}^2,$$

the sub-Laplacian associated to the ith particle. We note that

$$\mathcal{L} = \sum_{i=1}^{N} \mathcal{L}_i.$$

We now recall a simple but crucial inequality on $\mathbb{R}^m$.

Lemma 6.12.1. *Let $m \geq 1$, and let*

$$\mathcal{A} = (\mathcal{A}_1(x), \ldots, \mathcal{A}_m(x))$$

be a mapping $\mathcal{A} : \mathbb{R}^m \to \mathbb{R}^m$ whose components and their first derivatives are uniformly bounded on $\mathbb{R}^m$. Then for every non-trivial $u \in C_0^1(\mathbb{R}^m)$ we have

$$\int_{\mathbb{R}^m} |\nabla u|^2 dx \geq \frac{1}{4} \frac{\left(\int_{\mathbb{R}^m} \operatorname{div}\mathcal{A}|u|^2 dx\right)^2}{\int_{\mathbb{R}^m} |\mathcal{A}|^2 |u|^2 dx}. \tag{6.109}$$

Proof of Lemma 6.12.1. We have

$$\begin{aligned}\left|\int_{\mathbb{R}^m} \operatorname{div}\mathcal{A}|u|^2 dx\right| &= 2\left|\operatorname{Re}\int_{\mathbb{R}^m} \langle \mathcal{A}, \nabla u\rangle \overline{u} dx\right| \\ &\leq 2\left(\int_{\mathbb{R}^m} |\mathcal{A}|^2|u|^2 dx\right)^{1/2} \left(\int_{\mathbb{R}^m} |\nabla u|^2 dx\right)^{1/2},\end{aligned}$$

using the Cauchy–Schwarz inequality in the last line. This implies (6.109). □

Theorem 6.12.2 (Horizontal many-particle Hardy inequality). *Let $\mathbb{G}$ be a stratified group with N being the dimension of its first stratum, and let $\Omega \subset \mathbb{G}^n$ be an open set. Let $N \geq 2$ and $n \geq 3$. Let $r_{ij} = |(x_i x_j^{-1})'| = |x_i' - x_j'|$. Then we have*

$$\int_\Omega |\nabla_{H^n} u|^2 dx \geq \frac{(N-2)^2}{n} \int_\Omega \sum_{1 \leq i < j \leq n} \frac{|u|^2}{r_{ij}^2} dx, \tag{6.110}$$

for all $u \in C^1(\Omega)$.

Remark 6.12.3. The Euclidean case of the inequality (6.110) was obtained by M. Hoffmann-Ostenhof, T. Hoffmann-Ostenhof, A. Laptev, and J. Tidblom in [HOHOLT08]. The Euclidean case of the subsequent Theorem 6.12.4 was obtained by D. Lundholm [Lun15]. Theorem 6.12.2 and Theorem 6.12.4 were obtained in [RSS18a] and our presentation here follows the arguments there.

Proof of Theorem 6.12.2. Let us define a mapping $\mathcal{B}_1$ by the formula

$$\mathcal{B}_1(x_i', x_j') := \frac{(x_i x_j^{-1})'}{r_{ij}^2}, \quad 1 \leq i < j \leq n.$$

In the subsequent arguments we denote by $\operatorname{div}_{\mathbb{G}_i}$ the horizontal divergence on $\mathbb{G}_i$. Applying inequality (6.109) to the mapping $\mathcal{B}_1$ we have

$$\begin{aligned}
\int_\Omega |(\nabla_{H_i} - \nabla_{H_j}) u|^2 dx &\geq \frac{1}{4} \frac{\left(\int_\Omega \left((\operatorname{div}_{H_i} - \operatorname{div}_{H_j})\mathcal{B}_1\right) |u|^2 dx\right)^2}{\int_\Omega |\mathcal{B}_1|^2 |u|^2 dx} \\
&= \frac{1}{4} \frac{\left(\int_\Omega \frac{2(N-2)}{|(x_i x_j^{-1})'|^2} |u|^2 dx\right)^2}{\int_\Omega \frac{|u|^2}{|(x_i x_j^{-1})'|^2} dx} \\
&= (N-2)^2 \int_\Omega \frac{|u|^2}{r_{ij}^2} dx. \qquad (6.111)
\end{aligned}$$

Also, we define another mapping $\mathcal{B}_2$ by

$$\mathcal{B}_2(x) := \frac{\sum_{j=1}^n x_j'}{\left|\sum_{j=1}^n x_j'\right|^2}.$$

We can calculate

$$\begin{aligned}
\nabla_{H_i} \cdot \mathcal{B}_2 &= \sum_{k=1}^N X_{ik} \left(\frac{\sum_{j=1}^n x_{jk}'}{|\sum_{j=1}^n x_j'|^2} \right) \\
&= \frac{Nn |\sum_{j=1}^n x_j'|^2 - 2n \left((\sum_{j=1}^n x_{j1}')^2 + \cdots + (\sum_{j=1}^n x_{jN}')^2 \right)}{|\sum_{j=1}^n x_j'|^4} = \frac{Nn - 2n}{|\sum_{j=1}^n x_j'|^2}.
\end{aligned}$$

Applying inequality (6.109) to the mapping $\mathcal{B}_2$ we obtain

$$\begin{aligned}\int_\Omega \left|\sum_{i=1}^n \nabla_{H_i} u\right|^2 dx &\geq \frac{1}{4} \frac{\left(\int_\Omega (\sum_{i=1}^n \operatorname{div}_{H_i} \mathcal{B}_2)|u|^2 dx\right)^2}{\int_\Omega |\mathcal{B}_2|^2 |u|^2 dx} \\ &= \frac{1}{4} \frac{\left(\int_\Omega \sum_{i=1}^n \frac{Nn-2n}{|\sum_{j=1}^n x_j'|^2} |u|^2 dx\right)^2}{\int_\Omega \frac{|u|^2}{|\sum_{j=1}^n x_j'|^2} dx} \\ &= \frac{(N-2)^2 n^4}{4} \int_\Omega \frac{|u|^2}{\left|\sum_{j=1}^n x_j'\right|^2} dx. \end{aligned} \tag{6.112}$$

Adding inequalities (6.111) and (6.112) and using the identity

$$n \sum_{i=1}^n |\nabla_{H_i} u|^2 = \sum_{1 \leq i < j \leq n} \left|\nabla_{H_i} u - \nabla_{H_j} u\right|^2 + \left|\sum_{i=1}^n \nabla_{H_i} u\right|^2,$$

we arrive at

$$\sum_{i=1}^n \int_\Omega |\nabla_{H_i} u|^2 dx \geq \frac{(N-2)^2}{n} \int_\Omega \sum_{i<j} \frac{|u|^2}{r_{ij}^2} dx + \frac{(N-2)^2 n^3}{4} \int_\Omega \frac{|u|^2}{\left|\sum_{j=1}^n x_j'\right|^2} dx.$$

Because the last term on right-hand side is positive, we get

$$\sum_{i=1}^n \int_\Omega |\nabla_{H_i} u|^2 dx \geq \frac{(N-2)^2}{n} \int_\Omega \sum_{i<j} \frac{|u|^2}{r_{ij}^2} dx.$$

Also we have $\sum_{i=1}^n |\nabla_{H_i} u|^2 = |\nabla_{H^n} u|^2$. Putting everything together, the proof of Theorem 6.12.2 is complete. □

The following theorem deals with the total separation of $n \geq 2$ particles.

Theorem 6.12.4 (Total separation of many-particles). *Let $\mathbb{G}$ be a stratified group with N being the dimension of its first stratum, and let $\Omega \subset \mathbb{G}^n$ be an open set. Let $\rho^2 := \sum_{i<j} |(x_i x_j^{-1})'|^2 = \sum_{i<j} |x_i' - x_j'|^2$ with $x_i' \neq x_j'$. Then we have*

$$\int_\Omega |\nabla_H u|^2 dx = n \left(\frac{(n-1)}{2} N - 1\right)^2 \int_\Omega \frac{|u|^2}{\rho^2} dx + \int_\Omega |\nabla_H \rho^{-2\alpha} u|^2 \rho^{4\alpha} dx \tag{6.113}$$

for all $u \in C_0^\infty(\Omega)$ with $\alpha = \frac{2-(n-1)N}{4}$.

The proof of Theorem 6.12.4 will rely on the following identity.

Proposition 6.12.5. *Let $\mathbb{G}$ be a stratified group with N being the dimension of its first stratum, and let $\Omega \subset \mathbb{G}^n$ be an open set. Let $f : \Omega \to (0,\infty)$ be twice differentiable. Then for any function $u \in C_0^\infty(\Omega)$ and $\alpha \in \mathbb{R}$, we have*

$$\int_\Omega |\nabla_H u|^2 dx = \int_\Omega \left(\alpha(1-\alpha)\frac{|\nabla_H f|^2}{f^2} - \alpha\frac{\mathcal{L}f}{f}\right)|u|^2 dx + \int_\Omega |\nabla_H v|^2 f^{2\alpha} dx,$$

where $v := f^{-\alpha}u$.

Proof of Proposition 6.12.5. Let us first observe that for $u = f^\alpha v$, we have

$$\nabla_H u = \alpha f^{\alpha-1}(\nabla_H f)v + f^\alpha \nabla_H v.$$

By squaring the above expression we get

$$\begin{aligned}|\nabla_H u|^2 &= \alpha^2 f^{2(\alpha-1)}|\nabla_H f|^2|v|^2 + \mathrm{Re}(2\alpha v f^{2\alpha-1}(\nabla_H f)\cdot(\nabla_H v)) + f^{2\alpha}|\nabla_H v|^2 \\ &= \alpha^2 f^{2(\alpha-1)}|\nabla_H f|^2|v|^2 + \alpha f^{2\alpha-1}(\nabla_H f)\cdot\nabla_H|v|^2 + f^{2\alpha}|\nabla_H v|^2.\end{aligned}$$

By integrating this expression over Ω, we obtain

$$\begin{aligned}\int_\Omega |\nabla_H u|^2 dx &= \int_\Omega \alpha^2 f^{2(\alpha-1)}|\nabla_H f|^2|v|^2 dx \\ &\quad + \int_\Omega \mathrm{Re}(\alpha f^{2\alpha-1}(\nabla_H f)\cdot\nabla_H|v|^2)dx + \int_\Omega f^{2\alpha}|\nabla_H v|^2 dx \\ &= \int_\Omega \alpha^2 f^{2(\alpha-1)}|\nabla_H f|^2|v|^2 dx \\ &\quad - \alpha\int_\Omega \nabla_H\cdot(f^{2\alpha-1}\nabla_H f)|v|^2 dx + \int_\Omega f^{2\alpha}|\nabla_H v|^2 dx.\end{aligned}$$

We have used integration by parts to the middle term on the right-hand side. Since

$$\nabla_H\cdot(f^{2\alpha-1}\nabla_H f) = (2\alpha-1)f^{2\alpha-2}|\nabla_H f|^2 + f^{2\alpha-1}\mathcal{L}f,$$

we get

$$\begin{aligned}\int_\Omega |\nabla_H u|^2 dx &= \int_\Omega \alpha^2 f^{2(\alpha-1)}|\nabla_H f|^2|v|^2 dx - \int_\Omega \alpha f^{2\alpha-1}\mathcal{L}f|v|^2 dx \\ &\quad - \int_\Omega \alpha(2\alpha-1)f^{2\alpha-2}|\nabla_H f|^2|v|^2 dx + \int_\Omega f^{2\alpha}|\nabla_H v|^2 dx.\end{aligned}$$

Putting back $v = f^{-\alpha}u$ and collecting the terms we arrive at the equality of Proposition 6.12.5. □

Proof of Theorem 6.12.4. With $\nabla_{H_k} = (X_{k1},\ldots,X_{kN})$, using the definition of ρ we have

$$\nabla_{H_k}\rho^2 = (X_{k1}\rho^2,\ldots,X_{kN}\rho^2) = 2\sum_{k\neq j}^{n}(x_k x_j^{-1})'.$$

Hence

$$\mathcal{L}\rho^2 = 2\sum_{k=1}^{n}\sum_{k\neq j}^{n}\nabla_{H_k}\cdot(x_k x_j^{-1})' = 2n(n-1)N, \tag{6.114}$$

$$\begin{aligned}|\nabla_H \rho^2|^2 &= 8\sum_{1\leq i<j\leq n}|(x_k x_j^{-1})'|^2\\ &+8\sum_{k=1}^{n}\sum_{1\leq i<j\leq n}(x_k x_i^{-1})'\cdot(x_k x_j^{-1})' = 4n\rho^2,\end{aligned} \tag{6.115}$$

where in the last step we used the identity

$$\sum_{k=1}^{n}\sum_{1\leq i<j\leq n}(x_k x_i^{-1})'\cdot(x_k x_j^{-1})' = \frac{n-2}{2}\sum_{1\leq i<j\leq n}|(x_i x_j^{-1})'|^2.$$

By putting (6.114) and (6.115) in the identity of Proposition 6.12.5 with $f=\rho^2$, we obtain

$$\int_\Omega |\nabla_H u|^2 dx = 4n\alpha\left(\frac{2-(n-1)N}{2}-\alpha\right)\int_\Omega \frac{|u|^2}{\rho^2}dx + \int_\Omega |\nabla_H \rho^{-2\alpha}u|^2\rho^{4\alpha}dx.$$

To optimize we differentiate the integral

$$4n\alpha\left(\frac{2-(n-1)N}{2}-\alpha\right)\int_\Omega \frac{|u|^2}{\rho^2}dx$$

with respect to α, then we have

$$\frac{2-(n-1)N}{2}-2\alpha=0 \quad\text{and}\quad \alpha=\frac{2-(n-1)N}{4},$$

which completes the proof of Theorem 6.12.4. □

6.13 Hardy inequality with exponential weights

In this section, we discuss a horizontal Hardy inequality with exponential weights. In the Euclidean case such a type of inequalities is sometimes called two parabolic type Hardy inequalities, see Zhang [Zha17]. The following statement was obtained in [RSS18a].

Theorem 6.13.1 (Hardy inequality with exponential horizontal weights). *Let $\mathbb{G}$ be a stratified group with $N\geq 3$ being the dimension of its first stratum, and let $\Omega\subset\mathbb{G}$ be an open set. Let $x_0\in\Omega$. Then we have*

$$\int_\Omega e^{-\frac{|(xx_0^{-1})'|^2}{4\lambda}}\left(\frac{(N-2)^2}{4|x'|^2}-\frac{N}{4\alpha}+\frac{|(xx_0^{-1})'|^2}{16\lambda^2}\right)|u|^2dx \leq \int_\Omega e^{-\frac{|(xx_0^{-1})'|^2}{4\lambda}}|\nabla_H u|^2 dx$$

for all $u\in C^1(\Omega)$ and for all $\lambda>0$.

Proof of Theorem 6.13.1. We will use the horizontal Hardy inequality

$$\frac{(N-2)^2}{4}\int_\Omega \frac{|v|^2}{|x'|^2}dx \leq \int_\Omega |\nabla_H v|^2 dx, \tag{6.116}$$

see (6.6), valid for all $v \in C^1(\Omega)$, with the choice of $v = e^{-\frac{|(xx_0^{-1})'|^2}{8\lambda}}u$. We note that

$$\nabla_H v = e^{-\frac{|(xx_0^{-1})'|^2}{8\lambda}}\nabla_H u - \frac{(xx_0^{-1})'}{4\lambda}e^{-\frac{|(xx_0^{-1})'|^2}{8\lambda}}u,$$

for all $v \in C^1(\Omega)$. Then by inequality (6.116) we have

$$\begin{aligned}\frac{(N-2)^2}{4}&\int_\Omega e^{-\frac{|(xx_0^{-1})'|^2}{4\lambda}}\frac{|u|^2}{|x'|^2}dx\\ &\leq \int_\Omega e^{-\frac{|(xx_0^{-1})'|^2}{4\lambda}}|\nabla_H u|^2 + \frac{|(xx_0^{-1})'|^2}{16\lambda^2}e^{-\frac{|(xx_0^{-1})'|^2}{4\lambda}}|u|^2 dx\\ &\quad - \frac{1}{2\lambda}\mathrm{Re}\int_\Omega (xx_0^{-1})'\cdot(\nabla_H u)ue^{-\frac{|(xx_0^{-1})'|^2}{4\lambda}}dx.\end{aligned} \tag{6.117}$$

Integration by parts in the last term of the right-hand side of this inequality yields

$$\begin{aligned}\mathrm{Re}&\int_\Omega (xx_0^{-1})'\cdot(\nabla_H u)ue^{-\frac{|(xx_0^{-1})'|^2}{4\lambda}}dx\\ &= -\frac{1}{2}\int_\Omega\left(N - \frac{|(xx_0^{-1})'|^2}{2\lambda}\right)e^{-\frac{|(xx_0^{-1})'|^2}{4\lambda}}|u|^2dx.\end{aligned}$$

By using this in (6.117) and rearranging the terms, we complete the proof of Theorem 6.13.1. □

Chapter 7

Hardy–Rellich Inequalities and Fundamental Solutions

In this chapter, we describe the Hardy and other inequalities on stratified groups with the $\mathcal{L}$-gauge weights. The appearance of such weights has been discussed in the beginning of Chapter 6. The literature on inequalities with such weights is rather substantial. Apart from describing new results and methods we will be making relevant references to the results existing in the earlier literature.

While horizontal estimates in Chapter 6 can be established on general stratified groups, the picture is not so complete if one is working with the $\mathcal{L}$-gauge weights. We recall that the $\mathcal{L}$-gauge $d(x)$ is a homogeneous quasi-norm arising from the fundamental solution of the sub-Laplacian $\mathcal{L}$ by the condition (1.75), namely, that $d(x)^{2-Q}$ is a constant multiple of Folland's [Fol75] fundamental solution of the sub-Laplacian $\mathcal{L}$, with Q being the homogeneous dimension of the stratified group $\mathbb{G}$.

Using the $\mathcal{L}$-gauge as a weight, the classical Hardy inequality on the Euclidean space $\mathbb{R}^n$,

$$\left(\frac{n-p}{p}\right)^p \int_{\mathbb{R}^n} \frac{|\phi(x)|^p}{|x|_E^p} dx \leq \int_{\mathbb{R}^n} |\nabla\phi(x)|^p dx, \tag{7.1}$$

for all $\phi \in C_0^\infty(\mathbb{R}^n)$ if $1 \leq p < n$, and for all $\phi \in C_0^\infty(\mathbb{R}^n\backslash\{0\})$ if $n < p < \infty$, is replaced by inequalities involving powers of $d(x)$. For instance, D'Ambrosio in [D'A05] and Goldstein and Kombe in [GK08] established the following L^p-Hardy type inequality on polarizable Carnot groups $\mathbb{G}$,

$$\left(\frac{Q-p}{p}\right)^p \int_{\mathbb{G}} \frac{|\nabla_H d|^p}{d^p} |\phi|^p dx \leq \int_{\mathbb{G}} |\nabla_H \phi|^p dx, \tag{7.2}$$

for all $\phi \in C_0^\infty(\mathbb{G}\backslash\{0\})$, provided that $Q \geq 3$ and $1 < p < Q$. Here, as usual, Q is the homogeneous dimension of $\mathbb{G}$.

In such inequalities the explicit formula (1.103) relating the $\mathcal{L}$-gauge to the fundamental solution of the p-sub-Laplacian often plays an important role. In

M. Ruzhansky, D. Suragan, *Hardy Inequalities on Homogeneous Groups*, Progress in Mathematics 327, https://doi.org/10.1007/978-3-030-02895-4_8

the case $p = 2$, since the p-sub-Laplacian is the usual sub-Laplacian, formula (1.103) just reduces to the definition of the fundamental solution holding on general stratified groups. Consequently, in the case $p = 2$ the version of (7.2) holds on any stratified group $\mathbb{G}$ of homogeneous dimension $Q \geq 3$ and all $\phi \in C_0^\infty(\mathbb{G}\backslash\{0\})$:

$$\left(\frac{Q-2}{2}\right)^2 \int_{\mathbb{G}} \frac{|\nabla_H d|^2}{d^2}|\phi|^2 dx \leq \int_{\mathbb{G}} |\nabla_H \phi|^2 dx. \tag{7.3}$$

It was shown in [Kom10] (see also [GK08]) that the Hardy inequality (7.3) on general stratified groups of homogeneous dimension $Q \geq 3$ also holds in its weighted form

$$\int_{\mathbb{G}} d^\alpha |\nabla_H \phi|^2 dx \geq \left(\frac{Q+\alpha-2}{2}\right) \int_{\mathbb{G}} d^\alpha \frac{|\nabla_H d|^2}{d^2}|\phi|^2 dx, \quad \alpha > 2-Q, \tag{7.4}$$

for all $\phi \in C_0^\infty(\mathbb{G}\backslash\{0\})$. It can be noted that the constants appearing in (7.2) and (7.4) are sharp but are never achieved.

The aim of this chapter is to discuss these and other related inequalities, and their further extensions. In Remark 7.1.2 we provide a more extensive historical perspective on these inequalities.

7.1 Weighted L^p-Hardy inequalities

We start with a general version of a weighted Hardy inequality on general stratified groups. Subsequently, in the following sections, we consider further extensions from the point of view of the weights in the setting of polarizable Carnot groups. Here we will be mostly working with the $\mathcal{L}$-gauge defined in (1.75), namely, with

$$d(x) := \begin{cases} \varepsilon(x)^{\frac{1}{2-Q}}, & \text{for } x \neq 0, \\ 0, & \text{for } x = 0, \end{cases} \tag{7.5}$$

where ε is the fundamental solution of the sub-Laplacian $\mathcal{L}$ on $\mathbb{G}$.

Theorem 7.1.1 (Weighted L^p-Hardy inequalities with $\mathcal{L}$-gauge). *Let $\mathbb{G}$ be a stratified group of homogeneous dimension $Q \geq 3$. Let $\alpha \in \mathbb{R}$ and let $1 < p < Q - \alpha$. Then for all complex-valued functions $u \in C_0^\infty(\mathbb{G}\backslash\{0\})$ we have*

$$\int_{\mathbb{G}} \frac{1}{d^\alpha |\nabla_H d|^{p-2}} |\nabla_H u|^p dx \geq \left(\frac{Q-p-\alpha}{p}\right)^p \int_{\mathbb{G}} \frac{|\nabla_H d|^2}{d^{\alpha+p}} |u|^p dx, \tag{7.6}$$

and the constant $\left(\frac{Q-p-\alpha}{p}\right)^p$ in inequality (7.6) is sharp.

Remark 7.1.2.

1. The inequality (7.6) in the setting of stratified groups of different types has a long history. For $p = 2$ and $\alpha = 0$, on the Heisenberg group it was proved by Garofalo and Lanconelli in [GL90] with an explicit expression for d being the Koranyi norm. Still on the Heisenberg group, it was shown in [NZW01] for $\alpha = 0$ and $1 < p < Q$. The weighted inequality for $p = 2$ was obtained by Kombe in [Kom10]. Different further unweighted versions for $p \neq 2$ in the settings related to those of polarizable Carnot groups were obtained by D'Ambrosio [D'A05], Goldstein and Kombe [GK08], and Danielli, Garofalo and Phuc [DGP11]. The weighted L^p inequality on general stratified groups by using a special class of weighted p-sub-Laplacians and the corresponding fundamental solutions was obtained by Jin and Shen [JS11]. More recently, in [Lia13] Lian has also obtained a similar result but with a sharp constant. In the proof below we follow Lian's arguments, as well as Lian's proof [Lia13] of Theorem 7.2.1.
2. Different formulations are also possible in the setting of polarizable Carnot groups. We present them in Theorem 7.1.3 and in Theorem 7.2.2 following [Kom10, Theorem 3.1] and [Kom10, Theorem 4.1] or [GKY17, Corollary 3.1], respectively. Further improved remainder terms have been also analysed in [Kom10].
3. There are other versions of Hardy inequalities that one can find in the literature, such as multi-particle inequalities (see, e.g., [Lun15] and references therein) or Besov space versions of Hardy inequalities, see [BCG06] and [BFKG12] for the settings of the Heisenberg group and on graded groups, respectively.

Proof of Theorem 7.1.1. Since for some constant C_Q we have that $C_Q d^{2-Q}$ is the fundamental solution of $\mathcal{L}$, for all $u \in C_0^\infty(\mathbb{G}\backslash\{0\})$, it follows that

$$\int_{\mathbb{G}} \langle \nabla_H d^{2-Q}, \nabla_H u\rangle dx = -C_Q^{-1} u(0) = 0. \tag{7.7}$$

For $\epsilon > 0$, let us define

$$u_\epsilon := (|u|^2 + \epsilon^2)^{p/2} - \epsilon^p.$$

Then $u_\epsilon \geq 0$, $u_\epsilon \in C_0^\infty(\mathbb{G}\backslash\{0\})$, and it has the same support as u. Replacing u by $u_\epsilon d^{Q-p-\alpha}$ in inequality (7.7), we obtain

$$\int_{\mathbb{G}} \frac{\langle \nabla_H d, \nabla_H u_\epsilon\rangle}{d^{p+\alpha-1}} dx + (Q-p-\alpha)\int_{\mathbb{G}} \frac{u_\epsilon}{d^{p+\alpha}} |\nabla_H d|^2 dx = 0.$$

Then we can estimate

$$(Q-p-\alpha)\int_{\mathbb{G}} \frac{u_\epsilon}{d^{p+\alpha}} |\nabla_H d|^2 dx = -p\int_{\mathbb{G}} (|u|^2+\epsilon^2)^{(p-2)/2} u \langle \nabla_H u, \nabla_H d\rangle \frac{1}{d^{p+\alpha-1}} dx$$

$$\leq p \int_{\mathbb{G}} \frac{(|u|^2+\epsilon^2)^{(p-2)/2}|u||\nabla_H u||\nabla_H d|}{d^{p+\alpha-1}} dx$$
$$\leq p \int_{\mathbb{G}} \frac{(|u|^2+\epsilon^2)^{(p-1)/2}|\nabla_H u||\nabla_H d|}{d^{p+\alpha-1}} dx.$$

Letting $\epsilon \to 0$, by the dominated convergence theorem we obtain the estimate

$$(Q-p-\alpha)\int_{\mathbb{G}} \frac{|u|^p}{d^{p+\alpha}}|\nabla_H d|^2 dx \leq p \int_{\mathbb{G}} \frac{|u|^{p-1}|\nabla_H u||\nabla_H d|}{d^{p+\alpha-1}} dx.$$

By Hölder's inequality, this implies

$$(Q-p-\alpha)\int_{\mathbb{G}} \frac{|u|^p}{d^{p+\alpha}}|\nabla_H d|^2 dx \leq p\left(\int_{\mathbb{G}} \frac{|u|^p}{d^{p+\alpha}}|\nabla_H d|^2 dx\right)^{\frac{p-1}{p}} \left(\int_{\mathbb{G}} \frac{|\nabla_H u|^p}{|\nabla_H d|^{p-2} d^{\alpha}} dx\right)^{\frac{1}{p}},$$

which gives (7.6).

Let us now show that the constant $\left(\frac{Q-p-\alpha}{p}\right)^p$ in inequality (7.6) is sharp. Let $f \in C_0^\infty(0,+\infty)$. Since $f(d) \in C_0^\infty(\mathbb{G}\backslash\{0\})$, using the polar decomposition in Proposition 1.2.10 with respect to d, we have

$$\inf_{u\in C_0^\infty(\mathbb{G}\backslash\{0\})\backslash\{0\}} \frac{\int_{\mathbb{G}} \frac{|\nabla_H u|^p}{|\nabla_H d|^{p-2}d^\alpha}dx}{\int_{\mathbb{G}} \frac{|u|^p}{d^{p+\alpha}}|\nabla_H d|^2 dx}$$
$$\leq \inf_{f\in C_0^\infty(0,+\infty)\backslash\{0\}} \frac{\int_{\mathbb{G}} \frac{|\nabla_H f|^p}{|\nabla_H d|^{p-2}d^\alpha}dx}{\int_{\mathbb{G}} \frac{|f|^p}{d^{p+\alpha}}|\nabla_H d|^2 dx}$$
$$= \inf_{f\in C_0^\infty(0,+\infty)\backslash\{0\}} \frac{\int_0^\infty |f'(d)|^p d^{Q-\alpha-1} dd \cdot \int_{\wp} |\nabla d|^2 d\sigma}{\int_0^\infty |f(d)|^p d^{Q-p-\alpha-1} dd \cdot \int_{\wp} |\nabla d|^2 d\sigma}$$
$$= \inf_{f\in C_0^\infty(0,+\infty)\backslash\{0\}} \frac{\int_0^\infty |f'(d)|^p d^{Q-\alpha-1} dd}{\int_0^\infty |f(d)|^p d^{Q-p-\alpha-1} dd}$$
$$= \inf_{f\in C_0^\infty(0,+\infty)\backslash\{0\}} \frac{\int_{\mathbb{R}^Q} \frac{|\nabla f(|x|)|^p}{|x|^\alpha}dx}{\int_{\mathbb{R}^Q} \frac{|f(|x|)|^p}{|x|^{p+\alpha}}dx} = \left(\frac{Q-p-\alpha}{p}\right)^p,$$

where we abuse the notation by writing dd for the integration with respect to the radial variable determined by d. The last equality follows from the fact that the Euclidean weighted Hardy inequalities

$$\int_{\mathbb{R}^Q} \frac{|\nabla f(|x|)|^p}{|x|^\alpha}dx \geq \left(\frac{Q-p-\alpha}{p}\right)^p \int_{\mathbb{R}^Q} \frac{|f(|x|)|^p}{|x|^{p+\alpha}}dx$$

hold for all $f \in C_0^\infty(\mathbb{R}^Q\backslash\{0\})$ and the constant here sharp and is attained as a limit of radial functions, as it was shown by Davies and Hinz [DH98]. This completes the proof. $\square$

Another type of Hardy inequality is also known on polarizable Carnot groups. We give it next following [Kom10, Theorem 3.1] and its proof.

Theorem 7.1.3 (Another type of weighted Hardy inequalities with $\mathcal{L}$-gauge). *Let $\mathbb{G}$ be a polarizable Carnot group of homogeneous dimension $Q \geq 3$. Let $1 < p < Q$ and let $\alpha \in \mathbb{R}$ be such that $\alpha > -Q$. Then for all $f \in C_0^\infty(\mathbb{G}\backslash\{0\})$ we have the inequality*

$$\int_{\mathbb{G}} d^{\alpha+p} \frac{|\nabla_H d \cdot \nabla_H f|^p}{|\nabla_H d|^{2p}} dx \geq \left(\frac{Q+\alpha}{p}\right)^p \int_{\mathbb{G}} d^\alpha |f|^p dx, \tag{7.8}$$

where the constant $\left(\frac{Q+\alpha}{p}\right)^p$ is sharp.

Proof of Theorem 7.1.3. Let us first recall the formula (1.105), that is,

$$\nabla_H \left(\frac{d}{|\nabla_H d|^2} \nabla_H d\right) = Q$$

in $\mathbb{G}\backslash\mathcal{Z}$, where $\mathcal{Z} := \{0\} \bigcup \{x \in \mathbb{G}\backslash\{0\} : \nabla_H d = 0\}$ has Haar measure zero, and $\nabla_H d \neq 0$ for a.e. $x \in \mathbb{G}$. By using this formula as well as Green's formula (see Theorem 1.4.6) we obtain

$$(Q+\alpha) \int_{\mathbb{G}} d^\alpha |f|^p dx = -p \int_{\mathbb{G}} \frac{|f|^{p-2} f d^{\alpha+1}}{|\nabla_H d|^2} \nabla_H d \cdot \nabla_H f dx.$$

Moreover, by using Hölder's and Young's inequalities we can estimate

$$\begin{aligned}(Q+\alpha) \int_{\mathbb{G}} d^\alpha |f|^p dx &\leq p \left(\int_{\mathbb{G}} d^\alpha |f|^p dx\right)^{\frac{p-1}{p}} \left(\int_{\mathbb{G}} \frac{d^{\alpha+p} |\nabla_H d \cdot \nabla_H f|^p}{|\nabla_H d|^{2p}} dx\right)^{1/p} \\ &\leq (p-1)\epsilon^{-p/(p-1)} \int_{\mathbb{G}} d^\alpha |f|^p dx + \epsilon^p \int_{\mathbb{G}} \frac{d^{\alpha+p} |\nabla_H d \cdot \nabla_H f|^p}{|\nabla_H d|^{2p}} dx\end{aligned}$$

for any $\epsilon > 0$, that is,

$$\epsilon^{-p}(Q+\alpha-(p-1)\epsilon^{-p/(p-1)}) \int_{\mathbb{G}} d^\alpha |f|^p dx \leq \int_{\mathbb{G}} \frac{d^{\alpha+p} |\nabla_H d \cdot \nabla_H f|^p}{|\nabla_H d|^{2p}} dx.$$

Since the function $\epsilon \to \epsilon^{-p}(Q+\alpha-(p-1)\epsilon^{-p/(p-1)})$ attains its maximum $\left(\frac{Q+\alpha}{p}\right)^p$ at $\epsilon^{p/(p-1)} = \frac{p}{Q+\alpha}$, we obtain the inequality

$$\left(\frac{Q+\alpha}{p}\right)^p \int_{\mathbb{G}} d^\alpha |f|^p dx \leq \int_{\mathbb{G}} \frac{d^{\alpha+p} |\nabla_H d \cdot \nabla_H f|^p}{|\nabla_H d|^{2p}} dx.$$

Now let us show that $\left(\frac{Q+\alpha}{p}\right)^p$ is the best constant, that is, we show that we have

$$C_H := \inf_{0 \neq f \in C_0^\infty(\mathbb{G})} \frac{\int_{\mathbb{G}} \frac{d^{\alpha+p} |\nabla_H d \cdot \nabla_H f|^p}{|\nabla_H d|^{2p}} dx}{\int_{\mathbb{G}} d^\alpha |f|^p dx} = \left(\frac{Q+\alpha}{p}\right)^p.$$

Obviously, one has

$$\left(\frac{Q+\alpha}{p}\right)^p \leq \frac{\int_{\mathbb{G}} \frac{d^{\alpha+p}|\nabla_H d\cdot\nabla_H f|^p}{|\nabla_H d|^{2p}}dx}{\int_{\mathbb{G}} d^\alpha |f|^p dx}$$

for all $f \in C_0^\infty(\mathbb{G}\setminus\{0\})$, that is, $\left(\frac{Q+\alpha}{p}\right)^p \leq C_H$. So, we need to show the converse, namely, that $C_H \geq \left(\frac{Q+\alpha}{p}\right)^p$. For this, consider the following family of d-radial functions

$$f_\epsilon(d) := \begin{cases} d^{\frac{Q+\alpha}{p}+\epsilon} & \text{if} \quad d\in[0,1], \\ d^{-\left(\frac{Q+\alpha}{p}+\epsilon\right)} & \text{if} \quad d>1, \end{cases}$$

with $\epsilon > 0$. Note that $f_\epsilon(d)$ can be also approximated by smooth functions with compact support in $\mathbb{G}$. We can also readily calculate that

$$\frac{d^{\alpha+p}|\nabla_H d\cdot\nabla_H f_\epsilon|^p}{|\nabla_H d|^{2p}} = \begin{cases} \left(\frac{Q+\alpha}{p}+\epsilon\right)^p d^{Q+2\alpha-p\epsilon} & \text{if} \quad d\in[0,1], \\ \left(\frac{Q+\alpha}{p}+\epsilon\right)^p d^{-Q-\epsilon} & \text{if} \quad d>1. \end{cases}$$

Denoting by $\mathbb{B}_1 = \{x\in\mathbb{G} : d(x)\leq 1\}$ the unit d-ball, we have

$$\int_{\mathbb{G}} d^\alpha |f_\epsilon|^p dx = \int_{\mathbb{B}_1} d^{Q+2\alpha-p\epsilon}dx + \int_{\mathbb{G}\setminus\mathbb{B}_1} d^{-Q-\epsilon}dx.$$

For every $\epsilon > 0$, the weights $d^{Q+2\alpha+p\epsilon}$ and $d^{-Q-p\epsilon}$ are integrable at 0 and ∞, respectively. Thus, the integral $\int_{\mathbb{G}} d^\alpha|f_\epsilon|^p dx$ is finite. Therefore, we get

$$\begin{aligned}\left(\frac{Q+\alpha}{p}+\epsilon\right)^p \int_{\mathbb{G}} d^\alpha|f_\epsilon|^p dx &= \left(\frac{Q+\alpha}{p}+\epsilon\right)^p \left[\int_{\mathbb{B}_1} d^{Q+2\alpha-p\epsilon}dx + \int_{\mathbb{G}\setminus\mathbb{B}_1} d^{-Q-\epsilon}dx\right] \\ &= \int_{\mathbb{G}} d^{\alpha+p}\frac{|\nabla_H d\cdot\nabla_H f|^p}{|\nabla_H d|^{2p}}dx.\end{aligned}$$

Moreover, we have

$$\begin{aligned}\left(\frac{Q+\alpha}{p}+\epsilon\right)^p \int_{\mathbb{G}} d^{\alpha+p}\frac{|\nabla_H d\cdot\nabla_H f|^p}{|\nabla_H d|^{2p}}dx &\geq \left(\frac{Q+\alpha}{p}+\epsilon\right)^p \int_{\mathbb{G}} d^\alpha|f_\epsilon|^p dx \\ &= \int_{\mathbb{G}} d^{\alpha+p}\frac{|\nabla_H d\cdot\nabla_H f|^p}{|\nabla_H d|^{2p}}dx.\end{aligned}$$

That is, $C_H \leq \left(\frac{Q+\alpha}{p}+\epsilon\right)^p$ and letting $\epsilon\to 0$ we obtain $\left(\frac{Q+\alpha}{p}\right)^p \leq C_H$. This yields $C_H = \left(\frac{Q+\alpha}{p}\right)^p$, showing the sharpness of the constant. $\square$

7.2 Weighted L^p-Rellich inequalities

In this section we discuss the Rellich inequality with weights given in terms of the $\mathcal{L}$-gauge.

Theorem 7.2.1 (Weighted L^p-Rellich inequalities with $\mathcal{L}$-gauge). *Let $\mathbb{G}$ be a stratified group of homogeneous dimension $Q \geq 3$. Let $\alpha \in \mathbb{R}$ and let $1 < p < (Q-\alpha)/2$. Then for all $u \in C_0^\infty(\mathbb{G}\backslash\{0\})$ we have*

$$\int_{\mathbb{G}} \frac{|\mathcal{L}u|^p}{|\nabla_H d|^{2(p-1)} d^\alpha} dx \geq \left(\frac{[(p-1)Q+\alpha](Q-2p-\alpha)}{p^2}\right)^p \int_{\mathbb{G}} \frac{|\nabla_H d|^2}{d^{2p+\alpha}} |u|^p dx, \quad (7.9)$$

where the constant $\left(\frac{[(p-1)Q+\alpha](Q-2p-\alpha)}{p^2}\right)^p$ in inequality (7.9) is sharp.

Proof of Theorem 7.2.1. For $\epsilon > 0$, let us set

$$u_\epsilon := (|u|^2+\epsilon^2)^{p/2} - \epsilon^p \qquad \text{and} \qquad \omega_\epsilon := (|u|^2+\epsilon^2)^{p/4} - \epsilon^{p/2}.$$

Then we can calculate

$$\begin{aligned} \mathcal{L}u_\epsilon &= p(|u|^2+\epsilon^2)^{p/2-1}|\nabla_H u|^2 + p(p-2)(|u|^2+\epsilon^2)^{p/2-2}|u|^2|\nabla_H u|^2 \\ &\quad + p(|u|^2+\epsilon^2)^{p/2-1}u\mathcal{L}u \\ &\geq p(p-1)(|u|^2+\epsilon^2)^{p/2-2}|u|^2|\nabla_H u|^2 + p(|u|^2+\epsilon^2)^{p/2-1}u\mathcal{L}u \\ &= \frac{4(p-1)}{p}|\nabla_H \omega_\epsilon|^2 + p(|u|^2+\epsilon^2)^{p/2-1}u\mathcal{L}u. \end{aligned}$$

Therefore, we have the estimate

$$-p\int_{\mathbb{G}} \frac{(|u|^2+\epsilon^2)^{p/2-1}u\mathcal{L}u}{d^{\alpha+2(p-1)}}dx \geq \frac{4(p-1)}{p}\int_{\mathbb{G}} \frac{|\nabla_H \omega_\epsilon|^2}{d^{\alpha+2(p-1)}}dx - \int_{\mathbb{G}} \frac{\mathcal{L}u_\epsilon}{d^{\alpha+2(p-1)}}dx.$$

The integration by parts in the last term, using (1.78), yields

$$-\int_{\mathbb{G}} \frac{\mathcal{L}u_\epsilon}{d^{\alpha+2(p-1)}}dx = (\alpha+2p-2)(Q-\alpha-2p)\int_{\mathbb{G}} \frac{u_\epsilon}{d^{\alpha+2p}}|\nabla_H d|^2 dx.$$

Using this and Theorem 7.1.1 we obtain

$$\begin{aligned} -p\int_{\mathbb{G}} \frac{(|u|^2+\epsilon^2)^{p/2-1}u\mathcal{L}u}{d^{\alpha+2(p-1)}}dx &\geq \frac{4(p-1)}{p}\int_{\mathbb{G}} \frac{|\nabla_H \omega_\epsilon|^2}{d^{\alpha+2(p-1)}}dx - \int_{\mathbb{G}} \frac{\mathcal{L}u_\epsilon}{d^{\alpha+2(p-1)}}dx \\ &\geq \frac{(p-1)(Q-2p-\alpha)^2}{p}\int_{\mathbb{G}} \frac{\omega_\epsilon^2}{d^{\alpha+2p}}|\nabla_H d|^2 dx \\ &\quad + (\alpha+2p-2)(Q-\alpha-2p)\int_{\mathbb{G}} \frac{u_\epsilon}{d^{\alpha+2p}}|\nabla_H d|^2 dx. \end{aligned}$$

Hence we have

$$\frac{(p-1)(Q-2p-\alpha)^2}{p}\int_{\mathbb{G}}\frac{\omega_\epsilon^2}{d^{\alpha+2p}}|\nabla_H d|^2dx$$
$$+(\alpha+2p-2)(Q-\alpha-2p)\int_{\mathbb{G}}\frac{u_\epsilon}{d^{\alpha+2p}}|\nabla_H d|^2dx$$
$$\leq p\int_{\mathbb{G}}\frac{(|u|^2+\epsilon^2)^{p/2-1}|u||\mathcal{L}u|}{d^{\alpha+2(p-1)}}dx\leq p\int_{\mathbb{G}}\frac{(|u|^2+\epsilon^2)^{(p-1)/2}|\mathcal{L}u|}{d^{\alpha+2(p-1)}}dx.$$

Letting $\epsilon\to 0+$, the dominated convergence theorem implies that

$$\frac{(Q-2p-\alpha)\left((p-1)Q+\alpha\right)}{p}\int_{\mathbb{G}}\frac{|u|^p}{d^{\alpha+2p}}|\nabla_H d|^2dx\leq p\int_{\mathbb{G}}\frac{|u|^{p-1}|\mathcal{L}u|}{d^{\alpha+2(p-1)}}dx.$$

By Hölder's inequality we can estimate

$$\frac{(Q-2p-\alpha)\left((p-1)Q+\alpha\right)}{p}\int_{\mathbb{G}}\frac{|u|^p}{d^{\alpha+2p}}|\nabla_H d|^2dx$$
$$\leq p\left(\int_{\mathbb{G}}\frac{|u|^p}{d^{\alpha+2p}}|\nabla_H d|^2dx\right)^{\frac{p-1}{p}}\left(\int_{\mathbb{G}}\frac{|\mathcal{L}u|^p}{|\nabla_H d|^{2(p-1)}d^\alpha}dx\right)^{\frac{1}{p}},$$

which implies inequality (7.9).

The argument for the sharpness of the constant $\left(\frac{(Q-2p-\alpha)((p-1)Q+\alpha)}{p^2}\right)^p$ in (7.9) is similar to the sharpness argument in the proof of Theorem 7.1.1 (with the similar explanation for the notation dd). Namely, for functions $f\in C_0^\infty(0,+\infty)$, by using Proposition 1.2.10 we can estimate

$$\inf_{u\in C_0^\infty(\mathbb{G}\backslash\{0\})\backslash\{0\}}\frac{\int_{\mathbb{G}}\frac{|\mathcal{L}u|^p}{|\nabla_H d|^{2(p-1)}d^\alpha}dx}{\int_{\mathbb{G}}\frac{|u|^p}{d^{\alpha+2p}}|\nabla_H d|^2dx}$$
$$\leq\inf_{f\in C_0^\infty(0,+\infty)\backslash\{0\}}\frac{\int_{\mathbb{G}}\frac{|\mathcal{L}f(d)|^p}{|\nabla_H d|^{2(p-1)}d^\alpha}dx}{\int_{\mathbb{G}}\frac{|f(d)|^p}{d^{\alpha+2p}}|\nabla_H d|^2dx}$$
$$=\inf_{f\in C_0^\infty(0,+\infty)\backslash\{0\}}\frac{\int_0^\infty|f''(d)+(Q-1)f'(d)/d|^pd^{Q-\alpha-1}dd\cdot\int_\wp|\nabla d|^2d\sigma}{\int_0^\infty|f(d)|^pd^{Q-2p-\alpha-1}dd\cdot\int_\wp|\nabla d|^2d\sigma}$$
$$=\inf_{f\in C_0^\infty(0,+\infty)\backslash\{0\}}\frac{\int_0^\infty|f''(d)+(Q-1)f'(d)/d|^pd^{Q-\alpha-1}dd}{\int_0^\infty|f(d)|^pd^{Q-2p-\alpha-1}dd}$$
$$=\inf_{f\in C_0^\infty(0,+\infty)\backslash\{0\}}\frac{\int_{\mathbb{R}^Q}\frac{|\mathcal{L}f(|x|)|^p}{|x|^\alpha}dx}{\int_{\mathbb{R}^Q}\frac{|f(|x|)|^p}{|x|^{2p+\alpha}}dx}$$
$$=\left(\frac{((p-1)Q+\alpha)(Q-2p-\alpha)}{p^2}\right)^p,$$

where the last equality follows from the weighted Rellich inequalities

$$\int_{\mathbb{R}^Q} \frac{|\mathcal{L}f(|x|)|^p}{|x|^\alpha} dx \geq \left(\frac{((p-1)Q+\alpha)(Q-2p-\alpha)}{p^2} \right)^p \int_{\mathbb{R}^Q} \frac{|f(|x|)|^p}{|x|^{2p+\alpha}} dx$$

for all $f \in C_0^\infty(\mathbb{R}^Q \backslash \{0\})$, and the fact that the constant here is sharp and is attained in the limit of radial functions, [DH98]. This completes the proof. □

Another Rellich inequality is also possible, see Remark 7.1.2, Part 2. We present it in the next statement following [Kom10, Theorem 4.1] and its proof.

Theorem 7.2.2 (Another type of weighted L^2-Rellich inequalities with $\mathcal{L}$-gauge). *Let $\mathbb{G}$ be a stratified group of homogeneous dimension $Q \geq 3$, with the homogeneous $\mathcal{L}$-gauge norm d on $\mathbb{G}$. Let $\alpha \in \mathbb{R}$ be such that $Q + \alpha - 4 > 0$. Then for all $f \in C_0^\infty(\mathbb{G}\backslash\{0\})$ we have*

$$\int_{\mathbb{G}} \frac{d^\alpha}{|\nabla_H d|^2} |\mathcal{L}f|^2 dx \geq \frac{(Q+\alpha-4)^2(Q-\alpha)^2}{16} \int_{\mathbb{G}} d^\alpha \frac{|\nabla_H d|^2}{d^4} |f|^2 dx, \tag{7.10}$$

where the constant $\frac{(Q+\alpha-4)^2(Q-\alpha)^2}{16}$ *is sharp.*

Proof of Theorem 7.2.2. Recalling formula (7.5) for the $\mathcal{L}$-gauge, a direct calculation gives that

$$\mathcal{L}d^{\alpha-2} = (Q+\alpha-4)(\alpha-2)d^{\alpha-4}|\nabla_H d|^2 + \frac{\alpha-2}{2-Q} d^{Q+\alpha-4}\mathcal{L}\varepsilon. \tag{7.11}$$

As before, we can assume without loss of generality that f is real-valued. Then (7.11) implies

$$\int_{\mathbb{G}} f^2 \mathcal{L} d^{\alpha-2} dx = \int_{\mathbb{G}} d^{\alpha-2}(2f\mathcal{L}f + 2|\nabla_H f|^2) dx.$$

On the other hand, since ε is the fundamental solution of $\mathcal{L}$ we have

$$\int_{\mathbb{G}} f^2 \mathcal{L} d^{\alpha-2} dx = (Q+\alpha-4)(\alpha-2) \int_{\mathbb{G}} d^{\alpha-4}|\nabla_H d|^2 f^2 dx,$$

with $Q + \alpha - 4 > 0$. Thus, we have

$$\begin{aligned}(Q+\alpha-4)(\alpha-2)\int_{\mathbb{G}} d^{\alpha-4}|\nabla_H d|^2 f^2 dx - 2\int_{\mathbb{G}} d^{\alpha-2} f\mathcal{L}f dx \\ = 2\int_{\mathbb{G}} d^{\alpha-2}|\nabla_H f|^2 dx.\end{aligned} \tag{7.12}$$

Further, using the following weighted Hardy inequality (see Corollary 7.3.2, Part 1, related to Theorem 7.1.1 for $p = 2$)

$$\left(\frac{Q+\alpha-p}{p}\right)^p \int_{\mathbb{G}} d^\alpha \frac{|\nabla_H d|^p}{d^p} |f|^p dx \leq \int_{\mathbb{G}} d^\alpha |\nabla_H f|^p dx,$$

we arrive at

$$2\left(\frac{Q+\alpha-4}{2}\right)^2\int_{\mathbb{G}} d^{\alpha-4}|\nabla_H d|^2 f^2 dx$$
$$\leq (Q+\alpha-4)(\alpha-2)\int_{\mathbb{G}} d^{\alpha-4}|\nabla_H d|^2 f^2 dx - 2\int_{\mathbb{G}} d^{\alpha-2} f\mathcal{L}f dx.$$

It follows that

$$\left(\frac{Q+\alpha-4}{2}\right)\left(\frac{Q-\alpha}{2}\right)\int_{\mathbb{G}} d^{\alpha-4}|\nabla_H d|^2 f^2 dx \leq -\int_{\mathbb{G}} d^{\alpha-2} f\mathcal{L}f dx. \tag{7.13}$$

By the Cauchy–Schwarz inequality we have

$$-\int_{\mathbb{G}} d^{\alpha-2} f\mathcal{L}f dx \leq \left(\int_{\mathbb{G}} d^{\alpha-4}|\nabla_H d|^2 f^2 dx\right)^{1/2}\left(\int_{\mathbb{G}} \frac{|\mathcal{L}f|^2}{|\nabla_H d|^2} d^\alpha dx\right)^{1/2}. \tag{7.14}$$

Now combination of the inequalities (7.14) and (7.13) yields (7.10).

Let us now show that the constant $C_R = \frac{(Q+\alpha-4)^2(Q-\alpha)^2}{16}$ is sharp, that is, we have the equality

$$C_R := \inf_{0\neq f\in C_0^\infty(\mathbb{G})} \frac{\int_{\mathbb{G}} \frac{d^\alpha}{|\nabla_H d|^2}|\mathcal{L}f|^2 dx}{\int_{\mathbb{G}} d^\alpha \frac{|\nabla_H d|^2}{d^4} f^2 dx} = \frac{(Q+\alpha-4)^2(Q-\alpha)^2}{16}.$$

Obviously, we have

$$\frac{\int_{\mathbb{G}} \frac{d^\alpha}{|\nabla_H d|^2}|\mathcal{L}f|^2 dx}{\int_{\mathbb{G}} d^\alpha \frac{|\nabla_H d|^2}{d^4} f^2 dx} \leq \frac{(Q+\alpha-4)^2(Q-\alpha)^2}{16},$$

that is, $\frac{(Q+\alpha-4)^2(Q-\alpha)^2}{16} \leq C_R$. So, we need to show the converse, namely, that $C_R \leq \frac{(Q+\alpha-4)^2(Q-\alpha)^2}{16}$. To do this we define a family of d-radial functions by

$$f_\epsilon(d) := \begin{cases} \left(\frac{Q+\alpha-4}{2}+\epsilon\right)^2 |\nabla_H d|^4 \frac{(Q-1)^2}{d^2} & \text{if } \ d \leq 1, \\ \left(\frac{Q+\alpha-4}{2}+\epsilon\right)^2\left(\frac{Q+\alpha-4}{2}-\epsilon\right)^2 d^{-Q-\alpha-2\epsilon}|\nabla_H d|^4 & \text{if } \ d > 1, \end{cases}$$

for some $\epsilon > 0$. Denoting by $\mathbb{B}_1 = \{x \in \mathbb{G} : d(x) \leq 1\}$ the unit d-ball, we have

$$\int_{\mathbb{G}} d^\alpha \frac{|\mathcal{L}f_\epsilon|^2}{|\nabla_H d|^2} dx = A\int_{\mathbb{B}_1} d^{\alpha-2}|\nabla_H d|^2 dx + B\int_{\mathbb{G}\setminus\mathbb{B}_1} d^{-Q-2\epsilon}|\nabla_H d|^2 dx,$$

where

$$A = (Q-1)^2\left(\frac{Q+\alpha-4}{2}+\epsilon\right)^2$$

and

$$B = \left(\frac{Q+\alpha-4}{2}+\epsilon\right)^2\left(\frac{Q+\alpha-4}{2}-\epsilon\right)^2.$$

Since $|\nabla_H d|$ is uniformly bounded and $Q+\alpha-4>0$, the integral $\int_{\mathbb{B}_1} d^{\alpha-2}|\nabla_H d|^2 dx$ is finite. It implies that we have

$$\int_{\mathbb{G}} d^{\alpha} \frac{|\mathcal{L} f_\epsilon|^2}{|\nabla_H d|^2} dx = B \int_{\mathbb{G}\setminus\mathbb{B}_1} d^{-Q-2\epsilon} |\nabla_H d|^2 dx + O(1).$$

Moreover, we have

$$\int_{\mathbb{G}} d^{\alpha} \frac{|\nabla_H d|^2}{d^4} f_\epsilon^2 dx = \int_{\mathbb{B}_1} d^{\alpha} \frac{|\nabla_H d|^2}{d^4} f_\epsilon^2 dx + \int_{\mathbb{G}\setminus\mathbb{B}_1} d^{\alpha} \frac{|\nabla_H d|^2}{d^4} f_\epsilon^2 dx.$$

Since the first integral is finite we obtain

$$\int_{\mathbb{G}} d^{\alpha} \frac{|\nabla_H d|^2}{d^4} f_\epsilon^2 dx = \int_{\mathbb{G}\setminus\mathbb{B}_1} d^{-Q-2\epsilon} |\nabla_H d|^2 dx + O(1).$$

Taking $\epsilon \to 0$ and noting that

$$\int_{\mathbb{G}\setminus\mathbb{B}_1} d^{-Q-2\epsilon} |\nabla_H d|^2 dx \to \infty,$$

we arrive at

$$\frac{\int_{\mathbb{G}} \frac{d^{\alpha}}{|\nabla_H d|^2} |\mathcal{L} f_\epsilon|^2 dx}{\int_{\mathbb{G}} d^{\alpha} \frac{|\nabla_H d|^2}{d^4} f_\epsilon^2 dx} \leq \frac{(Q+\alpha-4)^2 (Q-\alpha)^2}{16}.$$

This means that $C_R = \frac{(Q+\alpha-4)^2(Q-\alpha)^2}{16}$, so that the constant is sharp. □

7.3 Two-weight Hardy inequalities and uncertainty principles

In this section we consider Hardy inequalities with more general weights, presenting the approach of Goldstein, Kombe and Yener [GKY17]. This can be also extended further to Rellich inequalities, see [GKY18]. Other types of two-weight inequalities are known in the classical Euclidean setting, see, e.g., [GM11], and a more extensive exposition in [GM13].

Another general two-weight inequality on general homogeneous groups was given in Theorem 2.1.14, without making any assumptions on the weights ϕ, ψ there. However, in the following result, the weights V and W will be assumed to satisfy relation (7.15).

Theorem 7.3.1 (Two-weight L^p-Hardy inequality). *Let $\mathbb{G}$ be a stratified group. Let $V \in C^1(\mathbb{G})$ and $W \in L^1_{\rm loc}(\mathbb{G})$ be non-negative functions, and let $\Phi \in C^\infty(\mathbb{G})$ be a positive function such that*

$$-\nabla_H \cdot (V(x) |\nabla_H \Phi|^{p-2} \nabla_H \Phi) \geq W(x) \Phi^{p-1} \tag{7.15}$$

holds almost everywhere.

Then there exists a positive constant $c_p > 0$ depending only on p such that for all $\phi \in C_0^\infty(\mathbb{G})$ we have:
if $p \geq 2$, then

$$\int_{\mathbb{G}} V(x)|\nabla_H \phi|^p dx \geq \int_{\mathbb{G}} W(x)|\phi|^p dx + c_p \int_{\mathbb{G}} V(x) \left|\nabla_H \frac{\phi}{\Phi}\right|^p \Phi^p dx, \tag{7.16}$$

and if $1 < p < 2$, then

$$\int_{\mathbb{G}} V(x)|\nabla_H \phi|^p dx \geq \int_{\mathbb{G}} W(x)|\phi|^p dx + c_p \int_{\mathbb{G}} V(x) \frac{\left|\nabla_H \frac{\phi}{\Phi}\right|^2 \Phi^2}{\left(\left|\frac{\phi}{\Phi}\nabla_H \Phi\right| + \left|\nabla_H \frac{\phi}{\Phi}\right| \Phi\right)^{2-p}} dx. \tag{7.17}$$

For $p = 2$ we have the equality in (7.16) with $c_2 = 1$.

Proof of Theorem 7.3.1. For the proof we follow [GKY17], relying on the following inequalities (see, for example, [Lin90, Appendix]): For any $1 < p < \infty$ there exists a positive constant $c_p > 0$ depending only on p such that for all $a, b \in \mathbb{R}^n$ we have

$$|a + b|^p \geq |a|^p + p|a|^{p-2} a \cdot b + c_p |b|^p, \quad \text{for} \quad p \geq 2, \tag{7.18}$$

and

$$|a + b|^p \geq |a|^p + p|a|^{p-2} a \cdot b + c_p \frac{|b|^2}{(|a| + |b|)^{2-p}}, \quad \text{for} \quad 1 < p < 2. \tag{7.19}$$

Let $\varphi := \frac{\phi}{\Phi}$, where $0 < \Phi \in C^\infty(\mathbb{G})$ and $\phi \in C_0^\infty(\mathbb{G})$. Applying the inequality (7.18) with $a = \varphi \nabla_H \Phi$ and $b = \Phi \nabla_H \varphi$, for $p \geq 2$ we get

$$\begin{aligned} |\nabla_H \phi|^p &= |\varphi \nabla_H \Phi + \Phi \nabla_H \varphi|^p \\ &\geq |\nabla_H \Phi|^p |\varphi|^p + \Phi |\nabla_H \Phi|^{p-2} \nabla_H \Phi \cdot \nabla_H (|\varphi|^p) + c_p |\nabla_H \varphi|^p \Phi^p. \end{aligned} \tag{7.20}$$

Multiplying this by $V(x)$ on both sides and integrating by parts yields

$$\begin{aligned} \int_{\mathbb{G}} V(x)|\nabla_H \phi|^p dx &\geq \int_{\mathbb{G}} V(x)|\nabla_H \Phi|^p |\varphi|^p dx + c_p \int_{\mathbb{G}} V(x)|\nabla_H \varphi|^p \Phi^p dx \\ &\quad - \int_{\mathbb{G}} \nabla_H \cdot \left(V(x)\Phi|\nabla_H \Phi|^{p-2} \nabla_H \Phi\right) |\varphi|^p dx \\ &= - \int_{\mathbb{G}} \nabla_H \cdot \left(V(x)\Phi|\nabla_H \Phi|^{p-2} \nabla_H \Phi\right) \Phi |\varphi|^p dx \\ &\quad + c_p \int_{\mathbb{G}} V(x)|\nabla_H \varphi|^p \Phi^p dx. \end{aligned}$$

Consequently, assumption (7.15) implies that

$$\int_{\mathbb{G}} V(x)|\nabla_H \phi|^p dx \geq \int_{\mathbb{G}} W(x)|\varphi|^p \Phi^p dx + c_p \int_{\mathbb{G}} V(x)|\nabla_H \varphi|^p \Phi^p dx.$$

Recalling that $\varphi = \frac{\phi}{\Phi}$ one gets (7.16).

For the case $1 < p < 2$ one can use inequality (7.19) with the same choice of a and b as above, and we leave the details to the reader. Also, the above arguments show that if $p = 2$, then (7.16) is an equality with $c_2 = 1$. □

Let us now collect some consequences of Theorem 7.3.1 on polarizable Carnot groups, following [GKY17]. As before, we fix d to be the $\mathcal{L}$-gauge on a stratified group $\mathbb{G}$. Consequently, we denote by

$$B_R := \{x \in \mathbb{G} : d(x) < R\} \tag{7.21}$$

the ball of radius R with respect to the quasi-norm d.

Corollary 7.3.2 (Special cases of two-weight inequalities). *Let $\mathbb{G}$ be a polarizable Carnot group. Then we have the following inequalities:*

1. *Let $\alpha \in \mathbb{R}$, $1 < p < Q + \alpha$, $\gamma > -1$. Then we have*

$$\int_{\mathbb{G}} d^{\alpha}|\nabla_H d|^{\gamma}|\nabla_H \phi|^p dx \geq \left(\frac{Q+\alpha-p}{p}\right)^p \int_{\mathbb{G}} d^{\alpha}\frac{|\nabla_H d|^{p+\gamma}}{d^p}|\phi|^p dx$$

for all $\phi \in C_0^{\infty}(\mathbb{G}\backslash\{0\})$.

2. *Let $Q = p > 1$ and $\alpha < -1$. Then we have*

$$\int_{B_R} \left(\log\frac{R}{d}\right)^{\alpha+p} |\nabla_H \phi|^p dx \geq \left(\frac{|\alpha+1|}{p}\right)^p \int_{B_R} \left(\log\frac{R}{d}\right)^{\alpha} \frac{|\nabla_H d|}{d^p}|\phi|^p dx$$

for all $\phi \in C_0^{\infty}(B_R)$.

3. *Let $\alpha \in \mathbb{R}$ and $Q + \alpha > p > 1$. Then we have*

$$\int_{\mathbb{G}} d^{\alpha}|\nabla_H \phi|^p dx \geq \left(\frac{Q+\alpha-p}{p-1}\right)^{p-1} (Q+\alpha)\int_{\mathbb{G}} d^{\alpha}\frac{|\nabla_H d|^p}{(1+d^{\frac{p}{p-1}})^p}|\phi|^p dx$$

for all $\phi \in C_0^{\infty}(\mathbb{G})$.

4. *Let $1 < p < Q$ and $\alpha > 1$. Then we have*

$$\int_{\mathbb{G}} (1+d^{\frac{p}{p-1}})^{\alpha(p-1)}|\nabla_H \phi|^p dx$$
$$\geq Q\left(\frac{p(\alpha-1)}{p-1}\right)^{p-1} \int_{\mathbb{G}} \frac{|\nabla_H d|^p}{(1+d^{\frac{p}{p-1}})^{(1-\alpha)(p-1)}}|\phi|^p dx$$

for all $\phi \in C_0^{\infty}(\mathbb{G})$.

5. *Let $a, b > 0$ and $\alpha, \beta, m \in \mathbb{R}$. If $\alpha\beta > 0$ and $m \leq \frac{Q-2}{2}$, then we have*

$$\int_{\mathbb{G}} \frac{(a+bd^{\alpha})^{\beta}}{d^{2m}}|\nabla_H \phi|^2 dx \geq C(Q,m)^2 \int_{\mathbb{G}} \frac{(a+bd^{\alpha})^{\beta}}{d^{2m+2}}|\nabla_H d|^2\phi^2 dx$$
$$+ C(Q,m)\alpha\beta b\int_{\mathbb{G}} \frac{(a+bd^{\alpha})^{\beta-1}}{d^{2m-\alpha+2}}|\nabla_H d|^2\phi^2 dx,$$

for all $\phi \in C_0^{\infty}(\mathbb{G})$, where $C(Q,m) = \frac{Q-2m-2}{2}$.

6. *Let $Q = p > 1$. Then we have*

$$\int_{B_R} |\nabla_H \phi|^p dx \geq \left(\frac{p-1}{p}\right)^p \int_{B_R} \frac{|\nabla_H d|^p}{(R-d)^p} |\phi|^p dx$$

for all $\phi \in C_0^\infty(B_R)$.

Proof of Corollary 7.3.2. To make the application of Theorem 7.3.1 rigorous one can replace the function d with its regularization $d_\epsilon := (\Gamma + \epsilon)^{\frac{1}{2-Q}}$ for $\epsilon > 0$, where Γ is the fundamental solution for $\mathcal{L}$, and after the application of Theorem 7.3.1 take the limit as $\epsilon \to 0$.

1. The inequality follows from Theorem 7.3.1 with the choice

$$V = d^\alpha |\nabla_H d|^\gamma \quad \text{with} \quad \Phi = d^{-\left(\frac{Q+\alpha-p}{p}\right)}.$$

2. This part follows by taking

$$V = \left(\log \frac{R}{d}\right)^{\alpha+p} \quad \text{and} \quad \Phi = \left(\log \frac{R}{d}\right)^{\frac{|\alpha+1|}{p}}.$$

3. This part follows by taking

$$V = d^\alpha \quad \text{and} \quad \Phi = \left(1 + d^{\frac{p}{p-1}}\right)^{-\left(\frac{Q+\alpha-p}{p},\right)}.$$

4. This part follows by taking

$$V = \left(1 + d^{\frac{p}{p-1}}\right)^{\alpha(p-1)} \quad \text{and} \quad \Phi = \left(1 + d^{\frac{p}{p-1}}\right)^{1-\alpha}.$$

5. This part follows by taking

$$V = \frac{(a + bd^\alpha)^\beta}{d^{2m}} \quad \text{and} \quad \Phi = d^{-\left(\frac{Q-2m-2}{2}\right)}.$$

6. This part follows by taking

$$V \equiv 1 \quad \text{and} \quad \Phi = (R-d)^{\frac{p-1}{p}}.$$

The proof is complete. □

Remark 7.3.3. The statement of Corollary 7.3.2, Part 1, was first shown by Wang and Niu [WN08]. Part 2 was shown in [D'A05, (3.40)]. Part 4 is a version of the Euclidean estimate [Skr13, (5.1)]. Part 5 is a version of the Euclidean estimate [GM11, (42)]. Part 6 was shown on the Heisenberg group in [HN03] and then for polarizable Carnot groups in [D'A05].

Theorem 7.3.1 also yields several versions of the uncertainty principles.

Corollary 7.3.4 (Special cases of two-weight uncertainty principles). *Let $\mathbb{G}$ be a polarizable Carnot group. Then we have the following inequalities:*

1. *We have*

$$\left(\int_{\mathbb{G}} \frac{|\nabla_H \phi|^2}{|\nabla_H d|^2} dx\right)\left(\int_{\mathbb{G}} d^2|\phi|^2 dx\right) \geq \frac{Q^2}{4}\left(\int_{\mathbb{G}} |\phi|^2 dx\right)^2$$

for all $\phi \in C_0^\infty(\mathbb{G})$.

2. *We have*

$$\left(\int_{\mathbb{G}} |\nabla_H \phi|^2 dx\right)\left(\int_{\mathbb{G}} d^2|\nabla_H d|^2|\phi|^2 dx\right) \geq \frac{Q^2}{4}\left(\int_{\mathbb{G}} |\nabla_H d|^2|\phi|^2 dx\right)^2$$

for all $\phi \in C_0^\infty(\mathbb{G})$.

3. *We have*

$$\left(\int_{\mathbb{G}} |\nabla_H \phi|^2 dx\right)\left(\int_{\mathbb{G}} |\nabla_H d|^2|\phi|^2 dx\right) \geq \frac{(Q-1)^2}{4}\left(\int_{\mathbb{G}} \frac{|\nabla_H d|^2}{d}|\phi|^2 dx\right)^2$$

for all $\phi \in C_0^\infty(\mathbb{G})$.

Proof of Corollary 7.3.4. 1. This inequality was first shown by Kombe in [Kom10], extending the Euclidean uncertainty principle (2). Considering

$$V = \frac{1}{|\nabla_H d|^2} \quad \text{and} \quad \Phi = e^{-\alpha d^2},$$

for $\alpha > 0$, Theorem 7.3.1 implies

$$\int_{\mathbb{G}} \frac{1}{|\nabla_H d|^2}|\nabla_H \phi|^2 dx \geq 2\alpha Q \int_{\mathbb{G}} |\phi|^2 dx - 4\alpha^2 \int_{\mathbb{G}} d^2|\phi|^2 dx.$$

Let now $A := -4\int_{\mathbb{G}} d^2\phi^2 dx$, $B := 2Q\int_{\mathbb{G}} \phi^2 dx$ and $C := -\int_{\mathbb{G}} \frac{|\nabla_H \phi|^2}{|\nabla_H d|^2} dx$. Then the above inequality can be expressed as $A\alpha^2 + B\alpha + C \leq 0$ for all $\alpha \in \mathbb{R}$. But this implies that $B^2 - 4AC \leq 0$, which proves the statement.

2. Let us take

$$V \equiv 1 \qquad \text{and} \qquad \Phi = e^{-\alpha d},$$

where $\alpha > 0$. Then by Theorem 7.3.1 we have

$$\int_{\mathbb{G}} |\nabla_H \phi|^2 dx \geq 2\alpha Q \int_{\mathbb{G}} |\nabla_H d|^2|\phi|^2 dx - 4\alpha^2 \int_{\mathbb{G}} d^2|\nabla_H d|^2|\phi|^2 dx.$$

The same argument as in Part 1 implies the statement.

3. The statement follows from Theorem 7.3.1 with

$$V \equiv 1 \qquad \text{and} \qquad \Phi = e^{-\alpha d},$$

for $\alpha > 0$, and the same argument as in Part 1. □

In [GKY17] the authors showed that on polarizable Carnot groups the statement of Theorem 7.3.1 can be refined to give also the remainder estimates. Following [GKY17] we recapture this statement, its proof, and its consequences. In the following theorem we consider the case $p \geq 2$ noting that a similar result can be shown also for $1 < p < 2$, with a different reminder term, if one uses in the proof (7.19) instead of (7.18).

Theorem 7.3.5 (Two-weight L^p-Hardy inequalities with remainder estimates)**.** *Let $\mathbb{G}$ be a polarizable Carnot group and let Ω be a bounded domain in $\mathbb{G}$ with smooth boundary $\partial\Omega$. Assume that V is a non-negative C^1-function and that δ is a positive C^∞-function such that*

$$-\nabla_H \cdot \left(V(x) d^{p-Q} \frac{|\nabla_H \delta|^{p-2}}{\delta^{p-2}} \nabla_H \delta \right) \geq 0 \tag{7.22}$$

holds almost everywhere in Ω. Then for any $\phi \in C_0^\infty(\Omega)$ we have

$$\begin{aligned} \int_\Omega V(x) d^\alpha |\nabla_H \phi|^p dx \geq & \left(\frac{Q+\alpha-p}{p} \right)^p \int_\Omega V(x) d^\alpha \frac{|\nabla_H d|^p}{d^p} |\phi|^p dx \\ & + \left(\frac{Q+\alpha-p}{p} \right)^{p-1} \int_\Omega V(x) d^\alpha \frac{|\nabla_H d|^{p-2}}{d^{p-1}} \nabla_H d \cdot \nabla_H V |\phi|^p dx \\ & + \frac{c_p}{p^p} \int_\Omega V(x) d^\alpha \frac{|\nabla_H \delta|^p}{\delta^p} |\phi|^p dx, \end{aligned} \tag{7.23}$$

where $Q+\alpha > p \geq 2$, $\alpha \in \mathbb{R}$ and $c_p = c(p) > 0$.

Proof. For any $\phi \in C_0^\infty(\Omega)$ we set $\varphi := d^{-\gamma}\phi$ with $\gamma < 0$, a constant that will be chosen later. By a direct computation we have

$$\nabla_H (d^\gamma \phi) = \gamma d^{\gamma-1} \varphi \nabla_H d + d^\gamma \nabla_H \varphi.$$

Applying inequality (7.18) with $a = \gamma d^{\gamma-1}\varphi\nabla_H d$ and $b = d^\gamma \nabla_H \varphi$ we get

$$\begin{aligned} |\nabla_H \phi|^p \geq & |\gamma|^p d^{p(\gamma-1)} |\nabla_H d|^p |\phi|^p \\ & + \gamma|\gamma|^{p-2} d^{p(\gamma-1)+1} |\nabla_H d|^{p-2} \nabla_H d \cdot \nabla_H (|\varphi|^p) + c_p d^{p\gamma} |\nabla_H \varphi|^p. \end{aligned} \tag{7.24}$$

Multiplying both sides of (7.24) by $V(x)d^\alpha$ and integrating by parts we get

$$\begin{aligned} \int_\Omega V(x) d^\alpha |\nabla_H \phi|^p dx \geq & |\gamma|^p \int_\Omega V(x) d^{\alpha+p(\gamma-1)} |\nabla_H d|^p |\varphi|^p dx \\ & - \gamma|\gamma|^{p-2} \int_\Omega \nabla_H \cdot (V(x) d^{\alpha+p(\gamma-1)+1} |\nabla_H d|^{p-2} \nabla_H d) |\varphi|^p dx \\ & + c_p \int_\Omega V(x) d^{\alpha+p\gamma} |\nabla_H \varphi|^p dx. \end{aligned} \tag{7.25}$$

Using (1.77) and (1.102) we have

$$\begin{aligned}\nabla_H \cdot (V(x)d^{\alpha+p(\gamma-1)+1}|\nabla_H d|^{p-2}\nabla_H d)\\ = d^{\alpha+p(\gamma-1)+1}|\nabla_H d|^{p-2}\nabla_H d\cdot\nabla_H V\\ + [Q+\alpha+p(\gamma-1)]V(x)d^{\alpha+p(\gamma-1)}|\nabla_H d|^p.\end{aligned} \tag{7.26}$$

Using (7.26) we can rewrite (7.25) as

$$\begin{aligned}\int_\Omega V(x)d^\alpha|\nabla_H\phi|^p dx &\geq \zeta(Q,\alpha,p;\gamma)\int_\Omega V(x)d^{\alpha+p(\gamma-1)}|\nabla_H d|^p|\varphi|^p dx\\ &- \gamma|\gamma|^{p-2}\int_\Omega d^{\alpha+p(\gamma-1)+1}|\nabla_H d|^{p-2}\nabla_H d\cdot\nabla_H V|\varphi|^p dx\\ &+ c_p\int_\Omega V(x)d^{\alpha+p\gamma}|\nabla_H\varphi|^p dx,\end{aligned}$$

where $\zeta(Q,\alpha,p;\gamma) = |\gamma|^p - \gamma|\gamma|^{p-2}(Q+\alpha+\gamma p-p)$. Since $\gamma<0$ we can choose $\gamma = (p-\alpha-Q)/p$. Therefore, we have

$$\begin{aligned}\int_\Omega V(x)d^\alpha|\nabla_H\phi|^p dx \geq \left(\frac{Q+\alpha-p}{p}\right)^p\int_\Omega V(x)\frac{|\nabla_H d|^p}{d^Q}|\varphi|^p dx\\ + \left(\frac{Q+\alpha-p}{p}\right)^{p-1}\int_\Omega \frac{|\nabla_H d|^{p-2}}{d^{Q-1}}\nabla_H d\cdot\nabla_H V|\varphi|^p dx\\ + c_p\int_\Omega V(x)d^{p-Q}|\nabla_H\varphi|^p dx.\end{aligned} \tag{7.27}$$

Let us analyse the last term in (7.27). Let us define $\vartheta := \delta^{-1/p}\varphi$, where $0<\delta\in C^\infty(\Omega)$ and $\varphi\in C_0^\infty(\Omega)$. It follows from (7.18) that

$$\begin{aligned}|\nabla_H\phi|^p &= |\frac{1}{p}\delta^{\frac{1-p}{p}}\vartheta\nabla_H\delta + \delta^{\frac{1}{p}}\nabla_H\vartheta|^p\\ &\geq \frac{1}{p^p}\frac{|\nabla_H\delta|^p}{\delta^{p-1}}|\vartheta|^p + \frac{1}{p^{p-1}}\frac{|\nabla_H\delta|^{p-2}}{\delta^{p-2}}\nabla_H\delta\cdot\nabla_H(|\vartheta|^p) + c_p\delta^p|\nabla_H\vartheta|^p.\end{aligned} \tag{7.28}$$

Since $c_p\delta^p|\nabla_H\vartheta|^p\geq 0$, integrating by parts in (7.28) we get

$$\begin{aligned}c_p\int_\Omega V(x)d^{p-Q}|\nabla_H\varphi|^p dx \geq \frac{c_p}{p^p}\int_\Omega V(x)d^{p-Q}\frac{|\nabla_H\delta|^p}{\delta^{p-1}}|\vartheta|^p dx\\ - \frac{c_p}{p^{p-1}}\int_\Omega \nabla_H\cdot\left(V(x)d^{p-Q}\frac{|\nabla_H\delta|^{p-2}}{\delta^{p-2}}\nabla_H\delta\right)|\vartheta|^p dx.\end{aligned}$$

Using (7.22) and the substitution $\vartheta := \delta^{-1/p}d^{\frac{Q+\alpha-p}{p}}\phi$ we obtain

$$c_p\int_\Omega V(x)d^{p-Q}|\nabla_H\varphi|^p dx \geq \frac{c_p}{p^p}\frac{C_p}{p^p}\int_\Omega V(x)d^{p-Q}\frac{|\nabla_H\delta|^p}{\delta^p}|\phi|^p dx. \tag{7.29}$$

Combining (7.27) and (7.29), and using $\varphi = d^{\frac{Q+\alpha-p}{p}}\phi$ we obtain (7.23). □

Let us now list several consequences of Theorem 7.3.5 for some specific choices of V and δ. We recall the notation

$$B_R := \{x \in \mathbb{G} : d(x) < R\}$$

for the ball of radius R with respect to the quasi-norm d, already used in (7.21).

Corollary 7.3.6 (A collection of L^p-Hardy inequalities with remainders). *Let $\mathbb{G}$ be a polarizable Carnot group and let Ω be a bounded domain in $\mathbb{G}$ with smooth boundary $\partial\Omega$. Then we have the following statements.*

1. *For all $\phi \in C_0^\infty(\Omega)$ we have*

$$\int_\Omega d^\alpha |\nabla_H \phi|^p dx \geq \left(\frac{Q+\alpha-p}{p}\right)^p \int_\Omega d^\alpha \frac{|\nabla_H d|^p}{d^p}|\phi|^p dx + \frac{c_p}{p^p}\int_\Omega d^\alpha \frac{|\nabla_H d|^p}{(d\log(\frac{R}{d}))}|\phi|^p dx,$$

where $Q+\alpha > p \geq 2$, $\alpha \in \mathbb{R}$, $c_p > 0$ and $R > \sup_{x\in\Omega} d(x)$.

2. *For all $\phi \in C_0^\infty(\Omega)$ we have*

$$\int_\Omega d^\alpha |\nabla_H \phi|^p dx \geq \left(\frac{Q+\alpha-p}{p}\right)^p \int_\Omega d^\alpha \frac{|\nabla_H d|^p}{d^p}|\phi|^p dx + \frac{c_p}{p^p}\int_\Omega d^\alpha \frac{|\nabla_H d|^p}{d^p(\log\frac{R}{d})^p(\log(\log\frac{R}{d}))^p}|\phi|^p dx,$$

where $Q+\alpha > p \geq 2$, $\alpha \in \mathbb{R}$, $c_p > 0$ and $R > e\sup_{x\in\Omega} d(x)$.

3. *For all $\phi \in C_0^\infty(\Omega)$ we have*

$$\int_\Omega e^d d^\alpha |\nabla_H \phi|^p dx \geq \left(\frac{Q+\alpha-p}{p}\right)^p \int_\Omega e^d d^\alpha \frac{|\nabla_H d|^p}{d^p}|\phi|^p dx + \left(\frac{Q+\alpha-p}{p}\right)^{p-1}\int_\Omega e^d d^\alpha \frac{|\nabla_H d|^p}{d^{p-1}}|\phi|^p dx + \frac{c_p}{p^p}\int_\Omega e^d d^\alpha |\nabla_H d|^p |\phi|^p dx,$$

where $Q+\alpha > p \geq 2$, $\alpha \in \mathbb{R}$, $c_p > 0$.

4. *For all $\phi \in C_0^\infty(B_R)$ we have*

$$\int_{B_R} d^\alpha |\nabla_H \phi|^p dx \geq \left(\frac{Q+\alpha-p}{p}\right)^p \int_{B_R} d^\alpha \frac{|\nabla_H d|^p}{d^p}|\phi|^p dx + \frac{c_p}{p^p}\int_{B_R} d^\alpha \frac{|\nabla_H d|^p}{(R-d)^p}|\phi|^p dx,$$

where $Q+\alpha > p \geq 2$, $\alpha \in \mathbb{R}$, $c_p > 0$.

Proof of Corollary 7.3.6. 1. The statement follows from Theorem 7.3.5 with

$$V \equiv 1 \quad \text{and} \quad \delta = \log\left(\frac{R}{d}\right).$$

This inequality is the stratified group version of the Euclidean inequality in [ACR02, (1.4)].

2. The statement follows from Theorem 7.3.5 with

$$V \equiv 1 \quad \text{and} \quad \delta = \log\left(\log\frac{R}{d}\right), \quad R > e\sup_{x\in\Omega} d(x).$$

3. The statement follows from Theorem 7.3.5 with

$$V = e^{d} \quad \text{and} \quad \delta = e^{-d}.$$

4. The statement follows from Theorem 7.3.5 with

$$V \equiv 1 \quad \text{and} \quad \delta = R - d.$$

As in the proof of Corollary 7.3.2, the above applications of Theorem 7.3.5 can be justified by considering the regularization $d_\epsilon := (\Gamma+\epsilon)^{\frac{1}{2-Q}}$ for $\epsilon > 0$, where Γ is the fundamental solution for $\mathcal{L}$, and after the application of Theorem 7.3.5 taking the limit as $\epsilon \to 0$. □

7.4 Rellich inequalities for sub-Laplacians with drift

In this section, we show the weighted Rellich inequality for sub-Laplacians with drift on polarizable Carnot groups expressing the weights in terms of the fundamental solution of the sub-Laplacian.

We recall that in Section 6.9 we already showed Rellich inequalities for sub-Laplacians with drift with weights expressed in terms of the variable x' from the first stratum.

In this section, we assume all the notation of Section 1.4.6 where sub-Laplacians with drift have been discussed.

Theorem 7.4.1 (Rellich inequality for sub-Laplacian with drift with $\mathcal{L}$-gauge weights on polarizable Carnot groups)**.** *Let $\mathbb{G}$ be a polarizable Carnot group of homogeneous dimension $Q \geq 3$ and let $\theta \in \mathbb{R}$ with $Q + 2\theta - 4 > 0$. Then for all functions $f \in C_0^\infty(\mathbb{G}\backslash\{0\})$ we have*

$$\left\|\frac{d^\theta}{|\nabla_H d|}\mathcal{L}_X f\right\|^2_{L^2(\mathbb{G},\mu_X)} \geq \frac{(Q+2\theta-4)^2(Q-2\theta)^2}{16}\left\|d^{\theta-2}|\nabla_H d| f\right\|^2_{L^2(\mathbb{G},\mu_X)}$$

$$+\gamma^4 b_X^4\left\|\frac{d^\theta}{|\nabla_H d|}f\right\|^2_{L^2(\mathbb{G},\mu_X)} + \gamma^2 b_X^2\left(\frac{(Q+2\theta-2)(Q-2\theta-2)}{2}\right)\left\|d^{\theta-1}f\right\|^2_{L^2(\mathbb{G},\mu_X)}$$

$$+ 2\gamma^2 b_X^2 (Q-1)(3Q-4) \left\| d^{\theta-1} f \right\|^2_{L^2(\mathbb{G},\mu_X)}$$
$$+ 2\gamma^2 b_X^2 \int_{\mathbb{G}} \frac{d^{2\theta+2Q-2}}{|\nabla_H d|^4} \left(d^{1-Q}|\nabla_H d| \mathcal{L}(d^{1-Q}|\nabla_H d|) - 3|\nabla_H (d^{1-Q}|\nabla_H d|)|^2 \right)$$
$$\times |f(x)|^2 d\mu_X(x), \tag{7.30}$$

where $\mathcal{L}_X$ and b_X are defined in (1.93) and (1.95), respectively.

Remark 7.4.2. In the Abelian case $\mathbb{G} = (\mathbb{R}^n, +)$, we have $N = n$, $\nabla_H = \nabla = (\partial_{x_1}, \ldots, \partial_{x_n})$ is the usual full gradient, $d = |x|_E$ is the Euclidean distance, hence $|\nabla_H d| = 1$, and setting $X = \sum_{i=1}^n a_i \partial_{x_i}$, the last two terms in (7.30) cancel each other, so that we obtain the same estimate as in (6.83).

Proof of Theorem 7.4.1. The proof follows [RY18b]. Let $g = g(x) \in C_0^\infty(\mathbb{G}\backslash\{0\})$ be such that $f = \chi^{-1/2} g$. By (1.101) we know that $L^2(\mathbb{G},\mu) \in g \mapsto \chi^{-1/2} g \in L^2(\mathbb{G},\mu_X)$. Then, we have

$$\left\| \frac{d^\theta}{|\nabla_H d|} \mathcal{L}_X f \right\|_{L^2(\mathbb{G},\mu_X)} = \left\| \frac{d^\theta}{|\nabla_H d|} \chi^{1/2} \mathcal{L}_X f \right\|_{L^2(\mathbb{G},\mu)}$$
$$= \left\| \frac{d^\theta}{|\nabla_H d|} \chi^{1/2} \mathcal{L}_X (\chi^{-1/2} g) \right\|_{L^2(\mathbb{G},\mu)},$$

where μ is the Haar (i.e., Lebesgue) measure on $\mathbb{G}$. By (1.100) and integration by parts, we calculate

$$\left\| \frac{d^\theta}{|\nabla_H d|} \mathcal{L}_X f \right\|^2_{L^2(\mathbb{G},\mu_X)} = \left\| \frac{d^\theta}{|\nabla_H d|} (\mathcal{L}_0 + \gamma^2 b_X^2) g \right\|^2_{L^2(\mathbb{G},\mu)}$$
$$= \left\| \frac{d^\theta}{|\nabla_H d|} \mathcal{L}_0 g \right\|^2_{L^2(\mathbb{G},\mu)} + 2\gamma^2 b_X^2 \mathrm{Re} \int_{\mathbb{G}} \frac{d^{2\theta}}{|\nabla_H d|^2} \mathcal{L}_0 g(x) \overline{g(x)} dx + \gamma^4 b_X^4 \left\| \frac{d^\theta}{|\nabla_H d|} g \right\|^2_{L^2(\mathbb{G},\mu)}$$
$$= \left\| \frac{d^\theta}{|\nabla_H d|} \mathcal{L}_0 g \right\|^2_{L^2(\mathbb{G},\mu)} - 2\gamma^2 b_X^2 \mathrm{Re} \sum_{j=1}^N \int_{\mathbb{G}} \frac{d^{2\theta}}{|\nabla_H d|^2} X_j^2 g(x) \overline{g(x)} dx$$
$$+ \gamma^4 b_X^4 \left\| \frac{d^\theta}{|\nabla_H d|} g \right\|^2_{L^2(\mathbb{G},\mu)}$$
$$= \left\| \frac{d^\theta}{|\nabla_H d|} \mathcal{L}_0 g \right\|^2_{L^2(\mathbb{G},\mu)} + 2\gamma^2 b_X^2 \left\| \frac{d^\theta}{|\nabla_H d|} \nabla_H g \right\|^2_{L^2(\mathbb{G},\mu)} + \gamma^4 b_X^4 \left\| \frac{d^\theta}{|\nabla_H d|} g \right\|^2_{L^2(\mathbb{G},\mu)}$$
$$+ 2\gamma^2 b_X^2 \mathrm{Re} \sum_{j=1}^N \int_{\mathbb{G}} X_j g(x) \overline{g(x)} X_j \left(\frac{d^{2\theta}}{|\nabla_H d|^2} \right) dx$$
$$= \left\| \frac{d^\theta}{|\nabla_H d|} \mathcal{L}_0 g \right\|^2_{L^2(\mathbb{G},\mu)} + 2\gamma^2 b_X^2 \left\| \frac{d^\theta}{|\nabla_H d|} \nabla_H g \right\|^2_{L^2(\mathbb{G},\mu)} + \gamma^4 b_X^4 \left\| \frac{d^\theta}{|\nabla_H d|} g \right\|^2_{L^2(\mathbb{G},\mu)}$$

$$
\begin{aligned}
&+ 4\gamma^2 b_X^2 \theta \mathrm{Re} \sum_{j=1}^{N} \int_{\mathbb{G}} X_j g(x) \overline{g(x) \frac{d^{2\theta-1} X_j d}{|\nabla_H d|^2}} dx \\
&- 4\gamma^2 b_X^2 \mathrm{Re} \sum_{j=1}^{N} \int_{\mathbb{G}} X_j g(x) \overline{g(x) \frac{d^{2\theta} X_j |\nabla_H d|}{|\nabla_H d|^3}} dx \\
&=: \left\| \frac{d^\theta}{|\nabla_H d|} \mathcal{L}_0 g \right\|^2_{L^2(\mathbb{G},\mu)} + 2\gamma^2 b_X^2 \left\| \frac{d^\theta}{|\nabla_H d|} \nabla_H g \right\|^2_{L^2(\mathbb{G},\mu)} + \gamma^4 b_X^4 \left\| \frac{d^\theta}{|\nabla_H d|} g \right\|^2_{L^2(\mathbb{G},\mu)} \\
&+ I_1 + I_2.
\end{aligned} \tag{7.31}
$$

Then, by Theorem 7.1.3 one has for $Q + 2\theta - 2 > 0$ that

$$
2\gamma^2 b_X^2 \left\| \frac{d^\theta}{|\nabla_H d|} \nabla_H g \right\|^2_{L^2(\mathbb{G},\mu)} \geq 2\gamma^2 b_X^2 \left(\frac{Q + 2\theta - 2}{2} \right)^2 \left\| d^{\theta-1} g \right\|^2_{L^2(\mathbb{G},\mu)}. \tag{7.32}
$$

On the other hand by Theorem 7.2.2 we get for $Q + 2\theta - 4 > 0$ that

$$
\left\| \frac{d^\theta}{|\nabla_H d|} \mathcal{L}_0 g \right\|^2_{L^2(\mathbb{G},\mu)} \geq \frac{(Q + 2\theta - 4)^2 (Q - 2\theta)^2}{16} \left\| d^{\theta-2} |\nabla_H d| g \right\|^2_{L^2(\mathbb{G},\mu)}. \tag{7.33}
$$

Putting (7.32) and (7.33) into (7.31) we obtain for $Q + 2\theta - 4 > 0$ that

$$
\begin{aligned}
&\left\| \frac{d^\theta}{|\nabla_H d|} \mathcal{L}_X f \right\|^2_{L^2(\mathbb{G},\mu_X)} \\
&\geq \frac{(Q + 2\theta - 4)^2 (Q - 2\theta)^2}{16} \left\| d^{\theta-2} |\nabla_H d| g \right\|^2_{L^2(\mathbb{G},\mu)} \\
&+ 2\gamma^2 b_X^2 \left(\frac{Q + 2\theta - 2}{2} \right)^2 \left\| d^{\theta-1} g \right\|^2_{L^2(\mathbb{G},\mu)} \\
&+ \gamma^4 b_X^4 \left\| \frac{d^\theta}{|\nabla_H d|} g \right\|^2_{L^2(\mathbb{G},\mu)} + I_1 + I_2.
\end{aligned} \tag{7.34}
$$

Let us calculate I_1 from (7.31):

$$
\begin{aligned}
I_1 &= 4\gamma^2 b_X^2 \theta \mathrm{Re} \sum_{j=1}^{N} \int_{\mathbb{G}} X_j g(x) \overline{g(x) \frac{d^{2\theta-1} X_j d}{|\nabla_H d|^2}} dx \\
&= -4\gamma^2 b_X^2 \theta \mathrm{Re} \sum_{j=1}^{N} \int_{\mathbb{G}} g(x) \overline{X_j g(x) \frac{d^{2\theta-1} X_j d}{|\nabla_H d|^2}} dx \\
&- 4\gamma^2 b_X^2 \theta \mathrm{Re} \sum_{j=1}^{N} \int_{\mathbb{G}} |g(x)|^2 X_j \left(\frac{d^{2\theta-1} X_j d}{|\nabla_H d|^2} \right) dx.
\end{aligned}
$$

It follows that

$$I_1 = 4\gamma^2 b_X^2 \theta \mathrm{Re} \sum_{j=1}^N \int_{\mathbb{G}} X_j g(x) \overline{g(x)} \frac{d^{2\theta-1} X_j d}{|\nabla_H d|^2} dx$$
$$= -2\gamma^2 b_X^2 \theta \sum_{j=1}^N \int_{\mathbb{G}} |g(x)|^2 X_j \left(\frac{d^{2\theta-1} X_j d}{|\nabla_H d|^2} \right) dx.$$

Putting $d = u^{\frac{1}{2-Q}}$ we calculate

$$\sum_{j=1}^N X_j \left(\frac{d^{2\theta-1} X_j d}{|\nabla_H d|^2} \right) = (2-Q) \sum_{j=1}^N X_j \left(\frac{u^{\frac{2\theta-Q}{2-Q}}}{|\nabla_H u|^2} X_j u \right)$$
$$= (2-Q) \sum_{j=1}^N \frac{2\theta - Q}{2-Q} u^{\frac{2\theta-2}{2-Q}} \frac{(X_j u)^2}{|\nabla_H u|^2} + (2-Q) \sum_{j=1}^N \frac{u^{\frac{2\theta-Q}{2-Q}} X_j^2 u}{|\nabla_H u|^2}$$
$$- 2(2-Q) \sum_{j=1}^N \frac{u^{\frac{2\theta-Q}{2-Q}} X_j u}{|\nabla_H u|^3} X_j |\nabla_H u|,$$

which implies, using (1.104), that

$$I_1 = -2\gamma^2 b_X^2 \theta \int_{\mathbb{G}} \left(\sum_{j=1}^N X_j \left(\frac{d^{2\theta-1} X_j d}{|\nabla_H d|^2} \right) \right) |g(x)|^2 dx$$
$$= -2\gamma^2 b_X^2 \theta (2\theta + Q - 2) \int_{\mathbb{G}} u^{\frac{2\theta-2}{2-Q}} |g(x)|^2 dx$$
$$= -2\gamma^2 b_X^2 \theta (2\theta + Q - 2) \int_{\mathbb{G}} |g(x)|^2 d^{2\theta-2} dx. \tag{7.35}$$

Now for I_2 by integration by parts we get

$$I_2 = -4\gamma^2 b_X^2 \mathrm{Re} \sum_{j=1}^N \int_{\mathbb{G}} X_j g(x) \overline{g(x)} \frac{d^{2\theta} X_j |\nabla_H d|}{|\nabla_H d|^3} dx$$
$$= 2\gamma^2 b_X^2 \sum_{j=1}^N \int_{\mathbb{G}} |g(x)|^2 X_j \left(\frac{d^{2\theta} X_j |\nabla_H d|}{|\nabla_H d|^3} \right) dx.$$

Using $d = u^{\frac{1}{2-Q}}$ one has

$$\sum_{j=1}^N X_j \left(\frac{d^{2\theta} X_j |\nabla_H d|}{|\nabla_H d|^3} \right) = (2-Q)^2 \sum_{j=1}^N X_j \left(\frac{u^{\frac{2\theta-3Q+3}{2-Q}}}{|\nabla_H u|^3} X_j \left(u^{\frac{Q-1}{2-Q}} |\nabla_H u| \right) \right)$$
$$= (2-Q)^2 \sum_{j=1}^N X_j \left(\frac{Q-1}{2-Q} u^{\frac{2\theta-Q}{2-Q}} \frac{X_j u}{|\nabla_H u|^2} + u^{\frac{2\theta-2Q+2}{2-Q}} \frac{X_j |\nabla_H u|}{|\nabla_H u|^3} \right)$$
$$=: (2-Q)^2 J_1 + (2-Q)^2 J_2. \tag{7.36}$$

Then, taking into account (1.104) we have for J_1 that

$$J_1 = \sum_{j=1}^{N} X_j \left(\frac{Q-1}{2-Q} u^{\frac{2\theta-Q}{2-Q}} \frac{X_j u}{|\nabla_H u|^2} \right) = \frac{(Q-1)(2\theta-Q)}{(Q-2)^2} u^{\frac{2\theta-2}{2-Q}} \sum_{j=1}^{N} \frac{(X_j u)^2}{|\nabla_H u|^2}$$
$$+ \frac{Q-1}{2-Q} u^{\frac{2\theta-Q}{2-Q}} \sum_{j=1}^{N} \frac{X_j^2 u}{|\nabla_H u|^2} - \frac{2(Q-1)}{(2-Q)} u^{\frac{2\theta-Q}{2-Q}} \sum_{j=1}^{N} \frac{X_j u X_j |\nabla_H u|}{|\nabla_H u|^3}$$
$$= \frac{(Q-1)(2\theta-Q)}{(Q-2)^2} u^{\frac{2\theta-2}{2-Q}} + \frac{Q-1}{2-Q} u^{\frac{2\theta-Q}{2-Q}} \frac{\mathcal{L}u}{|\nabla_H u|^2} + \frac{2(Q-1)^2}{(Q-2)^2} u^{\frac{2\theta-2}{2-Q}}. \tag{7.37}$$

Now we calculate for J_2 that

$$J_2 = \sum_{j=1}^{N} X_j \left(u^{\frac{2\theta-2Q+2}{2-Q}} \frac{X_j |\nabla_H u|}{|\nabla_H u|^3} \right)$$
$$= \frac{2\theta-2Q+2}{2-Q} u^{\frac{2\theta-Q}{2-Q}} \sum_{j=1}^{N} \frac{X_j u X_j |\nabla_H u|}{|\nabla_H u|^3}$$
$$+ u^{\frac{2\theta-2Q+2}{2-Q}} \sum_{j=1}^{N} \frac{X_j^2 |\nabla_H u|}{|\nabla_H u|^3} - 3u^{\frac{2\theta-2Q+2}{2-Q}} \sum_{j=1}^{N} \frac{(X_j |\nabla_H u|)^2}{|\nabla_H u|^4}$$
$$= \frac{2\theta-2Q+2}{2-Q} \left(\frac{Q-1}{Q-2} \right) u^{\frac{2\theta-2}{2-Q}} + u^{\frac{2\theta-2Q+2}{2-Q}} \frac{\mathcal{L}|\nabla_H u|}{|\nabla_H u|^3}$$
$$- 3u^{\frac{2\theta-2Q+2}{2-Q}} \frac{|\nabla_H |\nabla_H u||^2}{|\nabla_H u|^4}, \tag{7.38}$$

where we have used (1.104) in the last equality. Plugging (7.37) and (7.38) into (7.36), we obtain

$$\sum_{j=1}^{N} X_j \left(\frac{d^{2\theta} X_j |\nabla_H d|}{|\nabla_H d|^3} \right)$$
$$= (Q-1)(3Q-4) u^{\frac{2\theta-2}{2-Q}} + (2-Q)(Q-1) \frac{\mathcal{L}u}{|\nabla_H u|^2}$$
$$+ (Q-2)^2 u^{\frac{2\theta-2Q+2}{2-Q}} |\nabla_H u|^{-4} (|\nabla_H u| \mathcal{L} |\nabla_H u| - 3 |\nabla_H |\nabla_H u||^2).$$

Then we get for I_2 the expression

$$I_2 = 2\gamma^2 b_X^2 \sum_{j=1}^{N} \int_{\mathbb{G}} |g(x)|^2 X_j \left(\frac{d^{2\theta} X_j |\nabla_H d|}{|\nabla_H d|^3} \right) dx$$
$$= 2\gamma^2 b_X^2 (Q-1)(3Q-4) \int_{\mathbb{G}} u^{\frac{2\theta-2}{2-Q}} |g(x)|^2 dx$$
$$+ 2\gamma^2 b_X^2 (Q-2)^2 \int_{\mathbb{G}} u^{\frac{2\theta-2Q+2}{2-Q}} |\nabla_H u|^{-4} (|\nabla_H u| \mathcal{L} |\nabla_H u| - 3 |\nabla_H |\nabla_H u||^2) |g(x)|^2 dx.$$

Setting here $u = d^{2-Q}$, we get for I_2 that

$$I_2 = 2\gamma^2 b_X^2 (Q-1)(3Q-4) \int_{\mathbb{G}} d^{2\theta-2} |g(x)|^2 dx$$
$$+ 2\gamma^2 b_X^2 \int_{\mathbb{G}} \frac{d^{2\theta+2Q-2}}{|\nabla_H d|^4} \left(d^{1-Q} |\nabla_H d| \mathcal{L}(d^{1-Q} |\nabla_H d|) - 3|\nabla_H (d^{1-Q} |\nabla_H d|)|^2 \right) |g(x)|^2 dx.$$

Thus, by this and (7.35) we have

$$I_1 + I_2 = 2\gamma^2 b_X^2 ((Q-1)(3Q-4) - \theta(2\theta + Q - 2)) \left\| d^{\theta-1} g \right\|^2_{L^2(\mathbb{G},\mu)}$$
$$+ 2\gamma^2 b_X^2 \int_{\mathbb{G}} \frac{d^{2\theta+2Q-2}}{|\nabla_H d|^4} \left(d^{1-Q} |\nabla_H d| \mathcal{L}(d^{1-Q} |\nabla_H d|) - 3|\nabla_H (d^{1-Q} |\nabla_H d|)|^2 \right) |g(x)|^2 dx.$$

Combining this with (7.34) and taking into account (1.101) we obtain Theorem 7.4.1. □

7.5 Hardy inequalities on the complex affine group

The aim of this section is to show that some of the above techniques are also applicable for non-unimodular Lie groups. For example, consider the complex affine groups:

Definition 7.5.1 (Complex affine group). The complex affine group is the semi-direct product

$$\mathbb{G} = \mathbb{C} \rtimes \mathbb{C}^*,$$

where $\mathbb{C}^*$ is the multiplicative group of nonzero complex numbers. This means that $\mathbb{G}$ is equal to $\mathbb{C} \times \mathbb{C}^*$ as a set, with the group composition law of the complex affine group $\mathbb{G}$ given by

$$(x, y) \circ (x', y') = (x + yx', yy')$$

for all $x, x' \in \mathbb{C}$ and $y, y' \in \mathbb{C}^*$. We will be also using the notation $x := t + is$ and $y := \tau + i\varsigma$. The complex affine group is a Lie group, with its Lie algebra denoted by $\mathfrak{g}$.

We now fix a basis $\{X_1, X_2, X_3, X_4\}$ of $\mathfrak{g}$ given by

$$X_1 = \frac{\partial}{\partial t}, \quad X_3 = t\frac{\partial}{\partial t} + s\frac{\partial}{\partial s} + \tau\frac{\partial}{\partial \tau} + \varsigma\frac{\partial}{\partial \varsigma},$$
$$X_2 = \frac{\partial}{\partial s}, \quad X_4 = -s\frac{\partial}{\partial t} + t\frac{\partial}{\partial s} - \varsigma\frac{\partial}{\partial \tau} + \tau\frac{\partial}{\partial \varsigma}.$$

These right invariant vector fields correspond to the canonical basis elements of $\mathfrak{g}$, and it will be convenient to work with right invariant vector fields here. Therefore, the positive (sub-)Laplacian

$$\Delta_X = -\sum_{j=1}^{4} X_j^2 \tag{7.39}$$

is called a *right invariant canonical Laplacian of the complex affine group* $\mathbb{G}$. The fundamental solution of the Laplacian Δ_X was computed explicitly by Gaudry and Sjögren [GS98] in the following form

$$\varepsilon = \frac{1}{4\pi^2} \frac{|y|^2}{|x|^2 + |1-y|^2}.$$

We will also use the notation

$$\nabla_X = (X_1, X_2, X_3, X_4)$$

for the right invariant (canonical) gradient on $\mathbb{G}$. The right invariant and the left invariant Haar measures on $\mathbb{G}$ are defined by

$$d\mu_r = dx\frac{dy}{|y|^2}, \quad d\mu_l = dx\frac{dy}{|y|^4},$$

with the modular function $m(x,y) = |y|^2$, respectively. In addition, one has the following integration rules with respect to the modular function

$$\int_{\mathbb{G}} f(\eta\zeta)d\mu_l(\eta) = m^{-1}(\zeta)\int_{\mathbb{G}} f(\eta)d\mu_l(\eta),$$
$$\int_{\mathbb{G}} f(\eta^{-1})m^{-1}(\eta)d\mu_l(\eta) = \int_{\mathbb{G}} f(\eta)d\mu_l(\eta).$$

We now present a Hardy type inequality on $\mathbb{G}$ with the proof relying on properties of the fundamental solution of the right invariant canonical Laplacian Δ_X on the complex affine group $\mathbb{G}$ given in (7.39).

Theorem 7.5.2 (Hardy inequalities on the complex affine group). *Let $\mathbb{G}$ be the complex affine group. Let $\alpha \in \mathbb{R}$, $\alpha > 2-\beta$, $\beta > 2$. Then we have*

$$\int_{\mathbb{G}} \varepsilon^{\frac{\alpha}{2-\beta}} |\nabla_X u|^2 \, d\mu_l \geq \left(\frac{\beta+\alpha-2}{2}\right)^2 \int_{\mathbb{G}} \varepsilon^{\frac{\alpha-2}{2-\beta}} |\nabla_X \varepsilon^{\frac{1}{2-\beta}}|^2 |u|^2 \, d\mu_l, \tag{7.40}$$

for all $u \in C_0^\infty(\mathbb{G})$, where $\nabla_X = (X_1, X_2, X_3, X_4)$.

Proof of Theorem 7.5.2. By using formula (2.8) we can assume without loss of generality that u is real-valued. Then let us set $u = d^\gamma q$ for some real-valued functions $d > 0$, q, and a constant $\gamma \neq 0$ to be chosen later. We use our usual notation for the potential theory considerations:

$$\widetilde{\nabla} u := \sum_{k=1}^{4} (X_k u)\, X_k.$$

Then we can calculate

$$\begin{aligned}
(\widetilde{\nabla} u)u &= (\widetilde{\nabla} d^\gamma q) d^\gamma q \\
&= \sum_{k=1}^{4} X_k(d^\gamma q) X_k(d^\gamma q) \\
&= \gamma^2 d^{2\gamma-2} \sum_{k=1}^{4} (X_k d)^2 q^2 + 2\gamma d^{2\gamma-1} q \sum_{k=1}^{4} X_k d\, X_k q + d^{2\gamma} \sum_{k=1}^{4} (X_k q)^2 \\
&= \gamma^2 d^{2\gamma-2}((\widetilde{\nabla} d) d) q^2 + 2\gamma d^{2\gamma-1} q (\widetilde{\nabla} d) q + d^{2\gamma}(\widetilde{\nabla} q) q.
\end{aligned}$$

Integrating by parts we observe that

$$\begin{aligned}
2\gamma \int_{\mathbb{G}} d^{\alpha+2\gamma-1} q (\widetilde{\nabla} d) q d\mu_l &= \frac{\gamma}{\alpha+2\gamma} \int_{\mathbb{G}} (\widetilde{\nabla} d^{\alpha+2\gamma}) q^2 d\mu_l \\
&= \frac{\gamma}{\alpha+2\gamma} \int_{\mathbb{G}} (\widetilde{\nabla} q^2) d^{\alpha+2\gamma} d\mu_l \\
&= -\frac{\gamma}{\alpha+2\gamma} \int_{\mathbb{G}} q^2 \Delta_X d^{\alpha+2\gamma} d\mu_l.
\end{aligned}$$

In particular, because of this, we will later choose γ so that $d^{\alpha+2\gamma} = \varepsilon$. Consequently, we have

$$\begin{aligned}
\int_{\mathbb{G}} d^\alpha (\widetilde{\nabla} u) u d\mu_l &= \gamma^2 \int_{\mathbb{G}} d^{\alpha+2\gamma-2}((\widetilde{\nabla} d) d)\, q^2 d\mu_l + \frac{\gamma}{\alpha+2\gamma} \int_{\mathbb{G}} (\widetilde{\nabla} d^{\alpha+2\gamma}) q^2 d\mu_l \\
&\quad + \int_{\mathbb{G}} d^{\alpha+2\gamma} (\widetilde{\nabla} q) q d\mu_l \\
&= \gamma^2 \int_{\mathbb{G}} d^{\alpha+2\gamma-2}((\widetilde{\nabla} d) d)\, q^2 d\mu_l \\
&\quad - \frac{\gamma}{\alpha+2\gamma} \int_{\mathbb{G}} q^2 \Delta_X d^{\alpha+2\gamma} d\mu_l + \int_{\mathbb{G}} d^{\alpha+2\gamma} (\widetilde{\nabla} q) q d\mu_l \\
&\geq \gamma^2 \int_{\mathbb{G}} d^{\alpha+2\gamma-2}((\widetilde{\nabla} d) d)\, q^2 d\mu_l - \frac{\gamma}{\alpha+2\gamma} \int_{\mathbb{G}} q^2 \Delta_X d^{\alpha+2\gamma} d\mu_l,
\end{aligned} \tag{7.41}$$

since $d > 0$ and $(\widetilde{\nabla} q) q = |\nabla_X q|^2 \geq 0$. On the other hand, it can be readily checked that for a vector field X we have

$$\begin{aligned}
&\frac{\gamma}{\alpha+2\gamma} X^2(d^{\alpha+2\gamma}) = \gamma X(d^{\alpha+2\gamma-1} X d) = \frac{\gamma}{2-\beta} X(d^{\alpha+2\gamma+\beta-2} X(d^{2-\beta})) \\
&= \frac{\gamma}{2-\beta}(\alpha+2\gamma+\beta-2) d^{\alpha+2\gamma+\beta-3}(Xd) X(d^{2-\beta}) + \frac{\gamma}{2-\beta} d^{\alpha+2\gamma+\beta-2} X^2(d^{2-\beta}) \\
&= \gamma(\alpha+2\gamma+\beta-2) d^{\alpha+2\gamma-2}(Xd)^2 + \frac{\gamma}{2-\beta} d^{\alpha+2\gamma+\beta-2} X^2(d^{2-\beta}).
\end{aligned}$$

Consequently, we get the equality

$$-\frac{\gamma}{\alpha+2\gamma}\Delta_X d^{\alpha+2\gamma} = -\gamma(\alpha+2\gamma+\beta-2)d^{\alpha+2\gamma-2}(\widetilde{\nabla}d)d - \frac{\gamma}{2-\beta}d^{\alpha+2\gamma+\beta-2}\Delta_X d^{2-\beta}. \tag{7.42}$$

We now substitute (7.42) into (7.41) and use that $q^2 = d^{-2\gamma}u^2$, so that

$$\int_{\mathbb{G}} d^{\alpha}(\widetilde{\nabla}u)u d\mu_l \geq (-\gamma^2 - \gamma(\alpha+\beta-2))\int_{\mathbb{G}} d^{\alpha-2}((\widetilde{\nabla}d)d)u^2 d\mu_l$$
$$-\frac{\gamma}{2-\beta}\int_{\mathbb{G}}(\Delta_X d^{2-\beta})d^{\alpha+\beta-2}u^2 dx.$$

We now take $d := \varepsilon^{\frac{1}{2-\beta}}$, and since $\beta > 2$ and ε is the fundamental solution to Δ_X we have

$$\int_{\mathbb{G}}(\Delta_X\varepsilon)\varepsilon^{\frac{\alpha+\beta-2}{2-\beta}}u^2 dx = 0,\ \alpha > 2-\beta,\ \beta > 2.$$

Thus, we obtain

$$\int_{\mathbb{G}} \varepsilon^{\frac{\alpha}{2-\beta}}(\widetilde{\nabla}u)u\, d\mu_l \geq (-\gamma^2 - \gamma(\alpha+\beta-2))\int_{\mathbb{G}} \varepsilon^{\frac{\alpha-2}{2-\beta}}(\widetilde{\nabla}\varepsilon^{\frac{1}{2-\beta}})\varepsilon^{\frac{1}{2-\beta}}u^2\, d\mu_l.$$

Taking $\gamma = \frac{2-\beta-\alpha}{2}$, we obtain (7.40). □

As usual, a Hardy inequality, such as the one in Theorem 7.5.2, implies uncertainty principles:

Corollary 7.5.3 (Uncertainty principles on the complex affine group). *Let $\mathbb{G}$ be the complex affine group and let $\beta > 2$. Then for all $u \in C_0^\infty(\mathbb{G})$ we have*

$$\int_{\mathbb{G}} \varepsilon^{\frac{2}{2-\beta}}|\nabla_X\varepsilon^{\frac{1}{2-\beta}}|^2|u|^2 d\mu_l \int_{\mathbb{G}}|\nabla_X u|^2 d\mu_l \geq \left(\frac{\beta-2}{2}\right)^2\left(\int_{\mathbb{G}}|\nabla_X\varepsilon^{\frac{1}{2-\beta}}|^2|u|^2 d\mu_l\right)^2, \tag{7.43}$$

as well as

$$\int_{\mathbb{G}} \frac{\varepsilon^{\frac{2}{2-\beta}}}{|\nabla_X\varepsilon^{\frac{1}{2-\beta}}|^2}|u|^2 d\mu_l \int_{\mathbb{G}}|\nabla_X u|^2 d\mu_l \geq \left(\frac{\beta-2}{2}\right)^2\left(\int_{\mathbb{G}}|u|^2 d\mu_l\right)^2. \tag{7.44}$$

Proof of Corollary 7.5.3. Taking $\alpha = 0$ in the inequality (7.40) and using Hardy inequality in Theorem 7.5.2, we get

$$\int_{\mathbb{G}} \varepsilon^{\frac{2}{2-\beta}}|\nabla_X\varepsilon^{\frac{1}{2-\beta}}|^2|u|^2 d\mu_l \int_{\mathbb{G}}|\nabla_X u|^2 d\mu_l$$
$$\geq \left(\frac{\beta-2}{2}\right)^2\int_{\mathbb{G}} \varepsilon^{\frac{2}{2-\beta}}|\nabla_X\varepsilon^{\frac{1}{2-\beta}}|^2|u|^2 d\mu_l \int_{\mathbb{G}}\frac{|\nabla_X\varepsilon^{\frac{1}{2-\beta}}|^2}{\varepsilon^{\frac{2}{2-\beta}}}|u|^2\, d\mu_l$$
$$\geq \left(\frac{\beta-2}{2}\right)^2\left(\int_{\mathbb{G}}|\nabla_X\varepsilon^{\frac{1}{2-\beta}}|^2|u|^2 d\mu_l\right)^2,$$

which shows (7.43). The proof of (7.44) is similar. □

7.6 Hardy inequalities for Baouendi–Grushin operators

In this section, we describe a special case of the Hardy inequalities on homogeneous groups, namely, inequalities associated to the Baouendi–Grushin vector fields on $\mathbb{R}^n$. However, the described methods also work for non-smooth vector fields allowing a singularity at the origin, so we include such cases in our exposition as well.

Definition 7.6.1 (Baouendi–Grushin operator and vector fields)**.** Let

$$z = (x_1, \ldots, x_m, y_1, \ldots, y_k) = (x, y) \in \mathbb{R}^m \times \mathbb{R}^k$$

with $k, m \geq 1$, $k + m = n$. Let $\gamma \geq 0$. Let us consider the (Baouendi–Grushin) vector fields

$$X_i = \frac{\partial}{\partial x_i}, \; i = 1, \ldots, m, \quad Y_j = |x|^\gamma \frac{\partial}{\partial y_j}, \; j = 1, \ldots, k.$$

The corresponding subelliptic gradient, which is the n-dimensional vector field, is then defined as

$$\nabla_\gamma := (X_1, \ldots, X_m, Y_1, \ldots, Y_k) = (\nabla_x, |x|^\gamma \nabla_y). \tag{7.45}$$

The *Baouendi–Grushin operator* on $\mathbb{R}^{m+k}$ is defined by

$$\Delta_\gamma = \sum_{i=1}^{m} X_i^2 + \sum_{j=1}^{k} Y_j^2 = \Delta_x + |x|^{2\gamma} \Delta_y = \nabla_\gamma \cdot \nabla_\gamma, \tag{7.46}$$

where Δ_x and Δ_y are the Laplace operators in the variables $x \in \mathbb{R}^m$ and $y \in \mathbb{R}^k$, respectively.

If γ is an even positive integer then the vector fields X_i, Y_j are smooth, and Δ_γ is hypoelliptic as a sum of squares of C^∞ vector fields satisfying Hörmander's condition

$$\text{rank Lie}[X_1, \ldots, X_m, Y_1, \ldots, Y_k] = n.$$

For any $\gamma \geq 0$ the dilation structure on $\mathbb{R}^{m+k}$ associated to Δ_γ is

$$\delta_\lambda(x, y) := (\lambda x, \lambda^{1+\gamma} y)$$

for $\lambda > 0$. Indeed, it is easy to check that this dilation structure makes the vector fields homogeneous,

$$X_i(\delta_\lambda) = \lambda \delta_\lambda(X_i), \quad Y_i(\delta_\lambda) = \lambda \delta_\lambda(Y_i),$$

and hence also

$$\nabla_\gamma \circ \delta_\lambda = \lambda \delta_\lambda \nabla_\gamma.$$

The homogeneous dimension of $\mathbb{R}^m \times \mathbb{R}^k$ with respect to this dilation is

$$Q = m + (1 + \gamma)k. \tag{7.47}$$

Definition 7.6.2 (Baouendi–Grushin distance)**.** Let $\rho(z)$ be the distance function, for $z=(x,y)\in\mathbb{R}^m\times\mathbb{R}^k$ defined by

$$\rho=\rho(z):=\left(|x|^{2(1+\gamma)}+(1+\gamma)^2|y|^2\right)^{\frac{1}{2(1+\gamma)}}. \tag{7.48}$$

It is easy to check that it satisfies

$$|\nabla_\gamma\rho|=\frac{|x|^\gamma}{\rho^\gamma}. \tag{7.49}$$

We will formulate a refined version of the Hardy inequality for Baouendi–Grushin vector fields, and then in Remark 7.6.4 we will put it in the context of the existing rich literature on this subject.

Theorem 7.6.3 (Refined Hardy inequality for Baouendi–Grushin vector fields)**.** *Let $(x,y)=(x_1,\ldots,x_m,y_1,\ldots,y_k)\in\mathbb{R}^m\times\mathbb{R}^k$ with $k,m\geq 1$, $k+m=n$. Let $\alpha_1,\alpha_2\in\mathbb{R}$ be such that*

$$Q+\alpha_1-2>0 \text{ and } m+\gamma\alpha_2>0.$$

Then for all complex-valued functions $f\in C_0^\infty(\mathbb{R}^n\backslash\{0\})$ we have

$$\begin{aligned}&\int_{\mathbb{R}^n}\rho^{\alpha_1}|\nabla_\gamma\rho|^{\alpha_2}\left(\left|\frac{d}{d|x|}f\right|^2+|x|^{2\gamma}|\nabla_y f|^2\right)dxdy\\&\geq\left(\frac{Q+\alpha_1-2}{2}\right)^2\int_{\mathbb{R}^n}\rho^{\alpha_1}|\nabla_\gamma\rho|^{\alpha_2}\frac{|\nabla_\gamma\rho|^2}{\rho^2}|f|^2dxdy,\end{aligned} \tag{7.50}$$

with sharp constant $\left(\frac{Q+\alpha_1-2}{2}\right)^2$.

Remark 7.6.4.

1. First, a Hardy inequality for Grushin operators was obtained by Garofalo [Gar93], who has shown the inequality

$$\begin{aligned}&\int_{\mathbb{R}^n}(|\nabla_x f|^2+|x|^{2\gamma}|\nabla_y f|^2)dxdy\\&\geq\left(\frac{Q-2}{2}\right)^2\int_{\mathbb{R}^n}\left(\frac{|x|^{2\gamma}}{|x|^{2+2\gamma}+(1+\gamma)^2|y|^2}\right)|f|^2dxdy,\end{aligned} \tag{7.51}$$

where $x\in\mathbb{R}^m$, $y\in\mathbb{R}^k$ with $n=m+k$, $m,k\geq 1$, $\gamma\geq 0$, $Q=m+(1+\gamma)k$ and $f\in C_0^\infty(\mathbb{R}^m\times\mathbb{R}^k\backslash\{(0,0)\})$. Theorem 7.6.3 gives (7.51) when $\alpha_1=\alpha_2=0$ in view of the inequality

$$\left|\frac{d}{d|x|}f\right|\leq|\nabla_x f|. \tag{7.52}$$

2. Weighted L^p-versions of (7.51) were investigated by D'Ambrosio in [D'A04a] who has obtained the following estimate: Let $\Omega \subset \mathbb{R}^n$ be an open set. Let $p > 1$, $k, m \geq 1$, $\alpha, \beta \in \mathbb{R}$ be such that $m + (1+\gamma)k > \alpha - \beta$ and $m > \gamma p - \beta$. Then for every $f \in D^{1,p}_{\gamma}(\Omega, |x|^{\beta-\gamma p}\rho^{(1+\gamma)p-\alpha})$ we have

$$\int_{\Omega} |\nabla_{\gamma} f|^p |x|^{\beta-\gamma p} \rho^{(1+\gamma)p-\alpha} dxdy \geq \left(\frac{Q+\beta-\alpha}{p}\right)^p \int_{\Omega} |f|^p \frac{|x|^{\beta}}{\rho^{\alpha}} dxdy, \quad (7.53)$$

where $D^{1,p}_{\gamma}(\Omega, \omega)$ stands for the closure of $C_0^{\infty}(\Omega)$ in the norm $\left(\int_{\Omega} |\nabla_{\gamma} f|^p \omega dz dy\right)^{1/p}$ for a weight $\omega \in L^1_{\mathrm{loc}}(\Omega)$ with $\omega > 0$ a.e. on Ω.

If $0 \in \Omega$, then the constant $\left(\frac{Q+\beta-\alpha}{p}\right)^p$ in (7.53) is sharp. The inequality (7.53) has also been obtained in [Kom15], and in [SJ12] for $\Omega = \mathbb{R}^n$ with sharp constant.

In view of (7.52), Theorem 7.6.3 refines (7.53) when $p = 2$ and $\Omega = \mathbb{R}^n$. We also mention that in the case $p = 2$ inequality (7.53) has been also shown in [Kom15] and [SJ12] by different methods.

3. In [SJ12], a Hardy–Rellich type inequality for the Baouendi–Grushin operator was obtained in L^2 with sharp constant:

$$\left(\frac{Q-\alpha-2}{2}\right)^2 \int_{\mathbb{R}^n} |\nabla_{\gamma} f|^2 \rho^{\alpha} dxdy \leq \int_{\mathbb{R}^n} |\Delta_{\gamma} f|^2 \rho^{\alpha+2} |\nabla_{\gamma}\rho|^{-2} dxdy,$$

where $p > 1$, $\frac{2-Q}{3} \leq \alpha \leq Q-2$, $f \in C_0^{\infty}(\mathbb{R}^n \backslash \{0\})$.

4. Inequalities of the above types have been also studied for subelliptic operators of different types, see, e.g., [Gar93], [GL90], [D'A04b], [D'A04a] and [DGN06], and also with remainder estimates, see, e.g., [DGN10] and references therein.

5. Magnetic Hardy inequalities for the Baouendi–Grushin operators have been obtained in [LRY17]. There, the authors also obtained Hardy inequalities for the magnetic Landau Hamiltonian.

Proof of Theorem 7.6.3. For the proof we follow [LRY17]. We denote

$$r := |x| \text{ and } F(r, y) := \rho^{\alpha_1} |\nabla_{\gamma}\rho|^{\alpha_2}.$$

Then, using (7.48) and (7.49) we can write

$$F(r, y) = \rho^{\alpha_1} |\nabla_{\gamma}\rho|^{\alpha_2} = r^{\alpha_2\gamma} \rho^{\alpha_1-\alpha_2\gamma} = r^{\alpha_2\gamma} (r^{2(1+\gamma)} + (1+\gamma)^2 |y|^2)^{\frac{\alpha_1-\alpha_2\gamma}{2(1+\gamma)}}. \quad (7.54)$$

Let us first calculate the following expression

$$\int_{\mathbb{R}^k} \int_0^{\infty} \left(\left| \left(\partial_r + \alpha \frac{\partial_r \rho}{\rho}\right) f \right|^2 + r^{2\gamma} \left| \left(\nabla_y + \alpha \frac{\nabla_y \rho}{\rho}\right) f \right|^2 \right) r^{m-1} F(r, y) drdy$$

$$= \int_{\mathbb{R}^k} \int_0^{\infty} \left(|\partial_r f|^2 + r^{2\gamma} |\nabla_y f|^2 \right) r^{m-1} F(r, y) drdy$$

$$
\begin{aligned}
&+ \alpha^2 \int_{\mathbb{R}^k} \int_0^\infty \left(\left| \frac{\partial_r \rho}{\rho} \right|^2 + r^{2\gamma} \left| \frac{\nabla_y \rho}{\rho} \right|^2 \right) |f|^2 r^{m-1} F(r,y) dr dy \\
&+ 2\alpha \mathrm{Re} \int_{\mathbb{R}^k} \int_0^\infty \frac{\partial_r \rho}{\rho} r^{m-1} F(r,y) \overline{\partial_r f} \cdot f dr dy \\
&+ 2\alpha \mathrm{Re} \int_{\mathbb{R}^k} \int_0^\infty \frac{\nabla_y \rho}{\rho} \cdot \overline{\nabla_y f} r^{2\gamma+m-1} F(r,y) f dr dy \\
&=: I_1 + I_2 + I_3 + I_4. \qquad (7.55)
\end{aligned}
$$

We now calculate the terms I_2, I_3, I_4. Using the expressions

$$
\frac{\partial_r \rho}{\rho} = \frac{r^{2\gamma+1}}{\rho^{2\gamma+2}} \quad \text{and} \quad \frac{\nabla_y \rho}{\rho} = \frac{(\gamma+1)y}{\rho^{2\gamma+2}},
$$

we calculate

$$
\left| \frac{\partial_r \rho}{\rho} \right|^2 + r^{2\gamma} \left| \frac{\nabla_y \rho}{\rho} \right|^2 = \frac{r^{4\gamma+2} + r^{2\gamma}(\gamma+1)^2 |y|^2}{\rho^{4\gamma+4}} = \frac{r^{2\gamma}}{\rho^{2\gamma+2}} = \frac{|\nabla_\gamma \rho|^2}{\rho^2}. \qquad (7.56)
$$

Thus, we obtain

$$
I_2 = \alpha^2 \int_{-\infty}^{\infty} \int_0^\infty \frac{|\nabla_\gamma \rho|^2}{\rho^2} |f|^2 r^{m-1} F(r,y) dr dy. \qquad (7.57)
$$

For I_3, we integrate by parts to get

$$
\begin{aligned}
I_3 = &-\alpha \int_{\mathbb{R}^k} \int_0^\infty (2\gamma + m + \gamma\alpha_2) \rho^{\alpha_1 - \alpha_2\gamma - 2\gamma - 2} r^{2\gamma+m-1+\gamma\alpha_2} |f|^2 dr dy \\
&- \alpha \int_{\mathbb{R}^k} \int_0^\infty (\alpha_1 - \alpha_2\gamma - 2\gamma - 2) \rho^{\alpha_1 - \alpha_2\gamma - 4\gamma - 4} r^{4\gamma+m+\gamma\alpha_2+1} |f|^2 dr dy.
\end{aligned}
$$

Since $F(r,y) = r^{\alpha_2\gamma} \rho^{\alpha_1 - \alpha_2\gamma}$ by (7.54), we obtain

$$
\begin{aligned}
I_3 = &-\alpha \int_{\mathbb{R}^k} \int_0^\infty \left((2\gamma + m + \gamma\alpha_2) \frac{r^{2\gamma}}{\rho^{2\gamma+2}} + (\alpha_1 - \alpha_2\gamma - 2\gamma - 2) \frac{r^{4\gamma+2}}{\rho^{4\gamma+4}} \right) \\
&\qquad \times r^{m-1} F(r,y) |f|^2 dr dy \\
= &-\alpha \int_{\mathbb{R}^k} \int_0^\infty \left(2\gamma + m + \gamma\alpha_2 + (\alpha_1 - \alpha_2\gamma - 2\gamma - 2) \frac{r^{2\gamma+2}}{\rho^{2\gamma+2}} \right) \\
&\qquad \times \frac{|\nabla_\gamma \rho|^2}{\rho^2} |f|^2 r^{m-1} F(r,y) dr dy.
\end{aligned}
$$

Similarly, we have for I_4 that

$$
\begin{aligned}
I_4 = &-\alpha \int_{\mathbb{R}^k} \int_0^\infty \mathrm{div}_y \left(F(r,y) \frac{\nabla_y \rho}{\rho} \right) r^{2\gamma+m-1} |f|^2 dr dy \\
= &-\alpha(\gamma+1) \int_{\mathbb{R}^k} \int_0^\infty \mathrm{div}_y \left(\rho^{\alpha_1 - \alpha_2\gamma - 2\gamma - 2} y \right) r^{\alpha_2\gamma + 2\gamma + m - 1} |f|^2 dr dy
\end{aligned}
$$

$$= -\alpha \int_{\mathbb{R}^k}\int_0^\infty \left((\alpha_1-\alpha_2\gamma-2\gamma-2)\rho^{\alpha_1-\alpha_2\gamma-2\gamma-3}\frac{(\gamma+1)^2|y|^2}{\rho^{2\gamma+1}}\right)$$
$$\times r^{\alpha_2\gamma+2\gamma+m-1}|f|^2 drdy$$
$$-\alpha \int_{\mathbb{R}^k}\int_0^\infty k(\gamma+1)\rho^{\alpha_1-\alpha_2\gamma-2\gamma-2}r^{\alpha_2\gamma+2\gamma+m-1}|f|^2 drdy.$$

Since $\frac{r^{2\gamma}}{\rho^{2\gamma+2}}=\frac{|\nabla_\gamma\rho|^2}{\rho^2}$ and $F(r,y)=r^{\alpha_2\gamma}\rho^{\alpha_1-\alpha_2\gamma}$ by (7.56) and (7.54), respectively, we have

$$I_4 = -\alpha \int_{\mathbb{R}^k}\int_0^\infty \left((\alpha_1-\alpha_2\gamma-2\gamma-2)\frac{(\gamma+1)^2|y|^2}{\rho^{2\gamma+2}}+k(\gamma+1)\right)$$
$$\times \frac{|\nabla_\gamma\rho|^2}{\rho^2}|f|^2 r^{m-1}F(r,y)drdy.$$

Then, taking into account (7.48) we get

$$I_3+I_4 = -\alpha \int_{\mathbb{R}^k}\int_0^\infty (\alpha_1-\alpha_2\gamma-2\gamma-2+2\gamma+m+\gamma\alpha_2+k(\gamma+1))$$
$$\times \frac{|\nabla_\gamma\rho|^2}{\rho^2}|f|^2 r^{m-1}F(r,y)drdy.$$

Finally, using that $Q=m+(1+\gamma)k$ we obtain

$$I_3+I_4=-\alpha\int_{\mathbb{R}^k}\int_0^\infty (Q+\alpha_1-2)\frac{|\nabla_\gamma\rho|^2}{\rho^2}|f|^2r^{m-1}F(r,y)drdy. \tag{7.58}$$

Putting (7.57) and (7.58) in (7.55) we get

$$\int_{\mathbb{R}^k}\int_0^\infty\left(\left|\left(\partial_r+\alpha\frac{\partial_r\rho}{\rho}\right)f\right|^2+r^{2\gamma}\left|\left(\nabla_y+\alpha\frac{\nabla_y\rho}{\rho}\right)f\right|^2\right)r^{m-1}F(r,y)drdy$$
$$=\int_{\mathbb{R}^k}\int_0^\infty\left(|\partial_r f|^2+r^{2\gamma}|\nabla_y f|^2\right)r^{m-1}F(r,y)drdy$$
$$-\left((Q+\alpha_1-2)\alpha-\alpha^2\right)\int_{\mathbb{R}^k}\int_0^\infty\frac{|\nabla_\gamma\rho|^2}{\rho^2}|f|^2r^{m-1}F(r,y)drdy.$$

By substituting $\alpha=\frac{Q+\alpha_1-2}{2}$ and taking into account (7.54), we obtain (7.50).

The sharpness of the constant $\left(\frac{Q+\alpha_1-2}{2}\right)^2$ in (7.50) follows from the inequalities

$$\int_{\mathbb{R}^n}\rho^{\alpha_1}|\nabla_\gamma\rho|^{\alpha_2}\left(|\nabla_x f|^2+|x|^{2\gamma}|\nabla_y f|^2\right)dxdy$$
$$\geq\int_{\mathbb{R}^n}\rho^{\alpha_1}|\nabla_\gamma\rho|^{\alpha_2}\left(\left|\frac{d}{d|x|}f\right|^2+|x|^{2\gamma}|\nabla_y f|^2\right)dxdy$$
$$\geq\left(\frac{Q+\alpha_1-2}{2}\right)^2\int_{\mathbb{R}^n}\rho^{\alpha_1}|\nabla_\gamma\rho|^{\alpha_2}\frac{|\nabla_\gamma\rho|^2}{\rho^2}|f|^2dxdy,$$

since it is known that this inequality is sharp in (7.53), see Remark 7.6.4, Part 2.

□

7.7 Weighted L^p-inequalities with boundary terms

In this section, we present a generalization of the weighted L^p-Hardy, L^p-Caffarelli–Kohn–Nirenberg, and L^p-Rellich inequalities with respect to the inclusion of boundary terms in the setting of stratified Lie groups. The appearing weights are controlled by a real-valued function V with the property that $\mathcal{L}V$ does not change sign. In addition to the inequalities themselves we will also give their refined versions involving expressions appearing due to the boundary of the domain in which these inequalities are derived. As a consequence, one can recover many of the Hardy type inequalities and Heisenberg–Pauli–Weyl type uncertainty principles on stratified groups by choosing special cases of the real-valued function V and working with functions vanishing at the boundary. The exposition of this section follows [RSS18d]. In Section 11.4 we will discuss boundary terms again but emphasizing the use of the $\mathcal{L}$-gauge in that discussion.

Setting of this section

Thus, throughout this section Ω is an admissible domain in the stratified group $\mathbb{G}$, and V is a real-valued function in $L^1_{\rm loc}(\Omega)$ with partial derivatives of order up to two in $L^1_{\rm loc}(\Omega)$, and such that $\mathcal{L}V$ is of one sign. Also, as usual, N denotes the dimension of the first stratum of the group $\mathbb{G}$ and ∇_H the horizontal gradient on $\mathbb{G}$. Then, as in (1.87), the vector field $\widetilde{\nabla} u$ is defined by

$$\widetilde{\nabla} u := \sum_{k=1}^{N} (X_k u)\, X_k. \tag{7.59}$$

7.7.1 Hardy and Caffarelli–Kohn–Nirenberg inequalities

We start with Hardy and Caffarelli–Kohn–Nirenberg inequalities with generalized weights.

Theorem 7.7.1 (L^p-Hardy inequality with generalized weight and boundary term). *Let $1<p<\infty$. Let V be a real-valued function such that $\mathcal{L}V<0$ holds a.e. in Ω. Then for all complex-valued functions $u\in C^2(\Omega)\cap C^1(\overline{\Omega})$ we have the inequality*

$$\left\| |\mathcal{L}V|^{\frac{1}{p}} u \right\|^p_{L^p(\Omega)} \leq p \left\| \frac{|\nabla_H V|}{|\mathcal{L}V|^{\frac{p-1}{p}}} |\nabla_H u| \right\|_{L^p(\Omega)} \left\| |\mathcal{L}V|^{\frac{1}{p}} u \right\|^{p-1}_{L^p(\Omega)} - \int_{\partial\Omega} |u|^p \langle \widetilde{\nabla} V, dx\rangle. \tag{7.60}$$

Proof of Theorem 7.7.1. Let us denote

$$v_\epsilon := (|u|^2+\epsilon^2)^{\frac{1}{2}} - \epsilon.$$

Then $v_\epsilon^p \in C^2(\Omega)\cap C^1(\overline{\Omega})$ and using Green's first formula in Theorem 1.4.6 and the fact that $\mathcal{L}V<0$ we get

$$\int_\Omega |\mathcal{L}V| v_\epsilon^p dx = -\int_\Omega \mathcal{L}V v_\epsilon^p dx = \int_\Omega (\widetilde{\nabla} V) v_\epsilon^p dx - \int_{\partial\Omega} v_\epsilon^p \langle \widetilde{\nabla} V, dx\rangle$$

$$
\begin{aligned}
&= \int_\Omega \nabla_H V \cdot \nabla_H \upsilon_\epsilon^p dx - \int_{\partial\Omega} \upsilon_\epsilon^p \langle \widetilde{\nabla} V, dx\rangle \\
&\leq \int_\Omega |\nabla_H V||\nabla_H \upsilon_\epsilon^p| dx - \int_{\partial\Omega} \upsilon_\epsilon^p \langle \widetilde{\nabla} V, dx\rangle \\
&= p \int_\Omega \left(\frac{|\nabla_H V|}{|\mathcal{L}V|^{\frac{p-1}{p}}} \right) |\mathcal{L}V|^{\frac{p-1}{p}} \upsilon_\epsilon^{p-1} |\nabla_H \upsilon_\epsilon| dx - \int_{\partial\Omega} \upsilon_\epsilon^p \langle \widetilde{\nabla} V, dx\rangle,
\end{aligned}
$$

where, as usual, $(\widetilde{\nabla} u)v = \nabla_H u \cdot \nabla_H v$. We then have

$$
\nabla_H \upsilon_\epsilon = (|u|^2 + \epsilon^2)^{-\frac{1}{2}} |u| \nabla_H |u|,
$$

since $0 \leq \upsilon_\epsilon \leq |u|$. Thus, we also have

$$
\upsilon_\epsilon^{p-1} |\nabla_H \upsilon_\epsilon| \leq |u|^{p-1} |\nabla_H |u||.
$$

On the other hand, let us write

$$
u(x) = R(x) + iI(x),
$$

where $R(x)$ and $I(x)$ denote the real and imaginary parts of u. We can restrict to the set where $u \neq 0$. Then we have

$$
(\nabla_H |u|)(x) = \frac{1}{|u|} (R(x) \nabla_H R(x) + I(x) \nabla_H I(x)) \quad \text{if} \quad u \neq 0.
$$

Since

$$
\left| \frac{1}{|u|} (R \nabla_H R + I \nabla_H I) \right|^2 \leq |\nabla_H R|^2 + |\nabla_H I|^2,
$$

we get that $|\nabla_H |u|| \leq |\nabla_H u|$ a.e. in Ω. Therefore,

$$
\begin{aligned}
\int_\Omega |\mathcal{L}V| \upsilon_\epsilon^p dx &\leq p \int_\Omega \left(\frac{|\nabla_H V|}{|\mathcal{L}V|^{\frac{p-1}{p}}} |\nabla_H u| \right) |\mathcal{L}V|^{\frac{p-1}{p}} |u|^{p-1} dx - \int_{\partial\Omega} \upsilon_\epsilon^p \langle \widetilde{\nabla} V, dx\rangle \\
&\leq p \left(\int_\Omega \left(\frac{|\nabla_H V|^p}{|\mathcal{L}V|^{(p-1)}} |\nabla_H u|^p \right) dx \right)^{\frac{1}{p}} \left(\int_\Omega |\mathcal{L}V||u|^p dx \right)^{\frac{p-1}{p}} - \int_{\partial\Omega} \upsilon_\epsilon^p \langle \widetilde{\nabla} V, dx\rangle,
\end{aligned}
$$

where we have used Hölder's inequality in the last line. Thus, when $\epsilon \to 0$, we obtain (7.60). □

Remark 7.7.2.

1. If u vanishes on the boundary $\partial\Omega$, then (7.60) extends the Davies and Hinz result [DH98] to the following weighted L^p-Hardy type inequality on stratified groups:

$$
\left\| |\mathcal{L}V|^{\frac{1}{p}} u \right\|_{L^p(\Omega)} \leq p \left\| \frac{|\nabla_H V|}{|\mathcal{L}V|^{\frac{p-1}{p}}} |\nabla_H u| \right\|_{L^p(\Omega)}, \quad 1 < p < \infty. \tag{7.61}
$$

2. There are a number of other interesting consequences of Theorem 7.7.1 that we can record. We now discuss several such statements. First we present a horizontal L^p-Caffarelli–Kohn–Nirenberg type inequality with the boundary term on the stratified group $\mathbb{G}$. Incidentally, this also gives another proof of the horizontal L^p-Hardy type inequality (such as that in Theorem 6.2.1).

Corollary 7.7.3 (Horizontal L^p-Caffarelli–Kohn–Nirenberg inequality with boundary term). *Let $1 < p < \infty$ and let $\alpha, \beta \in \mathbb{R}$. Let Ω be an admissible domain in a stratified group $\mathbb{G}$ with $N \geq 3$ being the dimension of the first stratum. Let $|\cdot|_E$ be the Euclidean norm on $\mathbb{R}^N$. Then for all $u \in C^2(\Omega\backslash\{x'=0\}) \cap C^1(\overline{\Omega}\backslash\{x'=0\})$ we have*

$$\frac{|N-\gamma|}{p}\left\|\frac{u}{|x'|_E^{\frac{\gamma}{p}}}\right\|^p_{L^p(\Omega)} \leq \left\|\frac{\nabla_H u}{|x'|_E^{\alpha}}\right\|_{L^p(\Omega)}\left\|\frac{u}{|x'|_E^{\frac{\beta}{p-1}}}\right\|^{p-1}_{L^p(\Omega)} - \frac{1}{p}\int_{\partial\Omega}|u|^p\langle\widetilde{\nabla}|x'|_E^{2-\gamma}, dx\rangle, \tag{7.62}$$

for $2 < \gamma < N$ with $\gamma = \alpha+\beta+1$. In particular, if u vanishes on the boundary $\partial\Omega$, we have (6.3).

Proof of Corollary 7.7.3. We will show that (7.62) follows as a special case of (7.60). Let us take

$$V(x) := |x'|_E^{2-\gamma}.$$

Then we have

$$|\nabla_H V| = |2-\gamma||x'|_E^{1-\gamma}, \qquad |\mathcal{L}V| = |(2-\gamma)(N-\gamma)||x'|_E^{-\gamma},$$

and observe that $\mathcal{L}V = (2-\gamma)(N-\gamma)|x'|_E^{-\gamma} < 0$. To use (7.60) we calculate the following expressions:

$$\left\||\mathcal{L}V|^{\frac{1}{p}}u\right\|^p_{L^p(\Omega)} = |(2-\gamma)(N-\gamma)|\left\|\frac{u}{|x'|_E^{\frac{\gamma}{p}}}\right\|^p_{L^p(\Omega)},$$

$$\left\|\frac{|\nabla_H V|}{|\mathcal{L}V|^{\frac{p-1}{p}}}\nabla_H u\right\|_{L^p(\Omega)} = \frac{|2-\gamma|}{|(2-\gamma)(N-\gamma)|^{\frac{p-1}{p}}}\left\|\frac{|\nabla_H u|}{|x'|_E^{\frac{\gamma-p}{p}}}\right\|_{L^p(\Omega)},$$

$$\left\||\mathcal{L}V|^{\frac{1}{p}}u\right\|^{p-1}_{L^p(\Omega)} = |(2-\gamma)(N-\gamma)|^{\frac{p-1}{p}}\left\|\frac{u}{|x'|_E^{\frac{\gamma}{p}}}\right\|^{p-1}_{L^p(\Omega)}.$$

Thus, (7.60) implies the inequality

$$\frac{|N-\gamma|}{p}\left\|\frac{u}{|x'|_E^{\frac{\gamma}{p}}}\right\|^p_{L^p(\Omega)} \leq \left\|\frac{\nabla_H u}{|x'|_E^{\frac{\gamma-p}{p}}}\right\|_{L^p(\Omega)}\left\|\frac{u}{|x'|_E^{\frac{\gamma}{p}}}\right\|^{p-1}_{L^p(\Omega)} - \frac{1}{p}\int_{\partial\Omega}|u|^p\langle\widetilde{\nabla}|x'|_E^{2-\gamma}, dx\rangle.$$

If we denote $\alpha = \frac{\gamma-p}{p}$ and $\frac{\beta}{p-1} = \frac{\gamma}{p}$, we obtain (7.62). □

Another interesting feature of Theorem 7.7.1 is that it also allows one to obtain inequalities with the $\mathcal{L}$-gauge d.

Let us give an example.

Corollary 7.7.4 (Hardy inequality with $\mathcal{L}$-gauge weights and boundary term). *Let $\Omega \subset \mathbb{G}$ be an admissible domain in a stratified group $\mathbb{G}$ of homogeneous dimension $Q \geq 3$, and assume that $0 \notin \partial\Omega$. Let $2-Q < \alpha < 0$. Let $u \in C^1(\Omega\backslash\{0\}) \cap C(\overline{\Omega}\backslash\{0\})$. Then we have*

$$\frac{|Q+\alpha-2|}{p}\left\|d^{\frac{\alpha-2}{p}}|\nabla_H d|^{\frac{2}{p}}u\right\|_{L^p(\Omega)} \tag{7.63}$$
$$\leq \left\|d^{\frac{p+\alpha-2}{p}}|\nabla_H d|^{\frac{2-p}{p}}|\nabla_H u|\right\|_{L^p(\Omega)} - \frac{1}{p}\left\|d^{\frac{\alpha-2}{p}}|\nabla_H d|^{\frac{2}{p}}u\right\|_{L^p(\Omega)}^{1-p}\int_{\partial\Omega} d^{\alpha-1}|u|^p\langle\widetilde{\nabla}d, dx\rangle.$$

Proof of Corollary 7.7.4. First, we can multiply both sides of the inequality (7.60) by $\left\||\mathcal{L}V|^{\frac{1}{p}}u\right\|_{L^p(\Omega)}^{1-p}$, so that we have the inequality

$$\left\||\mathcal{L}V|^{\frac{1}{p}}u\right\|_{L^p(\Omega)} \leq p\left\|\frac{|\nabla_H V|}{|\mathcal{L}V|^{\frac{p-1}{p}}}|\nabla_H u|\right\|_{L^p(\Omega)} - \left\||\mathcal{L}V|^{\frac{1}{p}}u\right\|_{L^p(\Omega)}^{1-p}\int_{\partial\Omega}|u|^p\langle\widetilde{\nabla}V, dx\rangle. \tag{7.64}$$

Now, let us take $V := d^\alpha$. Since $d = \varepsilon^{\frac{1}{2-Q}}$ for the fundamental solution ε of $\mathcal{L}$, we have

$$\mathcal{L}d^\alpha = \nabla_H(\nabla_H\varepsilon^{\frac{\alpha}{2-Q}}) = \nabla_H\left(\frac{\alpha}{2-Q}\varepsilon^{\frac{\alpha+Q-2}{2-Q}}\nabla_H\varepsilon\right)$$
$$= \frac{\alpha(\alpha+Q-2)}{(2-Q)^2}\varepsilon^{\frac{\alpha-4+2Q}{2-Q}}|\nabla_H\varepsilon|^2 + \frac{\alpha}{2-Q}\varepsilon^{\frac{\alpha+Q-2}{2-Q}}\mathcal{L}\varepsilon.$$

Since ε is the fundamental solution of $\mathcal{L}$, it follows that

$$\mathcal{L}d^\alpha = \frac{\alpha(\alpha+Q-2)}{(2-Q)^2}\varepsilon^{\frac{\alpha-4+2Q}{2-Q}}|\nabla_H\varepsilon|^2 = \alpha(\alpha+Q-2)d^{\alpha-2}|\nabla_H d|^2.$$

From this we can observe that $\mathcal{L}d^\alpha < 0$, and also a direct calculation yields the identities

$$\left\||\mathcal{L}d^\alpha|^{\frac{1}{p}}u\right\|_{L^p(\Omega)} = \alpha^{\frac{1}{p}}|Q+\alpha-2|^{\frac{1}{p}}\left\|d^{\frac{\alpha-2}{p}}|\nabla_H d|^{\frac{2}{p}}u\right\|_{L^p(\Omega)},$$
$$\left\|\frac{|\nabla_H d^\alpha|}{|\mathcal{L}d^\alpha|^{\frac{p-1}{p}}}|\nabla_H u|\right\|_{L^p(\Omega)} = \alpha^{\frac{1}{p}}|Q+\alpha-2|^{\frac{1-p}{p}}\left\|d^{\frac{\alpha-2+p}{p}}|\nabla_H d|^{\frac{2-p}{p}}|\nabla_H u|\right\|_{L^p(\Omega)},$$
$$\left\||\mathcal{L}d^\alpha|^{\frac{1}{p}}u\right\|_{L^p(\Omega)}^{1-p}\int_{\partial\Omega}|u|^p\langle\widetilde{\nabla}d^\alpha, dx\rangle$$
$$= \alpha^{\frac{1}{p}}|Q+\alpha-2|^{\frac{1-p}{p}}\left\|d^{\frac{\alpha-2}{p}}|\nabla_H d|^{\frac{2}{p}}u\right\|_{L^p(\Omega)}^{1-p}\int_{\partial\Omega}d^{\alpha-1}|u|^p\langle\widetilde{\nabla}d, dx\rangle.$$

Using (7.64) we arrive at

$$\frac{|Q+\alpha-2|}{p}\left\|d^{\frac{\alpha-2}{p}}|\nabla_H d|^{\frac{2}{p}}u\right\|_{L^p(\Omega)}$$
$$\leq \left\|d^{\frac{p+\alpha-2}{p}}|\nabla_H d|^{\frac{2-p}{p}}|\nabla_H u|\right\|_{L^p(\Omega)} - \frac{1}{p}\left\|d^{\frac{\alpha-2}{p}}|\nabla_H d|^{\frac{2}{p}}u\right\|^{1-p}_{L^p(\Omega)}\int_{\partial\Omega} d^{\alpha-1}|u|^p\langle\widetilde{\nabla} d, dx\rangle,$$

which implies (7.63). □

The inequality (7.64) implies the following generalized Heisenberg–Pauli–Weyl type uncertainty principle on stratified groups.

Corollary 7.7.5 (Weighted Heisenberg–Pauli–Weyl uncertainty principle with boundary term). *Let $\Omega \subset \mathbb{G}$ be an admissible domain in a stratified group $\mathbb{G}$ and let $V \in C^2(\Omega)$ be real-valued. Then for all complex-valued functions $u \in C^2(\Omega)\cap C^1(\overline{\Omega})$ we have*

$$\left\||\mathcal{L}V|^{-\frac{1}{p}}u\right\|_{L^p(\Omega)}\left\|\frac{|\nabla_H V|}{|\mathcal{L}V|^{\frac{p-1}{p}}}|\nabla_H u|\right\|_{L^p(\Omega)}$$
$$\geq \frac{1}{p}\|u\|^2_{L^p(\Omega)} + \frac{1}{p}\left\||\mathcal{L}V|^{-\frac{1}{p}}u\right\|_{L^p(\Omega)}\left\||\mathcal{L}V|^{\frac{1}{p}}u\right\|^{1-p}_{L^p(\Omega)}\int_{\partial\Omega}|u|^p\langle\widetilde{\nabla} V, dx\rangle. \tag{7.65}$$

In particular, if u vanishes on the boundary $\partial\Omega$, then we have

$$\left\||\mathcal{L}V|^{-\frac{1}{p}}u\right\|_{L^p(\Omega)}\left\|\frac{|\nabla_H V|}{|\mathcal{L}V|^{\frac{p-1}{p}}}|\nabla_H u|\right\|_{L^p(\Omega)} \geq \frac{1}{p}\|u\|^2_{L^p(\Omega)}. \tag{7.66}$$

By setting $V = |x'|^\alpha$ in the inequality (7.66), we recover the Heisenberg–Pauli–Weyl type uncertainty principle on stratified groups.

Proof of Corollary 7.7.5. By using the extended Hölder inequality and (7.64) we have

$$\left\||\mathcal{L}V|^{-\frac{1}{p}}u\right\|_{L^p(\Omega)}\left\|\frac{|\nabla_H V|}{|\mathcal{L}V|^{\frac{p-1}{p}}}|\nabla_H u|\right\|_{L^p(\Omega)}$$
$$\geq \frac{1}{p}\left\||\mathcal{L}V|^{-\frac{1}{p}}u\right\|_{L^p(\Omega)}\left\||\mathcal{L}V|^{\frac{1}{p}}u\right\|_{L^p(\Omega)}$$
$$+\frac{1}{p}\left\||\mathcal{L}V|^{-\frac{1}{p}}u\right\|_{L^p(\Omega)}\left\||\mathcal{L}V|^{\frac{1}{p}}u\right\|^{1-p}_{L^p(\Omega)}\int_{\partial\Omega}|u|^p\langle\widetilde{\nabla} V, dx\rangle,$$
$$\geq \frac{1}{p}\left\||u|^2\right\|_{L^{\frac{p}{2}}(\Omega)} + \frac{1}{p}\left\||\mathcal{L}V|^{-\frac{1}{p}}u\right\|_{L^p(\Omega)}\left\||\mathcal{L}V|^{\frac{1}{p}}u\right\|^{1-p}_{L^p(\Omega)}\int_{\partial\Omega}|u|^p\langle\widetilde{\nabla} V, dx\rangle.$$
$$=\frac{1}{p}\|u\|^2_{L^p(\Omega)} + \frac{1}{p}\left\||\mathcal{L}V|^{-\frac{1}{p}}u\right\|_{L^p(\Omega)}\left\||\mathcal{L}V|^{\frac{1}{p}}u\right\|^{1-p}_{L^p(\Omega)}\int_{\partial\Omega}|u|^p\langle\widetilde{\nabla} V, dx\rangle,$$

proving (7.65). □

7.7.2 Rellich inequalities

In this section, we describe weighted Rellich inequalities with boundary terms. We consider first the L^2 and then the L^p case.

Theorem 7.7.6 (L^2-Rellich inequality with generalized weight and boundary term). *Let $V \in C^2(\Omega)$ be a real-valued function such that $\mathcal{L}V(x) < 0$ for all $x \in \Omega \subset \mathbb{G}$. Then for every $\epsilon > 0$ we have*

$$\begin{aligned}\left\| \frac{|V|}{|\mathcal{L}V|^{\frac{1}{2}}} \mathcal{L}u \right\|^2_{L^2(\Omega)} &\geq 2\epsilon \left\| V^{\frac{1}{2}} |\nabla_H u| \right\|^2_{L^2(\Omega)} + \epsilon(1-\epsilon) \left\| |\mathcal{L}V|^{\frac{1}{2}} u \right\|^2_{L^2(\Omega)} \\ &\quad - \epsilon \int_{\partial\Omega} (|u|^2 \langle \widetilde{\nabla} V, dx \rangle - V \langle \widetilde{\nabla} |u|^2, dx \rangle),\end{aligned} \tag{7.67}$$

for all complex-valued functions $u \in C^2(\Omega) \cap C^1(\overline{\Omega})$. In particular, if u vanishes on the boundary $\partial\Omega$, we have

$$\left\| \frac{|V|}{|\mathcal{L}V|^{\frac{1}{2}}} \mathcal{L}u \right\|^2_{L^2(\Omega)} \geq 2\epsilon \left\| V^{\frac{1}{2}} |\nabla_H u| \right\|^2_{L^2(\Omega)} + \epsilon(1-\epsilon) \left\| |\mathcal{L}V|^{\frac{1}{2}} u \right\|^2_{L^2(\Omega)}.$$

Remark 7.7.7. In the case of $\mathbb{R}^n$, an analogous L^2-Rellich inequality was proved by Schmincke [Sch72] and generalized further by Bennett [Ben89]. In the setting of stratified group this and other results of this section were obtained in [RSS18d].

Proof of Theorem 7.7.6. Using Green's second identity from Theorem 1.4.6 and the condition that $\mathcal{L}V(x) < 0$ in Ω, we obtain

$$\begin{aligned}\int_\Omega |\mathcal{L}V||u|^2 dx &= -\int_\Omega V \mathcal{L}|u|^2 dx - \int_{\partial\Omega} (|u|^2 \langle \widetilde{\nabla} V, dx \rangle - V \langle \widetilde{\nabla} |u|^2, dx \rangle) \\ &= -2 \int_\Omega V \left(\mathrm{Re}(\overline{u} \mathcal{L}u) + |\nabla_H u|^2 \right) dx - \int_{\partial\Omega} (|u|^2 \langle \widetilde{\nabla} V, dx \rangle - V \langle \widetilde{\nabla} |u|^2, dx \rangle).\end{aligned}$$

Using the Cauchy–Schwarz inequality this implies

$$\begin{aligned}\int_\Omega |\mathcal{L}V||u|^2 dx &\leq 2 \left(\frac{1}{\epsilon} \int_\Omega \frac{|V|^2}{|\mathcal{L}V|} |\mathcal{L}u|^2 dx \right)^{1/2} \left(\epsilon \int_\Omega |\mathcal{L}V||u|^2 dx \right)^{1/2} \\ &\quad - 2 \int_\Omega V |\nabla_H u|^2 dx - \int_{\partial\Omega} (|u|^2 \langle \widetilde{\nabla} V, dx \rangle - V \langle \widetilde{\nabla} |u|^2, dx \rangle) \\ &\leq \frac{1}{\epsilon} \int_\Omega \frac{|V|^2}{|\mathcal{L}V|} |\mathcal{L}u|^2 dx + \epsilon \int_\Omega |\mathcal{L}V||u|^2 dx \\ &\quad - 2 \int_\Omega V |\nabla_H u|^2 dx - \int_{\partial\Omega} (|u|^2 \langle \widetilde{\nabla} V, dx \rangle - V \langle \widetilde{\nabla} |u|^2, dx \rangle),\end{aligned}$$

yielding (7.67). ∎

We now move to the case of L^p-estimates.

Theorem 7.7.8 (L^p-Rellich inequality with generalized weight and boundary term). *Let $1 \leq p < \infty$. Let Ω be an admissible domain in a stratified group $\mathbb{G}$. If*

$$0 < V \in C(\Omega),\ \mathcal{L}V < 0 \ \text{ and } \mathcal{L}(V^{\sigma}) \leq 0$$

on Ω for some $\sigma > 1$, then for all $u \in C_0^\infty(\Omega)$ we have

$$\left\| |\mathcal{L}V|^{\frac{1}{p}} u \right\|_{L^p(\Omega)} \leq \frac{p^2}{(p-1)\sigma + 1} \left\| \frac{V}{|\mathcal{L}V|^{\frac{p-1}{p}}} \mathcal{L}u \right\|_{L^p(\Omega)}. \tag{7.68}$$

Before proving Theorem 7.7.8, let us make some remarks and establish some preliminary properties needed for its proof.

Remark 7.7.9.

1. Choosing $V = |x'|_E^{-(\alpha-2)}$ in Theorem 7.7.8, with the Euclidean distance $|\cdot|_E$ in the first stratum of $\mathbb{G}$, we obtain for any $2 < \alpha < N$ and all $u \in C_0^\infty(\mathbb{G}\backslash\{x' = 0\})$, the inequality

$$\int_{\mathbb{G}} \frac{|u|^p}{|x'|_E^{\alpha}} dx \leq C^p_{(N,p,\alpha)} \int_{\mathbb{G}} \frac{|\mathcal{L}u|^p}{|x'|_E^{\alpha-2p}} dx, \tag{7.69}$$

where

$$C_{(N,p,\alpha)} = \frac{p^2}{(N-\alpha)\left((p-1)N + \alpha - 2p\right)}. \tag{7.70}$$

2. Let $d = \varepsilon^{\frac{1}{2-Q}}$, where ε is the fundamental solution of the sub-Laplacian $\mathcal{L}$. Assume that $Q \geq 3$, $\alpha < 2$, and $Q + \alpha - 4 > 0$. Choosing $V = d^{\alpha-2}$ in Theorem 7.7.8, with d being the $\mathcal{L}$-gauge as above, we obtain

$$\frac{(Q+\alpha-4)^2(Q-\alpha)^2}{16} \int_{\mathbb{G}} d^{\alpha-4} |\nabla_H d|^2 |u|^2 dx \leq \int_{\mathbb{G}} \frac{d^{\alpha}}{|\nabla_H d|^2} |\mathcal{L}u|^2 dx. \tag{7.71}$$

Theorem 7.7.8 will be proved as follows: it is a consequence of Lemma 7.7.11, by putting $C = \frac{(p-1)(\sigma-1)}{p}$ in Lemma 7.7.10.

Lemma 7.7.10. *Let Ω be an admissible domain in a stratified group $\mathbb{G}$. If $V \geq 0$, $\mathcal{L}V < 0$, and there exists a constant $C \geq 0$ such that*

$$C \left\| |\mathcal{L}V|^{\frac{1}{p}} u \right\|^p_{L^p(\Omega)} \leq p(p-1) \left\| V^{\frac{1}{p}} |u|^{\frac{p-2}{p}} |\nabla_H u|^{\frac{2}{p}} \right\|^p_{L^p(\Omega)}, \quad 1 < p < \infty, \tag{7.72}$$

for all $u \in C_0^\infty(\Omega)$, then we have

$$(1+C) \left\| |\mathcal{L}V|^{\frac{1}{p}} u \right\|_{L^p(\Omega)} \leq p \left\| \frac{V}{|\mathcal{L}V|^{\frac{p-1}{p}}} \mathcal{L}u \right\|_{L^p(\Omega)}, \tag{7.73}$$

for all $u \in C_0^\infty(\Omega)$. If $p = 1$ then the statement holds for $C = 0$.

Proof of Lemma 7.7.10. In view of (2.8) we can assume that u is real-valued. Let $\epsilon > 0$ and set

$$u_\epsilon := (|u|^2 + \epsilon^2)^{p/2} - \epsilon^p.$$

Then $0 \le u_\epsilon \in C_0^\infty$ and

$$\int_\Omega |\mathcal{L}V| u_\epsilon dx = -\int_\Omega (\mathcal{L}V) u_\epsilon dx = -\int_\Omega V \mathcal{L} u_\epsilon dx,$$

where

$$\begin{aligned}\mathcal{L}u_\epsilon = \mathcal{L}\left((|u|^2+\epsilon^2)^{\frac{p}{2}} - \epsilon^p\right) &= \nabla_H \cdot (\nabla_H((|u|^2+\epsilon^2)^{\frac{p}{2}} - \epsilon^p)) \\ &= \nabla_H(p(|u|^2+\epsilon^2)^{\frac{p-2}{2}} u \nabla_H u) \\ &= p(p-2)(|u|^2+\epsilon^2)^{\frac{p-4}{2}} u^2 |\nabla_H u|^2 \\ &\quad + p(|u|^2+\epsilon^2)^{\frac{p-2}{2}} |\nabla_H u|^2 + p(|u|^2+\epsilon^2)^{\frac{p-2}{2}} u \mathcal{L} u.\end{aligned}$$

Then we have

$$\begin{aligned}\int_\Omega |\mathcal{L}V| u_\epsilon dx = &-\int_\Omega \left(p(p-2)u^2(u^2+\epsilon^2)^{\frac{p-4}{2}} + p(u^2+\epsilon^2)^{\frac{p-2}{2}}\right) V |\nabla_H u|^2 dx \\ &- p\int_\Omega V u (u^2+\epsilon^2)^{\frac{p-2}{2}} \mathcal{L} u dx.\end{aligned}$$

Hence we have the inequality

$$\begin{aligned}\int_\Omega |\mathcal{L}V| u_\epsilon + \left(p(p-2)u^2(u^2+\epsilon^2)^{\frac{p-4}{2}} + p(u^2+\epsilon^2)^{\frac{p-2}{2}}\right) V |\nabla_H u|^2 dx \\ \le p\int_\Omega V |u| (u^2+\epsilon^2)^{\frac{p-2}{2}} |\mathcal{L} u| dx.\end{aligned}$$

When $\epsilon \to 0$, the integrand on the left-hand side is non-negative and tends to

$$|\mathcal{L}V||u|^p + p(p-1)V|u|^{p-2}|\nabla_H u|^2$$

pointwise, only for $u \neq 0$ when $p < 2$, otherwise for any x. On the other hand, the integrand on the right-hand side is bounded by

$$V(\max|u|^2+1)^{(p-1)/2} \max|\mathcal{L}u|$$

and it is integrable because $u \in C_0^\infty(\Omega)$, and so the integral tends to

$$\int_\Omega V|u|^{p-1}|\mathcal{L}u| dx$$

by the dominated convergence theorem. It then follows by Fatou's lemma that

$$\left\| |\mathcal{L}V|^{\frac{1}{p}} u \right\|^p_{L^p(\Omega)} + p(p-1)\left\| V^{\frac{1}{p}} |u|^{\frac{p-2}{p}} |\nabla_H u|^{\frac{2}{p}} \right\|^p_{L^p(\Omega)} \le p \left\| V^{\frac{1}{p}} |u|^{\frac{p-1}{p}} |\mathcal{L}u|^{\frac{1}{p}} \right\|^p_{L^p(\Omega)}.$$

By using (7.72), followed by Hölder's inequality, we obtain

$$(1+C)\left\||\mathcal{L}V|^{\frac{1}{p}}u\right\|^p_{L^p(\Omega)} \leq p\left\||\mathcal{L}V|^{(p-1)}V^{\frac{1}{p}}|u|^{\frac{p-1}{p}}|\mathcal{L}V|^{-(p-1)}|\mathcal{L}u|^{\frac{1}{p}}\right\|^p_{L^p(\Omega)}$$
$$\leq p\left\||\mathcal{L}V|^{\frac{1}{p}}u\right\|^{p-1}_{L^p(\Omega)}\left\|\frac{|V|}{|\mathcal{L}V|^{\frac{p-1}{p}}}\mathcal{L}u\right\|_{L^p(\Omega)}.$$

This implies (7.73). □

Lemma 7.7.11. *Let $1 < p < \infty$. Let Ω be an admissible domain in a stratified group $\mathbb{G}$. If*

$$0 < V \in C(\Omega),\ \mathcal{L}V < 0 \ \text{ and } \ \mathcal{L}V^\sigma \leq 0$$

on Ω for some $\sigma > 1$, then we have

$$(\sigma - 1)\int_\Omega |\mathcal{L}V||u|^p dx \leq p^2 \int_{\{x\in\Omega, u(x)\neq 0\}} V|u|^{p-2}|\nabla_H u|^2 dx < \infty, \tag{7.74}$$

for all $u \in C_0^\infty(\Omega)$.

Proof of Lemma 7.7.11. We shall use that

$$0 \geq \mathcal{L}(V^\sigma) = \sigma V^{\sigma-2}\left((\sigma-1)|\nabla_H V|^2 + V\mathcal{L}V\right), \tag{7.75}$$

and hence

$$(\sigma - 1)|\nabla_H V|^2 \leq V|\mathcal{L}V|.$$

First we consider the case $p = 2$: we use the inequality (7.61) to get

$$(\sigma-1)\int_\Omega |\mathcal{L}V||u|^2 dx \leq 4(\sigma-1)\int_\Omega \frac{|\nabla_H V|^2}{|\mathcal{L}V|}|\nabla_H u|^2 dx$$
$$\leq 4\int_\Omega V|\nabla_H u|^2 dx$$
$$= 4\int_{\{x\in\Omega; u(x)\neq 0, |\nabla_H u|\neq 0\}} V|\nabla_H u|^2 dx, \tag{7.76}$$

the last equality valid since $|\{x \in \Omega; u(x) = 0, |\nabla_H u| \neq 0\}| = 0$. This proves Lemma 7.7.11 for $p = 2$.

For $p \neq 2$, denote

$$v_\epsilon := (u^2 + \epsilon^2)^{p/4} - \epsilon^{p/2},$$

and let $\epsilon \to 0$. Since

$$0 \leq v_\epsilon \leq |u|^{\frac{p}{2}},$$

the left-hand side of (7.76), with u replaced by v_ϵ, tends to

$$(\sigma - 1)\int_\Omega |\mathcal{L}V||u|^p dx$$

by the dominated convergence theorem. If $u \neq 0$, then

$$|\nabla_H v_\epsilon|^2 V = \left|\frac{p}{2} u(u^2+\epsilon^2)^{\frac{p-4}{4}} \nabla_H u\right|^2 V.$$

For $\epsilon \to 0$ we obtain

$$|\nabla_H u|^p V = \frac{p^2}{4} |u|^{p-2} |\nabla_H u|^2 V.$$

It follows as in the proof of Lemma 7.7.10, by using Fatou's lemma, that the right-hand side of (7.76) tends to

$$p^2 \int_{\{x\in\Omega; u(x)\neq 0, |\nabla_H u|\neq 0\}} V |u|^{p-2} |\nabla_H u|^2 dx,$$

and this completes the proof. □

Chapter 8

Geometric Hardy Inequalities on Stratified Groups

Given a domain in the space, the 'geometric' version of Hardy inequalities usually refers to the Hardy type inequalities where the weight is given in terms of the distance to the boundary of the domain. In this chapter we discuss L^2 and L^p versions of the geometric Hardy inequality on the stratified group $\mathbb{G}$. For the clarity of the exposition, we first deal with the half-space domains, and then with more general convex domains.

The results presented in this chapter have been obtained in [RSS18b], and our exposition here follows this paper. In particular, we discuss L^2 and L^p versions of the (subelliptic) geometric Hardy inequalities in half-spaces and convex domains on general stratified groups. As usual, these imply the geometric versions of the uncertainty principles. A certain current drawback of the methods in the case of convex domains is that the convexity is understood in the Euclidean sense.

8.1 L^2-Hardy inequality on the half-space

In this section, we discuss an L^2-version of the geometric Hardy inequality on the half-space of the stratified group $\mathbb{G}$. We start by recalling a few known results and by putting the further analysis in perspective.

Remark 8.1.1.

1. If Ω is a convex open set of the Euclidean space, then the geometric version of the Hardy inequality is well understood and given by

$$\int_{\Omega} |\nabla u|^2 dx \geq \frac{1}{4} \int_{\Omega} \frac{|u|^2}{\operatorname{dist}(x, \partial\Omega)^2} dx,$$

for $u \in C_0^{\infty}(\Omega)$, with the sharp constant $1/4$. Nowadays, there are many studies related to this subject, here we can mention, for example, [Anc86], [D'A04b], [AL10], [AW07], [Dav99] and [OK90].

M. Ruzhansky, D. Suragan, *Hardy Inequalities on Homogeneous Groups*, Progress in Mathematics 327, https://doi.org/10.1007/978-3-030-02895-4_9

2. In the setting of the Heisenberg group $\mathbb{H}$, the geometric Hardy inequality on the half-space
$$\mathbb{H}^+ := \{(x_1, x_2, x_3) \in \mathbb{H} \mid x_3 > 0\}$$
takes the form
$$\int_{\mathbb{H}^+} |\nabla_{\mathbb{H}} u|^2 dx \geq \int_{\mathbb{H}^+} \frac{|x_1|^2 + |x_2|^2}{x_3^2} |u|^2 dx,$$
for all $u \in C_0^\infty(\mathbb{H}^+)$. This inequality was obtained in [LY08], and we can also recapture it as a consequence in Corollary 8.1.5. There are further extensions to geometric L^p-Hardy inequalities as well as to the convex domains of the Heisenberg group obtained in [Lar16].

The following construction can be traced back to Garofalo [Gar08].

Definition 8.1.2 (Half-space and angle function)**.** Let $\mathbb{G}$ be a stratified group. In this section the half-space of $\mathbb{G}$ will be defined by
$$\mathbb{G}^+ := \{x \in \mathbb{G} : \langle x, \nu \rangle > d\},$$
where $d \in \mathbb{R}$, and $\nu := (\nu_1, \ldots, \nu_r)$ with $\nu_j \in \mathbb{R}^{N_j}$, $j = 1, \ldots, r$, is the Riemannian outer unit normal to $\partial\mathbb{G}^+$. The Euclidean distance to the boundary $\partial\mathbb{G}^+$ will be denoted by $\operatorname{dist}(x, \partial\mathbb{G}^+)$ and given by the formula
$$\operatorname{dist}(x, \partial\mathbb{G}^+) = \langle x, \nu \rangle - d.$$
The *angle function* on $\partial\mathbb{G}^+$ is defined by
$$\mathcal{W}(x) := \sqrt{\sum_{i=1}^{N} \langle X_i(x), \nu \rangle^2}. \tag{8.1}$$

In what follows we will be working in the setting of Definition 8.1.2.

Theorem 8.1.3 (Geometric L^2-Hardy inequality on half-space)**.** *Let $\mathbb{G}^+$ be a half-space of a stratified group $\mathbb{G}$.*

(1) *Let $\beta \in \mathbb{R}$ and set $C_1(\beta) := -(\beta^2 + \beta)$. Then we have*
$$\begin{aligned}\int_{\mathbb{G}^+} |\nabla_H u|^2 dx \geq C_1(\beta) \int_{\mathbb{G}^+} \frac{\mathcal{W}(x)^2}{\operatorname{dist}(x, \partial\mathbb{G}^+)^2} |u|^2 dx \\ + \beta \int_{\mathbb{G}^+} \sum_{i=1}^{N} \frac{X_i \langle X_i(x), \nu \rangle}{\operatorname{dist}(x, \partial\mathbb{G}^+)} |u|^2 dx,\end{aligned} \tag{8.2}$$
for all $u \in C_0^\infty(\mathbb{G}^+)$.

(2) *We have*

$$\int_{\mathbb{G}^+} |\nabla_H u|^2 dx \geq \frac{1}{4} \int_{\mathbb{G}^+} \frac{|u|^2}{\operatorname{dist}(x, \partial\mathbb{G}^+)^2} dx, \tag{8.3}$$

for all $u \in C_0^\infty(\mathbb{G}^+)$.

Remark 8.1.4 (Uncertainty principle and step 2 case). In the step 2 case we have the following simplification of Part (1) of Theorem 8.1.3.

1. Note that for the stratified groups of step 2 it follows from Proposition 1.2.19 that one can use the following basis of the left invariant vector fields

$$X_i = \frac{\partial}{\partial x_i'} + \sum_{s=1}^{N_2} \sum_{m=1}^{N} a_{m,i}^s x_m' \frac{\partial}{\partial x_s''}, \tag{8.4}$$

where $i = 1, \ldots, N$ and $a_{m,i}^s$ are the constants depending on the group. In addition, we can also write $x = (x', x'')$ with

$$x' = (x_1', \ldots, x_N'),\ x'' = (x_1'', \ldots, x_{N_2}''),$$

and also $\nu = (\nu', \nu'')$ with

$$\nu' = (\nu_1', \ldots, \nu_N'),\ \nu'' = (\nu_1'', \ldots, \nu_{N_2}'').$$

Then the statement of Theorem 8.1.3, Part (1), can be simplified as follows: for all $u \in C_0^\infty(\mathbb{G}^+)$ and $\beta \in \mathbb{R}$ we have

$$\begin{aligned} \int_{\mathbb{G}^+} |\nabla_H u|^2 dx \geq C_1(\beta) \int_{\mathbb{G}^+} \frac{\mathcal{W}(x)^2}{\operatorname{dist}(x, \partial\mathbb{G}^+)^2} |u|^2 dx \\ + K(a, \nu, \beta) \int_{\mathbb{G}^+} \frac{|u|^2}{\operatorname{dist}(x, \partial\mathbb{G}^+)} dx, \end{aligned} \tag{8.5}$$

where $C_1(\beta) := -(\beta^2 + \beta)$ and $K(a, \nu, \beta) = \beta \sum_{s=1}^{N_2} \sum_{i=1}^{N} a_{i,i}^s \nu_s''$.

2. In the standard way Theorem 8.1.3, Part (2), implies the geometric uncertainty principle on the half-space $\mathbb{G}^+$ for general stratified groups $\mathbb{G}$. Indeed, (8.3) and the Cauchy–Schwarz inequality imply

$$\begin{aligned} &\int_{\mathbb{G}^+} |\nabla_H u|^2 dx \int_{\mathbb{G}^+} \operatorname{dist}(x, \partial\mathbb{G}^+)^2 |u|^2 dx \\ &\geq \frac{1}{4} \int_{\mathbb{G}^+} \frac{1}{\operatorname{dist}(x, \partial\mathbb{G}^+)^2} |u|^2 dx \int_{\mathbb{G}^+} \operatorname{dist}(x, \partial\mathbb{G}^+)^2 |u|^2 dx \\ &\geq \frac{1}{4} \left(\int_{\mathbb{G}^+} |u|^2 dx \right)^2. \end{aligned}$$

That is, we have

$$\left(\int_{\mathbb{G}^+} |\nabla_H u|^2 dx \right)^{\frac{1}{2}} \left(\int_{\mathbb{G}^+} \operatorname{dist}(x, \partial\mathbb{G}^+)^2 |u|^2 dx \right)^{\frac{1}{2}} \geq \frac{1}{2} \int_{\mathbb{G}^+} |u|^2 dx$$

for all $u \in C_0^\infty(\mathbb{G}^+)$.

Proof of Theorem 8.1.3. Proof of Part (1). For the proof we apply the method of factorisation. So, for any real-valued $W := (W_1, \ldots, W_N)$, $W_i \in C^1(\mathbb{G}^+)$, which will be chosen later, a direct calculation gives

$$\begin{aligned} 0 \le \int_{\mathbb{G}^+} |\nabla_H u + \beta W u|^2 dx &= \int_{\mathbb{G}^+} |(X_1 u, \ldots, X_N u) + \beta(W_1, \ldots, W_N)u|^2 dx \\ &= \int_{\mathbb{G}^+} |(X_1 u + \beta W_1 u, \ldots, X_N u + \beta W_N u)|^2 dx \\ &= \int_{\mathbb{G}^+} \sum_{i=1}^N |X_i u + \beta W_i u|^2 dx \\ &= \int_{\mathbb{G}^+} \sum_{i=1}^N \left[|X_i u|^2 + 2\mathrm{Re}\beta W_i u X_i u + \beta^2 W_i^2 |u|^2 \right] dx \\ &= \int_{\mathbb{G}^+} \sum_{i=1}^N \left[|X_i u|^2 + \beta W_i X_i |u|^2 + \beta^2 W_i^2 |u|^2 \right] dx \\ &= \int_{\mathbb{G}^+} \sum_{i=1}^N \left[|X_i u|^2 - \beta (X_i W_i) |u|^2 + \beta^2 W_i^2 |u|^2 \right] dx. \end{aligned}$$

From the above expression we get the inequality

$$\int_{\mathbb{G}^+} |\nabla_H u|^2 dx \ge \int_{\mathbb{G}^+} \sum_{i=1}^N \left[(\beta(X_i W_i) - \beta^2 W_i^2)|u|^2 \right] dx. \tag{8.6}$$

Let us now take W_i in the form

$$W_i(x) = \frac{\langle X_i(x), \nu\rangle}{\mathrm{dist}(x, \partial\mathbb{G}^+)} = \frac{\langle X_i(x), \nu\rangle}{\langle x, \nu\rangle - d}, \tag{8.7}$$

where

$$X_i(x) = (\overbrace{0, \ldots, 1}^{i}, \ldots, 0, a_{i,1}^{(2)}(x'), \ldots, a_{i,N_r}^{(r)}(x', x^{(2)}, \ldots, x^{(r-1)})),$$

and

$$\nu = (\nu_1, \nu_2, \ldots, \nu_r),\ \nu_j \in \mathbb{R}^{N_j}.$$

Now $W_i(x)$ can be written as

$$W_i(x) = \frac{\nu_{1,i} + \sum_{l=2}^r \sum_{m=1}^{N_l} a_{i,m}^{(l)}(x', \ldots, x^{(l-1)})\nu_{l,m}}{\sum_{l=1}^r x^{(l)} \cdot \nu_l - d}.$$

By a direct computation we have

$$\begin{aligned} X_i W_i(x) &= \frac{X_i \langle X_i(x), \nu\rangle \,\mathrm{dist}(x, \partial\mathbb{G}^+) - \langle X_i(x), \nu\rangle X_i(\mathrm{dist}(x, \partial\mathbb{G}^+))}{\mathrm{dist}(x, \partial\mathbb{G}^+)^2} \\ &= \frac{X_i\langle X_i(x), \nu\rangle}{\mathrm{dist}(x, \partial\mathbb{G}^+)} - \frac{\langle X_i(x), \nu\rangle^2}{\mathrm{dist}(x, \partial\mathbb{G}^+)^2}, \end{aligned} \tag{8.8}$$

where

$$\begin{aligned} X_i(\mathrm{dist}(x,\partial\mathbb{G}^+)) &= X_i\left(\sum_{k=1}^{N} x'_k\nu_{1,k} + \sum_{l=2}^{r}\sum_{m=1}^{N_l} x_m^{(l)}\nu_{l,m} - d\right) \\ &= \nu_{1,i} + \sum_{l=2}^{r}\sum_{m=1}^{N_l} a_{i,m}^{(l)}(x',\dots,x^{(l-1)})\nu_{l,m} \\ &= \langle X_i(x),\nu\rangle. \end{aligned}$$

Now combining (8.8) with (8.6) we arrive at the inequality

$$\begin{aligned} \int_{\mathbb{G}^+} |\nabla_H u|^2 dx \geq &- (\beta^2+\beta)\int_{\mathbb{G}^+}\sum_{i=1}^{N}\frac{\langle X_i(x),\nu\rangle^2}{\mathrm{dist}(x,\partial\mathbb{G}^+)^2}|u|^2dx \\ &+ \beta\int_{\mathbb{G}^+}\sum_{i=1}^{N}\frac{X_i\langle X_i(x),\nu\rangle}{\mathrm{dist}(x,\partial\mathbb{G}^+)}|u|^2dx, \end{aligned}$$

which completes the proof of Part (1).

Proof of Part (2). Let $x := (x', x^{(2)}, \dots, x^{(r)}) \in \mathbb{G}$ with $x' = (x'_1, \dots, x'_N)$ and $x^{(j)} \in \mathbb{R}^{N_j}$, $j = 2, \dots, r$. By taking $\nu := (\nu', 0, \dots, 0)$ with $\nu' = (\nu'_1, \dots, \nu'_N)$, we have that

$$X_i(x) = (\overbrace{0,\dots,1}^{i},\dots,0,a_{i,1}^{(2)}(x'),\dots,a_{i,N_r}^{(r)}(x',x^{(2)},\dots,x^{(r-1)})),$$

so that

$$\sum_{i=1}^{N}\langle X_i(x),\nu\rangle^2 = \sum_{i=1}^{N}(\nu'_i)^2 = |\nu'|^2 = 1$$

and

$$X_i\langle X_i(x),\nu\rangle = X_i\nu'_i = 0.$$

Substituting this in (8.2) we get

$$\int_{\mathbb{G}^+}|\nabla_H u|^2dx \geq -(\beta^2+\beta)\int_{\mathbb{G}^+}\frac{|u|^2}{\mathrm{dist}(x,\partial\mathbb{G}^+)^2}dx.$$

To optimize we differentiate the right-hand side expression with respect to β, that is, we put $-2\beta - 1 = 0$, or $\beta = -\frac{1}{2}$ in this inequality, implying (8.3). □

8.1.1 Examples of Heisenberg and Engel groups

Let us give examples of the geometric L^2-Hardy inequality on half-spaces from Theorem 8.1.3 in the cases of groups of steps 2 and 3. The example of general stratified groups of step 2 was considered in Remark 8.1.4, Part 1, and now we look at the special case of the Heisenberg group. In particular, it yields the estimate that was given in Remark 8.1.1, Part 2.

Corollary 8.1.5 (Geometric L^2-Hardy inequality on half-space of the Heisenberg group). *Let $\mathbb{H}^+ = \{(x_1, x_2, x_3) \in \mathbb{H} \mid x_3 > 0\}$ be a half-space of the Heisenberg group $\mathbb{H}$. Then for all $u \in C_0^\infty(\mathbb{H}^+)$ we have*

$$\int_{\mathbb{H}^+} |\nabla_{\mathbb{H}} u|^2 dx \geq \int_{\mathbb{H}^+} \frac{|x_1|^2 + |x_2|^2}{x_3^2} |u|^2 dx,$$

where $\nabla_{\mathbb{H}} = (X_1, X_2)$.

Proof of Corollary 8.1.5. Since the left invariant vector fields on the Heisenberg group can be given by

$$X_1 = \frac{\partial}{\partial x_1} + 2x_2 \frac{\partial}{\partial x_3}, \qquad X_2 = \frac{\partial}{\partial x_2} - 2x_1 \frac{\partial}{\partial x_3},$$

with the commutator

$$[X_1, X_2] = -4\frac{\partial}{\partial x_3},$$

choosing $\nu = (0, 0, 1)$ as the unit vector in the direction of x_3 and taking $d = 0$ in inequality (8.2), we get

$$X_1(x) = (1, 0, 2x_2) \quad \text{and} \quad X_2(x) = (0, 1, -2x_1),$$

and

$$\begin{aligned} \langle X_1(x), \nu\rangle &= 2x_2, \quad &\text{and} \quad \langle X_2(x), \nu\rangle &= -2x_1, \\ X_1\langle X_1(x), \nu\rangle &= 0, \quad &\text{and} \quad X_2\langle X_2(x), \nu\rangle &= 0, \end{aligned}$$

where $x = (x_1, x_2, x_3)$. Thus, with $\mathcal{W}(x)$ as in (8.1), we get

$$\frac{\mathcal{W}(x)^2}{\operatorname{dist}(x, \partial\mathbb{G}^+)^2} = 4\frac{|x_1|^2 + |x_2|^2}{x_3^2}.$$

Inserting these to (8.2) with $\beta = -\frac{1}{2}$ we obtain

$$\int_{\mathbb{H}^+} |\nabla_{\mathbb{H}} u|^2 dx \geq \int_{\mathbb{H}^+} \frac{|x_1|^2 + |x_2|^2}{x_3^2} |u|^2 dx,$$

completing the proof. □

Next, let us give an example for a class of stratified groups of step $r = 3$, namely, the case of the Engel group.

Definition 8.1.6 (Engel group). The *Engel group* $\mathcal{E}$ is the space $\mathbb{R}^4$ with the group law given by

$$x \circ y = (x_1 + y_1, x_2 + y_2, x_3 + y_3 + P_1, x_4 + y_4 + P_2),$$

where

$$P_1 = \frac{1}{2}(x_1 y_2 - x_2 y_1),$$
$$P_2 = \frac{1}{2}(x_1 y_3 - x_3 y_1) + \frac{1}{12}(x_1^2 y_2 - x_1 y_1 (x_2 + y_2) + x_2 y_1^2).$$

The left invariant vector fields of $\mathcal{E}$ are generated by (the basis)

$$X_1 = \frac{\partial}{\partial x_1} - \frac{x_2}{2}\frac{\partial}{\partial x_3} - \left(\frac{x_3}{2} - \frac{x_1 x_2}{12}\right)\frac{\partial}{\partial x_4}, \quad X_3 = \frac{\partial}{\partial x_3} + \frac{x_1}{2}\frac{\partial}{\partial x_4},$$
$$X_2 = \frac{\partial}{\partial x_2} + \frac{x_1}{2}\frac{\partial}{\partial x_3} + \frac{x_1^2}{12}\frac{\partial}{\partial x_4}, \qquad X_4 = \frac{\partial}{\partial x_4}.$$

The group $\mathcal{E}$ is stratified of step 3, with the nonzero commutation relations given by

$$[X_1, X_2] = X_3, \; [X_1, X_3] = X_4.$$

So we have

Corollary 8.1.7 (Geometric L^2-Hardy inequality on half-space of the Engel group). *Let $\mathcal{E}^+ = \{x := (x_1, x_2, x_3, x_4) \in \mathcal{E} \mid \langle x, \nu \rangle > 0\}$ be a half-space of the Engel group $\mathcal{E}$. Then for all $\beta \in \mathbb{R}$ and $u \in C_0^\infty(\mathcal{E}^+)$ we have*

$$\begin{aligned}\int_{\mathcal{E}^+} |\nabla_{\mathcal{E}} u|^2 dx \geq C_1(\beta) \int_{\mathcal{E}^+} &\frac{\langle X_1(x), \nu\rangle^2 + \langle X_2(x), \nu\rangle^2}{\operatorname{dist}(x, \partial\mathcal{E}^+)^2}|u|^2 dx \\ &+ \frac{\beta}{3}\int_{\mathcal{E}^+} \frac{x_2 \nu_4}{\operatorname{dist}(x, \partial\mathcal{E}^+)}|u|^2 dx,\end{aligned} \tag{8.9}$$

where $\nabla_{\mathcal{E}} = (X_1, X_2)$, $\nu := (\nu_1, \nu_2, \nu_3, \nu_4)$, and $C_1(\beta) = -(\beta^2 + \beta)$.

In particular, if we take $\nu_4 = 0$ in (8.9), then by taking $\beta = -\frac{1}{2}$, we get the following inequality on such $\mathcal{E}^+$:

$$\int_{\mathcal{E}^+} |\nabla_{\mathcal{E}} u|^2 dx \geq \frac{1}{4}\int_{\mathcal{E}^+} \frac{\langle X_1(x), \nu\rangle^2 + \langle X_2(x), \nu\rangle^2}{\operatorname{dist}(x, \partial\mathcal{E}^+)^2}|u|^2 dx.$$

Proof of Corollary 8.1.7. Using the above basis of the left invariant vector fields, we have

$$X_1(x) = \left(1, 0, -\frac{x_2}{2}, -\left(\frac{x_3}{2} - \frac{x_1 x_2}{12}\right)\right),$$
$$X_2(x) = \left(0, 1, \frac{x_1}{2}, \frac{x_1^2}{12}\right).$$

It is then straightforward to see that

$$\begin{aligned}
\langle X_1(x), \nu\rangle &= \nu_1 - \frac{x_2}{2}\nu_3 - \left(\frac{x_3}{2} - \frac{x_1x_2}{12}\right)\nu_4,\\
\langle X_2(x), \nu\rangle &= \nu_2 + \frac{x_1}{2}\nu_3 + \frac{x_1^2}{12}\nu_4,\\
X_1\langle X_1(x), \nu\rangle &= \frac{x_2}{12}\nu_4 + \frac{x_2}{4}\nu_4 = \frac{x_2\nu_4}{3},\\
X_2\langle X_2(x), \nu\rangle &= 0.
\end{aligned}$$

Now plugging these in inequality (8.2) we get the desired inequality (8.9). □

8.2 L^p-Hardy inequality on the half-space

Now we discuss an L^p version of the geometric Hardy inequality on the half-space of $\mathbb{G}$ as an extension of the previous L^2 arguments. We recall that the p-version of Garofalo's angle function from Definition 8.1.2 can be defined by the formula

$$\mathcal{W}_p(x) = \left(\sum_{i=1}^{N} |\langle X_i(x), \nu\rangle|^p\right)^{1/p}, \tag{8.10}$$

with $\mathcal{W}(x) := \mathcal{W}_2(x)$, and where N denotes the dimension of the first stratum of $\mathbb{G}$. As before let $\mathbb{G}^+$ be a half-space of a stratified group $\mathbb{G}$. The L^p version of the geometric Hardy inequality from Theorem 8.1.3 can be written in the following form.

Theorem 8.2.1 (Geometric L^p-Hardy inequality on half-space). *Let $\mathbb{G}^+$ be a half-space of a stratified group $\mathbb{G}$ and let $1 < p < \infty$. Then for all $u \in C_0^\infty(\mathbb{G}^+)$ and all $\beta \in \mathbb{R}$ we have*

$$\begin{aligned}
\int_{\mathbb{G}^+} \sum_{i=1}^{N} |X_i u|^p dx \geq C_2(\beta, p) \int_{\mathbb{G}^+} \frac{\mathcal{W}_p(x)^p}{\operatorname{dist}(x, \partial\mathbb{G}^+)^p} |u|^p dx \\
+ \beta(p-1) \int_{\mathbb{G}^+} \sum_{i=1}^{N} \left(\frac{|\langle X_i(x), \nu\rangle|}{\operatorname{dist}(x, \partial\mathbb{G}^+)}\right)^{p-2} \frac{X_i\langle X_i(x), \nu\rangle}{\operatorname{dist}(x, \partial\mathbb{G}^+)} |u|^p dx,
\end{aligned} \tag{8.11}$$

where $C_2(\beta, p) := -(p-1)(|\beta|^{\frac{p}{p-1}} + \beta)$.

Remark 8.2.2. Note that for $p \geq 2$, since

$$|\nabla_H u|^p = \left(\sum_{i=1}^{N} |X_i u|^2\right)^{p/2} \geq \sum_{i=1}^{N} \left(|X_i u|^2\right)^{p/2},$$

the proof will also yield the inequality

$$\int_{\mathbb{G}^+} |\nabla_H u|^p dx \geq C_2(\beta,p)\int_{\mathbb{G}^+} \frac{\mathcal{W}_p(x)^p}{\operatorname{dist}(x,\partial\mathbb{G}^+)^p}|u|^p dx \tag{8.12}$$
$$+\beta(p-1)\int_{\mathbb{G}^+}\sum_{i=1}^{N}\left(\frac{|\langle X_i(x),\nu\rangle|}{\operatorname{dist}(x,\partial\mathbb{G}^+)}\right)^{p-2}\frac{X_i\langle X_i(x),\nu\rangle}{\operatorname{dist}(x,\partial\mathbb{G}^+)}|u|^p dx.$$

Proof of Theorem 8.2.1. For $W \in C^\infty(\mathbb{G}^+)$ and $f \in C^1(\mathbb{G}^+)$, a direct computation with Hölder's inequality gives

$$\int_{\mathbb{G}^+} \operatorname{div}_{\mathrm{H}}(fW)|u|^p dx = -\int_{\mathbb{G}^+} fW\cdot\nabla_H|u|^p dx = -p\int_{\mathbb{G}^+} f\langle W,\nabla_H u\rangle|u|^{p-1}dx$$
$$\leq p\left(\int_{\mathbb{G}^+}|\langle W,\nabla_H u\rangle|^p dx\right)^{1/p}\left(\int_{\mathbb{G}^+}|f|^{\frac{p}{p-1}}|u|^p dx\right)^{\frac{p-1}{p}}. \tag{8.13}$$

For $p>1$ and $q>1$ with $\frac{1}{p}+\frac{1}{q}=1$, we will use Young's inequality

$$ab \leq \frac{a^p}{p}+\frac{b^q}{q}, \quad \text{for} \quad a\geq 0,\, b\geq 0,$$

with

$$a := \left(\int_{\mathbb{G}^+}|\langle W,\nabla_H u\rangle|^p dx\right)^{1/p} \qquad \text{and} \qquad b := \left(\int_{\mathbb{G}^+}|f|^{\frac{p}{p-1}}|u|^p dx\right)^{\frac{p-1}{p}}.$$

Using this Young inequality in (8.13) and rearranging the terms, we get

$$\int_{\mathbb{G}^+}|\langle W,\nabla_H u\rangle|^p dx \geq \int_{\mathbb{G}^+}\left(\operatorname{div}_{\mathrm{H}}(fW)-(p-1)|f|^{\frac{p}{p-1}}\right)|u|^p dx. \tag{8.14}$$

Now choosing $W := I_i$, which has the following form $I_i = (\overbrace{0,\ldots,1}^{i},\ldots,0)$ and setting

$$f = \beta\frac{|\langle X_i(x),\nu\rangle|^{p-1}}{\operatorname{dist}(x,\partial\mathbb{G}^+)^{p-1}},$$

we calculate

$$\operatorname{div}_{\mathrm{H}}(Wf) = (\nabla_H\cdot I_i)f = X_i f = \beta X_i\left(\frac{|\langle X_i(x),\nu\rangle|}{\operatorname{dist}(x,\partial\mathbb{G}^+)}\right)^{p-1}$$
$$= \beta(p-1)\left(\frac{|\langle X_i(x),\nu\rangle|}{\operatorname{dist}(x,\partial\mathbb{G}^+)}\right)^{p-2}X_i\left(\frac{\langle X_i(x),\nu\rangle}{\operatorname{dist}(x,\partial\mathbb{G}^+)}\right)$$
$$= \beta(p-1)\left(\frac{|\langle X_i(x),\nu\rangle|}{\operatorname{dist}(x,\partial\mathbb{G}^+)}\right)^{p-2}\left(\frac{X_i\langle X_i(x),\nu\rangle}{\operatorname{dist}(x,\partial\mathbb{G}^+)}-\frac{|\langle X_i(x),\nu\rangle|^2}{\operatorname{dist}(x,\partial\mathbb{G}^+)^2}\right)$$
$$= \beta(p-1)\left[\left(\frac{|\langle X_i(x),\nu\rangle|}{\operatorname{dist}(x,\partial\mathbb{G}^+)}\right)^{p-2}\left(\frac{X_i\langle X_i(x),\nu\rangle}{\operatorname{dist}(x,\partial\mathbb{G}^+)}\right)-\frac{|\langle X_i(x),\nu\rangle|^p}{\operatorname{dist}(x,\partial\mathbb{G}^+)^p}\right],$$

and

$$|f|^{\frac{p}{p-1}} = |\beta|^{\frac{p}{p-1}} \frac{|\langle X_i(x), \nu\rangle|^p}{\operatorname{dist}(x, \partial\mathbb{G}^+)^p}.$$

Moreover, we have

$$\langle W, \nabla_H u\rangle = (\overbrace{0, \dots, 1}^{i}, \dots, 0) \cdot (X_1 u, \dots, X_i u, \dots, X_N u)^T = X_i u.$$

Substituting these in (8.14) and summing over $i = 1, \dots, N$, we obtain

$$\begin{aligned}\int_{\mathbb{G}^+} \sum_{i=1}^{N} |X_i u|^p dx \geq & -(p-1)(|\beta|^{\frac{p}{p-1}} + \beta) \int_{\mathbb{G}^+} \sum_{i=1}^{N} \frac{|\langle X_i(x), \nu\rangle|^p}{\operatorname{dist}(x, \partial\mathbb{G}^+)^p} |u|^p dx \\ & + \beta(p-1) \int_{\mathbb{G}^+} \sum_{i=1}^{N} \left(\frac{|\langle X_i(x), \nu\rangle|}{\operatorname{dist}(x, \partial\mathbb{G}^+)}\right)^{p-2} \frac{X_i \langle X_i(x), \nu\rangle}{\operatorname{dist}(x, \partial\mathbb{G}^+)} |u|^p dx.\end{aligned} \tag{8.15}$$

This completes the proof. □

8.3 L^2-Hardy inequality on convex domains

In this and the following sections we extend the proceeding arguments from half-spaces to convex domains in the stratified groups. Here, however, the convex domain is understood in the sense of the Euclidean space. Thus, let Ω be a convex domain of a stratified group $\mathbb{G}$ and let $\partial\Omega$ be its boundary. Here for $x \in \Omega$ we denote by $\nu(x)$ the unit normal for $\partial\Omega$ at a point $\hat{x} \in \partial\Omega$, determined by the condition

$$\operatorname{dist}(x, \partial\Omega) = \operatorname{dist}(x, \hat{x}).$$

For the half-space, we have the distance from the boundary $\operatorname{dist}(x, \partial\Omega) = \langle x, \nu\rangle - d$. As it was already defined in (8.10), we will use the p-version of the angle function

$$\mathcal{W}_p(x) = \left(\sum_{i=1}^{N} |\langle X_i(x), \nu\rangle|^p\right)^{1/p},$$

with $\mathcal{W}(x) := \mathcal{W}_2(x)$. We have the following extension of Theorem 8.1.3.

Theorem 8.3.1 (Geometric L^2-Hardy inequality on convex domains)**.** *Let Ω be a convex domain of a stratified group $\mathbb{G}$. Then for all $u \in C_0^\infty(\Omega)$ and all $\beta < 0$ we have*

$$\int_\Omega |\nabla_H u|^2 dx \geq C_1(\beta) \int_\Omega \frac{\mathcal{W}(x)^2}{\operatorname{dist}(x, \partial\Omega)^2} |u|^2 dx + \beta \int_\Omega \sum_{i=1}^{N} \frac{X_i \langle X_i(x), \nu\rangle}{\operatorname{dist}(x, \partial\Omega)} |u|^2 dx, \tag{8.16}$$

where $C_1(\beta) := -(\beta^2 + \beta)$.

Proof of Theorem 8.3.1. As elsewhere in this chapter, we follow the proof for general stratified groups of [RSS18b], based on the convex polytope approach used by Larson [Lar16] in the case of the Heisenberg group.

We denote the facets of Ω by $\{\mathcal{F}_j\}_j$ and unit normals of these facets by $\{\nu_j\}_j$, which are directed inward. So, Ω can be viewed as the union of the disjoint sets

$$\Omega_j := \{x \in \Omega : \operatorname{dist}(x, \partial\Omega) = \operatorname{dist}(x, \mathcal{F}_j)\}.$$

Now we follow the method as in the case of the half-space $\mathbb{G}^+$ for each element Ω_j with one exception that not all the boundary values are zero when we use the partial integration. As before we calculate

$$\begin{aligned} 0 \leq \int_{\Omega_j} |\nabla_H u + \beta W u|^2 dx &= \int_{\Omega_j} \sum_{i=1}^{N} |X_i u + \beta W_i u|^2 dx \\ &= \int_{\Omega_j} \sum_{i=1}^{N} \left[|X_i u|^2 + 2\mathrm{Re}\beta W_i u X_i u + \beta^2 W_i^2 |u|^2 \right] dx \\ &= \int_{\Omega_j} \sum_{i=1}^{N} \left[|X_i u|^2 + \beta W_i X_i |u|^2 + \beta^2 W_i^2 |u|^2 \right] dx \\ &= \int_{\Omega_j} \sum_{i=1}^{N} \left[|X_i u|^2 - \beta (X_i W_i) |u|^2 + \beta^2 W_i^2 |u|^2 \right] dx \\ &\quad + \beta \int_{\partial\Omega_j} \sum_{i=1}^{N} W_i \langle X_i(x), n_j(x) \rangle |u|^2 d\Gamma_{\partial\Omega_j}(x), \end{aligned}$$

where n_j is the unit normal of $\partial\Omega_j$ which is directed outward. Since $\mathcal{F}_j \subset \partial\Omega_j$ we have $n_j = -\nu_j$. That is, we have

$$\begin{aligned} \int_{\Omega_j} |\nabla_H u|^2 dx \geq & \int_{\Omega_j} \sum_{i=1}^{N} \left[(\beta(X_i W_i) - \beta^2 W_i^2) |u|^2 \right] dx \\ & - \beta \int_{\partial\Omega_j} \sum_{i=1}^{N} W_i \langle X_i(x), n_j(x) \rangle |u|^2 d\Gamma_{\partial\Omega_j}(x). \end{aligned} \tag{8.17}$$

The boundary terms on $\partial\Omega$ disappears since u is compactly supported in Ω. Thus, we only need to deal with the parts of $\partial\Omega_j$ in Ω. Note that for every facet of $\partial\Omega_j$ there exists some $\partial\Omega_l$ which shares this facet. Denote by Γ_{jl} the common facet of $\partial\Omega_j$ and $\partial\Omega_l$, with $n_k|_{\Gamma_{jl}} = -n_l|_{\Gamma_{jl}}$.

Now we choose W_i in the form

$$W_i(x) = \frac{\langle X_i(x), \nu_j \rangle}{\operatorname{dist}(x, \partial\Omega_j)} = \frac{\langle X_i(x), \nu_j \rangle}{\langle x, \nu_j \rangle - d},$$

and a direct computation shows that

$$X_i W_i(x) = \frac{X_i \langle X_i(x), \nu_j \rangle}{\operatorname{dist}(x, \partial\Omega_j)} - \frac{\langle X_i(x), \nu_j \rangle^2}{\operatorname{dist}(x, \partial\Omega_j)^2}. \tag{8.18}$$

Substituting (8.18) into (8.17) we get

$$\int_{\Omega_j} |\nabla_H u|^2 dx \geq -(\beta^2+\beta)\int_{\Omega_j}\sum_{i=1}^{N}\frac{\langle X_i(x),\nu_j\rangle^2}{\operatorname{dist}(x,\partial\Omega_j)^2}|u|^2dx \tag{8.19}$$
$$+\beta\int_{\Omega_j}\sum_{i=1}^{N}\frac{X_i\langle X_i(x),\nu_j\rangle}{\operatorname{dist}(x,\partial\Omega_j)}|u|^2dx-\beta\int_{\Gamma_{jl}}\sum_{i=1}^{N}\frac{\langle X_i(x),\nu_j\rangle\langle X_i(x),n_{jl}\rangle}{\operatorname{dist}(x,\mathcal{F}_j)}|u|^2d\Gamma_{jl}.$$

Now we sum over all partition elements Ω_j and let $n_{jl}=n_k|_{\Gamma_{jl}}$, i.e., the unit normal of Γ_{jl} pointing from Ω_j into Ω_l. Then we have

$$\begin{aligned}\int_{\Omega}|\nabla_H u|^2dx \geq & -(\beta^2+\beta)\int_{\Omega}\sum_{i=1}^{N}\frac{\langle X_i(x),\nu\rangle^2}{\operatorname{dist}(x,\partial\Omega)^2}|u|^2dx\\ &+\beta\int_{\Omega}\sum_{i=1}^{N}\frac{X_i\langle X_i(x),\nu\rangle}{\operatorname{dist}(x,\partial\Omega)}|u|^2dx\\ &-\beta\sum_{j\neq l}\int_{\Gamma_{jl}}\sum_{i=1}^{N}\frac{\langle X_i(x),\nu_j\rangle\langle X_i(x),n_{jl}\rangle}{\operatorname{dist}(x,\mathcal{F}_j)}|u|^2d\Gamma_{jl}\\ = & -(\beta^2+\beta)\int_{\Omega}\sum_{i=1}^{N}\frac{\langle X_i(x),\nu\rangle^2}{\operatorname{dist}(x,\partial\Omega)^2}|u|^2dx\\ &+\beta\int_{\Omega}\sum_{i=1}^{N}\frac{X_i\langle X_i(x),\nu\rangle}{\operatorname{dist}(x,\partial\Omega)}|u|^2dx\\ &-\beta\sum_{j<l}\int_{\Gamma_{jl}}\sum_{i=1}^{N}\frac{\langle X_i(x),\nu_j-\nu_l\rangle\langle X_i(x),n_{jl}\rangle}{\operatorname{dist}(x,\mathcal{F}_j)}|u|^2d\Gamma_{jl}.\end{aligned}$$

Here we have used the fact that (by the definition) Γ_{jl} is a set with

$$\operatorname{dist}(x,\mathcal{F}_j)=\operatorname{dist}(x,\mathcal{F}_l).$$

From

$$\Gamma_{jl}=\{x : x\cdot\nu_j-d_j=x\cdot\nu_l-d_l\}$$

rearranging $x\cdot(\nu_j-\nu_l)-d_j+d_l=0$ we see that Γ_{jl} is a hyperplane with a normal $\nu_j-\nu_l$. So, $\nu_j-\nu_l$ is parallel to n_{jl} and one only needs to check that $(\nu_j-\nu_l)\cdot n_{jl}>0$. Since n_{jl} points out and ν_j points into jth partition element, $\nu_j\cdot n_{jl}$ is non-negative. Similarly, we see that $\nu_l\cdot n_{jl}$ is non-positive. That is, $(\nu_j-\nu_l)\cdot n_{jl}>0$. On the other hand, it is easy to see that

$$\begin{aligned}|\nu_j-\nu_l|^2=(\nu_j-\nu_l)\cdot(\nu_j-\nu_l)&=2-2\nu_j\cdot\nu_l\\ &=2-2\cos(\alpha_{jl}),\end{aligned}$$

which implies that

$$(\nu_j - \nu_l) \cdot n_{jl} = \sqrt{2 - 2\cos(\alpha_{jl})},$$

where α_{jl} is the angle between ν_j and ν_l. So we obtain

$$\begin{aligned}\int_\Omega |\nabla_H u|^2 dx \geq & -(\beta^2+\beta)\int_\Omega \sum_{i=1}^N \frac{\langle X_i(x), \nu\rangle^2}{\text{dist}(x,\partial\Omega)^2}|u|^2 dx \\ & + \beta \int_\Omega \sum_{i=1}^N \frac{X_i\langle X_i(x),\nu\rangle}{\text{dist}(x,\partial\Omega)}|u|^2 dx \\ & - \beta \sum_{j<l}\sum_{i=1}^N \int_{\Gamma_{jl}} \sqrt{1-\cos(\alpha_{jl})}\frac{\langle X_i(x), n_{jl}\rangle^2}{\text{dist}(x,\mathcal{F}_j)}|u|^2 d\Gamma_{jl}.\end{aligned}$$

Here with $\beta < 0$ and due to the boundary term signs we prove the desired inequality for the polytope convex domains.

Now we are ready to consider the general case, that is, when Ω is an arbitrary convex domain. For each $u \in C_0^\infty(\Omega)$ one can always choose an increasing sequence of convex polytopes $\{\Omega_j\}_{j=1}^\infty$ such that $u \in C_0^\infty(\Omega_1)$, $\Omega_j \subset \Omega$ and $\Omega_j \to \Omega$ as $j \to \infty$. Assume that $\nu_j(x)$ is the above map ν (corresponding to Ω_j), and then we can calculate

$$\begin{aligned}\int_\Omega |\nabla_H u|^2 dx &= \int_{\Omega_j} |\nabla_H u|^2 dx \\ &\geq -(\beta^2+\beta)\int_{\Omega_j}\sum_{i=1}^N \frac{\langle X_i(x),\nu_j\rangle^2}{\text{dist}(x,\partial\Omega_j)^2}|u|^2dx + \beta\int_{\Omega_j}\sum_{i=1}^N \frac{X_i\langle X_i(x),\nu_j\rangle}{\text{dist}(x,\partial\Omega_j)}|u|^2dx \\ &= -(\beta^2+\beta)\int_{\Omega}\sum_{i=1}^N \frac{\langle X_i(x),\nu_j\rangle^2}{\text{dist}(x,\partial\Omega_j)^2}|u|^2dx + \beta\int_{\Omega}\sum_{i=1}^N \frac{X_i\langle X_i(x),\nu_j\rangle}{\text{dist}(x,\partial\Omega_j)}|u|^2dx \\ &\geq -(\beta^2+\beta)\int_{\Omega}\sum_{i=1}^N \frac{\langle X_i(x),\nu_j\rangle^2}{\text{dist}(x,\partial\Omega)^2}|u|^2dx + \beta\int_{\Omega}\sum_{i=1}^N \frac{X_i\langle X_i(x),\nu_j\rangle}{\text{dist}(x,\partial\Omega)}|u|^2dx\,.\end{aligned}$$

Now we obtain the desired result by letting $j \to \infty$. □

8.4 L^p-Hardy inequality on convex domains

The same arguments as in the previous section give the general L^p-version of Theorem 8.3.1.

Theorem 8.4.1 (Geometric L^p-Hardy inequality on convex domains). *Let Ω be a convex domain of a stratified group $\mathbb{G}$. Then for all $u \in C_0^\infty(\Omega)$ and all $\beta < 0$*

we have

$$\int_\Omega \sum_{i=1}^N |X_i u|^p dx \geq C_2(\beta,p) \int_\Omega \frac{\mathcal{W}_p(x)^p}{\operatorname{dist}(x,\partial\Omega)^p}|u|^p dx + \beta(p-1)\int_\Omega \sum_{i=1}^N \left(\frac{|\langle X_i(x),\nu\rangle|}{\operatorname{dist}(x,\partial\Omega)}\right)^{p-2}\left(\frac{X_i\langle X_i(x),\nu\rangle}{\operatorname{dist}(x,\partial\Omega)}\right)|u|^p dx, \tag{8.20}$$

where $C_2(\beta,p) := -(p-1)(|\beta|^{\frac{p}{p-1}}+\beta)$.

Remark 8.4.2. Note that for $p \geq 2$, since

$$|\nabla_H u|^p = \left(\sum_{i=1}^N |X_i u|^2\right)^{p/2} \geq \sum_{i=1}^N \left(|X_i u|^2\right)^{p/2}, \tag{8.21}$$

instead of (8.20) we have the inequality

$$\int_\Omega |\nabla_H u|^p dx \geq C_2(\beta,p)\int_\Omega \frac{\mathcal{W}_p(x)^p}{\operatorname{dist}(x,\partial\Omega)^p}|u|^p dx + \beta(p-1)\int_\Omega \sum_{i=1}^N \left(\frac{|\langle X_i(x),\nu\rangle|}{\operatorname{dist}(x,\partial\Omega)}\right)^{p-2}\left(\frac{X_i\langle X_i(x),\nu\rangle}{\operatorname{dist}(x,\partial\Omega)}\right)|u|^p dx. \tag{8.22}$$

Proof of Theorem 8.4.1. As in the proof of Theorem 8.3.1, let us first assume that Ω is the convex polytope. Thus, for $f \in C^1(\Omega_j)$ and $W \in C^\infty(\Omega_j)$, we calculate

$$\begin{aligned}\int_{\Omega_j} \operatorname{div}_{\mathbb{G}}(fW)|u|^p dx &= -p\int_{\Omega_j} f\langle W,\nabla_H u\rangle |u|^{p-1}dx + \int_{\partial\Omega_j} f\langle W, n_j(x)\rangle |u|^p d\Gamma_{\partial\Omega_j}(x) \\ &\leq p\left(\int_\Omega |\langle W,\nabla_H u\rangle|^p dx\right)^{\frac{1}{p}}\left(\int_{\Omega_j}|f|^{\frac{p}{p-1}}|u|^p dx\right)^{\frac{p-1}{p}} + \int_{\partial\Omega_j} f\langle W,n_j(x)\rangle|u|^p d\Gamma_{\partial\Omega_j}(x),\end{aligned} \tag{8.23}$$

where Ω_j is the partition as in the proof of Theorem 8.3.1. In the last line we have used the Hölder inequality. By using Young's inequality in (8.23) and rearranging the terms, we get

$$\int_{\Omega_j} |\langle W,\nabla_H u\rangle|^p dx \geq \int_\Omega \left(\operatorname{div}_{\mathbb{G}}(fW) - (p-1)|f|^{\frac{p}{p-1}}\right)|u|^p dx - \int_{\partial\Omega_j} f\langle W,n_j(x)\rangle|u|^p d\Gamma_{\partial\Omega_j}(x). \tag{8.24}$$

Choosing $W := I_i$ as a unit vector of the ith component and letting

$$f = \beta\frac{|\langle X_i(x),\nu_j\rangle|^{p-1}}{\operatorname{dist}(x,\mathcal{F}_j)^{p-1}},$$

we calculate

$$\operatorname{div}_{\mathbb{G}}(Wf) = X_i f = \beta X_i \left(\frac{|\langle X_i(x),\nu_j\rangle|}{\operatorname{dist}(x,\mathcal{F}_j)}\right)^{p-1}$$

$$= \beta(p-1)\left(\frac{|\langle X_i(x),\nu_j\rangle|}{\operatorname{dist}(x,\mathcal{F}_j)}\right)^{p-2} X_i\left(\frac{\langle X_i(x),\nu_j\rangle}{\operatorname{dist}(x,\mathcal{F}_j)}\right)$$

$$= \beta(p-1)\left(\frac{|\langle X_i(x),\nu_j\rangle|}{\operatorname{dist}(x,\mathcal{F}_j)}\right)^{p-2}\left(\frac{X_i\langle X_i(x),\nu_j\rangle}{\operatorname{dist}(x,\mathcal{F}_j)} - \frac{|\langle X_i(x),\nu_j\rangle|^2}{\operatorname{dist}(x,\mathcal{F}_j)^2}\right)$$

$$= \beta(p-1)\left[\left(\frac{|\langle X_i(x),\nu_j\rangle|}{\operatorname{dist}(x,\mathcal{F}_j)}\right)^{p-2}\left(\frac{X_i\langle X_i(x),\nu_j\rangle}{\operatorname{dist}(x,\mathcal{F}_j)}\right) - \frac{|\langle X_i(x),\nu_j\rangle|^p}{\operatorname{dist}(x,\mathcal{F}_j)^p}\right],$$

and

$$|f|^{\frac{p}{p-1}} = |\beta|^{\frac{p}{p-1}}\frac{|\langle X_i(x),\nu_j\rangle|^p}{\operatorname{dist}(x,\mathcal{F}_j)^p}.$$

Moreover, we have

$$\langle W,\nabla_H u\rangle = (\overbrace{0,\ldots,1}^{i},\ldots,0)\cdot(X_1u,\ldots,X_iu,\ldots,X_Nu)^T = X_iu.$$

Substituting these into (8.24) and summing over all $i=1,\ldots,N$, we obtain

$$\begin{aligned}\int_{\Omega_j}\sum_{i=1}^N |X_iu|^p dx \geq &-(p-1)(|\beta|^{\frac{p}{p-1}}+\beta)\int_{\Omega_j}\sum_{i=1}^N\frac{|\langle X_i(x),\nu_j\rangle|^p}{\operatorname{dist}(x,\mathcal{F}_j)^p}|u|^p dx \\ &+\beta(p-1)\int_{\Omega_j}\sum_{i=1}^N\left(\frac{|\langle X_i(x),\nu_j\rangle|}{\operatorname{dist}(x,\mathcal{F}_j)}\right)^{p-2}\left(\frac{X_i\langle X_i(x),\nu_j\rangle}{\operatorname{dist}(x,\mathcal{F}_j)}\right)|u|^p dx \\ &-\beta\int_{\partial\Omega_j}\sum_{i=1}^N\left(\frac{|\langle X_i(x),\nu_j\rangle|}{\operatorname{dist}(x,\mathcal{F}_j)}\right)^{p-1}\langle X_i(x),n_j(x)\rangle|u|^p d\Gamma_{\partial\Omega_j}(x).\end{aligned} \tag{8.25}$$

Now summing up over Ω_j, and with the interior boundary terms we get

$$\begin{aligned}\int_{\Omega}\sum_{i=1}^N |X_iu|^p dx \geq &-(p-1)(|\beta|^{\frac{p}{p-1}}+\beta)\sum_{i=1}^N\int_{\Omega}\frac{|\langle X_i(x),\nu\rangle|^p}{\operatorname{dist}(x,\partial\Omega)^p}|u|^p dx \\ &+\beta(p-1)\sum_{i=1}^N\int_{\Omega}\left(\frac{|\langle X_i(x),\nu\rangle|}{\operatorname{dist}(x,\partial\Omega)}\right)^{p-2}\left(\frac{X_i\langle X_i(x),\nu\rangle}{\operatorname{dist}(x,\partial\Omega)}\right)|u|^p dx \\ &-\beta\sum_{j\neq l}\sum_{i=1}^N\int_{\Gamma_{jl}}\left(\frac{|\langle X_i(x),\nu_j\rangle|}{\operatorname{dist}(x,\mathcal{F}_j)}\right)^{p-1}\langle X_i(x),n_{jl}(x)\rangle|u|^p d\Gamma_{jl} \\ = &-(p-1)(|\beta|^{\frac{p}{p-1}}+\beta)\sum_{i=1}^N\int_{\Omega}\frac{|\langle X_i(x),\nu\rangle|^p}{\operatorname{dist}(x,\partial\Omega)^p}|u|^p dx \\ &+\beta(p-1)\sum_{i=1}^N\int_{\Omega}\left(\frac{|\langle X_i(x),\nu\rangle|}{\operatorname{dist}(x,\partial\Omega)}\right)^{p-2}\left(\frac{X_i\langle X_i(x),\nu\rangle}{\operatorname{dist}(x,\partial\Omega)}\right)|u|^p dx \\ &-\beta\sum_{j<l}\sum_{i=1}^N\int_{\Gamma_{jl}}\left[\left(\frac{|\langle X_i(x),\nu_j\rangle|}{\operatorname{dist}(x,\mathcal{F}_j)}\right)^{p-1}\langle X_i(x),n_{jl}(x)\rangle\right.\end{aligned}$$

$$
\left. - \left(\frac{|\langle X_i(x), \nu_l \rangle|}{\operatorname{dist}(x, \mathcal{F}_l)} \right)^{p-1} \langle X_i(x), n_{jl}(x) \rangle \right] |u|^p d\Gamma_{jl}.
$$

As in the proof of Theorem 8.3.1, if the boundary term is positive we can discard it, so we need to show that

$$
\left[\left(\frac{|\langle X_i(x), \nu_j \rangle|}{\operatorname{dist}(x, \mathcal{F}_j)} \right)^{p-1} \langle X_i(x), n_{jl}(x) \rangle - \left(\frac{|\langle X_i(x), \nu_l \rangle|}{\operatorname{dist}(x, \mathcal{F}_l)} \right)^{p-1} \langle X_i(x), n_{jl}(x) \rangle \right] \geq 0.
$$

Since $n_{jl} = \frac{\nu_j - \nu_l}{\sqrt{2 - 2\cos(\alpha_{jl})}}$ and $\operatorname{dist}(x, \mathcal{F}_j) = \operatorname{dist}(x, \mathcal{F}_l)$ on Γ_{jl}, we have

$$
\begin{aligned}
&\frac{1}{2 - 2\cos(\alpha_{jl})} \left[\left(\frac{|\langle X_i(x), \nu_j \rangle|}{\operatorname{dist}(x, \mathcal{F}_j)} \right)^{p-1} \langle X_i(x), \nu_j - \nu_l \rangle \right. \\
&\qquad \left. - \left(\frac{|\langle X_i(x), \nu_l \rangle|}{\operatorname{dist}(x, \mathcal{F}_l)} \right)^{p-1} \langle X_i(x), \nu_j - \nu_l \rangle \right] \\
&= \frac{|\langle X_i(x), \nu_j \rangle|^p - |\langle X_i(x), \nu_j \rangle|^{p-1} \langle X_i(x), \nu_l \rangle}{(2 - 2\cos(\alpha_{jl})) \operatorname{dist}(x, \mathcal{F}_j)^{p-1}} \\
&\qquad + \frac{-|\langle X_i(x), \nu_l \rangle|^{p-1} \langle X_i(x), \nu_j \rangle + |\langle X_i(x), \nu_l \rangle|^p}{(2 - 2\cos(\alpha_{jl})) \operatorname{dist}(x, \mathcal{F}_j)^{p-1}} \\
&= \frac{(|\langle X_i(x), \nu_j \rangle| - |\langle X_i(x), \nu_l \rangle|) (|\langle X_i(x), \nu_j \rangle|^{p-1} - |\langle X_i(x), \nu_l \rangle|^{p-1})}{(2 - 2\cos(\alpha_{jl})) \operatorname{dist}(x, \mathcal{F}_j)^{p-1}} \geq 0.
\end{aligned}
$$

Here we have used the equality

$$
(a - b)(a^{p-1} - b^{p-1}) = a^p - a^{p-1}b - b^{p-1}a + b^{p-1}
$$

with $a = |\langle X_i(x), \nu_j \rangle|$ and $b = |\langle X_i(x), \nu_l \rangle|$. Thus, for $\beta < 0$ by discarding the above boundary term (integral) we complete the proof. □

Chapter 9

Uncertainty Relations on Homogeneous Groups

In this chapter we discuss relations between main operators of quantum mechanics, that is, relations between momentum and position operators as well as Euler and Coulomb potential operators on homogeneous groups as well as their consequences. Since in most uncertainty relations and in these operators the appearing weights are radially symmetric, it turns out that these relations can be extended to also hold on general homogeneous groups. In particular, we obtain both isotropic and anisotropic uncertainty principles in a refined form, where the radial derivative operators are used instead of the elliptic or hypoelliptic differential operators.

Throughout this book, most of the inequalities imply the corresponding uncertainty principles. An example of such an uncertainty principle was given, for example, in Corollary 2.1.3 as a consequence of the Hardy inequality, and also in Corollary 3.3.5. However, in this chapter we aim at presenting an independent treatment of inequalities following from certain identities involving the appearing operators. In this respect such uncertainty relations can be sometimes obtained independently from Hardy inequalities in alternative ways, see, e.g., also Ciatti, Ricci and Sundari [CRS07].

In general, the uncertainty principles in different form have attracted a lot of attention due to their physical applications. For example, a fundamental element of the quantum mechanics is the uncertainty principle of Werner Heisenberg [Hei27]. It is worth observing that his original argument, while conceptually enlightening, was experiential.

Then Wolfgang Pauli and Hermann Weyl provided the mathematical aspects of uncertainty relations involving position and momentum operators, but the first rigorous proof was given by Earle Kennard [Ken27]. Charles Fefferman's work [Fef83] and [FP81] was a starting point of studies to widely present the interpretation of uncertainty inequalities as spectral properties of differential operators. Nowadays there is vast literature on uncertainty relations and their applications. Since we do not aim here at presenting a survey of the uncertainty relations on the

M. Ruzhansky, D. Suragan, *Hardy Inequalities on Homogeneous Groups*, Progress in Mathematics 327, https://doi.org/10.1007/978-3-030-02895-4_10

Euclidean space $\mathbb{R}^n$, we only refer to relevant works. In general, we can refer to a recent survey [CBTW15] for further discussions and references on this subject in the Euclidean setting, as well as to [FS97] for an overview of the history and the relevance of this type of inequalities from a purely mathematical point of view. The link between the uncertainty principles and the adapted Fourier analysis has been explored in [Tha04]. The exposition of the present chapter is based on [RS17f].

9.1 Abstract position and momentum operators

The idea for our presentation is to introduce abstract position and momentum operators $\mathcal{P}$ and $\mathcal{M}$ that satisfy certain relations. In particular, the classical position and momentum operators of quantum physics satisfy these assumptions. However, this abstract point of view allows one to take different versions of position-momentum pairs depending on the setting. We will exemplify such a possibility of different choices in the case of the Heisenberg group, see Example 9.1.4.

9.1.1 Definition and assumptions

Throughout this chapter, the abstract position and momentum operators $\mathcal{P}$ and $\mathcal{M}$ will be assumed to satisfy the following properties.

Definition 9.1.1 (Abstract position and momentum operators)**.** Let $\mathcal{P}$ and $\mathcal{M}$ be linear operators which are densely defined from $L^2(\mathbb{G})$ to $L^2(\mathbb{G})$, with their domains containing $C_0^\infty(\mathbb{G})$, and such that $C_0^\infty(\mathbb{G})$ is an invariant subspace for them, that is,

$$\mathcal{P}(C_0^\infty(\mathbb{G})) \subset C_0^\infty(\mathbb{G}) \text{ and } \mathcal{M}(C_0^\infty(\mathbb{G})) \subset C_0^\infty(\mathbb{G}).$$

We will say that such operators $\mathcal{P}$ and $\mathcal{M}$ are *abstract position and momentum operators* if they satisfy the relations

$$2\,\mathrm{Re}\left(\mathcal{P}f\overline{(i\mathcal{M})f}\right) = (\mathcal{P}\circ(i\mathcal{M}))|f|^2 = \mathbb{E}|f|^2 \tag{9.1}$$

for all $f \in C_0^\infty(\mathbb{G})$. We will denote by $D(\mathcal{P})$ and $D(\mathcal{M})$ the domains of operators $\mathcal{P}$ and $\mathcal{M}$, respectively.

Before giving examples, let us make some remarks concerning the meaning of the equalities in (9.1).

Remark 9.1.2.

1. The first equality in (9.1) gives a relation between the position and momentum operator; it will be clear from Example 9.1.3 that it is satisfied by the classical position and momentum operators of the Euclidean quantum mechanics. This condition is instrumental in establishing several further properties of these operators.

2. The second equality in (9.1) relates to position and momentum operators to the Euler operator $\mathbb{E}$ on $\mathbb{G}$, which was defined by
$$\mathbb{E} := |x|\mathcal{R}, \tag{9.2}$$
where $\mathcal{R}$ is the radial derivative operator, see Section 1.3.2 for a discussion of its properties. The main characterizing feature of the Euler operator is its responsibility for the homogeneity property on $\mathbb{G}$ given in Proposition 1.3.1, namely, that
$$f(\lambda x) = \lambda^\nu f(x) \text{ for all } \lambda > 0 \text{ if and only if } \mathbb{E}f = \nu f,$$
for any differentiable function f on $\mathbb{G}$. Thus, the assumption (9.1) for the abstract position and momentum operators says that they have to give the factorisation of the Euler operator $\mathbb{E}$ as in the second equality in (9.1). In this sense, the second equality in (9.1) relates position and momentum operators to the homogeneous structure of the group $\mathbb{G}$.
3. We can note that already in the anisotropic and even isotropic $\mathbb{R}^n$ the results of this chapter give some new insights in view of an arbitrary choice of a homogeneous quasi-norm $|\cdot|$ and the abstract nature of these operators.
4. It is rather curious that equalities (9.1) already imply uncertainty relations of several types, such as the Heisenberg–Kennard and Heisenberg–Pauli–Weyl type uncertainty inequalities. Moreover, the property that the operators $\mathcal{P}$ and $i\mathcal{M}$ factorise the Euler operator allows one to establish further relations between them and other operators such as the radial operator, the dilations generating operator, and the Coulomb potential operator, and prove some equalities and inequalities among them. Such relations are presented in this chapter.

9.1.2 Examples

If the group $\mathbb{G}$ is the Euclidean $\mathbb{R}^n$ with isotropic (standard) dilations and the usual Abelian structure, then the operators
$$\mathcal{P} := x \quad \text{and} \quad \mathcal{M} := -i\nabla, \tag{9.3}$$
i.e., the multiplication and the gradient (multiplied by $-i$), satisfy assumptions (9.1). The same will hold on general homogeneous groups, as we show in Example 9.1.3. Thus, we now give several other examples extending this to general homogeneous groups. In particular, we give an example of a choice of abstract position and momentum operators on general homogeneous groups. Furthermore, we show that other choices are possible, which we exemplify in the case of the Heisenberg group.

Example 9.1.3 (Position and momentum on general homogeneous groups). Let $\mathbb{G}$ be a homogeneous group. Let us define the operators

$$\mathcal{P} := x,\ x \in \mathbb{G}, \quad \text{and} \quad \mathcal{M} := -i\nabla_E, \tag{9.4}$$

where

$$\nabla_E = \left(\frac{\partial}{\partial x_1}, \ldots, \frac{\partial}{\partial x_n}\right)$$

is an anisotropic gradient on $\mathbb{G}$ consisting of partial derivatives with respect to coordinate functions. We understand the operator $\mathcal{P}$ as the scalar multiplication operator by the coordinates of the variable x, i.e.,

$$\mathcal{P}v = \sum x_j v_j,$$

where x_j are the coordinate functions of $x \in \mathbb{G}$, see Section 1.2.4 for a discussion of these functions on homogeneous groups.

These operators $\mathcal{P}$ and $\mathcal{M}$ are the position and momentum operators in the sense of Definition 9.1.1. Indeed, first we observe that by elementary properties of derivatives the first equality in the following relations is satisfied:

$$2\,\mathrm{Re}\,(xf \cdot \nabla_E f) = x \cdot \nabla_E |f|^2 = \mathbb{E}|f|^2. \tag{9.5}$$

The second equality in (9.5) follows if we recall that $\mathbb{E}$ is the Euler operator from (1.37), that is, we have the relations

$$\mathbb{E} = x \cdot \nabla_E \text{ and } \mathcal{R} = \frac{x \cdot \nabla_E}{|x|} = \frac{d}{d|x|},$$

see (1.35). In the notation (9.4) the relations (9.5) can be expressed as

$$2\,\mathrm{Re}\left(\mathcal{P}f\overline{i\mathcal{M}f}\right) = (\mathcal{P} \circ (i\mathcal{M}))|f|^2 = \mathbb{E}|f|^2, \tag{9.6}$$

showing that (9.1) is satisfied.

We note that the left invariant gradient $\nabla = \nabla_X = (X_1, \ldots, X_n)$ and the anisotropic (Euclidean) gradient ∇_E are related and can be expressed in terms of each other. For example, we can recall the relations

$$\frac{\partial}{\partial x_j} = X_j + \sum_{\substack{1 \leq k \leq n \\ \nu_j < \nu_k}} p_{j,k} X_k,$$

for some homogeneous polynomials $p_{j,k}$ on $\mathbb{G}$ of homogeneous degree $\nu_k - \nu_j > 0$, see Section 1.2.4.

Example 9.1.4 (Another choice of position and momentum operators on the Heisenberg group). Let us consider the Heisenberg group $\mathbb{H} = \mathbb{H}^1$, topologically equivalent to $\mathbb{R}^3$. As for general nilpotent Lie groups (see Proposition 1.1.1), the exponential map of $\mathbb{H}$ is globally invertible and its inverse map is given by the formula

$$\exp_{\mathbb{H}}^{-1}(x) = e(x) \cdot \nabla_X \equiv \sum_{j=1}^{3} e_j(x) X_j, \tag{9.7}$$

where $\nabla_X = (X_1, X_2, X_3)$ is the full gradient of $\mathbb{H}$ with

$$\begin{aligned} X_1 &= \frac{\partial}{\partial x_1} + 2x_2 \frac{\partial}{\partial x_3}, \\ X_2 &= \frac{\partial}{\partial x_2} - 2x_1 \frac{\partial}{\partial x_3}, \\ X_3 &= -4 \frac{\partial}{\partial x_3}, \end{aligned}$$

as well as

$$e(x) = (e_1(x), e_2(x), e_3(x)),$$

where

$$\begin{aligned} e_1(x) &= x_1, \\ e_2(x) &= x_2, \\ e_3(x) &= -\frac{1}{4} x_3. \end{aligned}$$

We define the position and momentum operators for this case to be

$$\mathcal{P} := e(x),\ x \in \mathbb{G}, \text{ and } \mathcal{M} := -i\nabla_X. \tag{9.8}$$

One can readily see that these operators satisfy the relations (9.6). Now let us check the relation (1.37) between the Euler operator $E_{\mathbb{H}} := e(x) \cdot \nabla_X$ and the radial operator $\mathcal{R}_{\mathbb{H}} = \frac{d}{d|x|}$:

$$\begin{aligned} E_{\mathbb{H}} &= e(x) \cdot \nabla_X \\ &= x_1 \left(\frac{\partial}{\partial x_1} + 2x_2 \frac{\partial}{\partial x_3} \right) + x_2 \left(\frac{\partial}{\partial x_2} - 2x_1 \frac{\partial}{\partial x_3} \right) - \frac{1}{4} x_3 \left(-4 \frac{\partial}{\partial x_3} \right) \\ &= x_1 \frac{\partial}{\partial x_1} + x_2 \frac{\partial}{\partial x_2} + x_3 \frac{\partial}{\partial x_3} \\ &= |x| \left(\frac{x_1}{|x|} \frac{\partial}{\partial x_1} + \frac{x_2}{|x|} \frac{\partial}{\partial x_2} + \frac{x_3}{|x|} \frac{\partial}{\partial x_3} \right) \\ &= |x| \frac{d}{d|x|} = |x| \mathcal{R}_{\mathbb{H}}. \end{aligned}$$

9.2 Position-momentum relations

In this section, we show further relations between abstract position $\mathcal{P}$ and momentum $\mathcal{M}$ operators on homogeneous groups. The obtained relations are the consequences of the homogeneous group's structure and the equalities (9.1).

9.2.1 Further position-momentum identities

We start with certain further identities involving the abstract position and momentum operators as a consequence of equalities (9.1).

Theorem 9.2.1 (Position-momentum identities). *Let $\mathbb{G}$ be a homogeneous group of homogeneous dimension $Q \geq 2$. Then for every $f \in D(\mathcal{P}) \bigcap D(\mathcal{M})$ with $\mathcal{P}f \not\equiv 0$ and $\mathcal{M}f \not\equiv 0$, we have the identity*

$$\begin{aligned}\|\mathcal{P}f\|^2_{L^2(\mathbb{G})} + \|\mathcal{M}f\|^2_{L^2(\mathbb{G})} &= Q\|f\|^2_{L^2(\mathbb{G})} + \|\mathcal{P}f - i\mathcal{M}f\|^2_{L^2(\mathbb{G})} \\ &= \|\mathcal{P}f\|_{L^2(\mathbb{G})}\|\mathcal{M}f\|_{L^2(\mathbb{G})}\left(2 - \left\|\frac{\mathcal{P}f}{\|\mathcal{P}f\|_{L^2(\mathbb{G})}} + \frac{i\mathcal{M}f}{\|\mathcal{M}f\|_{L^2(\mathbb{G})}}\right\|^2_{L^2(\mathbb{G})}\right) \\ &\quad + \|\mathcal{P}f + i\mathcal{M}f\|^2_{L^2(\mathbb{G})}. \end{aligned} \tag{9.9}$$

Proof of Theorem 9.2.1. It is enough to show (9.9) for functions $f \in C_0^\infty(\mathbb{G})$. Indeed, in this case, because $C_0^\infty(\mathbb{G})$ is dense in $L^2(\mathbb{G})$, it is also true on $D(\mathcal{P}) \bigcap D(\mathcal{M})$ by density. Using the polar decomposition from Proposition 1.2.10, the definition (1.30) of the radial operator, and equality (9.1), we calculate

$$\begin{aligned}-2\,\mathrm{Re}\int_{\mathbb{G}} \mathcal{P}f\overline{i\mathcal{M}f}dx &= -\int_{\mathbb{G}} \mathcal{P}i\mathcal{M}|f|^2 dx \\ &= -\int_0^\infty \int_{\wp} r^Q \frac{1}{r}\mathbb{E}|f|^2 d\sigma(y)dr \\ &= -\int_0^\infty \int_{\wp} r^Q \frac{d|f|^2}{dr} d\sigma(y)dr \\ &= Q\int_0^\infty \int_{\wp} r^{Q-1}|f|^2 d\sigma(y)dr \\ &= Q\int_{\mathbb{G}} |f|^2 dx \\ &= Q\|f\|^2_{L^2(\mathbb{G})}.\end{aligned}$$

Combining this with the equality

$$\|\mathcal{P}f\|^2_{L^2(\mathbb{G})} + \|\mathcal{M}f\|^2_{L^2(\mathbb{G})} = \|\mathcal{P}f + i\mathcal{M}f\|^2_{L^2(\mathbb{G})} - 2\,\mathrm{Re}\int_{\mathbb{G}} \mathcal{P}f\overline{i\mathcal{M}f}dx$$

we obtain the first equality in (9.9). On the other hand, we have

$$-2\,\mathrm{Re}\int_{\mathbb{G}} \mathcal{P}f\overline{i\mathcal{M}f}dx$$
$$= \|\mathcal{M}f\|_{L^2(\mathbb{G})}\|\mathcal{P}f\|_{L^2(\mathbb{G})}\left(2-\left\|\frac{\mathcal{P}f}{\|\mathcal{P}f\|_{L^2(\mathbb{G})}}+\frac{i\mathcal{M}f}{\|\mathcal{M}f\|_{L^2(\mathbb{G})}}\right\|^2_{L^2(\mathbb{G})}\right),$$

yielding the second equality in (9.9). □

9.2.2 Heisenberg–Kennard and Pythagorean inequalities

Immediately from (9.9) we observe the Heisenberg–Kennard type inequality as its consequence. In the Abelian case (see, e.g., [SZ97] and [WM08]) it is also sometimes called the Kennard uncertainty inequality.

Corollary 9.2.2 (Heisenberg–Kennard uncertainty principle). *We have*

$$\frac{Q}{2}\|f\|^2_{L^2(\mathbb{G})} \leq \|\mathcal{P}f\|_{L^2(\mathbb{G})}\|\mathcal{M}f\|_{L^2(\mathbb{G})}. \tag{9.10}$$

The first equality in (9.9) also implies the following Pythagorean type inequality:

Corollary 9.2.3 (Pythagorean type inequality). *We have*

$$\|\sqrt{Q}f\|^2_{L^2(\mathbb{G})} \leq \|\mathcal{P}f\|^2_{L^2(\mathbb{G})} + \|\mathcal{M}f\|^2_{L^2(\mathbb{G})}. \tag{9.11}$$

Equalities (9.9) also imply the following conditions for reaching the equalities in Heisenberg–Kennard and Pythagorean inequalities:

Corollary 9.2.4 (Equalities in Heisenberg–Kennard and Pythagorean inequalities). *Let $f \in D(\mathcal{P})\bigcap D(\mathcal{M})$ be such that $\mathcal{P}f \not\equiv 0$ and $\mathcal{M}f \not\equiv 0$.*

(i) *The equality case in the Heisenberg–Kennard uncertainty inequality* (9.10) *holds, that is,*

$$\frac{Q}{2}\|f\|^2_{L^2(\mathbb{G})} = \|\mathcal{P}f\|_{L^2(\mathbb{G})}\|\mathcal{M}f\|_{L^2(\mathbb{G})}$$

if and only if

$$\|\mathcal{P}f\|_{L^2(\mathbb{G})}i\mathcal{M}f = \|\mathcal{M}f\|_{L^2(\mathbb{G})}\mathcal{P}f.$$

(ii) *For $f \in D(\mathcal{P})\bigcap D(\mathcal{M})$ we have the Pythagorean equality*

$$\|\sqrt{Q}f\|^2_{L^2(\mathbb{G})} = \|\mathcal{P}f\|^2_{L^2(\mathbb{G})} + \|\mathcal{M}f\|^2_{L^2(\mathbb{G})}$$

if and only if

$$\mathcal{P}f = i\mathcal{M}f.$$

9.3 Euler–Coulomb relations

In addition to the Euler operator that was defined by

$$\mathbb{E}f := |x|\mathcal{R}f, \tag{9.12}$$

we also define the *Coulomb potential operator* as

$$\mathcal{C}f := \frac{1}{|x|}f. \tag{9.13}$$

The domains of these operators are given, respectively, by

$$D(\mathbb{E}) = \{f \in L^2(\mathbb{G}) : \ \mathbb{E}f \in L^2(\mathbb{G})\} \tag{9.14}$$

and

$$D(\mathcal{C}) = \{f \in L^2(\mathbb{G}) : \frac{1}{|x|}f \in L^2(\mathbb{G})\}. \tag{9.15}$$

It is also immediate to observe from (1.30) that the composition of the Euler operator and Coulomb operators gives the radial derivative operator $\mathcal{R}$:

$$\mathcal{R} := \mathcal{C}\mathbb{E}. \tag{9.16}$$

Recall that in Theorem 2.1.5 it was shown that for each $f \in C_0^\infty(\mathbb{G}\backslash\{0\})$ one has the identity

$$\left\|\frac{1}{|x|^\alpha}\mathcal{R}f\right\|^2_{L^2(\mathbb{G})} = \left(\frac{Q-2}{2}-\alpha\right)^2\left\|\frac{f}{|x|^{\alpha+1}}\right\|^2_{L^2(\mathbb{G})} + \left\|\frac{1}{|x|^\alpha}\mathcal{R}f + \frac{Q-2-2\alpha}{2|x|^{\alpha+1}}f\right\|^2_{L^2(\mathbb{G})}, \tag{9.17}$$

for all $\alpha \in \mathbb{R}$. If $\alpha = 0$ from (9.17) we obtain the equality

$$\|\mathcal{R}f\|^2_{L^2(\mathbb{G})} = \left(\frac{Q-2}{2}\right)^2\left\|\frac{1}{|x|}f\right\|^2_{L^2(\mathbb{G})} + \left\|\mathcal{R}f + \frac{Q-2}{2|x|}f\right\|^2_{L^2(\mathbb{G})}. \tag{9.18}$$

As it was already shown before, by dropping the non-negative last term in (9.18) we immediately obtain a version of L^2-Hardy's inequality on $\mathbb{G}$:

$$\left\|\frac{f}{|x|}\right\|_{L^2(\mathbb{G})} \leq \frac{2}{Q-2}\|\mathcal{R}f\|_{L^2(\mathbb{G})}, \quad Q \geq 3, \tag{9.19}$$

with the constant being sharp for any quasi-norm $|\cdot|$.

9.3.1 Heisenberg–Pauli–Weyl uncertainty principle

An L^p-version of the Heisenberg–Pauli–Weyl uncertainty principle was given in Corollary 3.3.5 for a particular choice of position and momentum operators. In this section we show its abstract version for abstract position and momentum operators, although restricting the consideration, as usual in this chapter, to the case of L^2-spaces.

Thus, by a standard argument the inequality (9.19) implies the following Heisenberg–Pauli–Weyl type uncertainty principle on homogeneous groups:

Proposition 9.3.1 (Heisenberg–Pauli–Weyl type uncertainty principle)**.** *Let $\mathbb{G}$ be a homogeneous group of homogeneous dimension $Q \geq 3$. Then for each $f \in C_0^\infty(\mathbb{G}\backslash\{0\})$ and any homogeneous quasi-norm $|\cdot|$ on $\mathbb{G}$ we have*

$$\|f\|^2_{L^2(\mathbb{G})} \leq \frac{2}{Q-2} \|\mathcal{R}f\|_{L^2(\mathbb{G})} \||x|f\|_{L^2(\mathbb{G})}. \tag{9.20}$$

Proof of Corollary 9.3.1. From the inequality (9.19) we get

$$\begin{aligned} &\left(\int_{\mathbb{G}} |\mathcal{R}f|^2\, dx\right)^{1/2} \left(\int_{\mathbb{G}} |x|^2|f|^2 dx\right)^{1/2} \\ &\geq \frac{Q-2}{2}\left(\int_{\mathbb{G}} \frac{|f|^2}{|x|^2}\, dx\right)^{1/2} \left(\int_{\mathbb{G}} |x|^2|f|^2 dx\right)^{1/2} \geq \frac{Q-2}{2}\int_{\mathbb{G}} |f|^2 dx, \end{aligned}$$

where we have used the Hölder inequality in the last line. This shows (9.20). ☐

Remark 9.3.2.

1. In the Abelian case $\mathbb{G} = (\mathbb{R}^n, +)$, we have $Q = n$, so that (9.20) implies the uncertainty principle with any homogeneous quasi-norm $|\cdot|$:

$$\left(\int_{\mathbb{R}^n} |f(x)|^2 dx\right)^2 \leq \left(\frac{2}{n-2}\right)^2 \int_{\mathbb{R}^n} \left|\frac{x}{|x|}\cdot \nabla f(x)\right|^2 dx \int_{\mathbb{R}^n} |x|^2|f(x)|^2 dx, \tag{9.21}$$

   which in turn implies the classical uncertainty principle for $\mathbb{G} \equiv \mathbb{R}^n$ with the standard Euclidean distance $|x|_E$:

$$\left(\int_{\mathbb{R}^n} |f(x)|^2 dx\right)^2 \leq \left(\frac{2}{n-2}\right)^2 \int_{\mathbb{R}^n} |\nabla f(x)|^2 dx \int_{\mathbb{R}^n} |x|_E^2|f(x)|^2 dx, \tag{9.22}$$

   which is the classical *Heisenberg–Pauli–Weyl uncertainty principle* on $\mathbb{R}^n$. For the improved constant in (9.22) see (9.39).

2. Different versions of this uncertainty principle have been considered in different settings, for example in those of stratified groups. We can refer to [GL90], [CRS07], [CCR15] for some results, and further estimates will be shown in Section 12.4.

Moreover, we have the following Pythagorean relation for the Euler operator:

Proposition 9.3.3 (Pythagorean relation for Euler operator). *Let $\mathbb{G}$ be a homogeneous group of homogeneous dimension $Q \geq 3$. Then we have*

$$\|\mathbb{E}f\|^2_{L^2(\mathbb{G})} = \left\|\frac{Q}{2}f\right\|^2_{L^2(\mathbb{G})} + \left\|\mathbb{E}f + \frac{Q}{2}f\right\|^2_{L^2(\mathbb{G})} \tag{9.23}$$

for any $f \in D(\mathbb{E})$.

Proof of Proposition 9.3.3. Taking $\alpha = -1$, from (9.17) we obtain (9.23) for any $f \in C_0^\infty(\mathbb{G}\backslash\{0\})$. Since $D(\mathbb{E}) \subset L^2(\mathbb{G})$ and $C_0^\infty(\mathbb{G}\backslash\{0\})$ is dense in $L^2(\mathbb{G})$, this implies that (9.23) is also true on $D(\mathbb{E})$ by density. □

Simply by dropping the positive term in the right-hand side, (9.23) implies

Corollary 9.3.4 (Lower bound for Euler operator). *Let $\mathbb{G}$ be a homogeneous group of homogeneous dimension $Q \geq 3$. Then we have*

$$\|f\|_{L^2(\mathbb{G})} \leq \frac{2}{Q}\|\mathbb{E}f\|_{L^2(\mathbb{G})}, \tag{9.24}$$

for any $f \in D(\mathbb{E})$.

9.4 Radial dilations – Coulomb relations

Using the radial derivative and Coulomb operators we can define the generator of dilations operator by

$$\mathcal{R}_g := -i\left(\mathcal{R} + \frac{Q-1}{2}\mathcal{C}\right) \tag{9.25}$$

with the domain

$$D(\mathcal{R}_g) = \{f \in L^2(\mathbb{G}) : \mathcal{R}f \in L^2(\mathbb{G}), \mathcal{C}f \in L^2(\mathbb{G})\}. \tag{9.26}$$

First we record a commutator relation between this generator of dilations operator $\mathcal{R}_g$ and the Coulomb potential operator:

Lemma 9.4.1 (Commutator relation between generator of dilations and Coulomb operators). *Let $\mathbb{G}$ be a homogeneous group of dimension $Q \geq 1$. Then for any $f \in C_0^\infty(\mathbb{G}\backslash\{0\})$ we have*

$$[\mathcal{R}_g, \mathcal{C}]f = i\mathcal{C}^2 f, \tag{9.27}$$

where $[\mathcal{R}_g, \mathcal{C}] = \mathcal{R}_g\mathcal{C} - \mathcal{C}\mathcal{R}_g$.

Proof of Lemma 9.4.1. Denoting $r := |x|$ we have $\mathcal{C} = \frac{1}{r}$, and from (1.30) it follows that $\mathcal{R}_g = -i\left(\frac{d}{dr} + \frac{Q-1}{2r}\right)$. Thus, a direct calculation shows

$$\begin{aligned}\mathcal{R}_g, \mathcal{C}tf &= \mathcal{R}_g\mathcal{C}f - \mathcal{C}\mathcal{R}_g f \\ &= -i\left(-\frac{1}{r^2} + \frac{1}{r}\frac{d}{dr} + \frac{Q-1}{2r^2} - \frac{1}{r}\frac{d}{dr} - \frac{Q-1}{2r^2}\right)f = i\frac{1}{r^2}f = i\mathcal{C}^2 f,\end{aligned}$$

establishing (9.27). □

We now analyse further properties of these operators.

Lemma 9.4.2 (Operators $\mathcal{R}_g$ and $\mathcal{C}$ are symmetric)**.** *Operators $\mathcal{R}_g$ and $\mathcal{C}$ are symmetric.*

Proof of Lemma 9.4.2. It is a straightforward to see that $\mathcal{C}$ is symmetric, that is,

$$\int_{\mathbb{G}} (\mathcal{C}f)\overline{f}dx = \int_{\mathbb{G}} f(\overline{\mathcal{C}f})dx.$$

Now we need to show that

$$\int_{\mathbb{G}} (\mathcal{R}_g f)\overline{f}dx = \int_{\mathbb{G}} f(\overline{\mathcal{R}_g f})dx \tag{9.28}$$

for any $f \in C_0^\infty(\mathbb{G}\backslash\{0\})$. Since $D(\mathcal{R}_g) \subset L^2(\mathbb{G})$ and $C_0^\infty(\mathbb{G}\backslash\{0\})$ are dense in $L^2(\mathbb{G})$ it follows that it is enough to show (9.28) on $C_0^\infty(\mathbb{G}\backslash\{0\})$ since it then follows also on $D(\mathcal{R}_g)$ by density. Using the polar decomposition from Proposition 1.2.10 and the expression $\mathcal{R}_g = -i\left(\frac{d}{dr} + \frac{Q-1}{2r}\right)$ we can calculate

$$\begin{aligned}
\int_{\mathbb{G}} (\mathcal{R}_g f)\overline{f}dx &= -i\int_0^\infty \int_{\wp} r^{Q-1}\left(\frac{df}{dr} + \frac{Q-1}{2r}f\right)\overline{f}d\sigma(y)dr \\
&= -i\int_0^\infty \int_{\wp} \frac{df}{dr}\overline{f}r^{Q-1}d\sigma(y)dr - i\frac{Q-1}{2}\int_0^\infty \int_{\wp} r^{Q-1}\frac{f}{r}\overline{f}d\sigma(y)dr \\
&= i\int_0^\infty \int_{\wp} f\frac{d\overline{f}}{dr}r^{Q-1}d\sigma(y)dr + i(Q-1)\int_0^\infty \int_{\wp} r^{Q-1}\frac{f}{r}\overline{f}d\sigma(y)dr \\
&\quad - i\frac{Q-1}{2}\int_0^\infty \int_{\wp} r^{Q-1}\frac{f}{r}\overline{f}d\sigma(y)dr \\
&= \int_0^\infty \int_{\wp} r^{Q-1}f\overline{\left(-i\frac{df}{dr} - i\frac{Q-1}{2r}f\right)}d\sigma(y)dr = \int_{\mathbb{G}} f\overline{\mathcal{R}_g f}d\nu,
\end{aligned}$$

proving that $\mathcal{R}_g$ is also symmetric. □

For any symmetric operators A and B in L^2 with domains $D(A)$ and $D(B)$, respectively, a straightforward calculation (see, e.g., [OY17, Theorem 2.1]) shows the equality

$$\begin{aligned}
&-i\int_{\mathbb{G}} ([A,B]f)\overline{f}d\nu \\
&= \|Af\|_{L^2(\mathbb{G})}\|Bf\|_{L^2(\mathbb{G})}\left(2 - \left\|\frac{Af}{\|Af\|_{L^2(\mathbb{G})}} + i\frac{Bf}{\|Bf\|_{L^2(\mathbb{G})}}\right\|^2_{L^2(\mathbb{G})}\right),
\end{aligned} \tag{9.29}$$

for $f \in D(A) \cap D(B)$ with $Af \not\equiv 0$ and $Bf \not\equiv 0$, which will be useful in our next proof.

Theorem 9.4.3 (Identities involving $\mathcal{R}$, $\mathcal{R}_g$, and $\mathcal{C}$)**.** *Let $\mathbb{G}$ be a homogeneous group of homogeneous dimension $Q \geq 3$. Then for every $f \in D(\mathcal{R}) \cap D(\mathcal{C})$ such that $f \not\equiv 0$ and $\mathcal{R}_g f \not\equiv 0$ we have*

$$\|\mathcal{R}f\|^2_{L^2(\mathbb{G})} = \|\mathcal{R}_g f\|^2_{L^2(\mathbb{G})} + \frac{(Q-1)(Q-3)}{4}\|\mathcal{C}f\|^2_{L^2(\mathbb{G})}, \tag{9.30}$$

and

$$\|\mathcal{C}f\|_{L^2(\mathbb{G})} = \|\mathcal{R}_g f\|_{L^2(\mathbb{G})}\left(2 - \left\|\frac{\mathcal{R}_g f}{\|\mathcal{R}_g f\|_{L^2(\mathbb{G})}} + i\frac{\mathcal{C}f}{\|\mathcal{C}f\|_{L^2(\mathbb{G})}}\right\|^2_{L^2(\mathbb{G})}\right). \tag{9.31}$$

Proof of Theorem 9.4.3. As in the proof of Theorem 9.2.1 we can calculate

$$\begin{aligned}
&\|\mathcal{R}_g f\|^2_{L^2(\mathbb{G})} \\
&= \left\|\mathcal{R}f + \frac{Q-1}{2|x|}f\right\|^2_{L^2(\mathbb{G})} \\
&= \|\mathcal{R}f\|^2_{L^2(\mathbb{G})} + (Q-1)\operatorname{Re}\int_{\mathbb{G}} (\mathcal{R}f)\overline{\frac{1}{|x|}f}dx + \left\|\frac{Q-1}{2|x|}f\right\|^2_{L^2(\mathbb{G})} \\
&= \|\mathcal{R}f\|^2_{L^2(\mathbb{G})} + (Q-1)\operatorname{Re}\int_0^\infty\int_{\wp} r^{Q-1}\left(\frac{d}{dr}f\right)\overline{\frac{1}{r}f}d\sigma(y)dr + \left\|\frac{Q-1}{2|x|}f\right\|^2_{L^2(\mathbb{G})} \\
&= \|\mathcal{R}f\|^2_{L^2(\mathbb{G})} + \frac{Q-1}{2}\int_0^\infty\int_{\wp} r^{Q-2}\frac{d}{dr}|f|^2 d\sigma(y)dr + \frac{(Q-1)^2}{4}\|\mathcal{C}f\|^2_{L^2(\mathbb{G})} \\
&= \|\mathcal{R}f\|^2_{L^2(\mathbb{G})} - \frac{(Q-1)(Q-2)}{2}\int_0^\infty\int_{\wp} r^{Q-1}\frac{1}{r^2}|f|^2 d\sigma(y)dr + \frac{(Q-1)^2}{4}\|\mathcal{C}f\|^2_{L^2(\mathbb{G})} \\
&= \|\mathcal{R}f\|^2_{L^2(\mathbb{G})} - \frac{(Q-1)(Q-2)}{2}\int_{\mathbb{G}}|\mathcal{C}f|^2dx + \frac{(Q-1)^2}{4}\|\mathcal{C}f\|^2_{L^2(\mathbb{G})} \\
&= \|\mathcal{R}f\|^2_{L^2(\mathbb{G})} - \frac{(Q-1)(Q-3)}{4}\|\mathcal{C}f\|^2_{L^2(\mathbb{G})}.
\end{aligned}$$

This proves (9.30). Using (9.27) and Lemma 9.4.2, in view of (9.29) we obtain

$$\begin{aligned}
\|\mathcal{C}f\|^2_{L^2(\mathbb{G})} &= -i\int_{\mathbb{G}}[\mathcal{R}_g, \mathcal{C}]f\overline{f}dx \\
&= \|\mathcal{R}_g f\|_{L^2(\mathbb{G})}\|\mathcal{C}f\|_{L^2(\mathbb{G})}\left(2 - \left\|\frac{\mathcal{R}_g f}{\|\mathcal{R}_g f\|_{L^2(\mathbb{G})}} + i\frac{\mathcal{C}f}{\|\mathcal{C}f\|_{L^2(\mathbb{G})}}\right\|^2_{L^2(\mathbb{G})}\right).
\end{aligned}$$

Since $C_0^\infty(\mathbb{G})$ is dense in $L^2(\mathbb{G})$, it implies that this equality is also true on $D(\mathcal{R}) \cap D(\mathcal{C})$ by density. □

The equality (9.30) implies the following estimates:

Corollary 9.4.4 (Estimates for $\mathcal{R}_g$ and $\mathcal{C}$). *Let $\mathbb{G}$ be a homogeneous group of homogeneous dimension $Q \geq 3$. The generator of dilations and Coulomb potential operator are bounded by the radial operator $\mathcal{R}$, that is, we have the estimates*

$$\|\mathcal{R}_g f\|_{L^2(\mathbb{G})} \leq \|\mathcal{R} f\|_{L^2(\mathbb{G})}, \tag{9.32}$$

and

$$\frac{\sqrt{(Q-1)(Q-3)}}{2}\|\mathcal{C} f\|_{L^2(\mathbb{G})} \leq \|\mathcal{R} f\|_{L^2(\mathbb{G})}, \tag{9.33}$$

for all $f \in D(\mathcal{R}) \cap D(\mathcal{C})$.

The equality (9.31) implies the following bound with an explicit constant, independent on the choice of a homogeneous norm on $\mathbb{G}$.

Corollary 9.4.5 (Bound of Coulomb operator by generator of dilations). *Let $\mathbb{G}$ be a homogeneous group of homogeneous dimension $Q \geq 3$. The Coulomb potential operator is bounded by the generator of dilations operator with relative bound 2, that is,*

$$\|\mathcal{C} f\|_{L^2(\mathbb{G})} \leq 2\|\mathcal{R}_g f\|_{L^2(\mathbb{G})}, \tag{9.34}$$

for all $f \in D(\mathcal{R}) \cap D(\mathcal{C})$ such that $\mathcal{R}_g f \not\equiv 0$.

9.5 Further weighted uncertainty type inequalities

In this section, we give an overview of a number of further uncertainty type inequalities.

Theorem 9.5.1. *For any quasi-norm $|\cdot|$, all differentiable $|\cdot|$-radial functions ϕ, all $p > 1$, $Q \geq 2$ with $\frac{1}{p} + \frac{1}{q} = 1$, and all $f \in C_0^1(\mathbb{G})$ we have*

$$\int_{\mathbb{G}} \frac{\phi'(|x|)}{|x|^{Q-1}}|f|^p dx \leq \int_{\mathbb{G}} |\mathcal{R} f|^p \, dx + \frac{p}{q}\int_{\mathbb{G}} \frac{|\phi(|x|)|^q}{|x|^{q(Q-1)}}|f|^p dx, \tag{9.35}$$

and

$$\int_{\mathbb{G}} \frac{\phi'(|x|)}{|x|^{Q-1}}|f|^p dx \leq p\left(\int_{\mathbb{G}} |\mathcal{R} f|^p \, dx\right)^{1/p}\left(\int_{\mathbb{G}} \frac{|\phi(|x|)|^q}{|x|^{q(Q-1)}}|f|^p dx\right)^{1/q}. \tag{9.36}$$

Before proving this theorem, let us point out several of its consequences.

Remark 9.5.2.

1. In (9.35) taking $\phi = \log|x|$ in the Euclidean (Abelian) case $\mathbb{G} = (\mathbb{R}^n, +)$, $n \geq 2$, we have $Q = n$, and taking $p = n \geq 2$, for any quasi-norm $|\cdot|$ on $\mathbb{R}^n$, it implies the new inequality

$$\int_{\mathbb{R}^n} \frac{|f|^n}{|x|^n} dx \leq \int_{\mathbb{R}^n} |\mathcal{R} f|^n \, dx + (n-1)\int_{\mathbb{R}^n} \frac{\left|\log\frac{1}{|x|}\right|^{\frac{n}{n-1}}}{|x|^n}|f|^n \, dx.$$

In turn, by using Schwarz' inequality with the standard Euclidean distance $|x|_E = \sqrt{x_1^2 + \cdots + x_n^2}$, it implies the 'critical' Hardy inequality

$$\int_{\mathbb{R}^n} \frac{|f|^n}{|x|_E^n} dx \leq \int_{\mathbb{R}^n} |\nabla f|^n \, dx + (n-1) \int_{\mathbb{R}^n} \frac{\left|\log \frac{1}{|x|_E}\right|^{\frac{n}{n-1}}}{|x|_E^n} |f|^n \, dx, \tag{9.37}$$

where ∇ is the standard gradient on $\mathbb{R}^n$. It is known that there is no positive constant C such that

$$\int_{\mathbb{R}^n} \frac{|f|^n}{|x|_E^n} dx \leq C \int_{\mathbb{R}^n} |\nabla f|^n \, dx$$

for all $f \in C_0^1(\mathbb{R}^n)$. Therefore, the appearance of a positive additional term (the second term) on the right-hand side of (9.37) seems essential. Critical inequalities for different versions of critical Hardy–Sobolev type inequalities have been investigated in [RS16a].

2. Note that this type of inequalities (Hardy–Sobolev type inequalities with an additional term on the right-hand side) can be applied, for example in the Euclidean case, to establish the existence and nonexistence of positive exponentially bounded weak solutions to a parabolic type operator perturbed by a critical singular potential (see, e.g., [ST18b]).
3. In (9.36), taking $\phi = |x|^n$ in the Euclidean (Abelian) case $\mathbb{G} = (\mathbb{R}^n, +)$, $n \geq 2$, we have $Q = n$, so for any quasi-norm $|\cdot|$ on $\mathbb{R}^n$ it implies the following uncertainty principle

$$\int_{\mathbb{R}^n} |f|^p dx \leq \frac{p}{n} \left(\int_{\mathbb{R}^n} |\mathcal{R} f|^p \, dx \right)^{\frac{1}{p}} \left(\int_{\mathbb{R}^n} |x|^{\frac{p}{p-1}} |f|^p dx \right)^{\frac{p-1}{p}}. \tag{9.38}$$

In turn, by using Schwarz' inequality with the standard Euclidean distance $|x|_E = \sqrt{x_1^2 + \cdots + x_n^2}$, it implies that

$$\int_{\mathbb{R}^n} |f|^p dx \leq \frac{p}{n} \left(\int_{\mathbb{R}^n} |\nabla f|^p \, dx \right)^{\frac{1}{p}} \left(\int_{\mathbb{R}^n} |x|_E^{\frac{p}{p-1}} |f|^p dx \right)^{\frac{p-1}{p}}, \tag{9.39}$$

where ∇ is the standard gradient on $\mathbb{R}^n$. In the case when $p = 2$ we have

$$\left(\int_{\mathbb{R}^n} |f|^2 dx \right)^2 \leq \left(\frac{2}{n} \right)^2 \int_{\mathbb{R}^n} |\nabla f|^2 \, dx \int_{\mathbb{R}^n} |x|_E^2 |f|^2 dx, \quad n \geq 2, \tag{9.40}$$

for all $f \in C_0^1(\mathbb{R}^n)$. Thus, when $n = 2$ inequality (9.40) gives the critical case of the *Heisenberg–Pauli–Weyl uncertainty principle* (9.22). Moreover, since $\frac{2}{n-2} \geq \frac{2}{n}$, $n \geq 3$, inequality (9.40) is an improved version of (9.22).

Note that equality case in (9.40) holds for the family of functions $f = C \exp(-b|x|_E)$, $b > 0$.
4. Uncertainty inequalities have been extended to many settings such as more general Lie groups and manifolds; see Folland and Sitaram [FS97] for more

information about older studies. On the general topic of uncertainty principles on groups and manifolds we can refer to, e.g., [CRS07], [VSCC92], [Tao05], among many others.

Proof of Theorem 9.5.1. By applying the polar decomposition formula in Proposition 1.2.10, and integration by parts, we obtain

$$\begin{aligned}\int_{\mathbb{G}} \frac{\phi'(|x|)}{|x|^{Q-1}}|f|^p dx &= \int_0^\infty \int_\wp |f|^p \frac{\phi'(r)}{r^{Q-1}} r^{Q-1} d\sigma(y) dr \\ &= \int_0^\infty \int_\wp |f|^p \frac{d}{dr}\phi(r) d\sigma(y) dr = -\int_0^\infty \int_\wp \phi(r) \frac{d}{dr}|f|^p d\sigma(y) dr \\ &= -\int_{\mathbb{G}} \frac{\phi(|x|)}{|x|^{Q-1}} \mathcal{R}|f|^p dx = -p\mathrm{Re} \int_{\mathbb{G}} \frac{\phi(|x|)|f|^{p-2}f}{|x|^{Q-1}} \overline{\mathcal{R}f} dx.\end{aligned}$$

Now by using Young's inequality for $p > 1$ and $\frac{1}{p} + \frac{1}{q} = 1$, we arrive at

$$\begin{aligned}\int_{\mathbb{G}} \frac{\phi'(|x|)}{|x|^{Q-1}}|f|^p dx &= -p\mathrm{Re} \int_{\mathbb{G}} \frac{\phi|f|^{p-2}f}{|x|^{Q-1}} \overline{\mathcal{R}f} dx \\ &\leq p \int_{\mathbb{G}} \frac{|\phi||f|^{p-1}}{|x|^{Q-1}} |\mathcal{R}f| dx \\ &\leq \int_{\mathbb{G}} |\mathcal{R}f|^p \, dx + \frac{p}{q} \int_{\mathbb{G}} \frac{|\phi(|x|)|^q}{|x|^{q(Q-1)}} |f|^p dx. \end{aligned} \tag{9.41}$$

This proves inequality (9.35). Furthermore, from (9.41) by using Hölder's inequality for $p > 1$ and $\frac{1}{p} + \frac{1}{q} = 1$, we establish

$$\begin{aligned}\int_{\mathbb{G}} \frac{\phi'(|x|)}{|x|^{Q-1}}|f|^p dx &\leq p \int_{\mathbb{G}} \frac{|\phi||f|^{p-1}}{|x|^{Q-1}} |\mathcal{R}f| dx \\ &\leq p \left(\int_{\mathbb{G}} |\mathcal{R}f|^p \, dx \right)^{1/p} \left(\int_{\mathbb{G}} \frac{|\phi(|x|)|^q}{|x|^{q(Q-1)}} |f|^p dx \right)^{1/q}.\end{aligned}$$

This completes the proof. □

Chapter 10

Function Spaces on Homogeneous Groups

In this chapter, we describe several function spaces on homogeneous groups. The origins of the extensive use of homogeneous groups in analysis go back to the book [FS82] of Folland and Stein where Hardy spaces on homogeneous groups have been thoroughly analysed. It turns out that several other function spaces can be defined on homogeneous groups since their main structural properties essentially depend only on the group and dilation structures. Thus, in this chapter we carry out such a construction for Morrey and Campanato spaces and analyse their main properties. Moreover, we describe a version of Sobolev spaces associated to the Euler operator. We call such spaces the Euler–Hilbert–Sobolev spaces.

The constructions of this chapter are based on the analysis in [RSY18d] and [RSY18c]. Since on general homogeneous groups we may not have a (hypoelliptic) differential operator to start with for the usual construction of Sobolev spaces, we develop a version of Sobolev spaces associated to the Euler operator. The development of such spaces is linked to relevant Hardy and Sobolev inequalities from Chapter 2 and Chapter 3. Consequently, we can also analyse properties of maximal operators and fractional integral operators in the constructed Morrey and Campanato spaces on homogeneous groups.

In Definition 6.5.4 and the subsequent analysis we have already considered a collection of (horizontal) weighted Sobolev type on stratified groups, but here we will concentrate on general homogeneous groups. Since horizontal gradients are not available in such a setting, its action will be replaced by that of the radial derivative combined with the corresponding Euler operator.

Thus, throughout this chapter $\mathbb{G}$ will denote a general homogeneous group of homogeneous dimension denoted by Q.

10.1 Euler–Hilbert–Sobolev spaces

In this section, we introduce an Euler–Hilbert–Sobolev space on a homogeneous group $\mathbb{G}$ of homogeneous dimension Q. We start with more general Euler–Sobolev

M. Ruzhansky, D. Suragan, *Hardy Inequalities on Homogeneous Groups*,
Progress in Mathematics 327, https://doi.org/10.1007/978-3-030-02895-4_11

function spaces. The definitions and further analysis are based on the Euler operator $\mathbb{E}$ discussed in Section 1.3.

Definition 10.1.1 (Euler-Sobolev function spaces). We define the *Euler–Sobolev function space* on $\mathbb{G}$ by

$$\mathfrak{L}^{k,p}(\mathbb{G}) := \overline{C_0^\infty(\mathbb{G}\backslash\{0\})}^{\|\cdot\|_{\mathfrak{L}^{k,p}(\mathbb{G})}}, \; k \in \mathbb{Z}, \tag{10.1}$$

i.e., as the completion of $C_0^\infty(\mathbb{G}\backslash\{0\})$ with respect to the semi-norm

$$\|f\|_{\mathfrak{L}^{k,p}(\mathbb{G})} := \|\mathbb{E}^k f\|_{L^p(\mathbb{G})}.$$

Let us recall a special case of inequality (3.87) with $\alpha = 0$, that is,

$$\|f\|_{L^p(\mathbb{G})} \leq \left(\frac{p}{Q}\right)^k \|\mathbb{E}^k f\|_{L^p(\mathbb{G})}, \; 1 < p < \infty, \; k \in \mathbb{N}. \tag{10.2}$$

From the definition of the Euler–Sobolev space it follows that this inequality extends to all functions $f \in \mathfrak{L}^{k,p}(\mathbb{G})$:

Corollary 10.1.2 (Embeddings of Euler–Sobolev spaces). *The semi-normed spaces* $(\mathfrak{L}^{k,p}, \|\cdot\|_{\mathfrak{L}^{k,p}})$, $k \in \mathbb{Z}$, *are complete spaces for any* $1 < p < \infty$. *The norm of the embedding operator* $\iota : (\mathfrak{L}^{k,p}, \|\cdot\|_{\mathfrak{L}^{k,p}}) \hookrightarrow (L^p, \|\cdot\|_{L^p})$ *satisfies*

$$\|\iota\|_{\mathfrak{L}^{k,p}\to L^p} \leq \left(\frac{p}{Q}\right)^k, \quad k \in \mathbb{N}, \; 1 < p < \infty, \tag{10.3}$$

where the embedding ι *is an embedding of a semi-normed subspace of* L^p.

Based on Lemma 1.3.2 we can use the general theory of fractional powers of operators as in [MS01, Chapter 5], to define fractional powers of the operator $\mathbb{A} = \mathbb{E}\mathbb{E}^*$, and we denote

$$|\mathbb{E}|^\beta := \mathbb{A}^{\frac{\beta}{2}}, \quad \beta \in \mathbb{C}.$$

For a brief and specific account of the relevant theory of fractional powers that is required for this construction we can refer the reader to the open access presentation in [FR16, Appendix A]. Consequently, we obtain the following fractional Hardy inequalities in $L^2(\mathbb{G})$.

Theorem 10.1.3 (Fractional Hardy inequalities in $L^2(\mathbb{G})$). *Let* $\mathbb{G}$ *be a homogeneous group of homogeneous dimension* $Q \geq 1$. *Let* $\beta \in \mathbb{C}_+$ *and let* $k > \frac{\mathrm{Re}\beta}{2}$ *be a positive integer. Then for all complex-valued functions* $f \in C_0^\infty(\mathbb{G}\backslash\{0\})$ *we have*

$$\|f\|_{L^2(\mathbb{G})} \leq C(k - \frac{\beta}{2}, k) \left(\frac{2}{Q}\right)^{\mathrm{Re}\beta} \left\| |\mathbb{E}|^\beta f \right\|_{L^2(\mathbb{G})}, \tag{10.4}$$

where

$$C(\beta, k) = \frac{\Gamma(k+1)}{|\Gamma(\beta)\Gamma(k-\beta)|} \frac{2^{k-\mathrm{Re}\beta}}{\mathrm{Re}\beta(k-\mathrm{Re}\beta)}. \tag{10.5}$$

Proof of Theorem 10.1.3. By using [MS01, Proposition 7.2.1, p. 176] we have the interpolation inequality

$$\left\| |\mathbb{E}|^{-\beta} f \right\|_{L^2(\mathbb{G})} \leq C(k - \frac{\beta}{2}, k) \, \|f\|_{L^2(\mathbb{G})}^{1-\frac{\mathrm{Re}\beta}{2k}} \, \|\mathbb{A}^{-k} f\|_{L^2(\mathbb{G})}^{\frac{\mathrm{Re}\beta}{2k}} . \tag{10.6}$$

By the equality (1.43) and (3.119) with $\alpha = 0$ it follows that

$$\begin{aligned} C(k - \frac{\beta}{2}, k) \, \|f\|_{L^2(\mathbb{G})}^{1-\frac{\mathrm{Re}\beta}{2k}} \, \|\mathbb{A}^{-k} f\|_{L^2(\mathbb{G})}^{\frac{\mathrm{Re}\beta}{2k}} &\leq C(k - \frac{\beta}{2}, k) \, \|f\|_{L^2(\mathbb{G})}^{1-\frac{\mathrm{Re}\beta}{2k}} \left(\frac{4}{Q^2}\right)^{\frac{\mathrm{Re}\beta}{2}} \|f\|_{L^2(\mathbb{G})}^{\frac{\mathrm{Re}\beta}{2k}}, \\ &= C(k - \frac{\beta}{2}, k) \left(\frac{4}{Q^2}\right)^{\frac{\mathrm{Re}\beta}{2}} \|f\|_{L^2(\mathbb{G})}, \end{aligned}$$

which combined with (10.6) implies (10.4). □

Definition 10.1.4 (Euler–Hilbert–Sobolev function spaces)**.** For $\beta \in \mathbb{C}_+$, we define the *Euler–Hilbert–Sobolev function space* on $\mathbb{G}$ by

$$\mathbb{H}^\beta(\mathbb{G}) := \overline{C_0^\infty(\mathbb{G}\backslash\{0\})}^{\|\cdot\|_{\mathbb{H}^\beta(\mathbb{G})}}, \tag{10.7}$$

that is, as the completion of $C_0^\infty(\mathbb{G}\backslash\{0\})$ with respect to the semi-norm

$$\|f\|_{\mathbb{H}^\beta(\mathbb{G})} := \||\mathbb{E}|^\beta f\|_{L^2(\mathbb{G})}.$$

In view of this definition we have inequality (10.4) for all $f \in \mathbb{H}^\beta(\mathbb{G})$:

$$\|f\|_{L^2(\mathbb{G})} \leq C(k - \frac{\beta}{2}, k) \left(\frac{2}{Q}\right)^{\mathrm{Re}\beta} \left\| |\mathbb{E}|^\beta f \right\|_{L^2(\mathbb{G})}, \tag{10.8}$$

where $\beta \in \mathbb{C}_+$, $k > \frac{\mathrm{Re}\beta}{2}$, $k \in \mathbb{N}$, and $C(k - \frac{\beta}{2}, k)$ is given by (10.5). We can summarize these facts as follows:

Proposition 10.1.5 (Embeddings of Euler–Hilbert–Sobolev spaces)**.** *Let $\mathbb{G}$ be a homogeneous group of homogeneous dimension $Q \geq 1$. For any $\beta \in \mathbb{C}$ the semi-normed space $(\mathbb{H}^\beta, \|\cdot\|_{\mathbb{H}^\beta})$ is a complete space. Moreover, the norm of the embedding operator $\iota : (\mathbb{H}^\beta, \|\cdot\|_{\mathbb{H}^\beta}) \hookrightarrow (L^2, \|\cdot\|_{L^2})$ satisfies*

$$\|\iota\|_{\mathbb{H}^\beta \to L^2} \leq C\left(k - \frac{\beta}{2}, k\right) \left(\frac{2}{Q}\right)^{\mathrm{Re}\beta}, \quad \beta \in \mathbb{C}_+, \ k > \frac{\mathrm{Re}\beta}{2}, \ k \in \mathbb{N}, \tag{10.9}$$

where we understand the embedding ι as an embedding of a semi-normed subspace of L^2.

10.1.1 Poincaré type inequality

Let $\Omega \subset \mathbb{G}$ be an open set and let $\widehat{\mathfrak{L}}_0^{1,p}(\Omega)$ be the completion of $C_0^\infty(\Omega\backslash\{0\})$ with respect to

$$\|f\|_{\widehat{\mathfrak{L}}^{1,p}(\Omega)} := \|f\|_{L^p(\Omega)} + \|\mathbb{E}f\|_{L^p(\Omega)}, \quad 1 < p < \infty.$$

Then we have the following completion of Hardy inequalities:

Theorem 10.1.6 (Poincaré type inequality on homogeneous groups). *Let Ω be a bounded open subset of a homogeneous group $\mathbb{G}$ of homogeneous dimension Q. If $1 < p < \infty$, $f \in \widehat{\mathfrak{L}}_0^{1,p}(\Omega)$ and $\mathcal{R}f \equiv \frac{1}{|x|}\mathbb{E}f \in L^p(\Omega)$, then we have*

$$\|f\|_{L^p(\Omega)} \leq \frac{Rp}{Q}\|\mathcal{R}f\|_{L^p(\Omega)} = \frac{Rp}{Q}\left\|\frac{1}{|x|}\mathbb{E}f\right\|_{L^p(\Omega)}, \tag{10.10}$$

where $R = \sup_{x\in\Omega}|x|$.

In order to prove Theorem 10.1.6, we first show the following Hardy inequality on open sets.

Lemma 10.1.7 (Hardy inequality on open sets). *Let $\Omega \subset \mathbb{G}$ be an open set. If $1 < p < \infty$, $f \in \widehat{\mathfrak{L}}_0^{1,p}(\Omega)$ and $\mathbb{E}f \in L^p(\Omega)$, then we have*

$$\|f\|_{L^p(\Omega)} \leq \frac{p}{Q}\|\mathbb{E}f\|_{L^p(\Omega)}. \tag{10.11}$$

Proof of Lemma 10.1.7. Let $\zeta : \mathbb{R} \to \mathbb{R}$ be an even smooth function satisfying

- $0 \leq \zeta \leq 1$,
- $\zeta(r) = 1$ if $|r| \leq 1$,
- $\zeta(r) = 0$ if $|r| \geq 2$.

For $\lambda > 0$, we set

$$\zeta_\lambda(x) := \zeta(\lambda|x|).$$

By (3.70) we already have inequality (10.11) for $f \in C_0^\infty(\mathbb{G}\backslash\{0\})$. There exists some $\{f_\ell\}_{\ell=1}^\infty \in C_0^\infty(\Omega\backslash\{0\})$ such that $f_\ell \to f$ in $\widehat{\mathfrak{L}}_0^{1,p}(\Omega)$ as $\ell \to \infty$. Let $\lambda > 0$. From (3.70) we obtain

$$\|\zeta_\lambda f_\ell\|_{L^p(\Omega)} \leq \frac{p}{Q}\left(\|(\mathbb{E}\zeta_\lambda)f_\ell\|_{L^p(\Omega)} + \|\zeta_\lambda(\mathbb{E}f_\ell)\|_{L^p(\Omega)}\right)$$

for all $\ell \geq 1$. It is easy to see that

$$\lim_{\ell\to\infty} \zeta_\lambda f_\ell = \zeta_\lambda f,$$
$$\lim_{\ell\to\infty} (\mathbb{E}\zeta_\lambda) f_\ell = (\mathbb{E}\zeta_\lambda) f,$$
$$\lim_{\ell\to\infty} \zeta_\lambda(\mathbb{E}f_\ell) = \zeta_\lambda(\mathbb{E}f)$$

in $L^p(\Omega)$. These properties imply that

$$\|\zeta_\lambda f\|_{L^p(\Omega)} \le \frac{p}{Q}\left\{\|(\mathbb{E}\zeta_\lambda)f\|_{L^p(\Omega)} + \|\zeta_\lambda(\mathbb{E}f)\|_{L^p(\Omega)}\right\}.$$

Since

$$|(\mathbb{E}\zeta_\lambda)(x)| \le \begin{cases} \sup|\mathbb{E}\zeta|, & \text{if } \lambda^{-1} < |x| < 2\lambda^{-1}, \\ 0, & \text{otherwise,} \end{cases}$$

we obtain (10.11) in the limit as $\lambda \to 0$. □

Proof of Theorem 10.1.6. Since $R = \sup_{x\in\Omega}|x|$ using Proposition 10.1.7 we obtain

$$\|f\|_{L^p(\Omega)} \le \frac{p}{Q}\|\mathbb{E}f\|_{L^p(\Omega)} \le \frac{Rp}{Q}\|\mathcal{R}f\|_{L^p(\Omega)} = \frac{Rp}{Q}\left\|\frac{1}{|x|}\mathbb{E}f\right\|_{L^p(\Omega)},$$

which gives (10.10). □

10.2 Sobolev–Lorentz–Zygmund spaces

In this section, we define several families of Lorentz type spaces and analyse their basic properties.

Definition 10.2.1 (Lorentz, Lorentz–Zygmund, and Sobolev–Lorentz–Zygmund spaces)**.** Let $\mathbb{G}$ be a homogeneous group of homogeneous dimension Q with a homogeneous quasi-norm $|\cdot|$. We define the *Lorentz type spaces* on $\mathbb{G}$ by

$$L_{|\cdot|,Q,p,q}(\mathbb{G}) := \{f \in L^1_{\mathrm{loc}}(\mathbb{G}) : \|f\|_{L_{|\cdot|,Q,p,q}(\mathbb{G})} < \infty\}, \quad 0 \le p, q \le \infty,$$

where

$$\|f\|_{L_{|\cdot|,Q,p,q}(\mathbb{G})} := \left(\int_{\mathbb{G}} (|x|^{\frac{Q}{p}}|f(x)|)^q \frac{1}{|x|^Q}dx\right)^{1/q}.$$

In the sequel, if the quasi-norm $|\cdot|$ on $\mathbb{G}$ is fixed we will often abbreviate the notation by writing

$$L_{p,q}(\mathbb{G}) := L_{|\cdot|,Q,p,q}(\mathbb{G}).$$

Moreover, we define the *Lorentz–Zygmund spaces* on $\mathbb{G}$ by

$$L_{p,q,\lambda}(\mathbb{G}) := \{f \in L^1_{\mathrm{loc}}(\mathbb{G}) : \|f\|_{L_{p,q,\lambda}(\mathbb{G})} < \infty\}, \quad 0 \le p, q \le \infty, \lambda \in \mathbb{R},$$

where

$$\|f\|_{L_{p,q,\lambda}(\mathbb{G})} := \sup_{R>0}\left(\int_{\mathbb{G}}\left(|x|^{\frac{Q}{p}}\left|\log\frac{R}{|x|}\right|^{\lambda}|f(x)|\right)^q \frac{1}{|x|^Q}dx\right)^{1/q}.$$

Furthermore, we define the *Sobolev–Lorentz–Zygmund spaces* by

$$W^1L_{p,q,\lambda}(\mathbb{G}) := \left\{ f \in L_{p,q,\lambda}(\mathbb{G}) : \frac{1}{|x|}\mathbb{E}f \in L_{p,q,\lambda}(\mathbb{G}) \right\},$$

endowed with the norm

$$\|\cdot\|_{W^1L_{p,q,\lambda}(\mathbb{G})} := \|\cdot\|_{L_{p,q,\lambda}(\mathbb{G})} + \left\| \frac{1}{|x|}\mathbb{E}\cdot \right\|_{L_{p,q,\lambda}(\mathbb{G})}.$$

We also define

$$W_0^1L_{p,q,\lambda}(\mathbb{G}) := \overline{C_0^\infty(\mathbb{G})}^{\|\cdot\|_{W^1L_{p,q,\lambda}(\mathbb{G})}}$$

as the completion of $C_0^\infty(\mathbb{G})$ with respect to $\|\cdot\|_{W^1L_{p,q,\lambda}(\mathbb{G})}$.

In addition, for $\lambda_1, \lambda_2 \in \mathbb{R}$ we introduce the Lorentz–Zygmund spaces involving the double logarithmic weights by

$$L_{p,q,\lambda_1,\lambda_2}(\mathbb{G}) := \{f \in L^1_{\mathrm{loc}}(\mathbb{G}) : \|f\|_{L_{p,q,\lambda_1,\lambda_2}(\mathbb{G})} < \infty\},$$

where

$$\|f\|_{L_{p,q,\lambda_1,\lambda_2}(\mathbb{G})} := \sup_{R>0} \left(\int_{\mathbb{G}} \left(|x|^{\frac{Q}{p}} \left| \log \frac{R}{|x|} \right|^{\lambda_1} \left| \log \left| \log \frac{R}{|x|} \right| \right|^{\lambda_2} |f(x)| \right)^q \frac{dx}{|x|^Q} \right)^{1/q}.$$

Remark 10.2.2. The space $L_{p,q,\lambda_1,\lambda_2}(\mathbb{G})$ extends the scale of the spaces $L_{p,q,\lambda}(\mathbb{G})$ and $L_{p,q}(\mathbb{G})$ in the sense that $L_{p,q,\lambda,0}(\mathbb{G}) = L_{p,q,\lambda}(\mathbb{G})$ and $L_{p,q,0,0}(\mathbb{G}) = L_{p,q}(\mathbb{G})$.

Similarly, we define the Sobolev–Lorentz–Zygmund spaces $W^1L_{p,q,\lambda_1,\lambda_2}(\mathbb{G})$ by

$$W^1L_{p,q,\lambda_1,\lambda_2}(\mathbb{G}) := \left\{ f \in L_{p,q,\lambda_1,\lambda_2}(\mathbb{G}) : \frac{1}{|x|}\mathbb{E}f \in L_{p,q,\lambda_1,\lambda_2}(\mathbb{G}) \right\}, \tag{10.12}$$

endowed with the norm

$$\|\cdot\|_{W^1L_{p,q,\lambda_1,\lambda_2}(\mathbb{G})} := \|\cdot\|_{L_{p,q,\lambda_1,\lambda_2}(\mathbb{G})} + \left\| \frac{1}{|x|}\mathbb{E}\cdot \right\|_{L_{p,q,\lambda_1,\lambda_2}(\mathbb{G})},$$

and

$$W_0^1L_{p,q,\lambda_1,\lambda_2}(\mathbb{G}) := \overline{C_0^\infty(\mathbb{G})}^{\|\cdot\|_{W^1L_{p,q,\lambda_1,\lambda_2}(\mathbb{G})}}. \tag{10.13}$$

Taking into account the special behaviour of functions

$$f_R(x) := f\left(R\frac{x}{|x|}\right),$$

we introduce the Lorentz–Zygmund type spaces $\mathfrak{L}_{p,q,\lambda}$ by

$$\mathfrak{L}_{p,q,\lambda}(\mathbb{G}) := \{f \in L^1_{\mathrm{loc}}(\mathbb{G}) : \|f\|_{\mathfrak{L}_{p,q,\lambda}(\mathbb{G})} < \infty\}, \quad \lambda \in \mathbb{R},$$

where

$$\|f\|_{\mathfrak{L}_{p,q,\lambda}(\mathbb{G})} := \sup_{R>0}\left(\int_{\mathbb{G}}\left(|x|^{\frac{Q}{p}}\left|\log\frac{R}{|x|}\right|^{\lambda}|f-f_R|\right)^q\frac{dx}{|x|^Q}\right)^{1/q}.$$

For $p=\infty$ we define

$$\|f\|_{\mathfrak{L}_{\infty,q,\lambda}(\mathbb{G})} := \sup_{R>0}\left(\int_{\mathbb{G}}\left(\left|\log\frac{R}{|x|}\right|^{\lambda}|f-f_R|\right)^q\frac{dx}{|x|^Q}\right)^{1/q}.$$

Moreover, we define the Lorentz–Zygmund type spaces $\mathfrak{L}_{p,q,\lambda_1,\lambda_2}(\mathbb{G})$ by

$$\mathfrak{L}_{p,q,\lambda_1,\lambda_2}(\mathbb{G}) := \{f\in L^1_{\mathrm{loc}}(\mathbb{G}) : \|f\|_{\mathfrak{L}_{p,q,\lambda_1,\lambda_2}(\mathbb{G})}<\infty\}, \tag{10.14}$$

where

$$\begin{aligned}&\|f\|_{\mathfrak{L}_{p,q,\lambda_1,\lambda_2}(\mathbb{G})}\\ &:= \sup_{R>0}\Bigg(\int_{\mathbb{G}}\left(|x|^{\frac{Q}{p}}\left|\log\frac{eR}{|x|}\right|^{\lambda_1}\left|\log\left|\log\frac{eR}{|x|}\right|\right|^{\lambda_2}\right.\\ &\qquad\times\left(\chi_{B(0,eR)}(x)|f-f_R|+\chi_{B^c(0,eR)}(x)|f-f_{e^2R}|\right)\Bigg)^q\frac{dx}{|x|^Q}\Bigg)^{1/q},\end{aligned}$$

$$\chi_{B(0,eR)}(x)=\begin{cases}1, & x\in B(0,eR);\\ 0, & x\notin B(0,eR).\end{cases}$$

For $p=\infty$ we define

$$\begin{aligned}\|f\|_{\mathfrak{L}_{\infty,q,\lambda_1,\lambda_2}(\mathbb{G})} &:= \sup_{R>0}\Bigg(\int_{\mathbb{G}}\Bigg(\left|\log\frac{eR}{|x|}\right|^{\lambda_1}\left|\log\left|\log\frac{eR}{|x|}\right|\right|^{\lambda_2}\\ &\qquad\times\left(\chi_{B(0,eR)}(x)|f-f_R|+\chi_{B^c(0,eR)}(x)|f-f_{e^2R}|\right)\Bigg)^q\frac{dx}{|x|^Q}\Bigg)^{1/q}.\end{aligned}$$

We now show several embeddings between the Sobolev–Lorentz–Zygmund spaces.

Theorem 10.2.3 (Embeddings of Sobolev–Lorentz–Zygmund spaces)**.** *For all* $1<\gamma<\infty$ *and* $\max\{1,\gamma-1\}<q<\infty$ *we have the continuous embedding*

$$W_0^1L_{Q,q,\frac{q-1}{q},\frac{q-\gamma}{q}}(\mathbb{G})\hookrightarrow\mathfrak{L}_{\infty,q,-\frac{1}{q},-\frac{\gamma}{q}}(\mathbb{G}).$$

In particular, for all $f \in W_0^1 L_{Q,q,\frac{q-1}{q},\frac{q-\gamma}{q}}(\mathbb{G})$ *and for any* $R > 0$ *the following inequality holds*

$$\left(\int_{\mathbb{G}} \frac{\chi_{B(0,eR)}(x)|f-f_R|^q + \chi_{B^c(0,eR)}(x)|f-f_{e^2R}|^q}{\left|\log\left|\log\frac{eR}{|x|}\right|\right|^{\gamma}\left|\log\frac{eR}{|x|}\right|}\frac{dx}{|x|^Q}\right)^{1/q} \leq \frac{q}{\gamma-1}\left(\int_{\mathbb{G}} |x|^{q-Q}\left|\log\frac{eR}{|x|}\right|^{q-1}\left|\log\left|\log\frac{eR}{|x|}\right|\right|^{q-\gamma}\left|\frac{1}{|x|}\mathbb{E}f\right|^q dx\right)^{1/q}, \tag{10.15}$$

where the embedding constant $\frac{q}{\gamma-1}$ *is sharp and where we denote* $f_R(x) := f(R\frac{x}{|x|})$.

Remark 10.2.4.

1. The function spaces extend with respect to indices some known results for spaces in the Abelian case $\mathbb{R}^n$ analysed in [MOW15b]. Embeddings of such Euclidean spaces for some indices were considered in [Wad14], and logarithmic type Hardy inequalities in Euclidean Sobolev–Zygmund spaces were investigated in [MOW13a].
2. Despite the fact that the integrand on the right-hand side of (10.15) has singularities for $|x| = R$, $|x| = eR$ and $|x| = e^2R$ as it will be clear from the proof we do not need to subtract the boundary value of functions on $|x| = eR$ on the left-hand side.

In order to prove Theorem 10.2.3, let us first establish the following estimate.

Proposition 10.2.5. *Let* $Q \in \mathbb{N}$, $1 < \gamma < \infty$, *and assume that* $\max\{1, \gamma-1\} < q < \infty$. *Then for all* $f \in C_0^\infty(\mathbb{G})$ *and any* $R > 0$ *we have the inequality*

$$\left(\int_{B(0,eR)} \frac{|f-f_R|^q}{\left|\log\left|\log\frac{eR}{|x|}\right|\right|^{\gamma}\left|\log\frac{eR}{|x|}\right|}\frac{dx}{|x|^Q}\right)^{1/q} \leq \frac{q}{\gamma-1}\left(\int_{B(0,eR)} |x|^{q-Q}\left|\log\frac{eR}{|x|}\right|^{q-1}\left|\log\left|\log\frac{eR}{|x|}\right|\right|^{q-\gamma}\left|\frac{1}{|x|}\mathbb{E}f\right|^q dx\right)^{1/q}. \tag{10.16}$$

Proof of Proposition 10.2.5. Using the polar coordinates as in Proposition 1.2.10 and integrating by parts in a quasi-ball $B(0,R)$ we have

$$\int_{B(0,R)} \frac{|f-f_R|^q}{\left|\log\left|\log\frac{eR}{|x|}\right|\right|^{\gamma}\left|\log\frac{eR}{|x|}\right|}\frac{dx}{|x|^Q} = \int_0^R \frac{1}{r(\log\frac{eR}{r})(\log(\log\frac{eR}{r}))^{\gamma}}\int_{\wp}|f(ry)-f(Ry)|^q d\sigma(y)dr$$

$$
\begin{aligned}
&= \frac{1}{\gamma-1}\left[\left(\log\left(\log\frac{eR}{r}\right)\right)^{-\gamma+1}\int_{\wp}|f(ry)-f(Ry)|^q d\sigma(y)\right]_{r=0}^{r=R} \\
&\quad - \frac{1}{\gamma-1}\int_0^R\left(\log\left(\log\frac{eR}{r}\right)\right)^{-\gamma+1}\frac{d}{dr}\int_{\wp}|f(ry)-f(Ry)|^q d\sigma(y)dr \\
&= -\frac{q}{\gamma-1}\int_0^R\left(\log\left(\log\frac{eR}{r}\right)\right)^{-\gamma+1}\operatorname{Re}\int_{\wp}|f(ry)-f(Ry)|^{q-2} \\
&\quad \times (f(ry)-f(Ry))\overline{\frac{df(ry)}{dr}}d\sigma(y)dr.
\end{aligned}
$$

In the above calculation, since $q-\gamma+1>0$, the boundary term at $r=R$ vanishes due to inequalities

$$
\begin{aligned}
\log\left(\log\frac{eR}{r}\right) &= \int_1^{\log\frac{eR}{r}}\frac{dt}{t} \geq \frac{\log\frac{eR}{r}-1}{\log\frac{eR}{r}} = \frac{\log\frac{R}{r}}{\log\frac{eR}{r}} \\
&= \frac{1}{\log\frac{eR}{r}}\int_1^{\frac{R}{r}}\frac{dt}{t} \geq \frac{1}{\log\frac{eR}{r}}\frac{\frac{R}{r}-1}{\frac{R}{r}} = \frac{R-r}{R\log\frac{eR}{r}}
\end{aligned}
$$

and

$$
|f(ry)-f(Ry)| \leq C(R-r),
$$

for $0<r\leq R$. It follows that

$$
\begin{aligned}
&\int_0^R \frac{1}{r(\log\frac{eR}{r})(\log(\log\frac{eR}{r}))^{\gamma}}\int_{\wp}|f(ry)-f(Ry)|^q d\sigma(y)dr \\
&\leq \frac{q}{\gamma-1}\int_0^R\frac{1}{(\log(\log\frac{eR}{r}))^{\gamma-1}}\int_{\wp}|f(ry)-f(Ry)|^{q-1}\left|\frac{df(ry)}{dr}\right|d\sigma(y)dr \\
&= \frac{q}{\gamma-1} \\
&\quad\times\int_0^R \frac{1}{r^{\frac{q-1}{q}}(\log\frac{eR}{r})^{\frac{q-1}{q}}r^{-\frac{q-1}{q}}(\log\frac{eR}{r})^{-\frac{q-1}{q}}}\frac{1}{(\log(\log\frac{eR}{r}))^{\frac{(q-1)\gamma}{q}}(\log(\log\frac{eR}{r}))^{\frac{\gamma-q}{q}}} \\
&\quad\times\int_{\wp}|f(ry)-f(Ry)|^{q-1}\left|\frac{df(ry)}{dr}\right|d\sigma(y)dr.
\end{aligned}
$$

By the Hölder inequality, we get

$$
\begin{aligned}
&\int_0^R \frac{1}{r(\log\frac{eR}{r})(\log(\log\frac{eR}{r}))^{\gamma}}\int_{\wp}|f(ry)-f(Ry)|^q d\sigma(y)dr \\
&\leq \frac{q}{\gamma-1}\left(\int_0^R\int_{\wp}\frac{|f(ry)-f(Ry)|^q}{r(\log\frac{eR}{r})(\log(\log\frac{eR}{r}))^{\gamma}}d\sigma(y)dr\right)^{(q-1)/q} \\
&\quad\times\left(\int_0^R\int_{\wp}r^{q-1}\left(\log\frac{eR}{r}\right)^{q-1}\left(\log\left(\log\frac{eR}{r}\right)\right)^{q-\gamma}\left|\frac{df(ry)}{dr}\right|^q d\sigma(y)dr\right)^{1/q}.
\end{aligned}
$$

This gives

$$\left(\int_{B(0,R)} \frac{|f-f_R|^q}{\left|\log\left|\log\frac{eR}{|x|}\right|\right|^{\gamma}\left|\log\frac{eR}{|x|}\right|}\frac{dx}{|x|^Q}\right)^{1/q} \tag{10.17}$$
$$\leq \frac{q}{\gamma-1}\left(\int_{B(0,R)} |x|^{q-Q}\left|\log\frac{eR}{|x|}\right|^{q-1}\left|\log\left|\log\frac{eR}{|x|}\right|\right|^{q-\gamma}\left|\frac{1}{|x|}\mathbb{E}f\right|^q dx\right)^{1/q}.$$

Now we calculate the integrals in (10.16) restricted on $B(0,eR)\backslash B(0,R)$:

$$\int_{B(0,eR)\backslash B(0,R)} \frac{|f-f_R|^q}{\left|\log\left|\log\frac{eR}{|x|}\right|\right|^{\gamma}\left|\log\frac{eR}{|x|}\right|}\frac{dx}{|x|^Q}$$
$$= \int_R^{eR} \frac{1}{r(\log\frac{eR}{r})(\log((\log\frac{eR}{r})^{-1}))^{\gamma}}\int_{\wp}|f(ry)-f(Ry)|^q d\sigma(y)dr$$
$$= -\frac{1}{\gamma-1}\left[\left(\log\left(\left(\log\frac{eR}{r}\right)^{-1}\right)\right)^{-\gamma+1}\int_{\wp}|f(ry)-f(Ry)|^q d\sigma(y)\right]_{r=R}^{r=eR}$$
$$+\frac{1}{\gamma-1}\int_R^{eR}\left(\log\left(\left(\log\frac{eR}{r}\right)^{-1}\right)\right)^{-\gamma+1}\frac{d}{dr}\int_{\wp}|f(ry)-f(Ry)|^q d\sigma(y)dr$$
$$= \frac{q}{\gamma-1}\int_R^{eR}\left(\log\left(\left(\log\frac{eR}{r}\right)^{-1}\right)\right)^{-\gamma+1}\operatorname{Re}\int_{\wp}|f(ry)-f(Ry)|^{q-2}$$
$$\times (f(ry)-f(Ry))\overline{\frac{df(ry)}{dr}}d\sigma(y)dr.$$

Again, in the calculation above, since $q-\gamma+1>0$, the boundary term at $r=R$ vanishes due to inequalities

$$\log\left(\left(\log\frac{eR}{r}\right)^{-1}\right) = \int_1^{(\log\frac{eR}{r})^{-1}}\frac{dt}{t}$$
$$\geq \left(\log\frac{eR}{r}\right)\left(\left(\log\frac{eR}{r}\right)^{-1}-1\right)$$
$$= 1-\log\frac{eR}{r} \geq \frac{r-R}{R}$$

and

$$|f(ry)-f(Ry)| \leq C(R-r),$$

for $R \leq r \leq eR$. This implies

$$\int_R^{eR} \frac{1}{r(\log \frac{eR}{r})(\log((\log \frac{eR}{r})^{-1}))^{\gamma}} \int_{\wp} |f(ry) - f(Ry)|^q d\sigma(y) dr$$

$$\leq \frac{q}{\gamma - 1} \int_R^{eR} \left(\log \left(\left(\log \frac{eR}{r} \right)^{-1} \right) \right)^{-\gamma+1}$$

$$\times \int_{\wp} |f(ry) - f(Ry)|^{q-1} \left| \frac{df(ry)}{dr} \right| d\sigma(y) dr = \frac{q}{\gamma - 1}$$

$$\times \int_R^{eR} \frac{1}{r^{\frac{q-1}{q}} (\log \frac{eR}{r})^{\frac{q-1}{q}} r^{-\frac{q-1}{q}} (\log \frac{eR}{r})^{-\frac{q-1}{q}}}$$

$$\times \frac{1}{\left(\log\left((\log \frac{eR}{r})^{-1}\right)\right)^{\frac{(q-1)\gamma}{q}} \left(\log\left((\log \frac{eR}{r})^{-1}\right)\right)^{\frac{\gamma-q}{q}}}$$

$$\times \int_{\wp} |f(ry) - f(Ry)|^{q-1} \left| \frac{df(ry)}{dr} \right| d\sigma(y) dr.$$

By using the Hölder inequality, we get

$$\int_R^{eR} \frac{1}{r(\log \frac{eR}{r})(\log((\log \frac{eR}{r})^{-1}))^{\gamma}} \int_{\wp} |f(ry) - f(Ry)|^q d\sigma(y) dr$$

$$\leq \frac{q}{\gamma - 1} \left(\int_R^{eR} \int_{\wp} \frac{|f(ry) - f(Ry)|^q}{r(\log \frac{eR}{r})(\log((\log \frac{eR}{r})^{-1}))^{\gamma}} d\sigma(y) dr \right)^{(q-1)/q}$$

$$\times \left(\int_R^{eR} \int_{\wp} r^{q-1} \left(\log \frac{eR}{r} \right)^{q-1} \left(\log \left(\left(\log \frac{eR}{r} \right)^{-1} \right) \right)^{q-\gamma} \right.$$

$$\left. \times \left| \frac{df(ry)}{dr} \right|^q d\sigma(y) dr \right)^{1/q}.$$

Thus, we arrive at

$$\left(\int_{B(0,eR)\setminus B(0,R)} \frac{|f - f_R|^q}{\left|\log \left|\log \frac{eR}{|x|}\right|\right|^{\gamma} \left|\log \frac{eR}{|x|}\right|} \frac{dx}{|x|^Q} \right)^{1/q}$$

$$\leq \frac{q}{\gamma - 1} \left(\int_{B(0,eR)\setminus B(0,R)} |x|^{q-Q} \left|\log \frac{eR}{|x|}\right|^{q-1} \left|\log \left|\log \frac{eR}{|x|}\right|\right|^{q-\gamma} \left| \frac{1}{|x|} \mathbb{E} f \right|^q dx \right)^{1/q}.$$

This and (10.17) imply (10.16). □

By a similar argument one can prove a dual inequality to (10.16), that is, we have

Proposition 10.2.6. *Let* $1 < \gamma < \infty$ *and* $\max\{1, \gamma - 1\} < q < \infty$. *Then for all* $f \in C_0^\infty(\mathbb{G})$ *and for any* $R > 0$ *we have*

$$\left(\int_{B^c(0,R)} \frac{|f - f_{eR}|^q}{\left|\log\left|\log\frac{R}{|x|}\right|\right|^\gamma \left|\log\frac{R}{|x|}\right|} \frac{dx}{|x|^Q}\right)^{1/q} \tag{10.18}$$
$$\leq \frac{q}{\gamma - 1}\left(\int_{B^c(0,R)} |x|^{q-Q}\left|\log\frac{R}{|x|}\right|^{q-1}\left|\log\left|\log\frac{R}{|x|}\right|\right|^{q-\gamma}\left|\frac{1}{|x|}\mathbb{E}f\right|^q dx\right)^{1/q}.$$

Now we are ready to prove Theorem 10.2.3.

Proof of Theorem 10.2.3. By using (10.18) with R replaced by eR we have

$$\left(\int_{B^c(0,eR)} \frac{|f - f_{e^2R}|^q}{\left|\log\left|\log\frac{eR}{|x|}\right|\right|^\gamma \left|\log\frac{eR}{|x|}\right|} \frac{dx}{|x|^Q}\right)^{1/q} \tag{10.19}$$
$$\leq \frac{q}{\gamma - 1}\left(\int_{B^c(0,eR)} |x|^{q-Q}\left|\log\frac{eR}{|x|}\right|^{q-1}\left|\log\left|\log\frac{eR}{|x|}\right|\right|^{q-\gamma}\left|\frac{1}{|x|}\mathbb{E}f\right|^q dx\right)^{1/q}.$$

Then from (10.16) and (10.19) we get (10.15) for functions $f \in C_0^\infty(\mathbb{G})$.

Let us now show (10.15) for general functions $f \in W_0^1 L_{Q,q,\frac{q-1}{q},\frac{q-\gamma}{q}}(\mathbb{G})$. First let us verify that (10.16) holds for $f \in W_0^1 L_{Q,q,\frac{q-1}{q},\frac{q-\gamma}{q}}(\mathbb{G})$. Let $\{f_m\} \subset C_0^\infty(\mathbb{G})$ be a sequence such that $f_m \to f$ in $W_0^1 L_{Q,q,\frac{q-1}{q},\frac{q-\gamma}{q}}(\mathbb{G})$ as $m \to \infty$ and almost everywhere by the definition (10.13). If we define

$$f_{R,m}(x) := \frac{f_m(x) - f_m(R\frac{x}{|x|})}{\left|\log\left|\log\frac{eR}{|x|}\right|\right|^{\frac{\gamma}{q}}\left|\log\frac{eR}{|x|}\right|^{\frac{1}{q}}},$$

then $\{f_{R,m}\}_{m\in\mathbb{N}}$ is a Cauchy sequence in $L^q(\mathbb{G}; \frac{dx}{|x|^Q})$, which is a weighted Lebesgue space, since the inequality (10.16) holds for $f_m - f_k \in C_0^\infty(\mathbb{G})$. Consequently, there exists $g_R \in L^q(\mathbb{G}; \frac{dx}{|x|^Q})$ such that $f_{R,m} \to g_R$ in $L^q(G; \frac{dx}{|x|^Q})$ as $m \to \infty$. From the inclusion

$$\left\{x \in \mathbb{G}\backslash\{0\} : f_m\left(R\frac{x}{|x|}\right) \nrightarrow f\left(R\frac{x}{|x|}\right)\right\}$$
$$\subset \bigcup_{r>0}\left\{x \in \mathbb{G}\backslash\{0\} : f_m\left(r\frac{x}{|x|}\right) \nrightarrow f\left(r\frac{x}{|x|}\right)\right\} = \{x \in \mathbb{G}\backslash\{0\} : f_m(x) \nrightarrow f(x)\},$$

it follows that $f_m\left(R\frac{x}{|x|}\right) \to f\left(R\frac{x}{|x|}\right)$, that is, we obtain the equality

$$\frac{f(x) - f(R\frac{x}{|x|})}{\left|\log\left|\log\frac{eR}{|x|}\right|\right|^{\gamma/q}\left|\log\frac{eR}{|x|}\right|^{1/q}} = g_R(x)$$

almost everywhere.

That is, inequality (10.16) holds for all $f \in W_0^1 L_{Q,q,\frac{q-1}{q},\frac{q-\gamma}{q}}(\mathbb{G})$. In the same way we establish the inequality (10.19) for any $f \in W_0^1 L_{Q,q,\frac{q-1}{q},\frac{q-\gamma}{q}}(\mathbb{G})$. Since inequalities (10.16) and (10.19) hold for any $f \in W_0^1 L_{Q,q,\frac{q-1}{q},\frac{q-\gamma}{q}}(\mathbb{G})$, we get (10.15) for $f \in W_0^1 L_{Q,q,\frac{q-1}{q},\frac{q-\gamma}{q}}(\mathbb{G})$.

Now it remains to show the sharpness of the constant $\frac{q}{\gamma-1}$ in (10.15). By (10.15) for each $f \in W_0^1 L_{Q,q,\frac{q-1}{q},\frac{q-\gamma}{q}}(B(0,R))$, we have

$$\left(\int_{B(0,R)} \frac{|f(x)|^q}{\left|\log\left|\log\frac{eR}{|x|}\right|\right|^{\gamma}\left|\log\frac{eR}{|x|}\right|}\frac{dx}{|x|^Q}\right)^{1/q} \tag{10.20}$$

$$\leq \frac{q}{\gamma-1}\left(\int_{B(0,R)} |x|^{q-Q}\left|\log\frac{eR}{|x|}\right|^{q-1}\left|\log\left|\log\frac{eR}{|x|}\right|\right|^{q-\gamma}\left|\frac{1}{|x|}\mathbb{E}f\right|^q dx\right)^{1/q}.$$

Therefore, it is sufficient to show the sharpness of the constant $\frac{q}{\gamma-1}$ in (10.20). As in the Abelian case (see [MOW15b, Section 3]), we consider a sequence of functions $\{f_\ell\}$ for large $\ell \in \mathbb{N}$ defined by

$$f_\ell(x) := \begin{cases} (\log(\log(\ell eR)))^{\frac{\gamma-1}{q}}, & \text{when } |x| \leq \frac{1}{\ell}, \\ (\log(\log\frac{eR}{|x|}))^{\frac{\gamma-1}{q}}, & \text{when } \frac{1}{\ell} \leq |x| \leq \frac{R}{2}, \\ (\log(\log(2e)))^{\frac{\gamma-1}{q}}\frac{2}{R}(R-|x|), & \text{when } \frac{R}{2} \leq |x| \leq R. \end{cases}$$

It is clear that $f_\ell \in W_0^1 L_{Q,q,\frac{q-1}{q},\frac{q-\gamma}{q}}(B(0,R))$. Letting $\widetilde{f_\ell}(r) := f_\ell(x)$ with $r = |x| \geq 0$, we get

$$\frac{d}{dr}\widetilde{f_\ell}(r) = \begin{cases} 0, & \text{when } r < \frac{1}{\ell}, \\ -\frac{\gamma-1}{q}r^{-1}(\log(\log\frac{eR}{r}))^{\frac{\gamma-1}{q}-1}(\log\frac{eR}{r})^{-1}, & \text{when } \frac{1}{\ell} < r < \frac{R}{2}, \\ -\frac{2}{R}(\log(\log(2e)))^{\frac{\gamma-1}{q}}, & \text{when } \frac{R}{2} < r < R. \end{cases}$$

Denoting by $|\wp|$ the $Q-1$-dimensional surface measure of the unit sphere with respect to the quasi-norm $|\cdot|$, by a direct calculation we have

$$\int_{B(0,R)} |x|^{q-Q}\left|\log\frac{eR}{|x|}\right|^{q-1}\left|\log\left|\log\frac{eR}{|x|}\right|\right|^{q-\gamma}\left|\frac{1}{|x|}\mathbb{E}f\right|^q dx$$

$$= |\wp|\int_0^R r^{q-1}\left|\log\frac{eR}{r}\right|^{q-1}\left|\log\left|\log\frac{eR}{r}\right|\right|^{q-\gamma}\left|\frac{d}{dr}\widetilde{f_\ell}(r)\right|^q dr$$

$$= |\wp|\left(\frac{\gamma-1}{q}\right)^q\int_{\frac{1}{\ell}}^{\frac{R}{2}} r^{-1}\left(\log\frac{eR}{r}\right)^{-1}\left(\log\left(\log\frac{eR}{r}\right)\right)^{-1} dr$$

$$+ (\log(\log(2e)))^{\gamma-1}\left(\frac{2}{R}\right)^q|\wp|\int_{\frac{R}{2}}^R r^{q-1}\left(\log\frac{eR}{r}\right)^{q-1}\left(\log\left(\log\frac{eR}{r}\right)\right)^{q-\gamma} dr$$

$$= -|\wp|\left(\frac{\gamma-1}{q}\right)^q \int_{\frac{1}{\ell}}^{\frac{R}{2}} \frac{d}{dr}\left(\log\left(\log\left(\log\frac{eR}{r}\right)\right)\right)$$

$$+ (\log(\log(2e)))^{\gamma-1}\left(\frac{2}{R}\right)^q |\wp| \int_{\frac{R}{2}}^{R} r^{q-1}\left(\log\frac{eR}{r}\right)^{q-1}\left(\log\left(\log\frac{eR}{r}\right)\right)^{q-\gamma} dr$$

$$=: |\wp|\left(\frac{\gamma-1}{q}\right)^q (\log(\log(\log \ell eR)) - \log(\log(\log 2e))) + |\wp| C_{\gamma,q}, \qquad (10.21)$$

where

$$C_{\gamma,q} := (2e)^q (\log(\log(2e)))^{\gamma-1}\int_0^{(\log(\log(2e)))} s^{q-\gamma} e^{q(s-e^s)} ds.$$

The assumption $q-\gamma+1>0$ implies that $C_{\gamma,q}<+\infty$. Moreover, we have

$$\int_{B(0,R)} \frac{|f(x)|^q}{\left|\log\left|\log\frac{eR}{|x|}\right|\right|^{\gamma}\left|\log\frac{eR}{|x|}\right|}\frac{dx}{|x|^Q} = |\wp|\int_0^R \frac{|\widetilde{f}_\ell(r)|^q}{\left|\log\left|\log\frac{eR}{r}\right|\right|^{\gamma}\left|\log\frac{eR}{r}\right|}\frac{dr}{r}$$

$$= |\wp|(\log(\log(\ell eR)))^{\gamma-1}\int_0^{\frac{1}{\ell}} r^{-1}\left(\log\frac{eR}{r}\right)^{-1}\left(\log\left(\log\frac{eR}{r}\right)\right)^{-\gamma} dr$$

$$+ |\wp|\int_{\frac{1}{\ell}}^{\frac{R}{2}} r^{-1}\left(\log\frac{eR}{r}\right)^{-1}\left(\log\left(\log\frac{eR}{r}\right)\right)^{-1} dr$$

$$+ |\wp|(\log(\log(2e)))^{\gamma-1}\left(\frac{2}{R}\right)^q$$

$$\times \int_{\frac{R}{2}}^{R} r^{-1}(R-r)^q\left(\log\frac{eR}{r}\right)^{-1}\left(\log\left(\log\frac{eR}{r}\right)\right)^{-\gamma} dr$$

$$= \frac{|\wp|}{\gamma-1} + |\wp|(\log(\log(\log(\ell eR)) - \log(\log(\log(2e))) + |\wp| C_{R,\gamma,q}, \qquad (10.22)$$

where

$$C_{R,\gamma,q} := (\log(\log(2e)))^{\gamma-1}\left(\frac{2}{R}\right)^q$$

$$\times \int_{\frac{R}{2}}^{R} r^{-1}(R-r)^q\left(\log\frac{eR}{r}\right)^{-1}\left(\log\left(\log\frac{eR}{r}\right)\right)^{-\gamma} dr.$$

The inequality $\log(\log\frac{eR}{r}) \geq \frac{R-r}{R}$ for all $r\leq R$ and the assumption $q-\gamma>-1$ imply $C_{R,\gamma,q}<+\infty$. Then, by (10.21) and (10.22), we arrive at

$$\int_{B(0,R)} |x|^{q-Q}\left|\log\frac{eR}{|x|}\right|^{q-1}\left|\log\left|\log\frac{eR}{|x|}\right|\right|^{q-\gamma}\left|\frac{1}{|x|}\mathbb{E}f\right|^q dx$$

$$\times\left(\int_{B(0,R)} \frac{|f(x)|^q}{\left|\log\left|\log\frac{eR}{|x|}\right|\right|^{\gamma}\left|\log\frac{eR}{|x|}\right|}\frac{dx}{|x|^Q}\right)^{-1} \to \left(\frac{\gamma-1}{q}\right)^q$$

as $\ell\to\infty$, which implies that the constant $\frac{q}{\gamma-1}$ in (10.20) is sharp. □

10.3 Generalized Morrey spaces

In the next sections, we develop the theory of Morrey and Campanato spaces on general homogeneous groups. Moreover, we analyse properties of Bessel–Riesz operators, maximal operators, and fractional integral operators on these spaces.

A brief discussion of the general theory of Morrey and Campanato spaces was given in the introduction, and we can refer there also for references.

10.3.1 Bessel–Riesz kernels on homogeneous groups

The classical Bessel–Riesz operators on the Euclidean space $\mathbb{R}^n$ are the operators of the form

$$I_{\alpha,\gamma}f(x) = \int_{\mathbb{R}^n} K_{\alpha,\gamma}(x-y)f(y)dy = \int_{\mathbb{R}^n} \frac{|x-y|^{\alpha-n}}{(1+|x-y|)^\gamma} f(y)dy, \tag{10.23}$$

where $\gamma \geq 0$ and $0 < \alpha < n$. Here, $I_{\alpha,\gamma}$ and $K_{\alpha,\gamma}$ are called a *Bessel–Riesz operator* and a *Bessel–Riesz kernel*, respectively. The original works on these operators go back to Hardy and Littlewood in [HL27, HL32] and Sobolev in [Sob38]. We refer to Section 5.3 for the appearance of the related operators.

Definition 10.3.1 (Bessel–Riesz kernels)**.** A natural analogue of the operators (10.23) in the setting of homogeneous groups are operators of the form

$$I_{\alpha,\gamma}f(x) := \int_{\mathbb{G}} K_{\alpha,\gamma}(xy^{-1})f(y)dy = \int_{\mathbb{G}} \frac{|xy^{-1}|^{\alpha-Q}}{(1+|xy^{-1}|)^\gamma} f(y)dy, \tag{10.24}$$

with the Bessel–Riesz kernels defined by

$$K_{\alpha,\gamma}(z) = \frac{|z|^{\alpha-Q}}{(1+|z|)^\gamma}, \tag{10.25}$$

where $|\cdot|$ is a homogeneous quasi-norm on a homogeneous group $\mathbb{G}$.

First we calculate the L^p-norms of the Bessel–Riesz kernels.

Theorem 10.3.2 (L^p-norms of Bessel–Riesz kernels)**.** *Let $\mathbb{G}$ be a homogeneous group of homogeneous dimension Q with a homogeneous quasi-norm $|\cdot|$. If $0 < \alpha < Q$ and $\gamma > 0$ then $K_{\alpha,\gamma} \in L^{p_1}(\mathbb{G})$ and*

$$\|K_{\alpha,\gamma}\|_{L^{p_1}(\mathbb{G})} \sim \left(\sum_{k\in\mathbb{Z}} \frac{(2^kR)^{(\alpha-Q)p_1+Q}}{(1+2^kR)^{\gamma p_1}} \right)^{1/p_1},$$

for $\frac{Q}{Q+\gamma-\alpha} < p_1 < \frac{Q}{Q-\alpha}$.

We will use the following result in some proofs.

Lemma 10.3.3 ([IGLE15]). *If $b > a > 0$ then for every $u > 1$ and $R > 0$ we have*

$$\sum_{k\in\mathbb{Z}} \frac{(u^k R)^a}{(1+u^k R)^b} < \infty.$$

Proof of Theorem 10.3.2. By using the polar coordinates decomposition from Proposition 1.2.10, for any $R > 0$ we have

$$\begin{aligned}\int_{\mathbb{G}} |K_{\alpha,\gamma}(x)|^{p_1}\,dx &= \int_{\mathbb{G}} \frac{|x|^{(\alpha-Q)p_1}}{(1+|x|)^{\gamma p_1}}\,dx \\ &= \int_0^\infty \int_{\wp} \frac{r^{(\alpha-Q)p_1+Q-1}}{(1+r)^{\gamma p_1}}\,d\sigma(y)dr \\ &= |\wp| \sum_{k\in\mathbb{Z}} \int_{2^k R\le r<2^{k+1}R} \frac{r^{(\alpha-Q)p_1+Q-1}}{(1+r)^{\gamma p_1}}\,dr,\end{aligned}$$

where $|\wp|$ is the $Q-1$-dimensional surface measure of the unit sphere. This implies

$$\begin{aligned}\int_{\mathbb{G}} |K_{\alpha,\gamma}(x)|^{p_1}\,dx &\le |\wp| \sum_{k\in\mathbb{Z}} \frac{1}{(1+2^k R)^{\gamma p_1}} \int_{2^k R\le r<2^{k+1}R} r^{(\alpha-Q)p_1+Q-1}\,dr \\ &= \frac{|\wp|(2^{(\alpha-Q)p_1+Q}-1)}{(\alpha-Q)p_1+Q} \sum_{k\in\mathbb{Z}} \frac{(2^k R)^{(\alpha-Q)p_1+Q}}{(1+2^k R)^{\gamma p_1}}.\end{aligned}$$

Moreover, we have

$$\begin{aligned}\int_{\mathbb{G}} |K_{\alpha,\gamma}(x)|^{p_1}\,dx &\ge \frac{|\wp|}{2^{\gamma p_1}} \sum_{k\in\mathbb{Z}} \frac{1}{(1+2^k R)^{\gamma p_1}} \int_{2^k R\le r<2^{k+1}R} r^{(\alpha-Q)p_1+Q-1}\,dr \\ &= \frac{|\wp|(2^{(\alpha-Q)p_1+Q}-1)}{2^{\gamma p_1}((\alpha-Q)p_1+Q)} \sum_{k\in\mathbb{Z}} \frac{(2^k R)^{(\alpha-Q)p_1+Q}}{(1+2^k R)^{\gamma p_1}}.\end{aligned}$$

Combining these two inequalities, for each $R > 0$ we have

$$\int_{\mathbb{G}} |K_{\alpha,\gamma}(x)|^{p_1}\,dx \sim \sum_{k\in\mathbb{Z}} \frac{(2^k R)^{(\alpha-Q)p_1+Q}}{(1+2^k R)^{\gamma p_1}}.$$

For $p_1 \in \left(\frac{Q}{Q+\gamma-\alpha}, \frac{Q}{Q-\alpha}\right)$ using Lemma 10.3.3 with $u = 2$, $a = (\alpha-Q)p_1 + Q$ and $b = \gamma p_1$, we arrive at

$$\sum_{k\in\mathbb{Z}} \frac{(2^k R)^{(\alpha-Q)p_1+Q}}{(1+2^k R)^{\gamma p_1}} < \infty,$$

which also implies that $K_{\alpha,\gamma} \in L^{p_1}(\mathbb{G})$. $\square$

Since

$$I_{\alpha,\gamma} f = K_{\alpha,\gamma} * f,$$

the Young inequality in Proposition 1.2.13 immediately implies

Corollary 10.3.4 (L^p-L^q boundedness of Bessel–Riesz operators). *Let $0 < \alpha < Q$ and $\gamma > 0$. Assume that $1 \leq p, q, p_1 \leq \infty$, $\frac{1}{q}+1 = \frac{1}{p}+\frac{1}{p_1}$ and $\frac{Q}{Q+\gamma-\alpha} < p_1 < \frac{Q}{Q-\alpha}$. Then we have*

$$\|I_{\alpha,\gamma} f\|_{L^q(\mathbb{G})} \leq \|K_{\alpha,\gamma}\|_{L^{p_1}(\mathbb{G})} \|f\|_{L^p(\mathbb{G})}$$

for all $f \in L^p(\mathbb{G})$.

This means that $I_{\alpha,\gamma}$ is bounded from $L^p(\mathbb{G})$ to $L^q(\mathbb{G})$ and that its operator norm can be estimated as

$$\|I_{\alpha,\gamma}\|_{L^p(\mathbb{G})\to L^q(\mathbb{G})} \leq \|K_{\alpha,\gamma}\|_{L^{p_1}(\mathbb{G})}.$$

10.3.2 Hardy–Littlewood maximal operator in Morrey spaces

We now define Morrey and generalized Morrey spaces on homogeneous groups and show the boundedness of the Hardy–Littlewood maximal operator in these spaces.

As usual throughout this section $\mathbb{G}$ is a homogeneous group of homogeneous dimension Q.

Definition 10.3.5 (Local Morrey spaces). Let us define the *local Morrey spaces* $LM^{p,q}(\mathbb{G})$ by

$$LM^{p,q}(\mathbb{G}) := \{f \in L^p_{\mathrm{loc}}(\mathbb{G}) : \|f\|_{LM^{p,q}(\mathbb{G})} < \infty\}, \quad 1 \leq p \leq q, \tag{10.26}$$

where

$$\|f\|_{LM^{p,q}(\mathbb{G})} := \sup_{r>0} r^{Q(1/q-1/p)} \left(\int_{B(0,r)} |f(x)|^p dx\right)^{1/p}.$$

Sometimes these spaces are called central Morrey spaces in the literature.

Definition 10.3.6 (Generalized local Morrey spaces). Let $\phi : \mathbb{R}^+ \to \mathbb{R}^+$ and $1 \leq p < \infty$. We define the *generalized local Morrey space* $LM^{p,\phi}(\mathbb{G})$ by

$$LM^{p,\phi}(\mathbb{G}) := \{f \in L^p_{\mathrm{loc}}(\mathbb{G}) : \|f\|_{LM^{p,\phi}(\mathbb{G})} < \infty\}, \tag{10.27}$$

where

$$\|f\|_{LM^{p,\phi}(\mathbb{G})} := \sup_{r>0} \frac{1}{\phi(r)} \left(\frac{1}{r^Q}\int_{B(0,r)} |f(x)|^p dx\right)^{1/p}.$$

Let us first formulate the assumptions for the function ϕ in the above definition.

Assumptions on ϕ

From now on we will assume that ϕ is nonincreasing and $t^{Q/p}\phi(t)$ is nondecreasing, so that ϕ satisfies the doubling condition, i.e., there exists a constant $C_1 > 0$ such that we have

$$\frac{1}{2} \le \frac{r}{s} \le 2 \Longrightarrow \frac{1}{C_1} \le \frac{\phi(r)}{\phi(s)} \le C_1. \tag{10.28}$$

Definition 10.3.7 (Hardy–Littlewood maximal operator). For every $f \in L^p_{\mathrm{loc}}(\mathbb{G})$ we define the Hardy–Littlewood maximal operator $\mathscr{M}$ by

$$\mathscr{M}f(x) := \sup_{r>0} \frac{1}{|B(x,r)|} \int_{B(x,r)} |f(y)|dy, \tag{10.29}$$

where $|B(x,r)|$ denotes the Haar measure of the quasi-ball $B(x,r)$.

Using the definition (10.26) of local Morrey spaces one can readily obtain the following estimate:

Lemma 10.3.8. *For any* $1 \le p_2 \le p_1$ *and* $\frac{Q}{Q+\gamma-\alpha} < p_1 < \frac{Q}{Q-\alpha}$ *we have*

$$\|K_{\alpha,\gamma}\|_{LM^{p_2,p_1}(\mathbb{G})} \le \|K_{\alpha,\gamma}\|_{LM^{p_1,p_1}(\mathbb{G})} = \|K_{\alpha,\gamma}\|_{L^{p_1}(\mathbb{G})}. \tag{10.30}$$

We now prove the boundedness of the Hardy–Littlewood maximal operator on generalized local Morrey spaces.

Theorem 10.3.9 (Hardy–Littlewood maximal operator on generalized local Morrey spaces). *Let* $1 < p < \infty$. *Then there exists some* $C_p > 0$ *such that for all* $f \in LM^{p,\phi}(\mathbb{G})$ *we have*

$$\|\mathscr{M}f\|_{LM^{p,\phi}(\mathbb{G})} \le C_p\|f\|_{LM^{p,\phi}(\mathbb{G})}. \tag{10.31}$$

Remark 10.3.10. In the Euclidean space $\mathbb{R}^n$ this was shown by Nakai in [Nak94]. On stratified groups (or homogeneous Carnot groups) it was shown in [GAM13, Corollary 3.2]. For general homogeneous groups this and other results of this section were shown in [RSY18c].

Proof of Theorem 10.3.9. From the definition of the norm of the generalized local Morrey space (10.27) it follows that

$$\left(\int_{B(0,r)} |f(x)|^p dx\right)^{1/p} \le \phi(r) r^{\frac{Q}{p}} \|f\|_{LM^{p,\phi}(\mathbb{G})}, \tag{10.32}$$

for all $r > 0$.

On the other hand, using Corollary 2.5 (b) from Folland and Stein [FS82] we have the general property of the maximal function

$$\left(\int_{B(0,r)} |\mathscr{M}f(x)|^p dx\right)^{1/p} \le C_p \left(\int_{B(0,r)} |f(x)|^p dx\right)^{1/p}. \tag{10.33}$$

Combining (10.32) and (10.33) we arrive at

$$\frac{1}{\phi(r)} \left(\frac{1}{r^Q} \int_{B(0,r)} |\mathscr{M}f(x)|^p dx\right)^{1/p} \le C_p\|f\|_{LM^{p,\phi}(\mathbb{G})},$$

for all $r > 0$. But this is exactly (10.31). □

10.3.3 Bessel–Riesz operators in Morrey spaces

In this section, we show the boundedness of the Bessel–Riesz operators on the generalized local Morrey space from Definition 10.3.6. In the Abelian case $\mathbb{G} = (\mathbb{R}^n, +)$ and $Q = n$ with the standard Euclidean distance $|x|_E = \sqrt{x_1^2 + x_2^2 + \cdots + x_n^2}$ the results of this section have been obtained in [IGE16]. The exposition of this section follows the results obtained in [RSY18c].

Theorem 10.3.11 (Boundedness of Bessel–Riesz operators on Morrey spaces, I). *Let $\mathbb{G}$ be a homogeneous group of homogeneous dimension Q with a homogeneous quasi-norm $|\cdot|$. Let $\gamma > 0$ and $0 < \alpha < Q$. Let $\beta < -\alpha$, $1 < p < \infty$, and $\frac{Q}{Q+\gamma-\alpha} < p_1 < \frac{Q}{Q-\alpha}$. Assume that $0 < \phi(r) \leq Cr^{\beta}$ for all $r > 0$. Then for all $f \in LM^{p,\phi}(\mathbb{G})$ we have*

$$\|I_{\alpha,\gamma}f\|_{LM^{q,\psi}(\mathbb{G})} \leq C_{p,\phi,Q}\|K_{\alpha,\gamma}\|_{L^{p_1}(\mathbb{G})}\|f\|_{LM^{p,\phi}(\mathbb{G})}, \tag{10.34}$$

where $q = \frac{\beta p_1' p}{\beta p_1' + Q}$ and $\psi(r) = (\phi(r))^{p/q}$.

Proof of Theorem 10.3.11. For every $f \in LM^{p,\phi}(\mathbb{G})$, let us write $I_{\alpha,\gamma}f(x)$ in the form

$$I_{\alpha,\gamma}f(x) := I_1(x) + I_2(x),$$

where

$$I_1(x) := \int_{B(x,R)} \frac{|xy^{-1}|^{\alpha-Q}f(y)}{(1+|xy^{-1}|)^{\gamma}}dy$$

and

$$I_2(x) := \int_{B^c(x,R)} \frac{|xy^{-1}|^{\alpha-Q}f(y)}{(1+|xy^{-1}|)^{\gamma}}dy,$$

for some $R > 0$.

By using the dyadic decomposition for I_1 we obtain

$$\begin{aligned} |I_1(x)| &\leq \sum_{k=-\infty}^{-1} \int_{2^kR\leq|xy^{-1}|<2^{k+1}R} \frac{|xy^{-1}|^{\alpha-Q}|f(y)|}{(1+|xy^{-1}|)^{\gamma}}dy \\ &\leq \sum_{k=-\infty}^{-1} \frac{(2^kR)^{\alpha-Q}}{(1+2^kR)^{\gamma}} \int_{2^kR\leq|xy^{-1}|<2^{k+1}R} |f(y)|dy \\ &\leq C\mathscr{M}f(x) \sum_{k=-\infty}^{-1} \frac{(2^kR)^{\alpha-Q+Q/p_1}(2^kR)^{Q/p_1'}}{(1+2^kR)^{\gamma}}. \end{aligned}$$

From this using Hölder's inequality for $\frac{1}{p_1} + \frac{1}{p_1'} = 1$ we get

$$|I_1(x)| \leq C\mathscr{M}f(x)\left(\sum_{k=-\infty}^{-1} \frac{(2^kR)^{(\alpha-Q)p_1+Q}}{(1+2^kR)^{\gamma p_1}}\right)^{1/p_1} \left(\sum_{k=-\infty}^{-1} (2^kR)^{Q}\right)^{1/p_1'}.$$

Since

$$\left(\sum_{k=-\infty}^{-1}\frac{(2^kR)^{(\alpha-Q)p_1+Q}}{(1+2^kR)^{\gamma p_1}}\right)^{1/p_1}\le\left(\sum_{k\in\mathbb{Z}}\frac{(2^kR)^{(\alpha-Q)p_1+Q}}{(1+2^kR)^{\gamma p_1}}\right)^{1/p_1}\sim\|K_{\alpha,\gamma}\|_{L^{p_1}(\mathbb{G})},\tag{10.35}$$

we arrive at

$$|I_1(x)|\le C\|K_{\alpha,\gamma}\|_{L^{p_1}(\mathbb{G})}\mathscr{M}f(x)R^{Q/p_1'}.\tag{10.36}$$

For the second term I_2, by using Hölder's inequality for $\frac{1}{p}+\frac{1}{p'}=1$ we obtain that

$$\begin{aligned}|I_2(x)|&\le\sum_{k=0}^{\infty}\frac{(2^kR)^{\alpha-Q}}{(1+2^kR)^{\gamma}}\int_{2^kR\le|xy^{-1}|<2^{k+1}R}|f(y)|dy\\&\le\sum_{k=0}^{\infty}\frac{(2^kR)^{\alpha-Q}}{(1+2^kR)^{\gamma}}\left(\int_{2^kR\le|xy^{-1}|<2^{k+1}R}dy\right)^{\frac{1}{p'}}\left(\int_{2^kR\le|xy^{-1}|<2^{k+1}R}|f(y)|^pdy\right)^{\frac{1}{p}}\\&=\sum_{k=0}^{\infty}\frac{(2^kR)^{\alpha-Q}}{(1+2^kR)^{\gamma}}\left(\int_{2^kR}^{2^{k+1}R}\int_{\wp}r^{Q-1}d\sigma(y)dr\right)^{\frac{1}{p'}}\left(\int_{2^kR\le|xy^{-1}|<2^{k+1}R}|f(y)|^pdy\right)^{\frac{1}{p}}\\&\le C\sum_{k=0}^{\infty}\frac{(2^kR)^{\alpha-Q}}{(1+2^kR)^{\gamma}}(2^kR)^{Q/p'}\left(\int_{2^kR\le|xy^{-1}|<2^{k+1}R}|f(y)|^pdy\right)^{\frac{1}{p}}.\end{aligned}$$

This implies that

$$|I_2(x)|\le C\|f\|_{LM^{p,\phi}(\mathbb{G})}\sum_{k=0}^{\infty}\frac{(2^kR)^{\alpha-Q+Q/p_1}}{(1+2^kR)^{\gamma}}\phi(2^kR)(2^kR)^{Q/p_1'}.$$

Since $\phi(r)\le Cr^{\beta}$ by assumption, we can estimate

$$|I_2(x)|\le C\|f\|_{LM^{p,\phi}(\mathbb{G})}\sum_{k=0}^{\infty}\frac{(2^kR)^{\alpha-Q+Q/p_1}}{(1+2^kR)^{\gamma}}(2^kR)^{\beta+Q/p_1'}.$$

Applying Hölder's inequality again we get

$$|I_2(x)|\le C\|f\|_{LM^{p,\phi}(\mathbb{G})}\left(\sum_{k=0}^{\infty}\frac{(2^kR)^{(\alpha-Q)p_1+Q}}{(1+2^kR)^{\gamma p_1}}\right)^{1/p_1}\left(\sum_{k=0}^{\infty}(2^kR)^{\beta p_1'+Q}\right)^{1/p_1'}.$$

From the conditions $p_1<\frac{Q}{Q-\alpha}$ and $\beta<-\alpha$ we have $\beta p_1'+Q<0$. By Theorem 10.3.2, we also have

$$\left(\sum_{k=0}^{\infty}\frac{(2^kR)^{(\alpha-Q)p_1+Q}}{(1+2^kR)^{\gamma p_1}}\right)^{1/p_1}\le\left(\sum_{k\in\mathbb{Z}}\frac{(2^kR)^{(\alpha-Q)p_1+Q}}{(1+2^kR)^{\gamma p_1}}\right)^{1/p_1}\sim\|K_{\alpha,\gamma}\|_{L^{p_1}(\mathbb{G})}.$$

Using these, we arrive at

$$|I_2(x)| \le C\|K_{\alpha,\gamma}\|_{L^{p_1}(\mathbb{G})}\|f\|_{LM^{p,\phi}(\mathbb{G})}R^{Q/p_1'+\beta}. \tag{10.37}$$

Combining estimates (10.36) and (10.37) we get

$$|I_{\alpha,\gamma}f(x)| \le C\|K_{\alpha,\gamma}\|_{L^{p_1}(\mathbb{G})}(\mathscr{M}f(x)R^{Q/p_1'} + \|f\|_{LM^{p,\phi}(\mathbb{G})}R^{Q/p_1'+\beta}).$$

Assuming that f is not identically 0 and that $\mathscr{M}f$ is finite everywhere, we can choose $R>0$ such that $R^\beta = \frac{\mathscr{M}f(x)}{\|f\|_{LM^{p,\phi}(\mathbb{G})}}$, that is,

$$|I_{\alpha,\gamma}f(x)| \le C\|K_{\alpha,\gamma}\|_{L^{p_1}(\mathbb{G})}\|f\|_{LM^{p,\phi}(\mathbb{G})}^{-\frac{Q}{\beta p_1'}}(\mathscr{M}f(x))^{1+\frac{Q}{\beta p_1'}},$$

for every $x\in\mathbb{G}$. Setting $q=\frac{\beta p_1' p}{\beta p_1'+Q}$, for any $r>0$ we get

$$\left(\int_{|x|<r}|I_{\alpha,\gamma}f(x)|^q dx\right)^{1/q} \le C\|K_{\alpha,\gamma}\|_{L^{p_1}(\mathbb{G})}\|f\|_{LM^{p,\phi}(\mathbb{G})}^{1-p/q}\left(\int_{|x|<r}|\mathscr{M}f(x)|^p dx\right)^{1/q}.$$

Then we divide both sides by $(\phi(r))^{p/q}r^{Q/q}$ to get

$$\frac{\left(\int_{|x|<r}|I_{\alpha,\gamma}f(x)|^q dx\right)^{1/q}}{\psi(r)r^{Q/q}} \le C\|K_{\alpha,\gamma}\|_{L^{p_1}(\mathbb{G})}\|f\|_{LM^{p,\phi}(\mathbb{G})}^{1-p/q}\frac{\left(\int_{|x|<r}|\mathscr{M}f(x)|^p dx\right)^{1/q}}{(\phi(r))^{p/q}r^{Q/q}},$$

where $\psi(r)=(\phi(r))^{p/q}$. Now by taking the supremum over $r>0$ we obtain that

$$\|I_{\alpha,\gamma}f\|_{LM^{q,\psi}(\mathbb{G})} \le C\|K_{\alpha,\gamma}\|_{L^{p_1}(\mathbb{G})}\|f\|_{LM^{p,\phi}(\mathbb{G})}^{1-p/q}\|\mathscr{M}f\|_{LM^{p,\phi}(\mathbb{G})}^{p/q},$$

which gives (10.34) after applying estimate (10.31). □

Lemma 10.3.8 states, in particular, that the Bessel–Riesz kernel belongs to local Morrey spaces. This fact will be used in the following statement refining estimate (10.34) in Theorem 10.3.11.

Theorem 10.3.12 (Boundedness of Bessel–Riesz operators on Morrey spaces, II). *Let $\mathbb{G}$ be a homogeneous group of homogeneous dimension Q with a homogeneous quasi-norm $|\cdot|$. Let $\gamma>0$, $0<\alpha<Q$ and $1<p<\infty$. Let $\beta<-\alpha$, $\frac{Q}{Q+\gamma-\alpha}<p_2\le p_1<\frac{Q}{Q-\alpha}$ and $p_2\ge 1$. Assume that $0<\phi(r)\le Cr^\beta$ for all $r>0$. Then for all $f\in LM^{p,\phi}(\mathbb{G})$ we have*

$$\|I_{\alpha,\gamma}f\|_{LM^{q,\psi}(\mathbb{G})} \le C_{p,\phi,Q}\|K_{\alpha,\gamma}\|_{LM^{p_2,p_1}(\mathbb{G})}\|f\|_{LM^{p,\phi}(\mathbb{G})}, \tag{10.38}$$

where $q=\frac{\beta p_1' p}{\beta p_1'+Q}$ and $\psi(r)=(\phi(r))^{p/q}$.

Indeed, this refines Theorem 10.3.11 since by Lemma 10.3.8 we can estimate

$$\begin{aligned}\|I_{\alpha,\gamma}f\|_{LM^{q,\psi}(\mathbb{G})} &\le C\|K_{\alpha,\gamma}\|_{LM^{p_2,p_1}(\mathbb{G})}\|f\|_{LM^{p,\phi}(\mathbb{G})}\\ &\le C\|K_{\alpha,\gamma}\|_{L^{p_1}(\mathbb{G})}\|f\|_{LM^{p,\phi}(\mathbb{G})}.\end{aligned}$$

Proof of Theorem 10.3.12. Similarly to the proof of Theorem 10.3.11 we write $I_{\alpha,\gamma}f(x)$ in the form

$$I_{\alpha,\gamma}f(x) := I_1(x) + I_2(x),$$

where

$$I_1(x) := \int_{B(x,R)} \frac{|xy^{-1}|^{\alpha-Q}f(y)}{(1+|xy^{-1}|)^{\gamma}}dy$$

and

$$I_2(x) := \int_{B^c(x,R)} \frac{|xy^{-1}|^{\alpha-Q}f(y)}{(1+|xy^{-1}|)^{\gamma}}dy,$$

for some $R > 0$ to be chosen later.

As before, we estimate the first term I_1 by using the dyadic decomposition:

$$\begin{aligned}|I_1(x)| &\le \sum_{k=-\infty}^{-1} \int_{2^kR\le|xy^{-1}|<2^{k+1}R} \frac{|xy^{-1}|^{\alpha-Q}|f(y)|}{(1+|xy^{-1}|)^{\gamma}}dy\\ &\le \sum_{k=-\infty}^{-1} \frac{(2^kR)^{\alpha-Q}}{(1+2^kR)^{\gamma}} \int_{2^kR\le|xy^{-1}|<2^{k+1}R} |f(y)|dy\\ &\le CMf(x) \sum_{k=-\infty}^{-1} \frac{(2^kR)^{\alpha-Q+Q/p_2}(2^kR)^{Q/p_2'}}{(1+2^kR)^{\gamma}},\end{aligned}$$

where $1 \le p_2 \le p_1$. From this using Hölder's inequality for $\frac{1}{p_2} + \frac{1}{p_2'} = 1$, we get

$$|I_1(x)| \le C\mathscr{M}f(x) \left(\sum_{k=-\infty}^{-1} \frac{(2^kR)^{(\alpha-Q)p_2+Q}}{(1+2^kR)^{\gamma p_2}} \right)^{1/p_2} \left(\sum_{k=-\infty}^{-1} (2^kR)^{Q} \right)^{1/p_2'}.$$

In view of (10.35) we have

$$\begin{aligned}|I_1(x)| &\le C_2\mathscr{M}f(x) \left(\int_{0<|x|<R} K_{\alpha,\gamma}^{p_2}(x)dx \right)^{1/p_2} R^{Q/p_2'}\\ &\le C\|K_{\alpha,\gamma}\|_{LM^{p_2,p_1}(\mathbb{G})}\mathscr{M}f(x)R^{Q/p_1'}.\end{aligned} \tag{10.39}$$

Now for I_2 by using Hölder's inequality for $\frac{1}{p} + \frac{1}{p'} = 1$ we have

$$|I_2(x)| \le \sum_{k=0}^{\infty} \frac{(2^kR)^{\alpha-Q}}{(1+2^kR)^{\gamma}} (2^kR)^{Q/p'} \left(\int_{2^kR\le|xy^{-1}|<2^{k+1}R} |f(y)|^p dy \right)^{1/p},$$

that is,

$$|I_2(x)| \leq C\|f\|_{LM^{p,\phi}(\mathbb{G})} \sum_{k=0}^{\infty} \frac{(2^kR)^\alpha \phi(2^kR)}{(1+2^kR)^\gamma} \frac{\left(\int_{2^kR\leq|xy^{-1}|<2^{k+1}R} dy\right)^{1/p_2}}{(2^kR)^{Q/p_2}}$$

$$\leq C\|f\|_{LM^{p,\phi}(\mathbb{G})} \sum_{k=0}^{\infty} \phi(2^kR)(2^kR)^{Q/p_1'} \frac{\left(\int_{2^kR\leq|xy^{-1}|<2^{k+1}R} K_{\alpha,\gamma}^{p_2}(xy^{-1})dy\right)^{1/p_2}}{(2^kR)^{Q/p_2-Q/p_1}},$$

where we have used the inequality

$$\begin{aligned}&\left(\int_{2^kR\leq|xy^{-1}|<2^{k+1}R} K_{\alpha,\gamma}^{p_2}(xy^{-1})dy\right)^{1/p_2} \\ &\sim \frac{(2^kR)^{(\alpha-Q)+Q/p_2}}{(1+2^kR)^\gamma} \geq C\frac{(2^kR)^{(\alpha-Q)}}{(1+2^kR)^\gamma}\left(\int_{2^kR\leq|xy^{-1}|<2^{k+1}R} dy\right)^{1/p_2}.\end{aligned} \tag{10.40}$$

Since we have $\phi(r) \leq Cr^\beta$ by assumption and since

$$\frac{\left(\int_{2^kR\leq|xy^{-1}|<2^{k+1}R} K_{\alpha,\gamma}^{p_2}(xy^{-1})dy\right)^{1/p_2}}{(2^kR)^{Q/p_2-Q/p_1}} \lesssim \|K_{\alpha,\gamma}\|_{LM^{p_2,p_1}(\mathbb{G})}$$

for every $k = 0, 1, 2, \ldots$, we get

$$|I_2(x)| \leq C\|K_{\alpha,\gamma}\|_{LM^{p_2,p_1}(\mathbb{G})}\|f\|_{LM^{p,\phi}(\mathbb{G})} \sum_{k=0}^{\infty} (2^kR)^{\beta+Q/p_1'}.$$

Using that $\beta + Q/p_1' < 0$ we obtain

$$|I_2(x)| \leq C\|K_{\alpha,\gamma}\|_{LM^{p_2,p_1}(\mathbb{G})}\|f\|_{LM^{p,\phi}(\mathbb{G})} R^{\beta+Q/p_1'}. \tag{10.41}$$

Combining estimates (10.39) and (10.41) we get

$$|I_{\alpha,\gamma}f(x)| \leq C\|K_{\alpha,\gamma}\|_{LM^{p_2,p_1}(\mathbb{G})}(\mathscr{M}f(x)R^{Q/p_1'} + \|f\|_{LM^{p,\phi}(\mathbb{G})}R^{\beta+Q/p_1'}).$$

Assuming that f is not identically 0 and that $\mathscr{M}f$ is finite everywhere, we can choose $R > 0$ such that $R^\beta = \frac{\mathscr{M}f(x)}{\|f\|_{LM^{p,\phi}(\mathbb{G})}}$, which yields

$$|I_{\alpha,\gamma}f(x)| \leq C\|K_{\alpha,\gamma}\|_{LM^{p_2,p_1}(\mathbb{G})}\|f\|_{LM^{p,\phi}(\mathbb{G})}^{-\frac{Q}{\beta p_1'}}(\mathscr{M}f(x))^{1+\frac{Q}{\beta p_1'}}.$$

Now by using that $q = \frac{\beta p_1' p}{\beta p_1' + Q}$, for every $r > 0$ we obtain

$$\begin{aligned}&\left(\int_{|x|<r} |I_{\alpha,\gamma}f(x)|^q dx\right)^{1/q} \\ &\leq C\|K_{\alpha,\gamma}\|_{LM^{p_2,p_1}(\mathbb{G})}\|f\|_{LM^{p,\phi}(\mathbb{G})}^{1-p/q}\left(\int_{|x|<r} |\mathscr{M}f(x)|^p dx\right)^{1/q}.\end{aligned}$$

Then we divide both sides by $(\phi(r))^{p/q}r^{Q/q}$ to get

$$\frac{\left(\int_{|x|<r}|I_{\alpha,\gamma}f(x)|^q dx\right)^{1/q}}{\psi(r)r^{Q/q}} \le C\|K_{\alpha,\gamma}\|_{LM^{p_2,p_1}(\mathbb{G})}\|f\|^{1-p/q}_{LM^{p,\phi}(\mathbb{G})}\frac{\left(\int_{|x|<r}|\mathscr{M}f(x)|^p dx\right)^{1/q}}{(\phi(r))^{p/q}r^{Q/q}},$$

where $\psi(r) = (\phi(r))^{p/q}$. Taking the supremum over $r > 0$ and then using (10.31), we obtain the desired result

$$\begin{aligned}\|I_{\alpha,\gamma}f\|_{LM^{q,\psi}(\mathbb{G})} &\le C\|K_{\alpha,\gamma}\|_{LM^{p_2,p_1}(\mathbb{G})}\|f\|^{1-p/q}_{LM^{p,\phi}(\mathbb{G})}\|\mathscr{M}f\|^{p/q}_{LM^{p,\phi}(\mathbb{G})}\\ &\le C_{p,\phi,Q}\|K_{\alpha,\gamma}\|_{LM^{p_2,p_1}(\mathbb{G})}\|f\|_{LM^{p,\phi}(\mathbb{G})},\end{aligned}$$

proving (10.38). □

To refine and extend the obtained boundedness statements we first show that the Bessel–Riesz kernel $K_{\alpha,\gamma}$ belongs to the generalized local Morrey space $LM^{p_2,\omega}(\mathbb{G})$ for some $p_2 \ge 1$ and some function ω.

Lemma 10.3.13 (Bessel–Riesz kernel in generalized local Morrey space). *Let $\mathbb{G}$ be a homogeneous group of homogeneous dimension Q. Let $\gamma > 0$, $p_2 \ge 1$ and $Q - \frac{Q}{p_2} < \alpha < Q$. If $\omega : \mathbb{R}^+ \to \mathbb{R}^+$ satisfies $\omega(r) \ge Cr^{\alpha-Q}$ for all $r > 0$, then $K_{\alpha,\gamma} \in LM^{p_2,\omega}(\mathbb{G})$.*

Proof of Lemma 10.3.13. It is sufficient to evaluate the following integral around zero, and using polar decomposition from Proposition 1.2.10, we have

$$\begin{aligned}\int_{|x|\le R} K^{p_2}_{\alpha,\gamma}(x)dx &= \int_{|x|\le R}\frac{|x|^{(\alpha-Q)p_2}}{(1+|x|)^{\gamma p_2}}dx\\ &\le |\wp|\int_{0<r\le R} r^{(\alpha-Q)p_2+Q-1}dr \le C\omega^{p_2}(R)R^Q.\end{aligned}$$

By dividing both sides of this inequality by $\omega^{p_2}(R)R^Q$ and taking p_2th-root, we obtain

$$\frac{\left(\int_{|x|\le R}K^{p_2}_{\alpha,\gamma}(x)dx\right)^{1/p_2}}{\omega(R)R^{Q/p_2}} \le C^{1/p_2}.$$

Then, we take the supremum over $R > 0$ to get

$$\sup_{R>0}\frac{\left(\int_{|x|\le R}K^{p_2}_{\alpha,\gamma}(x)dx\right)^{1/p_2}}{\omega(R)R^{Q/p_2}} < \infty,$$

which implies that $K_{\alpha,\gamma} \in LM^{p_2,\omega}(\mathbb{G})$. □

Theorem 10.3.14 (Boundedness of Bessel–Riesz operators on Morrey spaces, III). *Let $\mathbb{G}$ be a homogeneous group of homogeneous dimension Q with a homogeneous quasi-norm $|\cdot|$. Let $0<\alpha<Q$, $\gamma>0$ and $1<p<\infty$. Let $\omega:\mathbb{R}^+\to\mathbb{R}^+$ satisfy the doubling condition (10.28) and assume that $0<\omega(r)\le Cr^{-\alpha}$ for all $r>0$, so that $K_{\alpha,\gamma}\in LM^{p_2,\omega}(\mathbb{G})$ for $\frac{Q}{Q+\gamma-\alpha}<p_2<\frac{Q}{Q-\alpha}$ and $p_2\ge 1$. Let $\beta<-\alpha<-Q-\beta$ and assume that $0<\phi(r)\le Cr^{\beta}$ for all $r>0$. Then for all $f\in LM^{p,\phi}(\mathbb{G})$ we have*

$$\|I_{\alpha,\gamma}f\|_{LM^{q,\psi}(\mathbb{G})}\le C_{p,\phi,Q}\|K_{\alpha,\gamma}\|_{LM^{p_2,\omega}(\mathbb{G})}\|f\|_{LM^{p,\phi}(\mathbb{G})}, \tag{10.42}$$

where $q=\frac{\beta p}{\beta+Q-\alpha}$ and $\psi(r)=(\phi(r))^{p/q}$.

Proof of Theorem 10.3.14. As in the proof of Theorem 10.3.11, we write

$$I_{\alpha,\gamma}f(x):=I_1(x)+I_2(x),$$

where

$$I_1(x):=\int_{B(x,R)}\frac{|xy^{-1}|^{\alpha-Q}f(y)}{(1+|xy^{-1}|)^{\gamma}}dy$$

and

$$I_2(x):=\int_{B^c(x,R)}\frac{|xy^{-1}|^{\alpha-Q}f(y)}{(1+|xy^{-1}|)^{\gamma}}dy,$$

for some $R>0$ to be chosen later.

First, we estimate I_1 by using the dyadic decomposition

$$\begin{aligned}|I_1(x)|&\le\sum_{k=-\infty}^{-1}\int_{2^kR\le|xy^{-1}|<2^{k+1}R}\frac{|xy^{-1}|^{\alpha-Q}|f(y)|}{(1+|xy^{-1}|)^{\gamma}}dy\\&\le\sum_{k=-\infty}^{-1}\frac{(2^kR)^{\alpha-Q}}{(1+2^kR)^{\gamma}}\int_{2^kR\le|xy^{-1}|<2^{k+1}R}|f(y)|dy\\&\le C\mathscr{M}f(x)\sum_{k=-\infty}^{-1}\frac{(2^kR)^{\alpha-Q+Q/p_2}(2^kR)^{Q/p_2'}}{(1+2^kR)^{\gamma}}.\end{aligned}$$

From this using Hölder's inequality for $\frac{1}{p_2}+\frac{1}{p_2'}=1$ we get

$$|I_1(x)|\le C\mathscr{M}f(x)\left(\sum_{k=-\infty}^{-1}\frac{(2^kR)^{(\alpha-Q)p_2+Q}}{(1+2^kR)^{\gamma p_2}}\right)^{1/p_2}\left(\sum_{k=-\infty}^{-1}(2^kR)^{Q}\right)^{1/p_2'}.$$

By (10.35) we have

$$\begin{aligned}|I_1(x)|&\le C\mathscr{M}f(x)\left(\int_{0<|x|<R}K_{\alpha,\gamma}^{p_2}(x)dx\right)^{1/p_2}R^{Q/p_2'}\\&\le C\|K_{\alpha,\gamma}\|_{LM^{p_2,\omega}(\mathbb{G})}\mathscr{M}f(x)\omega(R)R^{Q},\end{aligned}$$

and using that $\omega(r) \leq Cr^{-\alpha}$ by assumption, we arrive at

$$|I_1(x)| \leq C\|K_{\alpha,\gamma}\|_{LM^{p_2,\omega}(\mathbb{G})}\mathscr{M}f(x)R^{Q-\alpha}. \tag{10.43}$$

Now let us estimate the second term I_2 as follows

$$\begin{aligned}|I_2(x)| &\leq \sum_{k=0}^{\infty}\frac{(2^kR)^{\alpha-Q}}{(1+2^kR)^{\gamma}}\int_{2^kR\leq|xy^{-1}|<2^{k+1}R}|f(y)|dy\\ &\leq C\sum_{k=0}^{\infty}\frac{(2^kR)^{\alpha-Q}}{(1+2^kR)^{\gamma}}(2^kR)^{Q/p'}\left(\int_{2^kR\leq|xy^{-1}|<2^{k+1}R}|f(y)|^pdy\right)^{1/p}\\ &\leq C\|f\|_{LM^{p,\phi}(\mathbb{G})}\sum_{k=0}^{\infty}\frac{(2^kR)^{\alpha}\phi(2^kR)}{(1+2^kR)^{\gamma}}\frac{\left(\int_{2^kR\leq|xy^{-1}|<2^{k+1}R}dy\right)^{1/p_2}}{(2^kR)^{Q/p_2}},\end{aligned}$$

where we have used that $\left(\int_{2^kR\leq|xy^{-1}|<2^{k+1}R}dy\right)^{1/p_2}\sim(2^kR)^{Q/p_2}$. Using (10.40) we obtain

$$|I_2(x)| \leq C\|f\|_{LM^{p,\phi}(\mathbb{G})}\sum_{k=0}^{\infty}\frac{(2^kR)^{\alpha}\phi(2^kR)}{(2^kR)^{\alpha-Q}}\frac{\left(\int_{2^kR\leq|xy^{-1}|<2^{k+1}R}K^{p_2}_{\alpha,\gamma}(xy^{-1})dy\right)^{\frac{1}{p_2}}}{(2^kR)^{Q/p_2}}.$$

Taking into account that $\phi(r) \leq Cr^{\beta}$ and $\omega(r) \leq Cr^{-\alpha}$ for all $r > 0$ we have

$$|I_2(x)| \leq C\|f\|_{LM^{p,\phi}(\mathbb{G})}\sum_{k=0}^{\infty}(2^kR)^{Q-\alpha+\beta}\frac{\left(\int_{2^kR\leq|xy^{-1}|<2^{k+1}R}K^{p_2}_{\alpha,\gamma}(xy^{-1})dy\right)^{\frac{1}{p_2}}}{\omega(2^kR)(2^kR)^{Q/p_2}}.$$

Since we have

$$\frac{\left(\int_{2^kR\leq|xy^{-1}|<2^{k+1}R}K^{p_2}_{\alpha,\gamma}(xy^{-1})dy\right)^{1/p_2}}{\omega(2^kR)(2^kR)^{Q/p_2}} \lesssim \|K_{\alpha,\gamma}\|_{LM^{p_2,\omega}(\mathbb{G})}$$

for every $k = 0, 1, 2, \ldots$, it follows that

$$|I_2(x)| \leq C\|K_{\alpha,\gamma}\|_{LM^{p_2,\omega}(\mathbb{G})}\|f\|_{LM^{p,\phi}(\mathbb{G})}\sum_{k=0}^{\infty}(2^kR)^{Q-\alpha+\beta}.$$

Since $Q - \alpha + \beta < 0$, it implies that

$$|I_2(x)| \leq C\|K_{\alpha,\gamma}\|_{LM^{p_2,\omega}(\mathbb{G})}\|f\|_{LM^{p,\phi}(\mathbb{G})}R^{Q-\alpha+\beta}. \tag{10.44}$$

Combining estimates (10.43) and (10.44) we get

$$|I_{\alpha,\gamma}f(x)| \leq C\|K_{\alpha,\gamma}\|_{LM^{p_2,\omega}(\mathbb{G})}(\mathscr{M}f(x)R^{Q-\alpha} + \|f\|_{LM^{p,\phi}(\mathbb{G})}R^{Q-\alpha+\beta}).$$

Assuming that f is not identically 0 and that $\mathcal{M}f$ is finite everywhere, we can choose $R > 0$ such that $R^\beta = \frac{\mathcal{M}f(x)}{\|f\|_{LM^{p,\phi}(\mathbb{G})}}$, that is

$$|I_{\alpha,\gamma}f(x)| \leq C\|K_{\alpha,\gamma}\|_{LM^{p_2,\omega}(\mathbb{G})}\|f\|_{LM^{p,\phi}(\mathbb{G})}^{(\alpha-Q)/\beta}(\mathcal{M}f(x))^{1+(Q-\alpha)/\beta}.$$

Since $q = \frac{\beta p}{\beta+Q-\alpha}$, for all $r > 0$ we get

$$\begin{aligned}\left(\int_{|x|<r}|I_{\alpha,\gamma}f(x)|^q dx\right)^{1/q}& \\ \leq C\|K_{\alpha,\gamma}\|_{LM^{p_2,\omega}(\mathbb{G})}&\|f\|_{LM^{p,\phi}(\mathbb{G})}^{1-p/q}\left(\int_{|x|<r}|\mathcal{M}f(x)|^p dx\right)^{1/q}.\end{aligned}$$

Then we divide both sides by $(\phi(r))^{p/q}r^{Q/q}$ to get

$$\begin{aligned}&\frac{\left(\int_{|x|<r}|I_{\alpha,\gamma}f(x)|^q dx\right)^{1/q}}{\psi(r)r^{Q/q}} \\ &\leq C\|K_{\alpha,\gamma}\|_{LM^{p_2,\omega}(\mathbb{G})}\|f\|_{LM^{p,\phi}(\mathbb{G})}^{1-p/q}\frac{\left(\int_{|x|<r}|\mathcal{M}f(x)|^p dx\right)^{1/q}}{(\phi(r))^{p/q}r^{Q/q}},\end{aligned}$$

where $\psi(r) = (\phi(r))^{p/q}$. Finally, taking the supremum over $r > 0$ and using (10.31), we obtain the desired result

$$\begin{aligned}\|I_{\alpha,\gamma}f\|_{LM^{q,\psi}(\mathbb{G})} &\leq C\|K_{\alpha,\gamma}\|_{LM^{p_2,\omega}(\mathbb{G})}\|f\|_{LM^{p,\phi}(\mathbb{G})}^{1-p/q}\|Mf\|_{LM^{p,\phi}(\mathbb{G})}^{p/q} \\ &\leq C_{p,\phi,Q}\|K_{\alpha,\gamma}\|_{LM^{p_2,\omega}(\mathbb{G})}\|f\|_{LM^{p,\phi}(\mathbb{G})},\end{aligned}$$

which gives (10.42). □

Remark 10.3.15. We can make the following comparison between the obtained estimates, similarly to the Euclidean case [IGE16, Section 3], namely, that also in the case of general homogeneous groups, Theorem 10.3.14 gives the best estimate among the three. Indeed, if we take

$$\omega(R) := (1 + R^{Q/q_1})R^{-Q/p_1}$$

for some $q_1 > p_1$, then

$$\|K_{\alpha,\gamma}\|_{LM^{p_2,\omega}(\mathbb{G})} \leq \|K_{\alpha,\gamma}\|_{LM^{p_2,p_1}(\mathbb{G})}.$$

By Theorem 10.3.14 and Lemma 10.3.8 we then obtain

$$\begin{aligned}\|I_{\alpha,\gamma}f\|_{LM^{q,\psi}(\mathbb{G})} &\leq C\|K_{\alpha,\gamma}\|_{LM^{p_2,\omega}(\mathbb{G})}\|f\|_{LM^{p,\phi}(\mathbb{G})} \\ &\leq C\|K_{\alpha,\gamma}\|_{LM^{p_2,p_1}(\mathbb{G})}\|f\|_{LM^{p,\phi}(\mathbb{G})} \\ &\leq C\|K_{\alpha,\gamma}\|_{L^{p_1}(\mathbb{G})}\|f\|_{LM^{p,\phi}(\mathbb{G})}.\end{aligned}$$

10.3.4 Generalized Bessel–Riesz operators

In this section we investigate properties of some generalizations of Bessel–Riesz operators.

Definition 10.3.16 (Generalized Bessel–Riesz operators)**.** We define the *generalized Bessel–Riesz operator* $I_{\widetilde{\rho},\gamma}$ by

$$I_{\widetilde{\rho},\gamma}f(x) := \int_{\mathbb{G}} \frac{\widetilde{\rho}(|xy^{-1}|)}{(1+|xy^{-1}|)^{\gamma}} f(y)dy, \tag{10.45}$$

where $\gamma \geq 0$, $\widetilde{\rho} : \mathbb{R}^{+} \to \mathbb{R}^{+}$, $\widetilde{\rho}$ satisfies the doubling condition (10.28) and the condition

$$\int_{0}^{1} \frac{\widetilde{\rho}(t)}{t^{\gamma-Q+1}} dt < \infty. \tag{10.46}$$

We denote its kernel by

$$K_{\widetilde{\rho},\gamma}(z) = \frac{\widetilde{\rho}(|z|)}{(1+|z|)^{\gamma}}.$$

We recover the usual Bessel–Riesz kernels from Definition 10.3.1 by taking $\widetilde{\rho}(t) = t^{\alpha-Q}$ for $0 < \alpha < Q$, in which case we have

$$K_{\widetilde{\rho},\gamma}(z) = K_{\alpha,\gamma}(z) = \frac{|z|^{\alpha-Q}}{(1+|z|)^{\gamma}}.$$

Theorem 10.3.17 (Boundedness of generalized Bessel–Riesz operators on Morrey spaces)**.** *Let $\mathbb{G}$ be a homogeneous group of homogeneous dimension Q with a homogeneous quasi-norm $|\cdot|$. Let $\gamma > 0$. Let $\widetilde{\rho}$ and ϕ satisfy the doubling condition* (10.28)*. Assume that ϕ is surjective and for some $1 < p < q < \infty$ satisfies*

$$\int_{r}^{\infty} \frac{(\phi(t))^{p}}{t} dt \leq C_{1}(\phi(r))^{p}, \tag{10.47}$$

and

$$\phi(r)\int_{0}^{r} \frac{\widetilde{\rho}(t)}{t^{\gamma-Q+1}} dt + \int_{r}^{\infty} \frac{\widetilde{\rho}(t)\phi(t)}{t^{\gamma-Q+1}} dt \leq C_{2}(\phi(r))^{p/q}, \tag{10.48}$$

for all $r > 0$. Then we have

$$\|I_{\widetilde{\rho},\gamma}f\|_{LM^{q,\phi^{p/q}}(\mathbb{G})} \leq C_{p,q,\phi,Q}\|f\|_{LM^{p,\phi}(\mathbb{G})}. \tag{10.49}$$

Proof of Theorem 10.3.17*.* For every $R > 0$, let us write $I_{\widetilde{\rho},\gamma}f(x)$ in the form

$$I_{\widetilde{\rho},\gamma}f(x) = I_{1,\widetilde{\rho}}(x) + I_{2,\widetilde{\rho}}(x),$$

where

$$I_{1,\widetilde{\rho}}(x) := \int_{B(x,R)} \frac{\widetilde{\rho}(|xy^{-1}|)}{(1+|xy^{-1}|)^{\gamma}} f(y)dy$$

and

$$I_{2,\widetilde{\rho}}(x) := \int_{B^c(x,R)} \frac{\widetilde{\rho}(|xy^{-1}|)}{(1+|xy^{-1}|)^{\gamma}} f(y)dy.$$

For $I_{1,\widetilde{\rho}}(x)$, we have

$$\begin{aligned}|I_{1,\widetilde{\rho}}(x)| &\le \int_{|xy^{-1}|<R} \frac{\widetilde{\rho}(|xy^{-1}|)}{(1+|xy^{-1}|)^{\gamma}} |f(y)|dy \\ &\le \int_{|xy^{-1}|<R} \frac{\widetilde{\rho}(|xy^{-1}|)}{|xy^{-1}|^{\gamma}} |f(y)|dy \\ &= \sum_{k=-\infty}^{-1} \int_{2^k R\le |xy^{-1}|<2^{k+1}R} \frac{\widetilde{\rho}(|xy^{-1}|)}{|xy^{-1}|^{\gamma}} |f(y)|dy.\end{aligned}$$

In view of (10.28) we can estimate

$$\begin{aligned}|I_{1,\widetilde{\rho}}(x)| &\le C \sum_{k=-\infty}^{-1} \frac{\widetilde{\rho}(2^k R)}{(2^k R)^{\gamma}} \int_{|xy^{-1}|<2^{k+1}R} |f(y)|dy \\ &\le CMf(x) \sum_{k=-\infty}^{-1} \frac{\widetilde{\rho}(2^k R)}{(2^k R)^{\gamma-Q}} \\ &\le CMf(x) \sum_{k=-\infty}^{-1} \int_{2^k R}^{2^{k+1}R} \frac{\widetilde{\rho}(t)}{t^{\gamma-Q+1}} dt \\ &= CMf(x) \int_0^R \frac{\widetilde{\rho}(t)}{t^{\gamma-Q+1}} dt,\end{aligned}$$

where we have used the fact that

$$\int_{2^k R}^{2^{k+1}R} \frac{\widetilde{\rho}(t)}{t^{\gamma-Q+1}} dt \ge C \frac{\widetilde{\rho}(2^k R)}{(2^k R)^{\gamma-Q+1}} 2^k R \ge C \frac{\widetilde{\rho}(2^k R)}{(2^k R)^{\gamma-Q}}. \tag{10.50}$$

Now, using (10.48) we obtain

$$|I_{1,\widetilde{\rho}}(x)| \le CMf(x)(\phi(R))^{(p-q)/q}. \tag{10.51}$$

For $I_{2,\widetilde{\rho}}(x)$ can estimate

$$\begin{aligned}|I_{2,\widetilde{\rho}}(x)| &\le \int_{|xy^{-1}|\ge R} \frac{\widetilde{\rho}(|xy^{-1}|)}{(1+|xy^{-1}|)^{\gamma}} |f(y)|dy \\ &\le \int_{|xy^{-1}|\ge R} \frac{\widetilde{\rho}(|xy^{-1}|)}{|xy^{-1}|^{\gamma}} |f(y)|dy \\ &= \sum_{k=0}^{\infty} \int_{2^k R\le |xy^{-1}|<2^{k+1}R} \frac{\widetilde{\rho}(|xy^{-1}|)}{|xy^{-1}|^{\gamma}} |f(y)|dy.\end{aligned}$$

Applying (10.28), we get

$$|I_{2,\widetilde{\rho}}(x)| \leq C \sum_{k=0}^{\infty} \frac{\widetilde{\rho}(2^k R)}{(2^k R)^{\gamma}} \int_{|xy^{-1}|<2^{k+1}R} |f(y)|dy.$$

From this using the Hölder inequality, we obtain

$$\begin{aligned}
|I_{2,\widetilde{\rho}}(x)| &\leq C \sum_{k=0}^{\infty} \frac{\widetilde{\rho}(2^k R)}{(2^k R)^{\gamma}} \left(\int_{|xy^{-1}|<2^{k+1}R} dy \right)^{1-1/p} \left(\int_{|xy^{-1}|<2^{k+1}R} |f(y)|dy \right)^{1/p} \\
&\leq C \sum_{k=0}^{\infty} \frac{\widetilde{\rho}(2^k R)}{(2^k R)^{\gamma - Q + \frac{Q}{p}}} \left(\int_{|xy^{-1}|<2^{k+1}R} |f(y)|dy \right)^{1/p} \\
&\leq C \|f\|_{LM^{p,\phi}(\mathbb{G})} \sum_{k=0}^{\infty} \frac{\widetilde{\rho}(2^{k+1}R)\phi(2^{k+1}R)}{(2^k R)^{\gamma-Q}} \\
&\leq C \|f\|_{LM^{p,\phi}(\mathbb{G})} \sum_{k=0}^{\infty} \int_{2^k R}^{2^{k+1}R} \frac{\widetilde{\rho}(t)\phi(t)}{t^{\gamma-Q+1}} dt \\
&= C \|f\|_{LM^{p,\phi}(\mathbb{G})} \int_{R}^{\infty} \frac{\widetilde{\rho}(t)\phi(t)}{t^{\gamma-Q+1}} dt,
\end{aligned}$$

where we have used the fact that

$$\int_{2^k R}^{2^{k+1}R} \frac{\widetilde{\rho}(t)\phi(t)}{t^{\gamma-Q+1}} dt \geq C \frac{\widetilde{\rho}(2^{k+1}R)\phi(2^{k+1}R)}{(2^{k+1}R)^{\gamma-Q+1}} 2^k R \geq C \frac{\widetilde{\rho}(2^{k+1}R)\phi(2^{k+1}R)}{(2^k R)^{\gamma-Q}}.$$

Now, using (10.48) we obtain

$$|I_{2,\widetilde{\rho}}(x)| \leq C \|f\|_{LM^{p,\phi}(\mathbb{G})} (\phi(R))^{p/q}. \tag{10.52}$$

Combining estimates (10.51) and (10.52) we get

$$|I_{\widetilde{\rho},\gamma} f(x)| \leq C(Mf(x)(\phi(R))^{(p-q)/q} + \|f\|_{LM^{p,\phi}(\mathbb{G})} (\phi(R))^{p/q}).$$

Assuming that f is not identically 0 and that $\mathscr{M} f$ is finite everywhere and then using the fact that ϕ is surjective, we can choose $R > 0$ such that

$$\phi(R) = \mathscr{M} f(x) \cdot \|f\|^{-1}_{LM^{p,\phi}(\mathbb{G})}.$$

Thus, for every $x \in \mathbb{G}$, we have

$$|I_{\widetilde{\rho},\gamma} f(x)| \leq C (\mathscr{M} f(x))^{\frac{p}{q}} \|f\|^{\frac{q-p}{q}}_{LM^{p,\phi}(\mathbb{G})}.$$

It follows that

$$\left(\int_{B(0,r)} |I_{\widetilde{\rho},\gamma} f(x)|^q \right)^{1/q} \leq C \left(\int_{B(0,r)} |\mathscr{M} f(x)|^p \right)^{1/q} \|f\|^{\frac{q-p}{q}}_{LM^{p,\phi}(\mathbb{G})}.$$

Now dividing both sides by $(\phi(r))^{p/q} r^{Q/q}$ we get

$$\frac{1}{(\phi(r))^{p/q}}\left(\frac{1}{r^Q}\int_{B(0,r)}|I_{\widetilde{\rho},\gamma}f(x)|^q\right)^{1/q}$$

$$\leq C\frac{1}{(\phi(r))^{p/q}}\left(\frac{1}{r^Q}\int_{B(0,r)}|\mathscr{M}f(x)|^p\right)^{1/q}\|f\|_{LM^{p,\phi}(\mathbb{G})}^{\frac{q-p}{q}}.$$

Taking the supremum over $r>0$ and using the boundedness of the maximal operator $\mathscr{M}$ on $LM^{p,\phi}(\mathbb{G})$ from (10.31), we obtain

$$\|I_{\widetilde{\rho},\gamma}f\|_{LM^{q,\phi^{p/q}}(\mathbb{G})}\leq C_{p,q,\phi,Q}\|f\|_{LM^{p,\phi}(\mathbb{G})}.$$

This gives (10.49). □

10.3.5 Olsen type inequalities for Bessel–Riesz operator

In this section, we show the Olsen type inequalities for the generalized Bessel–Riesz operator $I_{\rho,\gamma}$. Another type of this inequality will be shown in Theorem 10.3.22.

Theorem 10.3.18 (Olsen type inequalities for generalized Bessel–Riesz operators). *Let $\mathbb{G}$ be a homogeneous group of homogeneous dimension Q with a homogeneous quasi-norm $|\cdot|$. Let $\gamma>0$. Let $\widetilde{\rho}$ and ϕ satisfy the doubling condition* (10.28). *Assume that ϕ is surjective and satisfies* (10.47) *and* (10.48). *Then we have*

$$\|WI_{\widetilde{\rho},\gamma}f\|_{LM^{p,\phi}(\mathbb{G})}\leq C_{p,\phi,Q}\|W\|_{LM^{p_2,\phi^{p/p_2}}(\mathbb{G})}\|f\|_{LM^{p,\phi}(\mathbb{G})}, \tag{10.53}$$

provided that $1<p<p_2<\infty$ and $W\in LM^{p_2,\phi^{p/p_2}}(\mathbb{G})$.

Proof of Theorem 10.3.18. By using Hölder's inequality, we have

$$\frac{1}{r^Q}\int_{B(0,r)}|W(x)I_{\widetilde{\rho},\gamma}f(x)|^p dx$$

$$\leq\left(\frac{1}{r^Q}\int_{B(0,r)}|W(x)|^{p_2}dx\right)^{p/p_2}\left(\frac{1}{r^Q}\int_{B(0,r)}|I_{\widetilde{\rho},\gamma}f(x)|^{\frac{pp_2}{p_2-p}}dx\right)^{(p_2-p)/p_2}.$$

Now let us take the pth roots and then divide both sides by $\phi(r)$, so that

$$\frac{1}{\phi(r)}\left(\frac{1}{r^Q}\int_{B(0,r)}|W(x)I_{\widetilde{\rho},\gamma}f(x)|^p dx\right)^{1/p}$$

$$\leq\frac{1}{(\phi(r))^{p/p_2}}\left(\frac{1}{r^Q}\int_{B(0,r)}|W(x)|^{p_2}dx\right)^{1/p_2}$$

$$\times\frac{1}{(\phi(r))^{\frac{p_2-p}{p_2}}}\left(\frac{1}{r^Q}\int_{B(0,r)}|I_{\widetilde{\rho},\gamma}f(x)|^{\frac{pp_2}{p_2-p}}dx\right)^{(p_2-p)/pp_2}.$$

By taking the supremum over $r > 0$ and using the inequality (10.49), we get

$$\|WI_{\widetilde{\rho},\gamma}f\|_{LM^{p,\phi}(\mathbb{G})} \leq C_{p,\phi,Q}\|W\|_{LM^{p_2,\phi^{p/p_2}}(\mathbb{G})}\|I_{\widetilde{\rho},\gamma}f\|_{L^{\frac{pp_2}{p_2-p},\phi^{\frac{p_2-p}{p_2}}}(\mathbb{G})}.$$

Taking into account that $1 < p < \frac{pp_2}{p_2-p} < \infty$ and putting $q = \frac{pp_2}{p_2-p}$ in (10.49), we obtain (10.53). □

10.3.6 Fractional integral operators in Morrey spaces

In this section we extend the previous results to more general fractional integral operators, in particular establishing their boundedness and the Olsen type inequality on generalized Morrey spaces on homogeneous groups.

Definition 10.3.19 (Generalized fractional integral operator). We define the *generalized fractional integral operator* T_ρ by

$$T_\rho f(x) := \int_{\mathbb{G}} \frac{\rho(|xy^{-1}|)}{|xy^{-1}|^Q} f(y)dy, \tag{10.54}$$

where $\rho : \mathbb{R}^+ \to \mathbb{R}^+$ satisfies the doubling condition (10.28) and the condition

$$\int_0^1 \frac{\rho(t)}{t} dt < \infty. \tag{10.55}$$

As in the standard Euclidean case, taking $\rho(t) = t^\alpha$ with $0 < \alpha < Q$, we have the classical Riesz transform

$$T_\rho f(x) = I_\alpha f(x) = \int_{\mathbb{G}} \frac{1}{|xy^{-1}|^{Q-\alpha}} f(y)dy.$$

Theorem 10.3.20 (Fractional integral operators on Morrey spaces). *Let $\mathbb{G}$ be a homogeneous group of homogeneous dimension Q with a homogeneous quasi-norm $|\cdot|$. Let ρ and ϕ satisfy the doubling condition (10.28). Assume that ϕ is surjective and for some $1 < p < q < \infty$ satisfies the inequalities*

$$\int_r^\infty \frac{\phi(t)^p}{t} dt \leq C_1(\phi(r))^p, \tag{10.56}$$

and

$$\phi(r)\int_0^r \frac{\rho(t)}{t} dt + \int_r^\infty \frac{\rho(t)\phi(t)}{t} dt \leq C_2(\phi(r))^{p/q}, \tag{10.57}$$

for all $r > 0$. Then we have

$$\|T_\rho f\|_{LM^{q,\phi^{p/q}}(\mathbb{G})} \leq C_{p,q,\phi,Q}\|f\|_{LM^{p,\phi}(\mathbb{G})}. \tag{10.58}$$

Proof of Theorem 10.3.20. For every $R > 0$, let us write $T_\rho f(x)$ in the form

$$T_\rho f(x) = T_1(x) + T_2(x),$$

where

$$T_1(x) := \int_{B(x,R)} \frac{\rho(|xy^{-1}|)}{(|xy^{-1}|)^Q} f(y)dy$$

and

$$T_2(x) := \int_{B^c(x,R)} \frac{\rho(|xy^{-1}|)}{(|xy^{-1}|)^Q} f(y)dy.$$

For $T_1(x)$, we have

$$\begin{aligned}|T_1(x)| &\le \int_{|xy^{-1}|<R} \frac{\rho(|xy^{-1}|)}{|xy^{-1}|^Q}|f(y)|dy \\ &= \sum_{k=-\infty}^{-1} \int_{2^kR\le|xy^{-1}|<2^{k+1}R} \frac{\rho(|xy^{-1}|)}{|xy^{-1}|^Q}|f(y)|dy.\end{aligned}$$

By view of (10.28) we can estimate

$$\begin{aligned}|T_1(x)| &\le C \sum_{k=-\infty}^{-1} \frac{\rho(2^kR)}{(2^kR)^Q} \int_{|xy^{-1}|<2^{k+1}R} |f(y)|dy \\ &\le C\mathscr{M}f(x) \sum_{k=-\infty}^{-1} \rho(2^kR) \\ &\le C\mathscr{M}f(x) \sum_{k=-\infty}^{-1} \int_{2^kR}^{2^{k+1}R} \frac{\rho(t)}{t}dt \\ &= C\mathscr{M}f(x) \int_0^R \frac{\rho(t)}{t}dt.\end{aligned}$$

Here we have used the fact that

$$\int_{2^kR}^{2^{k+1}R} \frac{\rho(t)}{t}dt \ge C\rho(2^kR) \int_{2^kR}^{2^{k+1}R} \frac{1}{t}dt = C\rho(2^kR)\ln 2. \tag{10.59}$$

Now, using (10.57), we obtain

$$|T_1(x)| \le C\mathscr{M}f(x)(\phi(R))^{(p-q)/q}. \tag{10.60}$$

For $T_2(x)$ we estimate it as

$$\begin{aligned}|T_2(x)| &\le \int_{|xy^{-1}|\ge R} \frac{\rho(|xy^{-1}|)}{|xy^{-1}|^Q}|f(y)|dy \\ &= \sum_{k=0}^{\infty} \int_{2^kR\le|xy^{-1}|<2^{k+1}R} \frac{\rho(|xy^{-1}|)}{|xy^{-1}|^Q}|f(y)|dy.\end{aligned}$$

Applying (10.28) we get

$$|T_2(x)| \leq C \sum_{k=0}^{\infty} \frac{\rho(2^k R)}{(2^k R)^Q} \int_{|xy^{-1}|<2^{k+1}R} |f(y)| dy.$$

From this using Hölder's inequality we obtain

$$\begin{aligned}
|T_2(x)| &\leq C \sum_{k=0}^{\infty} \frac{\rho(2^k R)}{(2^k R)^Q} \left(\int_{|xy^{-1}|<2^{k+1}R} dy \right)^{1-1/p} \left(\int_{|xy^{-1}|<2^{k+1}R} |f(y)| dy \right)^{1/p} \\
&\leq C \sum_{k=0}^{\infty} \frac{\rho(2^k R)}{(2^k R)^{Q/p}} \left(\int_{|xy^{-1}|<2^{k+1}R} |f(y)| dy \right)^{1/p} \\
&\leq C \|f\|_{LM^{p,\phi}(\mathbb{G})} \sum_{k=0}^{\infty} \rho(2^{k+1}R)\phi(2^{k+1}R) \\
&\leq C \|f\|_{LM^{p,\phi}(\mathbb{G})} \sum_{k=0}^{\infty} \int_{2^k R}^{2^{k+1}R} \frac{\rho(t)\phi(t)}{t} \\
&= C \|f\|_{LM^{p,\phi}(\mathbb{G})} \int_{R}^{\infty} \frac{\rho(t)\phi(t)}{t},
\end{aligned}$$

where we have used the fact that

$$\begin{aligned}
\int_{2^k R}^{2^{k+1}R} \frac{\rho(t)\phi(t)}{t} dt &\geq C\rho(2^{k+1}R)\phi(2^{k+1}R) \int_{2^k R}^{2^{k+1}R} \frac{1}{t} dt \\
&= C\rho(2^{k+1}R)\phi(2^{k+1}R) \ln 2.
\end{aligned}$$

Now, in view of (10.57) we obtain

$$|T_2(x)| \leq C \|f\|_{LM^{p,\phi}(\mathbb{G})} (\phi(R))^{p/q}. \tag{10.61}$$

Combining estimates (10.60) and (10.61) we get

$$|T_\rho f(x)| \leq C(\mathscr{M} f(x)(\phi(R))^{(p-q)/q} + \|f\|_{LM^{p,\phi}(\mathbb{G})} (\phi(R))^{p/q}).$$

Assuming that f is not identically 0 and that $\mathscr{M} f$ is finite everywhere and then using the fact that ϕ is surjective, we can choose $R > 0$ such that

$$\phi(R) = \mathscr{M} f(x) \cdot \|f\|^{-1}_{LM^{p,\phi}(\mathbb{G})}.$$

Thus, for every $x \in \mathbb{G}$ we have

$$|T_\rho f(x)| \leq C(\mathscr{M} f(x))^{\frac{p}{q}} \|f\|^{\frac{q-p}{q}}_{LM^{p,\phi}(\mathbb{G})}.$$

It follows that

$$\left(\int_{B(0,r)} |T_\rho f(x)|^q\right)^{1/q} \le C \left(\int_{B(0,r)} |\mathscr{M} f(x)|^p\right)^{1/q} \|f\|_{LM^{p,\phi}(\mathbb{G})}^{\frac{q-p}{q}}.$$

Dividing both sides by $(\phi(r))^{p/q} r^{Q/q}$ we obtain

$$\begin{aligned}&\frac{1}{(\phi(r))^{p/q}} \left(\frac{1}{r^Q}\int_{B(0,r)} |T_\rho f(x)|^q\right)^{1/q}\\ &\le C\frac{1}{(\phi(r))^{p/q}} \left(\frac{1}{r^Q}\int_{B(0,r)} |\mathscr{M} f(x)|^p\right)^{1/q} \|f\|_{LM^{p,\phi}(\mathbb{G})}^{\frac{q-p}{q}}.\end{aligned}$$

Taking the supremum over $r > 0$ and using the boundedness of the maximal operator $\mathscr{M}$ on $LM^{p,\phi}(\mathbb{G})$ shown in (10.31), we obtain

$$\|T_\rho f\|_{LM^{q,\phi^{p/q}}(\mathbb{G})} \le C_{p,q,\phi,Q} \|f\|_{LM^{p,\phi}(\mathbb{G})}.$$

This gives (10.58). □

10.3.7 Olsen type inequalities for fractional integral operators

We now show Olsen type inequalities for the generalized fractional integral operator T_ρ and the Bessel–Riesz operator $I_{\alpha,\gamma}$. This continues the analysis from Section 10.3.5.

Theorem 10.3.21 (Olsen type inequalities for fractional integral operators)**.** *Let $\mathbb{G}$ be a homogeneous group of homogeneous dimension Q with a homogeneous quasi-norm $|\cdot|$. Let ρ and ϕ satisfy the doubling condition* (10.28)*. Assume that ϕ is surjective and satisfies* (10.56) *and* (10.57)*. Then we have*

$$\|WT_\rho f\|_{LM^{p,\phi}(\mathbb{G})} \le C_{p,\phi,Q} \|W\|_{LM^{p_2,\phi^{p/p_2}}(\mathbb{G})} \|f\|_{LM^{p,\phi}(\mathbb{G})}, \tag{10.62}$$

provided that $1 < p < p_2 < \infty$ and $W \in LM^{p_2,\phi^{p/p_2}}(\mathbb{G})$.

Proof of Theorem 10.3.21. By using Hölder's inequality, we have

$$\begin{aligned}&\frac{1}{r^Q}\int_{B(0,r)} |W(x)T_\rho f(x)|^p dx\\ &\le \left(\frac{1}{r^Q}\int_{B(0,r)} |W(x)|^{p_2} dx\right)^{p/p_2} \left(\frac{1}{r^Q}\int_{B(0,r)} |T_\rho f(x)|^{\frac{pp_2}{p_2-p}} dx\right)^{(p_2-p)/p_2}.\end{aligned}$$

Now let us take the pth roots and then divide both sides by $\phi(r)$ to obtain

$$\frac{1}{\phi(r)}\left(\frac{1}{r^Q}\int_{B(0,r)}|W(x)T_\rho f(x)|^p dx\right)^{1/p}$$
$$\leq \frac{1}{(\phi(r))^{p/p_2}}\left(\frac{1}{r^Q}\int_{B(0,r)}|W(x)|^{p_2} dx\right)^{1/p_2}$$
$$\times \frac{1}{(\phi(r))^{\frac{p_2-p}{p_2}}}\left(\frac{1}{r^Q}\int_{B(0,r)}|T_\rho f(x)|^{\frac{pp_2}{p_2-p}} dx\right)^{(p_2-p)/pp_2}.$$

By taking the supremum over $r > 0$ and using the inequality (10.58) we get

$$\|WT_\rho f\|_{LM^{p,\phi}(\mathbb{G})} \leq C_{p,\phi,Q}\|W\|_{LM^{p_2,\phi^{p/p_2}}(\mathbb{G})}\|T_\rho f\|_{L^{\frac{pp_2}{p_2-p},\phi^{\frac{p_2-p}{p_2}}}(\mathbb{G})}.$$

Taking into account that $1 < p < \frac{pp_2}{p_2-p} < \infty$ and putting $q = \frac{pp_2}{p_2-p}$ in (10.58) we obtain (10.62). □

Theorem 10.3.22 (Olsen type inequalities for Bessel–Riesz operators). *Let $\mathbb{G}$ be a homogeneous group of homogeneous dimension Q with a homogeneous quasi-norm $|\cdot|$. Let $1 < p < \infty$, $0 < \alpha < Q$ and $\gamma > 0$. Let $\omega : \mathbb{R}^+ \to \mathbb{R}^+$ satisfy the doubling condition and assume that $0 < \omega(r) \leq Cr^{-\alpha}$ for every $r > 0$, so that $K_{\alpha,\gamma} \in LM^{p_2,\omega}(\mathbb{G})$ for $\frac{Q}{Q+\gamma-\alpha} < p_2 < \frac{Q}{Q-\alpha}$ and $p_2 \geq 1$, where $q = \frac{\beta p}{\beta+Q-\alpha}$. Assume that $0 < \phi(r) \leq Cr^\beta$ for all $r > 0$, where $\beta < -\alpha < -Q-\beta$. Then we have*

$$\|WI_{\alpha,\gamma}f\|_{LM^{p,\phi}(\mathbb{G})} \leq C_{p,\phi,Q}\|W\|_{LM^{p_2,\phi^{p/p_2}}(\mathbb{G})}\|f\|_{LM^{p,\phi}(\mathbb{G})}, \tag{10.63}$$

provided that $W \in LM^{p_2,\phi^{p/p_2}}(\mathbb{G})$, where $\frac{1}{p_2} = \frac{1}{p} - \frac{1}{q}$.

Proof of Theorem 10.3.22. As in Theorem 10.3.21, by using Hölder's inequality for $\frac{1}{p_2} + \frac{1}{q} = \frac{1}{p}$, we have

$$\frac{1}{r^Q}\int_{B(0,r)}|W(x)I_{\alpha,\gamma}f(x)|^p dx$$
$$\leq \left(\frac{1}{r^Q}\int_{B(0,r)}|W(x)|^{p_2} dx\right)^{p/p_2}\left(\frac{1}{r^Q}\int_{B(0,r)}|I_{\alpha,\gamma}f(x)|^q dx\right)^{p/q}.$$

Now we take the pth roots and then divide both sides by $\phi(r)$ to get

$$\begin{aligned}\frac{1}{\phi(r)}&\left(\frac{1}{r^Q}\int_{B(0,r)}|W(x)I_{\alpha,\gamma}f(x)|^p dx\right)^{1/p}\\&\leq\frac{1}{(\phi(r))^{p/p_2}}\left(\frac{1}{r^Q}\int_{B(0,r)}|W(x)|^{p_2}dx\right)^{1/p_2}\\&\times\frac{1}{(\phi(r))^{p/q}}\left(\frac{1}{r^Q}\int_{B(0,r)}|I_{\alpha,\gamma}f(x)|^q dx\right)^{1/q}.\end{aligned}$$

By taking the supremum over $r>0$ we have

$$\|WI_{\alpha,\gamma}f\|_{LM^{p,\phi}(\mathbb{G})}\leq C\|W\|_{LM^{p_2,\phi^{p/p_2}}(\mathbb{G})}\|I_{\alpha,\gamma}f\|_{LM^{q,\phi^{p/q}}(\mathbb{G})}.$$

Putting $\psi(r)=(\phi(r))^{p/q}$ and using Theorem 10.3.14 we obtain (10.63). □

10.3.8 Summary of results

Let us now briefly summarize and collect the main results shown in this section, for a clearer overview of the statements.

Corollary 10.3.23. *Let $\mathbb{G}$ be a homogeneous group of homogeneous dimension Q with a homogeneous quasi-norm $|\cdot|$. Then we have the following properties:*

1. *Let $K_{\alpha,\gamma}(x):=\frac{|x|^{\alpha-n}}{(1+|x|)^\gamma}$. If $0<\alpha<Q$ and $\gamma>0$, then $K_{\alpha,\gamma}\in L^{p_1}(\mathbb{G})$ for $\frac{Q}{Q+\gamma-\alpha}<p_1<\frac{Q}{Q-\alpha}$, and*

$$\|K_{\alpha,\gamma}\|_{L^{p_1}(\mathbb{G})}\sim\left(\sum_{k\in\mathbb{Z}}\frac{(2^kR)^{(\alpha-Q)p_1+Q}}{(1+2^kR)^{\gamma p_1}}\right)^{1/p_1}$$

   for any $R>0$.

2. *For any $f\in LM^{p,\phi}(\mathbb{G})$ and $1<p<\infty$, we have*

$$\|\mathscr{M}f\|_{LM^{p,\phi}(\mathbb{G})}\leq C_p\|f\|_{LM^{p,\phi}(\mathbb{G})},$$

   where generalized local Morrey space $LM^{p,\phi}(\mathbb{G})$ and the Hardy–Littlewood maximal operator $\mathscr{M}$ are defined in (10.27) and (10.29), respectively.

3. *Let $I_{\alpha,\gamma}$ be a Bessel–Riesz operator on a homogeneous group defined in (10.24). Let $\gamma>0$ and $0<\alpha<Q$. If $0<\phi(r)\leq Cr^\beta$ for every $r>0$, $\beta<-\alpha$, $1<p<\infty$, and $\frac{Q}{Q+\gamma-\alpha}<p_1<\frac{Q}{Q-\alpha}$, then for all $f\in LM^{p,\phi}(\mathbb{G})$ we have*

$$\|I_{\alpha,\gamma}f\|_{LM^{q,\psi}(\mathbb{G})}\leq C_{p,\phi,Q}\|K_{\alpha,\gamma}\|_{L^{p_1}(\mathbb{G})}\|f\|_{LM^{p,\phi}(\mathbb{G})},$$

   where $q=\frac{\beta p_1' p}{\beta p_1'+Q}$ and $\psi(r)=(\phi(r))^{p/q}$.

4. *Let $\gamma > 0$, $1 < p < \infty$ and $0 < \alpha < Q$. If $0 < \phi(r) \le Cr^{\beta}$ for all $r > 0$, $\beta < -\alpha$, $\frac{Q}{Q+\gamma-\alpha} < p_2 \le p_1 < \frac{Q}{Q-\alpha}$ and $p_2 \ge 1$, then for all $f \in LM^{p,\phi}(\mathbb{G})$ we have*

$$\|I_{\alpha,\gamma} f\|_{LM^{q,\psi}(\mathbb{G})} \le C_{p,\phi,Q} \|K_{\alpha,\gamma}\|_{LM^{p_2,p_1}(\mathbb{G})} \|f\|_{LM^{p,\phi}(\mathbb{G})},$$

where $q = \frac{\beta p_1' p}{\beta p_1' + Q}$ and $\psi(r) = (\phi(r))^{p/q}$.

5. *Let $1 < p < \infty$. Let $\omega : \mathbb{R}^+ \to \mathbb{R}^+$ satisfy the doubling condition and assume that $0 < \omega(r) \le Cr^{-\alpha}$ for all $r > 0$, so that $K_{\alpha,\gamma} \in LM^{p_2,\omega}(\mathbb{G})$ for $\frac{Q}{Q+\gamma-\alpha} < p_2 < \frac{Q}{Q-\alpha}$ and $p_2 \ge 1$, where $0 < \alpha < Q$ and $\gamma > 0$. If $0 < \phi(r) \le Cr^{\beta}$ for all $r > 0$, where $\beta < -\alpha < -Q - \beta$, then for all $f \in LM^{p,\phi}(\mathbb{G})$ we have*

$$\|I_{\alpha,\gamma} f\|_{LM^{q,\psi}(\mathbb{G})} \le C_{p,\phi,Q} \|K_{\alpha,\gamma}\|_{LM^{p_2,\omega}(\mathbb{G})} \|f\|_{LM^{p,\phi}(\mathbb{G})},$$

where $q = \frac{\beta p}{\beta + Q - \alpha}$ and $\psi(r) = (\phi(r))^{p/q}$.

6. *Let $I_{\rho,\gamma}$ be the generalized Bessel–Riesz operator defined in (10.45). Let $\gamma > 0$ and let ρ and ϕ satisfy the doubling condition (10.28). Let $1 < p < q < \infty$. Assume that ϕ is surjective and satisfies*

$$\int_r^{\infty} \frac{\phi(t)^p}{t} dt \le C_1(\phi(r))^p,$$

and

$$\phi(r) \int_0^r \frac{\rho(t)}{t^{\gamma-Q+1}} dt + \int_r^{\infty} \frac{\rho(t)\phi(t)}{t^{\gamma-Q+1}} dt \le C_2(\phi(r))^{p/q},$$

for all $r > 0$. Then we have

$$\|I_{\rho,\gamma} f\|_{LM^{q,\phi^{p/q}}(\mathbb{G})} \le C_{p,q,\phi,Q} \|f\|_{LM^{p,\phi}(\mathbb{G})}.$$

7. *Let ρ and ϕ satisfy the doubling condition (10.28). Let $\gamma > 0$ and assume that ϕ is surjective and satisfies (10.47) and (10.48). Then we have*

$$\|W I_{\rho,\gamma} f\|_{LM^{p,\phi}(\mathbb{G})} \le C_{p,\phi,Q} \|W\|_{LM^{p_2,\phi^{p/p_2}}(\mathbb{G})} \|f\|_{LM^{p,\phi}(\mathbb{G})}, \quad 1 < p < p_2 < \infty,$$

provided that $W \in LM^{p_2,\phi^{p/p_2}}(\mathbb{G})$.

8. *Let T_ρ be the generalized fractional integral operator defined in (10.54). Let ρ and ϕ satisfy the doubling condition (10.28). Let $1 < p < q < \infty$. Assume that ϕ is surjective and satisfies*

$$\int_r^{\infty} \frac{\phi(t)^p}{t} dt \le C_1(\phi(r))^p,$$

and

$$\phi(r) \int_0^r \frac{\rho(t)}{t} dt + \int_r^{\infty} \frac{\rho(t)\phi(t)}{t} dt \le C_2(\phi(r))^{p/q},$$

for all $r > 0$. Then we have

$$\|T_\rho f\|_{LM^{q,\phi^{p/q}}(\mathbb{G})} \le C_{p,q,\phi,Q}\|f\|_{LM^{p,\phi}(\mathbb{G})}.$$

9. *Let ρ and ϕ satisfy the doubling condition (10.28). Assume that ϕ is surjective and satisfies (10.56) and (10.57). Then we have*

$$\|WT_\rho f\|_{LM^{p,\phi}(\mathbb{G})} \le C_{p,\phi,Q}\|W\|_{LM^{p_2,\phi^{p/p_2}}(\mathbb{G})}\|f\|_{LM^{p,\phi}(\mathbb{G})},\quad 1 < p < p_2 < \infty,$$

provided that $W \in LM^{p_2,\phi^{p/p_2}}(\mathbb{G})$.

10. *Let $\omega : \mathbb{R}^+ \to \mathbb{R}^+$ satisfy the doubling condition and assume that $0 < \omega(r) \le Cr^{-\alpha}$ for all $r > 0$, so that $K_{\alpha,\gamma} \in LM^{p_2,\omega}(\mathbb{G})$ for $\frac{Q}{Q+\gamma-\alpha} < p_2 < \frac{Q}{Q-\alpha}$ and $p_2 \ge 1$, where $0 < \alpha < Q$, $1 < p < \infty$, $q = \frac{\beta p}{\beta+Q-\alpha}$ and $\gamma > 0$. If $0 < \phi(r) \le Cr^\beta$ for all $r > 0$, where $\beta < -\alpha < -Q-\beta$, then we have*

$$\|WI_{\alpha,\gamma} f\|_{LM^{p,\phi}(\mathbb{G})} \le C_{p,\phi,Q}\|W\|_{LM^{p_2,\phi^{p/p_2}}(\mathbb{G})}\|f\|_{LM^{p,\phi}(\mathbb{G})},$$

provided that $W \in LM^{p_2,\phi^{p/p_2}}(\mathbb{G})$, where $\frac{1}{p_2} = \frac{1}{p} - \frac{1}{q}$.

11. *Let the generalized local Campanato space $\mathcal{L}\mathcal{M}^{p,\psi}(\mathbb{G})$ and operator $\widetilde{T}_\rho$ be defined in (10.67) and (10.70), respectively. Let ρ satisfy (10.55), (10.28), (10.68), (10.69), and let ϕ satisfy the doubling condition (10.28) and $\int_1^\infty \frac{\phi(t)}{t}dt < \infty$. If*

$$\int_r^\infty \frac{\phi(t)}{t}dt \int_0^r \frac{\rho(t)}{t}dt + r\int_r^\infty \frac{\rho(t)\phi(t)}{t^2}dt \le C_3\psi(r) \quad \text{for all} \quad r > 0,$$

then we have

$$\|\widetilde{T}_\rho f\|_{\mathcal{L}\mathcal{M}^{p,\psi}(\mathbb{G})} \le C_{p,\phi,Q}\|f\|_{\mathcal{L}\mathcal{M}^{p,\phi}(\mathbb{G})},\quad 1 < p < \infty.$$

10.4 Besov type space: Gagliardo–Nirenberg inequalities

Now we discuss a family of Gagliardo–Nirenberg type inequalities on homogeneous groups. The formulation is based on the Besov type space $B^\alpha(\mathbb{R}\times\wp)$, which we define as the space of all tempered distributions f on $\mathbb{R}\times\wp$ with the norm

$$\|g\|_{B^\alpha(\mathbb{R}\times\wp)} := \sup_{t>0}\{t^{-\alpha/2}\|Fe^{-tA^2}F^{-1}f\|_{L^\infty(\mathbb{R}\times\wp)}\} < \infty, \tag{10.64}$$

where the operators F and A are defined in Section 1.3.4, and $\wp$ is the quasi-sphere from the polar decomposition in Proposition 1.2.10.

Now we state the Gagliardo–Nirenberg type inequalities:

Theorem 10.4.1 (Gagliardo–Nirenberg type inequalities with Besov norms). *Let $\mathbb{G}$ be a homogeneous group of homogeneous dimension Q. Let $1 \leq p < q < \infty$. Then there exists a positive constant $C = C(p,q) > 0$ such that we have*

$$\|f\|_{L^q(\mathbb{R}\times\wp)} \leq C\|\mathcal{R}f\|^{p/q}_{L^p(\mathbb{R}\times\wp)}\|f\|^{1-p/q}_{B^{p/(p-q)}(\mathbb{R}\times\wp)} \tag{10.65}$$

for all $f \in B^{p/(p-q)}(\mathbb{R}\times\wp)$ such that $\mathcal{R}f \in L^p(\mathbb{R}\times\wp)$.

The following one-dimensional result will be useful for the proof:

Theorem 10.4.2 ([Led03, Theorem 1]). *Let $1 \leq p < q < \infty$. Then for every function $f \in L^p_1(\mathbb{R}^n)$ there exists a positive constant $C = C(p,q,n) > 0$ such that*

$$\|f\|_{L^q(\mathbb{R}^n)} \leq C\|\nabla f\|^{p/q}_{L^p(\mathbb{R}^n)}\|f\|^{1-p/q}_{B^{p/(p-q)}_{\infty,\infty}(\mathbb{R}^n)}, \tag{10.66}$$

where

$$\|f\|_{B^{\alpha}_{\infty,\infty}(\mathbb{R}^n)} := \sup_{t>0,\ x\in\mathbb{R}^n}\left\{t^{-\alpha/2}\left|\frac{1}{(4\pi t)^{n/2}}\int_{\mathbb{R}^n} f(y)e^{-|x-y|^2/4t}dy\right|\right\}.$$

Proof of Theorem 10.4.1. Using Theorem 10.4.2 with $n=1$ and Theorem 1.3.5, we obtain

$$\begin{aligned}
\int_{\mathbb{R}}|f(r,y)|^q dr &\leq C^q\int_{\mathbb{R}}\left|\frac{\partial f(r,y)}{\partial r}\right|^p dr\\
&\quad\times\left(\sup_{t>0,r\in\mathbb{R}} t^{p/(2(q-p))}\left|\frac{1}{\sqrt{4\pi t}}\int_{\mathbb{R}} e^{-(r-s)^2/(4t)}f(s,y)ds\right|\right)^{q-p}\\
&= C^q\int_{\mathbb{R}}|\mathcal{R}f(r,y)|^p dr\left(\sup_{t>0,r\in\mathbb{R}} t^{p/(2(q-p))}\left|Fe^{-tA^2}F^{-1}f(r,y)\right|\right)^{q-p}\\
&\leq C^q\int_{\mathbb{R}}|\mathcal{R}f(r,y)|^p dr\left(\sup_{t>0} t^{p/(2(q-p))}\left\|Fe^{-tA^2}F^{-1}f\right\|_{L^\infty(\mathbb{R}\times\wp)}\right)^{q-p}\\
&= C^q\int_{\mathbb{R}}|\mathcal{R}f(r,y)|^p dr\|f\|^{q-p}_{B^{p/(p-q)}(\mathbb{R}\times\wp)},
\end{aligned}$$

for any $y\in\wp$, in view of (10.64). One obtains (10.65) after integrating the above inequality with respect to y over $\wp$. □

10.5 Generalized Campanato spaces

In this section we show the boundedness of the extended version of fractional integral operators in Campanato spaces on homogeneous groups.

As before, throughout this section $\mathbb{G}$ is a homogeneous group of homogeneous dimension Q with a homogeneous quasi-norm $|\cdot|$.

Definition 10.5.1 (Generalized local Campanato spaces). Assume that the function $\frac{\phi(r)}{r}$ is nonincreasing. Denote

$$f_B = f_{B(0,r)} := \frac{1}{r^Q}\int_{B(0,r)} f(y)dy,$$

where $B(0,r) := \{x \in \mathbb{G} : |x| < r\}$. We define the *generalized local Campanato space* by

$$\mathcal{LM}^{p,\phi}(\mathbb{G}) := \{f \in L^p_{\mathrm{loc}}(\mathbb{G}) : \|f\|_{\mathcal{LM}^{p,\phi}(\mathbb{G})} < \infty\}, \tag{10.67}$$

where

$$\|f\|_{\mathcal{LM}^{p,\phi}(\mathbb{G})} := \sup_{r>0}\frac{1}{\phi(r)}\left(\frac{1}{r^Q}\int_{B(0,r)}|f(x)-f_B|^p dx\right)^{1/p}.$$

Sometimes these spaces are called central Camponato spaces in the literature.

Definition 10.5.2 (Modified generalized fractional integral operators). Let $\rho : \mathbb{R}^+ \to \mathbb{R}^+$ satisfy (10.55), (10.28) and the following conditions:

$$\int_r^\infty \frac{\rho(t)}{t^2}dt \le C_1\frac{\rho(r)}{r} \quad \text{for all } r > 0; \tag{10.68}$$

$$\frac{1}{2} \le \frac{r}{s} \le 2 \Longrightarrow \left|\frac{\rho(r)}{r^Q} - \frac{\rho(s)}{s^Q}\right| \le C_2|r-s|\frac{\rho(s)}{s^{Q+1}}. \tag{10.69}$$

We define the *modified version of the generalized fractional integral operator* T_ρ by

$$\widetilde{T_\rho}f(x) := \int_{\mathbb{G}}\left(\frac{\rho(|xy^{-1}|)}{|xy^{-1}|^Q} - \frac{\rho(|y|)(1-\chi_{B(0,1)}(y))}{|y|^Q}\right)f(y)dy, \tag{10.70}$$

where $\chi_{B(0,1)}$ is the characteristic function of $B(0,1)$.

For instance, the function $\rho(r) = r^\alpha$ satisfies (10.55), (10.28) and (10.69) for $0 < \alpha < Q$, and also satisfies (10.68) for $0 < \alpha < 1$.

Theorem 10.5.3 (Fractional integral operators on Camponato spaces). *Let $\mathbb{G}$ be a homogeneous group of homogeneous dimension Q with a homogeneous quasi-norm $|\cdot|$. Let ρ satisfy* (10.55), (10.28), (10.68), (10.69), *and let ϕ satisfy the doubling condition* (10.28) *and $\int_1^\infty \frac{\phi(t)}{t}dt < \infty$. If*

$$\int_r^\infty \frac{\phi(t)}{t}dt\int_0^r \frac{\rho(t)}{t}dt + r\int_r^\infty \frac{\rho(t)\phi(t)}{t^2}dt \le C_3\psi(r) \quad \textit{for all } r > 0, \tag{10.71}$$

then we have

$$\|\widetilde{T}_\rho f\|_{\mathcal{LM}^{p,\psi}(\mathbb{G})} \leq C_{p,\phi,Q}\|f\|_{\mathcal{LM}^{p,\phi}(\mathbb{G})}, \quad 1<p<\infty. \tag{10.72}$$

Proof of Theorem 10.5.3. For every $x \in B(0,r)$ and $f \in \mathcal{LM}^{p,\phi}(\mathbb{G})$, let us write $\widetilde{T}_\rho f$ in the form

$$\begin{aligned}\widetilde{T}_\rho f(x) &= \widetilde{T}_{B(0,r)}(x) + C^1_{B(0,r)} + C^2_{B(0,r)}\\ &= \widetilde{T}^1_{B(0,r)}(x) + \widetilde{T}^2_{B(0,r)}(x) + C^1_{B(0,r)} + C^2_{B(0,r)},\end{aligned}$$

where

$$\widetilde{T}_{B(0,r)}(x) := \int_{\mathbb{G}} (f(y)-f_{B(0,2r)})\left(\frac{\rho(|xy^{-1}|)}{|xy^{-1}|^Q} - \frac{\rho(|y|)(1-\chi_{B(0,2r)}(y))}{|y|^Q}\right)dy,$$

$$C^1_{B(0,r)} := \int_{\mathbb{G}} (f(y)-f_{B(0,2r)})\left(\frac{\rho(|y|)(1-\chi_{B(0,2r)}(y))}{|y|^Q} - \frac{\rho(|y|)(1-\chi_{B(0,1)}(y))}{|y|^Q}\right)dy,$$

$$C^2_{B(0,r)} := \int_{\mathbb{G}} f_{B(0,2r)}\left(\frac{\rho(|xy^{-1}|)}{|xy^{-1}|^Q} - \frac{\rho(|y|)(1-\chi_{B(0,1)}(y))}{|y|^Q}\right)dy,$$

$$\widetilde{T}^1_{B(0,r)}(x) := \int_{B(0,2r)} (f(y)-f_{B(0,2r)})\frac{\rho(|xy^{-1}|)}{|xy^{-1}|^Q}dy,$$

$$\widetilde{T}^2_{B(0,r)}(x) := \int_{B^c(0,2r)} (f(y)-f_{B(0,2r)})\left(\frac{\rho(|xy^{-1}|)}{|xy^{-1}|^Q} - \frac{\rho(|y|)}{|y|^Q}\right)dy.$$

Since

$$\left|\frac{\rho(|y|)(1-\chi_{B(0,2r)}(y))}{|y|^Q} - \frac{\rho(|y|)(1-\chi_{B(0,1)}(y))}{|y|^Q}\right| \leq \begin{cases} 0, \ |y| < \min(1,2r) \text{ or } |y| \geq \max(1,2r);\\ \frac{\rho(|y|)}{|y|^Q} = \text{const}, \ \text{otherwise},\end{cases}$$

we have that $C^1_{B(0,r)}$ is finite.

Now let us show that $C^2_{B(0,r)}$ is finite. For this it is enough to prove that the following integral is finite:

$$\begin{aligned}&\int_{\mathbb{G}}\left(\frac{\rho(|xy^{-1}|)}{|xy^{-1}|^Q} - \frac{\rho(|y|)(1-\chi_{B(0,1)}(y))}{|y|^Q}\right)dy\\ &= \int_{\mathbb{G}}\left(\frac{\rho(|xy^{-1}|)}{|xy^{-1}|^Q} - \frac{\rho(|y|)}{|y|^Q}\right)dy + \int_{B(0,1)} \frac{\rho(|y|)}{|y|^Q}dy.\end{aligned}$$

Let us denote

$$A := \int_{\mathbb{G}}\left(\frac{\rho(|xy^{-1}|)}{|xy^{-1}|^Q} - \frac{\rho(|y|)}{|y|^Q}\right)dy.$$

For large $R>0$, we write A in the form

$$A = A_1 + A_2 + A_3,$$

where

$$A_1 = \int_{B(x,R)} \frac{\rho(|xy^{-1}|)}{|xy^{-1}|^Q} dy - \int_{B(0,R)} \frac{\rho(|y|)}{|y|^Q} dy,$$

$$A_2 = \int_{B(x,R+r)\setminus B(x,R)} \frac{\rho(|xy^{-1}|)}{|xy^{-1}|^Q} dy - \int_{B(x,R+r)\setminus B(0,R)} \frac{\rho(|y|)}{|y|^Q} dy,$$

$$A_3 = \int_{B^c(x,R+r)} \left(\frac{\rho(|xy^{-1}|)}{|xy^{-1}|^Q} - \frac{\rho(|y|)}{|y|^Q} \right) dy.$$

Since we have $\int_0^1 \frac{\rho(t)}{t} dt < +\infty$, it follows that

$$\frac{\rho(|xy^{-1}|)}{|xy^{-1}|^Q}, \frac{\rho(|y|)}{|y|^Q} \in L^1_{\mathrm{loc}}(\mathbb{G}),$$

and hence $A_1 = 0$.

For A_2, we have

$$\begin{aligned} |A_2| &\leq \int_{B(x,R+r)\setminus B(x,R-r)} \left(\frac{\rho(|xy^{-1}|)}{|xy^{-1}|^Q} + \frac{\rho(|y|)}{|y|^Q} \right) dy \\ &\sim ((R+r)^Q - (R-r)^Q) \frac{\rho(R)}{R^Q} \leq Cr \frac{\rho(R)}{R}. \end{aligned}$$

In view of conditions (10.28) and (10.68) we have that

$$|A_2| \leq Cr \frac{\rho(R)}{R} \to 0 \text{ as } R \to +\infty.$$

By (10.69) we have

$$\begin{aligned} |A_3| &\leq \int_{B^c(x,R+r)} \left| \frac{\rho(|xy^{-1}|)}{|xy^{-1}|^Q} - \frac{\rho(|y|)}{|y|^Q} \right| dy \\ &\leq C \int_{B^c(x,R+r)} ||xy^{-1}| - |y|| \frac{\rho(|xy^{-1}|)}{|xy^{-1}|^{Q+1}} dy. \end{aligned}$$

By using the triangle inequality from Proposition 1.2.4 and the symmetric property of homogeneous quasi-norms, we get

$$|A_3| \leq C||x| + |y^{-1}| - |y|| \int_{R+r}^{+\infty} \int_{\wp} \frac{\rho(t)}{t^{Q+1}} t^{Q-1} d\sigma(y) dt \leq C|\wp| r \int_{R+r}^{+\infty} \frac{\rho(t)}{t^2} dt.$$

The inequality (10.68) implies that the last integral is integrable and $|A_3| \to 0$ as $R \to +\infty$.

Summarizing these properties of A_1, A_2, A_3 we have that $A \to 0$ as $R \to +\infty$, so that $A = 0$ and hence

$$\int_{\mathbb{G}} \left(\frac{\rho(|xy^{-1}|)}{|xy^{-1}|^Q} - \frac{\rho(|y|)(1 - \chi_{B(0,1)}(y))}{|y|^Q} \right) dy = \int_{B(0,1)} \frac{\rho(|y|)}{|y|^Q} dy < \infty,$$

which implies that $C^2_{B(0,r)}$ is finite.

Now before estimating $\widetilde{T}^1_{B(0,r)}$, let us denote

$$\widetilde{f} := (f - f_{B(0,2r)})\chi_{B(0,2r)} \qquad \text{and} \qquad \widetilde{\phi}(r) := \int_r^\infty \frac{\phi(t)}{t} dt.$$

Then we have

$$\begin{aligned}|\widetilde{T}^1_{B(0,r)}(x)| &\leq \int_{B(0,2r)} |\widetilde{f}(y)| \frac{\rho(|xy^{-1}|)}{|xy^{-1}|^Q} dy \\ &= \sum_{k=-\infty}^{0} \int_{2^k r \leq |xy^{-1}| < 2^{k+1} r} \frac{\rho(|xy^{-1}|)}{|xy^{-1}|^Q} |\widetilde{f}(y)| dy.\end{aligned}$$

By using (10.28) and (10.59) we get

$$\begin{aligned}|\widetilde{T}^1_{B(0,r)}(x)| &\leq C \sum_{k=-\infty}^{0} \frac{\rho(2^k r)}{(2^k r)^Q} \int_{|xy^{-1}| < 2^{k+1} r} |\widetilde{f}(y)| dy \\ &\leq C\mathscr{M}\widetilde{f}(x) \sum_{k=-\infty}^{0} \rho(2^k r) \leq C\mathscr{M}\widetilde{f}(x) \sum_{k=-\infty}^{0} \rho(2^{k-1} r) \\ &\leq C\mathscr{M}\widetilde{f}(x) \sum_{k=-\infty}^{0} \int_{2^{k-1} r}^{2^k r} \frac{\rho(t)}{t} dt = C\mathscr{M}\widetilde{f}(x) \int_0^r \frac{\rho(t)}{t} dt.\end{aligned}$$

Now using (10.71) we have

$$|\widetilde{T}^1_{B(0,r)}(x)| \leq C \frac{\psi(r)}{\widetilde{\phi}(r)} \mathscr{M}\widetilde{f}(x).$$

Using (10.33) we have

$$\begin{aligned}&\frac{1}{\psi(r)} \left(\frac{1}{r^Q} \int_{B(0,r)} |\widetilde{T}^1_{B(0,r)}(x)|^p dx \right)^{1/p} \\ &\quad \leq C \frac{1}{\widetilde{\phi}(r) r^{Q/p}} \left(\int_{B(0,r)} |\mathscr{M}\widetilde{f}(x)|^p dx \right)^{1/p} \leq C \frac{1}{\widetilde{\phi}(r) r^{Q/p}} \|\widetilde{f}\|_{L^p(\mathbb{G})}.\end{aligned}$$

By the Minkowski inequality, we have

$$\begin{aligned}&\frac{1}{\widetilde{\phi}(r) r^{Q/p}} \|\widetilde{f}\|_{L^p(\mathbb{G})} = \frac{1}{\widetilde{\phi}(r) r^{Q/p}} \|(f - f_{B(0,2r)})\chi_{B(0,2r)}\|_{L^p(\mathbb{G})} \\ &\quad \leq C \frac{1}{\widetilde{\phi}(r) r^{Q/p}} (\|(f - \sigma(f))\chi_{B(0,2r)}\|_{L^p(\mathbb{G})} + (2r)^{Q/p} |f_{B(0,2r)} - \sigma(f)|),\end{aligned}$$

where $\sigma(f) = \lim_{r\to\infty} f_{B(0,r)}$.

Moreover, we obtain the following inequalities exactly in the same way as in the standard Euclidean case (see [EN04, Section 6]):

$$\|f-\sigma(f)\|_{LM^{p,\widetilde{\phi}}(\mathbb{G})} \le C_1\|f\|_{\mathcal{L}\mathcal{M}^{p,\phi}(\mathbb{G})},$$

and

$$|f_{B(0,r)}-\sigma(f)| \le C_2\|f\|_{\mathcal{L}\mathcal{M}^{p,\phi}(\mathbb{G})}\widetilde{\phi}(r).$$

Finally, using these inequalities we get the estimate for $\widetilde{T}^1_{B(0,r)}$ as

$$|\widetilde{T}^1_{B(0,r)}(x)| \le C\|f\|_{\mathcal{L}\mathcal{M}^{p,\phi}(\mathbb{G})}. \tag{10.73}$$

Now let us estimate $\widetilde{T}^2_{B(0,r)}$. By (10.28) and (10.69) we have

$$\begin{aligned}|\widetilde{T}^2_{B(0,r)}(x)| &\le \int_{B^c(0,2r)} |f(y)-f_{B(0,2r)}|\left|\frac{\rho(|xy^{-1}|)}{|xy^{-1}|^Q}-\frac{\rho(|y|)}{|y|^Q}\right|dy\\ &\le C\big||xy^{-1}|-|y|\big|\int_{|y|\ge 2r}|f(y)-f_{B(0,2r)}|\frac{\rho(|y|)}{|y|^{Q+1}}dy.\end{aligned}$$

By using the triangle inequality from Proposition 1.2.4 and symmetric property of homogeneous quasi-norms we get

$$\begin{aligned}|\widetilde{T}^2_{B(0,r)}(x)| &\le C\big||x|+|y^{-1}|-|y|\big|\int_{|y|\ge 2r}|f(y)-f_{B(0,2r)}|\frac{\rho(|y|)}{|y|^{Q+1}}dy\\ &\le C|x|\int_{|y|\ge 2r}|f(y)-f_{B(0,2r)}|\frac{\rho(|y|)}{|y|^{Q+1}}dy\\ &= C|x|\sum_{k=2}^{\infty}\int_{2^{k-1}r\le|y|<2^k r}\frac{\rho(|y|)|f(y)-f_{B(0,2r)}|}{|y|^{Q+1}}dy.\end{aligned}$$

By using (10.28) and Hölder's inequality we obtain

$$\begin{aligned}|\widetilde{T}^2_{B(0,r)}(x)| &\le C|x|\sum_{k=2}^{\infty}\frac{\rho(2^k r)}{(2^k r)^{Q+1}}\int_{|y|<2^k r}|f(y)-f_{B(0,2r)}|dy\\ &\le C|x|\sum_{k=2}^{\infty}\frac{\rho(2^k r)}{2^k r}\left(\frac{1}{(2^k r)^Q}\int_{|y|<2^k r}|f(y)-f_{B(0,2r)}|^p dy\right)^{1/p}.\end{aligned}$$

As in the Abelian case ([EN04]) we have

$$\left(\frac{1}{(2^k r)^Q}\int_{B(0,2^k r)}|f(y)-f_{B(0,2r)}|^p dy\right)^{1/p} \le C\|f\|_{\mathcal{L}\mathcal{M}^{p,\phi}(\mathbb{G})}\int_{2r}^{2^{k+1}r}\frac{\phi(s)}{s}ds,$$

for every $k\ge 2$. The inequality (10.59) implies that

$$\int_{2^k r}^{2^{k+1}r}\frac{\rho(t)}{t^2}dt \ge \frac{1}{2^{k+1}r}\int_{2^k r}^{2^{k+1}r}\frac{\rho(t)}{t}dt \ge C\frac{\rho(2^k r)}{2^k r}.$$

By using the last two inequalities we get

$$
\begin{aligned}
|\widetilde{T}^2_{B(0,r)}(x)| &\le C|x|\|f\|_{\mathcal{LM}^{p,\phi}(\mathbb{G})} \sum_{k=2}^{\infty} \frac{\rho(2^k r)}{2^k r} \int_{2r}^{2^{k+1}r} \frac{\phi(s)}{s} ds \\
&\le C|x|\|f\|_{\mathcal{LM}^{p,\phi}(\mathbb{G})} \sum_{k=2}^{\infty} \int_{2^k r}^{2^{k+1}r} \frac{\rho(t)}{t^2} \left(\int_{2r}^{t} \frac{\phi(s)}{s} ds \right) dt \\
&\le C|x|\|f\|_{\mathcal{LM}^{p,\phi}(\mathbb{G})} \int_{2r}^{\infty} \frac{\rho(t)}{t^2} \left(\int_{2r}^{t} \frac{\phi(s)}{s} ds \right) dt \\
&= C|x|\|f\|_{\mathcal{LM}^{p,\phi}(\mathbb{G})} \int_{2r}^{\infty} \left(\int_{s}^{\infty} \frac{\rho(t)}{t^2} dt \right) \frac{\phi(s)}{s} ds.
\end{aligned}
$$

Using (10.68) and then (10.71), it implies that

$$
|\widetilde{T}^2_{B(0,r)}(x)| \le Cr\|f\|_{\mathcal{LM}^{p,\phi}(\mathbb{G})} \int_{2r}^{\infty} \frac{\rho(s)\phi(s)}{s^2} ds \le C\psi(r)\|f\|_{\mathcal{LM}^{p,\phi}(\mathbb{G})}.
$$

Finally, it follows that

$$
\frac{1}{\psi(r)} \left(\frac{1}{r^Q} \int_{B(0,r)} |\widetilde{T}^2_{B(0,r)}(x)|^p dx \right)^{1/p} \le C\|f\|_{\mathcal{LM}^{p,\phi}(\mathbb{G})}. \tag{10.74}
$$

Combining estimates (10.73) and (10.74) we obtain (10.72). □

Chapter 11

Elements of Potential Theory on Stratified Groups

In this chapter, we discuss elements of the potential theory and the theory of boundary layer operators in the setting of stratified groups. The main tools for this analysis are the fundamental solution for the sub-Laplacian and Green's identities established in Section 1.4.4.

From a different perspective than ours, comparable problems have been considered by Folland and Stein [FS74], Geller [Gel90], Jerison [Jer81], Romero [Rom91], Capogna, Garofalo and Nhieu [CGN08], Bonfiglioli, Lanconelli and Uguzzoni [BLU07] and a number of other people. A general setting of degenerate elliptic operators was considered by Bony [Bon69]. However, it seems that the potential theory on stratified group based on the layer potentials is an ingredient missing in the literature, and here our presentation follows the development of such a theory in [RS17c] on general stratified groups and in [RS16c] on the Heisenberg group.

Elements of the developed potential theory and the constructed layer potentials are then used in this chapter to discuss several applications. In particular, we describe new well-posed (i.e., solvable in the classical sense) boundary value problems in addition to using it in solving the known problems such as Dirichlet and Neumann problems for the sub-Laplacian. Furthermore, the developed potential theory is used to derive trace formulae for the Newton potential of the sub-Laplacian to piecewise smooth surfaces, and using these conditions to construct the analogue of Kac's boundary value problem in the setting of stratified groups which is related to Kac's "principle of not feeling the boundary" for the sub-Laplacian.

In this section $\mathbb{G}$ will be a stratified group of homogeneous dimension $Q \geq 3$. Since our analysis is based on the fundamental solution for the sub-Laplacian we will be restricting our discussion to the case of $Q \geq 3$. This is not restrictive since it effectively rules out only the spaces $\mathbb{R}$ and $\mathbb{R}^2$ where the fundamental solution assumes a different form and where most things presented in this chapter are already known.

M. Ruzhansky, D. Suragan, *Hardy Inequalities on Homogeneous Groups*,
Progress in Mathematics 327, https://doi.org/10.1007/978-3-030-02895-4_12

11.1 Boundary value problems on stratified groups

In this section, we show how Green's identities established in Theorem 1.4.6 can be used to provide simple proofs for the well-posedness for a number of boundary value problems. In particular, we provide examples of different boundary conditions, such as Dirichlet, Neumann, Robin, mixed Dirichlet and Robin, or different types of conditions on different parts of the boundary. For simplicity we restrict the considerations in this section to zero boundary conditions only, otherwise these problems may become very delicate due to the presence of characteristic points, and we can refer to [DGN06] for a thorough analysis in this direction, and to Remark 11.1.7 for some discussion.

We also note that in the subsequently considered boundary value problems, we can assume without loss of generality (in the proofs) that functions are real-valued since otherwise we can always take real and imaginary parts which would then satisfy the same equations.

In this section Ω is always an *admissible domain* as given in Definition 1.4.4.

The following result is known by other methods but given Green's first formula in Theorem 1.4.6 its proof becomes elementary.

Proposition 11.1.1 (Dirichlet boundary value problem for sub-Laplacian). *Let Ω be an admissible domain in a stratified group $\mathbb{G}$. Then the Dirichlet boundary value problem*

$$\mathcal{L}u(x) = 0, \ \ x \in \Omega, \tag{11.1}$$

$$u(x) = 0, \ \ x \in \partial\Omega, \tag{11.2}$$

has the unique trivial solution $u \equiv 0$ in the class of functions $C^2(\Omega)\bigcap C^1(\overline{\Omega})$.

Proof of Proposition 11.1.1. Setting $v = u$ in (1.86), by (11.1) and (11.2) we get

$$\int_\Omega \widetilde{\nabla} u u d\nu = \int_\Omega \left(\widetilde{\nabla} u u + u\mathcal{L}u\right) d\nu = \int_{\partial\Omega} u\langle\widetilde{\nabla} u, d\nu\rangle = 0.$$

Therefore

$$\int_\Omega \sum_{k=1}^{N_1} |X_k u|^2 d\nu = 0,$$

that is, $X_k u = 0$, $k = 1, \ldots, N_1$. Since any element of a Jacobian basis of $\mathbb{G}$ is represented by Lie brackets of $\{X_1, \ldots, X_{N_1}\}$, we obtain that u is a constant, so $u \equiv 0$ on Ω by (11.2). ☐

This result has the following simple extension to stationary Schrödinger operators:

Proposition 11.1.2 (Dirichlet boundary value problem for Schrödinger operator). *Let Ω be an admissible domain in a stratified group $\mathbb{G}$. Let $q : \Omega \to \mathbb{R}$ be a non-negative bounded function: $q \in L^\infty(\Omega)$ and $q(x) \geq 0$ for all $x \in \Omega$. Then the Dirichlet boundary value problem for the Schrödinger equation*

$$-\mathcal{L}u(x) + q(x)u(x) = 0, \;\; x \in \Omega, \tag{11.3}$$

$$u(x) = 0, \;\; x \in \partial\Omega, \tag{11.4}$$

has the unique trivial solution $u \equiv 0$ in the class of functions $C^2(\Omega)\bigcap C^1(\overline{\Omega})$.

Proof of Proposition 11.1.2. Using Green's formula (1.86), from (11.3) and (11.4) we obtain

$$\begin{aligned}\int_\Omega \widetilde{\nabla} u u d\nu &= \int_\Omega \left(\widetilde{\nabla} u u + u\mathcal{L}u\right) d\nu - \int_\Omega q(y)|u(y)|^2 d\nu \\ &= \int_{\partial\Omega} u\langle\widetilde{\nabla} u, d\nu\rangle - \int_\Omega q(y)|u(y)|^2 d\nu \\ &= -\int_\Omega q(y)|u(y)|^2 d\nu.\end{aligned}$$

Therefore, we have

$$0 \leq \int_\Omega \sum_{k=1}^{N_1} |X_k u|^2 d\nu = -\int_\Omega q(y)|u(y)|^2 d\nu \leq 0,$$

that is, we must have $u \equiv 0$ because of (11.4). □

Similarly, we can treat von Neumann type boundary conditions.

Proposition 11.1.3 (Neumann boundary value problem for sub-Laplacian). *Let Ω be an admissible domain in a stratified group $\mathbb{G}$. Then the boundary value problem*

$$\mathcal{L}u(x) = 0, \;\; x \in \Omega \subset \mathbb{G}, \tag{11.5}$$

$$\sum_{j=1}^{N_1} X_j u\langle X_j, d\nu\rangle = 0 \;\; on \;\; \partial\Omega, \tag{11.6}$$

has only constant solutions $u \equiv$ const in the class of functions $C^2(\Omega)\bigcap C^1(\overline{\Omega})$.

Remark 11.1.4. Note that von Neumann type boundary value problems for the sub-Laplacian have been known and studied, see, e.g., [DGN06]. However, Proposition 11.1.3 provides a new measure type condition for the von Neumann type boundary value problem.

Proof of Proposition 11.1.3. Set $v = u$ in (1.86), then by (11.5) and (11.6) we get

$$\begin{aligned}\int_\Omega \widetilde{\nabla} u u d\nu &= \int_\Omega \left(\widetilde{\nabla} u u + u\mathcal{L}u\right) d\nu \\ &= \int_{\partial\Omega} u\langle \widetilde{\nabla} u, d\nu\rangle \\ &= \int_{\partial\Omega} u \sum_{j=1}^{N_1} X_j u \langle X_j, d\nu\rangle \\ &= 0.\end{aligned}$$

Therefore

$$\int_\Omega \sum_{k=1}^{N_1} |X_k u|^2 d\nu = 0,$$

that is, $X_k u = 0,\ k = 1, \ldots, N_1$. Since all vector fields in $\mathfrak{g}$ are represented by Lie brackets of $\{X_1, \ldots, X_{N_1}\}$, we obtain that u is a constant. □

Similarly, the Robin type boundary conditions can be also considered.

Proposition 11.1.5 (Robin boundary value problem for sub-Laplacian). *Let Ω be an admissible domain in a stratified group $\mathbb{G}$. Let $a_k : \partial\Omega \to \mathbb{R}$, $k = 1, \ldots, N_1$, be bounded functions such that the measure*

$$\sum_{j=1}^{N_1} a_j \langle X_j, d\nu\rangle \geq 0 \tag{11.7}$$

is non-negative on $\partial\Omega$. Then the boundary value problem

$$\mathcal{L}u(x) = 0,\ x \in \Omega \subset \mathbb{G}, \tag{11.8}$$

$$\sum_{j=1}^{N_1} (a_j u + X_j u)\langle X_j, d\nu\rangle = 0 \quad on \quad \partial\Omega, \tag{11.9}$$

has only constant solutions $u \equiv \mathrm{const}$ in the class of functions $C^2(\Omega)\bigcap C^1(\overline{\Omega})$.

Moreover, if the integral of the measure (11.7) is positive, i.e., if

$$\int_{\partial\Omega} \sum_{j=1}^{N_1} a_j \langle X_j, d\nu\rangle > 0, \tag{11.10}$$

then the boundary value problem (11.8)–(11.9) has the unique trivial solution $u \equiv 0$ in the class of functions $C^2(\Omega)\bigcap C^1(\overline{\Omega})$.

Proof of Proposition 11.1.5. Set $v = u$ in (1.86), then by (11.8) and (11.9) we get

$$\begin{aligned}\int_\Omega \widetilde{\nabla} u u d\nu &= \int_\Omega \left(\widetilde{\nabla} u u + u\mathcal{L}u\right) d\nu \\ &= \int_{\partial\Omega} u\langle \widetilde{\nabla} u, d\nu\rangle \\ &= \int_{\partial\Omega} u \sum_{j=1}^{N_1} X_j u\langle X_j, d\nu\rangle \\ &= -\int_{\partial\Omega} u^2 \sum_{j=1}^{N_1} a_j\langle X_j, d\nu\rangle.\end{aligned}$$

This means that

$$\int_\Omega \sum_{k=1}^{N_1} |X_k u|^2 d\nu = -\int_{\partial\Omega} u^2 \sum_{j=1}^{N_1} a_j\langle X_j, d\nu\rangle.$$

Therefore, we must have

$$\int_\Omega \sum_{k=1}^{N_1} |X_k u|^2 d\nu = 0 \qquad \text{and} \qquad \int_{\partial\Omega} u^2 \sum_{j=1}^{N_1} a_j\langle X_j, d\nu\rangle = 0.$$

As before the first equality implies that u is a constant and the first part of Proposition 11.1.5 is proved.

On the other hand, by the assumption (11.10) the second equality implies that $u = 0$ on $\partial\Omega$, and this implies that $u \equiv 0$ on Ω. □

The problems where Dirichlet or Robin conditions are imposed on different parts of the boundary can be also considered:

Proposition 11.1.6 (Dirichlet and Robin boundary value problem for sub-Laplacian)**.** *Let Ω be an admissible domain in a stratified group $\mathbb{G}$. Let $a_k : \partial\Omega \to \mathbb{R}$, $k = 1, \ldots, N_1$, be bounded functions such that the measure*

$$\sum_{j=1}^{N_1} a_j\langle X_j, d\nu\rangle \geq 0 \tag{11.11}$$

is non-negative on $\partial\Omega$. Let $\partial\Omega_1 \subset \partial\Omega$, $\partial\Omega_1 \neq \{\emptyset\}$ and $\partial\Omega_2 := \partial\Omega\backslash\partial\Omega_1$. Then the boundary value problem

$$\mathcal{L}u(x) = 0,\ x \in \Omega \subset \mathbb{G}, \tag{11.12}$$

$$u = 0 \ \ on \ \ \partial\Omega_1, \tag{11.13}$$

$$\sum_{j=1}^{N_1} (a_j u + X_j u)\langle X_j, d\nu\rangle = 0 \ \ on \ \ \partial\Omega_2, \tag{11.14}$$

has the unique trivial solution $u \equiv 0$ in the class of functions $C^2(\Omega)\bigcap C^1(\overline{\Omega})$.

Proposition 11.1.6 can be proved in the same way as Proposition 11.1.5.

Remark 11.1.7 (More general boundary value problems). In principle, one is certainly also interested in boundary value problems for non-zero boundary data or, more generally, for example in the Dirichlet case, for an admissible domain Ω, in

$$\begin{cases} \mathcal{L}u = f & \text{in } \Omega, \\ u = \phi & \text{on } \partial\Omega. \end{cases} \tag{11.15}$$

In the Euclidean case, when $\mathcal{L}$ is the Laplacian, which is elliptic, the boundary value problem (11.15) has a classical solution in $C^2(\Omega)\cap C^1(\overline{\Omega})$ for reasonably good functions f, ϕ, for example, for $f \in C^\alpha(\Omega)$ with $\alpha > 0$, and $\phi \in C(\partial\Omega)$.

However, in general, this fact fails completely for the hypoelliptic boundary value problem (11.15). Thus, already on the Heisenberg group $\mathbb{H}^n$, even if the domain Ω and the boundary datum ϕ are real analytic and $f \equiv 0$, D. Jerison [Jer81] gave an example when the solution of the Dirichlet problem (11.15) is not better than Hölder continuous near a characteristic boundary point, that is, the solution is not classical.

We recall that the *characteristic set of* Ω (related to vector fields $\{X_1, \ldots, X_{N_1}\}$ giving the first stratum of a stratified group $\mathbb{G}$) is the set

$$\{x \in \partial\Omega : X_k(x) \in T_x(\partial\Omega),\ k = 1, \ldots, N_1\},$$

with $T_x(\partial\Omega)$ being the tangent space to $\partial\Omega$ at the point x. We refer to Section 11.2 more discussions on characteristic points, and also to [CGN08, Section 4] for an extensive discussion on boundary value problems and additional conditions allowing one to partially handle the appearing characteristic points.

The appearance of characteristic points is hard to avoid, see the next section.

11.2 Layer potentials of the sub-Laplacian

Let $D \subset \mathbb{R}^N$ be an open set with boundary ∂D. Let D be a *domain of class* C^1, that is, a domain such that for every $x_0 \in \partial D$ there exist a neighbourhood U_{x_0} of x_0, and a function $\phi_{x_0} \in C^1(U_{x_0})$, with

$$|\nabla \phi_{x_0}| \geq \alpha > 0$$

in U_{x_0}, where ∇ is the standard gradient in $\mathbb{R}^N$, such that

$$D \cap U_{x_0} = \{x \in U_{x_0} \mid \phi_{x_0}(x) < 0\},$$
$$\partial D \cap U_{x_0} = \{x \in U_{x_0} \mid \phi_{x_0}(x) = 0\}.$$

A point $x_0 \in \partial D$ is called *characteristic* with respect to vector fields $\{X_1, \ldots, X_{N_1}\}$, if for U_{x_0}, ϕ_{x_0} as above we have

$$X_1\phi_{x_0}(x_0) = 0, \ldots, X_{N_1}\phi_{x_0}(x_0) = 0.$$

Usually, bounded domains have a non-empty collection of characteristic points. For instance, any bounded domain of class C^1 in the Heisenberg group $\mathbb{H}^n$ with the boundary homeomorphic to the $2n$-dimensional sphere $\mathbb{S}^{2n}$, has a non-empty characteristic set, see, e.g., [DGN06].

11.2.1 Single layer potentials

We start by defining single layer potentials for the sub-Laplacian and then analyse their basic properties.

Definition 11.2.1 (Single layer potentials)**.** Let the function $\varepsilon(y,x) = \varepsilon(x,y)$ be the fundamental solution of the sub-Laplacian as in (1.89). We define *single layer potentials* for an admissible domain Ω as the functionals

$$\mathcal{S}_j u(x) := \int_{\partial\Omega} u(y)\varepsilon(y,x)\langle X_j, d\nu(y)\rangle, \quad j = 1, \ldots, N_1, \tag{11.16}$$

where $\langle X_j, d\nu\rangle$ is the canonical pairing between vector fields and differential forms, and N_1 is the dimension of the first stratum of $\mathbb{G}$.

Remark 11.2.2. In [Jer81] Jerison used the single layer potential defined by

$$\mathcal{S}_0 u(x) = \int_{\partial\Omega} u(y)\varepsilon(y,x)dS(y).$$

However, it is not integrable over characteristic points of $\partial\Omega$, see [Rom91] for examples. On the contrary, as we will show in Lemma 11.2.3, the single layer potentials (11.16) are integrable over the whole boundary $\partial\Omega$ including the set of characteristic points.

We recall from (1.89) that

$$\varepsilon(x,y) = [d(x,y)]^{2-Q} = [d(x^{-1}y)]^{2-Q}, \tag{11.17}$$

with d being the $\mathcal{L}$-gauge as in (1.75).

Lemma 11.2.3 (Single layer potentials are well defined)**.** *Let $\partial\Omega$ be the boundary of an admissible domain Ω in a stratified group $\mathbb{G}$ of homogeneous dimension $Q \geq 3$. Then*

$$\int_{\partial\Omega} \varepsilon(x,y)\langle X_j, d\nu(y)\rangle$$

is a convergent integral for any $x \in \mathbb{G}$ such that $x \notin \partial\Omega$.

Proof of Lemma 11.2.3. Let $B_R := \{y : d(x,y) < R\}$ be a ball such that $\Omega \subset B_R$. In view of (11.17) we can estimate

$$\int_{\partial\Omega} [d(x,y)]^{2-Q}\langle X_j, d\nu(y)\rangle = \int_{\Omega} X_j[d(x,y)]^{2-Q} d\nu(y)$$

$$\leq \int_{\Omega} | X_j[d(x,y)]^{2-Q} | d\nu(y) \leq \int_{B_R} | X_j[d(x,y)]^{2-Q} | d\nu(y)$$
$$= C \int_0^R r^{1-Q} r^{Q-1} dr < \infty,$$

where we have used the polar decomposition from Proposition 1.2.10 with respect to the $\mathcal{L}$-gauge d. This proves Lemma 11.2.3. □

We now establish the main basic properties of the single layer potentials.

Theorem 11.2.4 (Continuity of single layer potentials). *Let $\partial\Omega$ be the boundary of an admissible domain Ω in a stratified group $\mathbb{G}$ of homogeneous dimension $Q \geq 3$. Let u be bounded on $\partial\Omega$, that is, $u \in L^\infty(\partial\Omega)$. Then the single layer potential $\mathcal{S}_j u$ is continuous on $\mathbb{G}$ for all $j = 1, \ldots, N_1$.*

Proof of Theorem 11.2.4. Let first $x_0 \in \mathbb{G}$ be such that $x_0 \notin \partial\Omega$. Then

$$|\mathcal{S}_j u(x) - \mathcal{S}_j u(x_0)| = \left| \int_{\partial\Omega} u(y)(\varepsilon(y,x) - \varepsilon(y,x_0))\langle X_j, d\nu(y)\rangle \right|$$
$$\leq \sup_{y\in\partial\Omega} |u(y)| \left| \int_{\partial\Omega} \varepsilon(y,x) - \varepsilon(y,x_0)\langle X_j, d\nu(y)\rangle \right| .$$

This means that

$$\lim_{x\to x_0} \mathcal{S}_j u(x) = \mathcal{S}_j u(x_0),$$

that is, the single layer potential $\mathcal{S}_j u$ is continuous on $\mathbb{G}\backslash\partial\Omega$.

Now let $x_0 \in \mathbb{G}$ be such that $x_0 \in \partial\Omega$. Let us denote

$$\Omega_\epsilon := \{y \in \Omega : d(x_0, y) < \epsilon\}.$$

Then we have

$$|\mathcal{S}_j u(x) - \mathcal{S}_j u(x_0)| = \left| \int_{\partial\Omega} u(y)(\varepsilon(y,x) - \varepsilon(y,x_0))\langle X_j, d\nu(y)\rangle \right|$$
$$= \sup_{y\in\partial\Omega} |u(y)| \left| \int_{\partial\Omega} (\varepsilon(y,x) - \varepsilon(y,x_0))\langle X_j, d\nu(y)\rangle \right|$$
$$= \sup_{y\in\partial\Omega} |u(y)| \left| \int_{\Omega} X_j(\varepsilon(y,x) - \varepsilon(y,x_0)) d\nu(y) \right|$$
$$\leq \sup_{y\in\partial\Omega} |u(y)| \lim_{\epsilon\to 0} \left(\left| \int_{\Omega\backslash\Omega_\epsilon} X_j(\varepsilon(y,x) - \varepsilon(y,x_0)) d\nu(y) \right| \right.$$
$$\left. + \left| \int_{\Omega_\epsilon} X_j(\varepsilon(y,x) - \varepsilon(y,x_0)) d\nu(y) \right| \right),$$

where the first term tends to zero when $x \to x_0$. Now what is left is to show that the second term tends to zero. This follows since

$$\begin{aligned}
&\lim_{\epsilon\to 0}\int_{\Omega_\epsilon} X_j(\varepsilon(y,x)-\varepsilon(y,x_0))d\nu(y)\\
&=\lim_{\epsilon\to 0}\left(\int_{\Omega_\epsilon} X_j\varepsilon(y,x)d\nu(y)-\int_{\Omega_\epsilon} X_j\varepsilon(y,x_0)d\nu(y)\right)\\
&=\lim_{\epsilon\to 0}(C\int_0^\epsilon r^{1-Q}r^{Q-1}dr)=C\lim_{\epsilon\to 0}\epsilon=0.
\end{aligned}$$

This completes the proof. □

11.2.2 Double layer potential

In this section, we discuss double layer potentials and establish Plemelj type jump relations for them.

Definition 11.2.5 (Double layer potential). We define the *double layer potential* as the operator

$$\mathcal{D}u(x) := \int_{\partial\Omega} u(y)\langle\widetilde{\nabla}\varepsilon(y,x), d\nu(y)\rangle, \tag{11.18}$$

where

$$\widetilde{\nabla}\varepsilon = \sum_{k=1}^{N_1}(X_k\varepsilon)X_k,$$

with vector fileds X_k acting on the y-variable.

We now describe the Plemelj type jump relations for the double layer potential $\mathcal{D}$.

Theorem 11.2.6 (Plemelj jump relations for double layer potential). *Let $\Omega \subset \mathbb{G}$ be an admissible domain in a stratified group $\mathbb{G}$ of homogeneous dimension $Q \geq 3$. Let $u \in C^1(\Omega)\bigcap C(\overline{\Omega})$. For $x_0 \in \partial\Omega$ define*

$$\mathcal{D}^0u(x_0) := \int_{\partial\Omega} u(y)\langle\widetilde{\nabla}\varepsilon(y,x_0), d\nu(y)\rangle,$$

$$\mathcal{D}^+u(x_0) := \lim_{x\to x_0,\, x\in\Omega}\int_{\partial\Omega} u(y)\langle\widetilde{\nabla}\varepsilon(y,x), d\nu(y)\rangle,$$

and

$$\mathcal{D}^-u(x_0) := \lim_{x\to x_0,\, x\notin\overline{\Omega}}\int_{\partial\Omega} u(y)\langle\widetilde{\nabla}\varepsilon(y,x), d\nu(y)\rangle.$$

Then $\mathcal{D}^+u(x_0), \mathcal{D}^-u(x_0)$ and $\mathcal{D}^0u(x_0)$ exist and verify the following jump relations:

$$\mathcal{D}^+u(x_0) - \mathcal{D}^-u(x_0) = u(x_0),$$

$$\mathcal{D}^0 u(x_0) - \mathcal{D}^- u(x_0) = \mathcal{J}(x_0)u(x_0),$$
$$\mathcal{D}^+ u(x_0) - \mathcal{D}^0 u(x_0) = (1 - \mathcal{J}(x_0))u(x_0),$$

where the jump value $\mathcal{J}(x_0)$ *is given by the formula*

$$\mathcal{J}(x_0) = \int_{\partial\Omega} \langle \widetilde{\nabla}\varepsilon(y, x_0), d\nu(y)\rangle,$$

in the sense of the (*Cauchy*) *principal value, and where*

$$\widetilde{\nabla}\varepsilon = \sum_{k=1}^{N_1} (X_k\varepsilon)\, X_k.$$

In order to prove Theorem 11.2.6 we will need the following property.

Lemma 11.2.7. *Let* $\Omega \subset \mathbb{G}$ *be an admissible domain with the boundary* $\partial\Omega$ *and let* $x_0 \in \partial\Omega$. *Let* $u \in C^1(\Omega)\bigcap C(\overline{\Omega})$. *Then*

$$\lim_{x\to x_0} \int_{\partial\Omega} [u(y) - u(x)]\langle\widetilde{\nabla}\varepsilon(y, x), d\nu(y)\rangle = \int_{\partial\Omega} [u(y) - u(x_0)]\langle\widetilde{\nabla}\varepsilon(y, x_0), d\nu(y)\rangle.$$

Proof of Lemma 11.2.7. To simplify the notation in this proof we will use the following Einstein type convention: if the index k is repeated in an integrand, it means that it is a sum over k from 1 to N_1. For example, we abbreviate it as follows:

$$\int_{\partial\Omega} [u(y)-u(x)]X_k\varepsilon(y,x)\langle X_k, d\nu(y)\rangle := \sum_{k=1}^{N_1}\int_{\partial\Omega} [u(y)-u(x)]X_k\varepsilon(y,x)\langle X_k, d\nu(y)\rangle.$$

First, let us show that

$$\lim_{\epsilon\to 0} \int_{d(x,y)<\epsilon} X_k \{(u(y) - u(x))X_k\varepsilon(y, x)\}\, d\nu(y) = 0.$$

By using the divergence formula in Theorem 1.4.5, we can estimate

$$\begin{aligned}
&\lim_{\epsilon\to 0}\left|\int_{d(x,y)<\epsilon} X_k \{(u(y) - u(x))X_k\varepsilon(y, x)\}\, d\nu(y)\right| \\
&\quad\le C_1 \lim_{\epsilon\to 0}\int_{d(x,y)<\epsilon} |X_k\varepsilon(y, x)|d\nu(y) \\
&\quad\quad + \lim_{\epsilon\to 0}\int_{d(x,y)<\epsilon} |(u(y) - u(x))X_kX_k\varepsilon(y, x)|d\nu(y) \\
&\quad\le C \lim_{\epsilon\to 0}\int_0^\epsilon dr = 0.
\end{aligned}$$

Therefore, we have

$$\begin{aligned}\int_{\Omega} & X_k \{[u(y)-u(x)]X_k\varepsilon(y,x)\}\, d\nu(y) \\ &= \int_{\Omega} X_k \{[u(y)-u(x)]X_k\varepsilon(y,x)\}\, d\nu(y) \\ &\quad + \lim_{\epsilon\to 0}\int_{d(x,y)<\epsilon} X_k \{(u(y)-u(x))X_k\varepsilon(y,x)\}\, d\nu(y).\end{aligned}$$

If we take $\Omega_\epsilon = \{y \in \mathbb{G} : d(x,y) < \epsilon\}$, then by the divergence formula in Theorem 1.4.5 we have

$$\begin{aligned}\lim_{x\to x_0} & \int_{\partial\Omega} [u(y)-u(x)]X_k\varepsilon(y,x)\langle X_k, d\nu(y)\rangle \\ &= \lim_{x\to x_0}\int_{\Omega} X_k \{[u(y)-u(x)]X_k\varepsilon(y,x)\}\, d\nu(y) \\ &= \lim_{x\to x_0}\lim_{\epsilon\to 0}\left\{\int_{\Omega\setminus\Omega_\epsilon} X_k \{[u(y)-u(x)]X_k\varepsilon(y,x)\}\, d\nu(y)\right. \\ &\quad \left. + \int_{\Omega_\epsilon} X_k \{[u(y)-u(x)]X_k\varepsilon(y,x)\}\, d\nu(y)\right\} \\ &= \int_{\Omega} X_k \{[u(y)-u(x_0)]X_k\varepsilon(y,x_0)\}\, d\nu(y).\end{aligned}$$

That is, we have

$$\begin{aligned}\lim_{x\to x_0} & \int_{\partial\Omega} [u(y)-u(x)]X_k\varepsilon(y,x)\langle X_k, d\nu(y)\rangle \\ &= \int_{\Omega} X_k \{[u(y)-u(x_0)]X_k\varepsilon(y,x_0)\}\, d\nu(y) \\ &= \int_{\partial\Omega} [u(y)-u(x_0)]X_k\varepsilon(y,x_0)\langle X_k, d\nu(y)\rangle.\end{aligned}$$

Recalling the summation convention over the repeated index k, we get

$$\lim_{x\to x_0}\int_{\partial\Omega}[u(y)-u(x)]\langle\widetilde{\nabla}\varepsilon(y,x), d\nu(y)\rangle = \int_{\partial\Omega}[u(y)-u(x_0)]\langle\widetilde{\nabla}\varepsilon(y,x_0), d\nu(y)\rangle.$$

This proves the statement of Lemma 11.2.7. □

Proof of Theorem 11.2.6. We have

$$\begin{aligned}\lim_{\substack{x\to x_0,\\ x\notin\partial\Omega}} & \int_{\partial\Omega} u(y)\langle\widetilde{\nabla}\varepsilon(y,x), d\nu(y)\rangle \\ &= \lim_{\substack{x\to x_0,\\ x\notin\partial\Omega}}\left(\int_{\partial\Omega}[u(y)-u(x)]\langle\widetilde{\nabla}\varepsilon(y,x), d\nu(y)\rangle + u(x)\int_{\partial\Omega}\langle\widetilde{\nabla}\varepsilon(y,x), d\nu(y)\rangle\right).\end{aligned}$$

Choosing $u = \varepsilon$ and $v = 1$ in Green's first formula (1.86) we get

$$\int_{\partial\Omega} \langle \widetilde{\nabla}\varepsilon(y,x), d\nu(y)\rangle = \begin{cases} 1, & x \in \Omega, \\ 0, & x \notin \bar{\Omega}, \end{cases}$$

see Corollary 1.4.9. Therefore, using Lemma 11.2.7 we obtain

$$\mathcal{D}^{+}u(x_0) = \int_{\partial\Omega} [u(y) - u(x_0)]\langle \widetilde{\nabla}\varepsilon(y,x_0), d\nu(y)\rangle + u(x_0) \tag{11.19}$$

and

$$\mathcal{D}^{-}u(x_0) = \int_{\partial\Omega} [u(y) - u(x_0)]\langle \widetilde{\nabla}\varepsilon(y,x_0), d\nu(y)\rangle. \tag{11.20}$$

This gives the first jump relation

$$\mathcal{D}^{+}u(x_0) - \mathcal{D}^{-}u(x_0) = u(x_0).$$

We also have

$$\begin{aligned} \mathcal{D}^{0}u(x_0) &= \int_{\partial\Omega} u(y)\langle \widetilde{\nabla}\varepsilon(y,x_0), d\nu(y)\rangle \\ &= \int_{\partial\Omega} [u(y) - u(x_0)]\langle \widetilde{\nabla}\varepsilon(y,x_0), d\nu(y)\rangle \\ &\quad + u(x_0)\int_{\partial\Omega} \langle \widetilde{\nabla}\varepsilon(y,x_0), d\nu(y)\rangle \\ &= \int_{\partial\Omega} [u(y) - u(x_0)]\left\langle \widetilde{\nabla}\varepsilon(y,x_0), d\nu(y)\right\rangle + \mathcal{J}(x_0)u(x_0). \end{aligned}$$

So we obtain

$$\mathcal{D}^{0}u(x_0) = \int_{\partial\Omega} [u(y) - u(x_0)]\left\langle \widetilde{\nabla}\varepsilon(y,x_0), d\nu(y)\right\rangle + \mathcal{J}(x_0)u(x_0). \tag{11.21}$$

Now we get the second jump relation by subtracting (11.20) from (11.21), and subtracting (11.21) from (11.19) we obtain the third one. □

11.3 Traces and Kac's problem for the sub-Laplacian

Let $\Omega \subset \mathbb{R}^d$, $d \geq 2$, be a bounded domain. Then it is well known that the solution to the Poisson equation in Ω,

$$\Delta u(x) = f(x), \quad x \in \Omega, \tag{11.22}$$

is given by the Green formula (or the Newton potential formula)

$$u(x) = \int_{\Omega} \varepsilon_d(x-y)f(y)dy, \quad x \in \Omega, \tag{11.23}$$

for appropriate functions f supported in Ω. Here ε_d is the fundamental solution to Δ in $\mathbb{R}^d$ given by

$$\varepsilon_d(x-y) = \begin{cases} \frac{1}{(2-d)s_d} \frac{1}{|x-y|^{d-2}}, & d \geq 3, \\ \frac{1}{2\pi} \log|x-y|, & d = 2, \end{cases} \tag{11.24}$$

where $s_d = \frac{2\pi^{\frac{d}{2}}}{\Gamma(\frac{d}{2})}$ is the surface area of the unit sphere in $\mathbb{R}^d$. A question related to the so-called Kac's boundary value problem in Ω is

> what boundary condition can be put on u on the (piecewise smooth) boundary $\partial\Omega$ so that equation (11.22) complemented by this boundary condition would have the solution in Ω still given by the same formula (11.23), with the same kernel ε_d given by (11.24)?

This question is equivalent to finding the trace of the Newton potential (11.23) on the boundary surface $\partial\Omega$.

It turns out that the answer to these questions is the integral boundary condition

$$-\frac{1}{2}u(x)+\int_{\partial\Omega} \frac{\partial\varepsilon_d(x-y)}{\partial n_y} u(y)dS_y - \int_{\partial\Omega} \varepsilon_d(x-y)\frac{\partial u(y)}{\partial n_y} dS_y = 0,\ x \in \partial\Omega, \tag{11.25}$$

where $\frac{\partial}{\partial n_y}$ is the outer normal derivative at a point y on $\partial\Omega$. Thus, the trace of the Newton potential (11.23) on the boundary surface $\partial\Omega$ is determined by (11.25).

The boundary condition (11.25) and the subsequent spectral analysis are often called "the principle of not feeling the boundary" and it has originally appeared in M. Kac's work [Kac51], with further extensions in Kac's book [Kac80]. Such an analysis has several applications to the spectral theory and the asymptotics of Weyl's eigenvalue counting function. Spectral problems related to the boundary value problem (11.22), (11.25) were considered in the papers [KS09], [KS11], [RS16b] and [RRS16]. In general, the boundary value problem (11.22), (11.25) has several interesting properties and applications, for example discussed by Kac [Kac51, Kac80] and Saito [Sai08].

11.3.1 Traces of Newton potential for the sub-Laplacian

The analysis of the trace of the Newton potential or of related Kac's boundary value problems will be relying to certain results of Folland and Stein established in the setting of anisotropic Hölder spaces. We now define these spaces following [FS74] and [Fol75].

Definition 11.3.1 (Hölder spaces on stratified groups). Let $0 < \alpha < 1$. Define the anisotropic Hölder spaces $\Gamma_\alpha(\Omega)$ on $\Omega \subset \mathbb{G}$ by

$$\Gamma_\alpha(\Omega) := \{f : \Omega \to \mathbb{C} : \sup_{\substack{x,y\in\Omega \\ x\neq y}} \frac{|f(x)-f(y)|}{[d(x,y)]^\alpha} < \infty\}.$$

For $k \in \mathbb{N}$ and $0 < \alpha < 1$, one defines $\Gamma_{k+\alpha}(\Omega)$ as the space of all $f : \Omega \to \mathbb{C}$ such that all derivatives of f of order k belong to $\Gamma_\alpha(\Omega)$. A bounded function f is called *α-Hölder continuous* in $\Omega \subset \mathbb{G}$ if $f \in \Gamma_\alpha(\Omega)$.

We now define the analogue of the Newton potential in the setting of stratified groups. Here and in the sequel we adopt all the notations of Section 1.4.3 and of the subsequent sections.

Definition 11.3.2 (Newton potential for sub-Laplacian). Let $\Omega \subset \mathbb{G}$ be an admissible domain in a stratified group $\mathbb{G}$ in the sense of Definition 1.4.4. The *sub-Laplacian Newton potential* is defined by

$$u(x) = \int_\Omega f(y)\varepsilon(y, x)d\nu(y), \quad x \in \Omega, \quad f \in \Gamma_\alpha(\Omega), \tag{11.26}$$

where

$$\varepsilon(y, x) = \varepsilon(x, y) = \varepsilon(y^{-1}x, 0) = \varepsilon(y^{-1}x)$$

is the fundamental solution (1.89) of the sub-Laplacian $\mathcal{L}$, i.e.,

$$\varepsilon(x, y) = [d(x, y)]^{2-Q},$$

where $d(x, y) = d(y^{-1}x)$ is the $\mathcal{L}$-gauge on $\mathbb{G}$.

Remark 11.3.3. It was shown by Folland in [Fol75] that if $f \in \Gamma_\alpha(\Omega)$ for $\alpha > 0$ then u defined by (11.26) is twice differentiable and satisfies the equation

$$\mathcal{L}u = f.$$

The following theorem describes the trace of the integral operator in (11.26) on $\partial\Omega$. This is equivalent to finding a boundary condition for u such that the equation $\mathcal{L}u = f$ has a unique solution in $C^2(\Omega)$ and this solution is the Newton potential (11.26).

Theorem 11.3.4 (Trace of Newton potential and Kac's boundary value problem). *Let $\Omega \subset \mathbb{G}$ be an admissible domain in a stratified group $\mathbb{G}$ of homogeneous dimension $Q \geq 3$. Let $\varepsilon(y, x) = \varepsilon(y^{-1}x)$ be the fundamental solution to $\mathcal{L}$, so that*

$$\mathcal{L}\varepsilon = \delta \quad \text{on } \mathbb{G}. \tag{11.27}$$

Then for all $f \in \Gamma_\alpha(\Omega)$, $0 < \alpha < 1$, $\operatorname{supp} f \subset \Omega$, the Newton potential (11.26) is the unique solution in $C^2(\Omega) \cap C^1(\overline{\Omega})$ of the equation

$$\mathcal{L}u = f \quad \text{in } \Omega, \tag{11.28}$$

with the boundary condition

$$\begin{aligned}(1 - \mathcal{J}(x))u(x) &+ \int_{\partial\Omega} u(y)\langle\widetilde{\nabla}\varepsilon(y, x), d\nu(y)\rangle \\ &- \int_{\partial\Omega} \varepsilon(y, x)\langle\widetilde{\nabla}u(y), d\nu(y)\rangle = 0 \quad \text{for} \quad x \in \partial\Omega,\end{aligned} \tag{11.29}$$

where the jump value is given by the formula

$$\mathcal{J}(x) = \int_{\partial\Omega} \langle \widetilde{\nabla}\varepsilon(y,x), d\nu(y)\rangle, \tag{11.30}$$

with $\widetilde{\nabla} = \widetilde{\nabla}_y$ *defined by*

$$\widetilde{\nabla}g = \sum_{k=1}^{N_1}(X_k g)X_k.$$

Proof of Theorem 11.3.4. Since the Newton potential (11.26) is a solution of (11.28) it follows from Remark 11.3.3 that u is locally in $\Gamma_{\alpha+2}(\Omega, \mathrm{loc})$ and that it is twice differentiable in Ω. In particular, it follows that $u \in C^2(\Omega) \cap C^1(\overline{\Omega})$.

We will use the following representation formula for functions $u \in C^2(\Omega) \cap C^1(\overline{\Omega})$ shown in Corollary 1.4.10, Part 1, for all $x \in \Omega$:

$$\begin{aligned} u(x) = \int_{\Omega} f(y)\varepsilon(y,x)d\nu(y) &+ \int_{\partial\Omega} u(y)\langle \widetilde{\nabla}\varepsilon(y,x), d\nu(y)\rangle \\ &- \int_{\partial\Omega} \varepsilon(y,x)\langle \widetilde{\nabla}u(y), d\nu(y)\rangle. \end{aligned} \tag{11.31}$$

Since u given by the Newton potential formula (11.26) is a solution of (11.28), using it in (11.31) we get, for all $x \in \Omega$,

$$\int_{\partial\Omega} u(y)\langle \widetilde{\nabla}\varepsilon(y,x), d\nu(y)\rangle - \int_{\partial\Omega} \varepsilon(y,x)\langle \widetilde{\nabla}u(y), d\nu(y)\rangle = 0. \tag{11.32}$$

When $x \in \Omega$ approaches the boundary $\partial\Omega$ from the interior we can use Theorem 11.2.4 and Theorem 11.2.6 yielding the identity

$$\begin{aligned} (1 - \mathcal{J}(x))u(x) &+ \int_{\partial\Omega} u(y)\langle \widetilde{\nabla}\varepsilon(y,x), d\nu(y)\rangle \\ &- \int_{\partial\Omega} \varepsilon(y,x)\langle \widetilde{\nabla}u(y), d\nu(y)\rangle = 0 \quad \text{for any } x \in \partial\Omega. \end{aligned} \tag{11.33}$$

It follows from (11.26) that u is then a solution of the boundary value problem (11.28) with the boundary condition (11.29).

Let us now prove its uniqueness. Assuming that u and u_1 are two solutions of the boundary value problem (11.28) and (11.29), their difference

$$w := u - u_1 \in C^2(\Omega) \cap C^1(\overline{\Omega})$$

satisfies the homogeneous equation

$$\mathcal{L}w = 0 \quad \text{in } \Omega, \tag{11.34}$$

and the boundary condition (11.29), i.e.,

$$\begin{aligned}(1-\mathcal{J}(x))w(x)+\int_{\partial\Omega} w(y)\langle\widetilde{\nabla}\varepsilon(y,x),d\nu(y)\rangle\\ -\int_{\partial\Omega}\varepsilon(y,x)\langle\widetilde{\nabla}w(y),d\nu(y)\rangle=0,\quad x\in\partial\Omega.\end{aligned}\tag{11.35}$$

The representation formula from Corollary 1.4.10, Part 2, yields the formula for w as

$$w(x)=\int_{\partial\Omega} w(y)\langle\widetilde{\nabla}\varepsilon(y,x),d\nu(y)\rangle-\int_{\partial\Omega}\varepsilon(y,x)\langle\widetilde{\nabla}w(y),d\nu(y)\rangle,\tag{11.36}$$

for any $x\in\Omega$. As above, when $x\in\Omega$ approaches the boundary $\partial\Omega$ from the interior we can use Theorem 11.2.4 and Theorem 11.2.6 by using the properties of the double and single layer potentials as $x\to\partial\Omega$ from interior, formula (11.36) implies that for every $x\in\partial\Omega$ we have

$$w(x)=(1-\mathcal{J}(x))w(x)+\int_{\partial\Omega} w(y)\langle\widetilde{\nabla}\varepsilon(y,x),d\nu(y)\rangle-\int_{\partial\Omega}\varepsilon(y,x)\langle\widetilde{\nabla}w,d\nu(y)\rangle.$$

Comparing this with (11.35) we obtain that w has to satisfy

$$w(x)=0,\ x\in\partial\Omega.\tag{11.37}$$

By Proposition 11.1.1 the boundary value problem (11.28) with the boundary condition (11.34) has a unique trivial solution $w\equiv 0$ in $C^2(\Omega)\cap C^1(\overline{\Omega})$ which shows that $u=u_1$ in Ω. □

11.3.2 Powers of the sub-Laplacian

We now extend the analysis of Section 11.3.1 to traces and boundary value problems for powers of the sub-Laplacian on stratified groups. Here we naturally understand the powers by

$$\mathcal{L}^m:=\mathcal{L}\mathcal{L}^{m-1},\tag{11.38}$$

for all $m\in\mathbb{N}$.

Let $\Omega\subset\mathbb{G}$ be an admissible domain in a stratified group $\mathbb{G}$ of homogeneous dimension $Q\geq 3$. Then for $m=1,2,\ldots$, for a given $f\in\Gamma_\alpha(\Omega)$, we consider the equations

$$\mathcal{L}^m u(x)=f(x),\ x\in\Omega,\tag{11.39}$$

Definition 11.3.5 (Generalized Newton potentials)**.** Let $\varepsilon(y,x)=\varepsilon(y^{-1}x)$ be the fundamental solution of the sub-Laplacian $\mathcal{L}$ as in (1.89). We define the *generalized Newton potentials* by

$$u(x)=\int_\Omega f(y)\varepsilon_m(y,x)d\nu(y),\tag{11.40}$$

where $\varepsilon_m(y,x)$ is the fundamental solution of (11.39) such that

$$\mathcal{L}^{m-1}\varepsilon_m = \varepsilon, \quad m = 1, 2, \ldots.$$

Namely, with a proper distributional interpretation for $m = 2, 3, \ldots$, we take

$$\varepsilon_m(y,x) := \int_\Omega \varepsilon_{m-1}(y,\zeta)\varepsilon(\zeta,x)d\nu(\zeta), \qquad y, x \in \Omega, \tag{11.41}$$

with

$$\varepsilon_1(y,x) = \varepsilon(y,x).$$

We note that in general higher-order hypoelliptic operators on stratified groups may not have unique fundamental solutions, see Geller [Gel83], or [FR16, Section 3.2.7] for a detailed discussion. However, in the case of the iterated sub-Laplacian $\mathcal{L}^m$ we have uniqueness in the sense of the next theorem and of the uniqueness argument in its proof.

Theorem 11.3.6 (Traces of generalized Newton potentials and Kac's boundary value problem). *Let $\Omega \subset \mathbb{G}$ be an admissible domain in a stratified group $\mathbb{G}$ of homogeneous dimension $Q \geq 3$. For any $f \in \Gamma_\alpha(\Omega)$, $0 < \alpha < 1$, $\operatorname{supp} f \subset \Omega$, the generalized Newton potential (11.40) is a unique solution of the equation (11.39) in $C^{2m}(\Omega) \cap C^{2m-1}(\overline{\Omega})$ with m boundary conditions*

$$\begin{aligned}(1 - \mathcal{J}(x))\mathcal{L}^i u(x) &+ \sum_{j=0}^{m-i-1} \int_{\partial\Omega} \mathcal{L}^{j+i}u(y)\langle \widetilde{\nabla}\mathcal{L}^{m-1-j}\varepsilon_m(y,x), d\nu(y)\rangle \\ &- \sum_{j=0}^{m-i-1} \int_{\partial\Omega} \mathcal{L}^{m-1-j}\varepsilon_m(y,x)\langle \widetilde{\nabla}\mathcal{L}^{j+i}u(y)d\nu(y)\rangle = 0, \quad x \in \partial\Omega,\end{aligned} \tag{11.42}$$

for all $i = 0, 1, \ldots, m-1$, where $\widetilde{\nabla}$ is given by

$$\widetilde{\nabla}g = \sum_{k=1}^{N_1}(X_k g)X_k$$

and $\mathcal{J}(x)$ is the jump function given by the formula (11.30).

Remark 11.3.7. From Theorem 11.3.6 we have that the kernel (11.41), which is a fundamental solution of the equation (11.39), is the Green function of the boundary value problem (11.39), (11.42) in Ω. This presents an example of an explicitly solvable boundary value problem for the iterated sub-Laplacian in any (admissible) domain Ω, which can be regarded as the higher order version of the original Kac's boundary value problem.

Proof of Theorem 11.3.6. For each $x \in \Omega$, applying Green's second formula from Theorem 1.4.6 we obtain

$$u(x) = \int_\Omega f(y)\varepsilon_m(y,x)d\nu(y) = \int_\Omega \mathcal{L}^m u(y)\varepsilon_m(y,x)d\nu(y)$$

$$
\begin{aligned}
&= \int_\Omega \mathcal{L}^{m-1}u(y)\mathcal{L}\varepsilon_m(y,x)d\nu(y) - \int_{\partial\Omega} \mathcal{L}^{m-1}u(y)\langle\widetilde{\nabla}\varepsilon_m(y,x), d\nu(y)\rangle \\
&\quad + \int_{\partial\Omega} \varepsilon_m(y,x)\langle\widetilde{\nabla}\mathcal{L}^{m-1}u(y), d\nu(y)\rangle \\
&= \int_\Omega \mathcal{L}^{m-2}u(y)\mathcal{L}^2\varepsilon_m(y,x)d\nu(y) - \int_{\partial\Omega} \mathcal{L}^{m-2}u(y)\langle\widetilde{\nabla}\mathcal{L}\varepsilon_m(y,x), d\nu(y)\rangle \\
&\quad + \int_{\partial\Omega} \mathcal{L}\varepsilon_m(y,x)\langle\widetilde{\nabla}\mathcal{L}^{m-2}u(y), d\nu(y)\rangle - \int_{\partial\Omega} \mathcal{L}^{m-1}u(y)\langle\widetilde{\nabla}\varepsilon_m(y,x), d\nu(y)\rangle \\
&\quad + \int_{\partial\Omega} \varepsilon_m(y,x)\langle\widetilde{\nabla}\mathcal{L}^{m-1}u(y), d\nu(y)\rangle \\
&= \cdots \\
&= u(x) - \sum_{j=0}^{m-1}\int_{\partial\Omega} \mathcal{L}^j u(y)\langle\widetilde{\nabla}\mathcal{L}^{m-1-j}\varepsilon_m(y,x), d\nu(y)\rangle \\
&\quad + \sum_{j=0}^{m-1}\int_{\partial\Omega} \mathcal{L}^{m-1-j}\varepsilon_m(y,x)\langle\widetilde{\nabla}\mathcal{L}^j u(y), d\nu(y)\rangle, \quad x \in \Omega.
\end{aligned}
$$

This implies the identity

$$
\begin{aligned}
&\sum_{j=0}^{m-1}\int_{\partial\Omega} \mathcal{L}^j u(y)\langle\widetilde{\nabla}\mathcal{L}^{m-1-j}\varepsilon_m(y,x), d\nu(y)\rangle \\
&- \sum_{j=0}^{m-1}\int_{\partial\Omega} \mathcal{L}^{m-1-j}\varepsilon_m(y,x)\langle\widetilde{\nabla}\mathcal{L}^j u(y), d\nu(y)\rangle = 0, \quad x \in \Omega.
\end{aligned} \tag{11.43}
$$

We note that the first term of the first summand, i.e., the term with $j = 0$ given by

$$
\int_{\partial\Omega} u(y)\langle\widetilde{\nabla}\varepsilon(y,x), d\nu(y)\rangle
$$

is the double layer potential analysed in Theorem 11.2.6. The other terms in the above sum are single layer type potentials. In particular, by Theorem 11.2.4 they are continuous functions on $\mathbb{G}$. By using the properties of the double and single layer potentials as x approaches the boundary $\partial\Omega$ from the interior, from (11.43) we obtain the equality

$$
\begin{aligned}
(1 - \mathcal{J}(x))u(x) &+ \sum_{j=0}^{m-1}\int_{\partial\Omega} \mathcal{L}^j u(y)\langle\widetilde{\nabla}\mathcal{L}^{m-1-j}\varepsilon_m(y,x), d\nu(y)\rangle \\
&- \sum_{j=0}^{m-1}\int_{\partial\Omega} \mathcal{L}^{m-1-j}\varepsilon_m(y,x)\langle\widetilde{\nabla}\mathcal{L}^j u(y), d\nu(y)\rangle = 0, \quad x \in \partial\Omega.
\end{aligned}
$$

From this relation we obtain the first boundary conditions for (11.40). The remaining boundary conditions can be derived by writing

$$\mathcal{L}^{m-i}\mathcal{L}^{i}u = f, \quad i = 0, 1, \ldots, m-1, \quad m = 1, 2, \ldots, \tag{11.44}$$

and carrying out similar arguments as above. Indeed, we have

$$\begin{aligned}
\mathcal{L}^{i}u(x) &= \int_{\Omega} f(y)\mathcal{L}^{i}\varepsilon_m(y,x)d\nu(y) = \int_{\Omega} \mathcal{L}^{m-i}\mathcal{L}^{i}u(y)\mathcal{L}^{i}\varepsilon_m(y,x)d\nu(y) \\
&= \int_{\Omega} \mathcal{L}^{m-i-1}\mathcal{L}^{i}u(y)\mathcal{L}\mathcal{L}^{i}\varepsilon_m(y,x)d\nu(y) \\
&\quad - \int_{\partial\Omega} \mathcal{L}^{m-i-1}\mathcal{L}^{i}u(y)\langle\widetilde{\nabla}\mathcal{L}^{i}\varepsilon_m(y,x), d\nu(y)\rangle \\
&\quad + \int_{\partial\Omega} \mathcal{L}^{i}\varepsilon_m(y,x)\langle\widetilde{\nabla}\mathcal{L}^{m-i-1}\mathcal{L}^{i}u(y), d\nu(y)\rangle \\
&= \int_{\Omega} \mathcal{L}^{m-i-2}\mathcal{L}^{i}u(y)\mathcal{L}^{2}\mathcal{L}^{i}\varepsilon_m(y,x)d\nu(y) \\
&\quad - \int_{\partial\Omega} \mathcal{L}^{m-i-2}\mathcal{L}^{i}u(y)\langle\widetilde{\nabla}\mathcal{L}\mathcal{L}^{i}\varepsilon_m(y,x), d\nu(y)\rangle \\
&\quad + \int_{\partial\Omega} \mathcal{L}\mathcal{L}^{i}\varepsilon_m(y,x)\langle\widetilde{\nabla}\mathcal{L}^{m-i-2}\mathcal{L}^{i}u(y), d\nu(y)\rangle \\
&\quad - \int_{\partial\Omega} \mathcal{L}^{m-i-1}\mathcal{L}^{i}u(y)\langle\widetilde{\nabla}\mathcal{L}^{i}\varepsilon_m(y,x), d\nu(y)\rangle \\
&\quad + \int_{\partial\Omega} \mathcal{L}^{i}\varepsilon_m(y,x)\langle\widetilde{\nabla}\mathcal{L}^{m-i-1}\mathcal{L}^{i}u(y), d\nu(y)\rangle \\
&= \cdots \\
&= \int_{\Omega} \mathcal{L}^{i}u(y)\mathcal{L}^{m-i}\mathcal{L}^{i}\varepsilon_m(y,x)d\nu(y) \\
&\quad - \sum_{j=0}^{m-i-1} \int_{\partial\Omega} \mathcal{L}^{j}\mathcal{L}^{i}u(y)\langle\widetilde{\nabla}\mathcal{L}^{m-i-1-j}\mathcal{L}^{i}\varepsilon_m(y,x), d\nu(y)\rangle \\
&\quad + \sum_{j=0}^{m-i-1} \int_{\partial\Omega} \mathcal{L}^{m-i-1-j}\mathcal{L}^{i}\varepsilon_m(y,x)\langle\widetilde{\nabla}\mathcal{L}^{j}\mathcal{L}^{i}u(y), d\nu(y)\rangle \\
&= \mathcal{L}^{i}u(x) - \sum_{j=0}^{m-i-1} \int_{\partial\Omega} \mathcal{L}^{j+i}u(y)\langle\widetilde{\nabla}\mathcal{L}^{m-1-j}\varepsilon_m(y,x), d\nu(y)\rangle \\
&\quad + \sum_{j=0}^{m-i-1} \int_{\partial\Omega} \mathcal{L}^{m-1-j}\varepsilon_m(y,x)\langle\widetilde{\nabla}\mathcal{L}^{j+i}u(y), d\nu(y)\rangle, \quad x \in \Omega,
\end{aligned}$$

where $\mathcal{L}^{i}\varepsilon_m$ is a fundamental solution of the equation (11.44), that is,

$$\mathcal{L}^{m-i}\mathcal{L}^{i}\varepsilon_m = \delta, \qquad i = 0, 1, \ldots, m-1.$$

From the previous relations, for all $x \in \Omega$ and $i = 0, 1, \ldots, m-1$, we get the identities

$$\sum_{j=0}^{m-i-1} \int_{\partial\Omega} \mathcal{L}^{j+i} u(y) \langle \widetilde{\nabla} \mathcal{L}^{m-1-j} \varepsilon_m(y,x), d\nu(y) \rangle - \sum_{j=0}^{m-i-1} \int_{\partial\Omega} \mathcal{L}^{m-1-j} \varepsilon_m(y,x) \langle \widetilde{\nabla} \mathcal{L}^{j+i} u(y), d\nu(y) \rangle = 0.$$

By using the properties of the double and single layer potentials in Theorem 11.2.6 and Theorem 11.2.4 as x approaches the boundary $\partial\Omega$ from the interior of Ω, we find that

$$(1 - \mathcal{J}(x)) \mathcal{L}^i u(x) + \sum_{j=0}^{m-i-1} \int_{\partial\Omega} \mathcal{L}^{j+i} u(y) \langle \widetilde{\nabla} \mathcal{L}^{m-1-j} \varepsilon_m(y,x), d\nu(y) \rangle - \sum_{j=0}^{m-i-1} \int_{\partial\Omega} \mathcal{L}^{m-1-j} \varepsilon_m(y,x) \langle \widetilde{\nabla} \mathcal{L}^{j+i} u(y), d\nu(y) \rangle = 0, \quad x \in \partial\Omega,$$

are all the boundary conditions of (11.40) for each $i = 0, 1, \ldots, m-1$.

Let us now show the uniqueness: if a function $w \in C^{2m}(\Omega) \cap C^{2m-1}(\overline{\Omega})$ satisfies the equation $\mathcal{L}^m w = f$ and the boundary conditions (11.42), then it must coincide with the solution (11.40). Indeed, the function

$$v := u - w \in C^{2m}(\Omega) \cap C^{2m-1}(\overline{\Omega}),$$

where u is the generalized Newton potential (11.40), satisfies the homogeneous equation

$$\mathcal{L}^m v = 0 \tag{11.45}$$

and the boundary conditions (11.42), i.e.,

$$I_i(v)(x) := (1 - \mathcal{J}(x)) \mathcal{L}^i v(x) + \sum_{j=0}^{m-i-1} \int_{\partial\Omega} \mathcal{L}^{j+i} v(y) \langle \widetilde{\nabla} \mathcal{L}^{m-1-j} \varepsilon_m(y,x), d\nu(y) \rangle - \sum_{j=0}^{m-i-1} \int_{\partial\Omega} \mathcal{L}^{m-1-j} \varepsilon_m(y,x) \langle \widetilde{\nabla} \mathcal{L}^{j+i} v(y), d\nu(y) \rangle = 0, \quad i = 0, 1, \ldots, m-1,$$

for $x \in \partial\Omega$. By applying the Green formula from Theorem 1.4.6 to the function $v \in C^{2m}(\Omega) \cap C^{2m-1}(\overline{\Omega})$ and by following the lines of the previous calculation we obtain

$$0 = \int_{\Omega} \mathcal{L}^m v(x) \mathcal{L}^i \varepsilon_m(y,x) d\nu(y) = \int_{\Omega} \mathcal{L}^{m-i} \mathcal{L}^i v(x) \mathcal{L}^i \varepsilon_m(y,x) d\nu(y)$$

$$
\begin{aligned}
&= \int_{\Omega} \mathcal{L}^{m-1} v(x) \mathcal{L}\mathcal{L}^i \varepsilon_m(y,x) d\nu(y) - \int_{\partial\Omega} \mathcal{L}^{m-1} v(x) \langle \widetilde{\nabla} \mathcal{L}^i \varepsilon_m(y,x), d\nu(y)\rangle \\
&\quad + \int_{\partial\Omega} \mathcal{L}^i \varepsilon_m(y,x) \langle \widetilde{\nabla} \mathcal{L}^{m-1} v(x), d\nu(y)\rangle \\
&= \int_{\Omega} \mathcal{L}^{m-2} v(x) \mathcal{L}^2 \mathcal{L}^i \varepsilon_m(y,x) d\nu(y) - \int_{\partial\Omega} \mathcal{L}^{m-2} v(x) \langle \widetilde{\nabla} \mathcal{L}^{i+1} \varepsilon_m(y,x), d\nu(y)\rangle \\
&\quad + \int_{\partial\Omega} \mathcal{L}^{i+1} \varepsilon_m(y,x) \langle \widetilde{\nabla} \mathcal{L}^{m-2} v(x), d\nu(y)\rangle - \int_{\partial\Omega} \mathcal{L}^{m-1} v(x) \langle \widetilde{\nabla} \mathcal{L}^i \varepsilon_m(y,x), d\nu(y)\rangle \\
&\quad + \int_{\partial\Omega} \mathcal{L}^i \varepsilon_m(y,x) \langle \widetilde{\nabla} \mathcal{L}^{m-1} v(x), d\nu(y)\rangle \\
&= \cdots \\
&= \mathcal{L}^i v(x) - \sum_{j=0}^{m-i-1} \int_{\partial\Omega} \mathcal{L}^{j+i} v(y) \langle \widetilde{\nabla} \mathcal{L}^{m-1-j} \varepsilon_m(y,x), d\nu(y)\rangle \\
&\quad + \sum_{j=0}^{m-i-1} \int_{\partial\Omega} \mathcal{L}^{m-1-j} \varepsilon_m(y,x) \langle \widetilde{\nabla} \mathcal{L}^{j+i} v(y), d\nu(y)\rangle, \quad i = 0, 1, \ldots, m-1.
\end{aligned}
$$

By passing to the limit as $x \to \partial\Omega$ from interior, we obtain the relations

$$
\mathcal{L}^i v(x) \mid_{x\in\partial\Omega} = I_i(v)(x) \mid_{x\in\partial\Omega} = 0, \quad i = 0, 1, \ldots, m-1. \tag{11.46}
$$

We claim that the boundary value problem

$$
\begin{aligned}
\mathcal{L}^m v &= 0, \\
\mathcal{L}^i v \mid_{\partial\Omega} &= 0, \quad i = 0, 1, \ldots, m-1,
\end{aligned} \tag{11.47}
$$

has uniqueness trivial solution $v \equiv 0$ in $C^{2m}(\Omega) \cap C^{2m-1}(\overline{\Omega})$.

Assuming this claim for the moment we get that $v = u - w \equiv 0$, for all $x \in \Omega$, i.e., w coincides with u in Ω. Thus (11.40) is the unique solution of the boundary value problem (11.39), (11.42) in Ω.

Let us now show the above claim, that is, that the boundary value problem (11.47) has a unique solution in $C^{2m}(\Omega) \cap C^{2m-1}(\overline{\Omega})$. Denoting

$$
\tilde{v} := \mathcal{L}^{m-1} v,
$$

the claim follows by induction from the uniqueness in $C^2(\Omega) \cap C^1(\overline{\Omega})$ of the boundary value problem

$$
\begin{aligned}
\mathcal{L}\tilde{v} &= 0, \\
\tilde{v} \mid_{\partial\Omega} &= 0.
\end{aligned}
$$

The proof of Theorem 11.3.6 is complete. □

11.3.3 Extended Kohn Laplacians on the Heisenberg group

In this section we describe more explicit formulae for the traces of the Newton potential presented in Section 11.3.1 in the setting of the Heisenberg group. In this analysis we adapt the complex description of the Heisenberg group as described in Section 1.4.8, so we follow all the notation introduced there. The presentation in this and the following sections follows [RS16c].

In particular, we will use the coordinates on the Heisenberg group $\mathbb{H}^n$ given by variables $z = (\zeta, t)$, so that a basis for the complex tangent space of $\mathbb{H}^n$ at the point z is given by the left invariant vector fields

$$X_j = \frac{\partial}{\partial z_j} + i\bar{z}\frac{\partial}{\partial t}, \quad j = 1, \ldots, n.$$

Denoting their conjugates by

$$X_{\bar{j}} := \overline{X}_j = \frac{\partial}{\partial \bar{z}_j} - iz\frac{\partial}{\partial t},$$

we will be dealing with the extended Kohn Laplacian

$$\mathcal{L}_{a,b} = \sum_{j=1}^{n}(aX_jX_{\bar{j}} + bX_{\bar{j}}X_j), \quad a + b = n. \tag{11.48}$$

As described in Section 1.4.8, this operator has a fundamental solution given by a constant multiple of (1.110). Moreover, in [FS74], Folland and Stein defined the Newton potential (volume potential) for a function f with compact support contained in a set $\Omega \subset \mathbb{H}_n$ by

$$u(z) = \int_{\Omega} f(\xi)\varepsilon(\xi^{-1}z)d\nu(\xi), \tag{11.49}$$

with $d\nu$ being the volume element: this is given by the Haar measure on $\mathbb{H}_n$ which in turn coincides with the Lebesgue measure on $\mathbb{C}^n \times \mathbb{R}$. More precisely, they proved that

$$\mathcal{L}_{a,b}u = c_{a,b}f,$$

where the constant $c_{a,b}$ is zero if a and $b = -1, -2, \ldots, n, n+1, \ldots$, and $c_{a,b} \neq 0$ if a or $b \neq -1, -2, \ldots, n, n+1, \ldots$, given by (1.112) for $a \notin \mathbb{Z}$. For different approaches to the fundamental solutions of $\mathcal{L}_{a,b}$ we can also refer to Greiner and Stein [GS77].

With the notation for $\varepsilon_{a,b} = \varepsilon$ as in (1.110) the distribution $\frac{1}{c_{a,b}}\varepsilon$ is the fundamental solution of $\mathcal{L}_{a,b}$, while ε satisfies the equation

$$\mathcal{L}_{a,b}\varepsilon = c_{a,b}\delta. \tag{11.50}$$

In order to make our presentation compatible with the accepted ones in [FS74] and [Rom91], in this section we will also work with the function ε although there is a constant $c_{a,b}$ appearing in the identity (11.50).

Therefore, throughout this and next sections we assume that $c_{a,b} \neq 0$, which means that a and b satisfy the condition

$$a \text{ and } b \neq -1, -2, \ldots, n, n+1, \ldots.$$

In addition, without loss of generality we may also assume that $a, b \geq 0$.

We recall the anisotropic Hölder space $\Gamma_\alpha(\Omega)$ from Definition 11.3.1. Since all homogeneous quasi-norms on homogeneous groups are equivalent for such a definition we can also use an explicitly defined distance on $\mathbb{H}^n$ given by

$$|z| := (|\zeta|^4 + |t|^2)^{1/4}.$$

Thus, for $0 < \alpha < 1$, the anisotropic Hölder spaces $\Gamma_\alpha(\Omega)$ from Definition 11.3.1 can be also described by

$$\Gamma_\alpha(\Omega) = \left\{ f : \Omega \to \mathbb{C} : \sup_{\substack{z_1, z_2 \in \Omega \\ z_1 \neq z_2}} \frac{|f(z_2) - f(z_1)|}{|z_2^{-1} z_1|^\alpha} < \infty \right\}.$$

Remark 11.3.8. In addition to Remark 11.3.3, in the setting of the Heisenberg groups we have that if $f \in \Gamma_\alpha(\Omega)$ for $\alpha > 0$ then u defined by (12.43) is twice differentiable in the complex directions and satisfies the equation

$$\mathcal{L}_{a,b} u = c_{a,b} f.$$

There are several different approaches to show this result by Folland and Stein [FS74], Greiner and Stein [GS77], and Romero [Rom91]. Moreover, Folland and Stein have shown that if $f \in \Gamma_\alpha(\Omega, \mathrm{loc})$ and $\mathcal{L}_{a,b} u = c_{a,b} f$, then $f \in \Gamma_{\alpha+2}(\Omega, \mathrm{loc})$, for naturally defined localisations $\Gamma_\alpha(\Omega, \mathrm{loc})$ of the anisotropic Hölder spaces $\Gamma_\alpha(\Omega)$.

We will be using the single layer potentials introduced in Definition 11.2.1, namely, the operators

$$S_j g(z) = \int_{\partial\Omega} g(\xi) \varepsilon(\xi, z) \langle X_j, d\nu(\xi) \rangle,$$

where $\langle X, d\nu \rangle$ is the canonical pairing between vector fields and differential forms, which are well defined by Theorem 11.2.3. Moreover, specifically in the setting of the Heisenberg group, it was shown in [Rom91, Theorem 2.3] that if the density of $g(\xi)\langle X_j, d\nu \rangle$ in the operator S_j is bounded then $Sg \in \Gamma_\alpha(\mathbb{H}_n)$ for all $\alpha < 1$.

The double layer potentials have been introduced in Definition 11.2.5 on general stratified groups. However, it will be now convenient to adapt its definition slightly to the extended Kohn Laplacians $\mathcal{L}_{a,b}$. Thus, for every differentiable function g we define the vector field

$$\nabla^{a,b} g = \sum_{j=1}^{n-1} (a X_j g X_{\bar{j}} + b X_{\bar{j}} g X_j). \tag{11.51}$$

Consequently, we will be working with the double layer potential

$$Wu(z) = \int_{\partial\Omega} u(\xi)\langle \nabla^{a,b}\varepsilon(\xi, z), d\nu(\xi)\rangle. \tag{11.52}$$

Let us now record the Plemelj jump relations for double layer potential which follow from Theorem 11.2.6.

Corollary 11.3.9. *The double layer potential Wu in (11.52) has two limits*

$$W^{+}u(z) = \lim_{\substack{z_0\to z\\ z_0\in\Omega}} \int_{\partial\Omega} u(\xi)\langle \nabla^{a,b}\varepsilon(\xi, z_0), d\nu(\xi)\rangle$$

and

$$W^{-}u(z) = \lim_{\substack{z_0\to z\\ z_0\notin\Omega}} \int_{\partial\Omega} u(\xi)\langle \nabla^{a,b}\varepsilon(\xi, z_0), d\nu(\xi)\rangle,$$

and the principal value

$$W^{0}u(z) = \text{p.v. } Wu(z) = \lim_{\delta\to 0} \int_{\partial\Omega\setminus\{|\xi^{-1}z|<\delta\}} u(\xi)\langle \nabla^{a,b}\varepsilon(\xi, z), d\nu(\xi)\rangle.$$

For $u \in \Gamma_\alpha(\Omega)$ and $z \in \partial\Omega$ the above limits exist and satisfy the jump relations

$$\begin{aligned} W^{+}u(z) - W^{-}u(z) &= c_{a,b}u(z),\\ W^{0}u(z) - W^{-}u(z) &= H.R.(z)u(z),\\ W^{+}u(z) - W^{0}u(z) &= (c_{a,b} - H.R.(z))u(z), \end{aligned} \tag{11.53}$$

where $H.R(z)$ is the so-called half-residue given by the formula

$$H.R(z) = \lim_{\delta\to 0} \int_{\partial\Omega\setminus\{|\xi^{-1}z|<\delta\}} \langle \nabla^{a,b}\varepsilon(\xi, z), d\nu(\xi)\rangle. \tag{11.54}$$

The first two jump relations in (11.53) follow by an adaptation of the argument in Theorem 11.2.6 keeping in mind the appearing constant $c_{a,b}$. The third jump relation in (11.53) follows by subtraction of the first two. The statement of Corollary 11.3.9 was shown in [Rom91, Theorem 2.4].

Then we have the following analogue of Theorem 11.3.4.

Theorem 11.3.10 (Trace of Newton potential and Kac's boundary value problem for $\mathcal{L}_{a,b}$)**.** *Let $\varepsilon(\xi, z) = \varepsilon(\xi^{-1}z)$ be the rescaled fundamental solution to $\mathcal{L}_{a,b}$, so that*

$$\mathcal{L}_{a,b}\varepsilon = c_{a,b}\delta \quad \text{on } \mathbb{H}_n. \tag{11.55}$$

For any $f \in \Gamma_\alpha(\Omega)$, $0 < \alpha < 1$, $\operatorname{supp} f \subset \Omega$, the Newton potential (12.43) is the unique solution in $C^2(\Omega) \cap C^1(\overline{\Omega})$ of the equation

$$\mathcal{L}_{a,b}u = c_{a,b}f \tag{11.56}$$

with the boundary condition

$$(c_{a,b} - H.R(z))u(z) + \lim_{\delta\to 0} \int_{\partial\Omega\setminus\{|\xi^{-1}z|<\delta\}} u(\xi)\langle \nabla^{a,b}\varepsilon(\xi,z), d\nu(\xi)\rangle \\ - \int_{\partial\Omega} \varepsilon(\xi,z)\langle \nabla^{b,a}u(\xi), d\nu(\xi)\rangle = 0, \qquad \text{for } z \in \partial\Omega, \tag{11.57}$$

where $H.R(z)$ is the so-called half-residue given by (11.54).

Proof of Theorem 11.3.10. Since the Newton potential

$$u(z) = \int_\Omega f(\xi)\varepsilon(\xi,z)d\nu(\xi) \tag{11.58}$$

is a solution of (11.56), from Remark 11.3.8 it follows that u is locally in $\Gamma_{\alpha+2}(\Omega, \mathrm{loc})$ and that it is twice complex differentiable in Ω. In particular, it follows that $u \in C^2(\Omega) \cap C^1(\overline{\Omega})$.

From Green's second formula in Theorem 1.4.6, similarly to Corollary 1.4.10, Part 1, by taking into account constants $c_{a,b}$, for every $u \in C^2(\Omega) \cap C^1(\overline{\Omega})$ and every $z \in \Omega$ we have the representation formula

$$c_{a,b}u(z) = c_{a,b}\int_\Omega f(\xi)\varepsilon(\xi,z)d\nu(\xi) + \int_{\partial\Omega} u(\xi)\langle \nabla^{a,b}\varepsilon(\xi,z), d\nu(\xi)\rangle \\ - \int_{\partial\Omega} \varepsilon(\xi,z)\langle \nabla^{b,a}u(\xi), d\nu(\xi)\rangle. \tag{11.59}$$

Since $u(z)$ given by (11.58) is a solution of (11.56), using it in (11.59), for every $z \in \Omega$ we have

$$\int_{\partial\Omega} u(\xi)\langle \nabla^{a,b}\varepsilon(\xi,z), d\nu(\xi)\rangle - \int_{\partial\Omega} \varepsilon(\xi,z)\langle \nabla^{b,a}u(\xi), d\nu(\xi)\rangle = 0. \tag{11.60}$$

The fundamental solution $\varepsilon(z)$ is homogeneous of degree $2 - Q = -2n$; this is a general fact (see, e.g., [FR16, Theorem 3.2.40]), or can be seen directly from formula (1.110) since

$$\varepsilon(\lambda z) = \lambda^{-2a-2b}\varepsilon(z) = \lambda^{-2n}\varepsilon(z) \quad \text{for any } \lambda > 0,$$

since $a + b = n$. It follows that ε and its first-order complex derivatives are locally integrable. Since $\varepsilon(\xi,z) = \varepsilon(\xi^{-1}z)$, we obtain that as z approaches the boundary, we can pass to the limit in the second term in (11.60).

By using this and the relation (11.53) as $z \in \Omega$ approaches the boundary $\partial\Omega$ from the interior, by passing to the limit, for every $z \in \partial\Omega$ we get

$$(c_{a,b} - H.R(z))u(z) + \lim_{\delta\to 0} \int_{\partial\Omega\setminus\{|\xi^{-1}z|<\delta\}} u(\xi)\langle \nabla^{a,b}\varepsilon(\xi,z), d\nu(\xi)\rangle \\ - \int_{\partial\Omega} \varepsilon(\xi,z)\langle \nabla^{b,a}u(\xi), d\nu(\xi)\rangle = 0. \tag{11.61}$$

This shows that (11.58) is a solution of the boundary value problem (11.56) with the boundary condition (11.57).

Now let us show the uniqueness which can be done similarly to the proof of the uniqueness in Theorem 11.3.4. If the boundary value problem (11.56)–(11.57) has two solutions u and u_1 then the function $w := u-u_1 \in C^2(\Omega)\cap C^1(\overline{\Omega})$ satisfies the homogeneous equation

$$\mathcal{L}_{a,b}w = 0 \quad \text{in } \Omega, \tag{11.62}$$

and the boundary condition (11.57), that is,

$$\begin{aligned}(c_{a,b} - H.R(z))w(z) + \lim_{\delta\to 0}\int_{\partial\Omega\setminus\{|\xi^{-1}z|<\delta\}} & w(\xi)\langle\nabla^{a,b}\varepsilon(\xi,z), d\nu(\xi)\rangle \\ - \int_{\partial\Omega} & \varepsilon(\xi,z)\langle\nabla^{b,a}w(\xi), d\nu(\xi)\rangle = 0,\end{aligned} \tag{11.63}$$

for all $z \in \partial\Omega$.

Using the representation formula (11.59) for solutions with $f \equiv 0$ we have the following representation formula for w:

$$c_{a,b}w(z) = \int_{\partial\Omega} w(\xi)\langle\nabla^{a,b}\varepsilon(\xi,z), d\nu(\xi)\rangle - \int_{\partial\Omega} \varepsilon(\xi,z)\langle\nabla^{b,a}w(\xi), d\nu(\xi)\rangle$$

for all $z \in \Omega$. As in the existence proof, by using the properties of the double and single layer potentials as $z \to \partial\Omega$, we obtain that

$$\begin{aligned}c_{a,b}w(z) = (c_{a,b} - H.R(z))w(z) + \lim_{\delta\to 0}\int_{\partial\Omega\setminus\{|\xi^{-1}z|<\delta\}} & w(\xi)\langle\nabla^{a,b}\varepsilon(\xi,z), d\nu(\xi)\rangle \\ - \int_{\partial\Omega} & \varepsilon(\xi,z)\langle\nabla^{b,a}w, d\nu(\xi)\rangle\end{aligned}$$

for all $z \in \partial\Omega$. Comparing this with (11.63) we see that w must satisfy

$$w(z) = 0, \ \ z \in \partial\Omega. \tag{11.64}$$

By an argument similar to that in Proposition 11.1.1, the homogeneous equation (11.62) with the Dirichlet boundary condition (11.64) has a unique trivial solution $w \equiv 0$ in Ω. Therefore, we must have $u = u_1$ in Ω. This completes the proof of Theorem 11.3.10. □

11.3.4 Powers of the Kohn Laplacian

In this section, we carry out the analysis similar to that in Section 11.3.2 for the powers of the extended Kohn Laplacian $\mathcal{L}_{a,b}$.

As before, let Ω be an admissible domain in $\mathbb{H}^n$. For $m \in \mathbb{N}$, we denote

$$\mathcal{L}^m_{a,b} := \mathcal{L}_{a,b}\mathcal{L}^{m-1}_{a,b}.$$

Then for $m \in \mathbb{N}$, we consider the equation

$$\mathcal{L}_{a,b}^{m} u(z) = c_{a,b} f(z), \;\; z \in \Omega. \tag{11.65}$$

Let $\varepsilon(\xi, z) = \varepsilon(\xi^{-1} z)$ be the rescaled fundamental solution of the Kohn Laplacian as in (11.55). Let us now define the Newton potential for the operator $\mathcal{L}_{a,b}^{m}$ by

$$u(z) = \int_{\Omega} f(\xi) \varepsilon_m(\xi, z) d\nu(\xi), \tag{11.66}$$

where $\varepsilon_m(\xi, z)$ is a rescaled fundamental solution of (11.65) satisfying

$$\mathcal{L}_{a,b}^{m-1} \varepsilon_m = \varepsilon.$$

With a suitable distributional interpretation, for $m = 2, 3, \ldots$, we can write

$$\varepsilon_m(\xi, z) = \int_{\Omega} \varepsilon_{m-1}(\xi, \zeta) \varepsilon(\zeta, z) d\nu(\zeta), \qquad \xi, z \in \Omega, \tag{11.67}$$

with

$$\varepsilon_1(\xi, z) = \varepsilon(\xi, z).$$

Theorem 11.3.11 (Traces and Kac's boundary value problem for powers of Kohn Laplacians $\mathcal{L}_{a,b}$)**.** *For any $f \in \Gamma_\alpha(\Omega)$, $0 < \alpha < 1$, $\mathrm{supp} f \subset \Omega$, the generalized Newton potential (11.66) is a unique solution of the equation (11.65) in $C^{2m}(\Omega) \cap C^{2m-1}(\overline{\Omega})$ with m boundary conditions*

$$\begin{aligned} &(c_{a,b} - H.R(z)) \mathcal{L}_{a,b}^{i} u(z) \\ &\quad + \sum_{j=0}^{m-i-1} \lim_{\delta \to 0} \int_{\partial\Omega \setminus \{|\xi^{-1} z| < \delta\}} \mathcal{L}_{a,b}^{j+i} u(\xi) \langle \nabla^{a,b} \mathcal{L}_{a,b}^{m-1-j} \varepsilon_m(\xi, z), d\nu(\xi) \rangle \\ &\quad - \sum_{j=0}^{m-i-1} \int_{\partial\Omega} \mathcal{L}_{a,b}^{m-1-j} \varepsilon_m(\xi, z) \langle \nabla^{b,a} \mathcal{L}_{a,b}^{j+i} u(\xi) d\nu(\xi) \rangle = 0, \quad z \in \partial\Omega, \end{aligned} \tag{11.68}$$

for $i = 0, 1, \ldots, m-1$, where $H.R(z)$ is the half-residue given by (11.54).

Proof of Theorem 11.3.11. Applying Green's second formula from Theorem 1.4.6 for each $z \in \Omega$, similarly to (11.59) we get

$$\begin{aligned} &c_{a,b} u(z) = c_{a,b} \int_{\Omega} f(\xi) \varepsilon_m(\xi, z) d\nu(\xi) \\ &= \int_{\Omega} \mathcal{L}_{a,b}^{m} u(\xi) \varepsilon_m(\xi, z) d\nu(\xi) = \int_{\Omega} \mathcal{L}_{a,b}^{m-1} u(\xi) \mathcal{L}_{a,b} \varepsilon_m(\xi, z) d\nu(\xi) \\ &\quad - \int_{\partial\Omega} \mathcal{L}_{a,b}^{m-1} u(\xi) \langle \nabla^{a,b} \varepsilon_m(\xi, z), d\nu(\xi) \rangle + \int_{\partial\Omega} \varepsilon_m(\xi, z) \langle \nabla^{b,a} \mathcal{L}_{a,b}^{m-1} u(\xi), d\nu(\xi) \rangle \end{aligned}$$

$$= \int_{\Omega} \mathcal{L}_{a,b}^{m-2} u(\xi) \mathcal{L}_{a,b}^{2} \varepsilon_m(\xi, z) d\nu(\xi) - \int_{\partial\Omega} \mathcal{L}_{a,b}^{m-2} u(\xi) \langle \nabla^{a,b} \mathcal{L}_{a,b} \varepsilon_m(\xi, z), d\nu(\xi) \rangle$$
$$+ \int_{\partial\Omega} \mathcal{L}_{a,b} \varepsilon_m(\xi, z) \langle \nabla^{b,a} \mathcal{L}_{a,b}^{m-2} u(\xi), d\nu(\xi) \rangle - \int_{\partial\Omega} \mathcal{L}_{a,b}^{m-1} u(\xi) \langle \nabla^{a,b} \varepsilon_m(\xi, z), d\nu(\xi) \rangle$$
$$+ \int_{\partial\Omega} \varepsilon_m(\xi, z) \langle \nabla^{b,a} \mathcal{L}_{a,b}^{m-1} u(\xi), d\nu(\xi) \rangle = \cdots$$
$$= c_{a,b} u(z) - \sum_{j=0}^{m-1} \int_{\partial\Omega} \mathcal{L}_{a,b}^{j} u(\xi) \langle \nabla^{a,b} \mathcal{L}_{a,b}^{m-1-j} \varepsilon_m(\xi, z), d\nu(\xi) \rangle$$
$$+ \sum_{j=0}^{m-1} \int_{\partial\Omega} \mathcal{L}_{a,b}^{m-1-j} \varepsilon_m(\xi, z) \langle \nabla^{b,a} \mathcal{L}_{a,b}^{j} u(\xi), d\nu(\xi) \rangle, \quad z \in \Omega.$$

It follows that

$$\sum_{j=0}^{m-1} \int_{\partial\Omega} \mathcal{L}_{a,b}^{j} u(\xi) \langle \nabla^{a,b} \mathcal{L}_{a,b}^{m-1-j} \varepsilon_m(\xi, z), d\nu(\xi) \rangle$$
$$- \sum_{j=0}^{m-1} \int_{\partial\Omega} \mathcal{L}_{a,b}^{m-1-j} \varepsilon_m(\xi, z) \langle \nabla^{b,a} \mathcal{L}_{a,b}^{j} u(\xi), d\nu(\xi) \rangle = 0, \quad z \in \Omega. \tag{11.69}$$

By using the continuity of the single layer potential and Corollary 11.3.9 for the double layer potential as z approaches the boundary $\partial\Omega$ from the interior, from (11.69) we obtain

$$(c_{a,b} - H.R(z))u(z) + \sum_{j=0}^{m-1} \lim_{\delta \to 0} \int_{\partial\Omega \setminus \{|\xi^{-1} z| < \delta\}} \mathcal{L}_{a,b}^{j} u(\xi) \langle \nabla^{a,b} \mathcal{L}_{a,b}^{m-1-j} \varepsilon_m(\xi, z), d\nu(\xi) \rangle$$
$$- \sum_{j=0}^{m-1} \int_{\partial\Omega} \mathcal{L}_{a,b}^{m-1-j} \varepsilon_m(\xi, z) \langle \nabla^{b,a} \mathcal{L}_{a,b}^{j} u(\xi), d\nu(\xi) \rangle = 0, \ z \in \partial\Omega.$$

This relation is one of the boundary conditions (11.68). Let us now derive the remaining boundary conditions. For this, we will rewrite the equation as

$$\mathcal{L}_{a,b}^{m-i} \mathcal{L}_{a,b}^{i} u = c_{a,b} f, \quad i = 0, 1, \ldots, m-1, \quad m = 1, 2, \ldots, \tag{11.70}$$

and then use a similar argument. Thus,

$$c_{a,b} \mathcal{L}_{a,b}^{i} u(z) = c_{a,b} \int_{\Omega} f(\xi) \mathcal{L}_{a,b}^{i} \varepsilon_m(\xi, z) d\nu(\xi) = \int_{\Omega} \mathcal{L}_{a,b}^{m-i} \mathcal{L}_{a,b}^{i} u(\xi) \mathcal{L}_{a,b}^{i} \varepsilon_m(\xi, z) d\nu(\xi)$$
$$= \int_{\Omega} \mathcal{L}_{a,b}^{m-i-1} \mathcal{L}_{a,b}^{i} u(\xi) \mathcal{L}_{a,b} \mathcal{L}_{a,b}^{i} \varepsilon_m(\xi, z) d\nu(\xi)$$
$$- \int_{\partial\Omega} \mathcal{L}_{a,b}^{m-i-1} \mathcal{L}_{a,b}^{i} u(\xi) \langle \nabla^{a,b} \mathcal{L}_{a,b}^{i} \varepsilon_m(\xi, z), d\nu(\xi) \rangle$$
$$+ \int_{\partial\Omega} \mathcal{L}_{a,b}^{i} \varepsilon_m(\xi, z) \langle \nabla^{b,a} \mathcal{L}_{a,b}^{m-i-1} \mathcal{L}_{a,b}^{i} u(\xi), d\nu(\xi) \rangle$$

$$
\begin{aligned}
&= \int_\Omega \mathcal{L}_{a,b}^{m-i-2}\mathcal{L}_{a,b}^{i}u(\xi)\mathcal{L}_{a,b}^{2}\mathcal{L}_{a,b}^{i}\varepsilon_m(\xi,z)d\nu(\xi) \\
&\quad - \int_{\partial\Omega} \mathcal{L}_{a,b}^{m-i-2}\mathcal{L}_{a,b}^{i}u(\xi)\langle\nabla^{a,b}\mathcal{L}_{a,b}\mathcal{L}_{a,b}^{i}\varepsilon_m(\xi,z), d\nu(\xi)\rangle \\
&\quad + \int_{\partial\Omega} \mathcal{L}_{a,b}\mathcal{L}_{a,b}^{i}\varepsilon_m(\xi,z)\langle\nabla^{b,a}\mathcal{L}_{a,b}^{m-i-2}\mathcal{L}_{a,b}^{i}u(\xi), d\nu(\xi)\rangle \\
&\quad - \int_{\partial\Omega} \mathcal{L}_{a,b}^{m-i-1}\mathcal{L}_{a,b}^{i}u(\xi)\langle\nabla^{a,b}\mathcal{L}_{a,b}^{i}\varepsilon_m(\xi,z), d\nu(\xi)\rangle \\
&\quad + \int_{\partial\Omega} \mathcal{L}_{a,b}^{i}\varepsilon_m(\xi,z)\langle\nabla^{b,a}\mathcal{L}_{a,b}^{m-i-1}\mathcal{L}_{a,b}^{i}u(\xi), d\nu(\xi)\rangle \\
&= \cdots \\
&= \int_\Omega \mathcal{L}_{a,b}^{i}u(\xi)\mathcal{L}_{a,b}^{m-i}\mathcal{L}_{a,b}^{i}\varepsilon_m(\xi,z)d\nu(\xi) \\
&\quad - \sum_{j=0}^{m-i-1}\int_{\partial\Omega} \mathcal{L}_{a,b}^{j}\mathcal{L}_{a,b}^{i}u(\xi)\langle\nabla^{a,b}\mathcal{L}_{a,b}^{m-i-1-j}\mathcal{L}_{a,b}^{i}\varepsilon_m(\xi,z), d\nu(\xi)\rangle \\
&\quad + \sum_{j=0}^{m-i-1}\int_{\partial\Omega} \mathcal{L}_{a,b}^{m-i-1-j}\mathcal{L}_{a,b}^{i}\varepsilon_m(\xi,z)\langle\nabla^{b,a}\mathcal{L}_{a,b}^{j}\mathcal{L}_{a,b}^{i}u(\xi), d\nu(\xi)\rangle \\
&= c_{a,b}\mathcal{L}_{a,b}^{i}u(z) - \sum_{j=0}^{m-i-1}\int_{\partial\Omega} \mathcal{L}_{a,b}^{j+i}u(\xi)\langle\nabla^{a,b}\mathcal{L}_{a,b}^{m-1-j}\varepsilon_m(\xi,z), d\nu(\xi)\rangle \\
&\quad + \sum_{j=0}^{m-i-1}\int_{\partial\Omega} \mathcal{L}_{a,b}^{m-1-j}\varepsilon_m(\xi,z)\langle\nabla^{b,a}\mathcal{L}_{a,b}^{j+i}u(\xi), d\nu(\xi)\rangle, \quad z\in\Omega,
\end{aligned}
$$

where we used that $\mathcal{L}_{a,b}^{i}\varepsilon_m$ is a rescaled fundamental solution of $\mathcal{L}_{a,b}^{m-i}$ as in (11.70).

This implies the identities

$$
\begin{aligned}
&\sum_{j=0}^{m-i-1}\int_{\partial\Omega} \mathcal{L}_{a,b}^{j+i}u(\xi)\langle\nabla^{a,b}\mathcal{L}_{a,b}^{m-1-j}\varepsilon_m(\xi,z), d\nu(\xi)\rangle \\
&\quad - \sum_{j=0}^{m-i-1}\int_{\partial\Omega} \mathcal{L}_{a,b}^{m-1-j}\varepsilon_m(\xi,z)\langle\nabla^{b,a}\mathcal{L}_{a,b}^{j+i}u(\xi), d\nu(\xi)\rangle = 0,
\end{aligned}
$$

for all $z\in\Omega$ and $i=0,1,\ldots,m-1$. As before, by using the continuity of the single layer potential and Corollary 11.3.9 for the double layer potential as z approaches the boundary $\partial\Omega$ from the interior, we find that for all $z\in\partial\Omega$ we have

$$
\begin{aligned}
&(c_{a,b}-H.R(z))\mathcal{L}_{a,b}^{i}u(z) \\
&\quad + \sum_{j=0}^{m-i-1}\lim_{\delta\to 0}\int_{\partial\Omega\setminus\{|\xi^{-1}z|<\delta\}} \mathcal{L}_{a,b}^{j+i}u(\xi)\langle\nabla^{a,b}\mathcal{L}_{a,b}^{m-1-j}\varepsilon_m(\xi,z), d\nu(\xi)\rangle \\
&\quad - \sum_{j=0}^{m-i-1}\int_{\partial\Omega} \mathcal{L}_{a,b}^{m-1-j}\varepsilon_m(\xi,z)\langle\nabla^{b,a}\mathcal{L}_{a,b}^{j+i}u(\xi), d\nu(\xi)\rangle = 0,
\end{aligned}
$$

which are the boundary conditions (11.68) for all $i=0,1,\ldots,m-1$.

Let us now show the uniqueness, that is, if a function $w \in C^{2m}(\Omega) \cap C^{2m-1}(\overline{\Omega})$ satisfies the equation $\mathcal{L}_{a,b}^m w = f$ and the boundary conditions (11.68), then it must be given by (11.66). Indeed, with u given by (11.66), the function

$$v := u - w \in C^{2m}(\Omega) \cap C^{2m-1}(\overline{\Omega}),$$

satisfies the homogeneous equation

$$\mathcal{L}_{a,b}^m v = 0$$

and the boundary conditions for all $i = 0, 1, \ldots, m-1$,

$$\begin{aligned} I_i(v)(z) &:= (c_{a,b} - H.R(z))\mathcal{L}_{a,b}^i v(z) \\ &+ \sum_{j=0}^{m-i-1} \lim_{\delta \to 0} \int_{\partial\Omega \backslash \{|\xi^{-1}z|<\delta\}} \mathcal{L}_{a,b}^{j+i} v(\xi) \langle \nabla^{a,b} \mathcal{L}_{a,b}^{m-1-j} \varepsilon_m(\xi, z), d\nu(\xi)\rangle \\ &- \sum_{j=0}^{m-i-1} \int_{\partial\Omega} \mathcal{L}_{a,b}^{m-1-j} \varepsilon_m(\xi, z) \langle \nabla^{b,a} \mathcal{L}_{a,b}^{j+i} v(\xi), d\nu(\xi)\rangle = 0, \end{aligned}$$

for $z \in \partial\Omega$. Applying the Green formula from Theorem 1.4.6 to the function $v \in C^{2m}(\Omega) \cap C^{2m-1}(\overline{\Omega})$ we obtain

$$\begin{aligned} 0 &= \int_\Omega \mathcal{L}_{a,b}^m v(z) \mathcal{L}_{a,b}^i \varepsilon_m(\xi, z) d\nu(\xi) = \int_\Omega \mathcal{L}_{a,b}^{m-i} \mathcal{L}_{a,b}^i v(z) \mathcal{L}_{a,b}^i \varepsilon_m(\xi, z) d\nu(\xi) \\ &= \int_\Omega \mathcal{L}_{a,b}^{m-1} v(z) \mathcal{L}_{a,b} \mathcal{L}_{a,b}^i \varepsilon_m(\xi, z) d\nu(\xi) - \int_{\partial\Omega} \mathcal{L}_{a,b}^{m-1} v(z) \langle \nabla^{a,b} \mathcal{L}_{a,b}^i \varepsilon_m(\xi, z), d\nu(\xi)\rangle \\ &+ \int_{\partial\Omega} \mathcal{L}_{a,b}^i \varepsilon_m(\xi, z) \langle \nabla^{a,b} \mathcal{L}_{a,b}^{m-1} v(z), d\nu(\xi)\rangle \\ &= \cdots \\ &= c_{a,b} \mathcal{L}_{a,b}^i v(z) - \sum_{j=0}^{m-i-1} \int_{\partial\Omega} \mathcal{L}_{a,b}^{j+i} v(\xi) \langle \nabla^{a,b} \mathcal{L}_{a,b}^{m-1-j} \varepsilon_m(\xi, z), d\nu(\xi)\rangle \\ &+ \sum_{j=0}^{m-i-1} \int_{\partial\Omega} \mathcal{L}_{a,b}^{m-1-j} \varepsilon_m(\xi, z) \langle \nabla^{b,a} \mathcal{L}_{a,b}^{j+i} v(\xi), d\nu(\xi)\rangle, \quad i = 0, 1, \ldots, m-1. \end{aligned}$$

By passing to the limit as $z \to \partial\Omega$, we obtain the additional relations

$$\mathcal{L}_{a,b}^i v(z) \mid_{z\in\partial\Omega} = I_i(v)(z) \mid_{z\in\partial\Omega} = 0, \quad i = 0, 1, \ldots, m-1.$$

We claim that the boundary value problem

$$\begin{aligned} &\mathcal{L}_{a,b}^m v = 0, \\ &\mathcal{L}_{a,b}^i v \mid_{\partial\Omega} = 0, \quad i = 0, 1, \ldots, m-1, \end{aligned} \tag{11.71}$$

has a unique trivial solution. Assuming this for a moment, we would get that $v = u - w \equiv 0$, which means that w coincides with u in Ω, so that (11.66) is the unique solution of the boundary value problem (11.65), (11.68) in Ω.

To show that the boundary value problem (11.71) has a unique solution in $C^{2m}(\Omega) \cap C^{2m-1}(\overline{\Omega})$, we denote $\tilde{v} := \mathcal{L}_{a,b}^{m-1} v$, and then the statement follows by induction from the uniqueness in $C^2(\Omega) \cap C^1(\overline{\Omega})$ of the problem

$$\mathcal{L}_{a,b}\tilde{v} = 0, \quad \tilde{v}\,|_{\partial\Omega} = 0.$$

The proof of Theorem 11.3.11 is complete. □

11.4 Hardy inequalities with boundary terms on stratified groups

In this section we demonstrate how the developed potential theory for the sub-Laplacian can be used to derive a refinement to Hardy inequalities. The basic inequality that we are referring to here is inequality (7.4) which says that on a stratified group $\mathbb{G}$ of homogeneous dimension $Q \geq 3$, for all $u \in C_0^\infty(\mathbb{G}\backslash\{0\})$ and $\alpha > 2 - Q$ we have

$$\int_{\mathbb{G}} d^\alpha |\nabla_{\mathbb{G}} u|^2 \, d\nu \geq \left(\frac{Q+\alpha-2}{2}\right)^2 \int_{\mathbb{G}} d^\alpha \frac{|\nabla_{\mathbb{G}} d|^2}{d^2} |u|^2 \, d\nu, \tag{11.72}$$

and the constant $\left(\frac{Q+\alpha-2}{2}\right)^2$ is sharp. We refer to Remark 7.1.2 for a historic discussion of this and other related inequalities.

We will now derive a version of (11.72) over domains in $\mathbb{G}$ in such a way that the boundary term appears as well, in the case the function does not vanish on the boundary. Certainly, the following inequality implies (11.72) if we take the domain Ω in such a way that it contains the support of a compactly supported function u. As before, d stands for the $\mathcal{L}$-gauge and we also write

$$\nabla_H = (X_1, \ldots, X_{N_1})$$

for the horizontal gradient. A discussion from a point of view of more general weights was also done in Section 7.7.

Theorem 11.4.1 (Hardy inequality with boundary term). *Let $\Omega \subset \mathbb{G}$ be an admissible domain in a stratified group of homogeneous dimension $Q \geq 3$ with $0 \notin \partial\Omega$. Let $\alpha \in \mathbb{R}$ and $\alpha > 2 - Q$. Let $u \in C^1(\Omega) \bigcap C(\overline{\Omega})$. Then we have*

$$\begin{aligned} \int_\Omega d^\alpha |\nabla_H u|^2 \, d\nu &\geq \left(\frac{Q+\alpha-2}{2}\right)^2 \int_\Omega d^\alpha \frac{|\nabla_H d|^2}{d^2} |u|^2 \, d\nu \\ &\quad + \frac{Q+\alpha-2}{2(Q-2)} \int_{\partial\Omega} d^{Q+\alpha-2} |u|^2 \langle \widetilde{\nabla} d^{2-Q}, d\nu \rangle. \end{aligned} \tag{11.73}$$

Remark 11.4.2.

1. If $u = 0$ on $\partial\Omega$ then (11.73) reduces to (11.72).
2. The boundary term in (11.73) sometimes can be of different sign. To show this, take $u = e^{-\frac{R}{2}d}$ for $R > 0$. Green's first formula in Theorem 1.4.6 allows us to calculate

$$\begin{aligned}\int_{\partial\Omega} & d^{Q+\alpha-2}e^{-Rd}\langle\widetilde{\nabla}d^{2-Q}, d\nu\rangle \\ &= \int_{\Omega}\widetilde{\nabla}(d^{Q+\alpha-2}e^{-Rd})d^{2-Q}d\nu + \frac{1}{\beta_d}\int_{\Omega} d^{Q+\alpha-2}e^{-Rd}\mathcal{L}\beta_d d^{2-Q}d\nu \\ &= \int_{\Omega}\widetilde{\nabla}(d^{Q+\alpha-2}e^{-Rd})d^{2-Q}d\nu \\ &= \sum_{k=1}^{N_1}\int_{\Omega} X_k(d^{Q+\alpha-2}e^{-Rd})X_k d^{2-Q}d\nu \\ &= \sum_{k=1}^{N_1}\int_{\Omega}((Q+\alpha-2)d^{Q+\alpha-2-1}e^{-Rd}X_k d - Rd^{Q+\alpha-2}e^{-Rd}X_k d) \\ &\quad \times(2-Q)d^{2-Q-1}X_k d\, d\nu.\end{aligned}$$

Let $\alpha = 0$, $Q = 3$, and $\Omega\bigcap B_{\frac{1}{R}} = \{\emptyset\}$, where $B_{\frac{1}{R}} = \{x \in \mathbb{G} : d(x) < \frac{1}{R}\}$. Then we get

$$\int_{\partial\Omega} de^{-Rd}\langle\widetilde{\nabla}d^{-1}, d\nu\rangle = \sum_{k=1}^{N_1}\int_{\Omega}(Rd^{-1} - d^{-2})e^{-Rd}(X_k d)^2\, d\nu > 0$$

is positive. On the other hand, if $u := C = \text{const}$, then we get

$$\begin{aligned}\int_{\partial\Omega} & d^{Q+\alpha-2}C^2\langle\widetilde{\nabla}d^{2-Q}, d\nu\rangle \\ &= C^2\sum_{k=1}^{N_1}\int_{\Omega}((Q+\alpha-2)d^{Q+\alpha-2-1}X_k d)\,(2-Q)d^{2-Q-1}X_k d\, d\nu \\ &= -C^2(Q+\alpha-2)(Q-2)\sum_{k=1}^{N_1}\int_{\Omega} d^{\alpha-2}(X_k d)^2 d\nu < 0,\end{aligned}$$

which shows that the boundary term can also be negative.

3. Since it is known that the constant $\left(\frac{Q+\alpha-2}{2}\right)^2$ in (11.73) (or rather in (11.72)) is sharp, the local inequality (11.73) gives a refinement to (11.72), and even more so if the boundary term is positive.
4. Note that in comparison to (11.72), we do not assume in Theorem 11.4.1 that 0 is not in the support of the function u since for $\alpha > 2 - Q$ all the integrals in (11.73) are convergent.

5. Even if $0 \in \partial\Omega$, the statement of Theorem 11.4.1 remains true if $0 \notin \partial\Omega \cap \operatorname{supp} u$.
6. Without the second (boundary) term inequality (11.73) was studied on the Heisenberg group in [GL90] for a particular choice of $d(x)$, or on Carnot groups [GK08] for $\mathcal{L}$-gauges $d(x)$. We can refer to these papers as well as to [GZ01] for other references on this subject. From the point of view of the boundary term the inequality (11.73) can be thought of as a refinement of the usual Hardy because, also as it was pointed out in Part 2 of this Remark, this boundary term in (11.73) can be positive. We call these inequalities local due to the presence of a contribution from the boundary.

Proof of Theorem 11.4.1. By an argument similar to that in Remark 2.6, Part 3, without loss of generality we can assume that u is real-valued. In this case, recalling that

$$(\widetilde{\nabla} u)u = \sum_{k=1}^{N_1} (X_k u) X_k u = |\nabla_H u|^2,$$

the estimate (11.73) reduces to

$$\begin{aligned}\int_\Omega d^\alpha (\widetilde{\nabla} u) u\, d\nu \geq \left(\frac{Q+\alpha-2}{2}\right)^2 \int_\Omega d^\alpha \frac{(\widetilde{\nabla} d) d}{d^2} u^2\, d\nu \\ + \frac{Q+\alpha-2}{2(Q-2)} \int_{\partial\Omega} d^{Q+\alpha-2} u^2 \langle \widetilde{\nabla} d^{2-Q}, d\nu\rangle,\end{aligned} \tag{11.74}$$

which we will now prove. Let us set $u = d^\gamma q$ for some $\gamma \neq 0$ to be chosen later. Then we can calculate

$$\begin{aligned}(\widetilde{\nabla} u)u &= (\widetilde{\nabla} d^\gamma q) d^\gamma q = \sum_{k=1}^{N_1} X_k(d^\gamma q) X_k(d^\gamma q) \\ &= \gamma^2 d^{2\gamma-2} \sum_{k=1}^{N_1} (X_k d)^2 q^2 + 2\gamma d^{2\gamma-1} q \sum_{k=1}^{N_1} X_k d\, X_k q + d^{2\gamma} \sum_{k=1}^{N_1} (X_k q)^2 \\ &= \gamma^2 d^{2\gamma-2} ((\widetilde{\nabla} d) d) q^2 + 2\gamma d^{2\gamma-1} q (\widetilde{\nabla} d) q + d^{2\gamma} (\widetilde{\nabla} q) q.\end{aligned}$$

Multiplying both sides of this equality by d^α and applying Green's first formula from Theorem 1.4.6, we get

$$\begin{aligned}\int_\Omega d^\alpha (\widetilde{\nabla} u) u d\nu &= \gamma^2 \int_\Omega d^{\alpha+2\gamma-2} ((\widetilde{\nabla} d) d)\, q^2 d\nu + \frac{\gamma}{\alpha+2\gamma} \int_{\partial\Omega} q^2 \langle \widetilde{\nabla} d^{\alpha+2\gamma}, d\nu\rangle \\ &\quad - \frac{\gamma}{\alpha+2\gamma} \int_\Omega (\mathcal{L} d^{\alpha+2\gamma}) q^2 d\nu + \int_\Omega d^{\alpha+2\gamma} (\widetilde{\nabla} q) q d\nu \\ &\geq \int_\Omega \gamma^2 d^{\alpha+2\gamma-2} ((\widetilde{\nabla} d) d)\, q^2 d\nu + \frac{\gamma}{\alpha+2\gamma} \int_{\partial\Omega} q^2 \langle \widetilde{\nabla} d^{\alpha+2\gamma}, d\nu\rangle \\ &\quad - \frac{\gamma}{\alpha+2\gamma} \int_\Omega (\mathcal{L} d^{\alpha+2\gamma}) q^2 d\nu.\end{aligned} \tag{11.75}$$

On the other hand, it can be readily checked that we have

$$-\frac{\gamma}{\alpha+2\gamma}\mathcal{L}d^{\alpha+2\gamma} = -\gamma(\alpha+2\gamma+Q-2)d^{\alpha+2\gamma-2}(\widetilde{\nabla} d)d - \frac{\gamma}{2-Q}d^{\alpha+2\gamma+Q-2}\mathcal{L}d^{2-Q}.$$

Since $q^2 = d^{-2\gamma}u^2$, substituting this into (11.75) we obtain

$$\int_\Omega d^\alpha(\widetilde{\nabla} u)ud\nu \geq (-\gamma^2 - \gamma(\alpha+Q-2))\int_\Omega d^\alpha \frac{(\widetilde{\nabla} d)d}{d^2}u^2 d\nu$$
$$-\frac{\gamma}{2-Q}\int_\Omega (\mathcal{L}d^{2-Q})d^{\alpha+Q-2}u^2 dx + \frac{\gamma}{\alpha+2\gamma}\int_{\partial\Omega} d^{-2\gamma}u^2\langle \widetilde{\nabla} d^{\alpha+2\gamma}, d\nu\rangle.$$

Recalling that for $Q \geq 3$, $\varepsilon = \beta_d d^{2-Q}$ is the fundamental solution of the sub-Laplacian $\mathcal{L}$, we have

$$\int_\Omega (\mathcal{L}d^{2-Q})d^{\alpha+Q-2}u^2 dx = 0, \ \alpha > 2-Q,$$

independent of whether 0 belongs to Ω or not, since $\mathcal{L}d^{2-Q} = \frac{1}{\beta_d}\delta$ in $\mathbb{G}$. It follows that

$$\int_\Omega d^\alpha(\widetilde{\nabla} u)ud\nu$$
$$\geq (-\gamma^2 - \gamma(\alpha+Q-2))\int_\Omega d^\alpha \frac{(\widetilde{\nabla} d)d}{d^2}u^2 d\nu + \frac{\gamma}{\alpha+2\gamma}\int_{\partial\Omega} d^{-2\gamma}u^2\langle \widetilde{\nabla} d^{\alpha+2\gamma}, d\nu\rangle.$$

Taking $\gamma = \frac{2-Q-\alpha}{2}$, we obtain (11.74). □

As usual, a Hardy inequality implies several uncertainty principles.

Corollary 11.4.3 (Local uncertainty principle with boundary terms). *Let $\Omega \subset \mathbb{G}$ be an admissible domain in a stratified group $\mathbb{G}$ of homogeneous dimension $Q \geq 3$, with $0 \notin \partial\Omega$. Then for all $u \in C^1(\Omega)\bigcap C(\overline{\Omega})$ we have*

$$\left(\int_\Omega d^2|\nabla_H d|^2|u|^2 d\nu\right)\left(\int_\Omega |\nabla_H u|^2 d\nu\right)$$
$$\geq \left(\frac{Q-2}{2}\right)^2\left(\int_\Omega |\nabla_H d|^2|u|^2 d\nu\right)^2 \tag{11.76}$$
$$+\frac{1}{2}\int_{\partial\Omega} d^{Q-2}|u|^2\langle \widetilde{\nabla} d^{2-Q}, d\nu\rangle\left(\int_\Omega d^2|\nabla_H d|^2|u|^2 d\nu\right)$$

and

$$\left(\int_\Omega \frac{d^2}{|\nabla_H d|^2}|u|^2 d\nu\right)\left(\int_\Omega |\nabla_H u|^2 d\nu\right) \tag{11.77}$$
$$\geq \left(\frac{Q-2}{2}\right)^2\left(\int_\Omega |u|^2 d\nu\right)^2 \frac{1}{2}\int_{\partial\Omega} d^{Q-2}|u|^2\langle \widetilde{\nabla} d^{2-Q}, d\nu\rangle\left(\int_\Omega \frac{d^2}{|\nabla_H d|^2}|u|^2 d\nu\right).$$

Proof of Corollary 11.4.3. Again, assuming u is real-valued, and taking $\alpha = 0$ in the inequality (11.74) we get

$$\begin{aligned}
&\left(\int_\Omega d^2((\widetilde{\nabla} d)d)|u|^2 d\nu\right)\left(\int_\Omega (\widetilde{\nabla} u)u d\nu\right) \\
&\geq \left(\frac{Q-2}{2}\right)^2 \left(\int_\Omega d^2((\widetilde{\nabla} d)d)|u|^2 d\nu\right) \int_\Omega \frac{(\widetilde{\nabla} d)d}{d^2}|u|^2\, d\nu \\
&\quad + \frac{1}{2}\int_{\partial\Omega} d^{Q-2}|u|^2 \langle \widetilde{\nabla} d^{2-Q}, d\nu\rangle \left(\int_\Omega d^2|\nabla_H d|^2|u|^2 d\nu\right) \\
&\geq \left(\frac{Q-2}{2}\right)^2 \left(\int_\Omega ((\widetilde{\nabla} d)d)|u|^2 d\nu\right)^2 \\
&\quad + \frac{1}{2}\int_{\partial\Omega} d^{Q-2}|u|^2 \langle \widetilde{\nabla} d^{2-Q}, d\nu\rangle \left(\int_\Omega d^2|\nabla_H d|^2|u|^2 d\nu\right),
\end{aligned}$$

where we have used the Hölder inequality in the last line. This shows (11.76). The proof of (11.77) is similar. □

11.5 Green functions on H-type groups

In this section we discuss applications of the described potential theory to the construction of solutions of boundary value problems. We restrict our attention to the prototype H-type groups discussed in Section 1.4.10. The examples of the prototype H-type groups are the Abelian group $(\mathbb{R}^d; +)$ and the Heisenberg group $\mathbb{H}^d$. The fact that will be important for our purposes in this section is that the fundamental solution to the sub-Laplacian in this setting is explicitly known, see Theorem 1.4.20. Our presentation in this section follows [GRS17].

Thus, in this section, let $\mathbb{G} \simeq \mathbb{R}^m \times \mathbb{R}^n$ be a prototype H-type group and let $\mathcal{L}$ be the sub-Laplacian on $\mathbb{G}$ as defined in (1.119). We consider an open set $\Omega \subset \mathbb{G}$ with piecewise smooth boundary $\partial\Omega$, and study the Dirichlet problem for the sub-Laplacian

$$\begin{cases} \mathcal{L}u = f & \text{in } \Omega, \\ u = \phi & \text{on } \partial\Omega. \end{cases} \tag{11.78}$$

As discussed in Remark 11.1.7, there are several difficulties in solving such boundary problems in view of the usually present characteristic points.

However, it turns out that there are still a number of cases of interest when the Dirichlet boundary value problem (11.78) can be solved in the classical sense. This is achieved by constructing a well-behaved Green function for the problem (11.78). In view of Jerison's example discussed in Remark 11.1.7 one wants to avoid the characteristic points which is not always possible, since even ball-like bounded domains on (non-Abelian) H-type groups have non-empty collection of characteristic points. However, domains that are unbounded may have no characteristic points. Thus, in this section we present such an analysis for so-called l-wedge and l-strip domains. This includes domains such as half-spaces, quadrant-spaces and so on.

An important tool for this analysis will be the fundamental solution for the sub-Laplacian $\mathcal{L}$. More precisely, recalling Theorem 1.4.20, the function

$$\Gamma(\xi) := c\left(|x|^4 + 16|t|^2\right)^{(2-Q)/4} \tag{11.79}$$

is the fundamental solution of the sub-Laplacian, that is,

$$\mathcal{L}\Gamma_\zeta = -\delta_\zeta, \tag{11.80}$$

where $\Gamma_\zeta(\xi) = \Gamma(\zeta^{-1} \circ \xi)$ and δ_ζ is the Dirac distribution at $\zeta \equiv (y, \tau) \in \mathbb{G}$.

Definition 11.5.1 (Green function for Dirichlet sub-Laplacian). We define the *Green function for the Dirichlet sub-Laplacian* in Ω by the formula

$$G_\Omega(\xi, \zeta) := \Gamma(\zeta^{-1} \circ \xi) - h_\zeta(\xi),$$

where $h_\zeta(\xi)$ is a harmonic function, that is,

$$\mathcal{L}h_\zeta(\xi) = 0 \quad \text{in } \Omega, \tag{11.81}$$

having the same boundary values on $\partial\Omega$ as the fundamental solution Γ_ζ with pole at $\zeta \in \Omega$.

In particular, we have

$$G_\Omega(\xi, \zeta) = 0, \quad \xi \in \partial\Omega. \tag{11.82}$$

11.5.1 Green functions and Dirichlet problem in wedge domains

We first define the class of domains that we will be working in.

Definition 11.5.2 (l-wedge domains). Let $1 \le l \le m$ and let $\mathbb{G}^\ddagger$ be the *l-wedge space*

$$\mathbb{G}^\ddagger := \{\xi = (x_1, \ldots, x_m, t_1, \ldots, t_n) : \, x_1, \ldots, x_l > 0\}.$$

Let the point $\zeta = (y, \tau) = (y_1, y_2, \ldots, y_m, \tau_1, \ldots, \tau_n)$ lie in this l-wedge space, so that $y_1 > 0, \ldots, y_l > 0$. The point ζ_{x_k} defined by

$$\zeta_{x_k} := (y_1, \ldots, -y_k, \ldots, y_m, \tau_1, \ldots, \tau_n)$$

is said to be *symmetric for the point ζ with respect to the hyperplane* $x_k = 0$. Consecutively, the point

$$\zeta_{x_k x_s} := (y_1, \ldots, -y_k, \ldots, -y_s, \ldots, y_m, \tau_1, \ldots, \tau_n)$$

is said to be symmetric for the point ζ_{x_k} with respect to the hyperplane $x_s = 0$, and this process can be continued further.

It is clear that the symmetry indices are invariant under permutations. We will also use the notation

$$\Gamma((\zeta_{(j,l)})^{-1}\circ\xi) \text{ for } j\le l,$$

meaning that we take the sum of the functions $\Gamma((\zeta^{-1}_{x_{k_1}\cdots x_{k_j}}\circ\xi)$, $j\le l$, over all possible combinations of the symmetry arguments $x_{k_1}\cdots x_{k_j}$: here in order to reduce the number of subindices we write $(\zeta_{(j,l)})^{-1}\circ\xi$ for $\zeta^{-1}_{x_{k_1}\cdots x_{k_j}}\circ\xi$. For example, if $l=3$, $j=2$, then

$$\Gamma((\zeta_{(2,3)})^{-1}\circ\xi)=\Gamma((\zeta_{x_1x_2})^{-1}\circ\xi)+\Gamma((\zeta_{x_1x_3})^{-1}\circ\xi)+\Gamma((\zeta_{x_3x_2})^{-1}\circ\xi),$$

and if $l=3$, $j=3$, then

$$\Gamma((\zeta_{(3,3)})^{-1}\circ\xi)=\Gamma((\zeta_{x_1x_2x_3})^{-1}\circ\xi).$$

In this notation the Green function for the Dirichlet sub-Laplacian in wedge domains takes the following form, for any $1\le l\le m$.

Proposition 11.5.3 (Green function for Dirichlet sub-Laplacian in wedge domains). *Let $\mathbb{G}^{\ddagger}$ be an l-wedge domain for $1\le l\le m$. Then the function*

$$G_{\mathbb{G}^{\ddagger}}(\xi,\zeta)=\Gamma(\zeta^{-1}\circ\xi)+\sum_{j=1}^{l}(-1)^j\Gamma((\zeta_{(j,l)})^{-1}\circ\xi) \tag{11.83}$$

is the Green function for the Dirichlet sub-Laplacian in $\mathbb{G}^{\ddagger}$.

Proof of Proposition 11.5.3. Since any symmetric point $\zeta_{(j,l)}$ is not in $\mathbb{G}^{\ddagger}$, for any $j=1,\dots,l$, it follows from (11.80) that

$$\mathcal{L}\Gamma((\zeta_{(j,l)})^{-1}\circ\xi)=-\delta_{\zeta_{(j,l)}}=0 \text{ in } \mathbb{G}^{\ddagger},$$

for any $\xi\in\mathbb{G}^{\ddagger}$ and $j=1,\dots,l$. Thus, the function

$$\sum_{j=1}^{l}(-1)^j\Gamma((\zeta_{(j,l)})^{-1}\circ\xi)$$

satisfies condition (11.81), i.e., it is harmonic in $\mathbb{G}^{\ddagger}$. Now let us verify boundary condition for the domain $\mathbb{G}^{\ddagger}$, that is, the function $G_{\mathbb{G}^{\ddagger}}$ should become zero at $x_1=0$ and at infinity. Recall that

$$d(\xi,\zeta):=\left(\Gamma(\zeta^{-1}\circ\xi)\right)^{\frac{1}{2-Q}}$$

is an actual distance on $\mathbb{G}$ defining a norm (and not only a quasi-norm, see, e.g., [Cyg81]). Now it is easy to see that the d-distance from any point of the hyperplane $x_k = 0$ to the points ζ and ζ_{x_k} is the same, that is, $G_{\mathbb{G}^\ddagger}$ satisfies the Dirichlet condition at the hyperplanes $x_1 = 0, \ldots, x_l = 0$ and it is also clear (by the construction) that the function $G_{\mathbb{G}^\ddagger}$ is zero at the infinity. It proves that

$$G_{\mathbb{G}^\ddagger}(\xi, \zeta) = 0, \quad \xi \in \partial\mathbb{G}^\ddagger,$$

that is, $G_{\mathbb{G}^\ddagger}$ is a Green function according to Definition 11.5.1. □

In the next theorem we will use the anisotropic Hölder space Γ_α from Definition 11.3.1. We consider the Dirichlet problem for the sub-Laplacian

$$\begin{cases} \mathcal{L}u = f & \text{in } \mathbb{G}^\ddagger, \\ \quad u = \phi & \text{on } \partial\mathbb{G}^\ddagger. \end{cases} \tag{11.84}$$

Theorem 11.5.4 (Dirichlet problem for sub-Laplacian in wedge domain)**.** *Let $f \in \Gamma_\alpha(\mathbb{G}^\ddagger)$ for $0 < \alpha < 1$, and assume that* $\operatorname{supp} f \subset \mathbb{G}^\ddagger$, *and that $\phi \in C^\infty(\partial\mathbb{G}^\ddagger)$. Then the boundary value problem* (11.84) *has a unique solution $u \in C^2(\mathbb{G}^\ddagger) \cap C^1(\overline{\mathbb{G}^\ddagger})$ and it can be represented by the formula*

$$u(\xi) = \int_{\mathbb{G}^\ddagger} G_{\mathbb{G}^\ddagger}(\xi, \zeta) f(\zeta) d\nu(\zeta) - \int_{\partial\mathbb{G}^\ddagger} \phi(\zeta) \langle \widetilde{\nabla} G_{\mathbb{G}^\ddagger}(\xi, \zeta), d\nu(\zeta) \rangle, \quad \xi \in \mathbb{G}^\ddagger, \tag{11.85}$$

where

$$\widetilde{\nabla} G_{\mathbb{G}^\ddagger} = \sum_{k=1}^{m} (X_k G_{\mathbb{G}^\ddagger}) X_k.$$

Proof of Theorem 11.5.4. Let us take $u \in C^2(\mathbb{G}^\ddagger) \cap C^1(\overline{\mathbb{G}^\ddagger})$ and assume that u tends to zero at infinity. By Remark 1.4.7, Part 2, we can apply Green's identities in $\mathbb{G}^\ddagger$. Thus, Green's second formula (1.88) applied to the functions u and $v(\zeta) = G_{\mathbb{G}^\ddagger}(\xi, \zeta)$ yields

$$u(\xi) = \int_{\mathbb{G}^\ddagger} G_{\mathbb{G}^\ddagger}(\xi, \zeta) f(\zeta) d\nu(\zeta) - \int_{\partial\mathbb{G}^\ddagger} \phi(\zeta) \langle \widetilde{\nabla} G_{\mathbb{G}^\ddagger}(\xi, \zeta), d\nu(\zeta) \rangle.$$

Here by using the properties of the Green function, we have

$$G_{\mathbb{G}^\ddagger}(\xi, \zeta) = 0, \quad \zeta \in \partial\mathbb{G}^\ddagger,$$

and, by construction the function $G_{\mathbb{G}^\ddagger}$ is symmetric, that is,

$$G_{\mathbb{G}^\ddagger}(\xi, \zeta) = G_{\mathbb{G}^\ddagger}(\zeta, \xi)$$

in $\mathbb{G}^\ddagger$, so

$$\mathcal{L}_\zeta G_{\mathbb{G}^\ddagger}(\xi, \zeta) = -\delta_\xi,$$

where δ_ξ is the Dirac distribution at $\xi \in \mathbb{G}^\ddagger$.

Let us now show that the function defined by (11.85) belongs to $C^2(\mathbb{G}^\ddagger) \cap C^1(\overline{\mathbb{G}^\ddagger})$. Since $f \in \Gamma_\alpha(\mathbb{G}^\ddagger)$ and $\operatorname{supp} f \subset \mathbb{G}^\ddagger$, the volume potential (i.e., the first term of the right-hand side in (11.85)) belongs to $C^2(\overline{\mathbb{G}^\ddagger})$ by Folland's theorem (see [Fol75, Theorem 6.1], see also [FS74]). Hörmander's hypoellipticity theorem (see [Hör67]) guarantees that every harmonic function is C^∞, hence the Dirichlet double layer potential (the second term of the right-hand side in (11.85)) is in $C^2(\mathbb{G}^\ddagger)$. On the other hand, since $\phi \in C^\infty(\partial\mathbb{G}^\ddagger)$, $\operatorname{supp}\phi \subset \{x_1 = 0, \ldots, x_l = 0\}$ and the boundary hyperplanes $\{x_1 = 0\}, \ldots, \{x_l = 0\}$ have no characteristic points, that is, recalling that the characteristic set of $\mathbb{G}^\ddagger$ is the set

$$\{x \in \partial\mathbb{G}^\ddagger : X_k(x) \in T_x(\partial\mathbb{G}^\ddagger),\ k = 1, \ldots, m\},$$

with $T_x(\partial\mathbb{G}^\ddagger)$ being the tangent space to $\partial\mathbb{G}^\ddagger$ at the point x, so we see that $X_k(x_0) \notin T_x\{\partial\mathbb{G}^\ddagger\}$ for all $x_0 \in \{x_1 = 0\}, \ldots, \{x_l = 0\}$

(see [GV00, Section 8] for more discussions on the non-characteristic hyperplanes in $\mathbb{G}$). Thus, the Dirichlet double layer potential is continuous on the boundary by the Kohn–Nirenberg theorem (see [CGN08, Theorem 3.12], which is a consequence of [KN65, Theorem 4], see also [Der71]-[Der72]). □

Remark 11.5.5. Let us point out some special cases and extensions of Theorem 11.5.4 and Proposition 11.5.3.

1. Let $\mathbb{G}^+$ be the *half-space*.

$$\mathbb{G}^+ = \{\xi = (x_1, \ldots, x_m, t_1, \ldots, t_n) : x_1 > 0\}.$$

Let the point $\zeta = (y, \tau) = (y_1, y_2, \ldots, y_m, \tau_1, \ldots, \tau_n)$ lie in this half-space, so that $y_1 > 0$. The point

$$\zeta^* = (y^*, \tau) := (-y_1, y_2, \ldots, y_m, \tau_1, \ldots, \tau_n)$$

is said to be symmetric for the point ζ with respect to the hyperplane $x_1 = 0$. As a direct consequence of Proposition 11.5.3 we have that the function

$$G_{\mathbb{G}^+}(\xi, \zeta) = \Gamma(\zeta^{-1} \circ \xi) - \Gamma((\zeta^*)^{-1} \circ \xi)$$

is the Green function for the Dirichlet sub-Laplacian in $\mathbb{G}^+$.

Let now $f \in \Gamma_\alpha(\mathbb{G}^+)$, $0 < \alpha < 1$, with $\operatorname{supp} f \subset \mathbb{G}^+$, and let $\phi \in C^\infty(\partial\mathbb{G}^+)$ with $\operatorname{supp}\phi \subset \{x_1 = 0\}$, and consider the Dirichlet sub-Laplacian problem

$$\begin{cases} \mathcal{L}u = f & \text{in } \mathbb{G}^+, \\ u = \phi & \text{on } \partial\mathbb{G}^+. \end{cases} \tag{11.86}$$

As a consequence of Theorem 11.5.4 we get that the boundary value problem (11.86) has a unique solution $u \in C^2(\mathbb{G}^+) \cap C^1(\overline{\mathbb{G}^+})$ and it can be represented by the formula

$$u(\xi) = \int_{\mathbb{G}^+} G_{\mathbb{G}^+}(\xi, \zeta) f(\zeta) d\nu(\zeta) - \int_{\partial\mathbb{G}^+} \phi(\zeta) \langle \widetilde{\nabla} G_{\mathbb{G}^+}(\xi, \zeta), d\nu(\zeta) \rangle, \quad \xi \in \mathbb{G}^+.$$

2. Let us point out another consequence for the quadrant-space in $\mathbb{G}$. Let $\mathbb{G}^{\oplus}$ be the *quadrant-space*

$$\mathbb{G}^{\oplus} := \{\xi = (x_1, x_2, \ldots, x_m, t_1, \ldots, t_n) : \ x_1 > 0, \ x_2 > 0\}.$$

Let the point $\zeta = (y, \tau) = (y_1, y_2, \ldots, y_m, \tau_1, \ldots, \tau_n)$ lie in this quadrant-space, so that $y_1 > 0$, $y_2 > 0$. Denote by

$$\zeta^* = (y^*, \tau) := (-y_1, y_2, \ldots, y_m, \tau_1, \ldots, \tau_n)$$

and

$$\overline{\zeta} = (\overline{y}, \tau) := (y_1, -y_2, \ldots, y_m, \tau_1, \ldots, \tau_n)$$

the symmetric points for ζ with respect to the hyperplanes $x_1 = 0$ and $x_2 = 0$, respectively. The point

$$\overline{\zeta}^* = (\overline{y}^*, \tau) = (-y_1, -y_2, \ldots, y_m, \tau_1, \ldots, \tau_n)$$

is the symmetric point for ζ^* with respect to the hyperplane $x_2 = 0$ and the symmetric point for $\overline{\zeta}$ with respect to the hyperplane $x_1 = 0$. Then as a direct consequence of Proposition 11.5.3 we have that the function

$$G_{\mathbb{G}^{\oplus}}(\xi, \zeta) = \Gamma(\zeta^{-1} \circ \xi) + \Gamma((\overline{\zeta}^*)^{-1} \circ \xi) - \Gamma((\zeta^*)^{-1} \circ \xi) - \Gamma((\overline{\zeta})^{-1} \circ \xi)$$

is the Green function for the Dirichlet sub-Laplacian in $\mathbb{G}^{\oplus}$.

Furthermore, if $f \in \Gamma_\alpha(\mathbb{G}^{\oplus})$, $0 < \alpha < 1$, with $\operatorname{supp} f \subset \mathbb{G}^{\oplus}$, and $\phi \in C^\infty(\partial \mathbb{G}^{\oplus})$ with $\operatorname{supp} \phi \subset \{x_1 = 0\}$, then the boundary value problem

$$\begin{cases} \mathcal{L}u = f & \text{in } \mathbb{G}^{\oplus}, \\ u = \phi & \text{on } \partial\mathbb{G}^{\oplus}, \end{cases}$$

has a unique solution $u \in C^2(\mathbb{G}^{\oplus}) \cap C^1(\overline{\mathbb{G}^{\oplus}})$ and it can be represented by the formula

$$u(\xi) = \int_{\mathbb{G}^{\oplus}} G_{\mathbb{G}^{\oplus}}(\xi, \zeta) f(\zeta) d\nu(\zeta) - \int_{\partial \mathbb{G}^{\oplus}} \phi(\zeta) \langle \widetilde{\nabla} G_{\mathbb{G}^{\oplus}}(\xi, \zeta), d\nu(\zeta) \rangle, \quad \xi \in \mathbb{G}^{\oplus}.$$

3. The statements of Theorem 11.5.4 and Proposition 11.5.3 can be extended in a straightforward way to shifted l-wedge like spaces, for any $a = (a_1, \ldots, a_l) \in \mathbb{R}^l$ defined by

$$\mathbb{G}^{\ddagger}_a := \{\xi = (x_1, \ldots, x_m, t_1, \ldots, t_n) : \ x_1 > a_1, \ldots, x_l > a_l\}.$$

In this space the Green function $G_{\mathbb{G}^{\ddagger}_a}$ has the same formula as in (11.83) if we choose the symmetry points now with respect to the hyperplanes $\{x_1 = a_1\}, \ldots, \{x_l = a_l\}$. In this case Theorem 11.5.4 remains true and can be obtained by the same argument.

11.5.2 Green functions and Dirichlet problem in strip domains

Similarly to wedge domain we can carry out a similar analysis in strip domains.

Definition 11.5.6 (l-strip domains). Let $1 \le l \le m$. We define $\mathbb{G}^{\models}$ to be the *l-strip space* if it is given by

$$\mathbb{G}^{\models} = \{\xi = (x_1, \ldots, x_m, t_1, \ldots, t_n) : \ a > x_l > 0\}.$$

Let the point $\zeta = (y, \tau) = (y_1, \ldots, y_m, \tau_1, \ldots, \tau_n)$ lie in this l-strip space, so that $a > y_l > 0$. We will use the notations

$$\zeta_{+,j} := (y_1, \ldots, y_l - 2aj, \ldots, y_m, \tau_1, \ldots, \tau_n),$$

and

$$\zeta_{-,j} := (y_1, \ldots, -y_l + 2aj, \ldots, y_m, \tau_1, \ldots, \tau_n),$$

for all $j = 0, 1, 2, \ldots$.

The Green function for the Dirichlet sub-Laplacian in the strip domains takes the following form, where compared to Proposition 11.5.3 in wedge domains, the formula is now given by an infinite series.

Proposition 11.5.7 (Green function for Dirichlet sub-Laplacian in strip domains). *Let $\mathbb{G}^{\ddagger}$ be an l-strip domain for $1 \le l \le m$. The function*

$$G_{\mathbb{G}^{\models}}(\xi, \zeta) = \sum_{j=-\infty}^{\infty} \left(\Gamma(\zeta_{+,j}^{-1} \circ \xi) - \Gamma(\zeta_{-,j}^{-1} \circ \xi) \right) \tag{11.87}$$

is the Green function for the Dirichlet sub-Laplacian in $\mathbb{G}^{\models}$.

Proof of Proposition 11.5.7. The formula (11.87) consists in the $j = 0$ term, i.e., the term $\Gamma(\zeta_{+,0}^{-1} \circ \xi)$, which corresponds to the fundamental solution, and all the other terms which are subharmonic functions in $\mathbb{G}^{\models}$. Let us check that traces of (11.87) vanish on hyperplanes $x_l = 0$ and $x_l = a$. If $x_l = 0$, then (11.87) gives

$$\begin{aligned}
&G_{\mathbb{G}^{\models}}(\xi,\zeta)|_{x_l=0} \\
&= c \sum_{j=-\infty}^{\infty} \Big(\big(((x_1-y_1)^2 + \cdots + (-y_l+2aj)^2 + \cdots + (x_m-y_m)^2)^2 + 16|t-\tau|^2 \big)^{(2-Q)/4} \\
&\quad - \big(((x_1-y_1)^2 + \cdots + (y_l-2aj)^2 + \cdots + (x_m-y_m)^2)^2 + 16|t-\tau|^2 \big)^{(2-Q)/4} \Big) = 0.
\end{aligned}$$

If $x_l = a$, then (11.87) gives

$$\begin{aligned}
&G_{\mathbb{G}^{\models}}(\xi,\zeta)|_{x_l=a} \\
&= c \sum_{j=-\infty}^{\infty} \Big(\big(((x_1-y_1)^2 + \cdots + (a-y_l+2aj)^2 + \cdots + (x_m-y_m)^2)^2 + 16|t-\tau|^2 \big)^{(2-Q)/4} \\
&\quad - \big(((x_1-y_1)^2 + \cdots + (a+y_l-2aj)^2 + \cdots + (x_m-y_m)^2)^2 + 16|t-\tau|^2 \big)^{(2-Q)/4} \Big)
\end{aligned}$$

$$
\begin{aligned}
&= c\sum_{j=0}^{\infty}\left(((x_1-y_1)^2+\cdots+(a-y_l+2aj)^2+\cdots+(x_m-y_m)^2)^2+16|t-\tau|^2\right)^{(2-Q)/4}\\
&\quad - c\sum_{j=1}^{\infty}\left(((x_1-y_1)^2+\cdots+(a+y_l-2aj)^2+\cdots+(x_m-y_m)^2)^2+16|t-\tau|^2\right)^{(2-Q)/4}\\
&\quad + c\sum_{j=-1}^{-\infty}\left(((x_1-y_1)^2+\cdots+(a-y_l+2aj)^2+\cdots+(x_m-y_m)^2)^2+16|t-\tau|^2\right)^{(2-Q)/4}\\
&\quad - c\sum_{j=0}^{-\infty}\left(((x_1-y_1)^2+\cdots+(a+y_l-2aj)^2+\cdots+(x_m-y_m)^2)^2+16|t-\tau|^2\right)^{(2-Q)/4}\\
&= 0.
\end{aligned}
$$

Here the first term ($j = 0$ term) of the first sum is cancelled with the first term ($j = 1$ term) of the second sum and the second terms of the first sum is cancelled with the second term of the second sum and so on, that is, the first two sums give zero. Similarly, the first term of the third sum is cancelled with the first term of the last sum and the second term of the third sum is cancelled with the second term of the last sum and so on, that is, the last two sums also give zero. As a result, the trace vanishes at $x_l = a$. $\square$

For $f \in \Gamma_\alpha(\mathbb{G}^{\models})$, $0 < \alpha < 1$, with $\operatorname{supp} f \subset \mathbb{G}^{\models}$, and for $\phi \in C^\infty(\partial\mathbb{G}^{\models})$ with $\operatorname{supp}\phi \subset \{x_l = 0\}\bigcup\{x_l = a\}$, we now consider the Dirichlet problem for the sub-Laplacian

$$
\begin{cases} \mathcal{L}u = f & \text{in } \mathbb{G}^{\models},\\ u = \phi & \text{on } \partial\mathbb{G}^{\models}. \end{cases} \tag{11.88}
$$

Theorem 11.5.8 (Dirichlet problem for sub-Laplacian in strip domain). *Let $f \in \Gamma_\alpha(\mathbb{G}^{\models})$, $0 < \alpha < 1$, with $\operatorname{supp} f \subset \mathbb{G}^{\models}$, and let $\phi \in C^\infty(\partial\mathbb{G}^{\models})$ with $\operatorname{supp}\phi \subset \{x_l = 0\}\bigcup\{x_l = a\}$. Then the boundary value problem (11.88) has a unique solution $u \in C^2(\mathbb{G}^{\models}) \cap C^1(\overline{\mathbb{G}^{\models}})$ and it can be represented by the formula*

$$
u(\xi) = \int_{\mathbb{G}^{\models}} G_{\mathbb{G}^{\models}}(\xi,\zeta)f(\zeta)d\nu(\zeta) - \int_{\partial\mathbb{G}^{\models}} \phi(\zeta)\langle\widetilde{\nabla} G_{\mathbb{G}^{\models}}(\xi,\zeta), d\nu(\zeta)\rangle, \quad \xi \in \mathbb{G}^{\models},
$$

where

$$
\widetilde{\nabla} G_{\mathbb{G}^{\models}} = \sum_{k=1}^{m} (X_k G_{\mathbb{G}^{\models}}) X_k.
$$

Proof of Theorem 11.5.8. The proof is almost the same as the proof of Theorem 11.5.4, so we omit it. $\square$

11.6 p-sub-Laplacian Picone's inequality and consequences

In this section, we present Picone's identities for the p-sub-Laplacian on stratified Lie groups, which implies a generalized Díaz–Saá inequality for the p-sub-Laplacian on stratified Lie groups. As consequences, a comparison principle and uniqueness of a positive solution to nonlinear p-sub-Laplacian equations are derived. The presentation of this section follows the results of [RS17a].

First let us introduce the functional spaces $S^{1,p}(\Omega)$.

Definition 11.6.1 (Functional spaces $S^{1,p}(\Omega)$ and $\overset{\circ}{S}{}^{1,p}(\Omega)$). We define

$$S^{1,p}(\Omega) := \{u : \Omega \to \mathbb{R};\ u,\ |\nabla_H u| \in L^p(\Omega)\}.$$

Consider the functional

$$J_p(u) := \left(\int_\Omega |\nabla_H u|^p dx\right)^{1/p},$$

then we define the functional class $\overset{\circ}{S}{}^{1,p}(\Omega)$ to be the completion of $C_0^1(\Omega)$ in the norm generated by J_p (see, e.g., [CDG93]).

In an admissible domain $\Omega \subset \mathbb{G}$ with smooth boundary $\partial\Omega$ we study the p-sub-Laplacian Dirichlet problem:

$$\begin{cases} -\mathcal{L}_p u = F(x,u), & \text{in } \Omega, \\ u = 0, & \text{on } \partial\Omega. \end{cases} \tag{11.89}$$

Here we assume that

(a) The function $F : \Omega \times \mathbb{R} \to \mathbb{R}$ is a positive, bounded and measurable, and there exists a positive constant $C > 0$ such that

$$F(x,\rho) \le C(\rho^{p-1} + 1)$$

holds for a.e. $x \in \Omega$.

(b) The function

$$\rho \mapsto \frac{F(x,\rho)}{\rho^{p-1}}$$

is strictly decreasing on $(0,\infty)$ for a.e. $x \in \Omega$.

As usual, if a function $u \in \overset{\circ}{S}{}^{1,p}(\Omega) \cap L^\infty(\Omega)$ satisfies

$$\int_\Omega |\nabla_H u|^{p-2}(\widetilde{\nabla} u)\phi d\nu = \int_\Omega F(x,u)\phi d\nu$$

for all $\phi \in C_0^\infty(\Omega)$, then it is called a *weak solution* of (11.89).

Keeping in mind the stratified group discussions in previous sections for any set $\Omega \subset \mathbb{G}$ we denote, for any $1 < p < \infty$,

$$L(u,v) := |\nabla_H u|^p - p\frac{|u|^{p-2}u}{f(v)}\nabla_H u \cdot \nabla_H v|\nabla_H v|^{p-2} + \frac{f'(v)|u|^p}{f^2(v)}|\nabla_H v|^p. \quad (11.90)$$

We also denote

$$R(u,v) := |\nabla_H u|^p - \nabla_H \left(\frac{|u|^p}{f(v)}\right)|\nabla_H v|^{p-2}\nabla_H v, \quad 1 < p < \infty, \quad (11.91)$$

a.e. in Ω. Here $f : \mathbb{R}^+ \to \mathbb{R}^+$ is a locally Lipschitz function such that

$$(p-1)|f(t)|^{\frac{p-2}{p-1}} \leq f'(t) \quad (11.92)$$

holds a.e. in $\mathbb{R}^+$. Then we have the following Picone identity on a stratified Lie group $\mathbb{G}$.

Lemma 11.6.2 (Picone identity). *Under the above assumptions we have*

$$L(u,v) = R(u,v) \geq 0$$

a.e. in Ω, where u and v are differentiable real-valued functions and $\Omega \subset \mathbb{G}$ is any set.

Proof of Lemma 11.6.2. We have the equality

$$\begin{aligned}\nabla_H \left(\frac{|u|^p}{f(v)}\right) &= \frac{pf(v)|u|^{p-2}u\nabla_H u - f'(v)|u|^p\nabla_H v}{f^2(v)}\\ &= \frac{p|u|^{p-2}u\nabla_H u}{f(v)} - \frac{f'(v)|u|^p\nabla_H v}{f^2(v)}.\end{aligned}$$

It implies that

$$\begin{aligned}R(u,v) &= |\nabla_H u|^p - \nabla_H \left(\frac{|u|^p}{f(v)}\right)|\nabla_H v|^{p-2}\nabla_H v\\ &= |\nabla_H u|^p - \frac{p|u|^{p-2}u}{f(v)}|\nabla_H v|^{p-2}\nabla_H u \cdot \nabla_H v + \frac{f'(v)|u|^p}{f^2(v)}|\nabla_H v|^p\\ &= L(u,v).\end{aligned}$$

Now it remains to show the non-negativity of $R(u,v)$ (that is, alternatively, of $L(u,v)$). Since

$$\frac{p|u|^{p-2}u}{f(v)}|\nabla_H v|^{p-2}\nabla_H u \cdot \nabla_H v \leq \frac{p|u|^{p-1}}{f(v)}|\nabla_H v|^{p-1}|\nabla_H u|,$$

by applying the Young inequality to the right-hand side of this inequality we obtain

$$\frac{p|u|^{p-2}u}{f(v)}|\nabla_H v|^{p-2}\nabla_H u\cdot\nabla_H v\leq|\nabla_H u|^p+(p-1)\frac{|u|^p|\nabla_H v|^p}{f^{\frac{p}{p-1}}(v)}.$$

Thus, we have

$$\frac{f'(v)|u|^p|\nabla_H v|^p}{f^2(v)}-(p-1)\frac{|u|^p|\nabla_H v|^p}{f^{\frac{p}{p-1}}(v)}\leq R(u,v).$$

By assumption (11.92) we have $(p-1)|f(t)|^{\frac{p-2}{p-1}}\leq f'(t)$, which means that $0\leq R(u,v)$, completing the proof. □

Lemma 11.6.3. *Let $1<p<\infty$ and let $f:\mathbb{R}^+\to\mathbb{R}^+$ be a locally Lipschitz function such that*

$$(p-1)|f(t)|^{\frac{p-2}{p-1}}\leq f'(t)$$

holds a.e. in $\mathbb{R}^+$. Let $\Omega\subset\mathbb{G}$ be a bounded open set and let $v\in\overset{\circ}{S}{}^{1,p}(\Omega)$ be such that $v\geq\epsilon>0$. Then for any $u\in C_0^\infty(\Omega)$ we have

$$\int_\Omega\frac{|u|^p}{f(v)}(-\mathcal{L}_p v)dx\leq\int_\Omega|\nabla_H u|^p dx. \tag{11.93}$$

Proof of Lemma 11.6.3. Let $v\in\overset{\circ}{S}{}^{1,p}(\Omega)$ be such as in the assumptions, that is, $v\geq\epsilon>0$. Then by the density argument we can choose $v_k\in C_0^1(\Omega)$, $k=1,2,\ldots$, such that $v_k>\frac{\epsilon}{2}$ in Ω and $v_k\to v$ a.e. in Ω. By using Lemma 11.6.2 we have

$$0\leq\int_\Omega R(u,v_k)dx,$$

for each k. That is,

$$\int_\Omega\frac{|u|^p}{f(v_k)}(-\mathcal{L}_p v_k)dx\leq\int_\Omega|\nabla_H u|^p dx.$$

Since $\mathcal{L}_p$ is a continuous operator from $\overset{\circ}{S}{}^{1,p}(\Omega)$ to $S^{-1,p'}(\Omega)$, $p'=\frac{p}{p-1}$, we have $\mathcal{L}_p v_k\to\mathcal{L}_p v$ in $S^{-1,p'}(\Omega)$ and $f(v_k)\to f(v)$ pointwise since f is a locally Lipschitz continuous function on $(0,\infty)$. Thus, by the Lebesque dominated convergence theorem and using the fact that f is an increasing function on $(0,\infty)$, for any $u\in C_0^\infty(\Omega)$ we arrive at the inequality

$$\int_\Omega\frac{|u|^p}{f(v)}(-\mathcal{L}_p v)dx\leq\int_\Omega|\nabla_H u|^p dx,$$

proving (11.93). □

As a consequence of the Harnack inequality for the general hypoelliptic operator (see [CDG93, Theorem 3.1]) one has the following strong maximum principle for the p-sub-Laplacian.

Lemma 11.6.4 (Strong maximum principle for the p-sub-Laplacian). *Let $1 < p \leq Q$, let $\Omega \subset \mathbb{G}$ be a bounded open set and let $F : \Omega \times \mathbb{R} \to \mathbb{R}$ be a measurable function such that*

$$|F(x, \rho)| \leq C(\rho^{p-1} + 1)$$

for all $\rho > 0$. Let $u \in \overset{\circ}{S}{}^{1,p}(\Omega)$ be a non-negative solution of

$$\begin{cases} -\mathcal{L}_p u = F(x, u), & \text{in } \Omega, \\ u = 0, & \text{on } \partial\Omega. \end{cases}$$

Then $u \equiv 0$ or $u > 0$ in Ω.

Proof of Lemma 11.6.4. Since $u \in \overset{\circ}{S}{}^{1,p}(\Omega)$ by using the Harnack inequality [CDG93, Theorem 3.1] for $1 < p \leq Q$ there exists a constant C_R such that

$$\operatorname*{ess\,sup}_{B(x,R)} u \leq C_R \operatorname*{ess\,inf}_{B(x,R)} u$$

holds for any $x \in \Omega$ and quasi-ball $B(0, R)$. This means that $u \equiv 0$ or $u > 0$ in Ω. □

Combining Lemma 11.6.3 and Lemma 11.6.4 one has the following generalized Picone inequality on $\mathbb{G}$, which is a key ingredient for proofs of both a comparison principle and uniqueness of a positive solution to nonlinear p-sub-Laplacian equations.

Lemma 11.6.5 (Generalized Picone inequality). *Let $\Omega \subset \mathbb{G}$ be a bounded open set and let $g : \Omega \times \mathbb{R} \to \mathbb{R}$ be positive, bounded and measurable function such that*

$$g(x, \rho) \leq C(\rho^{p-1} + 1)$$

for all $\rho > 0$. Then we have

$$\int_\Omega \frac{|u|^p}{f(v)}(-\mathcal{L}_p v)dx \leq \int_\Omega |\nabla_H u|^p dx, \quad 1 < p < \infty, \tag{11.94}$$

for all $v, u \in \overset{\circ}{S}{}^{1,p}(\Omega)$ and $v(\not\equiv 0) \geq 0$ a.e. $\Omega \in \mathbb{G}$ such that

$$-\mathcal{L}_p v = g(x, v).$$

Proof of Lemma 11.6.5. By Lemma 11.6.4 we have $v > 0$ in Ω. Let

$$v_k(x) := v(x) + \frac{1}{k}, \ k = 1, 2, \ldots.$$

Then we have $\mathcal{L}_p v_k = \mathcal{L}_p v$ in $S^{-1,p'}(\Omega)$, $v_k \to v$ a.e. in Ω and also $f(v_k) \to f(v)$ pointwise in Ω. Let $u_k \in C_0^\infty(\Omega)$ be such that $u_k \to u$ in $\overset{\circ}{S}{}^{1,p}(\Omega)$. For the functions u_k and v_k Lemma 11.6.3 gives

$$\int_\Omega \frac{|u_k|^p}{f(v_k)}(-\mathcal{L}_p v_k)dx \le \int_\Omega |\nabla_H u_k|^p dx.$$

Now since $f(v_k) \to f(v)$ pointwise, by the Fatou lemma we arrive at

$$\int_\Omega \frac{|u|^p}{f(v)}(-\mathcal{L}_p v)dx \le \int_\Omega |\nabla_H u|^p dx.$$

This completes the proof. □

Lemma 11.6.5 implies the following comparison type principle:

Theorem 11.6.6 (Comparison principle for p-sub-Laplacian). *Let $\Omega \subset \mathbb{G}$ be an admissible domain. Let $0 < q < p-1$ and let F be a non-negative function such that $F \not\equiv 0$. Let $u, v \in \overset{\circ}{S}{}^{1,p}(\Omega)$ be real-valued functions such that*

$$\begin{cases} -\mathcal{L}_p u \ge F(x)u^q, & u > 0 \quad \text{in } \Omega, \\ -\mathcal{L}_p v \le F(x)v^q, & v > 0 \quad \text{in } \Omega. \end{cases} \tag{11.95}$$

Then we have $v \le u$ a.e. in Ω.

Proof of Theorem 11.6.6. It follows from (11.95) that

$$F(x)\left(\frac{u^q}{u^{p-1}} - \frac{v^q}{v^{p-1}}\right) \le \frac{-\mathcal{L}_p u}{u^{p-1}} + \frac{\mathcal{L}_p v}{v^{p-1}}.$$

Multiplying both sides by $w = (v^p - u^p)_+$ and integrating over Ω we have

$$\begin{aligned} \int_{[v>u]} F(x)\left(\frac{u^q}{u^{p-1}} - \frac{v^q}{v^{p-1}}\right) w dx &= \int_\Omega F(x)\left(\frac{u^q}{u^{p-1}} - \frac{v^q}{v^{p-1}}\right) w dx \\ &\le \int_\Omega \left(\frac{-\mathcal{L}_p u}{u^{p-1}} + \frac{\mathcal{L}_p v}{v^{p-1}}\right) w dx. \end{aligned} \tag{11.96}$$

Moreover, we have

$$\begin{aligned} &\int_\Omega \left(\frac{-\mathcal{L}_p u}{u^{p-1}} + \frac{\mathcal{L}_p v}{v^{p-1}}\right) w dx \\ &= \int_\Omega |\nabla_H u|^{p-2}\nabla_H u \cdot \nabla_H \left(\frac{w}{u^{p-1}}\right) dx \\ &- \int_\Omega |\nabla_H v|^{p-2}\nabla_H v \cdot \nabla_H \left(\frac{w}{v^{p-1}}\right) dx \end{aligned}$$

$$
\begin{aligned}
&= \int_{\Omega\cap[v>u]} |\nabla_H u|^{p-2}\nabla_H u \cdot \nabla_H\left(\frac{v^p-u^p}{u^{p-1}}\right)dx \\
&\quad - \int_{\Omega\cap[v>u]} |\nabla_H v|^{p-2}\nabla_H v \cdot \nabla_H\left(\frac{v^p-u^p}{v^{p-1}}\right)dx \\
&= \int_{\Omega\cap[v>u]} \left(|\nabla_H u|^{p-2}\nabla_H u \cdot \nabla_H\left(\frac{v^p}{u^{p-1}}\right) - |\nabla_H v|^p\right)dx \\
&\quad + \int_{\Omega\cap[v>u]} \left(|\nabla_H v|^{p-2}\nabla_H v \cdot \nabla_H\left(\frac{u^p}{v^{p-1}}\right) - |\nabla_H u|^p\right)dx \\
&= I_1 + I_2,
\end{aligned}
$$

where

$$
I_1 := \int_{\Omega\cap[v>u]} \left(|\nabla_H u|^{p-2}\nabla_H u \cdot \nabla_H\left(\frac{v^p}{u^{p-1}}\right) - |\nabla_H v|^p\right)dx
$$

and

$$
I_2 := \int_{\Omega\cap[v>u]} \left(|\nabla_H v|^{p-2}\nabla_H v \cdot \nabla_H\left(\frac{u^p}{v^{p-1}}\right) - |\nabla_H u|^p\right)dx.
$$

We notice that

$$
\begin{aligned}
I_1 &= \int_{\Omega\cap[v>u]} |\nabla_H u|^{p-2}\nabla_H u \cdot \nabla_H\left(\frac{v^p}{u^{p-1}}\right)dx - \int_{\Omega\cap[v>u]} |\nabla_H v|^p dx \\
&= -\int_{\Omega\cap[v>u]} \frac{v^p}{u^{p-1}}\mathcal{L}_p u dx - \int_{\Omega\cap[v>u]} |\nabla_H v|^p dx \le 0.
\end{aligned}
$$

In the last line we have used Green's first identity (1.90) and the Picone inequality (11.94). Similarly, we see that $I_2 \le 0$. Thus, we obtain

$$
\int_\Omega \left(\frac{-\mathcal{L}_p u}{u^{p-1}} + \frac{\mathcal{L}_p v}{v^{p-1}}\right) w dx \le 0.
$$

Consequently, (11.96) implies that

$$
\int_{\Omega\cap[v>u]} F(x)\left(\frac{u^q}{u^{p-1}} + \frac{v^q}{v^{p-1}}\right)(v^p - u^p)dx \le 0.
$$

On the other hand, we have

$$
0 \le F(x)\left(\frac{u^q}{u^{p-1}} + \frac{v^q}{v^{p-1}}\right)
$$

on the set $[v>u]$. This means that $|[v>u]| = 0$. □

As another consequence of Lemma 11.6.5 we have the following Díaz–Saá inequality on stratified Lie groups.

Lemma 11.6.7 (Díaz–Saá inequality). *Let Ω be an admissible domain. Let functions g_1 and g_2 satisfy the assumption of Theorem 11.6.5. If functions u_1, $u_2 \in \overset{\circ}{S}{}^{1,p}(\Omega)$ with u_1, $u_2(\not\equiv 0) \geq 0$ a.e. $\Omega \in \mathbb{G}$ are such that*

$$-\mathcal{L}_p u_1 = g_1(x, u_1) \quad \textit{and} \quad -\mathcal{L}_p u_2 = g_2(x, u_2),$$

then we have

$$0 \leq \int_\Omega \left(\frac{-\mathcal{L}_p u_1}{u_1^{p-1}} + \frac{\mathcal{L}_p u_2}{u_2^{p-1}} \right) (u_1^p - u_2^p) dx.$$

Proof of Lemma 11.6.7. Let functions u_1 and u_2 satisfy the assumptions. Then by the inequality (11.94) with $f(u) = u^{p-1}$ as well as for u_1 and u_2 we have

$$\int_\Omega \frac{|u_1|^p}{u_2^{p-1}} (-\mathcal{L}_p u_2) dx \leq \int_\Omega |\nabla_H u_1|^p dx.$$

Using Green's first identity (1.90) we get

$$0 \leq \int_\Omega \left(\frac{-\mathcal{L}_p u_1}{u_1^{p-1}} + \frac{\mathcal{L}_p u_2}{u_2^{p-1}} \right) u_1^p dx. \tag{11.97}$$

Again, by the inequality (11.94) we have

$$\int_\Omega \frac{|u_2|^p}{u_1^{p-1}} (-\mathcal{L}_p u_1) dx \leq \int_\Omega |\nabla_H u_2|^p dx.$$

As above, this implies

$$0 \leq \int_\Omega \left(\frac{\mathcal{L}_p u_1}{u_1^{p-1}} - \frac{\mathcal{L}_p u_2}{u_2^{p-1}} \right) u_2^p dx. \tag{11.98}$$

Now the combination of (11.97) and (11.98) completes the proof. □

The established properties allow one to show the uniqueness of a positive solution for the equation

$$\begin{cases} -\mathcal{L}_p u = F(x, u), & \text{in } \Omega, \\ u = 0, & \text{on } \partial\Omega, \end{cases} \tag{11.99}$$

where Ω is an admissible domain. For convenience, we recall once again the assumptions on $F(x, u)$:

(a) The function $F : \Omega \times \mathbb{R} \to \mathbb{R}$ is a positive, bounded and measurable function and there exists a positive constant $C > 0$ such that

$$F(x, \rho) \leq C(\rho^{p-1} + 1)$$

holds for a.e. $x \in \Omega$.

(b) The function

$$\rho \mapsto \frac{F(x,\rho)}{\rho^{p-1}}$$

is strictly decreasing on $(0,\infty)$ for a.e. $x \in \Omega$.

Theorem 11.6.8 (Uniqueness of positive solutions). *There exists at most one positive weak solution to* (11.99) *for* $1 < p \leq Q$.

Proof of Theorem 11.6.8. First, assuming that u_1 and u_2 are two different ($u_1 \not\equiv u_2$) non-negative solutions of (11.99) and using Lemma 11.6.4 for these functions we conclude that $u_1 > 0$ and $u_2 > 0$ in Ω. Then Lemma 11.6.7 implies that

$$0 \leq \int_\Omega \left(\frac{-\mathcal{L}_p u_1}{u_1^{p-1}} + \frac{\mathcal{L}_p u_2}{u_2^{p-1}} \right) (u_1^p - u_2^p) dx.$$

Moreover, from the assumption (b) it follows that

$$\int_\Omega \left(\frac{F(x,u_1)}{u_1^{p-1}} - \frac{F(x,u_2)}{u_2^{p-1}} \right) (u_1^p - u_2^p) dx < 0.$$

On the other hand, we have

$$\int_\Omega \left(\frac{-\mathcal{L}_p u_1}{u_1^{p-1}} + \frac{\mathcal{L}_p u_2}{u_2^{p-1}} \right) (u_1^p - u_2^p) dx = \int_\Omega \left(\frac{F(x,u_1)}{u_1^{p-1}} - \frac{F(x,u_2)}{u_2^{p-1}} \right) (u_1^p - u_2^p) dx,$$

and this contradicts that both u_1 and u_2 ($u_1 \not\equiv u_2$) are non-negative solutions of (11.99). □

Chapter 12

Hardy and Rellich Inequalities for Sums of Squares of Vector Fields

In this chapter, we demonstrate how some ideas originating in the analysis on groups can be applied in related settings without the group structure. In particular, in Chapter 7 we showed a number of Hardy and Rellich inequalities with weights expressed in terms of the so-called $\mathcal{L}$-gauge. There, the $\mathcal{L}$-gauge is a homogeneous quasi-norm on a stratified group which is obtained from the fundamental solution to the sub-Laplacian. At the same time, in Chapter 11 we used the fundamental solutions of the sub-Laplacian for the advancement of the potential theory on stratified groups, and in Section 7.3 fundamental solutions for the p-sub-Laplacian and their properties were used on polarizable Carnot groups for the derivation of further Hardy estimates in that setting.

The aim of this chapter is to show that given the existence of a fundamental solution one can use the ideas from the analysis on groups to establish a number of Hardy inequalities on spaces without group structure.

Thus, let M be a smooth manifold of dimension n with a volume form $d\nu$. Let $\{X_k\}_{k=1}^N$ be a family of real vector fields on M, and denote by $\mathcal{L}$ the sum of their squares:

$$\mathcal{L} := \sum_{k=1}^{N} X_k^2. \tag{12.1}$$

Identifying each vector field X with the derivative in its direction, second-order differential operators in the form (12.1) have been widely studied in the literature. For instance, by the well-known Hörmander sums of the squares theorem from [Hör67], the operator $\mathcal{L}$ is locally hypoelliptic if the iterated commutators of the vector fields $\{X_k\}_{k=1}^N$ generate the tangent space at each point. Such operators have been also investigated under weaker conditions or without the hypoellipticity property. There are many geometric considerations related to such operators, see, e.g., the seminal papers of Rothschild and Stein [RS76] and of Nagel, Stein and Wainger [NSW85].

M. Ruzhansky, D. Suragan, *Hardy Inequalities on Homogeneous Groups*,
Progress in Mathematics 327, https://doi.org/10.1007/978-3-030-02895-4_13

In this chapter our main assumption on the operator $\mathcal{L}$ in (12.1) will be that it has a local fundamental solution. In particular, this is the case when M is a stratified or a graded group, but the group assumption is, in principle, not necessary. In what follows we will give other examples of naturally appearing operators in other contexts having local or global fundamental solutions. The presentation of this chapters is based on the results obtained in [RS17d].

12.1 Assumptions

We start by formulating assumptions for the presentation in this chapter. Then we discuss several settings where these assumptions are satisfied. This will include stratified Lie groups and operators on $\mathbb{R}^n$ satisfying the Hörmander commutator condition.

Let M be a smooth manifold of dimension n with a volume form $d\nu$, and let $\mathcal{L}$ be an operator as in (12.1). At a point $y \in M$ we will be making the following assumption that we call (A_y), asking for the existence of a local fundamental solution at y:

(A_y) For $y \in M$, assume that there is an open set $T_y \subset M$ containing y such that the operator $-\mathcal{L}$ has a fundamental solution in T_y, that is, there exists a function $\Gamma_y \in C^2(T_y \setminus \{y\})$ such that

$$-\mathcal{L}\Gamma_y = \delta_y \text{ in } T_y, \tag{12.2}$$

where δ_y is the Dirac δ-distribution at y.

When the point y is fixed, we will often use the notation $\Gamma(x, y) = \Gamma_y(x)$ or simply $\Gamma(x)$. Here C^2 stands for the space of functions with continuous second derivatives with respect to $\{X_k\}_{k=1}^N$. We note that among other things the existence of a fundamental solution implies that $\mathcal{L}$ is hypoelliptic.

Sometimes we will strengthen Assumption (A_y) to the following assumption that we call (A_y^+) asking for the local positivity of the fundamental solution:

(A_y^+) For $y \in M$, assume that (A_y) holds and, moreover, we have

$$\Gamma_y(x) > 0 \text{ in } T_y \setminus \{y\}, \text{ and } \frac{1}{\Gamma_y}(y) = 0.$$

The second part of the assumption is usually naturally satisfied since for a fundamental solution Γ_y, the quotient $\frac{1}{\Gamma_y}$ is usually well-defined and is equal to 0 at y since Γ_y normally blows up at y.

As before, we will be using the notation $\langle X_k, d\nu \rangle$ for the duality product of the vector field X_k with the volume form $d\nu$, that is, since $d\nu$ is an n-form, $\langle X_k, d\nu \rangle$ is an $(n-1)$-form on M.

It will be convenient to use the following notion of admissible domains in this chapter. We note that this notion here differs from the one in Definition 1.4.4.

However, there should be no confusion since the following definition will be used in this chapter only.

Definition 12.1.1 (Admissible domains). We will say (in this chapter) that an open bounded set $\Omega \subset M$ is an *admissible domain* if its boundary $\partial\Omega$ has no self-intersections, and if the vector fields $\{X_k\}_{k=1}^N$ satisfy the equality

$$\sum_{k=1}^{N} \int_{\Omega} X_k f_k d\nu = \sum_{k=1}^{N} \int_{\partial\Omega} f_k \langle X_k, d\nu \rangle, \tag{12.3}$$

for all $f_k \in C^1(\Omega) \bigcap C(\overline{\Omega})$, $k = 1, \ldots, N$.

We will also say that an admissible domain Ω is *strongly admissible with* $y \in M$ if assumption (A_y) is satisfied, $\Omega \subset T_y$, and (12.3) holds for $f_k = vX_k\Gamma_y$ for all $v \in C^1(\Omega) \bigcap C(\overline{\Omega})$.

Although there are several conditions incorporated in the notion of a strongly admissible domain the examples below will actually show that in a number of natural settings, any open bounded set with a piecewise smooth boundary without self-intersections is strongly admissible, see Proposition 12.2.1. The condition that the boundary $\partial\Omega$ has no self-intersections implies that $\partial\Omega$ is orientable. For brevity, we will say that such boundaries are *simple*.

12.1.1 Examples

Let us now describe several rather general settings when bounded domains with simple boundaries are strongly admissible in the sense of Definition 12.1.1. Moreover, we discuss also the validity of assumptions (A_y) and (A_y^+).

For the examples (E2) and (E3) below we will need the following definition.

Definition 12.1.2 (Control distance and Hölder spaces). The *control distance* $d_c(x, y)$ associated to the vector fields X_k is defined as the infimum of $T > 0$ such that there is a piecewise continuous integral curve γ of $X_1, \ldots, X_N$ such that $\gamma(0) = x$ and $\gamma(T) = y$.

The *Hölder space* $C^\alpha(\Omega)$ *with respect to the control distance* is then defined for $0 < \alpha \le 1$ as the space of all functions u for which there is $C > 0$ such that

$$|u(x) - u(y)| \le C d_c^\alpha(x, y)$$

holds for all $x, y \in \Omega$. Then, $u \in C^{1,\alpha}$ if $X_k u \in C^\alpha$ for all $k = 1, \ldots, N$, and the spaces $C^{r,\alpha}$ are defined inductively.

Example 12.1.3 (Examples of strongly admissible domains). Let us give several examples.

(E1) Let M be a stratified Lie group, $n \ge 3$, and let $\{X_k\}_{k=1}^N$ be left invariant vector fields giving the first stratum of M. Then for any $y \in M$ the assumption (A_y^+) is satisfied with $T_y = M$. Moreover, any open bounded set $\Omega \subset M$ with a piecewise smooth simple boundary is strongly admissible.

(E2) Let $M = \mathbb{R}^n$, $n \geq 3$, and let the vector fields X_k, $k = 1, \ldots, N$, $N \leq n$, be of the form

$$X_k = \frac{\partial}{\partial x_k} + \sum_{m=N+1}^{n} a_{k,m}(x) \frac{\partial}{\partial x_m}, \tag{12.4}$$

where $a_{k,m}(x)$ are locally $C^{1,\alpha}$-regular for some $0 < \alpha \leq 1$, where $C^{1,\alpha}$ stands for the space of functions with X_k-derivatives in the Hölder space C^α with respect to the control distance defined by these vector fields. Assume also

$$\frac{\partial}{\partial x_k} = \sum_{1 \leq i < j \leq N} \lambda_k^{i,j}(x)[X_i, X_j] \tag{12.5}$$

for all $k = N+1, \ldots, n$, with $\lambda_k^{i,j} \in L^\infty_{\text{loc}}(M)$. Then for any $y \in M$ the assumption (A_y^+) is satisfied. Moreover, any open bounded set $\Omega \subset M$ with a piecewise smooth simple boundary is strongly admissible.

(E3) More generally, let $M = \mathbb{R}^n$, $n \geq 3$, and let the vector fields X_k, $k = 1, \ldots, N$, $N \leq n$, satisfy the Hörmander commutator condition of step $r \geq 2$. Assume that all X_k, $k = 1, \ldots, N$, belong to $C^{r,\alpha}(U)$ for some $0 < \alpha \leq 1$ and $U \subset M$, and if $r = 2$ we assume $\alpha = 1$. Then for any $y \in M$ the assumption (A_y^+) is satisfied. Moreover, if X_k's are in the form (12.4), then any open bounded set $\Omega \subset M$ with a piecewise smooth simple boundary is strongly admissible.

Some remarks are in order.

Remark 12.1.4.

1. In Example (E1), the validity of Assumption (A_y^+) for any y follows from (1.74) and (1.75). The equality of (12.3) for (E1) and the strong admissibility for any domain with piecewise smooth simple boundary follows from Theorem 1.4.5.
2. In Example (E2), the existence of a local fundamental solution, that is (A_y) for any $y \in M$ was shown by Manfredini [Man12]. While the positivity of Γ_y does not seem to be explicitly stated there, see Sánchez-Calle [SC84], or Fefferman and Sánchez-Calle [FSC86] for the positivity, thus assuring that Assumption (A_y^+) holds. The validity of (12.3) and the strong admissibility for any domain with piecewise smooth simple boundary will follow from Theorem 12.2.1.
3. Condition (12.5) implies that the collection of vector fields $\{X_k\}_{k=1}^N$ satisfies Hörmander's commutator condition of step two.
4. The condition (12.4) on the vector fields in (E2) and (E3) is not restrictive. In fact, by a change of variables one can show that any collection of linearly independent vector fields which are locally $C^{r,\alpha}$-regular ($r \in \mathbb{N}$) can be transformed to a collection of the same regularity which satisfies condition (12.4), see Manfredini [Man12, page 975].

5. In Example (E3), the validity of condition (A_y^+) was studied by Bramanti, Brandolini, Manfredini and Pedroni [BBMP17, Theorem 4.8 and Theorem 5.9]. The validity of (12.3) and the strong admissibility for any domain with piecewise smooth simple boundary will follow from Theorem 12.2.1.
6. Assumptions (A_y) or (A_y^+) hold also in some other settings. The subject of the existence of local and global fundamental solutions for $\mathcal{L}$ is well studied when $\mathcal{L}$ is a hypoelliptic operator, see, e.g., [Man12, BLU04, FSC86, SC84, OR73] for more general and detailed discussions.
7. For both Examples (E2) and (E3) let us give the following explicit example: In $\mathbb{R}^3$ let $N = 2$ and let

$$X_1 = \frac{\partial}{\partial x_1} + a(x)\frac{\partial}{\partial x_3},$$

$$X_2 = \frac{\partial}{\partial x_2} + b(x)\frac{\partial}{\partial x_3},$$

be vector fields with coefficients

$$a(x) = x_2(1 + |x_2|),\ b(x) = -x_1(1 + |x_1|).$$

Clearly, these coefficients are not smooth. Then

$$[X_1, X_2] = -2(1 + |x_1| + |x_2|)\frac{\partial}{\partial x_3}.$$

The vector fields X_1, X_2 are $C^{1,1}$ and satisfy Hörmander's commutator condition of step two, so that assumptions of Example (E2) hold. Replacing $|x_1|$, $|x_2|$ with $x_1|x_1|$, $x_2|x_2|$ we get $C^{2,1}$ vector fields, satisfying assumptions of Example (E3).

These examples and the corresponding sub-Laplacian $\mathcal{L} = X_1^2 + X_2^2$ were studied in [BBMP17, Section 6]. Other explicit examples can be built from the so-called Δ_λ-Laplacians, see, e.g., [KS16].

12.2 Divergence formula

For this, there is no need to make any assumptions on the step to which Hörmander's commutator condition is satisfied, whether it is satisfied or not, or on the existence of fundamental solutions as in (A_y). Thus, let us formulate this property as a general statement which shall be of interest on its own. The assumption for smoothness on X_k can be reduced here, e.g., to $a_{k,m} \in C^1$.

Theorem 12.2.1 (Divergence formula). *Let $\Omega \subset \mathbb{R}^n$ be an open bounded domain with a piecewise smooth boundary that has no self-intersections. Let X_k, $k = 1, \dots, N$, be C^1 vector fields in the form*

$$X_k = \frac{\partial}{\partial x_k} + \sum_{m=N+1}^{n} a_{k,m}(x)\frac{\partial}{\partial x_m}. \tag{12.6}$$

Let $f_k \in C^1(\Omega)\bigcap C(\overline{\Omega})$, $k = 1, \dots, N$. Then for each $k = 1, \dots, N$, we have

$$\int_\Omega X_k f_k d\nu = \int_{\partial\Omega} f_k \langle X_k, d\nu\rangle. \tag{12.7}$$

Consequently, we also have the divergence type formula

$$\int_\Omega \sum_{k=1}^N X_k f_k d\nu = \int_{\partial\Omega} \sum_{k=1}^N f_k \langle X_k, d\nu\rangle. \tag{12.8}$$

If $y \in \mathbb{R}^n$ is such that (A_y) is satisfied, then we can also take $f_k = vX_k\Gamma_y$ in formulae above, for all $v \in C^1(\Omega)\bigcap C(\overline{\Omega})$.

Formula (12.8) is exactly the one needed for the admissibility of a domain in Definition 12.1.1. For a discussion of other related versions of the divergence formula in the literature see Remark 1.4.7. The proof of Theorem 12.2.1 is similar to that of Theorem 1.4.5.

Proof of Theorem 12.2.1. For any function f we calculate the following differentiation formula

$$\begin{aligned} df &= \sum_{k=1}^N \frac{\partial f}{\partial x_k} dx_k + \sum_{m=N+1}^n \frac{\partial f}{\partial x_m} dx_m \\ &= \sum_{k=1}^N X_k f dx_k - \sum_{k=1}^N \sum_{m=N+1}^n a_{k,m}(x) \frac{\partial f}{\partial x_m} dx_k + \sum_{m=N+1}^n \frac{\partial f}{\partial x_m} dx_m \\ &= \sum_{k=1}^N X_k f dx_k + \sum_{m=N+1}^n \frac{\partial f}{\partial x_m} \Big(-\sum_{k=1}^N a_{k,m}(x) dx_k + dx_m\Big) \\ &= \sum_{k=1}^N X_k f dx_k + \sum_{m=N+1}^n \frac{\partial f}{\partial x_m} \theta_m, \end{aligned}$$

where we denote

$$\theta_m := -\sum_{k=1}^N a_{k,m}(x) dx_k + dx_m, \ m = N+1, \dots, n. \tag{12.9}$$

That is, we have

$$df = \sum_{k=1}^N X_k f dx_k + \sum_{m=N+1}^n \frac{\partial f}{\partial x_m} \theta_m. \tag{12.10}$$

It is simple to see that

$$\langle X_s, dx_j\rangle = \frac{\partial}{\partial x_s} dx_j = \delta_{sj}, \ 1 \le s \le N, \ 1 \le j \le n,$$

where δ_{sj} is the Kronecker delta. Moreover, we have

$$
\begin{aligned}
\langle X_s, \theta_m \rangle &= \left\langle \frac{\partial}{\partial x_s} + \sum_{g=N+1}^{n} a_{s,g}(x)\frac{\partial}{\partial x_g}, -\sum_{k=1}^{N} a_{k,m}(x)dx_k + dx_m \right\rangle \\
&= -\sum_{k=1}^{N}\left(\frac{\partial}{\partial x_s}a_{k,m}(x)\right)dx_k - \sum_{k=1}^{N} a_{k,m}(x)\frac{\partial}{\partial x_s}dx_k + \frac{\partial}{\partial x_s}dx_m \\
&\quad -\sum_{k=1}^{N}\sum_{g=N+1}^{n} a_{s,g}(x)\left(\frac{\partial}{\partial x_g}a_{k,m}(x)\right)dx_k - \sum_{k=1}^{N}\sum_{g=N+1}^{n} a_{s,g}(x)a_{k,m}(x)\frac{\partial}{\partial x_g}dx_k \\
&\quad + \sum_{g=N+1}^{n} a_{s,g}(x)\frac{\partial}{\partial x_g}dx_m \\
&= -\sum_{k=1}^{N}\left(\frac{\partial}{\partial x_s}a_{k,m}(x)\right)dx_k - \sum_{k=1}^{N} a_{k,m}(x)\delta_{sk} \\
&\quad -\sum_{k=1}^{N}\sum_{g=N+1}^{n} a_{s,g}(x)\left(\frac{\partial}{\partial x_g}a_{k,m}(x)\right)dx_k + \sum_{g=N+1}^{n} a_{s,g}(x)\delta_{gm} \\
&= -\sum_{k=1}^{N}\sum_{g=N+1}^{n} a_{s,g}(x)\left(\frac{\partial}{\partial x_g}a_{k,m}(x)\right)dx_k - \sum_{k=1}^{N}\left(\frac{\partial}{\partial x_s}a_{k,m}(x)\right)dx_k \\
&= -\sum_{k=1}^{N}\left[\sum_{g=N+1}^{n} a_{s,g}(x)\left(\frac{\partial}{\partial x_g}a_{k,m}(x)\right) + \frac{\partial}{\partial x_s}a_{k,m}(x)\right]dx_k.
\end{aligned}
$$

That is, we have

$$\langle X_s, dx_j \rangle = \delta_{sj},$$

for $s = 1, \ldots, N$, $j = 1, \ldots, n$, and

$$\langle X_s, \theta_m \rangle = \sum_{k=1}^{N} \mathcal{C}_k(s,m)dx_k,$$

for $s = 1, \ldots, N$, $m = N+1, \ldots, n$, where we denote

$$\mathcal{C}_k(s,m) := -\sum_{g=N+1}^{n} a_{s,g}(x)\frac{\partial}{\partial x_g}a_{k,m}(x) - \frac{\partial}{\partial x_s}a_{k,m}(x).$$

We have

$$d\nu := d\nu(x) = \bigwedge_{j=1}^{N} dx_j = \bigwedge_{j=1}^{N} dx_j \bigwedge_{m=N+1}^{n} dx_m = \bigwedge_{j=1}^{N} dx_j \bigwedge_{m=N+1}^{n} \theta_m,$$

so that

$$\langle X_k, d\nu(x)\rangle = \bigwedge_{j=1, j\neq k}^{N} dx_j \bigwedge_{m=N+1}^{n} \theta_m. \tag{12.11}$$

Therefore, by using formula (12.10) we get

$$\begin{aligned} &d(f_s\langle X_s, d\nu(x)\rangle) \\ &\quad = df_s \wedge \langle X_s, d\nu(x)\rangle \\ &\quad = \sum_{k=1}^{N} X_k f_s dx_k \wedge \langle X_s, d\nu(x)\rangle + \sum_{m=N+1}^{n} \frac{\partial f_s}{\partial x_m}\theta_m \wedge \langle X_s, d\nu(x)\rangle \\ &\quad = \sum_{k=1}^{N} X_k f_s dx_k \wedge \bigwedge_{\substack{j=1,\\ j\neq k}}^{N} dx_j \bigwedge_{m=N+1}^{n} \theta_m + \sum_{m=N+1}^{n} \frac{\partial f_s}{\partial x_m}\theta_m \wedge \bigwedge_{\substack{j=1,\\ j\neq k}}^{N} dx_j \bigwedge_{m=N+1}^{n} \theta_m. \end{aligned}$$

The first term in the last line is equal to $X_s f_s d\nu(x)$ and the second term is zero by the wedge product rules. Therefore, we obtain

$$d(\langle f_s X_s, d\nu(x)\rangle) = X_s f_s d\nu(x), \quad s = 1, \ldots, N. \tag{12.12}$$

Now using the Stokes theorem (see, e.g., [DFN84, Theorem 26.3.1]) we obtain (12.7). Taking a sum over k we also obtain (12.8) for all $f_k \in C^1(\Omega)\bigcap C(\overline{\Omega})$.

As in the classical case, the formula (12.7) is still valid for the fundamental solution of $\mathcal{L}$ since Γ can be estimated by a distance function associated to $\{X_k\}$ (see, e.g., [Man12, Proposition 4.8]), or [FSC86, SC84] for such estimates in a more general setting. □

12.3 Green's identities for sums of squares

Similar to Theorem 1.4.6 the divergence formula in Theorem 12.2.1 implies the corresponding Green identities.

Theorem 12.3.1 (Green's identities)**.** *Let M be a smooth manifold of dimension n with a volume form $d\nu$ and let $\mathcal{L}$ be an operator as in (12.1). Let $\Omega \subset M$ be an admissible domain.*

1. *Green's first identity: If $v \in C^1(\Omega)\bigcap C(\overline{\Omega})$ and $u \in C^2(\Omega)\bigcap C^1(\overline{\Omega})$ then we have*

$$\int_\Omega \left((\widetilde{\nabla} v)u + v\mathcal{L}u\right) d\nu = \int_{\partial\Omega} v\langle \widetilde{\nabla} u, d\nu\rangle, \tag{12.13}$$

where

$$\widetilde{\nabla} u = \sum_{k=1}^{N} (X_k u)\, X_k. \tag{12.14}$$

2. *Green's second identity: If* $u, v \in C^2(\Omega)\bigcap C^1(\overline{\Omega})$ *then we have*

$$\int_{\Omega}(u\mathcal{L}v - v\mathcal{L}u)d\nu = \int_{\partial\Omega}(u\langle\widetilde{\nabla}v, d\nu\rangle - v\langle\widetilde{\nabla}u, d\nu\rangle). \tag{12.15}$$

Moreover, if Ω *is strongly admissible, we can put* $u = \Gamma$ *in* (12.13), *and* $u = \Gamma$ *or* $v = \Gamma$ *in* (12.15).

As in Remark 1.4.7, Part 1, the notation (12.14) implies that for functions u and v we have

$$\left(\widetilde{\nabla}v\right)u = \widetilde{\nabla}vu = \sum_{k=1}^{N}(X_k v)(X_k u) = \sum_{k=1}^{N} X_k v X_k u = \left(\widetilde{\nabla}u\right)v \tag{12.16}$$

is a scalar.

Proof of Theorem 12.3.1. Taking $f_k = vX_k u$, we get

$$\sum_{k=1}^{N} X_k f_k = (\widetilde{\nabla}v)u + v\mathcal{L}u.$$

Since Ω is admissible we can use (12.3), so that we obtain

$$\begin{aligned}\int_{\Omega}\left(\widetilde{\nabla}vu + v\mathcal{L}u\right)d\nu &= \int_{\Omega}\sum_{k=1}^{N} X_k f_k d\nu\\ &= \int_{\partial\Omega}\sum_{k=1}^{N}\langle f_k X_k, d\nu\rangle\\ &= \int_{\partial\Omega}\sum_{k=1}^{N}\langle vX_k u X_k, d\nu\rangle\\ &= \int_{\partial\Omega} v\langle\widetilde{\nabla}u, d\nu\rangle.\end{aligned}$$

This proves (12.13). Then by rewriting (12.13) for interchanged functions u and v we have

$$\int_{\Omega}\left((\widetilde{\nabla}u)v + u\mathcal{L}v\right)d\nu = \int_{\partial\Omega} u\langle\widetilde{\nabla}v, d\nu\rangle,$$
$$\int_{\Omega}\left((\widetilde{\nabla}v)u + v\mathcal{L}u\right)d\nu = \int_{\partial\Omega} v\langle\widetilde{\nabla}u, d\nu\rangle.$$

By subtracting the second identity from the first one and using $(\widetilde{\nabla}u)v = (\widetilde{\nabla}v)u$ in view of (12.16), we obtain (12.15).

If Ω is strongly admissible, we can put Γ for u or v as stated since (12.3) holds in these cases as well. □

Remark 12.3.2. It is crucial that Green's identities are valid for the fundamental solution Γ. In the classical (Euclidean) case Green's identities are valid for the fundamental solution of the Laplacian and this fact is of fundamental importance in the classical theory as well.

12.3.1 Consequences of Green's identities

Let us now record several useful consequences of Theorem 12.3.1. Setting $v = 1$ we obtain the following analogue of Gauss' mean value type formulae:

Corollary 12.3.3 (Gauss' mean value formulae). *Let $\Omega \subset M$ be an admissible domain. Then we have*

$$\mathcal{L}u \geq 0 \text{ in } \Omega \implies \int_{\partial\Omega} \langle \widetilde{\nabla} u, d\nu \rangle \geq 0$$

and

$$\mathcal{L}u \leq 0 \text{ in } \Omega \implies \int_{\partial\Omega} \langle \widetilde{\nabla} u, d\nu \rangle \leq 0.$$

Consequently, we also have

$$\mathcal{L}u = 0 \text{ in } \Omega \implies \int_{\partial\Omega} \langle \widetilde{\nabla} u, d\nu \rangle = 0.$$

Also, for a fixed $x \in \Omega$, taking $v = 1$ and $u(y) = \Gamma(x, y)$ in (12.13) we obtain:

Corollary 12.3.4. *Let $\Omega \subset M$ be a strongly admissible domain such that $\Omega \subset T_y$ for all $y \in \Omega$, and let $x \in \Omega$. Then we have*

$$\int_{\partial\Omega} \langle \widetilde{\nabla}\Gamma(x, y), d\nu(y) \rangle = -1,$$

where $\widetilde{\nabla}\Gamma(x, y) = \widetilde{\nabla}_y \Gamma(x, y)$ refers to the notation (12.14) with derivatives taken with respect to the variable y.

The assumption of $\Omega \subset T_y$ for all $y \in \Omega$ in Corollary 12.3.4 just assures that the family of Γ_y is defined over $y \in \Omega$.

Corollary 12.3.5 (Representation formulae). *Let us assume the conditions of Corollary 12.3.4. Taking v in (12.15) to be the fundamental solution Γ we obtain the following representation formulae.*

1. *Let $u \in C^2(\Omega) \bigcap C^1(\overline{\Omega})$. Then for all $x \in \Omega$ we have*

$$\begin{aligned} u(x) = &- \int_{\Omega} \Gamma(x, y) \mathcal{L}u(y) d\nu(y) \\ &- \int_{\partial\Omega} u(y) \langle \widetilde{\nabla}\Gamma(x, y), d\nu(y) \rangle + \int_{\partial\Omega} \Gamma(x, y) \langle \widetilde{\nabla} u(y), d\nu(y) \rangle. \end{aligned}$$

2. *Let* $u \in C^2(\Omega)\bigcap C^1(\overline{\Omega})$ *and* $\mathcal{L}u = 0$ *on* Ω. *Then for all* $x \in \Omega$ *we have*

$$u(x) = -\int_{\partial\Omega} u(y)\langle\widetilde{\nabla}\Gamma(x,y), d\nu(y)\rangle + \int_{\partial\Omega} \Gamma(x,y)\langle\widetilde{\nabla}u(y), d\nu(y)\rangle.$$

3. *Let* $u \in C^2(\Omega)\bigcap C^1(\overline{\Omega})$ *and*

$$u(x) = 0, \ x \in \partial\Omega.$$

Then for all $x \in \Omega$ *we have*

$$u(x) = -\int_{\Omega} \Gamma(x,y)\mathcal{L}u(y)d\nu(y) + \int_{\partial\Omega} \Gamma(x,y)\langle\widetilde{\nabla}u(y), d\nu(y)\rangle.$$

4. *Let* $u \in C^2(\Omega)\bigcap C^1(\overline{\Omega})$ *and*

$$\sum_{j=1}^{N} X_j u\langle X_j, d\nu\rangle = 0 \quad on \ \partial\Omega.$$

Then for all $x \in \Omega$ *we have*

$$u(x) = -\int_{\Omega} \Gamma(x,y)\mathcal{L}u(y)d\nu(y) - \int_{\partial\Omega} u(y)\langle\widetilde{\nabla}\Gamma(x,y), d\nu(y)\rangle.$$

12.3.2 Differential forms, perimeter and surface measures

In this section we briefly describe the relation between the forms $\langle X_j, d\nu\rangle$, perimeter measure, and the surface measure on the boundary $\partial\Omega$. In this we follow [RS17c] where this topic was discussed in the setting of stratified groups, and we would like to thank Nicola Garofalo and Valentino Magnani for discussions.

Definition 12.3.6 (Perimeter measure). Let $\Omega \subset M$ be an open set with a piecewise smooth boundary. The *perimeter measure* on $\partial\Omega$ is defined by

$$\sigma_H(\partial\Omega) = \sup\left\{\sum_{i=1}^{N}\int_{\partial\Omega} \psi_i\langle X_i, d\nu\rangle : \ \psi = (\psi_1, \ldots, \psi_{N_1}), \ |\psi| \le 1, \ \psi \in C^1\right\}.$$

Then we have the following simple proof of the divergence formula in Theorem 12.2.1.

Proposition 12.3.7 (Divergence formula). *Let* X *be a vector field and let* $\langle X, d\nu\rangle$ *be the contraction of the volume form* $d\nu = dx_1 \wedge \cdots \wedge dx_n$ *by* X. *Then we have*

$$\int_{\Omega} X\varphi \, d\nu = \int_{\partial\Omega} \varphi\langle X, d\nu\rangle. \tag{12.17}$$

Proof of Proposition 12.3.7. Let L_X denote the Lie derivative with respect to the vector field X. The Cartan formula for L_X gives

$$L_X = d\imath_X + \imath_X\, d, \qquad \text{where} \qquad \imath_X d\nu = \langle X, d\nu\rangle.$$

Then we have

$$\int_\Omega X\varphi\, d\nu = \int_\Omega \operatorname{div}(\varphi X) d\nu = \int_\Omega L_{\varphi X} d\nu = \int_\Omega d(\imath_{\varphi X} d\nu) = \int_{\partial\Omega} \varphi\langle X, d\nu\rangle,$$

showing (12.17). □

Proposition 12.3.8 (Relation between forms, perimeter and surface measures). *Let $\langle\cdot,\cdot\rangle_E$ denote the Euclidean scalar product. Then the perimeter measure $d\sigma_H$ and the surface measure dS on $\partial\Omega$ are related by*

$$\int_{\partial\Omega} \varphi\langle v, X_j\rangle_E dS = \int_{\partial\Omega} \varphi \frac{\langle v, X_j\rangle_E}{\left(\sum_{j=1}^N \langle v, X_j\rangle_E^2\right)^{\frac{1}{2}}} d\sigma_H = \int_{\partial\Omega} \varphi\langle X_j, d\nu\rangle, \tag{12.18}$$

for all outer unit vectors v and all $\varphi \in C^\infty(\partial\Omega)$.

Moreover, if $\mathfrak{g}$ denotes the vector space spanned by $\{X_j\}_{j=1}^N$ and X_j are orthonormal on $\mathfrak{g}$, then for any $f_j \in C^\infty(\partial\Omega)$ we have

$$\int_{\partial\Omega} \sum_{j=1}^N f_j\langle X_j, d\nu\rangle = \int_{\partial\Omega} \langle X, v_H\rangle_{\mathfrak{g}}\, d\sigma_H, \tag{12.19}$$

where $X = \sum_{j=1}^N f_j X_j$ and $v_H = \sum_{j=1}^N \langle v, X_j\rangle_E X_j$.

Proof of Proposition 12.3.8. For an outer unit vector v on $\partial\Omega$ let us write

$$|v_H| := \left(\sum_{j=1}^N \langle v, X_j\rangle_E^2\right)^{1/2} \qquad \text{and} \qquad |v_H|_j := \frac{\langle v, X_j\rangle_E}{|v_H|}.$$

If dS is the surface measure on $\partial\Omega$, we have

$$d\sigma_H = |v_H| dS,$$

and all these relations are well defined because the perimeter measure of the set of characteristic points of a smooth domain Ω is zero. We can now calculate

$$\begin{aligned}\int_\Omega X_j\varphi\, d\nu &= \int_\Omega \operatorname{div}(\varphi X_j) d\nu = \int_{\partial\Omega} \varphi\,\imath_{X_j}(d\nu) = \int_{\partial\Omega} \varphi\langle v, X_j\rangle_E dS \\ &= \int_{\partial\Omega} \varphi\frac{\langle v, X_j\rangle_E}{|v_H|}|v_H| dS = \int_{\partial\Omega} \varphi|v_H|_j\, d\sigma_H,\end{aligned}$$

giving one equality in (12.18). Combining this with (12.17) we obtain

$$\int_{\partial\Omega} \varphi\langle X_j, d\nu\rangle = \int_{\partial\Omega} \varphi|v_H|_j \, d\sigma_H, \tag{12.20}$$

the other equality in (12.18). Let us now assume that X_j are orthonormal on $\mathfrak{g}$, and let

$$X = \sum_{j=1}^{N} f_j X_j.$$

We write

$$v_H = \sum_{j=1}^{N} \langle v, X_j\rangle_E X_j$$

for a vector v with $|v_H| = 1$. Then we have

$$\langle X, v_H\rangle_{\mathfrak{g}} = \sum_{j=1}^{N} f_j |v_H|_j.$$

Now, applying (12.20) with $\varphi = f_j$ and summing over j, we get

$$\int_{\partial\Omega} \sum_{j=1}^{N} f_j \langle X_j, d\nu\rangle = \int_{\partial\Omega} \langle X, v_H\rangle_{\mathfrak{g}} \, d\sigma_H, \quad X = \sum_{j=1}^{N} f_j X_j,$$

which gives (12.19). □

12.4 Local Hardy inequalities

In this section we describe local versions of the Hardy inequality including boundary terms. The weights are formulated in terms of the fundamental solution and the proof relies on Green's first formula from Theorem 12.3.1. As usual, we denote

$$\nabla_X = (X_1, \ldots, X_N).$$

Theorem 12.4.1 (Local Hardy inequality with boundary terms). *Let $y \in M$ be such that (A_y^+) holds with the fundamental solution $\Gamma = \Gamma_y$ in T_y. Let $\Omega \subset T_y$ be a strongly admissible domain such that $y \notin \partial\Omega$. Let $\alpha \in \mathbb{R}$, $\alpha > 2 - \beta$, $\beta > 2$ and $R \geq e\sup_\Omega \Gamma^{\frac{1}{2-\beta}}$. Then for all $u \in C^1(\Omega) \bigcap C(\overline{\Omega})$ we have*

$$\begin{aligned} \int_\Omega \Gamma^{\frac{\alpha}{2-\beta}} |\nabla_X u|^2 \, d\nu \geq & \left(\frac{\beta+\alpha-2}{2}\right)^2 \int_\Omega \Gamma^{\frac{\alpha-2}{2-\beta}} |\nabla_X \Gamma^{\frac{1}{2-\beta}}|^2 |u|^2 \, d\nu \\ & + \frac{\beta+\alpha-2}{2(\beta-2)} \int_{\partial\Omega} \Gamma^{\frac{\alpha}{2-\beta}-1} |u|^2 \langle \widetilde{\nabla}\Gamma, d\nu\rangle, \end{aligned} \tag{12.21}$$

as well as its further refinement

$$\begin{aligned}\int_\Omega \Gamma^{\frac{\alpha}{2-\beta}}|\nabla_X u|^2\,d\nu \geq \left(\frac{\beta+\alpha-2}{2}\right)^2 \int_\Omega \Gamma^{\frac{\alpha-2}{2-\beta}}|\nabla_X \Gamma^{\frac{1}{2-\beta}}|^2|u|^2\,d\nu \\ + \frac{1}{4}\int_\Omega \Gamma^{\frac{\alpha-2}{2-\beta}}|\nabla_X \Gamma^{\frac{1}{2-\beta}}|^2 \left(\ln\frac{R}{\Gamma^{\frac{1}{2-\beta}}}\right)^{-2}|u|^2 d\nu \\ + \frac{1}{2(\beta-2)}\int_{\partial\Omega} \Gamma^{\frac{\alpha}{2-\beta}-1}\left(\ln\frac{R}{\Gamma^{\frac{1}{2-\beta}}}\right)^{-1}|u|^2\langle\widetilde{\nabla}\Gamma, d\nu\rangle \\ + \frac{\beta+\alpha-2}{2(\beta-2)}\int_{\partial\Omega}\Gamma^{\frac{\alpha}{2-\beta}-1}|u|^2\langle\widetilde{\nabla}\Gamma, d\nu\rangle.\end{aligned} \tag{12.22}$$

Remark 12.4.2.

1. If $u=0$ on the boundary $\partial\Omega$, for example when $\operatorname{supp} u \subset \Omega$, then (12.21) can be regarded as a usual Hardy inequality (without boundary term). Inequality (12.22) can be regarded as a further refinement of (12.21) since it includes further positive interior terms as well as further boundary terms.
2. Even if $y \in \partial\Omega$, the estimates (12.21) and (12.22) of Theorem 12.4.1 remain true if $y \notin \partial\Omega \cap \operatorname{supp} u$.
3. In (12.21) the boundary term can be positive, see Remark 11.4.2, Part 2, i.e., we sometimes have

$$\frac{\beta+\alpha-2}{2(\beta-2)}\int_{\partial\Omega}\Gamma^{\frac{\alpha}{2-\beta}-1}|u|^2\langle\widetilde{\nabla}\Gamma, d\nu\rangle \geq 0, \tag{12.23}$$

for some u.

4. In the setting of Example (E1), i.e., when M is a stratified group, and X_j's are the vectors from the first stratum, then (12.21) is equivalent to (11.73) in Theorem 11.4.1, where this inequality was expressed in terms of the $\mathcal{L}$-gauge d, taking $\beta = Q \geq 3$, and $d(x) = \Gamma(x,0)^{\frac{1}{2-Q}}$, where Q is the homogeneous dimension of the group. For example, with $\alpha=0$ we get

$$\begin{aligned}\int_\Omega |\nabla_X u|^2\,d\nu \geq \left(\frac{Q-2}{2}\right)^2\int_\Omega \frac{|\nabla_X d|^2}{d^2}|u|^2\,d\nu \\ + \frac{1}{2}\int_{\partial\Omega} d^{Q-2}|u|^2\langle\widetilde{\nabla} d^{2-Q}, d\nu\rangle,\end{aligned} \tag{12.24}$$

with the sharp constant $\left(\frac{Q-2}{2}\right)^2$.

Proof of Theorem 12.4.1. In the proof and in the subsequent analysis we follow [RS17d].

First, let us prove (12.21). By an argument of Remark 2.1.2, Part 3, we can assume that u is real-valued. In this case, recalling that

$$(\widetilde{\nabla}u)u = \sum_{k=1}^{N}(X_k u)X_k u = |\nabla_X u|^2,$$

inequality (12.21) reduces to

$$\begin{aligned}\int_\Omega \Gamma^{\frac{\alpha}{2-\beta}}(\widetilde{\nabla}u)u\,d\nu \geq \left(\frac{\beta+\alpha-2}{2}\right)^2 \int_\Omega \Gamma^{\frac{\alpha-2}{2-\beta}}(\widetilde{\nabla}\Gamma^{\frac{1}{2-\beta}})\Gamma^{\frac{1}{2-\beta}}u^2\,d\nu \\ + \frac{\beta+\alpha-2}{2(\beta-2)}\int_{\partial\Omega}\Gamma^{\frac{\alpha}{2-\beta}-1}u^2\langle\widetilde{\nabla}\Gamma, d\nu\rangle,\end{aligned} \tag{12.25}$$

which we will now prove. Setting

$$u = d^\gamma q \tag{12.26}$$

for some real-valued functions $d>0$, q, and a constant $\gamma \neq 0$ to be chosen later, we have

$$\begin{aligned}(\widetilde{\nabla}u)u &= (\widetilde{\nabla}d^\gamma q)d^\gamma q \\ &= \sum_{k=1}^{N} X_k(d^\gamma q)X_k(d^\gamma q) \\ &= \gamma^2 d^{2\gamma-2}\sum_{k=1}^{N}(X_k d)^2 q^2 + 2\gamma d^{2\gamma-1}q\sum_{k=1}^{N}X_k d\,X_k q + d^{2\gamma}\sum_{k=1}^{N}(X_k q)^2 \\ &= \gamma^2 d^{2\gamma-2}((\widetilde{\nabla}d)d)q^2 + 2\gamma d^{2\gamma-1}q(\widetilde{\nabla}d)q + d^{2\gamma}(\widetilde{\nabla}q)q.\end{aligned}$$

Multiplying both sides of this equality by d^α and applying Green's first formula from Theorem 12.3.1 to the second term in the last line we observe that

$$\begin{aligned}2\gamma\int_\Omega d^{\alpha+2\gamma-1}q(\widetilde{\nabla}d)qd\nu &= \frac{\gamma}{\alpha+2\gamma}\int_\Omega(\widetilde{\nabla}d^{\alpha+2\gamma})q^2 d\nu = \frac{\gamma}{\alpha+2\gamma}\int_\Omega(\widetilde{\nabla}q^2)d^{\alpha+2\gamma}d\nu \\ &= -\frac{\gamma}{\alpha+2\gamma}\int_\Omega q^2\mathcal{L}d^{\alpha+2\gamma}d\nu + \frac{\gamma}{\alpha+2\gamma}\int_{\partial\Omega}q^2\langle\widetilde{\nabla}d^{\alpha+2\gamma}, d\nu\rangle,\end{aligned}$$

where we note that later on we will choose γ so that $d^{\alpha+2\gamma} = \Gamma$, and hence Theorem 12.3.1 is applicable. Consequently, we get

$$\begin{aligned}\int_\Omega d^\alpha(\widetilde{\nabla}u)ud\nu &= \gamma^2\int_\Omega d^{\alpha+2\gamma-2}((\widetilde{\nabla}d)d)\,q^2 d\nu + \frac{\gamma}{\alpha+2\gamma}\int_\Omega(\widetilde{\nabla}d^{\alpha+2\gamma})q^2 d\nu \\ &\quad + \int_\Omega d^{\alpha+2\gamma}(\widetilde{\nabla}q)qd\nu \\ &= \gamma^2\int_\Omega d^{\alpha+2\gamma-2}((\widetilde{\nabla}d)d)\,q^2 d\nu + \frac{\gamma}{\alpha+2\gamma}\int_{\partial\Omega}q^2\langle\widetilde{\nabla}d^{\alpha+2\gamma}, d\nu\rangle \\ &\quad - \frac{\gamma}{\alpha+2\gamma}\int_\Omega q^2\mathcal{L}d^{\alpha+2\gamma}d\nu + \int_\Omega d^{\alpha+2\gamma}(\widetilde{\nabla}q)qd\nu\end{aligned}$$

$$\geq \gamma^2 \int_\Omega d^{\alpha+2\gamma-2}((\widetilde{\nabla} d)d)\, q^2 d\nu + \frac{\gamma}{\alpha+2\gamma}\int_{\partial\Omega} q^2 \langle \widetilde{\nabla} d^{\alpha+2\gamma}, d\nu\rangle$$
$$- \frac{\gamma}{\alpha+2\gamma}\int_\Omega q^2 \mathcal{L} d^{\alpha+2\gamma} d\nu, \tag{12.27}$$

since $d > 0$ and $(\widetilde{\nabla} q)q = |\nabla_X q|^2 \geq 0$. On the other hand, it can be readily checked that for a vector field X we have

$$\frac{\gamma}{\alpha+2\gamma} X^2(d^{\alpha+2\gamma}) = \gamma X(d^{\alpha+2\gamma-1} Xd) = \frac{\gamma}{2-\beta} X(d^{\alpha+2\gamma+\beta-2} X(d^{2-\beta}))$$
$$= \frac{\gamma}{2-\beta}(\alpha+2\gamma+\beta-2) d^{\alpha+2\gamma+\beta-3}(Xd)X(d^{2-\beta}) + \frac{\gamma}{2-\beta} d^{\alpha+2\gamma+\beta-2} X^2(d^{2-\beta})$$
$$= \gamma(\alpha+2\gamma+\beta-2) d^{\alpha+2\gamma-2}(Xd)^2 + \frac{\gamma}{2-\beta} d^{\alpha+2\gamma+\beta-2} X^2(d^{2-\beta}).$$

Consequently, we get the equality

$$-\frac{\gamma}{\alpha+2\gamma}\mathcal{L} d^{\alpha+2\gamma} = -\gamma(\alpha+2\gamma+\beta-2) d^{\alpha+2\gamma-2}(\widetilde{\nabla} d)d - \frac{\gamma}{2-\beta} d^{\alpha+2\gamma+\beta-2}\mathcal{L} d^{2-\beta}. \tag{12.28}$$

Since $q^2 = d^{-2\gamma}u^2$ in view of (12.26), substituting (12.28) into (12.27) we obtain

$$\int_\Omega d^\alpha (\widetilde{\nabla} u) u d\nu \geq (-\gamma^2 - \gamma(\alpha+\beta-2)) \int_\Omega d^{\alpha-2}((\widetilde{\nabla} d)d) u^2 d\nu$$
$$-\frac{\gamma}{2-\beta}\int_\Omega (\mathcal{L} d^{2-\beta}) d^{\alpha+\beta-2} u^2 dx + \frac{\gamma}{\alpha+2\gamma}\int_{\partial\Omega} d^{-2\gamma} u^2 \langle \widetilde{\nabla} d^{\alpha+2\gamma}, d\nu\rangle.$$

Taking $d = \Gamma^{\frac{1}{2-\beta}}$, $\beta > 2$, concerning the second term we observe that for $\alpha > 2-\beta$ and $\beta > 2$ we have

$$\int_\Omega (\mathcal{L}\Gamma)\Gamma^{\frac{\alpha+\beta-2}{2-\beta}} u^2 dx = 0, \tag{12.29}$$

since $\Gamma = \Gamma_y$ is the fundamental solution to $\mathcal{L}$. Indeed, the above equality is clear when y is outside of Ω. If y belongs to Ω we have

$$\int_\Omega (\mathcal{L}\Gamma)\Gamma^{\frac{\alpha+\beta-2}{2-\beta}} u^2 dx = \Gamma^{\frac{\alpha+\beta-2}{2-\beta}}(y) u^2(y) = 0,$$

since conditions $\alpha > 2-\beta$ and $\beta > 2$ imply that $\frac{\alpha+\beta-2}{2-\beta} < 0$, and since $\frac{1}{\Gamma}(y) = 0$ by (A_y^+). Thus, with $d = \Gamma^{\frac{1}{2-\beta}}$, $\beta > 2$, we get

$$\int_\Omega \Gamma^{\frac{\alpha}{2-\beta}}(\widetilde{\nabla} u) u\, d\nu \geq (-\gamma^2 - \gamma(\alpha+\beta-2)) \int_\Omega \Gamma^{\frac{\alpha-2}{2-\beta}}(\widetilde{\nabla}\Gamma^{\frac{1}{2-\beta}})\Gamma^{\frac{1}{2-\beta}} u^2\, d\nu$$
$$+ \frac{\gamma}{\alpha+2\gamma}\int_{\partial\Omega} \Gamma^{-\frac{2\gamma}{2-\beta}} u^2 \langle \widetilde{\nabla}\Gamma^{\frac{\alpha+2\gamma}{2-\beta}}, d\nu\rangle.$$

Taking $\gamma = \frac{2-\beta-\alpha}{2}$, we obtain (12.25). Finally, we note that with this γ, we have $d^{\alpha+2\gamma} = \Gamma$, so that the use of Theorem 12.3.1 is justified.

Let us now prove (12.22), with the proof similar to the above proof of (12.21). Recalling that

$$(\widetilde{\nabla} u)u = \sum_{k=1}^{N} (X_k u) X_k u = |\nabla_X u|^2,$$

inequality (12.22) reduces to

$$\begin{aligned}\int_\Omega \Gamma^{\frac{\alpha}{2-\beta}} (\widetilde{\nabla} u) u \, d\nu \geq & \left(\frac{\beta+\alpha-2}{2}\right)^2 \int_\Omega \Gamma^{\frac{\alpha-2}{2-\beta}} (\widetilde{\nabla} \Gamma^{\frac{1}{2-\beta}}) \Gamma^{\frac{1}{2-\beta}} u^2 \, d\nu \\ & + \frac{1}{4} \int_\Omega \Gamma^{\frac{\alpha-2}{2-\beta}} (\widetilde{\nabla} \Gamma^{\frac{1}{2-\beta}}) \Gamma^{\frac{1}{2-\beta}} \left(\ln \frac{R}{\Gamma^{\frac{1}{2-\beta}}}\right)^{-2} u^2 d\nu \\ & + \frac{1}{2(\beta-2)} \int_{\partial\Omega} \Gamma^{\frac{\alpha}{2-\beta}-1} \left(\ln \frac{R}{\Gamma^{\frac{1}{2-\beta}}}\right)^{-1} u^2 \langle \widetilde{\nabla} \Gamma, d\nu \rangle \\ & + \frac{\beta+\alpha-2}{2(\beta-2)} \int_{\partial\Omega} \Gamma^{\frac{\alpha}{2-\beta}-1} u^2 \langle \widetilde{\nabla} \Gamma, d\nu \rangle,\end{aligned} \tag{12.30}$$

which we will now prove. Let us recall the first part of (12.27) as

$$\begin{aligned}\int_\Omega & d^\alpha (\widetilde{\nabla} u) u d\nu \\ &= \gamma^2 \int_\Omega d^{\alpha+2\gamma-2} ((\widetilde{\nabla} d) d) \, q^2 d\nu + \frac{\gamma}{\alpha+2\gamma} \int_\Omega (\widetilde{\nabla} d^{\alpha+2\gamma}) q^2 d\nu + \int_\Omega d^{\alpha+2\gamma} (\widetilde{\nabla} q) q d\nu \\ &= \gamma^2 \int_\Omega d^{\alpha+2\gamma-2} ((\widetilde{\nabla} d) d) \, q^2 d\nu + \frac{\gamma}{\alpha+2\gamma} \int_{\partial\Omega} q^2 \langle \widetilde{\nabla} d^{\alpha+2\gamma}, d\nu \rangle \\ &\quad - \frac{\gamma}{\alpha+2\gamma} \int_\Omega q^2 \mathcal{L} d^{\alpha+2\gamma} d\nu + \int_\Omega d^{\alpha+2\gamma} (\widetilde{\nabla} q) q d\nu.\end{aligned} \tag{12.31}$$

Since $q^2 = d^{-2\gamma} u^2$, substituting (12.28) into (12.31) we obtain

$$\begin{aligned}\int_\Omega d^\alpha (\widetilde{\nabla} u) u d\nu = & \left(-\gamma^2 - \gamma(\alpha+\beta-2)\right) \int_\Omega d^{\alpha-2} ((\widetilde{\nabla} d) d) u^2 d\nu \\ & - \frac{\gamma}{2-\beta} \int_\Omega (\mathcal{L} d^{2-\beta}) d^{\alpha+\beta-2} u^2 dx \\ & + \frac{\gamma}{\alpha+2\gamma} \int_{\partial\Omega} d^{-2\gamma} u^2 \langle \widetilde{\nabla} d^{\alpha+2\gamma}, d\nu \rangle + \int_\Omega d^{\alpha+2\gamma} (\widetilde{\nabla} q) q d\nu.\end{aligned}$$

Using (12.29), with $d = \Gamma^{\frac{1}{2-\beta}}$, $\beta > 2$, we obtain

$$\begin{aligned}\int_\Omega \Gamma^{\frac{\alpha}{2-\beta}} (\widetilde{\nabla} u) u \, d\nu = & (-\gamma^2 - \gamma(\alpha+\beta-2)) \int_\Omega \Gamma^{\frac{\alpha-2}{2-\beta}} (\widetilde{\nabla} \Gamma^{\frac{1}{2-\beta}}) \Gamma^{\frac{1}{2-\beta}} u^2 \, d\nu \\ & + \frac{\gamma}{\alpha+2\gamma} \int_{\partial\Omega} \Gamma^{-\frac{2\gamma}{2-\beta}} u^2 \langle \widetilde{\nabla} \Gamma^{\frac{\alpha+2\gamma}{2-\beta}}, d\nu \rangle + \int_\Omega d^{\alpha+2\gamma} (\widetilde{\nabla} q) q d\nu.\end{aligned}$$

Taking $\gamma = \frac{2-\beta-\alpha}{2}$ we obtain

$$\int_\Omega \Gamma^{\frac{\alpha}{2-\beta}} (\widetilde{\nabla} u) u \, d\nu = \left(\frac{\beta+\alpha-2}{2}\right)^2 \int_\Omega \Gamma^{\frac{\alpha-2}{2-\beta}} (\widetilde{\nabla} \Gamma^{\frac{1}{2-\beta}}) \Gamma^{\frac{1}{2-\beta}} u^2 \, d\nu \\ + \frac{\beta+\alpha-2}{2(\beta-2)} \int_{\partial\Omega} \Gamma^{\frac{\alpha}{2-\beta}-1} u^2 \langle \widetilde{\nabla}\Gamma, d\nu\rangle + \int_\Omega \Gamma (\widetilde{\nabla} q) q d\nu. \tag{12.32}$$

Let us now take

$$q = \left(\ln \frac{R}{\Gamma^{\frac{1}{2-\beta}}}\right)^{1/2} \varphi,$$

that is,

$$\varphi = \left(\ln \frac{R}{\Gamma^{\frac{1}{2-\beta}}}\right)^{-\frac{1}{2}} \Gamma^{-\frac{2-\beta-\alpha}{2(2-\beta)}} u.$$

A straightforward computation shows that

$$\begin{aligned} \int_\Omega \Gamma(\widetilde{\nabla} q) q d\nu &= \sum_{j=1}^N \int_\Omega \Gamma \left(X_j \left(\ln \frac{R}{\Gamma^{\frac{1}{2-\beta}}}\right)^{\frac{1}{2}} \varphi + \left(\ln \frac{R}{\Gamma^{\frac{1}{2-\beta}}}\right)^{\frac{1}{2}} X_j \varphi \right)^2 d\nu \\ &= \frac{1}{4} \int_\Omega \Gamma^{\frac{-\beta}{2-\beta}} (\widetilde{\nabla} \Gamma^{\frac{1}{2-\beta}}) \Gamma^{\frac{1}{2-\beta}} \left(\ln \frac{R}{\Gamma^{\frac{1}{2-\beta}}}\right)^{-1} \varphi^2 d\nu \\ &\quad - \int_\Omega \Gamma^{1-\frac{1}{2-\beta}} \varphi (\widetilde{\nabla} \Gamma^{\frac{1}{2-\beta}}) \varphi d\nu + \int_\Omega \Gamma \ln \frac{R}{\Gamma^{\frac{1}{2-\beta}}} (\widetilde{\nabla} \varphi) \varphi d\nu \\ &= \frac{1}{4} \int_\Omega \Gamma^{\frac{-\beta}{2-\beta}} (\widetilde{\nabla} \Gamma^{\frac{1}{2-\beta}}) \Gamma^{\frac{1}{2-\beta}} \left(\ln \frac{R}{\Gamma^{\frac{1}{2-\beta}}}\right)^{-1} \varphi^2 d\nu \\ &\quad + \frac{1}{2(\beta-2)} \int_\Omega (\widetilde{\nabla}\Gamma) \varphi^2 d\nu + \int_\Omega \Gamma \ln \frac{R}{\Gamma^{\frac{1}{2-\beta}}} (\widetilde{\nabla} \varphi) \varphi d\nu \\ &= \frac{1}{4} \int_\Omega \Gamma^{\frac{-\beta}{2-\beta}} (\widetilde{\nabla} \Gamma^{\frac{1}{2-\beta}}) \Gamma^{\frac{1}{2-\beta}} \left(\ln \frac{R}{\Gamma^{\frac{1}{2-\beta}}}\right)^{-1} \varphi^2 d\nu \\ &\quad + \frac{1}{2(\beta-2)} \int_\Omega \mathcal{L}\Gamma \, \varphi^2 d\nu + \frac{1}{2(\beta-2)} \int_{\partial\Omega} \varphi^2 \langle \widetilde{\nabla}\Gamma, d\nu\rangle \\ &\quad + \int_\Omega \Gamma \ln \frac{R}{\Gamma^{\frac{1}{2-\beta}}} (\widetilde{\nabla} \varphi) \varphi d\nu. \end{aligned} \tag{12.33}$$

Since the second integral term of the right-hand side vanishes and the last integral term is positive from (12.33) we obtain that

$$\int_\Omega \Gamma(\widetilde{\nabla} q) q d\nu \geq \frac{1}{4} \int_\Omega \Gamma^{\frac{-\beta}{2-\beta}} (\widetilde{\nabla} \Gamma^{\frac{1}{2-\beta}}) \Gamma^{\frac{1}{2-\beta}} \left(\ln \frac{R}{\Gamma^{\frac{1}{2-\beta}}}\right)^{-1} \varphi^2 d\nu \\ + \frac{1}{2(\beta-2)} \int_{\partial\Omega} \varphi^2 \langle \widetilde{\nabla}\Gamma, d\nu\rangle$$

$$= \frac{1}{4}\int_\Omega \Gamma^{\frac{\alpha-2}{2-\beta}}(\widetilde{\nabla}\Gamma^{\frac{1}{2-\beta}})\Gamma^{\frac{1}{2-\beta}}\left(\ln\frac{R}{\Gamma^{\frac{1}{2-\beta}}}\right)^{-2}u^2 d\nu$$

$$+ \frac{1}{2(\beta-2)}\int_{\partial\Omega}\Gamma^{\frac{\alpha}{2-\beta}-1}\left(\ln\frac{R}{\Gamma^{\frac{1}{2-\beta}}}\right)^{-1}u^2\langle\widetilde{\nabla}\Gamma, d\nu\rangle. \tag{12.34}$$

Finally, (12.32) and (12.34) imply (12.30). □

12.5 Anisotropic Hardy inequalities via Picone identities

In this section we discuss the anisotropic versions of local Hardy inequalities for general (real-valued) vector fields as in the previous sections. As in the most of this chapter, such weighted anisotropic Hardy type inequalities will also include the boundary terms, which of course disappear if one works with functions supported in the interior of the considered domain. The analysis is based on the anisotropic Picone type identities, analogous to those described in Section 6.10.1. As consequences, we also recover some of the Hardy type inequalities of the Euclidean space described earlier in the setting of the stratified groups. The presentation of this section is based on [RSS18c].

Throughout this and further sections, let M be a smooth manifold of dimension n equipped with a volume form $d\nu$, and let $\{X_k\}_{k=1}^N$, $N \le n$, be a family of real vector fields.

We start with the following weighted anisotropic Hardy type inequalities in admissible domains in the sense of Definition 12.1.1.

Theorem 12.5.1 (Weighted anisotropic Hardy type inequality). *Let $\Omega \subset M$ be an admissible domain. Let $W_i(x)$, $H_i(x)$ be non-negative functions for $i = 1, \ldots, N$, such that for $v \in C^1(\Omega)\bigcap C(\overline{\Omega})$ satisfying $v > 0$ a.e. in Ω, we have*

$$-X_i(W_i(x)|X_i v|^{p_i-2}X_i v) \ge H_i(x)v^{p_i-1}, \quad i = 1, \ldots, N. \tag{12.35}$$

Then, for all non-negative functions $u \in C^2(\Omega)\bigcap C^1(\overline{\Omega})$ and the positive function $v \in C^1(\Omega)\bigcap C(\overline{\Omega})$ satisfying (12.35), we have

$$\sum_{i=1}^N \int_\Omega W_i(x)|X_i u|^{p_i} d\nu \ge \sum_{i=1}^N \int_\Omega H_i(x)|u|^{p_i} d\nu \tag{12.36}$$

$$+\sum_{i=1}^N \int_{\partial\Omega} \frac{u^{p_i}}{v^{p_i-1}}\langle\widetilde{\nabla}_i\left(W_i(x)|X_i v|^{p_i-2}X_i v\right), d\nu\rangle,$$

where $\widetilde{\nabla}_i f = X_i f X_i$ and $p_i > 1$, for $i = 1, \ldots, N$.

Before proving this inequality let us formulate several of its consequences, recovering and extending a number of known results, see Remark 12.5.3. In these examples of the weighted anisotropic Hardy type inequalities on M we express the weights in terms of the fundamental solution $\Gamma = \Gamma_y(x)$ in the assumption A_y. For brevity, we can just denote it by Γ, if we fix some $y \in M$ and the corresponding T_y and Γ_y.

Corollary 12.5.2 (Anisotropic Hardy inequalities and fundamental solutions). *Let $\Omega \subset M$ be an admissible domain. Then we have the following estimates.*

(1) *Let $\alpha \in \mathbb{R}$, $1 < p_i < \beta + \alpha$, $i = 1, \ldots, N$, and $\gamma > -1, \beta > 2$. Then for all $u \in C_0^\infty(\Omega \backslash \{0\})$ we have*

$$\sum_{i=1}^{N} \int_\Omega \Gamma^{\frac{\alpha}{2-\beta}} |X_i \Gamma^{\frac{1}{2-\beta}}|^\gamma |X_i u|^{p_i} d\nu \geq \sum_{i=1}^{N} \left(\frac{\beta+\alpha-p_i}{p_i}\right)^{p_i} \int_\Omega \Gamma^{\frac{\alpha-p_i}{2-\beta}} |X_i \Gamma^{\frac{1}{2-\beta}}|^{p_i+\gamma} |u|^{p_i} d\nu. \tag{12.37}$$

(2) *Let $\alpha, \gamma \in \mathbb{R}$ and $\alpha \neq 0, \beta > 2$. Then for any $u \in C_0^1(\Omega)$ we have*

$$\sum_{i=1}^{N} \int_\Omega \Gamma^{\frac{\gamma+p_i}{2-\beta}} |X_i u|^{p_i} d\nu \geq \sum_{i=1}^{N} C_i(\alpha, \gamma, p_i)^{p_i} \int_\Omega \Gamma^{\frac{\gamma}{2-\beta}} |X_i \Gamma^{\frac{1}{2-\beta}}|^{p_i} |u|^{p_i} d\nu, \tag{12.38}$$

where $C_i(\alpha, \gamma, p_i) := \frac{(\alpha-1)(p_i-1)-\gamma-1}{p_i}$, $p_i > 1$, and $i = 1, \ldots, N$.

(3) *Let $\alpha \in \mathbb{R}, \beta > 2$, $1 < p_i < \beta + \alpha$ for $i = 1, \ldots, N$. Then for all $u \in C_0^\infty(\Omega)$ we have*

$$\sum_{i=1}^{N} \int_\Omega \Gamma^{\frac{\alpha}{2-\beta}} |X_i u|^{p_i} d\nu \geq \sum_{i=1}^{N} C_i(\beta, \alpha, p_i) \int_\Omega \Gamma^{\frac{\alpha}{2-\beta}} \frac{|X_i \Gamma^{\frac{1}{2-\beta}}|^{p_i}}{\left(1 + \Gamma^{\frac{p_i}{(p_i-1)(2-\beta)}}\right)^{p_i}} |u|^{p_i} d\nu, \tag{12.39}$$

where $C_i(\beta, \alpha, p_i) := \left(\frac{\beta+\alpha-p_i}{p_i-1}\right)^{p_i-1} (\beta+\alpha)$.

(4) *Let $\alpha \in \mathbb{R}, \beta > 2$, $1 < p_i < \beta + \alpha$ for $i = 1, \ldots, N$. Then for all $u \in C_0^\infty(\Omega)$ we have*

$$\sum_{i=1}^{N} \int_\Omega \left(1 + \Gamma^{\frac{p_i}{(p_i-1)(2-\beta)}}\right)^{\alpha(p_i-1)} |X_i u|^{p_i} d\nu \geq \sum_{i=1}^{N} C_i(\beta, p_i, \alpha) \int_\Omega \frac{|X_i \Gamma^{\frac{1}{2-\beta}}|^{p_i}}{\left(1 + \Gamma^{\frac{p_i}{(p_i-1)(2-\beta)}}\right)^{(1-p_i)(1-\alpha)}} |u|^{p_i} d\nu, \tag{12.40}$$

where $C_i(\beta, p_i, \alpha) := \beta \left(\frac{p_i(\alpha-1)}{p_i-1}\right)^{p_i-1}$.

(5) *Let $\beta > 2$, $a, b > 0$ and $\alpha, \gamma, m \in \mathbb{R}$. If $\alpha\gamma > 0$ and $m \leq \frac{\beta-2}{2}$. Then for all $u \in C_0^\infty(\Omega)$ we have*

$$\begin{aligned}&\int_\Omega \frac{(a+b\Gamma^{\frac{\alpha}{2-\beta}})^\gamma}{\Gamma^{\frac{2m}{2-\beta}}}|\nabla_X u|^2 d\nu \\ &\geq C(\beta,m)^2 \int_\Omega \frac{(a+b\Gamma^{\frac{\alpha}{2-\beta}})^\gamma}{\Gamma^{\frac{2m+2}{2-\beta}}}|\nabla_X \Gamma^{\frac{1}{2-\beta}}|^2|u|^2 d\nu \\ &+ C(\beta,m)\alpha\gamma b \int_\Omega \frac{(a+b\Gamma^{\frac{\alpha}{2-\beta}})^{\gamma-1}}{\Gamma^{\frac{2m-\alpha+2}{2-\beta}}}|\nabla_X \Gamma^{\frac{1}{2-\beta}}|^2|u|^2 d\nu, \qquad (12.41)\end{aligned}$$

where $C(\beta,m) := \frac{\beta-2m-2}{2}$ and $\nabla_X = (X_1, \ldots, X_N)$.

Remark 12.5.3.

1. In Theorem 12.5.1, if u vanishes on the boundary $\partial\Omega$ and if $p_i = p$, then we have the two-weighted Hardy type inequalities for general vector fields of the form

$$\int_\Omega W(x)|\nabla_X u|^p d\nu \geq \int_\Omega H(x)|u|^p d\nu, \qquad (12.42)$$

   where $\nabla_X := (X_1, \ldots, X_N)$.

2. Inequality (12.37) is an analogue of the result of Wang and Niu [WN08], but now for general vector fields. Also, by taking $\gamma = 0$ and $p_i = 2$ we have the following inequality

$$\int_\Omega \Gamma^{\frac{\alpha}{2-\beta}}|\nabla_X u|^2 d\nu \geq \sum_{i=1}^N \left(\frac{\beta+\alpha-2}{2}\right)^2 \int_\Omega \Gamma^{\frac{\alpha-2}{2-\beta}}|\nabla_X \Gamma^{\frac{1}{2-\beta}}|^2|u|^2 d\nu, \quad (12.43)$$

   for all $u \in C_0^\infty(\Omega)$ and where $\nabla_X = (X_1, \ldots, X_N)$, which gives (12.21) without the boundary term.

3. Inequality (12.38) recovers the result of D'Ambrosio in [D'A05, Theorem 2.7].
4. A Carnot group version of inequality (12.39) was established by Goldstein, Kombe and Yener in [GKY17].
5. The Carnot and Euclidean versions of inequality (12.40) were established in [GKY17] and [Skr13], respectively.
6. The Carnot and Euclidean versions of inequality (12.41) were established in [GKY17] and [GM11], respectively.

As in the setting of stratified groups let us first present the anisotropic Picone type identity, now for general vector fields.

Lemma 12.5.4 (Anisotropic Picone identity for general vector fields). *Let $\Omega \subset M$ be an open set. Let u, v be differentiable a.e. in Ω, $v > 0$ a.e. in Ω and $u \geq 0$.*

Define

$$R(u,v) := \sum_{i=1}^{N} |X_i u|^{p_i} - \sum_{i=1}^{N} X_i \left(\frac{u^{p_i}}{v^{p_i-1}} \right) |X_i v|^{p_i-2} X_i v, \tag{12.44}$$

$$\begin{aligned} L(u,v) := \sum_{i=1}^{N} |X_i u|^{p_i} &- \sum_{i=1}^{N} p_i \frac{u^{p_i-1}}{v^{p_i-1}} |X_i v|^{p_i-2} X_i v X_i u \\ &+ \sum_{i=1}^{N} (p_i - 1) \frac{u^{p_i}}{v^{p_i}} |X_i v|^{p_i}, \end{aligned} \tag{12.45}$$

where $p_i > 1$, $i = 1, \dots, N$. Then

$$L(u,v) = R(u,v) \geq 0. \tag{12.46}$$

In addition, we have $L(u,v) = 0$ a.e. in Ω if and only if $u = cv$ a.e. in Ω with a positive constant $c > 0$.

Proof of Lemma 12.5.4. A direct calculation yields

$$\begin{aligned} R(u,v) &= \sum_{i=1}^{N} |X_i u|^{p_i} - \sum_{i=1}^{N} X_i \left(\frac{u^{p_i}}{v^{p_i-1}} \right) |X_i v|^{p_i-2} X_i v \\ &= \sum_{i=1}^{N} |X_i u|^{p_i} - \sum_{i=1}^{N} p_i \frac{u^{p_i-1}}{v^{p_i-1}} |X_i v|^{p_i-2} X_i v X_i u + \sum_{i=1}^{N} (p_i - 1) \frac{u^{p_i}}{v^{p_i}} |X_i v|^{p_i} \\ &= L(u,v), \end{aligned}$$

which gives the equality in (12.46). Now we restate $L(u,v)$ in a different form, with the aim to show that $L(u,v) \geq 0$. Thus, we write

$$\begin{aligned} L(u,v) = &\sum_{i=1}^{N} |X_i u|^{p_i} - \sum_{i=1}^{N} p_i \frac{u^{p_i-1}}{v^{p_i-1}} |X_i v|^{p_i-1} |X_i u| + \sum_{i=1}^{N} (p_i - 1) \frac{u^{p_i}}{v^{p_i}} |X_i v|^{p_i} \\ &+ \sum_{i=1}^{N} p_i \frac{u^{p_i-1}}{v^{p_i-1}} |X_i v|^{p_i-2} \left(|X_i v| |X_i u| - X_i v X_i u \right) \\ = &\, S_1 + S_2, \end{aligned}$$

where

$$\begin{aligned} S_1 := &\sum_{i=1}^{N} p_i \left[\frac{1}{p_i} |X_i u|^{p_i} + \frac{p_i - 1}{p_i} \left(\left(\frac{u}{v} |X_i v| \right)^{p_i-1} \right)^{\frac{p_i}{p_i-1}} \right] \\ &- \sum_{i=1}^{N} p_i \frac{u^{p_i-1}}{v^{p_i-1}} |X_i v|^{p_i-1} |X_i u|, \end{aligned}$$

and

$$S_2 := \sum_{i=1}^{N} p_i \frac{u^{p_i-1}}{v^{p_i-1}} |X_i v|^{p_i-2} \left(|X_i v||X_i u| - X_i v X_i u \right).$$

Since

$$|X_i v||X_i u| \geq X_i v X_i u,$$

we have $S_2 \geq 0$. To check that $S_1 \geq 0$ we will use Young's inequality for $a \geq 0$ and $b \geq 0$ stating that

$$ab \leq \frac{a^{p_i}}{p_i} + \frac{b^{q_i}}{q_i}, \tag{12.47}$$

where $p_i > 1, q_i > 1$, and $\frac{1}{p_i} + \frac{1}{q_i} = 1$ for $i = 1, \ldots, N$. The equality in (12.47) holds if and only if $a^{p_i} = b^{q_i}$, i.e., if $a = b^{\frac{1}{p_i-1}}$. By setting

$$a := |X_i u| \text{ and } b := \left(\frac{u}{v} |X_i v| \right)^{p_i-1}$$

in (12.47), we get

$$p_i |X_i u| \left(\frac{u}{v} |X_i v| \right)^{p_i-1} \leq p_i \left[\frac{1}{p_i} |X_i u|^{p_i} + \frac{p_i-1}{p_i} \left(\left(\frac{u}{v} |X_i v| \right)^{p_i-1} \right)^{\frac{p_i}{p_i-1}} \right]. \tag{12.48}$$

This implies $S_1 \geq 0$ which proves that $L(u,v) = S_1 + S_2 \geq 0$.

It is straightforward to see that $u = cv$ implies $R(u,v) = 0$.

Now let us show that $L(u,v) = 0$ implies $u = cv$. Due to $u(x) \geq 0$ and $L(u,v)(x_0) = 0$, $x_0 \in \Omega$, we consider the two cases $u(x_0) > 0$ and $u(x_0) = 0$. For the case $u(x_0) > 0$ we conclude from $L(u,v)(x_0) = 0$ that $S_1 = 0$ and $S_2 = 0$. Then $S_1 = 0$ yields

$$|X_i u| = \frac{u}{v} |X_i v|, \quad i = 1, \ldots, N, \tag{12.49}$$

and $S_2 = 0$ implies

$$|X_i v||X_i u| - X_i v X_i u = 0, \quad i = 1, \ldots, N. \tag{12.50}$$

The combination of (12.49) and (12.50) gives

$$\frac{X_i u}{X_i v} = \frac{u}{v} = c, \quad \text{with} \quad c \neq 0, \quad i = 1, \ldots, N. \tag{12.51}$$

Let us denote

$$\Omega^* := \{x \in \Omega : \ u(x) = 0\}.$$

If $\Omega^* \neq \Omega$, then suppose that $x_0 \in \partial\Omega^*$. So there exists a sequence $x_k \notin \Omega^*$ such that $x_k \to x_0$. In particular, $u(x_k) \neq 0$, and hence by the first case we have $u(x_k) = cv(x_k)$. Passing to the limit we get $u(x_0) = cv(x_0)$. Since $u(x_0) = 0$ and $v(x_0) \neq 0$, we get that $c = 0$. But then by the first case again, since $u = cv$ and $u \neq 0$ in $\Omega \backslash \Omega^*$, it is impossible that $c = 0$. This contradiction implies that $\Omega^* = \Omega$. It completes the proof of Lemma 12.5.4. □

The established anisotropic Picone identity can be used to prove Theorem 12.5.1.

Proof of Theorem 12.5.1. In the following calculation, we will use the following properties: anisotropic Picone type identity (12.46), then we apply the divergence theorem and the hypothesis (12.35), respectively, finally yielding (12.36). Thus, we obtain

$$\begin{aligned}
0 \leq \int_\Omega \sum_{i=1}^N W_i(x) L(u,v) d\nu &= \int_\Omega \sum_{i=1}^N W_i(x) R(u,v) d\nu \\
&= \sum_{i=1}^N \int_\Omega W_i(x) |X_i u|^{p_i} d\nu - \sum_{i=1}^N \int_\Omega X_i \left(\frac{u^{p_i}}{v^{p_i-1}} \right) W_i(x) |X_i v|^{p_i-2} X_i v d\nu \\
&= \sum_{i=1}^N \int_\Omega W_i(x) |X_i u|^{p_i} d\nu + \sum_{i=1}^N \int_\Omega \frac{u^{p_i}}{v^{p_i-1}} X_i \left(W_i(x) |X_i v|^{p_i-2} X_i v \right) d\nu \\
&\quad - \sum_{i=1}^N \int_{\partial\Omega} \frac{u^{p_i}}{v^{p_i-1}} \langle \widetilde{\nabla}_i \left(W_i(x) |X_i v|^{p_i-2} X_i v \right), d\nu \rangle \\
&\leq \sum_{i=1}^N \int_\Omega W_i(x) |X_i u|^{p_i} d\nu - \sum_{i=1}^N \int_\Omega H_i(x) u^{p_i} d\nu \\
&\quad - \sum_{i=1}^N \int_{\partial\Omega} \frac{u^{p_i}}{v^{p_i-1}} \langle \widetilde{\nabla}_i \left(W_i(x) |X_i v|^{p_i-2} X_i v \right), d\nu \rangle,
\end{aligned}$$

where $\widetilde{\nabla}_i f = X_i f X_i$. The proof of Theorem 12.5.1 is complete. □

Finally, we prove Corollary 12.5.2.

Proof of Corollary 12.5.2. Part (1). Consider the functions W_i and v such that

$$W_i = d^\alpha |X_i d|^\gamma \text{ and } v = \Gamma^{\frac{\psi}{2-\beta}} = d^\psi, \tag{12.52}$$

where, to abbreviate the calculation, we denote

$$d := \Gamma^{\frac{1}{2-\beta}} \text{ and } \psi := -\frac{\beta + \alpha - p_i}{p_i}.$$

Now we plug (12.52) in (12.35) to determine the candidate for the function H_i. For this, we first prepare several calculations. We can readily find

$$\begin{aligned}
X_i v &= \psi d^{\psi-1} X_i d, \\
|X_i v|^{p_i-2} &= |\psi|^{p_i-2} d^{(\psi-1)(p_i-2)} |X_i d|^{p_i-2}, \\
W_i |X_i v|^{p_i-2} X_i v &= |\psi|^{p_i-2} \psi d^{\alpha+(\psi-1)(p_i-1)} |X_i d|^{\gamma+p_i-2} X_i d.
\end{aligned}$$

Also, we get

$$\begin{aligned}\sum_{i=1}^{N} X_i^2 d^\alpha &= \sum_{i=1}^{N} X_i(X_i \Gamma^{\frac{\alpha}{2-\beta}}) = \sum_{i=1}^{N} X_i \left(\frac{\alpha}{2-\beta} \Gamma^{\frac{\alpha+\beta-2}{2-\beta}} X_i \Gamma \right) \\ &= \frac{\alpha(\alpha+\beta-2)}{(2-\beta)^2} \Gamma^{\frac{\alpha+2\beta-4}{2-\beta}} \sum_{i=1}^{N} |X_i \Gamma|^2 + \frac{\alpha}{2-\beta} \Gamma^{\frac{\alpha+\beta-2}{2-\beta}} \sum_{i=1}^{N} X_i^2 \Gamma \\ &= \frac{\alpha(\alpha+\beta-2)}{(2-\beta)^2} d^{\alpha+2\beta-4} \sum_{i=1}^{N} |X_i d^{2-\beta}|^2 \\ &= \alpha(\alpha+\beta-2) d^{\alpha-2} \sum_{i=1}^{N} |X_i d|^2. \end{aligned} \tag{12.53}$$

We observe that $\sum_{i=1}^{N} X_i^2 \Gamma = 0$ outside y, since $\Gamma = \Gamma_y$ is the fundamental solution for $\mathcal{L}$. Also, we have

$$\begin{aligned} X_i |X_i d|^\gamma &= X_i((X_i d)^2)^{\gamma/2} = \gamma |X_i d|^{\gamma-2} X_i d X_i^2 d \\ &= \gamma(\beta-1) d^{-1} |X_i d|^\gamma X_i d. \end{aligned} \tag{12.54}$$

In the last line, we have used (12.53) with $\alpha = 1$. Using (12.53) and (12.54), we compute

$$\begin{aligned} & X_i(W_i |X_i v|^{p_i-2} X_i v) \\ &= |\psi|^{p_i-2} \psi X_i \left(d^{\alpha+(\psi-1)(p_i-1)} |X_i d|^{\gamma+p_i-2} X_i d \right) \\ &= |\psi|^{p_i-2} \psi \left((\alpha + (\psi-1)(p_i-1)) d^{\alpha+(\psi-1)(p_i-1)-1} |X_i d|^{\gamma+p_i} \right) \\ &\quad + |\psi|^{p_i-2} \psi \left((\gamma+p_i-2)(\beta-1) d^{\alpha+(\psi-1)(p_i-1)-1} |X_i d|^{\gamma+p_i} \right) \\ &\quad + |\psi|^{p_i-2} \psi \left((\beta-1) d^{\alpha+(\psi-1)(p_i-1)-1} |X_i d|^{\gamma+p_i} \right) \\ &= |\psi|^{p_i-2} \psi \left(-\psi + (\gamma+p_i-2)(\beta-1) \right) d^{\alpha-p_i+\psi(p_i-1)} |X_i d|^{\gamma+p_i} \\ &= -|\psi|^{p_i} d^{\alpha-p_i} |X_i d|^{\gamma+p_i} v^{p_i-1} \\ &\quad + |\psi|^{p_i-2} \psi (\gamma+p_i-2)(\beta-1) d^{\alpha-p_i} |X_i d|^{\gamma+p_i} v^{p_i-1}. \end{aligned}$$

If we put back the value of ψ, we get

$$\begin{aligned} & -X_i(W_i |X_i v|^{p_i-2} X_i v) \\ &= \left| \frac{\beta+\alpha-p_i}{p_i} \right|^{p_i} d^{\alpha-p_i} |X_i d|^{\gamma+p_i} v^{p_i-1} \\ &\quad + \left| \frac{\beta+\alpha-p_i}{p_i} \right|^{p_i-2} \left(\frac{\beta+\alpha-p_i}{p_i} \right) (\gamma+p_i-2)(\beta-1) d^{\alpha-p_i} |X_i d|^{\gamma+p_i} v^{p_i-1} \end{aligned}$$

$$\geq \left|\frac{\beta+\alpha-p_i}{p_i}\right|^{p_i} d^{\alpha-p_i}|X_i d|^{\gamma+p_i} v^{p_i-1} \geq H_i(x)v^{p_i-1},$$

the last inequality being the desired one. So having satisfied the hypothesis, we plug the values of these functions W_i and

$$H_i = \left|\frac{\beta+\alpha-p_i}{p_i}\right|^{p_i} \Gamma^{\frac{\alpha-p_i}{2-\beta}} |X_i \Gamma^{\frac{1}{2-\beta}}|^{\gamma+p_i},$$

in (12.36), which completes the proof of Part (1).

Part (2) can be proved by the same approach as the previous case by considering the functions

$$W_i = \Gamma^{\frac{\gamma+p_i}{2-\beta}} \qquad \text{and} \qquad v = \Gamma^{-\frac{(\alpha-1)(p_i-1)-\gamma-1}{(2-\beta)p_i}}.$$

Part (3) can be proved by the same approach as the previous cases by considering the functions

$$W_i = \Gamma^{\frac{\alpha}{2-\beta}} \qquad \text{and} \qquad v = \left(1+\Gamma^{\frac{p_i}{(p_i-1)(2-\beta)}}\right)^{-\frac{\beta+\alpha-p_i}{p_i}}.$$

Part (4) can be proved by the same approach as the previous case by considering the functions

$$W_i = \left(1+\Gamma^{\frac{p_i}{(p_i-1)(2-\beta)}}\right)^{\alpha(p_i-1)} \qquad \text{and} \qquad v = \left(1+\Gamma^{\frac{p_i}{(p_i-1)(2-\beta)}}\right)^{1-\alpha}.$$

Part (5) can be proved by the same approach for $p_i = 2$, $i = 1, \ldots, N$, as the previous cases by considering the functions

$$W = \frac{(a+b\Gamma^{\frac{\alpha}{2-\beta}})^{\gamma}}{\Gamma^{\frac{2m}{2-\beta}}} \qquad \text{and} \qquad v = \Gamma^{-\frac{\beta-2m-2}{2(2-\beta)}}.$$

This completes the proof of Corollary 12.5.2. □

12.6 Local uncertainty principles

As usual, Hardy inequalities imply uncertainty principles, and we now formulate such consequences of Theorem 12.4.1 and Theorem 12.5.1.

Corollary 12.6.1 (Local uncertainty principles for sums of squares). *Let $y \in M$ be such that (A_y^+) holds with the fundamental solution $\Gamma = \Gamma_y$ in T_y. Let $\Omega \subset T_y$ be an admissible domain and let $\beta > 2$. Then for all $u \in C^1(\Omega)\bigcap C(\overline{\Omega})$ we have*

$$\begin{aligned}\int_\Omega \Gamma^{\frac{2}{2-\beta}} |\nabla_X \Gamma^{\frac{1}{2-\beta}}|^2 |u|^2 d\nu \int_\Omega |\nabla_X u|^2 d\nu \\ \geq \left(\frac{\beta-2}{2}\right)^2 \left(\int_\Omega |\nabla_X \Gamma^{\frac{1}{2-\beta}}|^2 |u|^2 d\nu\right)^2 \\ + \frac{1}{2}\int_{\partial\Omega} \Gamma^{-1}|u|^2 \langle \widetilde{\nabla}\Gamma, d\nu\rangle \int_\Omega \Gamma^{\frac{2}{2-\beta}} |\nabla_X \Gamma^{\frac{1}{2-\beta}}|^2 |u|^2 d\nu,\end{aligned} \tag{12.55}$$

and also

$$\int_\Omega \frac{\Gamma^{\frac{2}{2-\beta}}}{|\nabla_X \Gamma^{\frac{1}{2-\beta}}|^2}|u|^2 d\nu \int_\Omega |\nabla_X u|^2 d\nu \tag{12.56}$$
$$\geq \left(\frac{\beta-2}{2}\right)^2 \left(\int_\Omega |u|^2 d\nu\right)^2 + \frac{1}{2}\int_{\partial\Omega} \Gamma^{-1}|u|^2 \langle \widetilde{\nabla}\Gamma, d\nu\rangle \int_\Omega \frac{\Gamma^{\frac{2}{2-\beta}}}{|\nabla_X \Gamma^{\frac{1}{2-\beta}}|^2}|u|^2 d\nu.$$

Remark 12.6.2.

1. As in Remark 12.4.2, Part 3, the last (boundary) terms in (12.55) and (12.56) can also be positive, thus providing refined uncertainty principles with respect to the boundary conditions.
2. One can readily check that (12.55) extends the classical Hardy inequality. Indeed, in the case of $M = \mathbb{R}^n$ and $X_k = \frac{\partial}{\partial x_k}$, $k = 1, \ldots, n$, taking $\alpha = 0$ and $\beta = n \geq 3$, the fundamental solution for the Laplacian is given by $\Gamma(x) = C_n |x|^{2-n}$ for some constant C_n and $|x|_E$ being the Euclidean norm, so that (12.55) reduces to the classical Hardy inequality
$$\int_{\mathbb{R}^n} |\nabla u(x)|^2 dx \geq \left(\frac{n-2}{2}\right)^2 \int_{\mathbb{R}^n} \frac{|u(x)|^2}{|x|_E^2} dx, \quad n \geq 3, \tag{12.57}$$
where ∇ is the standard gradient in $\mathbb{R}^n$, $u \in C_0^\infty(\mathbb{R}^n \backslash \{0\})$, and the constant $\left(\frac{n-2}{2}\right)^2$ is known to be sharp. The constant C_n does not enter (12.57) due to the scaling invariance of the inequality (12.55) with respect to the multiplication of Γ by positive constants.
3. Further to the Euclidean example (12.57), with $\Gamma^{\frac{1}{2-\beta}}(x) = C|x|_E$ we have $|\nabla \Gamma^{\frac{1}{2-\beta}}| = C$, and hence both (12.55) and (12.56) reduce to the classical uncertainty principle for $\Omega \subset \mathbb{R}^n$ if $u = 0$ on $\partial\Omega$ (for example, for $u \in C_0^\infty(\Omega)$):
$$\int_\Omega |x|_E^2 |u(x)|^2 dx \int_\Omega |\nabla u(x)|^2 dx \geq \left(\frac{n-2}{2}\right)^2 \left(\int_\Omega |u(x)|^2 dx\right)^2, \quad n \geq 3.$$
4. In the example of stratified Lie groups with $\beta = Q \geq 3$ being the homogeneous dimension of the group, and $\Gamma^{\frac{1}{2-\beta}}(x) = d(x)$ being the $\mathcal{L}$-gauge, inequality (12.55) reduces to
$$\int_\Omega d^2 |\nabla_X d|^2 |u|^2 d\nu \int_\Omega |\nabla_X u|^2 d\nu$$
$$\geq \left(\frac{Q-2}{2}\right)^2 \left(\int_\Omega |\nabla_X d|^2 |u|^2 d\nu\right)^2$$
$$+\frac{1}{2}\int_{\partial\Omega} d^{Q-2}|u|^2 \langle \widetilde{\nabla} d^{2-Q}, d\nu\rangle \int_\Omega d^2 |\nabla_X d|^2 |u|^2 d\nu,$$
which gives inequality (12.24).

Proof of Corollary 12.6.1. Taking $\alpha = 0$ in inequality (12.21) we get

$$\begin{aligned}
&\int_\Omega \Gamma^{\frac{2}{2-\beta}} |\nabla_X \Gamma^{\frac{1}{2-\beta}}|^2 |u|^2 d\nu \int_\Omega |\nabla_X u|^2 d\nu \\
&\geq \left(\frac{\beta-2}{2}\right)^2 \int_\Omega \Gamma^{\frac{2}{2-\beta}} |\nabla_X \Gamma^{\frac{1}{2-\beta}}|^2 |u|^2 d\nu \int_\Omega \frac{|\nabla_X \Gamma^{\frac{1}{2-\beta}}|^2}{\Gamma^{\frac{2}{2-\beta}}} |u|^2\, d\nu \\
&\quad + \frac{1}{2} \int_{\partial\Omega} \Gamma^{-1} |u|^2 \langle \widetilde{\nabla}\Gamma, d\nu\rangle \int_\Omega \Gamma^{\frac{2}{2-\beta}} |\nabla_X \Gamma^{\frac{1}{2-\beta}}|^2 |u|^2 d\nu \\
&\geq \left(\frac{\beta-2}{2}\right)^2 \left(\int_\Omega |\nabla_X \Gamma^{\frac{1}{2-\beta}}|^2 |u|^2 d\nu\right)^2 \\
&\quad + \frac{1}{2} \int_{\partial\Omega} \Gamma^{-1} |u|^2 \langle \widetilde{\nabla}\Gamma, d\nu\rangle \int_\Omega \Gamma^{\frac{2}{2-\beta}} |\nabla_X \Gamma^{\frac{1}{2-\beta}}|^2 |u|^2 d\nu,
\end{aligned}$$

where we have used the Hölder inequality in the last line. This shows (12.55). The proof of (12.56) is similar. □

Inequality (12.22) also implies the following refinement of Corollary 12.6.1.

Corollary 12.6.3 (Refined local uncertainty principles for sums of squares). *Let $y \in M$ be such that (A_y^+) holds with the fundamental solution $\Gamma = \Gamma_y$ in T_y. Let $\Omega \subset T_y$, $y \notin \partial\Omega$, be an admissible domain and let $\beta > 2$. Then for all $u \in C^1(\Omega) \bigcap C(\overline{\Omega})$ we have*

$$\begin{aligned}
&\int_\Omega \Gamma^{\frac{2}{2-\beta}} |\nabla_X \Gamma^{\frac{1}{2-\beta}}|^2 |u|^2 d\nu \int_\Omega |\nabla_X u|^2 d\nu \\
&\geq \left(\frac{\beta-2}{2}\right)^2 \left(\int_\Omega |\nabla_X \Gamma^{\frac{1}{2-\beta}}|^2 |u|^2 d\nu\right)^2 \\
&\quad + \frac{1}{4} \int_\Omega \frac{|\nabla_X \Gamma^{\frac{1}{2-\beta}}|^2}{\Gamma^{\frac{2}{2-\beta}}} \left(\ln \frac{R}{\Gamma^{\frac{1}{2-\beta}}}\right)^{-2} |u|^2 d\nu \int_\Omega \Gamma^{\frac{2}{2-\beta}} |\nabla_X \Gamma^{\frac{1}{2-\beta}}|^2 |u|^2 d\nu \\
&\quad + \frac{1}{2(\beta-2)} \int_{\partial\Omega} \Gamma^{-1} \left(\ln \frac{R}{\Gamma^{\frac{1}{2-\beta}}}\right)^{-1} |u|^2 \langle \widetilde{\nabla}\Gamma, d\nu\rangle \int_\Omega \Gamma^{\frac{2}{2-\beta}} |\nabla_X \Gamma^{\frac{1}{2-\beta}}|^2 |u|^2 d\nu \\
&\quad + \frac{1}{2} \int_{\partial\Omega} \Gamma^{-1} |u|^2 \langle \widetilde{\nabla}\Gamma, d\nu\rangle \int_\Omega \Gamma^{\frac{2}{2-\beta}} |\nabla_X \Gamma^{\frac{1}{2-\beta}}|^2 |u|^2 d\nu, \qquad (12.58)
\end{aligned}$$

and also

$$\begin{aligned}
&\int_\Omega \frac{\Gamma^{\frac{2}{2-\beta}}}{|\nabla_X \Gamma^{\frac{1}{2-\beta}}|^2} |u|^2 d\nu \int_\Omega |\nabla_X u|^2 d\nu \\
&\geq \left(\frac{\beta-2}{2}\right)^2 \left(\int_\Omega |u|^2 d\nu\right)^2 \\
&\quad + \frac{1}{4} \int_\Omega \frac{|\nabla_X \Gamma^{\frac{1}{2-\beta}}|^2}{\Gamma^{\frac{2}{2-\beta}}} \left(\ln \frac{R}{\Gamma^{\frac{1}{2-\beta}}}\right)^{-2} |u|^2 d\nu \int_\Omega \frac{\Gamma^{\frac{2}{2-\beta}}}{|\nabla_X \Gamma^{\frac{1}{2-\beta}}|^2} |u|^2 d\nu
\end{aligned}$$

$$+\frac{1}{2(\beta-2)}\int_{\partial\Omega}\Gamma^{-1}\left(\ln\frac{R}{\Gamma^{\frac{1}{2-\beta}}}\right)^{-1}|u|^2\langle\widetilde{\nabla}\Gamma,d\nu\rangle\int_\Omega\frac{\Gamma^{\frac{2}{2-\beta}}}{|\nabla_X\Gamma^{\frac{1}{2-\beta}}|^2}|u|^2d\nu$$
$$+\frac{1}{2}\int_{\partial\Omega}\Gamma^{-1}|u|^2\langle\widetilde{\nabla}\Gamma,d\nu\rangle\int_\Omega\frac{\Gamma^{\frac{2}{2-\beta}}}{|\nabla_X\Gamma^{\frac{1}{2-\beta}}|^2}|u|^2d\nu. \tag{12.59}$$

Proof of Corollary 12.6.3. Taking $\alpha=0$ in inequality (12.22) we get

$$\int_\Omega\Gamma^{\frac{2}{2-\beta}}|\nabla_X\Gamma^{\frac{1}{2-\beta}}|^2|u|^2d\nu\int_\Omega|\nabla_Xu|^2d\nu$$
$$\geq\left(\frac{\beta-2}{2}\right)^2\int_\Omega\Gamma^{\frac{2}{2-\beta}}|\nabla_X\Gamma^{\frac{1}{2-\beta}}|^2|u|^2d\nu\int_\Omega\frac{|\nabla_X\Gamma^{\frac{1}{2-\beta}}|^2}{\Gamma^{\frac{2}{2-\beta}}}|u|^2\,d\nu$$
$$+\frac{1}{4}\int_\Omega\frac{|\nabla_X\Gamma^{\frac{1}{2-\beta}}|^2}{\Gamma^{\frac{2}{2-\beta}}}\left(\ln\frac{R}{\Gamma^{\frac{1}{2-\beta}}}\right)^{-2}|u|^2d\nu\int_\Omega\Gamma^{\frac{2}{2-\beta}}|\nabla_X\Gamma^{\frac{1}{2-\beta}}|^2|u|^2d\nu$$
$$+\frac{1}{2(\beta-2)}\int_{\partial\Omega}\Gamma^{-1}\left(\ln\frac{R}{\Gamma^{\frac{1}{2-\beta}}}\right)^{-1}|u|^2\langle\widetilde{\nabla}\Gamma,d\nu\rangle\int_\Omega\Gamma^{\frac{2}{2-\beta}}|\nabla_X\Gamma^{\frac{1}{2-\beta}}|^2|u|^2d\nu$$
$$+\frac{1}{2}\int_{\partial\Omega}\Gamma^{-1}|u|^2\langle\widetilde{\nabla}\Gamma,d\nu\rangle\int_\Omega\Gamma^{\frac{2}{2-\beta}}|\nabla_X\Gamma^{\frac{1}{2-\beta}}|^2|u|^2d\nu$$
$$\geq\left(\frac{\beta-2}{2}\right)^2\left(\int_\Omega|\nabla_X\Gamma^{\frac{1}{2-\beta}}|^2|u|^2d\nu\right)^2$$
$$+\frac{1}{4}\int_\Omega\frac{|\nabla_X\Gamma^{\frac{1}{2-\beta}}|^2}{\Gamma^{\frac{2}{2-\beta}}}\left(\ln\frac{R}{\Gamma^{\frac{1}{2-\beta}}}\right)^{-2}|u|^2d\nu\int_\Omega\Gamma^{\frac{2}{2-\beta}}|\nabla_X\Gamma^{\frac{1}{2-\beta}}|^2|u|^2d\nu$$
$$+\frac{1}{2(\beta-2)}\int_{\partial\Omega}\Gamma^{-1}\left(\ln\frac{R}{\Gamma^{\frac{1}{2-\beta}}}\right)^{-1}|u|^2\langle\widetilde{\nabla}\Gamma,d\nu\rangle\int_\Omega\Gamma^{\frac{2}{2-\beta}}|\nabla_X\Gamma^{\frac{1}{2-\beta}}|^2|u|^2d\nu$$
$$+\frac{1}{2}\int_{\partial\Omega}\Gamma^{-1}|u|^2\langle\widetilde{\nabla}\Gamma,d\nu\rangle\int_\Omega\Gamma^{\frac{2}{2-\beta}}|\nabla_X\Gamma^{\frac{1}{2-\beta}}|^2|u|^2d\nu,$$

where we have used the Hölder inequality. This shows (12.58). The proof of (12.59) is similar. □

Remark 12.6.4.

1. In the Euclidean case $M=\mathbb{R}^n$ with $\beta=n\geq 3$, we have $\Gamma^{\frac{1}{2-\beta}}(x)=C|x|_E$ is a constant multiple of the Euclidean distance, so that $|\nabla\Gamma^{\frac{1}{2-\beta}}|=C$. Consequently both (12.58) and (12.59) reduce to the improved uncertainty principle for $\Omega\subset\mathbb{R}^n$ if $u=0$ on $\partial\Omega$ (for example, usually one takes $u\in C_0^\infty(\Omega)$):

$$\int_\Omega|x|^2|u(x)|^2dx\int_\Omega|\nabla u(x)|^2dx$$
$$\geq\left(\frac{n-2}{2}\right)^2\left(\int_\Omega|u(x)|^2dx\right)^2$$
$$+\frac{1}{4}\int_\Omega\frac{1}{|x|^2}\left(\ln\frac{R}{|x|}\right)^{-2}|u(x)|^2d\nu\int_\Omega|x|^2|u(x)|^2d\nu,\quad n\geq 3.$$

2. In the Example of stratified Lie groups with $\beta = Q \geq 3$ being the homogeneous dimension of the group $\mathbb{G}$, and $\Gamma^{\frac{1}{2-\beta}}(x) = d(x)$ being the $\mathcal{L}$-gauge, inequality (12.58) reduces to

$$\begin{aligned}
&\int_\Omega d^2|\nabla_X d|^2|u|^2 d\nu \int_\Omega |\nabla_X u|^2 d\nu \\
&\geq \left(\frac{Q-2}{2}\right)^2 \left(\int_\Omega |\nabla_X d|^2|u|^2 d\nu\right)^2 \\
&+ \frac{1}{4}\int_\Omega \frac{|\nabla_X d|^2}{d^2}\left(\ln\frac{R}{d}\right)^{-2}|u|^2 d\nu \int_\Omega d^2|\nabla_X d|^2|u|^2 d\nu \\
&+ \frac{1}{2(Q-2)}\int_{\partial\Omega} d^{Q-2}\left(\ln\frac{R}{d}\right)^{-1}|u|^2\langle\widetilde{\nabla} d^{2-Q}, d\nu\rangle \int_\Omega d^2|\nabla_X d|^2|u|^2 d\nu \\
&+ \frac{1}{2}\int_{\partial\Omega} d^{Q-2}|u|^2\langle\widetilde{\nabla} d^{2-Q}, d\nu\rangle \int_\Omega d^2|\nabla_X d|^2|u|^2 d\nu.
\end{aligned}$$

Again, if $u \in C_0^\infty(\mathbb{G})$, the last terms disappear, and one obtains the improved uncertainty principle on stratified Lie groups compared to the statement of Corollary 11.4.3.

Theorem 12.5.1 also implies the following uncertainty principles:

Corollary 12.6.5 (Further uncertainty inequalities). *Let $\Omega \subset M$ be an admissible domain. Let $\beta > 2$. Then we have the following uncertainty inequalities:*

(1) *For all $u \in C_0^\infty(\Omega)$ we have*

$$\frac{\beta^2}{4}\left(\int_\Omega |u|^2 d\nu\right)^2 \leq \left(\int_\Omega |\nabla_X \Gamma^{\frac{1}{2-\beta}}|^{-2}|\nabla_X u|^2 d\nu\right)\left(\int_\Omega \Gamma^{\frac{2}{2-\beta}}|u|^2 d\nu\right). \tag{12.60}$$

(2) *For all $u \in C_0^\infty(\Omega)$ we have*

$$\left(\int_\Omega |\nabla_X u|^2 d\nu\right)\left(\int_\Omega \Gamma^{\frac{2}{2-\beta}}|\nabla_X \Gamma^{\frac{1}{2-\beta}}|^2|u|^2 d\nu\right) \geq \frac{\beta^2}{4}\left(\int_\Omega |\nabla_X \Gamma^{\frac{1}{2-\beta}}|^2|u|^2 d\nu\right)^2. \tag{12.61}$$

(3) *For all $u \in C_0^\infty(\Omega)$ we have*

$$\begin{aligned}
&\left(\int_\Omega |\nabla_X u|^2 d\nu\right)\left(\int_\Omega \Gamma^{\frac{2}{2-\beta}}|\nabla_X \Gamma^{\frac{1}{2-\beta}}|^2|u|^2 d\nu\right) \\
&\geq \frac{(\beta-1)^2}{4}\left(\int_\Omega \Gamma^{-\frac{1}{2-\beta}}|\nabla_X \Gamma^{\frac{1}{2-\beta}}|^2|u|^2 d\nu\right)^2.
\end{aligned} \tag{12.62}$$

The Carnot group versions of these uncertainty principles in were established in [Kom10] and [GKY17], and in our proof we follow [RSS18c].

Proof of Corollary 12.6.5. Part (1). In Theorem 12.5.1, by letting

$$W(x) = |\nabla_X \Gamma^{\frac{1}{2-\beta}}|^{-2} \quad \text{and} \quad v = e^{-\alpha\Gamma^{\frac{2}{2-\beta}}}$$

with $\alpha \in \mathbb{R}$, we obtain the inequality

$$-4\alpha^2 \int_\Omega \Gamma^{\frac{2}{2-\beta}} |u|^2 d\nu + 2\alpha\beta \int_\Omega |u|^2 d\nu - \int_\Omega |\nabla_X \Gamma^{\frac{1}{2-\beta}}|^{-2} |\nabla_X u|^2 d\nu \leq 0.$$

This inequality is of the form $a\alpha^2 + b\alpha + c \leq 0$, if we denote by

$$a := -4 \int_\Omega \Gamma^{\frac{2}{2-\beta}} |u|^2 d\nu, \qquad b := 2\beta \int_\Omega |u|^2 d\nu,$$

and

$$c := - \int_\Omega |\nabla_X \Gamma^{\frac{1}{2-\beta}}|^{-2} |\nabla_X u|^2 d\nu.$$

Thus, we must have $b^2 - 4ac \leq 0$ which proves (12.60).

Part (2). Setting

$$W = 1 \text{ and } v = e^{-\alpha\Gamma^{\frac{2}{2-\beta}}}$$

with $\alpha \in \mathbb{R}$, we obtain

$$\int_\Omega |\nabla_X u|^2 d\nu \geq 2\alpha\beta \int_\Omega |\nabla_X \Gamma^{\frac{1}{2-\beta}}|^2 |u|^2 d\nu - 4\alpha^2 \int_\Omega \Gamma^{\frac{2}{2-\beta}} |\nabla_X \Gamma^{\frac{1}{2-\beta}}|^2 |u|^2 d\nu.$$

Using the same technique as in Part (1) we prove (12.61).

We can prove Part (3) by the same approach, considering the pair

$$W = 1 \text{ and } v = e^{-\alpha\Gamma^{\frac{1}{2-\beta}}}.$$

The proof is complete. □

12.7 Local Rellich inequalities

In this section we present local refined versions of Rellich inequalities with additional boundary terms on the right-hand side, in the way analogous to Hardy inequalities and uncertainty principles in the previous sections. As before, we use the notation

$$\nabla_X = (X_1, \ldots, X_N).$$

Theorem 12.7.1 (Local Rellich inequalities for sums of squares). *Let $y \in M$ be such that* (A_y^+) *holds with the fundamental solution $\Gamma = \Gamma_y$ in T_y. Let $\Omega \subset T_y$ be* *a strongly admissible domain such that $y \notin \partial\Omega$. Let $\alpha \in \mathbb{R}$, $\beta > \alpha > 4-\beta$, $\beta > 2$ and $R \geq e\sup_\Omega \Gamma^{\frac{1}{2-\beta}}$. Then for all $u \in C^2(\Omega)\bigcap C^1(\overline{\Omega})$ we have*

$$\begin{aligned}\int_\Omega \frac{\Gamma^{\frac{\alpha}{2-\beta}}}{|\nabla_X \Gamma^{\frac{1}{2-\beta}}|^2}|\mathcal{L}u|^2 d\nu \geq{}& \frac{(\beta+\alpha-4)^2(\beta-\alpha)^2}{16}\int_\Omega \Gamma^{\frac{\alpha-4}{2-\beta}}|\nabla_X \Gamma^{\frac{1}{2-\beta}}|^2|u|^2\,d\nu \\ &+ \frac{(\beta+\alpha-4)^2(\beta-\alpha)}{4(\beta-2)}\int_{\partial\Omega}\Gamma^{\frac{\alpha-2}{2-\beta}-1}|u|^2\langle\widetilde{\nabla}\Gamma, d\nu\rangle \\ &+ \frac{(\beta+\alpha-4)(\beta-\alpha)}{4}\mathcal{C}(u), \end{aligned} \tag{12.63}$$

as well as its further refinement

$$\begin{aligned}&\int_\Omega \frac{\Gamma^{\frac{\alpha}{2-\beta}}}{|\nabla_X \Gamma^{\frac{1}{2-\beta}}|^2}|\mathcal{L}u|^2 d\nu \\ &\geq \frac{(\beta+\alpha-4)^2(\beta-\alpha)^2}{16}\int_\Omega \Gamma^{\frac{\alpha-4}{2-\beta}}|\nabla_X \Gamma^{\frac{1}{2-\beta}}|^2|u|^2\,d\nu \\ &+ \frac{(\beta+\alpha-4)(\beta-\alpha)}{8}\int_\Omega \Gamma^{\frac{\alpha-4}{2-\beta}}|\nabla_X \Gamma^{\frac{1}{2-\beta}}|^2\left(\ln\frac{R}{\Gamma^{\frac{1}{2-\beta}}}\right)^{-2}|u|^2 d\nu \\ &+ \frac{(\beta+\alpha-4)(\beta-\alpha)}{4(\beta-2)}\int_{\partial\Omega}\Gamma^{\frac{\alpha-2}{2-\beta}-1}\left(\ln\frac{R}{\Gamma^{\frac{1}{2-\beta}}}\right)^{-1}|u|^2\langle\widetilde{\nabla}\Gamma, d\nu\rangle \\ &+ \frac{(\beta+\alpha-4)^2(\beta-\alpha)}{4(\beta-2)}\int_{\partial\Omega}\Gamma^{\frac{\alpha-2}{2-\beta}-1}|u|^2\langle\widetilde{\nabla}\Gamma, d\nu\rangle \\ &+ \frac{(\beta+\alpha-4)(\beta-\alpha)}{4}\mathcal{C}(u), \end{aligned} \tag{12.64}$$

where

$$\mathcal{C}(u) := \frac{\alpha-2}{2-\beta}\int_{\partial\Omega} u^2\Gamma^{\frac{\alpha-2}{2-\beta}-1}\langle\widetilde{\nabla}\Gamma, d\nu\rangle - 2\int_{\partial\Omega}\Gamma^{\frac{\alpha-2}{2-\beta}}u\langle\widetilde{\nabla}u, d\nu\rangle.$$

Proof of Theorem 12.7.1. Let us prove (12.63) first. A direct calculation shows that

$$\begin{aligned}\mathcal{L}\Gamma^{\frac{\alpha-2}{2-\beta}} &= \sum_{k=1}^N X_k^2\Gamma^{\frac{\alpha-2}{2-\beta}} = (\alpha-2)\sum_{k=1}^N X_k\left(\Gamma^{\frac{\alpha-3}{2-\beta}}X_k\Gamma^{\frac{1}{2-\beta}}\right) \\ &= (\alpha-2)(\alpha-3)\Gamma^{\frac{\alpha-4}{2-\beta}}\sum_{k=1}^N\left|X_k\Gamma^{\frac{1}{2-\beta}}\right|^2 + (\alpha-2)\Gamma^{\frac{\alpha-3}{2-\beta}}\sum_{k=1}^N X_k\left(X_k\Gamma^{\frac{1}{2-\beta}}\right) \\ &= (\alpha-2)(\alpha-3)\Gamma^{\frac{\alpha-4}{2-\beta}}\sum_{k=1}^N\left|X_k\Gamma^{\frac{1}{2-\beta}}\right|^2 + \frac{\alpha-2}{2-\beta}\Gamma^{\frac{\alpha-3}{2-\beta}}\sum_{k=1}^N X_k\left(\Gamma^{\frac{\beta-1}{2-\beta}}X_k\Gamma\right)\end{aligned}$$

$$
\begin{aligned}
&= (\alpha-2)(\alpha-3)\Gamma^{\frac{\alpha-4}{2-\beta}}\sum_{k=1}^{N}\left|X_k\Gamma^{\frac{1}{2-\beta}}\right|^2 \\
&\quad+\frac{(\alpha-2)(\beta-1)}{2-\beta}\Gamma^{\frac{\alpha-3}{2-\beta}}\Gamma^{-1}\sum_{k=1}^{N}(X_k\Gamma^{\frac{1}{2-\beta}})(X_k\Gamma) \\
&\quad+\frac{\alpha-2}{2-\beta}\Gamma^{\frac{\beta+\alpha-4}{2-\beta}}\mathcal{L}\Gamma = (\alpha-2)(\alpha-3)\Gamma^{\frac{\alpha-4}{2-\beta}}\sum_{k=1}^{N}\left|X_k\Gamma^{\frac{1}{2-\beta}}\right|^2 \\
&\quad+(\alpha-2)(\beta-1)\Gamma^{\frac{\alpha-4}{2-\beta}}\sum_{k=1}^{N}(X_k\Gamma^{\frac{1}{2-\beta}})(X_k\Gamma^{\frac{1}{2-\beta}})+\frac{\alpha-2}{2-\beta}\Gamma^{\frac{\beta+\alpha-4}{2-\beta}}\mathcal{L}\Gamma \\
&= (\beta+\alpha-4)(\alpha-2)\Gamma^{\frac{\alpha-4}{2-\beta}}|\nabla_X\Gamma^{\frac{1}{2-\beta}}|^2+\frac{\alpha-2}{2-\beta}\Gamma^{\frac{\beta+\alpha-4}{2-\beta}}\mathcal{L}\Gamma,
\end{aligned}
$$

that is, we have

$$
\mathcal{L}\Gamma^{\frac{\alpha-2}{2-\beta}} = (\beta+\alpha-4)(\alpha-2)\Gamma^{\frac{\alpha-4}{2-\beta}}|\nabla_X\Gamma^{\frac{1}{2-\beta}}|^2+\frac{\alpha-2}{2-\beta}\Gamma^{\frac{\beta+\alpha-4}{2-\beta}}\mathcal{L}\Gamma. \tag{12.65}
$$

As in the proof of Theorem 12.4.1 we can assume that u is real-valued. Multiplying both sides of (12.65) by u^2 and integrating over Ω, since Γ is the fundamental solution of $\mathcal{L}$ and $\beta+\alpha-4>0$, we obtain

$$
\int_\Omega u^2\mathcal{L}\Gamma^{\frac{\alpha-2}{2-\beta}}\,d\nu = (\beta+\alpha-4)(\alpha-2)\int_\Omega \Gamma^{\frac{\alpha-4}{2-\beta}}|\nabla_X\Gamma^{\frac{1}{2-\beta}}|^2u^2\,d\nu. \tag{12.66}
$$

On the other hand, by using Green's second formula (12.15) we have

$$
\begin{aligned}
\int_\Omega u^2\mathcal{L}\Gamma^{\frac{\alpha-2}{2-\beta}}\,d\nu &= \int_\Omega \Gamma^{\frac{\alpha-2}{2-\beta}}\mathcal{L}u^2\,d\nu+\int_{\partial\Omega}u^2\langle\widetilde{\nabla}\Gamma^{\frac{\alpha-2}{2-\beta}},d\nu\rangle-\int_{\partial\Omega}\Gamma^{\frac{\alpha-2}{2-\beta}}\langle\widetilde{\nabla}u^2,d\nu\rangle \\
&= \int_\Omega \Gamma^{\frac{\alpha-2}{2-\beta}}(2u\mathcal{L}u+2|\nabla_Xu|^2)\,d\nu+\mathcal{C}(u),
\end{aligned} \tag{12.67}
$$

where

$$
\mathcal{C}(u) := \frac{\alpha-2}{2-\beta}\int_{\partial\Omega}u^2\Gamma^{\frac{\alpha-2}{2-\beta}-1}\langle\widetilde{\nabla}\Gamma,d\nu\rangle-\int_{\partial\Omega}2\Gamma^{\frac{\alpha-2}{2-\beta}}u\langle\widetilde{\nabla}u,d\nu\rangle.
$$

Combining (12.66) and (12.67) we obtain

$$
\begin{aligned}
-2\int_\Omega \Gamma^{\frac{\alpha-2}{2-\beta}}u\mathcal{L}u d\nu+(\beta+\alpha-4)(\alpha-2)\int_\Omega \Gamma^{\frac{\alpha-4}{2-\beta}}|\nabla_X\Gamma^{\frac{1}{2-\beta}}|^2\,u^2 d\nu& \\
= 2\int_\Omega \Gamma^{\frac{\alpha-2}{2-\beta}}|\nabla_Xu|^2 d\nu+\mathcal{C}(u).&
\end{aligned} \tag{12.68}
$$

By using (12.21) we have

$$
\begin{aligned}
&-2\int_{\Omega}\Gamma^{\frac{\alpha-2}{2-\beta}}u\mathcal{L}ud\nu+(\beta+\alpha-4)(\alpha-2)\int_{\Omega}\Gamma^{\frac{\alpha-4}{2-\beta}}|\nabla_X\Gamma^{\frac{1}{2-\beta}}|^2\,|u|^2d\nu\\
&\geq 2\left(\frac{\beta+\alpha-4}{2}\right)^2\int_{\Omega}\Gamma^{\frac{\alpha-4}{2-\beta}}|\nabla_X\Gamma^{\frac{1}{2-\beta}}|^2|u|^2\,d\nu\\
&\quad+\frac{\beta+\alpha-4}{\beta-2}\int_{\partial\Omega}\Gamma^{\frac{\alpha-2}{2-\beta}-1}|u|^2\langle\widetilde{\nabla}\Gamma,d\nu\rangle+\mathcal{C}(u).
\end{aligned}
$$

It follows that

$$
\begin{aligned}
-\int_{\Omega}\Gamma^{\frac{\alpha-2}{2-\beta}}u\mathcal{L}ud\nu\geq&\left(\frac{\beta+\alpha-4}{2}\right)\left(\frac{\beta-\alpha}{2}\right)\int_{\Omega}\Gamma^{\frac{\alpha-4}{2-\beta}}|\nabla_X\Gamma^{\frac{1}{2-\beta}}|^2|u|^2\,d\nu\\
&+\frac{\beta+\alpha-4}{2(\beta-2)}\int_{\partial\Omega}\Gamma^{\frac{\alpha-2}{2-\beta}-1}|u|^2\langle\widetilde{\nabla}\Gamma,d\nu\rangle+\frac{1}{2}\mathcal{C}(u).
\end{aligned}
\tag{12.69}
$$

On the other hand, for any $\epsilon>0$, Hölder's and Young's inequalities give

$$
\begin{aligned}
-\int_{\Omega}\Gamma^{\frac{\alpha-2}{2-\beta}}u\mathcal{L}ud\nu&\leq\left(\int_{\Omega}\Gamma^{\frac{\alpha-4}{2-\beta}}|\nabla_X\Gamma^{\frac{1}{2-\beta}}|^2|u|^2d\nu\right)^{1/2}\left(\int_{\Omega}\frac{\Gamma^{\frac{\alpha}{2-\beta}}}{|\nabla_X\Gamma^{\frac{1}{2-\beta}}|^2}|\mathcal{L}u|^2d\nu\right)^{1/2}\\
&\leq\epsilon\int_{\Omega}\Gamma^{\frac{\alpha-4}{2-\beta}}|\nabla_X\Gamma^{\frac{1}{2-\beta}}|^2|u|^2d\nu+\frac{1}{4\epsilon}\int_{\Omega}\frac{\Gamma^{\frac{\alpha}{2-\beta}}}{|\nabla_X\Gamma^{\frac{1}{2-\beta}}|^2}|\mathcal{L}u|^2d\nu.
\end{aligned}
\tag{12.70}
$$

Inequalities (12.70) and (12.69) imply that

$$
\begin{aligned}
\int_{\Omega}\frac{\Gamma^{\frac{\alpha}{2-\beta}}}{|\nabla_X\Gamma^{\frac{1}{2-\beta}}|^2}|\mathcal{L}u|^2d\nu\geq&\left(-4\epsilon^2+(\beta+\alpha-4)(\beta-\alpha)\epsilon\right)\int_{\Omega}\Gamma^{\frac{\alpha-4}{2-\beta}}|\nabla_X\Gamma^{\frac{1}{2-\beta}}|^2|u|^2\,d\nu\\
&+\frac{2(\beta+\alpha-4)\epsilon}{\beta-2}\int_{\partial\Omega}\Gamma^{\frac{\alpha-2}{2-\beta}-1}|u|^2\langle\widetilde{\nabla}\Gamma,d\nu\rangle+2\epsilon\mathcal{C}(u).
\end{aligned}
$$

Taking $\epsilon=\frac{(\beta+\alpha-4)(\beta-\alpha)}{8}$, we obtain

$$
\begin{aligned}
&\int_{\Omega}\frac{\Gamma^{\frac{\alpha}{2-\beta}}}{|\nabla_X\Gamma^{\frac{1}{2-\beta}}|^2}|\mathcal{L}u|^2d\nu\\
&\quad\geq\frac{(\beta+\alpha-4)^2(\beta-\alpha)^2}{16}\int_{\Omega}\Gamma^{\frac{\alpha-4}{2-\beta}}|\nabla_X\Gamma^{\frac{1}{2-\beta}}|^2|u|^2\,d\nu\\
&\quad\quad+\frac{(\beta+\alpha-4)^2(\beta-\alpha)}{4(\beta-2)}\int_{\partial\Omega}\Gamma^{\frac{\alpha-2}{2-\beta}-1}|u|^2\langle\widetilde{\nabla}\Gamma,d\nu\rangle+\frac{(\beta+\alpha-4)(\beta-\alpha)}{4}\mathcal{C}(u),
\end{aligned}
$$

which proves (12.63).

Let us now prove (12.64). From (12.68), if we use (12.22), we get

$$\begin{aligned}
&-2\int_\Omega \Gamma^{\frac{\alpha-2}{2-\beta}} u\mathcal{L}u d\nu + (\beta+\alpha-4)(\alpha-2)\int_\Omega \Gamma^{\frac{\alpha-4}{2-\beta}} |\nabla_X \Gamma^{\frac{1}{2-\beta}}|^2 |u|^2 d\nu \\
&\geq 2\left(\frac{\beta+\alpha-4}{2}\right)^2 \int_\Omega \Gamma^{\frac{\alpha-4}{2-\beta}} |\nabla_X \Gamma^{\frac{1}{2-\beta}}|^2 |u|^2\, d\nu \\
&\quad + \frac{1}{2}\int_\Omega \Gamma^{\frac{\alpha-4}{2-\beta}} |\nabla_X \Gamma^{\frac{1}{2-\beta}}|^2 \left(\ln \frac{R}{\Gamma^{\frac{1}{2-\beta}}}\right)^{-2} |u|^2 d\nu \\
&\quad + \frac{1}{(\beta-2)}\int_{\partial\Omega} \Gamma^{\frac{\alpha-2}{2-\beta}-1} \left(\ln \frac{R}{\Gamma^{\frac{1}{2-\beta}}}\right)^{-1} |u|^2 \langle \widetilde{\nabla}\Gamma, d\nu\rangle \\
&\quad + \frac{\beta+\alpha-4}{\beta-2}\int_{\partial\Omega} \Gamma^{\frac{\alpha-2}{2-\beta}-1} |u|^2 \langle \widetilde{\nabla}\Gamma, d\nu\rangle + \mathcal{C}(u).
\end{aligned}$$

It follows that

$$\begin{aligned}
-\int_\Omega \Gamma^{\frac{\alpha-2}{2-\beta}} u\mathcal{L}u d\nu \geq{}& \left(\frac{\beta+\alpha-4}{2}\right)\left(\frac{\beta-\alpha}{2}\right)\int_\Omega \Gamma^{\frac{\alpha-4}{2-\beta}} |\nabla_X \Gamma^{\frac{1}{2-\beta}}|^2 |u|^2\, d\nu \\
&+ \frac{1}{4}\int_\Omega \Gamma^{\frac{\alpha-4}{2-\beta}} |\nabla_X \Gamma^{\frac{1}{2-\beta}}|^2 \left(\ln \frac{R}{\Gamma^{\frac{1}{2-\beta}}}\right)^{-2} |u|^2 d\nu \\
&+ \frac{1}{2(\beta-2)}\int_{\partial\Omega} \Gamma^{\frac{\alpha-2}{2-\beta}-1} \left(\ln \frac{R}{\Gamma^{\frac{1}{2-\beta}}}\right)^{-1} |u|^2 \langle \widetilde{\nabla}\Gamma, d\nu\rangle \\
&+ \frac{\beta+\alpha-4}{2(\beta-2)}\int_{\partial\Omega} \Gamma^{\frac{\alpha-2}{2-\beta}-1} |u|^2 \langle \widetilde{\nabla}\Gamma, d\nu\rangle + \frac{1}{2}\mathcal{C}(u). \qquad (12.71)
\end{aligned}$$

Inequalities (12.70) and (12.71) imply that

$$\begin{aligned}
\int_\Omega \frac{\Gamma^{\frac{\alpha}{2-\beta}}}{|\nabla_X \Gamma^{\frac{1}{2-\beta}}|^2} |\mathcal{L}u|^2 d\nu \geq{}& (-4\epsilon^2 + (\beta+\alpha-4)(\beta-\alpha)\epsilon)\int_\Omega \Gamma^{\frac{\alpha-4}{2-\beta}} |\nabla_X \Gamma^{\frac{1}{2-\beta}}|^2 |u|^2\, d\nu \\
&+ \epsilon\int_\Omega \Gamma^{\frac{\alpha-4}{2-\beta}} |\nabla_X \Gamma^{\frac{1}{2-\beta}}|^2 \left(\ln \frac{R}{\Gamma^{\frac{1}{2-\beta}}}\right)^{-2} |u|^2 d\nu \\
&+ \frac{2\epsilon}{(\beta-2)}\int_{\partial\Omega} \Gamma^{\frac{\alpha-2}{2-\beta}-1} \left(\ln \frac{R}{\Gamma^{\frac{1}{2-\beta}}}\right)^{-1} |u|^2 \langle \widetilde{\nabla}\Gamma, d\nu\rangle \\
&+ \frac{2(\beta+\alpha-4)\epsilon}{\beta-2}\int_{\partial\Omega} \Gamma^{\frac{\alpha-2}{2-\beta}-1} |u|^2 \langle \widetilde{\nabla}\Gamma, d\nu\rangle + 2\epsilon\mathcal{C}(u).
\end{aligned}$$

Taking $\epsilon = \frac{(\beta+\alpha-4)(\beta-\alpha)}{8}$, we obtain

$$\begin{aligned}
&\int_\Omega \frac{\Gamma^{\frac{\alpha}{2-\beta}}}{|\nabla_X \Gamma^{\frac{1}{2-\beta}}|^2} |\mathcal{L}u|^2 d\nu \\
&\geq \frac{(\beta+\alpha-4)^2(\beta-\alpha)^2}{16}\int_\Omega \Gamma^{\frac{\alpha-4}{2-\beta}} |\nabla_X \Gamma^{\frac{1}{2-\beta}}|^2 |u|^2\, d\nu
\end{aligned}$$

$$+\frac{(\beta+\alpha-4)(\beta-\alpha)}{8}\int_\Omega \Gamma^{\frac{\alpha-4}{2-\beta}}|\nabla_X\Gamma^{\frac{1}{2-\beta}}|^2\left(\ln\frac{R}{\Gamma^{\frac{1}{2-\beta}}}\right)^{-2}|u|^2d\nu$$
$$+\frac{(\beta+\alpha-4)(\beta-\alpha)}{4(\beta-2)}\int_{\partial\Omega} \Gamma^{\frac{\alpha-2}{2-\beta}-1}\left(\ln\frac{R}{\Gamma^{\frac{1}{2-\beta}}}\right)^{-1}|u|^2\langle\widetilde{\nabla}\Gamma,d\nu\rangle$$
$$+\frac{(\beta+\alpha-4)^2(\beta-\alpha)}{4(\beta-2)}\int_{\partial\Omega} \Gamma^{\frac{\alpha-2}{2-\beta}-1}|u|^2\langle\widetilde{\nabla}\Gamma,d\nu\rangle+\frac{(\beta+\alpha-4)(\beta-\alpha)}{4}\mathcal{C}(u).$$

This completes the proof of Theorem 12.7.1. □

By a modification and refinement of the proof of Theorem 12.7.1 we can obtain another alternative of an improved Rellich inequality with boundary terms.

Theorem 12.7.2 (Refined local Rellich inequalities for sums of squares)**.** *Let $y\in M$ be such that (A_y^+) holds with the fundamental solution $\Gamma=\Gamma_y$ in T_y. Let $\Omega\subset T_y$ be a strongly admissible domain such that $y\notin\partial\Omega$. Let $\alpha\in\mathbb{R}$, $\beta>\alpha>\frac{8-\beta}{3}$, $\beta>2$ and $R\geq e\sup_\Omega\Gamma^{\frac{1}{2-\beta}}$. Then for all $u\in C^2(\Omega)\bigcap C^1(\overline{\Omega})$ we have*

$$\int_\Omega \frac{\Gamma^{\frac{\alpha}{2-\beta}}}{|\nabla_X\Gamma^{\frac{1}{2-\beta}}|^2}|\mathcal{L}u|^2d\nu\geq\frac{(\beta-\alpha)^2}{4}\int_\Omega \Gamma^{\frac{\alpha-2}{2-\beta}}|\nabla_Xu|^2d\nu$$
$$+\frac{(\beta+3\alpha-8)(\beta+\alpha-4)(\beta-\alpha)}{8(\beta-2)}\int_{\partial\Omega}\Gamma^{\frac{\alpha-2}{2-\beta}-1}|u|^2\langle\widetilde{\nabla}\Gamma,d\nu\rangle$$
$$+\frac{(\beta+\alpha-4)(\beta-\alpha)}{4}\mathcal{C}(u),\tag{12.72}$$

and its further refinement

$$\int_\Omega \frac{\Gamma^{\frac{\alpha}{2-\beta}}}{|\nabla_X\Gamma^{\frac{1}{2-\beta}}|^2}|\mathcal{L}u|^2d\nu$$
$$\geq\frac{(\beta-\alpha)^2}{4}\int_\Omega \Gamma^{\frac{\alpha-2}{2-\beta}}|\nabla_Xu|^2d\nu$$
$$+\frac{(\beta+3\alpha-8)(\beta-\alpha)}{16}\int_\Omega \Gamma^{\frac{\alpha-4}{2-\beta}}|\nabla_X\Gamma^{\frac{1}{2-\beta}}|^2\left(\ln\frac{R}{\Gamma^{\frac{1}{2-\beta}}}\right)^{-2}|u|^2d\nu$$
$$+\frac{(\beta+3\alpha-8)(\beta-\alpha)}{8(\beta-2)}\int_{\partial\Omega} \Gamma^{\frac{\alpha-2}{2-\beta}-1}\left(\ln\frac{R}{\Gamma^{\frac{1}{2-\beta}}}\right)^{-1}|u|^2\langle\widetilde{\nabla}\Gamma,d\nu\rangle$$
$$+\frac{(\beta+3\alpha-8)(\beta+\alpha-4)(\beta-\alpha)}{8(\beta-2)}\int_{\partial\Omega}\Gamma^{\frac{\alpha-2}{2-\beta}-1}|u|^2\langle\widetilde{\nabla}\Gamma,d\nu\rangle$$
$$+\frac{(\beta+\alpha-4)(\beta-\alpha)}{4}\mathcal{C}(u),\tag{12.73}$$

where

$$\mathcal{C}(u):=\frac{\alpha-2}{2-\beta}\int_{\partial\Omega}u^2\Gamma^{\frac{\alpha-2}{2-\beta}-1}\langle\widetilde{\nabla}\Gamma,d\nu\rangle-2\int_{\partial\Omega}\Gamma^{\frac{\alpha-2}{2-\beta}}u\langle\widetilde{\nabla}u,d\nu\rangle.$$

Proof of Theorem 12.7.2. Let us first prove (12.72). Let us rewrite (12.68) in the form

$$\begin{aligned}\frac{1}{2}\mathcal{C}(u)+\int_{\Omega}\Gamma^{\frac{\alpha-2}{2-\beta}}|\nabla_X u|^2 d\nu \\ = -\int_{\Omega}\Gamma^{\frac{\alpha-2}{2-\beta}}u\mathcal{L}u d\nu+\frac{(\beta+\alpha-4)(\alpha-2)}{2}\int_{\Omega}\Gamma^{\frac{\alpha-4}{2-\beta}}|\nabla_X\Gamma^{\frac{1}{2-\beta}}|^2\,|u|^2 d\nu.\end{aligned} \tag{12.74}$$

Also recalling (12.70) we have

$$-\int_{\Omega}\Gamma^{\frac{\alpha-2}{2-\beta}}u\mathcal{L}u d\nu\le\epsilon\int_{\Omega}\Gamma^{\frac{\alpha-4}{2-\beta}}|\nabla_X\Gamma^{\frac{1}{2-\beta}}|^2|u|^2 d\nu+\frac{1}{4\epsilon}\int_{\Omega}\frac{\Gamma^{\frac{\alpha}{2-\beta}}}{|\nabla_X\Gamma^{\frac{1}{2-\beta}}|^2}|\mathcal{L}u|^2 d\nu. \tag{12.75}$$

Inequalities (12.75) and (12.74) imply that

$$\begin{aligned}\frac{1}{2}\mathcal{C}(u)+\int_{\Omega}\Gamma^{\frac{\alpha-2}{2-\beta}}|\nabla_X u|^2 d\nu \\ \le\left(\frac{(\beta+\alpha-4)(\alpha-2)}{2}+\epsilon\right)\int_{\Omega}\Gamma^{\frac{\alpha-4}{2-\beta}}|\nabla_X\Gamma^{\frac{1}{2-\beta}}|^2|u|^2\,d\nu \\ +\frac{1}{4\epsilon}\int_{\Omega}\frac{\Gamma^{\frac{\alpha}{2-\beta}}}{|\nabla_X\Gamma^{\frac{1}{2-\beta}}|^2}|\mathcal{L}u|^2 d\nu.\end{aligned} \tag{12.76}$$

The already obtained inequality (12.63) can be rewritten as

$$\begin{aligned}\frac{16}{(\beta+\alpha-4)^2(\beta-\alpha)^2}\int_{\Omega}\frac{\Gamma^{\frac{\alpha}{2-\beta}}}{|\nabla_X\Gamma^{\frac{1}{2-\beta}}|^2}|\mathcal{L}u|^2 d\nu-\frac{4}{(\beta+\alpha-4)(\beta-\alpha)}\mathcal{C}(u) \\ -\frac{4}{(\beta-\alpha)(\beta-2)}\int_{\partial\Omega}\Gamma^{\frac{\alpha-2}{2-\beta}-1}|u|^2\langle\widetilde{\nabla}\Gamma,d\nu\rangle\ge\int_{\Omega}\Gamma^{\frac{\alpha-4}{2-\beta}}|\nabla_X\Gamma^{\frac{1}{2-\beta}}|^2|u|^2\,d\nu.\end{aligned}$$

Combining this with (12.76) we obtain

$$\begin{aligned}\frac{1}{2}\mathcal{C}(u)+\left(\frac{(\beta+\alpha-4)(\alpha-2)}{2}+\epsilon\right)\frac{4}{(\beta+\alpha-4)(\beta-\alpha)}\mathcal{C}(u) \\ +\left(\frac{(\beta+\alpha-4)(\alpha-2)}{2}+\epsilon\right)\frac{4}{(\beta-\alpha)(\beta-2)}\int_{\partial\Omega}\Gamma^{\frac{\alpha-2}{2-\beta}-1}|u|^2\langle\widetilde{\nabla}\Gamma,d\nu\rangle \\ +\int_{\Omega}\Gamma^{\frac{\alpha-2}{2-\beta}}|\nabla_X u|^2 d\nu \\ \le\left(\frac{16\epsilon}{(\beta+\alpha-4)^2(\beta-\alpha)^2}+\frac{8(\alpha-2)}{(\beta+\alpha-4)(\beta-\alpha)^2}+\frac{1}{4\epsilon}\right) \\ \times\int_{\Omega}\frac{\Gamma^{\frac{\alpha}{2-\beta}}}{|\nabla_X\Gamma^{\frac{1}{2-\beta}}|^2}|\mathcal{L}u|^2 d\nu.\end{aligned}$$

Taking $\epsilon = \frac{(\beta+\alpha-4)(\beta-\alpha)}{8}$ this implies

$$\frac{(\beta+\alpha-4)(\beta-\alpha)}{4}\mathcal{C}(u) + \frac{(\beta+3\alpha-8)(\beta+\alpha-4)(\beta-\alpha)}{8(\beta-2)} \int_{\partial\Omega} \Gamma^{\frac{\alpha-2}{2-\beta}-1}|u|^2\langle\widetilde{\nabla}\Gamma, d\nu\rangle$$
$$+ \frac{(\beta-\alpha)^2}{4}\int_\Omega \Gamma^{\frac{\alpha-2}{2-\beta}}|\nabla_X u|^2 d\nu \le \int_\Omega \frac{\Gamma^{\frac{\alpha}{2-\beta}}}{|\nabla_X \Gamma^{\frac{1}{2-\beta}}|^2}|\mathcal{L}u|^2 d\nu,$$

which shows (12.72).

Let us now prove (12.73). Inequality (12.64) can be rewritten as

$$\frac{16}{(\beta+\alpha-4)^2(\beta-\alpha)^2}\int_\Omega \frac{\Gamma^{\frac{\alpha}{2-\beta}}}{|\nabla_X \Gamma^{\frac{1}{2-\beta}}|^2}|\mathcal{L}u|^2 d\nu - \frac{16}{(\beta+\alpha-4)^2(\beta-\alpha)^2}\mathcal{D}$$
$$\ge \int_\Omega \Gamma^{\frac{\alpha-4}{2-\beta}}|\nabla_X \Gamma^{\frac{1}{2-\beta}}|^2|u|^2\, d\nu,$$

where

$$\mathcal{D} := \frac{(\beta+\alpha-4)(\beta-\alpha)}{8}\int_\Omega \Gamma^{\frac{\alpha-4}{2-\beta}}|\nabla_X \Gamma^{\frac{1}{2-\beta}}|^2 \left(\ln\frac{R}{\Gamma^{\frac{1}{2-\beta}}}\right)^{-2}|u|^2 d\nu$$
$$+ \frac{(\beta+\alpha-4)(\beta-\alpha)}{4(\beta-2)}\int_{\partial\Omega} \Gamma^{\frac{\alpha-2}{2-\beta}-1}\left(\ln\frac{R}{\Gamma^{\frac{1}{2-\beta}}}\right)^{-1}|u|^2\langle\widetilde{\nabla}\Gamma, d\nu\rangle$$
$$+ \frac{(\beta+\alpha-4)^2(\beta-\alpha)}{4(\beta-2)}\int_{\partial\Omega} \Gamma^{\frac{\alpha-2}{2-\beta}-1}|u|^2\langle\widetilde{\nabla}\Gamma, d\nu\rangle + \frac{(\beta+\alpha-4)(\beta-\alpha)}{4}\mathcal{C}(u).$$

Combining it with (12.76) we obtain

$$\frac{1}{2}\mathcal{C}(u) + \left(\frac{(\beta+\alpha-4)(\alpha-2)}{2} + \epsilon\right)\frac{16}{(\beta+\alpha-4)^2(\beta-\alpha)^2}\mathcal{D} + \int_\Omega \Gamma^{\frac{\alpha-2}{2-\beta}}|\nabla_X u|^2 d\nu$$
$$\le \left(\frac{16\epsilon}{(\beta+\alpha-4)^2(\beta-\alpha)^2} + \frac{8(\alpha-2)}{(\beta+\alpha-4)(\beta-\alpha)^2} + \frac{1}{4\epsilon}\right)\int_\Omega \frac{\Gamma^{\frac{\alpha}{2-\beta}}}{|\nabla_X \Gamma^{\frac{1}{2-\beta}}|^2}|\mathcal{L}u|^2 d\nu.$$

Taking $\epsilon = \frac{(\beta+\alpha-4)(\beta-\alpha)}{8}$ we obtain

$$\frac{(\beta-\alpha)^2}{8}\mathcal{C}(u) + \frac{\beta+3\alpha-8}{2(\beta+\alpha-4)}\mathcal{D} + \frac{(\beta-\alpha)^2}{4}\int_\Omega \Gamma^{\frac{\alpha-2}{2-\beta}}|\nabla_X u|^2 d\nu$$
$$\le \int_\Omega \frac{\Gamma^{\frac{\alpha}{2-\beta}}}{|\nabla_X \Gamma^{\frac{1}{2-\beta}}|^2}|\mathcal{L}u|^2 d\nu,$$

completing the proof. □

Remark 12.7.3. Let us formulate several consequences of the described estimates for the setting of functions $u \in C_0^\infty(\Omega)$, so that we have $\mathcal{C}(u) = 0$. Thus if $M = \mathbb{G}$ is a stratified group of homogeneous dimension $Q \ge 3$, we take $\beta = Q$, so that $\Gamma^{\frac{1}{2-\beta}}(x) = d(x)$ is the $\mathcal{L}$-gauge on the group $\mathbb{G}$. Then the above estimates give refinements compared to known estimates, as, e.g., in Kombe [Kom10], with respect to the inclusion of boundary terms. Thus, for all $u \in C_0^\infty(\Omega)$, we have that

1. Estimate (12.63) is reduced to

$$\int_\Omega \frac{d^\alpha}{|\nabla_X d|^2}|\mathcal{L}u|^2 d\nu \geq \frac{(Q+\alpha-4)^2(Q-\alpha)^2}{16}\int_\Omega d^{\alpha-4}|\nabla_X d|^2|u|^2\, d\nu,$$

for $Q > \alpha > 4-Q$.

2. Estimate (12.64) is reduced to

$$\int_\Omega \frac{d^\alpha}{|\nabla_X d|^2}|\mathcal{L}u|^2 d\nu \geq \frac{(Q+\alpha-4)^2(Q-\alpha)^2}{16}\int_\Omega d^{\alpha-4}|\nabla_X d|^2|u|^2\, d\nu + \frac{(Q+\alpha-4)(Q-\alpha)}{8}\int_\Omega d^{\alpha-4}|\nabla_X d|^2\left(\ln\frac{R}{d}\right)^{-2}|u|^2 d\nu,$$

for $Q > \alpha > 4-Q$.

3. Estimate (12.72) is reduced to

$$\int_\Omega \frac{d^\alpha}{|\nabla_X d|^2}|\mathcal{L}u|^2 d\nu \geq \frac{(Q-\alpha)^2}{4}\int_\Omega d^{\alpha-2}|\nabla_X u|^2 d\nu,$$

for $Q > \alpha > \frac{8-Q}{3}$.

4. Estimate (12.73) is reduced to

$$\int_\Omega \frac{d^\alpha}{|\nabla_X d|^2}|\mathcal{L}u|^2 d\nu \geq \frac{(Q-\alpha)^2}{4}\int_\Omega d^{\alpha-2}|\nabla_X u|^2 d\nu + \frac{(Q+3\alpha-8)(Q-\alpha)}{16}\int_\Omega d^{\alpha-4}|\nabla_X d|^2\left(\ln\frac{R}{d}\right)^{-2}|u|^2 d\nu,$$

for $Q > \alpha > \frac{8-Q}{3}$.

For unweighted versions (with $\alpha = 0$) inequalities (12.63)–(12.64) work under the condition $Q \geq 5$ which is usually appearing in Rellich inequalities, while (12.72)–(12.73) work for homogeneous dimensions $Q \geq 9$.

12.8 Rellich inequalities via Picone identities

In this section we discuss weighted Rellich inequalities in the spirit of Hardy inequalities from Section 12.5, relying on an appropriate version of Picone identities.

Theorem 12.8.1 (Weighted anisotropic Rellich type inequality). *Let $\Omega \subset M$ be an admissible domain. Let $W_i(x) \in C^2(\Omega)$ and $H_i(x) \in L^1_{\text{loc}}(\Omega)$ be the non-negative weight functions. Let $v > 0$, $v \in C^2(\Omega) \bigcap C^1(\overline{\Omega})$ with*

$$X_i^2 \left(W_i(x)|X_i^2 v|^{p_i-2} X_i^2 v\right) \geq H_i(x) v^{p-1} \quad \text{and} \quad - X_i^2 v > 0, \tag{12.77}$$

a.e. in Ω, for all $i = 1, \ldots, N$. Then for every $0 \leq u \in C^2(\Omega) \bigcap C^1(\overline{\Omega})$ we have the following inequality

$$\begin{aligned}\sum_{i=1}^{N} \int_{\Omega} H_i(x)|u|^{p_i} d\nu \leq & \sum_{i=1}^{N} \int_{\Omega} W_i(x)|X_i^2 u|^{p_i} d\nu \\ & + \sum_{i=1}^{N} \int_{\partial\Omega} W_i(x)|X_i^2 v|^{p_i-2} X_i^2 v \langle \widetilde{\nabla}_i \left(\frac{u^{p_i}}{v^{p_i-1}}\right), d\nu\rangle \\ & - \sum_{i=1}^{N} \int_{\partial\Omega} \left(\frac{u^{p_i}}{v^{p_i-1}}\right) \langle \widetilde{\nabla}_i (W_i(x)|X_i^2 v|^{p_i-2} X_i^2 v), d\nu\rangle,\end{aligned} \tag{12.78}$$

where $1 < p_i < N$ for $i = 1, \ldots, N$, and $\widetilde{\nabla}_i u = X_i u X_i$.

The proof of Theorem 12.8.1 will rely on first establishing an appropriate version of the second-order Picone identity. Before giving its formulation and the proofs, let us make some remarks and also formulate several of its consequences.

Remark 12.8.2.

1. A Carnot group version of Theorem 12.8.1 was obtained by Goldstein, Kombe and Yener in [GKY18]. In our exposition for general vector fields we follow [RSS18c], also allowing one to include boundary terms into the inequality.
2. Note that the function v from the assumption (12.77) appears in the boundary terms (12.78), which seems a new effect unlike known particular cases of Theorem 12.8.1.

As a consequence of Theorem 12.8.1 we can obtain several Rellich type inequalities involving the sum of squares operator.

Corollary 12.8.3 (Rellich inequalities for sums of squares). *Let $\Omega \subset M$ be an admissible domain, and let the operator $\mathcal{L}$ is the sum of squares of vector fields:*

$$\mathcal{L} := \sum_{i=1}^{N} X_i^2.$$

Then we have the following estimates:

(1) *Let* $\beta > 2$, $\alpha \in \mathbb{R}, \beta + \alpha > 4$ *and* $\beta > \alpha$. *Then we have*

$$\int_{\Omega} \frac{\Gamma^{\frac{\alpha}{2-\beta}}}{|\nabla_X \Gamma^{\frac{1}{2-\beta}}|^2} |\mathcal{L}u|^2 d\nu \geq \frac{(\beta + \alpha - 4)^2 (\beta - \alpha)^2}{16} \int_{\Omega} \Gamma^{\frac{\alpha-4}{2-\beta}} |\nabla_X \Gamma^{\frac{1}{2-\beta}}|^2 |u|^2 d\nu,$$

for all $u \in C_0^{\infty}(\Omega \backslash \{0\})$.

(2) *Let* $1 < p < \infty$ *and* $2 - \beta < \alpha < \min\{(\beta - 2)(p - 1), (\beta - 2)\}$. *Then for all* $u \in C_0^{\infty}(\Omega \backslash \{0\})$ *we have*

$$\int_{\Omega} \frac{\Gamma^{\frac{\alpha+2p-2}{2-\beta}}}{|\nabla_X \Gamma^{\frac{1}{2-\beta}}|^{2p-2}} |\mathcal{L}u|^p d\nu \geq C(\beta, \alpha, p)^p \int_{\Omega} \Gamma^{\frac{\alpha-2}{2-\beta}} |\nabla_X \Gamma^{\frac{1}{2-\beta}}|^2 |u|^p d\nu, \quad (12.79)$$

where $C(\beta, \alpha, p) := \frac{(\beta+\alpha-2)}{p} \frac{(\beta-2)(p-1)-\alpha}{p}$.

Proof of Corollary 12.8.3. To prove Part (1), we take $\gamma = -\frac{\beta+\alpha-4}{2}$, and choose the functions $W(x)$ and v such that

$$W(x) = \frac{\Gamma^{\frac{\alpha}{2-\beta}}}{|X_i \Gamma^{\frac{1}{2-\beta}}|^2} \quad \text{and} \quad v = \Gamma^{\frac{\gamma}{2-\beta}},$$

and apply Theorem 12.8.1.

To prove Part (2), we set

$$W(x) = \frac{\Gamma^{\frac{\alpha+2p-2}{2-\beta}}}{|\nabla_X \Gamma^{\frac{1}{2-\beta}}|^{2p-2}} \quad \text{and} \quad v = \gamma^{-\frac{\beta+\alpha-2}{p(2-\beta)}},$$

and apply Theorem 12.8.1. □

Now let us prove the following anisotropic (second-order) Picone type identity, extending its horizontal version in Lemma 6.10.4. Then, as a consequence, we obtain Theorem 12.8.1.

Lemma 12.8.4 (Second-order Picone identity). *Let* $\Omega \subset \mathbb{G}$ *be an open set. Let* u, v *be twice differentiable a.e. in* Ω *and satisfying the following conditions:*

$$u \geq 0, \ v > 0, \ X_i^2 v < 0 \ \text{ a.e. in } \Omega.$$

Let $p_i > 1$, $i = 1, \ldots, N$. *Then we have*

$$L_1(u, v) = R_1(u, v) \geq 0, \quad (12.80)$$

where

$$R_1(u, v) := \sum_{i=1}^{N} |X_i^2 u|^{p_i} - \sum_{i=1}^{N} X_i^2 \left(\frac{u^{p_i}}{v^{p_i - 1}} \right) |X_i^2 v|^{p_i - 2} X_i^2 v,$$

and

$$
\begin{aligned}
L_1(u,v) := & \sum_{i=1}^N |X_i^2 u|^{p_i} - \sum_{i=1}^N p_i \left(\frac{u}{v}\right)^{p_i-1} X_i^2 u X_i^2 v |X_i^2 v|^{p_i-2} \\
& + \sum_{i=1}^N (p_i - 1) \left(\frac{u}{v}\right)^{p_i} |X_i^2 v|^{p_i} \\
& - \sum_{i=1}^N p_i(p_i-1) \frac{u^{p_i-2}}{v^{p_i-1}} |X_i^2 v|^{p_i-2} X_i^2 v \left(X_i u - \frac{u}{v} X_i v\right)^2 .
\end{aligned}
$$

Proof of Lemma 12.8.4. A direct computation yields

$$
\begin{aligned}
X_i^2 \left(\frac{u^{p_i}}{v^{p_i-1}}\right) &= X_i \left(p_i \frac{u^{p_i-1}}{v^{p_i-1}} X_i u - (p_i-1) \frac{u^{p_i}}{v^{p_i}} X_i v\right) \\
&= p_i(p_i-1) \frac{u^{p_i-2}}{v^{p_i-2}} \left(\frac{(X_i u) v - u (X_i v)}{v^2}\right) X_i u + p_i \frac{u^{p_i-1}}{v^{p_i-1}} X_i^2 u \\
&\quad - p_i(p_i-1) \frac{u^{p_i-1}}{v^{p_i-1}} \left(\frac{(X_i u) v - u (X_i v)}{v^2}\right) X_i v - (p_i-1) \frac{u^{p_i}}{v^{p_i}} X_i^2 v \\
&= p_i(p_i-1) \left(\frac{u^{p_i-2}}{v^{p_i-1}} |X_i u|^2 - 2 \frac{u^{p_i-1}}{v^{p_i}} X_i v X_i u + \frac{u^{p_i}}{v^{p_i+1}} |X_i v|^2\right) \\
&\quad + p_i \frac{u^{p_i-1}}{v^{p_i-1}} X_i^2 u - (p_i-1) \frac{u^{p_i}}{v^{p_i}} X_i^2 v \\
&= p_i(p_i-1) \frac{u^{p_i-2}}{v^{p_i-1}} \left(X_i u - \frac{u}{v} X_i v\right)^2 + p_i \frac{u^{p_i-1}}{v^{p_i-1}} X_i^2 u - (p_i-1) \frac{u^{p_i}}{v^{p_i}} X_i^2 v,
\end{aligned}
$$

which gives the equality in (12.80). By Young's inequality we have

$$
\frac{u^{p_i-1}}{v^{p_i-1}} X_i^2 u X_i^2 v |X_i^2 v|^{p_i-2} \le \frac{|X_i^2 u|^{p_i}}{p_i} + \frac{1}{q_i} \frac{u^{p_i}}{v^{p_i}} |X_i^2 v|^{p_i}, \quad i = 1, \ldots, N,
$$

where $p_i > 1$ and $q_i > 1$ with $\frac{1}{p_i} + \frac{1}{q_i} = 1$. Since $X_i^2 v < 0$, $i = 1, \ldots, N$, we arrive at

$$
\begin{aligned}
L_1(u,v) \ge & \sum_{i=1}^N |X_i^2 u|^{p_i} + \sum_{i=1}^N (p_i-1) \frac{u^{p_i}}{v^{p_i}} |X_i^2 v|^{p_i} - \sum_{i=1}^N p_i \left(\frac{|X_i^2 u|^{p_i}}{p_i} + \frac{1}{q_i} \frac{u^{p_i}}{v^{p_i}} |X_i^2 v|^{p_i}\right) \\
& - \sum_{i=1}^N p_i(p_i-1) \frac{u^{p_i-2}}{v^{p_i-1}} |X_i^2 v|^{p_i-2} X_i^2 v \left|X_i u - \frac{u}{v} X_i v\right|^2 \\
= & \sum_{i=1}^N \left(p_i - 1 - \frac{p_i}{q_i}\right) \frac{u^{p_i}}{v^{p_i}} |X_i^2 v|^{p_i} \\
& - \sum_{i=1}^N p_i(p_i-1) \frac{u^{p_i-2}}{v^{p_i-1}} |X_i^2 v|^{p_i-2} X_i^2 v \left|X_i u - \frac{u}{v} X_i v\right|^2 \ge 0.
\end{aligned}
$$

This completes the proof of Lemma 12.8.4. □

Proof of Theorem 12.8.1. Let us give a brief outline of the following proof as it is similar to the proof of Theorem 12.5.1: we start by using the Picone type identity (12.80), then we apply an analogue of Green's second formula and the hypothesis (12.77), respectively. Finally, we arrive at (12.78) by using $H_i(x) \geq 0$. Summarizing, we have

$$\begin{aligned}
0 &\leq \int_\Omega W_i(x) L_1(u,v) d\nu = \int_\Omega W_i(x) R_1(u,v) d\nu \\
&= \int_\Omega W_i(x)|X_i^2 u|^{p_i} d\nu - \int_\Omega X_i^2 \left(\frac{u^{p_i}}{v^{p_i-1}}\right) W_i(x)|X_i^2 v|^{p_i-2} X_i^2 v d\nu \\
&= \int_\Omega W_i(x)|X_i^2 u|^{p_i} d\nu - \int_\Omega \frac{u^{p_i}}{v^{p_i-1}} X_i^2 \left(W_i(x)|X_i^2 v|^{p_i-2} X_i^2 v\right) d\nu \\
&\quad + \int_{\partial\Omega} \left(W_i(x)|X_i^2 v|^{p_i-2} X_i^2 v \langle \widetilde{\nabla}_i \left(\frac{u^{p_i}}{v^{p_i-1}}\right), d\nu \rangle \right. \\
&\quad \left. - \left(\frac{u^{p_i}}{v^{p_i-1}}\right) \langle \widetilde{\nabla}_i (W_i(x)|X_i^2 v|^{p_i-2} X_i^2 v), d\nu \rangle \right) \\
&\leq \int_\Omega W_i(x)|X_i^2 u|^{p_i} d\nu - \int_\Omega H_i(x)|u|^{p_i} d\nu \\
&\quad + \int_{\partial\Omega} \left(W_i(x)|X_i^2 v|^{p_i-2} X_i^2 v \langle \widetilde{\nabla}_i \left(\frac{u^{p_i}}{v^{p_i-1}}\right), d\nu \rangle \right. \\
&\quad \left. - \left(\frac{u^{p_i}}{v^{p_i-1}}\right) \langle \widetilde{\nabla}_i (W_i(x)|X_i^2 v|^{p_i-2} X_i^2 v), d\nu \rangle \right).
\end{aligned}$$

In the last line, we have used (12.77) which leads to (12.78). □

Bibliography

[AB17] B. Abdellaoui and R. Bentifour. Caffarelli–Kohn–Nirenberg type inequalities of fractional order with applications. *J. Funct. Anal.*, 272(10):3998–4029, 2017.

[ACP05] B. Abdellaoui, E. Colorado, and I. Peral. Some improved Caffarelli–Kohn–Nirenberg inequalities. *Calc. Var. Partial Differential Equations*, 23:327–345, 2005.

[ACR02] Adimurthi, N. Chaudhuri, and N. Ramaswamy. An improved Hardy Sobolev inequality and its applications. *Proc. Amer. Math. Soc.*, 130:489–505, 2002.

[Ada75] D.R. Adams. A note on Riesz potentials. *Duke Math. J.*, 42:765–778, 1975.

[AH98] W. Allegretto and Y.X. Huang. A Picone's identity for the p-Laplacian and applications. *Nonlinear Anal.*, 32(7):819–830, 1998.

[AL10] F. Avkhadiev and A. Laptev. Hardy inequalities for nonconvex domains. In *Around the research of Vladimir Maz'ya. I*, volume 11 of *Int. Math. Ser. (N.Y.)*, pages 1–12. Springer, New York, 2010.

[Anc86] A. Ancona. On strong barriers and an inequality of Hardy for domains in $\mathbf{R}^n$. *J. London Math. Soc.* (2), 34(2):274–290, 1986.

[AOS16] R.P. Agarwal, D. O'Regan, and S.H. Saker. *Hardy type inequalities on time scales.* Springer, Cham, 2016.

[AS06] Adimurthi and A. Sekar. Role of the fundamental solution in Hardy–Sobolev type inequalities. *Proc. Roy. Soc. Edinburgh Sect. A*, 136(6): 1111–1130, 2006.

[AW07] F.G. Avkhadiev and K.-J. Wirths. Unified Poincaré and Hardy inequalities with sharp constants for convex domains. *ZAMM Z. Angew. Math. Mech.*, 87(8-9):632–642, 2007.

[BBMP17] M. Bramanti, L. Brandolini, M. Manfredini, and M. Pedroni. Fundamental solutions and local solvability for nonsmooth Hörmander's operators. *Mem. Amer. Math. Soc.*, 249(1182):v + 79, 2017.

[BCG06] H. Bahouri, J.-Y. Chemin, and I. Gallagher. Refined Hardy inequalities. *Ann. Sc. Norm. Super. Pisa Cl. Sci.*, 5:375–391, 2006.

M. Ruzhansky, D. Suragan, *Hardy Inequalities on Homogeneous Groups*, Progress in Mathematics 327, https://doi.org/10.1007/978-3-030-02895-4

[BEHL08] A. Balinsky, W.D. Evans, D. Hundertmark, and R.T. Lewis. On inequalities of Hardy–Sobolev type. *Banach J. Math. Anal.*, 2(2):94–106, 2008.

[BEL15] A.A. Balinsky, W.D. Evans, and R.T. Lewis. *The analysis and geometry of Hardy's inequality.* Universitext. Springer, Cham, 2015.

[Ben89] D.M. Bennett. An extension of Rellich's inequality. *Proc. Amer. Math. Soc.*, 106(4):987–993, 1989.

[BFKG12] H. Bahouri, C. Fermanian-Kammerer, and I. Gallagher. Refined inequalities on graded Lie groups. *C. R. Math. Acad. Sci. Paris*, 350:393–397, 2012.

[Bir61] M.S. Birman. On the spectrum of singular boundary-value problems. *Mat. Sb. (N.S.)*, 55 (97):125–174, 1961.

[BL76] J. Bergh and J. Löfström. *Interpolation spaces. An introduction.* Springer-Verlag, Berlin-New York, 1976. Grundlehren der Mathematischen Wissenschaften, No. 223.

[BL85] H. Brezis and E. Lieb. Inequalities with remainder terms. *J. Funct. Anal.*, 62:73–86, 1985.

[Bli30] G.A. Bliss. An Integral Inequality. *J. London Math. Soc.*, 5(1):40–46, 1930.

[BLS04] A. Balinsky, A. Laptev, and A.V. Sobolev. Generalized Hardy inequality for the magnetic Dirichlet forms. *J. Statist. Phys.*, 116(1-4):507–521, 2004.

[BLU04] A. Bonfiglioli, E. Lanconelli, and F. Uguzzoni. Fundamental solutions for non-divergence form operators on stratified groups. *Trans. Amer. Math. Soc.*, 356(7):2709–2737, 2004.

[BLU07] A. Bonfiglioli, E. Lanconelli, and F. Uguzzoni. *Stratified Lie groups and potential theory for their sub-Laplacians.* Springer-Verlag, Berlin, Heidelberg, 2007.

[BM97] H. Brezis and M. Marcus. Hardy's inequalities revisited. *Ann. Scuola Norm. Sup. Pisa Cl. Sci.*, 25(4):217–237, 1997.

[BN83] H. Brezis and L. Nirenberg. Positive solutions of nonlinear elliptic equations involving critical Sobolev exponents. *Comm. Pure Appl. Math.*, 36:437–477, 1983.

[BNC14] V.I. Burenkov, E.D. Nursultanov, and D.K. Chigambayeva. Description of the interpolation spaces for a pair of local Morrey type spaces and their generalizations. *Proc. Steklov Inst. Math.*, 284:97–128, 2014.

[Bon69] J.-M. Bony. Principe du maximum, inégalité de Harnack et unicité du problème de Cauchy pour les opérateurs elliptiques dégénérés. *Ann. Inst. Fourier (Grenoble)*, 19:277–304, 1969.

[Bon09] A. Bonfiglioli. Taylor formula for homogeneous groups and applications. *Math. Z.*, 262(2):255–279, 2009.

[BT02a] N. Badiale and G. Tarantello. A Sobolev–Hardy inequality with applications to a nonlinear elliptic equation arising in astrophysics. *Arch. Ration. Mech. Anal.*, 163:259–293, 2002.

[BT02b] Z.M. Balogh and J.T. Tyson. Polar coordinates in Carnot groups. *Math. Z.*, 241(4):697–730, 2002.

[Bur13] V.I. Burenkov. Recent progress in studying the boundedness of classical operators of real analysis in general Morrey type spaces, I, II. *Eurasian Math. J.*, 3:11–32, 2013.

[BV97] H. Brezis and J.L. Vázquez. Blow-up solutions of some nonlinear elliptic problems. *Rev. Mat. Univ. Complut. Madrid*, 10(2):443–469, 1997.

[CBTW15] P.J. Coles, M. Berta, M. Tomamichel, and S. Wehner. Entropic uncertainty relations and their applications. *arXiv 1511.04857*, 2015.

[CCR15] P. Ciatti, M.G. Cowling, and F. Ricci. Hardy and uncertainty inequalities on stratified Lie groups. *Adv. Math.*, 277:365–387, 2015.

[CDG93] L. Capogna, D. Danielli, and N. Garofalo. An embedding theorem and the Harnack inequality for nonlinear subelliptic equations. *Comm. Partial Differential Equations*, 18(9-10):1765–1794, 1993.

[CDP18] J. Carrillo, M.G. Delgadino, and F.S. Patacchini. Existence of ground states for aggregation-diffusion equations. *arXiv:1803.01915*, 2018.

[CF87] F. Chiarenza and M. Frasca. Morrey spaces and Hardy–Littlewood maximal function. *Rend. Mat.*, 7:273–279, 1987.

[CF08] A. Cianci and A. Ferone. Hardy inequalities with non-standard remainder terms. *Ann. Inst. H. Poincaré. Anal. Nonlinéaire*, 25:889–906, 2008.

[CG90] L.J. Corwin and F.P. Greenleaf. *Representations of nilpotent Lie groups and their applications. Part I*, volume 18 of *Cambridge Studies in Advanced Mathematics.* Cambridge University Press, Cambridge, 1990. Basic theory and examples.

[CG95] M. Christ and L. Grafakos. Best constants for two nonconvolution inequalities. *Proc. Amer. Math. Soc.*, 123(6):1687–1693, 1995.

[CGL93] G. Citti, N. Garofalo, and E. Lanconelli. Harnack's inequality for sum of squares of vector fields plus a potential. *Amer. J. Math.*, 115(3):699–734, 1993.

[CGN08] L. Capogna, N. Garofalo, and D. Nhieu. Mutual absolute continuity of harmonic and surface measures for Hörmander type operators. In *Perspectives in partial differential equations*, pages 49–100, Proc. Sympos. Pure Math., 79, Amer. Math. Soc., Providence, RI, 2008.

[Chr91] M. Christ. L^p bounds for spectral multipliers on nilpotent groups. *Trans. Amer. Math. Soc.*, 328(1):73–81, 1991.

[Chr98] D. Christodoulou. On the geometry and dynamics of crystalline continua. *Ann. Inst. H. Poincaré Phys. Theor.*, 69:335–358, 1998.

[CKN84] L.A. Caffarelli, R. Kohn, and L. Nirenberg. First order interpolation inequalities with weights. *Composito Math.*, 53(3):259–275, 1984.

[CMS04] G. Citti, G. Manfredini, and A. Sarti. Neuronal oscillations in the visual cortex: Γ-convergence to the Riemannian Mumford–Shah functional. *SIAM J. Math. Anal.*, 35:1394–1419, 2004.

[Cos08] D.G. Costa. Some new and short proofs for a class of Caffarelli–Kohn–Nirenberg type inequalities. *J. Math. Anal. Appl.*, 337:311–317, 2008.

[CR16] D. Cardona and M. Ruzhansky. Littlewood–Paley theorem, Nikolskii inequality, Besov spaces, Fourier and spectral multipliers on graded Lie groups. *https://arxiv.org/abs/1610.04701*, 2016.

[CR17] D. Cardona and M. Ruzhansky. Multipliers for Besov spaces on graded Lie groups. *C. R. Math. Acad. Sci. Paris*, 355(4):400–405, 2017.

[CRS07] P. Ciatti, F. Ricci, and M. Sundari. Heisenberg–Pauli–Weyl uncertainty inequalities and polynomial volume growth. *Adv. Math.*, 215(2):616–625, 2007.

[CW01] F. Catrina and Z. Q. Wang. On the Caffarelli–Kohn–Nirenberg inequalities: sharp constants, existence (and non existence), and symmetry of extremals functions. *Comm. Pure Appl. Math.*, 54:229–258, 2001.

[Cyg81] J. Cygan. Subadditivity of homogeneous norms on certain nilpotent Lie groups. *Proc. Amer. Math. Soc.*, 83:69–70, 1981.

[D'A04a] L. D'Ambrosio. Hardy inequalities related to Grushin type operators. *Proc. Amer. Math. Soc.*, 132(3):725–734, 2004.

[D'A04b] L. D'Ambrosio. Some Hardy inequalities on the Heisenberg group. *Differential Equations*, 40(4):552–564, 2004.

[D'A05] L. D'Ambrosio. Hardy type inequalities related to degenerate elliptic differential operators. *Ann. Sc. Norm. Super. Pisa Cl. Sci.* (5), 4(3):451–486, 2005.

[Dav80] E.B. Davies. *One-parameter semigroups*, volume 15. London Mathematical Society Monographs. Academic Press, Inc. [Harcourt Brace Jovanovich, Publishers], London-New York, 1980.

[Dav99] E.B. Davies. A review of Hardy inequalities. In *The Maz'ya anniversary collection, Vol. 2 (Rostock, 1998)*, volume 110 of *Oper. Theory Adv. Appl.*, pages 55–67. Birkhäuser, Basel, 1999.

[Der71] M. Derridj. Un problème aux limites pour une classe d'opérateurs du second ordre hypoelliptiques. *Ann. Inst. Fourier (Grenoble)*, 21(4):99–148, 1971.

[Der72] M. Derridj. Sur un théorème de traces. *Ann. Inst. Fourier (Grenoble)*, 22(2):73–83, 1972.

[DFH18] J. Dolbeault, R. Frank, and F. Hoffmann. Reverse Hardy–Littlewood–Sobolev inequalities. *arXiv:1803.06151*, 2018.

[DFN84] B.A. Duvrovin, A.T. Fomenko, and S.P. Novikov. *Modern Geometry-Methods and Applications. Part I. The Geometry of Surfaces, Transformation Groups, and Fields.* Springer Science+Business Media, LLC, 1984.

[DGN06] D. Danielli, N. Garofalo, and D.-M. Nhieu. Non-doubling Ahlfors measures, perimeter measures, and the characterization of the trace spaces of Sobolev functions in Carnot–Carathéodory spaces. *Mem. Amer. Math. Soc.*, 182(857):x+119, 2006.

[DGN07] D. Danielli, N. Garofalo, and D. Nhieu. Sub-Riemannian calculus on hypersurfaces in Carnot groups. *Adv. Math.*, 215:292–378, 2007.

[DGN10] J. Dou, Q. Guo, and P. Niu. Hardy inequalities with remainder terms for the generalized Baouendi–Grushin vector fields. *Math. Inequal. Appl.*, 13(3):555–570, 2010.

[DGP11] D. Danielli, N. Garofalo, and N.C. Phuc. Hardy–Sobolev type inequalities with sharp constants in Carnot–Carathéodory spaces. *Potential Anal.*, 34:223–242, 2011.

[DH98] E.B. Davies and A.M. Hinz. Explicit constants for Rellich inequalities in $L_p(\Omega)$. *Math. Z.*, 227(3):511–523, 1998.

[DHK97] P. Drábek, H.P. Heinig, and A. Kufner. Higher-dimensional Hardy inequality. In *General inequalities,* 7 (*Oberwolfach,* 1995), volume 123 of *Internat. Ser. Numer. Math.*, pages 3–16. Birkhäuser, Basel, 1997.

[Dix77] J. Dixmier. *C^*-algebras.* North-Holland Publishing Co., Amsterdam, 1977. Translated from the French by Francis Jellett, North-Holland Mathematical Library, Vol. 15.

[Dix81] J. Dixmier. *von Neumann algebras*, volume 27 of *North-Holland Mathematical Library.* North-Holland Publishing Co., Amsterdam, 1981. With a preface by E.C. Lance, translated from the second French edition by F. Jellett.

[DJSJ13] Y. Di, L. Jiang, S. Shen, and Y. Jin. A note on a class of Hardy–Rellich type inequalities. *J. Inequal. Appl.*, pages 2013:84, 6, 2013.

[DL12] P. D'Ancona and R. Luca. Stein–Weiss and Caffarelli–Kohn–Nirenberg inequalities with angular integrability. *J. Math. Anal. Appl.*, 388(2): 1061–1079, 2012.

[DNPV12] E. Di Nezza, G. Palatucci, and E. Valdinoci. Hitchhiker's guide to the fractional Sobolev spaces. *Bull. Sci. Math.*, 136(5):521–573, 2012.

[Dye70] J.L. Dyer. A nilpotent Lie algebra with nilpotent automorphism group. *Bull. Amer. Math. Soc.*, 76:52–56, 1970.

[DZ15] J. Dou and M. Zhu. Reversed Hardy–Littewood–Sobolev inequality. *Int. Math. Res. Not. IMRN*, (19):9696–9726, 2015.

[EE04] D.E. Edmunds and W.D. Evans. *Hardy operators, function spaces and embeddings.* Springer Monographs in Mathematics. Springer-Verlag, Berlin, 2004.

[EKL15] T. Ekholm, H. Kovarík, and A. Laptev. Hardy inequalities for p-Laplacians with Robin boundary conditions. *Nonlinear Anal.*, 128:365–379, 2015.

[Elb81] A. Elbert. A half-linear second order differential equation. In *Qualitative theory of differential equations, Vol. I, II* (*Szeged,* 1979), volume 30 of *Colloq. Math. Soc. János Bolyai*, pages 153–180. North-Holland, Amsterdam-New York, 1981.

[EN04] H.G. Eridani and E. Nakai. On generalized fractional integral operators. *Scientiae Mathematicae Japanicae Online*, 10:307–318, 2004.

[Eri02] Eridani. On the boundedness of a generalized fractional integral on generalized Morrey spaces. *Tamkang J. Math.*, 33(4):335–340, 2002.

[ET99] D.E. Edmunds and H. Triebel. Sharp Sobolev embeddings and related Hardy inequalities: the critical case. *Math. Nachr.*, 207:79–92, 1999.

[FC17] T. Feng and X. Cui. Anisotropic Picone identities and anisotropic Hardy inequalities. *J. Inequal. Appl.*, pages Paper No. 16, 9, 2017.

[Fef83] C. Fefferman. The uncertainty principle. *Bull. Amer. Math. Soc.*, 9:129–206, 1983.

[FK72] G.B. Folland and J.J. Kohn. *The Neumann problem for the Cauchy–Riemann complex*, volume 75 of *Annals of Mathematics Studies.* Princeton University Press, Princeton, N.J.; University of Tokyo Press, Tokyo, 1972.

[FL12] R.L. Frank and E.H. Lieb. Sharp constants in several inequalities on the Heisenberg group. *Ann. of Math.*, 176:349–381, 2012.

[Fol75] G.B. Folland. Subelliptic estimates and function spaces on nilpotent Lie groups. *Ark. Mat.*, 13(2):161–207, 1975.

[Fol77] G.B. Folland. On the Rothschild–Stein lifting theorem. *Comm. Partial Differential Equations*, 2(2):165–191, 1977.

[Fol89] G.B. Folland. *Harmonic analysis in phase space*, volume 122 of *Annals of Mathematics Studies.* Princeton University Press, Princeton, NJ, 1989.

[Fol95] G.B. Folland. *A course in abstract harmonic analysis.* Studies in Advanced Mathematics. CRC Press, Boca Raton, FL, 1995.

[Fol16] G.B. Folland. *A course in abstract harmonic analysis.* Textbooks in Mathematics. CRC Press, Boca Raton, FL, second edition, 2016.

[FP81] C. Fefferman and D.H. Phong. The uncertainty principle and sharp Gårding inequality. *Comm. Pure Appl. Math.*, 34:285–331, 1981.

[FR16] V. Fischer and M. Ruzhansky. *Quantization on nilpotent Lie groups*, volume 314 of *Progress in Mathematics*. Birkhäuser. (open access book), 2016.

[FR17] V. Fischer and M. Ruzhansky. Sobolev spaces on graded groups. *Ann. Inst. Fourier*, 67(4):1671–1723, 2017.

[FR75] F. Forelli and W. Rudin. Projections on spaces of holomorphic functions in balls. *Indiana Univ. Math. J.*, 24:593–602, 1974/75.

[FS74] G.B. Folland and E.M. Stein. Estimates for the $\overline{\partial_b}$ complex and analysis on the Heisenberg group. *Comm. Pure Appl. Math.*, 27:429–522, 1974.

[FS82] G.B. Folland and E.M. Stein. *Hardy spaces on homogeneous groups*, volume 28 of *Mathematical Notes*. Princeton University Press, Princeton, N.J.; University of Tokyo Press, Tokyo, 1982.

[FS97] G. Folland and A. Sitaram. The uncertainty principle: A mathematical survey. *J. Fourier Anal. Appl.*, 3:207–238, 1997.

[FS08] R.L. Frank and R. Seiringer. Non-linear ground state representations and sharp Hardy inequalities. *J. Func. Anal.*, 255:3407–3430, 2008.

[FSC86] C.L. Fefferman and A. Sánchez-Calle. Fundamental solutions for second order subelliptic operators. *Ann. Math.*, 124(2):247–272, 1986.

[GAM13] V. Guliyev, A. Akbulut, and Y. Mammadov. Boundedness of fractional maximal operator and their higher order commutators in generalized Morrey spaces on Carnot groups. *Acta Math. Sci. Ser. B Engl. Ed.*, 33(5):1329–1346, 2013.

[Gar93] N. Garofalo. Unique continuation for a class of elliptic operators which degenerate on a manifold of arbitrary codimension. *J. Differential Equations*, 104(1):117–146, 1993.

[Gar08] N. Garofalo. Geometric second derivative estimates in Carnot groups and convexity. *Manuscripta Math.*, 126(3):353–373, 2008.

[Gar17] N. Garofalo. Fractional thoughts. *arXiv:1712.03347*, 2017.

[Gau77] C.F. Gauss. *Allgemeine Lehrsätze in Beziehung auf die im verkehrten Verhältnisse des Quadrats der Entfernung wirkenden Anziehungs- und Abstoßungs-Kräfte.* (1839), volume Werke 5:194–242, 2nd Ed. Göttingen, 1877.

[Gau29] C.F. Gauss. *Atlas des Erdmagnetismus* (1840), volume Werke 12:326-408. Göttingen, 1929.

[Gav77] B. Gaveau. Principle de moindre action, propagation de la chaleur et estimées sous elliptiques sur certains groupes nilpotents. *Acta Math*, 139:95–153, 1977.

[GE98] K.-G. Grosse-Erdmann. *The blocking technique, weighted mean operators and Hardy's inequality*, volume 1679 of *Lecture Notes in Mathematics*. Springer-Verlag, Berlin, 1998.

[GE09] H. Gunawan and Eridani. Fractional integrals and generalized Olsen inequalities. *Kyungpook Math. J.*, 49:31–39, 2009.

[Gel83] D. Geller. Liouville's theorem for homogeneous groups. *Comm. Partial Differential Equations*, 8:1665–1677, 1983.

[Gel90] D. Geller. *Analytic pseudodifferential operators for the Heisenberg group and local solvability*, volume 37 of *Mathematical Notes*. Princeton University Press, Princeton, NJ, 1990.

[Ges84] F. Gesztesy. On nondegenerate ground states for Schrödinger operators. *Rep. Math. Phys.*, 20(1):93–109, 1984.

[GK08] J.A. Goldstein and I. Kombe. The Hardy inequality and nonlinear parabolic equations on Carnot groups. *Nonlinear Anal.*, 69(12):4643–4653, 2008.

[GKY17] J.A. Goldstein, I. Kombe, and A. Yener. A unified approach to weighted Hardy type inequalities on Carnot groups. *Discrete Contin. Dyn. Syst.*, 37(4):2009–2021, 2017.

[GKY18] J.A. Goldstein, I. Kombe, and A. Yener. A general approach to weighted Rellich type inequalities on Carnot groups. *Monatsh. Math.*, 186(1):49–72, 2018.

[GL90] N. Garofalo and E. Lanconelli. Frequency functions on the Heisenberg group, the uncertainty principle and unique continuation. *Ann. Inst. Fourier (Grenoble)*, 40:313–356, 1990.

[GL03] C.E. Gutiérrez and E. Lanconelli. Maximum principle, nonhomogeneous Harnack inequality, and Liouville theorems for X-elliptic operators. *Comm. Partial Differential Equations*, 28:1833–1862, 2003.

[GL17] F. Geszstesy and L. Littlejohn. Factorizations and Hardy–Rellich type inequalities. In F. Gesztesy, H. Hanche-Olsen, E. Jakobsen, Y. Lyubarskii, N. Risebro, and K. Seip, editors, *Partial Differential Equations, Mathematical Physics, and Stochastic Analysis. A Volume in Honor of Helge Holden's* 60*th Birthday*. EMS Congress Reports, arXiv:1701.08929v2, 2017.

[GLMW17] F. Gesztesy, L.L. Littlejohn, I. Michael, and R. Wellman. On Birman's sequence of Hardy–Rellich type inequalities. *arXiv:1710.06955*, 2017.

[GM08] N. Ghoussoub and A. Moradifam. On the best possible remaining term in the Hardy inequality. *Proc. Natl. Acad. Sci. USA*, 105(37):13746–13751, 2008.

[GM11] N. Ghoussoub and A. Moradifam. Bessel pairs and optimal Hardy and Hardy–Rellich inequalities. *Math. Ann.*, 349(1):1–57, 2011.

[GM13] N. Ghoussoub and A. Moradifam. *Functional inequalities: new perspectives and new applications*, volume 187 of *Mathematical Surveys and Monographs*. American Mathematical Society, Providence, RI, 2013.

[GMS10] V.S. Guliyev, R.C. Mustafayev, and A. Serbetci. Stein–Weiss inequalities for the fractional integral operators in Carnot groups and applications. *Complex Var. Elliptic Equ.*, 55(8-10):847–863, 2010.

[GN88] C.E. Gutiérrez and G.S. Nelson. Bounds for the fundamental solution of degenerate parabolic equations. *Comm. Partial Differential Equations*, 13:635–649, 1988.

[Goo76] R.W. Goodman. *Nilpotent Lie groups: structure and applications to analysis.* Lecture Notes in Mathematics, Vol. 562. Springer-Verlag, Berlin, 1976.

[GP80] F. Gesztesy and L. Pittner. A generalization of the virial theorem for strongly singular potentials. *Rep. Math. Phys.*, 18(2):149–162 (1983), 1980.

[Gre28] G. Green. *An essay on the applicability of mathematical analysis of the theories of elasticity and magnetism.* Nottingham, U.K, 1828.

[Gri03] G. Grillo. Hardy and Rellich type inequalities for metrics defined by vector fields. *Potential Anal.*, 18(3):187–217, 2003.

[Gro96] M. Gromov. Carnot–Carathéodory spaces seen from within. In *Sub-Riemannian geometry*, volume 144 of *Progress in Mathematics*, pages 79–323. Birkhäuser, Basel, 1996.

[GRS17] N. Garofalo, M. Ruzhansky, and D. Suragan. On Green functions for Dirichlet sub-Laplacians on H-type groups. *J. Math. Anal. Appl.*, 452(2):896–905, 2017.

[GS77] P.C. Greiner and E.M. Stein. *Estimates for the $\overline{\partial}$-Neumann problem*, volume 19 of *Mathematical Notes.* Princeton University Press, Princeton, N.J., 1977.

[GS98] G. Gaudry and P. Sjögren. Singular integrals on the complex affine group. *Colloq. Math.*, 75(1):133–148, 1998.

[GS13] V.S. Guliyev and S.G. Samko. Maximal, potential, and singular operators in the generalized variable exponent Morrey spaces on unbounded sets. *J. Math. Sci. (N.Y.)*, 193(2):228–248, 2013. Problems in mathematical analysis. No. 71.

[GS15a] V. Guliyev and L. Softova. Generalized Morrey estimates for the gradient of divergence form parabolic operators with discontinuous coefficients. *J. Differential Equations*, 259:2368–2387, 2015.

[GS15b] V. Guliyev and L. Softova. Generalized Morrey regularity for parabolic equations with discontinuous data. *Proc. Edinb. Math. Soc.*, 58:199–218, 2015.

[GS16] V. Guliyev and S. Samko. Maximal operator in variable exponent generalized Morrey spaces on quasi-metric measure space. *Mediterr. J. Math.*, 13:1151–1165, 2016.

[GV00] N. Garofalo and D. Vassilev. Regularity near the characteristic set in the non-linear Dirichlet problem and conformal geometry of sub-Laplacians on Carnot groups. *Math. Ann.*, 318:453–516, 2000.

[GW92] C.E. Gutiérrez and R.L. Wheeden. Bounds for the fundamental solution of degenerate parabolic equations. *Comm. Partial Differential Equations*, 17:1287–1307, 1992.

[GZ01] J.A. Goldstein and Q.S. Zhang. On a degenerate heat equation with a singular potential. *J. Funct. Anal.*, 186:342–359, 2001.

[Han15] Y. Han. Weighted Caffarelli–Kohn–Nirenberg type inequality on the Heisenberg group. *Indian J. Pure Appl. Math.*, 46(2):147–161, 2015.

[Har19] G.H. Hardy. Notes on some points in the integral calculus. *Messenger Math.*, 48:107–112, 1919.

[Har20] G.H. Hardy. Note on a theorem of Hilbert. *Math. Z.*, 6(3-4):314–317, 1920.

[Hei27] W. Heisenberg. Über den anschaulichen Inhalt der quantentheoretischen Kinematik und Mechanik. *Zeitschrift fur Physik*, 43(3-4):172–198, 1927.

[Hei85] W. Heisenberg. *Über den anschaulichen Inhalt der quantentheoretischen Kinematik und Mechanik*, pages 478–504. Springer, Berlin, Heidelberg, 1985.

[HKK01] H.P. Heinig, R. Kerman, and M. Krbec. Weighted exponential inequalities. *Georgian Math. J.*, 8(1):69–86, 2001.

[HL27] G.H. Hardy and J.E. Littlewood. Some properties of fractional integrals. I. *Math. Z.*, 27:565–606, 1927.

[HL30] G.H. Hardy and J.E. Littlewood. Notes on the Theory of Series (XII): On Certain Inequalities Connected with the Calculus of Variations. *J. London Math. Soc.*, 5(1):34–39, 1930.

[HL32] G.H. Hardy and J.E. Littlewood. Some properties of fractional integrals. II. *Math. Z.*, 34:403–439, 1932.

[HLZ12] X. Han, G. Lu, and J. Zhu. Hardy–Littlewood–Sobolev and Stein–Weiss inequalities and integral systems on the Heisenberg group. *Nonlinear Anal.*, 75(11):4296–4314, 2012.

[HMM04] W. Hebisch, G. Mauceri, and S. Meda. Holomorphy of spectral multipliers of the Ornstein–Uhlenbeck operator. *J. Funct. Anal.*, 210(1):101–124, 2004.

[HMM05] W. Hebisch, G. Mauceri, and S. Meda. Spectral multipliers for sub-Laplacians with drift on Lie groups. *Math. Z.*, 251(4):899–927, 2005.

[HN03] Y. Han and P. Niu. Some Hardy type inequalities in the Heisenberg group. *JIPAM. J. Inequal. Pure Appl. Math.*, 4(5): Article 103, 5, 2003.

[HNZ11] Y.Z. Han, P.C. Niu, and T. Zhang. On first order interpolation inequalities with weights on the Heisenberg group. *Acta Mathematica Sinica, English series*, 27(12):2493–2506, 2011.

[HOHOL02] M. Hoffmann-Ostenhof, T. Hoffmann-Ostenhof, and A. Laptev. A geometrical version of Hardy's inequality. *J. Funct. Anal.*, 189(2):539–548, 2002.

[HOHOLT08] M. Hoffmann-Ostenhof, T. Hoffmann-Ostenhof, A. Laptev, and J. Tidblom. Many-particle Hardy inequalities. *J. Lond. Math. Soc. (2)*, 77(1):99–114, 2008.

[HOL15] T. Hoffmann-Ostenhof and A. Laptev. Hardy inequalities with homogeneous weights. *J. Funct. Anal.*, 268(11):3278–3289, 2015.

[Hör67] H. Hörmander. Hypoelliptic second-order differential equations. *Acta Math.*, 43:147–171, 1967.

[Hun56] G.A. Hunt. Semi-groups of measures on Lie groups. *Trans. Amer. Math. Soc.*, 81:264–293, 1956.

[HZ11] Y.Z. Han and Q. Zhao. A class of Caffarelli–Kohn–Nirenberg type inequalities for generalized Baouendi–Grushin vector fields. *Acta Math. Sci. Ser. A Chin. Ed.*, 31(5):1181–1189, 2011.

[HZD11] Y.Z. Han, S.T. Zhang, and J.B. Dou. On first order interpolation inequalities with weights on the H-type group. *Bull. Braz. Math. Soc. (N.S.)*, 42(2):185–202, 2011.

[IGE16] M. Idris, H. Gunawan, and Eridani. The boundedness of Bessel–Riesz operators on generalised Morrey spaces. *Aust. J. Math. Anal. Appl.*, 13(1):1–10, 2016.

[IGLE15] M. Idris, H. Gunawan, J. Lindiarni, and Eridani. In *The boundedness of Bessel–Riesz operators on Morrey spaces*, volume 1729. AIP Conf. Proc, http://dx.doi.org/10.1063/1.4946909, 2015.

[II15] N. Ioku and M. Ishiwata. A scale invariant form of a critical Hardy inequality. *Int. Math. Res. Not. IMRN*, (18):8830–8846, 2015.

[IIO16a] N. Ioku, M. Ishiwata, and T. Ozawa. Sharp remainder of a critical Hardy inequality. *Arch. Math. (Basel)*, 106(1):65–71, 2016.

[IIO16b] N. Ioku, M. Ishiwata, and T. Ozawa. Sharp remainder terms of Hardy type inequalities in L^p. *Preprint*, 2016.

[IIO17] N. Ioku, M. Ishiwata, and T. Ozawa. Hardy type inequalities in L^p with sharp remainders. *J. Inequal. Appl.*, pages 2017:5, 7, 2017.

[Jer81] D.S. Jerison. The Dirichlet problem for the Kohn Laplacian on the Heisenberg group, I and II. *J. Funct. Anal.*, 43:97–142, 224–257, 1981.

[JKS17] M. Jleli, M. Kirane, and B. Samet. Lyapunov type inequalities for fractional partial differential equations. *Appl. Math. Lett.*, 66:30–39, 2017.

[JS11] Y. Jin and S. Shen. Weighted Hardy and Rellich inequality on Carnot groups. *Arch. Math. (Basel)*, 96(3):263–271, 2011.

[Kac51] M. Kac. On some connections between probability theory and differential and integral equations. In *Proceedings of the Second Berkeley Symposium on Mathematical Statistics and Probability,* 1950, pages 189–215, Berkeley and Los Angeles, 1951. University of California Press.

[Kac80] M. Kac. *Integration in function spaces and some of its applications.* Accademia Nazionale dei Lincei, Pisa Lezioni Fermiane. [Fermi Lectures], 1980.

[Kap80] A. Kaplan. Fundamental solutions for a class of hypoelliptic PDE generated by composition of quadratic forms. *Trans. Amer. Math. Soc.*, 258:147–153, 1980.

[Ken27] E.H. Kennard. Zur Quantenmechanik einfacher Bewegungstypen. *Zeitschrift für Physik*, 44(4-5):326, 1927.

[KHPP13] K. Krulić Himmelreich, J. Pecarić, and D. Pokaz. *Inequalities of Hardy and Jensen*, volume 6 of *Monographs in Inequalities.* ELEMENT, Zagreb, 2013. New Hardy type inequalities with general kernels.

[Kir04] A.A. Kirillov. *Lectures on the orbit method*, volume 64 of *Graduate Studies in Mathematics.* American Mathematical Society, Providence, RI, 2004.

[KL12] H. Kovarík and A. Laptev. Hardy inequalities for Robin Laplacians. *J. Funct. Anal.*, 262(12):4972–4985, 2012.

[KL16] L. Kapitanski and A. Laptev. On continuous and discrete Hardy inequalities. *J. Spectr. Theory*, 6(4):837–858, 2016.

[KMP06] A. Kufner, L. Maligranda, and L.-E. Persson. The prehistory of the Hardy inequality. *Amer. Math. Monthly*, 113(8):715–732, 2006.

[KMP07] A. Kufner, L. Maligranda, and L.-E. Persson. *The Hardy inequality.* Vydavatelský Servis, Plzeň, 2007. About its history and some related results.

[KN65] J.J. Kohn and L. Nirenberg. Non-coercitive boundary value problems. *Comm. Pure and Appl. Math.*, 18(18):443–492, 1965.

[KNS99] K. Kurata, S. Nishigaki, and S. Sugano. Boundedness of integral operator on generalized Morrey spaces and its application to Schrödinger operator. *Proc. Amer. Math. Soc.*, 128:587–602, 1999.

[KÖ13] I. Kombe and M. Özaydin. Hardy–Poincaré, Rellich and uncertainty principle inequalities on Riemannian manifolds. *Trans. Amer. Math. Soc.*, 365(10):5035–5050, 2013.

[Kom10] I. Kombe. Sharp weighted Rellich and uncertainty principle inequalities on Carnot groups. *Commun. Appl. Anal.*, 14(2):251–271, 2010.

[Kom15] I. Kombe. Hardy and Rellich type inequalities with remainders for Baouendi–Grushin vector fields. *Houston J. Math.*, 41(3):849–874, 2015.

[KP03] A. Kufner and L.-E. Persson. *Weighted inequalities of Hardy type.* World Scientific Publishing Co., Inc., River Edge, NJ, 2003.

[KPS17] A. Kufner, L.-E. Persson, and N. Samko. *Weighted inequalities of Hardy type.* World Scientific Publishing Co. Pte. Ltd., Hackensack, NJ, second edition, 2017.

[KRS18a] A. Kassymov, M. Ruzhansky, and D. Suragan. Anisotropic fractional Gagliardo–Nirenberg, weighted Caffarelli–Kohn–Nirenberg and Lyapunov type inequalities, and applications to Riesz potentials and p-sub-Laplacian systems. *https://arxiv.org/abs/1806.08940*, 2018.

[KRS18b] A. Kassymov, M. Ruzhansky, and D. Suragan. Hardy–Littlewood–Sobolev and Stein–Weiss inequalities on homogeneous Lie groups. *https://arxiv.org/pdf/1810.11439*, 2018.

[KS09] T.S. Kal'menov and D. Suragan. On spectral problems for the volume potential. *Doklady Mathematics*, 80(2):646–649, 2009.

[KS11] T.S. Kalmenov and D. Suragan. A boundary condition and spectral problems for the Newton potential. In *Modern aspects of the theory of partial differential equations, volume* 216 *of* Oper. Theory Adv. Appl., pages 187–210. Birkhäuser, Basel AG, Basel, 2011.

[KS16] A.E. Kogoj and S. Sonner. Hardy type inequalities for Δ_λ-Laplacians. *Complex Var. Elliptic Equ.*, 61(3):422–442, 2016.

[Lar16] S. Larson. Geometric Hardy inequalities for the sub-elliptic Laplacian on convex domains in the Heisenberg group. *Bull. Math. Sci.*, 6(3):335–352, 2016.

[Led03] M. Ledoux. On improved Sobolev embedding theorems. *Math. Res. Lett.*, 10(5-6):659–669, 2003.

[Ler33] J. Leray. Étude de diverses équations intégrales non linéaires et de quelques problèmes que pose l'hydrodynamique. *J. Math. Pures Appl.*, 12:1–82, 1933.

[Lia13] B. Lian. Some sharp Rellich type inequalities on nilpotent groups and application. *Acta Math. Sci. Ser. B Engl. Ed.*, 33(1):59–74, 2013.

[Lie83] E.H. Lieb. Sharp constants in the Hardy–Littlewood–Sobolev and related inequalities. *Ann. of Math.* (2), 118(2):349–374, 1983.

[Lin90] P. Lindqvist. On the equation $\operatorname{div}(|\nabla u|^{p-2}\nabla u) + \lambda |u|^{p-2}u = 0$. *Proc. Amer. Math. Soc.*, 109(1):157–164, 1990.

[LRY17] A. Laptev, M. Ruzhansky, and N. Yessirkegenov. Hardy inequalities for Landau Hamiltonian and for Baouendi–Grushin operator with Aharonov–Bohm type magnetic field. Part I. *to appear in Math. Scand., https://arxiv.org/abs/1705.00062*, 2017.

[LU97] E. Lanconelli and F. Uguzzoni. On the Poisson kernel for the Kohn Laplacian. *Rend. Mat. Appl.*, 17(4):659–677, 1997.

[Lun15] D. Lundholm. Geometric extensions of many-particle Hardy inequalities. *J. Phys. A: Math. Theor.*, 48:175–203, 2015.

[LW99] A. Laptev and T. Weidl. Hardy inequalities for magnetic Dirichlet forms. In *Mathematical results in quantum mechanics* (*Prague,* 1998), volume 108 of *Oper. Theory Adv. Appl.*, pages 299–305. Birkhäuser, Basel, 1999.

[LY08] J.-W. Luan and Q.-H. Yang. A Hardy type inequality in the half-space on $\mathbb{R}^n$ and Heisenberg group. *J. Math. Anal. Appl.*, 347(2):645–651, 2008.

[Lya07] A.M. Lyapunov. Problème général de la stabilité du mouvement. *Ann. Fac. Sci. Univ. Toulouse*, 2:203–407, 1907.

[Man12] M. Manfredini. Fundamental solutions for sum of squares of vector fields operators with $C^{1,\alpha}$ coefficients. *Forum Math.*, 24:973–1011, 2012.

[Maz85] V.G. Maz'ja. *Sobolev spaces.* Springer Series in Soviet Mathematics. Springer-Verlag, Berlin, 1985. Translated from the Russian by T.O. Shaposhnikova.

[Maz11] V. Maz'ya. *Sobolev spaces with applications to elliptic partial differential equations*, volume 342 of *Grundlehren der Mathematischen Wissenschaften* [*Fundamental Principles of Mathematical Sciences*]. Springer, Heidelberg, augmented edition, 2011.

[MM90] G. Mauceri and S. Meda. Vector-valued multipliers on stratified groups. *Rev. Mat. Iberoamericana*, 6(3-4):141–154, 1990.

[MOV17] A. Martini, A. Ottazz, and M. Vallarino. A multiplier theorem for sub-Laplacians with drift on Lie groups. *arXiv:1705.04752*, 2017.

[MOW13a] S. Machihara, T. Ozawa, and H. Wadade. Generalizations of the logarithmic Hardy inequality in critical Sobolev–Lorentz spaces. *J. Inequal. Appl.*, pages 2013:381, 14, 2013.

[MOW13b] S. Machihara, T. Ozawa, and H. Wadade. Hardy type inequalities on balls. *Tohoku Math. J.* (2), 65(3):321–330, 2013.

[MOW15a] S. Machihara, T. Ozawa, and H. Wadade. On the Hardy type inequalities. *Preprint*, 2015.

[MOW15b] S. Machihara, T. Ozawa, and H. Wadade. Scaling invariant Hardy inequalities of multiple logarithmic type on the whole space. *J. Inequal. Appl.*, 281:1–13, 2015.

[MOW17a] S. Machihara, T. Ozawa, and H. Wadade. Remarks on the Hardy type inequalities with remainder terms in the framework of equalities. *to appear in Adv. Studies Pure Math.*, 2017.

[MOW17b] S. Machihara, T. Ozawa, and H. Wadade. Remarks on the Rellich inequality. *Math. Z.*, 286(3-4):1367–1373, 2017.

[MS01] C.C. Martinez and M.A. Sanz. *The theory of fractional powers of operators*, volume 187. North-Holland Mathematics Studies. North-Holland Publishing Co., Amsterdam, 2001.

[Nak94] E. Nakai. Hardy–Littlewood maximal operator, singular integral operator and Riesz potentials on generalized Morrey spaces. *Math. Nachr.*, 166:95–103, 1994.

[Nak01] E. Nakai. On generalized fractional integrals. *Taiwanese J. Math.*, 5:587–602, 2001.

[Nak02] E. Nakai. On generalized fractional integrals on the weak Orlicz spaces, BMO_ϕ, the Morrey spaces and the Campanato spaces. In *Function spaces, interpolation theory and related topics* (*Lund,* 2000), pages 389–401. de Gruyter, 2002.

[NDD12] P. Nápoli, I. Drelichman, and R. Durán. Improved Caffarelli–Kohn–Nirenberg and trace inequalities for radial functions. *Commun. Pure Appl. Anal.*, 11:1629–1642, 2012.

[Ngu17] V.H. Nguyen. The sharp higher order Hardy–Rellich type inequalities on the homogeneous groups. *https://arxiv.org/abs/1708.09311*, 2017.

[NN17] Q.A. Ngô and V.H. Nguyen. Sharp reversed Hardy–Littlewood–Sobolev inequality on $\mathbf{R}^n$. *Israel J. Math.*, 220(1):189–223, 2017.

[NS18a] H.-M. Nguyen and M. Squassina. Fractional Caffarelli–Kohn–Nirenberg inequalities. *J. Funct. Anal.*, 274(9):2661–2672, 2018.

[NS18b] H.-M. Nguyen and M. Squassina. On Hardy and Caffarelli–Kohn–Nirenberg inequalities. *arXiv:1801.06329*, 2018.

[NSW85] A. Nagel, E.M. Stein, and S. Wainger. Balls and metrics defined by vector fields. I. Basic properties. *Acta Math.*, 155(1-2):103–147, 1985.

[NZW01] P. Niu, H. Zhang, and Y. Wang. Hardy type and Rellich type inequalities on the Heisenberg group. *Proc. Amer. Math. Soc.*, 129(12):3623–3630, 2001.

[OK90] B. Opic and A. Kufner. *Hardy type inequalities*, volume 219 of *Pitman Research Notes in Mathematics Series.* Longman Scientific & Technical, Harlow, 1990.

[OR73] O. Oleinik and E. Radkevitch. *Second-order equations with nonnegative characteristic form.* AMS, Providence, R.I, 1973.

[ORS18] T. Ozawa, M. Ruzhansky, and D. Suragan. L^p-Caffarelli–Kohn–Nirenberg type inequalities on homogeneous groups. *to appear in Quart. J. Math., arXiv 1605.02520*, 2018.

[OS09] T. Ozawa and H. Sasaki. Inequalities associated with dilations. *Commun. Contemp. Math.*, 11(2):265–277, 2009.

[OY17] T. Ozawa and K. Yuasa. Uncertainty relations in the framework of equalities. *J. Math. Anal. Appl.*, 445:998–1012, 2017.

[PS12] L.-E. Persson and N. Samko. What should have happened if Hardy had discovered this? *J. Inequal. Appl.*, pages 2012:29, 11, 2012.

[Rel56] F. Rellich. Halbbeschränkte Differentialoperatoren höherer Ordnung. In *Proceedings of the International Congress of Mathematicians, 1954, Amsterdam, vol. III*, pages 243–250, Groningen; North-Holland Publishing Co., Amsterdam, 1956. Erven P. Noordhoff N.V.

[Ric] F. Ricci. Sub-Laplacians on nilpotent Lie groups. Unpublished lecture notes accessible on webpage http://homepage.sns.it/fricci/corsi.html.

[Rom91] C. Romero. *Potential theory for the Kohn Laplacian on the Heisenberg group*. PhD thesis, University of Minnesota, 1991.

[Rot83] L.P. Rothschild. A remark on hypoellipticity of homogeneous invariant differential operators on nilpotent Lie groups. *Comm. Partial Differential Equations*, 8(15):1679–1682, 1983.

[RRS16] G. Rozenblum, M. Ruzhansky, and D. Suragan. Isoperimetric inequalities for Schatten norms of Riesz potentials. *J. Funct. Anal.*, 271(1):224–239, 2016.

[RS76] L.P. Rothschild and E.M. Stein. Hypoelliptic differential operators and nilpotent groups. *Acta Math.*, 137:247–320, 1976.

[RS16a] M. Ruzhansky and D. Suragan. Critical Hardy inequalities. *https://arxiv.org/abs/1602.04809*, 2016.

[RS16b] M. Ruzhansky and D. Suragan. Isoperimetric inequalities for the logarithmic potential operator. *J. Math. Anal. Appl.*, 434(2):1676–1689, 2016.

[RS16c] M. Ruzhansky and D. Suragan. On Kac's principle of not feeling the boundary for the Kohn Laplacian on the Heisenberg group. *Proc. Amer. Math. Soc.*, 144(2):709–721, 2016.

[RS17a] M. Ruzhansky and D. Suragan. Green's identities, comparison principle and uniqueness of positive solutions for nonlinear p-sub-Laplacian equations on stratified Lie groups. *https://arxiv.org/abs/1706.09214*, 2017.

[RS17b] M. Ruzhansky and D. Suragan. Hardy and Rellich inequalities, identities, and sharp remainders on homogeneous groups. *Adv. Math.*, 317:799–822, 2017.

[RS17c] M. Ruzhansky and D. Suragan. Layer potentials, Kac's problem, and refined Hardy inequality on homogeneous Carnot groups. *Adv. Math.*, 308:483–528, 2017.

[RS17d] M. Ruzhansky and D. Suragan. Local Hardy and Rellich inequalities for sums of squares of vector fields. *Adv. Differential Equations*, 22(7-8):505–540, 2017.

[RS17e] M. Ruzhansky and D. Suragan. On horizontal Hardy, Rellich, Caffarelli–Kohn–Nirenberg and p-sub-Laplacian inequalities on stratified groups. *J. Differential Equations*, 262:1799–1821, 2017.

[RS17f] M. Ruzhansky and D. Suragan. Uncertainty relations on nilpotent Lie groups. *Proc. A.*, 473(2201):20170082, 12, 2017.

[RS18] M. Ruzhansky and D. Suragan. A note on stability of Hardy inequalities. *Ann. Funct. Anal.*, 9(4):451–462, 2018.

[RSS13] H. Rafeiro, N. Samko, and S. Samko. Morrey–Campanato spaces: an overview. In *Operator theory, pseudo-differential equations, and mathematical physics*, volume 228 of *Oper. Theory Adv. Appl.*, pages 293–323. Birkhäuser/Springer Basel AG, Basel, 2013.

[RSS18a] M. Ruzhansky, B. Sabitbek, and D. Suragan. Hardy and Rellich inequalities for anisotropic p-sub-Laplacians, and horizontal Hardy inequalities for multiple singularities and multi-particles on stratified groups. *https://arxiv.org/abs/1803.09996*, 2018.

[RSS18b] M. Ruzhansky, B. Sabitbek, and D. Suragan. Subelliptic geometric Hardy type inequalities on half-spaces and convex domains. *https://arxiv.org/abs/1806.06226*, 2018.

[RSS18c] M. Ruzhansky, B. Sabitbek, and D. Suragan. Weighted anisotropic Hardy and Rellich type inequalities for general vector fields. *https://arxiv.org/abs/1808.05532*, 2018.

[RSS18d] M. Ruzhansky, B. Sabitbek, and D. Suragan. Weighted L^p-Hardy and L^p-Rellich inequalities with boundary terms on stratified Lie groups. *Rev. Mat. Complutense, https://doi.org/10.1007/s13163-018-0268-3, https://arxiv.org/abs/1707.06754*, 2018.

[RSY17a] M. Ruzhansky, D. Suragan, and N. Yessirkegenov. Caffarelli–Kohn–Nirenberg and Sobolev type inequalities on stratified Lie groups. *NoDEA Nonlinear Differential Equations Appl.*, 24(5):24:56, 2017.

[RSY17b] M. Ruzhansky, D. Suragan, and N. Yessirkegenov. Extended Caffarelli–Kohn–Nirenberg inequalities and superweights for L^p-weighted Hardy inequalities. *C. R. Math. Acad. Sci. Paris*, 355(6):694–698, 2017.

[RSY18a] M. Ruzhansky, D. Suragan, and N. Yessirkegenov. Euler semigroup, Hardy–Sobolev and Gagliardo–Nirenberg type inequalities on homogeneous groups. *https://arxiv.org/abs/1805.01842*, 2018.

[RSY18b] M. Ruzhansky, D. Suragan, and N. Yessirkegenov. Extended Caffarelli–Kohn–Nirenberg inequalities, and remainders, stability, and superweights for L^p-weighted Hardy inequalities. *Trans. Amer. Math. Soc. Ser. B*, 5:32–62, 2018.

[RSY18c] M. Ruzhansky, D. Suragan, and N. Yessirkegenov. Hardy–Littlewood, Bessel–Riesz, and fractional integral operators in anisotropic Morrey and Campanato spaces. *Fract. Calc. Appl. Anal.*, 21(3):577–612, 2018.

[RSY18d] M. Ruzhansky, D. Suragan, and N. Yessirkegenov. Sobolev type inequalities, Euler–Hilbert–Sobolev and Sobolev–Lorentz–Zygmund spaces on homogeneous groups. *Integral Equations Operator Theory*, 90(1):Art. 10, 33, 2018.

[RT10] M. Ruzhansky and V. Turunen. *Pseudo-differential operators and symmetries. Background analysis and advanced topics*, volume 2 of *Pseudo-Differential Operators. Theory and Applications.* Birkhäuser Verlag, Basel, 2010.

[RT15] M. Rosenthal and H. Triebel. Morrey spaces, their duals and preduals. *Rev. Mat. Complut.*, 28(1):1–30, 2015.

[RV18] M. Ruzhansky and D. Verma. Hardy inequalities on metric measure spaces. *https://arxiv.org/abs/1806.03728*, 2018.

[RY17] M. Ruzhansky and N. Yessirkegenov. Factorizations and Hardy–Rellich inequalities on stratified groups. *https://arxiv.org/abs/1706.05108*, 2017.

[RY18a] M. Ruzhansky and N. Yessirkegenov. Hypoelliptic functional inequalities. *https://arxiv.org/abs/1805.01064*, 2018.

[RY18b] M. Ruzhansky and N. Yessirkegenov. Rellich inequalities for sub-Laplacians with drift. *To appear in Proc. Amer. Math. Soc., https://arxiv.org/abs/1707.03731*, 2018.

[Sai08] N. Saito. Data analysis and representation on a general domain using eigenfunctions of Laplacian. *Appl. Comput. Harmon. Anal.*, 25(1):68–97, 2008.

[Sak79] K. Saka. Besov spaces and Sobolev spaces on a nilpotent Lie group. *Tôhoku Math. J.* (2), 31(4):383–437, 1979.

[San18] M. Sano. Scaling invariant Hardy type inequalities with non-standard remainder terms. *Math. Inequal. Appl.*, 21(1):77–90, 2018.

[Saw84] E. Sawyer. Weighted Lebesgue and Lorentz norm inequalities for the Hardy operator. *Trans. Amer. Math. Soc.*, 281(1):329–337, 1984.

[SC84] A. Sánchez-Calle. Fundamental solutions and geometry of the sum of squares of vector fields. *Invent. Math.*, 78:143–160, 1984.

[Sch11] J. Schur. Bemerkungen zur Theorie der beschränkten Bilinearformen mit unendlich vielen Veränderlichen. *J. Reine Angew. Math.*, 140:1–28, 1911.

[Sch72] U.-W. Schmincke. Essential selfadjointness of a Schrödinger operator with strongly singular potential. *Math. Z.*, 124:47–50, 1972.

[SJ12] S.-F. Shen and Y.-Y. Jin. Rellich type inequalities related to Grushin type operator and Greiner type operator. *Appl. Math. J. Chinese Univ. Ser. B*, 27(3):353–362, 2012.

[Skr13] I. Skrzypczak. Hardy type inequalities derived from p-harmonic problems. *Nonlinear Anal.*, 93:30–50, 2013.

[Sob38] S.L. Sobolev. On a theorem in functional analysis (Russian). *Math. Sb.*, 46(471–497):39–68, 1938. [English translation in Amer. Math. Soc. Transl. 34(2):, 1963].

[SST12] Y. Sawano, S. Sugano, and H. Tanaka. Orlicz–Morrey spaces and fractional operators. *Potential Anal.*, 36:517–556, 2012.

[SSW03] S. Secchi, D. Smets, and M. Willen. Remarks on a Hardy–Sobolev inequality. *C. R. Acad. Sci. Paris, Ser. I*, 336:811–815, 2003.

[ST15] M. Sano and F. Takahashi. Improved Hardy inequality in a limiting case and their applications. *http://www.sci.osaka-cu.ac.jp/math/OCAMI/preprint/2015/15_03.pdf*, 2015.

[ST16] M. Sano and F. Takahashi. Improved Rellich type inequalities in $\mathbb{R}^N$. In *Geometric properties for parabolic and elliptic PDE's*, volume 176 of *Springer Proc. Math. Stat.*, pages 241–255. Springer, [Cham], 2016.

[ST17] M. Sano and F. Takahashi. Scale invariance structures of the critical and the subcritical Hardy inequalities and their improvements. *Calc. Var. Partial Differential Equations*, 56(3):Art. 69, 14, 2017.

[ST18a] M. Sano and F. Takahashi. Some improvements for a class of the Caffarelli–Kohn–Nirenberg inequalities. *Differential Integral Equations*, 31(1-2):57–74, 2018.

[ST18b] M. Sano and F. Takahashi. Weighted Hardy's inequality in a limiting case and the perturbed Kolmogorov equation. *arXiv 1803.02971*, 2018.

[Ste56] E.M. Stein. Interpolation of linear operators. *Trans. Amer. Math. Soc.*, 83:482–492, 1956.

[Stu90] J. Stubbe. Bounds on the number of bound states for potentials with critical decay at infinity. *J. Math. Phys.*, 31(5):1177–1180, 1990.

[SW58] E.M. Stein and G. Weiss. Fractional integrals on n-dimensional Euclidean space. *J. Math. Mech.*, 7:503–514, 1958.

[SZ97] M.O. Scully and M.S. Zubairy. *Quantum optics.* Cambridge University Press, Cambridge, 1997.

[Tak15] F. Takahashi. A simple proof of Hardy's inequality in a limiting case. *Arch. Math.* (*Basel*), 104:77–82, 2015.

[Tao05] T. Tao. An uncertainty principle for cyclic groups of prime order. *Math. Res. Lett.*, 12(1):121–127, 2005.

[Tha04] S. Thangavelu. *An introduction to the uncertainty principle*, volume 217 of *Progress in Mathematics.* Birkhäuser Boston, Inc., Boston, MA, 2004. Hardy's theorem on Lie groups, with a foreword by Gerald B. Folland.

[TW09] J. Tie and M.W. Wong. The heat kernel and Green functions of sub-Laplacians on the quaternion Heisenberg group. *J. Geom. Anal.*, 19(1):191–210, 2009.

[VS14] J. Van Schaftingen. Interpolation inequalities between Sobolev and Morrey–Campanato spaces: a common gateway to concentration-compactness and Gagliardo–Nirenberg interpolation inequalities. *Port. Math.*, 71(3-4):159–175, 2014.

[VSCC92] N.T. Varopoulos, L. Saloff-Coste, and T. Coulhon. *Analysis and geometry on groups*, volume 100 of *Cambridge Tracts in Mathematics*. Cambridge University Press, Cambridge, 1992.

[Wad14] H. Wadade. Optimal embeddings of critical Sobolev–Lorentz–Zygmund spaces. *Studia Math.*, 223(1):77–96, 2014.

[Wey08a] H. Weyl. *Singuläre Integralgleichungen.* Göttingen, 1908.

[Wey08b] H. Weyl. Singuläre Integralgleichungen. *Math. Ann.*, 66(3):273–324, 1908.

[Wey50] H. Weyl. *The theory of groups and quantum mechanics.* Dover Publications, Inc., New York, 1950. Translated from the second (revised) German edition by H.P. Roberton, reprint of the 1931 English translation.

[Wie10] F. Wiener. Elementarer Beweis eines Reihensatzes von Herrn Hilbert. *Math. Ann.*, 68(3):361–366, 1910.

[WM08] D.F. Walls and G.J. Milburn. *Quantum optics.* Springer, Berlin, 2nd edition, 2008.

[WN08] J. Wang and P. Niu. Sharp weighted Hardy type inequalities and Hardy–Sobolev type inequalities on polarizable Carnot groups. *C. R. Math. Acad. Sci. Paris Ser. I*, 346:1231–1234, 2008.

[WN16] L. Wu and P. Niu. Green functions for weighted subelliptic p-Laplace operator constructed by Hörmander's vector fields. *J. Math. Anal. Appl.*, 444:1565–1590, 2016.

[WW03] Z.-Q. Wang and M. Willem. Caffarelli–Kohn–Nirenberg inequalities with remainder terms. *J. Funct. Anal.*, 203(2):550–568, 2003.

[WZ03] Z. Wang and M. Zhu. Hardy inequalities with boundary terms. *EJDE*, 2003(43):1–8, 2003.

[Xia11] Y.-X. Xiao. Remarks on inequalities of Hardy–Sobolev type. *J. Inequal. Appl.*, pages 2011: 132, 8, 2011.

[Yac18] C. Yacoub. Caffarelli–Kohn–Nirenberg inequalities on Lie groups of polynomial growth. *Math. Nachr.*, 291(1):204–214, 2018.

[Yaf99] D. Yafaev. Sharp constants in the Hardy–Rellich inequalities. *J. Funct. Anal.*, 168(1):121–144, 1999.

[Yan13] Q.-H. Yang. Hardy type inequalities related to Carnot–Carathéodory distance on the Heisenberg group. *Proc. Amer. Math. Soc.*, 141(1):351–362, 2013.

[Yen16] A. Yener. Weighted Hardy type inequalities on the Heisenberg group $\mathbb{H}^n$. *Math. Ineq. Appl.*, 19(2):671–683, 2016.

[Zha17] C. Zhang. Quantitative unique continuation for the heat equation with Coulomb potentials. *arXiv:1707.07744v1*, 2017.

[ZHD14] S. Zhang, Y. Han, and J. Dou. A class of Caffarelli–Kohn–Nirenberg type inequalities on the H-type group. *Rend. Sem. Mat. Univ. Padova*, 132:249–266, 2014.

[ZHD15] S. Zhang, Y. Han, and J. Dou. Weighted Hardy–Sobolev type inequality for generalized Baouendi–Grushin vector fields and its application. *Advances in Mathematics* (*China*), 44(3):411–420, 2015.

Index

M. Ruzhansky, D. Suragan, *Hardy Inequalities on Homogeneous Groups*, Progress in Mathematics 327, https://doi.org/10.1007/978-3-030-02895-4